Studies in Economic Theory

Volume 29

More information about this series at http://www.springer.com/series/2584

Graciela Chichilnisky · Armon Rezai
Editors

The Economics of the Global Environment

Catastrophic Risks in Theory and Policy

58 €
WS
'18

Springer

Editors
Graciela Chichilnisky
Columbia University
New York, NY
USA

Armon Rezai
Department of Socio-Economics
Vienna University of Economics
and Business
Vienna
Austria

Partly previously published in the journal Economic Theory, Volume 49(2)

ISSN 1431-8849 ISSN 2196-9930 (electronic)
Studies in Economic Theory
ISBN 978-3-319-31941-4 ISBN 978-3-319-31943-8 (eBook)
DOI 10.1007/978-3-319-31943-8

Library of Congress Control Number: 2016940814

Printed on acid-free paper

This Springer imprint is published by Springer Nature
The registered company is Springer International Publishing AG
The registered company address is: Gewerbestrasse 11, 6330 Cham, Switzerland

Contents

The Economics of the Global Environment—Catastrophic Risks in Theory and Practice

Graciela Chichilnisky and Armon Rezai

1 Introduction

The world economy is changing fundamentally and irrevocably in front of our eyes. There is no disputing the fact. Yet the evolution of economics as a science does not match the sea of change we observe in the real world. This book attempts to address this somewhat remarkable and risky gap. It offers a collection of essays by leading economists who offer their vision on new foundations of economics, on experimental research testing the new theory and on its use in path breaking global policy. Some of the essays have been presented at the AFOSR workshop on Catastrophic Risk at SRI, Menlo Park, California, on May 31 and June 1, 2012, organized and sponsored by the Columbia Consortium for Risk Management at Columbia University.[1]

Global economic change is unfolding as we write this book. In December 2015, 200 hundred nations met in the most important climate change negotiations in decades, the Paris Convention of the Parties COP 21, to decide on the fate of the practical consequences of the world's single international agreement to achieve needed reductions of

[1]www.columbiariskmanagement.net.

G. Chichilnisky (✉)
Departments of Economics and Mathematical Statistics, Columbia University, 335 Riverside Drive, New York, NY 10027, USA
e-mail: gc9@columbia.edu

A. Rezai
Department of Socio-Economics, WU—Vienna University of Economics and Business, Welthandelsplatz 1, 1020 Vienna, Austria
e-mail: arezai@wu.ac.at

A. Rezai
IIASA, Schlossplatz 1, 2361 Laxenburg, Austria

A. Rezai
Vienna Institute for International Studies, Rahlgasse 3, 1060 Vienna, Austria

G. Chichilnisky and A. Rezai (eds.), *The Economics of the Global Environment*, Studies in Economic Theory 29, DOI 10.1007/978-3-319-31943-8_1

carbon in the atmosphere—the Kyoto Protocol and its path breaking Carbon Market. The debate centers on who should reduce carbon emissions—the poor or the rich nations of the world. Poor nations have been an unexpected engine of the growth of the world economy since 2000, and the G-20 was created as the first leading group of nations to include poor countries, the BRIC nations. The BRICS Bank they created can compete with the Bretton Woods' institutions such as the IMF and the World Bank that were created by the rich nations in 1945, at the end of WWII. These are the first global financial institutions in the world economy, largely responsible for an extraordinary period of success and globalization of capitalism, and a sea of change we observe today. The force of globalization since WWII led to a 300 % growth of the international economy over and above national growth, and joined all nations at the hip. At the same time the successful internationalization of capitalism led to the largest consumption and overexploitation of extractive resources the world ever saw. Minerals, metals, fuels, soil, and even the oceans that are 70 % of the planet's surface are disappearing in front of our eyes. The catastrophic risk of climate change from the overuse of fossil fuels is only matched by an equally catastrophic risk of biodiversity destruction and the vast and wasteful overexploitation the oceans. We know we have created the 6th largest destruction of life in the planet in the entire 4.5 billion history. We also know that the global environment is the trump card for the success or the demise of human societies. This is one of the key areas where economic science and economic reality are seriously out of step. And this is why the book is about the global environment and the catastrophic risks we face.

2 Part I. Catastrophic Risk in Economic Theory

The first part of the book goes to the foundations of economics and decision theory and expands existing theory to include rare events that are catastrophic and yet often neglected by existing theory. The articles introduce new breakthrough axioms, theorems that characterize the new probabilities that they characterizes and new criteria of optimality, updating the definition of probabilistic inference and formalizing sustainable preferences and behavior.

Peter Hammond's piece **Catastrophic Risk, Rare Events, and Black Swans: Toward an Alternative Synthesis** follows a series of articles (Chichilnisky 1996a, 2000, 2009, 2010) that set out an integrated decision theory intended to capture special properties of catastrophic risk, rare events, and black swans. Here these three are treated as separate extensions of orthodox decision theory. Precursors discussed include: (i) following Jones-Lee (1974), undefined willingness to pay in the case of catastrophic risk; (ii) following Hammond (1994) and others, rare events with infinitesimal probabilities. The author sketches a theory where enlivened decision trees can represent a dynamic process involving successive unforeseeable "true black swan" events and argues that a different integrated theory, yet to be developed, could perhaps include all these three features.

In **Preference Representations for Catastrophic Risk Analysis**, Richard E. Ericson and Jamie B. Kruse survey behavioral models of preference relations applicable for analysis of decisions in the face of catastrophic risk. The authors characterizes catastrophic risk and the implications of various representations are explored in simple, illustrative examples. They present an argument for the applicability of "variational preferences" for the analysis of behavior and decision making in the face of catastrophic risk. Louis Narens article provides a different formalization of risk. Almost all models of decision making assume an underlying Boolean space of events. This gives a logical structure to events that matches the structure of propositions of classical logic. In his article **Modeling Decisions Involving Ambiguous, Vague, or Rare Events** Narens takes a different approach, employing events that form a topology instead of a Boolean algebra. This allows for new modeling concepts for judgmental heuristics, rare events, and the influence of context on decisions.

Like Narens', Jun Zhang's article **Modeling Uncertainty, Context and Information Fusion via Lattice-Based Probability** investigates mathematical foundations for probabilistic inference, uncertainty representation, and information from disparate information sources. Probability measures are defined on an event space that is modeled as a bounded distributive lattice, including Boolean lattice as a special case, and standard probability theory is axiomatized. Following Narens (2009, 2011), Zhang invokes the relative pseudo-complementation operator on a distributed lattice, leading to Heyting algebra of event space that underlies intuitionistic logic. He considers basic probability assignment on a finite distributive lattice, leading to lower probability (belief function) and upper probability (plausibility function) on such lattices. Making use of the fact that topology on a set, that is, the collection of all open sets, forms a distributive lattice, pseudo-complementation can be addressed through the closure operation under an arbitrary topology introduced on the event space. He models contextual information for uncertainty as prescribing a topology on the event space. The totality of all topologies on an event space form a bounded complete lattice, ordered by coarse-grading, with the discrete topology as the top element (the finest/largest topology), and the indiscrete topology (consisting of only two elements, the null set and the full set) as the bottom element, the coarsest/smallest topology. By identifying topology with context, it provides a principle for defining "focal elements" (on open sets of the topology) combining them across different contexts in the lattice of topologies, more general than do current theories such as Dempster-Shafer belief function and Zadeh's fuzzy probability.

In **The foundations of uncertainty with black swans**, Chichilnisky extends the foundation of probability to include samples with rare events that are potentially catastrophic, called black swans. Examples are market crashes, catastrophic climate change, and species extinction. Such events are generally treated as ''outliers'' and disregarded. She proposes a new axiomatization of probability requiring equal treatment of rare and of frequent events—the Swan Axiom—and then characterizes all the probabilities that the axioms imply. These are new probabilities are neither finitely additive nor countably additive measures but rather a combination of both. They exclude countably additive probabilities as in De Groot (1970) and Arrow (1971) and Savage's (1954) finitely additive measures. The probabilities that satisfy the new axioms are standard distributions when the sample has no black swans, so the axioms are an extension of standard theory. The finitely additive part assigns however more weight to rare events than do standard distributions and in that sense explains the persistent observation of ''power laws'' and ''heavy tails'' that eludes classic theory. The axioms and the representation theorems extend earlier work by the author (Chichilnisky 1996a, 2000, 2002, 2009) to encompass and extend the foundation of probability (Villegas 1964; Dubins and Savage 1965; Dubins 1975; De Groot 1970; Purves and Sudderth 1976) and the theory of choice under uncertainty of Arrow (1971).

The Topology of Change seeks to overcome a bias in standard probability theory, which treats rare events as 'outliers' that are often disregarded and underestimated. The author argues that in a moment of change rare events can become frequent and frequent events rare. Chichilnisky therefore postulates new axioms for probability theory that require a balanced treatment for rare and frequent events, based on what she calls "the topology of change". The axioms extend the foundation of probability to integrate rare but potentially catastrophic events or black swans: such as natural hazards, market crashes, catastrophic climate change and major episodes of species extinction. New results presented in this article include a characterization of a family of purely finitely additive measures that are—somewhat surprisingly—absolutely continuous with respect to the Lebesgue measure. This is a new development from an earlier characterization of all the probabilities measures implied by the new axioms as a combination of purely finitely additive and countably additive measures that was first established in Chichilnisky (2000, 2002, 2009). The results are contrasted to the work of Kolmogorov (1933), De Groot (1970), Arrow (1971), Savage (1954), and Von Neumann and Morgernstern (1944).

In **Sustainable markets with short sales** Chichilnisky argues that market objectives may conflict with long-term goals, and that behind the conflict is the impatience axiom introduced by Koopmans to describe choices over time. The conflict is resolved here by introducing a new concept, "sustainable markets". These differ from Arrow-Debreu markets in that traders have sustainable preferences and have no bounds on short sales. Sustainable preferences were defined by Chichilnisky as preferences that are sensitive to the basic needs of the present without sacrificing the needs of future generations and embody the essence of sustainable development (Chichilnisky 1996b). In her theorems 1 and 2 Chichilnisky shows that limited arbitrage is a necessary and sufficient condition describing diversity and ensures the existence of a sustainable market equilibrium where the invisible hand delivers sustainable as well as efficient solutions (Chichilnisky 1995; Chichilnisky and Heal 1997). In sustainable markets prices have a new role: they reflect both the value of instantaneous consumption and the value of the long-run future. The latter are connected to the independence of the axiom of choice at the foundations of mathematics (Gödel 1940).

3 Part II. Ethical and Welfare Considerations

The second part of the book covers new fundamental thinking on the ethical and welfare considerations that should guide us when our actions today irrevocably change the future, and in particular the welfare of new generations that are yet to come. The problem requires a rewriting of neoclassical thinking that is dominated by the excellent work of Tjallings Koopmans in the 1960s who wrote when the pressing global environmental issues we face today were still to be understood (Koopmans 1960). He defined the neoclassical theory of economic choice based on "impatience," an axiom that he created which exponentially discounts the future by requiring fixes rates of discount from 1 year to the next and forever, thus obliterating the long term future implications from our actions today. This section seeks practical new axioms to replace Koopman's and criteria of optimality that define a sustainable economy. These authors and their articles are creative, polemic, theoretic and practical.

Geir Asheim, Tapan Mitra, and Bertil Tungodden ask what ethical criteria should be adopted when managing the global environment. In **Sustainable recursive social welfare functions** they study ethical criteria for intergenerational justice when faced with the task of managing the global environment. They argue that Koopmans' axiomatization of discounted utilitarianism is based on seemingly compelling conditions, yet this criterion leads to hard-to-justify outcomes. Their analysis considers a class of sustainable recursive social welfare functions within Koopmans' general framework. This class is axiomatized by means of a weak equity condition ("Hammond Equity for the Future") and general existence is established. Any member of the class satisfies the key axioms of Chichilnisky's "sustainable preferences" (Chichilnisky 1996b). The analysis singles out one of Koopmans' original separability conditions (his Postulate 3′a), that they call "Independent Present", as particularly questionable from an ethical perspective.

In **Intergenerational equity, efficiency, and constructability**, Luc Lauwers addresses global environmental issue such as biodiversity conservation or climate change, and argues that they are in reality long-term issues that are not properly taken into account with traditional models that incorporate the impatience axiom, which is manifested in fixed discount factors and in the use of present discounted utility criteria. When both the short and the very long run are important, he argues that one can appeal to overtaking criteria and Chichilnisky criteria. Unfortunately, overtaking criteria are highly incomplete. In order to decrease this incompleteness, stronger anonymity (or equity) axioms were developed. The author shows that a maximal anonymity axiom compatible with Pareto is a non-constructible object; its existence relies on the Axiom of Choice. The Chichilnisky criterion is based upon two axioms: non-dictatorship of the present and non-dictatorship of the future. Here, the very long run is captured by a finitely additive measure. Such a measure is a non-constructible object and has therefore no explicit description.

In **Sustainable exploitation of a natural resource: a satisfying use of Chichilnisky's criterion** Charles Figuières and Mabel Tidball analyze Chichilnisky's criterion for sustainability and argue that it has the merit to be, so far, the unique explicit, complete and continuous social welfare criterion that combines successfully the requirement of efficiency with an instrumental notion of intergenerational equity (Chichilnisky requires no dictatorship of the present and no dictatorship of the future). But they argue that it has one drawback: when applied in the context of renewable resources, and with a constant discount factor, there exists no exploitation path that maximizes this criterion. The present article suggests a way to cope with this problem. The idea is to restrict attention to the set of convex combinations between the optimal discounted utilitarian program and the stationary program leading to the green golden rule. It is shown that an optimal path in this set exists under rather weak sufficient conditions on the fundamentals of the problem. Some ethical properties of this approach are also discussed. In some cases, it turns out that the restricted solution implies no loss of efficiency and benefits intermediate and infinitely distant generations.

Finally Luc Lauwers studies the history of the axiomatic approach to the ranking of infinite streams that starts with Koopmans' (1960) characterization of the discounted utilitarian rule. In **The axiomatic approach to the ranking of infinite streams** the author argues that this rule, however, while it meets Chichilnisky's axiom of dictatorship of the present it sets asides future generations. Recently, Lauwers (2010) and Zame (2007) uncovered the impossibility to combine in a constructible way the requirements of equal treatment, sensitivity, and completeness. This contribution presents and discusses different axioms proposed to guide the ranking of infinite streams and the criteria they imply. The literature covered in this overview definitely points towards a set of meaningful alternatives to discounted utilitarianism.

4 Part III. The Environment in a Global Context

Thinking about policy in a context of global change is a challenging task. In one of her last articles, Elinor Ostrom's **Nested externalities and polycentric institutions: must we wait for global solutions to climate change before taking actions at other scales?** argues that the literature on global climate change has largely ignored the small but positive steps that many public and private actors are taking to reduce greenhouse gas emissions. She says that a global policy is frequently posited as the *only* strategy needed. The author argues that it is important to balance the major attention on global solutions as the only strategy for coping with climate change. Positive actions are underway at multiple, smaller scales to start the process of climate change mitigation, and researchers need to understand the strength of polycentric systems where enterprises at multiple levels may complement each other. Building a global regime is a necessity, but encouraging the emergence of a polycentric system starts the process of reducing greenhouse gas emissions and acts as a spur to international regimes to do their part.

In **Capital growth in a global warming model: will China and India sign a climate treaty?** Prajit Dutta and Roy Radner point out that global warming is now recognized as a significant threat to sustainable development on an international scale. They argue that one of the key challenges in mounting a global response to it is the seeming unwillingness of the fastest growing economies such as China and India to sign a treaty that limits their emissions. The aim of their paper is to examine the differential incentives of countries on different trajectories of capital growth. A benchmark dynamic game to study global warming, introduced in Dutta and Radner (2009), is generalized to allow for exogenous capital accumulation. It is shown that the presence of capital exacerbates the "tragedy of the common". Furthermore, even with high discount factors, the threat of reverting to the inefficient "tragedy" equilibrium is not sufficient to deter the emissions growth of the fastest growing economies—in contrast to standard folk theorem like results. However, foreign aid can help. If the slower growth economies—like the United States and Western Europe—are willing to make transfers to China and India, then the latter can be incentivized to cut emissions. Such an outcome is Pareto improving for both slower and faster growth economies.

The ethical foundations of climate change policy are also the topic of Franck Lecocq and Jean-Charles Hourcade's article **Unspoken ethical issues in the climate affair: Insights from a theoretical analysis of negotiation mandates**. Taking climate change as an example, their article provides new insights on the optimal provision of a long-term public good within and across generations. They consider the Bowen–Lindhal–Samuelson (BLS) conditions for the optimal provision of the public good in a world divided into N countries, with two periods, present and future, and we simultaneously determine the optimal response in the first and second periods for a given rate of pure time preference. However, the Negishi weights at second period cannot be determined unambiguously, even under a "no redistribution constraint" within each generation, because they depend on non-observable future incomes; and thus on the answers to two often-overlooked ethical questions: (i) Do rich countries agree on deals which recognize that developing countries may catch up with developed countries in the long run, or do they use their negotiating powers to preserve the current balance of power? And (ii) does each country consider only the welfare of its own future citizens (dynastic solidarity) or does it extend its concern to all future human beings (universal solidarity)? Answers to (i) and (ii)—critical in the debate about how to correct the market failures causing global warming—define four sets of Negishi weights and intertemporal welfare functions, which we interpret as four mandates that countries could give to the Chair of an international negotiation on climate change to find an optimal solution. Leqcocq and Hourcade find that in all mandates, public good provision expenditures are decreasing functions of income at first period. But each mandate leads to a different allocation of expenditures at second period and to different optimal levels of public good provision at both first and second periods. Finally, they show that only one of these four mandates defines a space for viable compromises. The effectiveness of international climate policy to curb carbon emissions has also been challenged because of possible "carbon leakages", which refers to the rise of emissions in non-participating countries.

In **Carbon leakages: a general equilibrium view** Jean-Marc Burniaux and Joaquim Oliveira offers a general equilibrium (GE) exploration of the key mechanisms and factors underlying the size of such carbon leakages. They develop a two-region, two-goods simplified GE framework, incorporating three types of fossil fuels (coal, oil and low-carbon energy), international trade and capital mobility. The model is designed to make tractable extensive multidimensional sensitivity analysis. The results suggest that the coal supply elasticity plays a critical role, while substitution elasticities between traded goods and international capital mobility appear relatively less influential. The shape of the production function also matters for the size of the leakages. Confirming the results obtained with large computable GE models, for a wide range of parameters' values, overall carbon leakages appear to be small. Therefore, the argument that unilateral carbon abatement action taken by a large group of countries (such as the Annex 1 group) is flawed by significant carbon leakages is not supported by Burniaux and Oliveira's sensitivity analysis. The likelihood of small leakage effects strengthens the call for the formation of a worldwide coalition to stabilize atmospheric carbon levels and halt climate change.

5 Part IV. The Case of Climate Change

The issue of climate change is the most pressing global policy issue today. For decades the world economy has been on an unsustainable trajectory with worldwide carbon emissions projected to increase beyond the mid of this century. The chapters of this section discuss the case of climate change from various points of view: natural science, social choice theory, and welfare economics. The IPCC (2014) recently projected that, due to years of inaction, reductions of carbon emissions are not enough anymore. The removal of carbon from the atmosphere is now needed in order to avert catastrophic climate change and has become a critical issue in climate policy. Peter Eisenberger tackles the issue in the article **Chaos Control—Climate Stabilization by Closing the Global Carbon Cycle**. The central idea behind the control of chaotic systems is that the same feedbacks that destabilize a complex system producing chaotic dynamics can be used to relatively easily stabilize it, drawing on an idea developed by Chichilnisky earlier. While many argue that the carbon cycle feedbacks are destabilizing the climate, Eisenberger argues that those same feedbacks can be used to stabilize the climate. The controlling variable is the amount of CO_2 in the atmosphere and the control strategy is to close the global carbon cycle of our planet, including human and planetary components, so the ambient concentration is fixed. The stabilization using CO_2 capture from or release to the atmosphere requires less energy per year than the amount used to stabilize the climate in our buildings and for less cost than 1 % of the global GDP. What is more, closing the carbon cycle by using carbon from the air and combining it with hydrogen from water to produce a new energy sources and thus removing the current negative feedback between our energy use and the planet, can enable us to use as much energy as we need for economic growth and well-being. Chaos control of our climate transforms the threat of climate change into an opportunity for our species and our planet to flourish in the Anthropocene era.

According to Norman Schofield, the enlightenment period was a philosophical project to construct a rational society without the need for a supreme being. It opened the way for the creation of market democracy and rapid economic growth. At the same time economic growth is the underlying cause of climate change, and we have become aware that this may destroy our civilization. In **Climate Change and Social Choice Theory** Schofield argues that the principal underpinning of the enlightenment project is the general equilibrium theorem (GET) of Arrow and Debreu (1954), asserting the existence of a Pareto optimal price equilibrium. Arrow's work in social choice can be interpreted as an attempt to construct a more general social equilibrium theorem. His article surveys recent results in social choice which suggests that chaos rather than equilibrium is generic. He also considers models of belief aggregation similar to Condorcet's Jury theorem and mentions Penn's Theorem on the existence of a belief equilibrium. However, Schofield argues that a belief equilibrium with regard to the appropriate response to climate change depends on the creation of a fundamental social principle of "guardianship of our planetary home." Schofield suggests that this will involve conflict between entrenched economic interests and ordinary people, as the effects of climate change make themselves felt in many countries.

Larry Karp's article studies the policy implications of standard economic theory in **Discounting and the evaluation of climate policy.** His essay discusses the relation between utility discounting and climate policy, returning to the problem of Koopmans' impatience axiom. Karp uses simple cost-benefit analysis to show that the planner's willingness to pay for the elimination of a climate event is greater, but less sensitive to discounting, when the event is random instead of deterministic. Examples in an optimizing setting show that policy may be less sensitive to discounting the more nonlinear is the underlying model. He then explains why, in general, there should be no presumption that the risk of catastrophe swamps discounting in the assessment of climate policy and concludes by pointing out that intertemporal transfers between the same agent at different points in their life, and transfers between different agents at different points in time, are qualitatively different, and should not be assessed using the same discount rate.

In **Global warming and economic externalities** Armon Rezai, Duncan Foley, and Lance Taylor return to the problem of the open carbon cycle and argue that despite important and worldwide policy efforts such as the Kyoto Protocol, the emission of greenhouse gases (GHG) remains a large negative externality. Rezai et al. point out that economic equilibrium paths in the presence of such an uncorrected externality are inefficient and that, as a consequence, there is no real economic opportunity cost to correcting this externality by mitigating global warming. The conclusion is that mitigation investment using resources diverted from conventional investments can be Pareto improving for present and future generations in the sense of raising the economic well-being of both current and future generations. The authors argue that the economic literature on GHG emissions misleadingly focuses attention on the intergenerational equity aspects of mitigation by using a hybrid constrained optimal path as the "business-as-usual" benchmark. In a simple, calibrated Keynes-Ramsey growth model they then illustrate the significant potential Pareto improvement from mitigation investment and the equilibrium concept appropriate to modeling an uncorrected negative externality.

6 Part V. Economic Policy and Regulation

Having discussed the foundations of catastrophic risks in economic theory, articles in this section discuss the implications for economic policy particularly with respect to the risks posed by climate change. The article **Detrimental externalities, pollution rights, and the "Coase theorem"** by John Chipman and Guoqiang Tian revisits the old question of taxes vs. quotas. Extending Chipman (1998), they analyze a simple model formulated by Hurwicz (1995) of two agents—a polluter and a pollutee—and two commodities—"money" (standing for an exchangeable private good desired by both agents) and "pollution" (a public commodity desired by the polluter but undesired by the pollutee). A government issues legal rights to the two agents to emit a certain amount of pollution, which can be bought and sold with money and it is assumed that both agents act as price-takers in the market for pollution rights, so that competitive equilibrium is possible. The "Coase theorem" (Stigler 1966) asserts that the equilibrium amount of pollution is independent of the allocation of pollution rights. A sufficient condition for this was in another context already obtained by Edgeworth (1925), namely that preferences of the two agents be "parallel" in the money commodity, whose marginal utility is constant. Hurwicz (1995) argued that this parallelism is also necessary and Chipman and Guoqiang critically discuss these results and provide an alternative set of necessary and sufficient conditions.

Larry Karp and Jiangfeng Zhang also discuss the emissions taxes or quotas choice in a setting where a (strategic) regulator and (non-strategic) firms have asymmetric information about abatement costs, and all agents use Markov perfect decision rules. In **Taxes versus quantities for a stock pollutant with endogenous abatement costs and asymmetric information** firms make investment decisions that affect their future abatement costs. For general functional forms, firms' investment policy is information-constrained efficient when the regulator uses a quota, but not when the regulator uses an emissions tax. This advantage of quotas over emissions taxes has not previously been recognized. For a special functional form (linear–quadratic) both policies are constrained efficient. Using numerical methods, Karp and Zhang find that there is no simple answer to the question of which instrument to use.

In **Walrasian prices in markets with tradable rights** Carlos Hervés-Beloso, Francisco Martínez, and Jorge Rivera consider an exchange economy where there is an external restriction for the consumption of goods. This restriction is defined by both, a cap on consumption of certain commodities and the requirement of an amount of rights for the consumption of these commodities. The caps for consumption are imposed exogenously due to the negative effects that the consumption may produce. The consumption rights or licenses are distributed among the agents. This fact leads to the possibility of establishing license markets. These licenses do not participate in agents' preferences, however, the individual's budgetary constraint may be modified, leading to a reassignment of resources. The authors' aim is to show the existence of a Walrasian equilibrium price system linking tradable rights prices with commodity prices.

7 Part VI. Catastrophic Risk in Economic Practice

Catastrophic risks have far-reaching implications for existing economic approaches and will force significant reconsideration of the fundamental assumptions. The chapters of this section present empirical and experimental evidence for this pressing necessity. In **Exploring the role of emotions in decisions involving catastrophic risks: Lessons from a double investigation** Olivier Chanel, Graciela Chichilnisky, Sébastien Massoni, and Jean-Christophe Vergnaud report on experimental results about how natural disasters due to climate change (like floods, hurricanes, heat waves or droughts) combine a risk of large losses and a low probability of occurrence, and require decisions to be made in highly uncertain universes. They highlight the inability of standard decision under uncertainty models to provide rankings when some outcomes are catastrophic impedes rational (public) decision-making. This paper examines the role of emotions in individuals' choices among alternatives involving catastrophic events, either in real life (flooding) or artificial (laboratory experiment) situations. The author report a survey on 599 respondents aimed at determining how people exposed to different levels of flood risk form beliefs and make decisions under uncertainty before and after emotion-generating events. Data on their emotions, the emotions they expect to experience, their personality and psychological determinants, their symptoms before and after emotion-generating events are collected and analyzed. In parallel with this survey, experimental protocols replicate the emotional experience of a catastrophe and measure its impact on behavior and formation of beliefs. Emotions are induced by framing effects and measured through a self-declared worry scale. The authors collect behavioral data (insurance choice, subjective beliefs, performance) and measure how they are affected by the emotions felt during the decision-making. These protocols test some assumptions in the survey using experimental paradigms from psychophysics that allow us to control the sources of uncertainty experienced by the subjects. The results reported confirm that emotions connected with the nature of the risk can significantly affect desire to reduce it. The survey provides valuable material for comparative analysis, revealing how actual experience of an anticipated event affects decisions. The experiments show that emotions affect the decision-making process and the forming of probabilistic beliefs.

In **How the change of risk announcement on catastrophic disaster affects property prices** Hayato Nakanishi presents evidence of the benefits of expected utility that is sensitive to rare events. Specifically, he estimate the treatment effect of reports on rare catastrophic tsunamis by using a land price hedonic approach. To identify this effect, he employs a difference in differences (DD) design. Ordinal expected utility predicts no effect since the probability of catastrophic events is sufficiently small to be ignored. The estimation results, however, reveal a significant effect. Just the results of Chanel et al., this implies that ordinal expected utility derived from objective risk distribution may be insufficient for the study of the economic implications of rare catastrophic events.

Finally in **Modeling US stock market volatility-return dependence using conditional Copula and quantile regression** Temisan Agbeyegbe examines the return-volatility relationship for some indices reported on exchanges in the United States of America. He utilizes both linear quantile regression and copula quantile regression to evaluate the asymmetric volatility-return relationship between changes in the volatility index (VXD, VIX, VXO and VXN) and the corresponding stock index return series (DJIA, S&P 500, the S&P 100 and NASDAQ), as the quantile copula models allow for inference at different quantiles of interest. Agbeyegbe finds, first, that the relationship between stock return and implied volatility depends on the quartile at which the relationship is being investigated. Second, he obtains results similar to those reported for European exchanges showing the existence of an inverted U-shaped relationship between stock return and implied volatility. This result was obtained even after controlling for changes in volatility of return using a GARCH (1,1) filter.

In consonance with the Introduction to this volume and Peter Eisenberger's article, Alan Kirman argues in **Economic Crises: Natural or Unnatural Catastrophes?** that whilst models of the environment and particularly of the climate represent evolving complex systems with non-linear dynamics and complicated feedbacks, macroeconomic models have remained essentially in an equilibrium framework in which the only major changes that can occur are the result of exogenous shocks. Kirman explains why this has been the route taken by macroeconomists and in his article he suggests that the economy shares many of the features of the environment and that it should also be viewed as a complex system which is prone to experience major, sudden and sometimes catastrophic, changes. These changes are largely due to the endogenous evolution of the system and not only to outside influence. In the Anthropocene we have to take account of the co-evolution of two complex systems, the environment and the economy, and the economic models that have been proposed as "integrated" models do not capture the complexity of the economy nor of its interactions with the environment. According to Kirman, successfully doing this will provide a better explanation of the evolution of the economy but will also imply that economists have to be much more modest their claims.

References

Arrow, K. (1971). *Essays in the theory of risk-bearing*. Amsterdam, The Netherlands: North-Holland.

Arrow, K. J., & Debreu, G. (1954). Existence of an equilibrium for a competitive economy. *Econometrica, 22*(3), 265–290.

Chichilnisky, G. (1995). Limited arbitrage is necessary and sufficient for the existence of competitive equilibrium with or without short sales. *Economic Theory, 95*, 79–108.

Chichilnisky, G. (1996a). Updating von Neumann Morgenstern Axioms for Choice under Uncertainty, *Proceedings of a Conference on Catastrophic Risks*. Toronto: The Fields Institute for Mathematical Sciences.

Chichilnisky, G. (1996b). An axiomatic approach to sustainable development. *Social Choice and Welfare, 13*(2), 231–257.

Chichilnisky, G. (2000). An axiomatic approach to choice under uncertainty with catastrophic risks. *Resource and Energy Economics, 22*, 221–231.

Chichilnisky, G. (2002). Catastrophic risk. In A. H. El-Shaarawi & W. W. Piegorsch (Eds.), *Encyclopedia of environmetrics* (vol. 1, pp. 274–279) Chichester, UK: Wiley.

Chichilnisky, G. (2009). The topology of fear. *Journal of Mathematical Economics, 45*, 807–816.

Chichilnisky, G. (2010). The foundations of statistics with black swans. *Mathematical Social Sciences, 59*, 184–192.

Chichilnisky, G., & Heal, G. (1997). Social choice with infinite populations. *Social Choice and Welfare, 14*, 303–319.

Chipman, J. S. (1998). A close look at the Coase theorem. In J. M. Buchanan & B. Monissen (Eds.), *The economists' vision. Essays in modern economic perspectives* (pp. 131–162). Frankfurt am Main: Campus Verlag.

De Groot, M. (1970). *Optimal statistical decisions*. New York, NY USA: McGraw-Hill.

Dubins, L. (1975). Finitely additive conditional probabilities, conglomerability and disintegration. *Annals of Probability, 3*, 89–99.

Dubins, L., & Savage, L. (1965). *How to Gamble if you Must*. New York: McGraw Hill.

Dutta, P., & Radner, R. (2009). A strategic model of global warming: theory and some numbers. *Journal of Economic Behaviour and Organization, 71*(2), 187–209.

Edgeworth, F. Y. (1925). On the determinateness of economic equilibrium. In: *Papers Relating to Political Economy*, London: Macmillan and Co., Limited, vol. II, 313–319.

Gödel, K. (1940). The Consistency of the Continuum Hypothesis, *Annals of Mathematics Studies* (vol. 3) Princeton, NJ, USA: Princeton University Press.

Hammond, P. (1994). Elementary non-archimedean representations of probability for decision theory and games. In: P. Humphreys (Ed.), *Patrick Suppes: Scientific Philosopher, Vol. I: Probability and Probabilistic Causality* (pp. 25–59). Kluwer Academic Publishers), ch. 2.

Hurwicz, L. (1995). What is the coase theorem? *Japan and the World Economy, 7*, 49–74.

IPCC (2014). Climate Change 2014, AR5 Synthesis Report. Accessed 14 Nov 2014 http://www.ipcc.ch/pdf/assessment-report/ar5/syr/SYR_AR5_SPM.pdf.

Jones-Lee, M. (1974). The Value of Changes in the Probability of Death or Injury. *Journal of Political Economy, 82*, 835–849.

Kolmogorov, A. N. (1933/1950). *Foundations of the theory of probability*, Chelsea Publishing Co., New York.

Koopmans, T. (1960). Stationary ordinal utility and impatience. *Econometrica, 28*, 287–309.

Lauwers, L. (2010). Ordering infinite utility streams comes at the cost of a non-Ramsey set. *Journal of Mathematical Economics, 46*, 32–37.

Narens, L. (2009). A foundation for support theory based on a non-Boolean event space. *Journal of Mathematical Psychology, 53*, 399–407.

Narens, L. (2011). Probabilistic lattices: theory with an application to decision theory. In: Dzhafarov et al. (Eds.), *Descriptive and normative approaches to human behavior* (pp. 161–202).

Purves, R. W., & Sudderth, W. D. (1976). Some finitely additive probability. *Annals of Probability, 4*, 259–276.
Savage, L. (1954). *The Foundations of Statistics*. New York: Wiley.
Stigler, G. J. (1966). *The Theory of Price* (3rd ed.). London: Collier-Macmillan Limited.
Villegas, C. (1964). On quantitative probability σ-algebras. *The Annals of Mathematical Statistics, 35*, 1789–1800.
Von Neumann, J., & Morgernstern, O. (1944). *Theory of Games and Economic Behavior*. New Jersey: Princeton University Press.
Zame, W. R. (2007). Can intergenerational equity be operationalised. *Theoretical Economics, 2*, 187–202.

Part I
Catastrophic Risk in Economic Theory

Catastrophic Risk, Rare Events, and Black Swans: Could There Be a Countably Additive Synthesis?

Peter J. Hammond

1 Introduction

1.1 Countably Additive Subjective Probability

In his classic *The Foundations of Statistics*, Savage (1954) provided an axiom system sufficient to imply that a decision maker's preferences could be represented by the expected value of a von Neumann–Morgenstern utility function, with personal or subjective probabilities attached to unknown events ranging over a sample space of states of the world. His axioms, however, implied that probabilities are only finitely rather than countably additive. Yet countable additivity is a key measure-theoretic property that probabilitists since the time of Kolmogorov (1933), at least, are accustomed to using. It allows, for instance, the probability of an interval of the real line to be found by integrating a density function over that interval.

1.2 Monotonicity

To bridge this gap between finite and countable additivity, Villegas (1964), Arrow (1965) and Fishburn (1982) all introduced an additional monotonicity axiom ensuring that subjective probabilities are countably additive. One version of the relevant axiom can be derived by combining slightly modified versions of two axioms set out in Sects. 7.2 and 7.3 of Hammond (1998b).

2015 May 1st, typeset from pjhForAFOSR2.tex.

P.J. Hammond (✉)
Department of Economics and CAGE (Centre for Competitive Advantage in the Global Economy), University of Warwick, Coventry CV4 7AL, UK
e-mail: p.j.hammond@warwick.ac.uk

G. Chichilnisky and A. Rezai (eds.), *The Economics of the Global Environment*, Studies in Economic Theory 29, DOI 10.1007/978-3-319-31943-8_2

Let $(Y, \mathcal{F})$ denote a measurable space of *consequences*, and $(S, \mathcal{S})$ a measurable space of *states of the world*. Following the evocative terminology introduced by Anscombe and Aumann (1963), let $\Delta(Y, \mathcal{F})$ denote the space of *roulette lotteries* in the form of probability measures over $(Y, \mathcal{F})$.

In the special case when S is a finite set, for each $E \subseteq S$, let Y^E denote the Cartesian product set $\prod_{s \in E} Y_s$, where each Y_s is a copy of Y, and let $\mathcal{F}^E$ denote the product σ-field $\bigotimes_{s \in E} \mathcal{F}_s$, where each $\mathcal{F}_s$ is a copy of $\mathcal{F}$. Consider then the space $\Delta(Y^E, \mathcal{F}^E)$ of roulette lotteries, in the form of probability measures over $(Y^E, \mathcal{F}^E)$ whose random outcomes are *horse lotteries* y^E in the space of measurable mappings from E to $(Y, \mathcal{F})$.

The key *reversal of order axiom* (RO) due to Anscombe and Aumann (1963) treats, for any event $E \subseteq S$, any pair $\lambda^E, \mu^E \in \Delta(Y^E, \mathcal{F}^E)$ as equivalent if and only if their marginal measures $\lambda_s, \mu_s \in \Delta(Y, \mathcal{F})$ are equal for each state $s \in E$. Then each $\pi^E \in \Delta(Y^E, \mathcal{F}^E)$ can be identified with the list $\langle \pi_s \rangle_{s \in E}$ of marginal probability measures $\pi_s \in \Delta(Y, \mathcal{F})$. In particular, this treats as irrelevant the extent of any correlation between consequences $y_s \in Y$ that arise in different states $s \in E$.

Next, we revert to the case of a general measurable space $(S, \mathcal{S})$. Then, for each measurable event $E \in \mathcal{S}$, define the conditional sub-σ-field

$$\mathcal{S}_{|E} := \{G \in \mathcal{S} \mid G \subseteq E\} \subseteq \mathcal{S}$$

Obviously, in case $E = S$, this definition implies that $\mathcal{S}_{|S} = \mathcal{S}$.

In the spirit of the case when S is a finite set, for each measurable event $E \in \mathcal{S}$, let $\Delta(Y^E, \mathcal{S}_{|E}, \mathcal{F})$ denote the space of functions $\pi^E : E \to \Delta(Y, \mathcal{F})$ with the property that, for each $K \in \mathcal{F}$, the mapping

$$E \ni s \mapsto \pi^E(s, K) \in \mathbb{R}_+$$

is measurable w.r.t. the σ-field $\mathcal{S}_{|E}$ on E and the Borel σ-field on $\mathbb{R}$.

The other axioms to be discussed here concern:

1. the preference ordering $\succsim^*$ on $\Delta(Y, \mathcal{F})$ having the property that for each $y \in Y$, in addition to the set $\{y\}$, the upper and lower contour sets

$$\{y' \in Y \mid \delta_{y'} \succsim^* \delta_y\} \quad \text{and} \quad \{y' \in Y \mid \delta_{y'} \precsim^* \delta_y\}$$

   are both $\mathcal{F}$-measurable;
2. for each measurable event $E \in \mathcal{S}$, the conditional preference ordering $\succsim^E$ on $\Delta(Y^E, \mathcal{S}_{|E}, \mathcal{F})$.

Definition 1 (*Event Dominance (ED)*)

Suppose that the event $E \in \mathcal{S}$, the list of probability measures $\pi^E = \langle \pi_s \rangle_{s \in E} \in \Delta(Y^E, \mathcal{S}_{|E}, \mathcal{F})$, and the simple lottery $\lambda \in \Delta(Y)$ are all given. Let $\lambda\, 1^E$ denote the particular list $\lambda^E = \langle \lambda_s \rangle_{s \in E} \in \Delta(Y^E, \mathcal{S}_{|E}, \mathcal{F})$ that satisfies $\lambda_s = \lambda$ for all $s \in E$. Then:

1. $\pi_s \succsim^* \lambda$ (all $s \in E$) implies $\pi^E \succsim^E \lambda\, 1^E$;
2. $\pi_s \precsim^* \lambda$ (all $s \in E$) implies $\pi^E \precsim^E \lambda\, 1^E$.

In case the set E is finite, condition (ED) is an obvious implication of Anscombe and Aumann's extension of Savage's sure thing principle. The force of (ED) comes in partially extending this principle to the case when E is any measurable subset of S.

Next, given any measurable event $E \in \mathcal{S}$ satisfying $\emptyset \neq E \neq S$, let $(\pi\, 1^E, \tilde{\pi}\, 1^{S\setminus E})$ denote the particular list of probability measures $\lambda^S = \langle \lambda_s \rangle_{s \in S} \in \Delta(Y^E, \mathcal{S}, \mathcal{F})$ whose marginal distribution $\lambda_s \in \Delta(Y, \mathcal{F})$ for each $s \in S$ is a roulette lottery that satisfies

$$\lambda_s = \begin{cases} \pi & \text{if } s \in E \\ \tilde{\pi} & \text{if } s \in S\setminus E \end{cases}$$

Definition 2 (*Event Continuity (EC)*)

Let $\succsim^*$ on $\Delta(Y, \mathcal{F})$ and $\succsim^S$ on $\Delta(Y^S, \mathcal{S}^S, \mathcal{F})$ be fixed preference orderings. Suppose that the two measurable events $E, E^* \subset S$, as well as the sequence of measurable events E_k $(k \in \mathbb{N})$, and the two probability measures $\pi, \tilde{\pi} \in \Delta(Y, \mathcal{F})$, together satisfy:

1. $E_1 \subset E_2 \subset \cdots \subset E_k \subset E_{k+1} \subset \cdots \subset S$;
2. $E^* = \cup_{k=1}^{\infty} E_k$;
3. $\pi \succ^* \tilde{\pi}$;
4. $(\pi\, 1^{E^*}, \tilde{\pi}\, 1^{S\setminus E^*}) \succ^S (\pi\, 1^E, \tilde{\pi}\, 1^{S\setminus E})$.

Then there must exist a finite k such that $(\pi\, 1^{E_k}, \tilde{\pi}\, 1^{S\setminus E_k}) \succ^S (\pi\, 1^E, \tilde{\pi}\, 1^{S\setminus E})$.

Equivalently,

$$(\pi\, 1^{E_k}, \tilde{\pi}\, 1^{S\setminus E_k}) \succsim^S (\pi\, 1^E, \tilde{\pi}\, 1^{S\setminus E}) \text{ (all } k \in \mathbb{N}) \\ \Longrightarrow (\pi\, 1^{E^*}, \tilde{\pi}\, 1^{S\setminus E^*}) \succsim^S (\pi\, 1^E, \tilde{\pi}\, 1^{S\setminus E})$$

1.3 Beyond Monotonicity

In several recent papers, Chichilnisky (1996, 2000, 2009, 2010) has explored a particular weakening of this kind of monotonicity axiom. This weakening allows a revised decision theory in which rare events, catastrophes, perhaps even "black swans", can all be given more prominence. Of course, the weakening comes at the cost of allowing probabilities that are only finitely additive. For this reason, ultimately it may be useful to investigate whether some alternative approach could allow for such phenomena while retaining probabilities that are countably additive measures.

1.4 Outline of Paper

The rest of this paper considers three different strands of literature. First, Sect. 2 considers some background on the use of the word “catastrophe”, in drama, mathematics, and finally decision theory. It goes on to formalize a notion of catastrophic risk in decision theory, based on pioneering work on the value of life due to Drèze (1962), followed by Jones-Lee (1974).

The second strand discussed in Sect. 3 concerns the use of infinitesimals to represent the subjective probability of events so rare that they should not be accorded any positive probability. Third, Sect. 4 offers a possible approach to modelling the “true black swans” that Taleb (2007) in particular regards as beyond any kind of systematic analysis. Finally, Sect. 5 combines a suggestion for an alternative synthesis of these three strands with some concluding remarks.

2 Catastrophic Risk

2.1 Etymology

According to http://www.etymonline.com/, the word “catastrophe” entered the English language during the 1530s with the meaning “‘reversal of what is expected’ (especially a fatal turning point in a drama)”. It is derived from the Greek “katastrophe”, meaning “overturning; a sudden end”, itself a compound of the prefix “kata” meaning “down” and “strephein” meaning “turn”.

The extension of the meaning of “catastrophe” to include “sudden disaster” is first recorded in 1748. In medicine, catastrophe is often taken to mean death related to what should have been routine surgery. In engineering, a “catastrophic failure” is the complete breakdown of a system from which recovery is impossible. A celebrated example is the Tay Bridge disaster of 1879 which, thanks to to the doggerel in McGonagall (1880), has become a classic of British folklore.

There is a branch of mathematics known as “catastrophe theory” that concerns the possible instability of the minimum of a non-linear potential function when that function depends on exogenous parameters which may be subject to sudden shocks. The monograph by Thom (1973) provided a systematic classification of different types of catastrophe. Zeeman (1976) did much to popularize the application of catastrophe theory to the study of many different dynamic phenomena where there is a sudden change. These applications include:

- in animal psychology, aggression in dogs;
- in medicine, the beating heart;
- in structural engineering, beams that first buckle and then collapse;
- in economics and finance, crashes in stock markets, as well as structural properties of the Walrasian equilibrium manifold as described by Balasko (1978).

2.2 Catastrophic Consequences

Standard decision theory considers acts whose consequences range over a specified *consequence domain* in the form of an abstract set Y equipped with a σ-algebra $\mathcal{F}$ of measurable sets. In principle, catastrophes can be described by letting the consequence domain Y be the union of the two disjoint measurable sets: (i) Y_0 of *non-catastropic* consequences; and (ii) Y_1 of *catastrophic* consequences.

Here, however, our concern will be to discuss how catastrophes can be modeled as events so extreme that a suitable money metric utility function becomes undefined whenever the probability of a catastrophe is sufficiently high. Accordingly, consider a *consequence domain* $K \times \mathbb{R}_+$ of pairs (κ, y) where:

1. $y \in \mathbb{R}_+$ is income or wealth (depending on context);
2. $\kappa \in K = \{0, 1\}$ is a binary indicator variable indicating whether a "catastrophe":
   - occurs, iff $\kappa = 1$;
   - or does not occur, iff $\kappa = 0$.

Hence $Y_0 = \{0\} \times \mathbb{R}_+$, whereas $Y_1 = \{1\} \times \mathbb{R}_+$,

Following Drèze (1962), consider too a consumer whose preference ordering $\succsim$ on the set $\Delta(K \times \mathbb{R}_+)$ of lotteries over $K \times \mathbb{R}_+$ is represented by the expected value $\mathbb{E}u$ of each real-valued von Neumann–Morgenstern utility function (or NMUF) $K \times \mathbb{R}_+ \mapsto (\kappa, y) \mapsto u(\kappa, y) \in \mathbb{R}$ in a unique cardinal equivalence class. The literature on decision theory inspired by Drèze often regards the mapping $y \mapsto u(\kappa, y)$ as a *state-dependent* utility function of income y, though it can perhaps be more usefully regarded as a *state-independent* utility function of the *fully specified consequence* (κ, y).

2.3 Assumptions

Within the framework of Sect. 2.2, we assume that:

1. for each fixed $\kappa \in K$, each NMUF $y \mapsto u(\kappa, y)$ is continuous, strictly increasing, and bounded above, with upper bound $\bar{u}_\kappa := \sup_y u(\kappa, y)$;
2. for each fixed $y \in \mathbb{R}_+$, one has $u(0, y) > u(1, y)$;
3. $\bar{u}_0 > \bar{u}_1$.

The second assumption, of course, is that the consumer is worse off with a catastrophe than without, *ceteris paribus*. Taking the limit as $y \to \infty$ implies that $\bar{u}_0 \geq \bar{u}_1$, obviously, but the third assumption that $\bar{u}_0 > \bar{u}_1$ strengthens this to a strict inequality. In particular, this third assumption holds if and only if there is a continuous extended utility function $\tilde{u} : K \times (\mathbb{R}_+ \cup \{\infty\}) \to \mathbb{R}$ for which there exists $y^* \in \mathbb{R}$ such that $\tilde{u}(0, y^*) = \tilde{u}(1, \infty)$ and so $\tilde{u}(0, y) > \tilde{u}(1, \infty)$ whenever $y > y^*$.

2.4 Money Metric Utility

Following Jones-Lee (1974), consider this consumer's willingness to pay for a reduction in the probability p of catastrophe. Specifically, consider any *reference* or *baseline* lottery

$$\lambda^R := (1 - p^R)\delta_{(0,y_0^R)} + p^R \delta_{(1,y_1^R)} \tag{1}$$

which is a mixture of the two degenerate lotteries $\delta_{(0,y_0^R)}$ and $\delta_{(1,y_1^R)}$ that attach probability one to the consequences $(0, y_0^R)$ and $(1, y_1^R)$ respectively. Thus, the consumer faces the probability p^R of a catastrophe, along with reference income levels y_κ^R ($\kappa \in \{1, 0\}$) with and without a catastrophe. Let

$$U^R := (1 - p^R)u(0, y_0^R) + p^R u(1, y_1^R) \tag{2}$$

denote expected utility in the reference situation. One can use these reference levels and the equation

$$(1 - p)u(0, m) + pu(1, y_1) = U^R \tag{3}$$

in an attempt to define implicitly a *money metric* utility function

$$\mathbb{R}_+ \times [0, 1] \ni (y_1; p) \mapsto m(y_1; p) \in \mathbb{R}_+ \tag{4}$$

Note that this function will be the same whenever u is replaced by an alternative NMUF that is cardinally equivalent.

Definition (4), when valid, implies that $m(y_1; p) - y_0^R$ is the consumer's *willingness to accept* the net increase $p - p^R$ in the risk of catastrophe, when compensation in the event of the catastrophe raises income from y_1^R to y_1. Alternatively, $y_0^R - m(y_1; p)$ is the consumer's (net) *willingness to pay*, in terms of foregone income in the absence of catastrophe, for the decrease in the probability of catastrophe from p^R to p.

2.5 A Critical Probability Level: Catastrophic Risk

The money metric utility function (4) really is defined by Eq. (3) for the pair $(y_1; p)$ if and only if

$$(1 - p)u(0, 0) + pu(1, y_1) \leq U^R.$$

Otherwise giving up all income is insufficient to compensate for the increase in p, which one could then regard as a *true* catastrophe.

In particular, the function (4) is defined iff $p \leq p_C$ for the *critical probability level* defined by

$$p_C := \frac{U^R - u(0,0)}{u(1,y_1) - u(0,0)} = \frac{(1-p^R)u(0,y_0^R) + p^R u(1,y_1^R) - u(0,0)}{u(1,y_1) - u(0,0)} \tag{5}$$

Thus, once p has reached p_C, no compensation is possible for any further increase in the probability of catastrophe.

Note that p_C, as the ratio of expected utility differences, is not only preserved under positive affine utility transformations. In addition, as discussed in Hammond (1998a), the formula (5) that expresses p_C as the ratio of utility differences implies that it must equal the constant marginal rate of substitution between shifts in probability away from (0, 0), the worst possible outcome without a catastrophe, toward respectively:

1. the reference lottery defined by (1);
2. the consequence $(1, y_1)$ that represents the occurrence of the catastrophe combined with the income level y_1.

2.6 Extreme Economic Catastrophes

One can also have an *extreme catastrophe* where p is large enough to satisfy

$$(1-p)u(0,0) + p\bar{u}_1 > U^R$$

This, of course, is equivalent to

$$p > \frac{U^R - u(0,0)}{\bar{u}_1 - u(0,0)} \tag{6}$$

Inequality (6) implies that the probability of catastrophe is so high that no matter how large y_1 may be, there is no value of m that satisfies (3). In this sense, compensation is completely impossible.

3 Rare Events

3.1 Standard Decision Theory

Standard decision theory uses the expected utility (EU) criterion. Traditionally, moreover, a distinction is made between *objective* and *subjective* EU theory, depending on whether one faces:

- *risk* or *roulette lotteries* described by *objective* probabilities, as in von Neumann and Morgenstern (1953) and then Jensen (1967);
- *uncertainty* or *horse lotteries* described by *subjective* probabilities, as in Savage (1954);
- combinations of roulette and horse lotteries, as in Anscombe and Aumann (1963).

3.2 *Infinitesimal Probability*

Recall that, by definition, an *infinitesimal* ϵ is some positive entity (not a real number) that is smaller than any positive real number in the sense that $0 < n\epsilon < 1$ for all natural numbers $n \in \mathbb{N}$. To accommodate rare events, one can follow the game-theoretic literature emanating from Selten (1975) by allowing "trembles" whose probability is taken to be some positive multiple of a particular *basic infinitesimal* ϵ. See Halpern (2009, 2010) for discussion of some recent developments.

3.3 *Rare Events and Infinitesimal Probabilities*

Probabilities must be:

1. *added* when calculating the probability of the union of two or more pairwise disjoint events;
2. *subtracted* when calculating the probability of the set-theoretic difference of any two events;
3. *multiplied* when compounding probabilities at successive stages of a stochastic process;
4. *divided* when calculating conditional probabilities.

This suggests that Selten's space of trembles should be enriched so that the extended probabilities we construct take values in an algebraic field, where all these four operations are well-defined—except, of course, when trying to divide by zero. This motivates the following definition:

Definition 3 A *polynomial function* of ϵ takes the form

$$P(\epsilon) \equiv \sum_{k \in K} p_k \epsilon^k = \sum_{j=1}^{r} p_{k_j} \epsilon^{k_j} \tag{7}$$

for some finite set $K = \{k_1, k_2, \ldots, k_r\} \subset \mathbb{Z}_+$, where $k_j < k_{j+1}$ for $j = 1, 2, \ldots, r-1$, and $p_k \neq 0$ for all $k \in K$. The *leading non-zero coefficient* of the polynomial (7) is p_{k_1}. The polynomial (7) is *positive* just in case $p_{k_1} > 0$.

A *rational function* of ϵ takes the form of a quotient $P(\epsilon)/Q(\epsilon)$ of two polynomial functions of ϵ, where the denominator $Q(\epsilon)$ is positive. Without loss of generality, the leading non-zero coefficient of $Q(\epsilon)$ can be normalized to 1.

Following Robinson (1973), define $\mathbb{R}(\epsilon)$ as the algebraic field whose members are rational functions of ϵ, equipped with the standard algebraic binary operations of addition and multiplication, as well as the additive identity 0 and the multiplicative identity 1. Define the *positive cone* $\mathbb{R}_+(\epsilon)$ of rational functions $P(\epsilon)/Q(\epsilon)$ as those where $P(\epsilon)$ as well as $Q(\epsilon)$ is a positive polynomial.

Following Rényi (1955, 1956) and associated ideas that were surveyed in Hammond (1994), rare events E in a finite set S of states of the world can be modelled formally as having infinitesimal probability $p(E;\epsilon)$ in an extended EU theory with "non-Archimedean" probabilities in the positive cone $\mathbb{R}_+(\epsilon)$ of the field $\mathbb{R}(\epsilon)$. That is, we must have $p(E;\epsilon) = P(\epsilon)/Q(\epsilon)$ where the coefficient of ϵ^0 in the polynomial (7) is zero. Obviously one requires the probability mapping $2^S \ni E \mapsto p(E;\epsilon) \in \mathbb{R}_+(\epsilon)$ to satisfy the *additivity condition* $p(E;\epsilon) \equiv p(E';\epsilon) + p(E'';\epsilon)$ whenever $E = E' \cup E''$ with $E' \cap E'' = \emptyset$, as well as the *normalization condition* $p(S;\epsilon) \equiv 1$.

3.4 A Metric Completion

As discussed in Hammond (1999), following an approach set out in Lightstone and Robinson (1975), the set $\mathbb{R}(\epsilon)$ of rational functions can be given a (real-valued) metric $d : \mathbb{R}(\epsilon) \times \mathbb{R}(\epsilon) \to \mathbb{R}_+$. This metric induces a very fine topology, according to which a sequence $r^{\mathbb{N}} = \langle r_n \rangle_{n\in\mathbb{N}}$ of real numbers converges to $r^* \in \mathbb{R}$ if and only if r_n is *eventually equal* to r^*—i.e., there exists $n^* \in \mathbb{N}$ such that $n \geq n^* \implies r_n = r^*$.

Let $\mathbb{R}^{\mathbb{N}}(\epsilon)$ denote the Cartesian product of countably many copies of the algebraic field $\mathbb{R}(\epsilon)$. The elements of $\mathbb{R}^{\mathbb{N}}(\epsilon)$ are infinite sequences $r^{\mathbb{N}}(\epsilon) = (r^n(\epsilon))_{n\in\mathbb{N}}$ of rational functions of ϵ. Following standard terminology in metric space theory, say that $r^{\mathbb{N}}(\epsilon) = (r^n(\epsilon))_{n\in\mathbb{N}}$ is a *Cauchy sequence* if for every small $\delta > 0$, there exists $n_\delta \in \mathbb{N}$ such that whenever $n', n'' \in \mathbb{N}$ with $n' > n_\delta$ and $n'' > n_\delta$, one has $d(r^{n'}(\epsilon), r^{n''}(\epsilon)) < \delta$.

Define the binary relation $\sim$ on the space of Cauchy sequences in $\mathbb{R}^{\mathbb{N}}(\epsilon)$ so that $r^{\mathbb{N}}(\epsilon) \sim \tilde{r}^{\mathbb{N}}(\epsilon)$ just in case, for every small $\delta > 0$, there exists $n_\delta \in \mathbb{N}$ such that whenever $n', n'' \in \mathbb{N}$ with $n' > n_\delta$ and $n'' > n_\delta$, one has $d(r^{n'}(\epsilon), \tilde{r}^{n''}(\epsilon)) < \delta$. It is easy to check that the relation $\sim$ is symmetric, reflexive, and transitive — i.e., it is an *equivalence relation*. Then the metric space $(\mathbb{R}(\epsilon), d)$, like any other, has a *metric completion* consisting of equivalence classes of Cauchy sequences. In Hammond (1997) it is shown that each member of this metric completion can be expressed uniquely as a *power series* $\sum_{k=0}^{\infty} a_k \epsilon^k$ of the basic infinitesimal ϵ, for an infinite sequence $a^{\mathbb{N}} = (a_k)_{k\in\mathbb{N}} \in \mathbb{R}^{\mathbb{N}}$ of real constants. We denote this metric completion by $(\mathbb{R}^\infty(\epsilon), d)$, where d denotes an obvious extension to the set $\mathbb{R}^\infty(\epsilon)$ of power series of the original metric d on the set $\mathbb{R}(\epsilon)$ of rational functions.

In the following, let $\mathbb{R}^\infty_+(\epsilon)$ denote the subset of power series that are *positive* in the sense that the leading non-zero coefficient is positive. We also introduce the *lexicographic strict ordering* $>_L$ on $\mathbb{R}^\infty(\epsilon)$, defined so that

$$\sum_{k=0}^{\infty} a_k \epsilon^k >_L \sum_{k=0}^{\infty} b_k \epsilon^k$$

if and only if the leading non-zero coefficient of the difference

$$\sum_{k=0}^{\infty} (a_k - b_k) \epsilon^k$$

is positive. Let $\geq_L$ denote the corresponding weak ordering defined so that

$$\sum_{k=0}^{\infty} a_k \epsilon^k \geq_L \sum_{k=0}^{\infty} b_k \epsilon^k \iff \sum_{k=0}^{\infty} b_k \epsilon^k \not>_L \sum_{k=0}^{\infty} a_k \epsilon^k$$

3.5 Extended Probability Measures

In order to treat compound lotteries in decision trees where branches at one or more successive chance nodes can have infinitesimal probabilities, and also to have a satisfactory theory of subjective probability, it seems desirable to allow probabilities to have values in $\mathbb{R}_+^\infty(\epsilon)$ rather than just in $\mathbb{R}_+$.

Definition 4 Let $(S, \mathcal{S})$ be any measurable state space S with σ -field $\mathcal{S}$. An *extended probability measure* on $(S, \mathcal{S})$ is a mapping

$$\mathcal{S} \ni E \mapsto \pi(E; \epsilon) = \sum_{k=0}^{\infty} \pi_k(E) \epsilon^k \in \mathbb{R}^\infty(\epsilon)$$

that satisfies:

1. $\pi(E; \epsilon) \in \mathbb{R}_+^\infty(\epsilon)$ for all $E \in \mathcal{S} \setminus \{\emptyset\}$;
2. $\pi(S; \epsilon) = 1$;
3. if the countable collection of sets E_n $(n \in \mathbb{N})$ is pairwise disjoint, then $\pi(\cup_n E_n; \epsilon) = \sum_n \pi(E_n; \epsilon)$ (countable additivity).

Let $\Delta(S, \mathcal{S}; \mathbb{R}_+^\infty(\epsilon))$ denote the family of all extended probability measures on $(S, \mathcal{S})$.

Note that, apart from having values in the algebraic field $\mathbb{R}^\infty(\epsilon)$, such probabilities are required to be positive for all possible events; a zero probability is attached only to the empty set.

3.6 Extended Subjective Expected Utility

For the case when S is finite, Hammond (1997) offers axioms which imply that a preference ordering $\succsim$ over the space $\Delta(Y^S)$ of all possible combination of roulette and horse lotteries can be represented by the lexicographic weak ordering $\geq_L$ applied to subjectively expected utility, in the form of a power series

$$\sum_{y^S \in Y^S} \lambda(y^S) \sum_{s \in S} \pi(s; \epsilon) v(y_s) \in \mathbb{R}^\infty(\epsilon)$$

Note in particular that the von Neumann–Morgenstern utility function (or NMUF) $v : Y \to \mathbb{R}$ is *real* valued; there is no need for any form of lexicographic utility, as opposed to lexicographic expected utility. The following is the main theorem of Hammond (1997):

Theorem 1 *Let S denote a finite set of unknown states of the world, and Y a consequence domain. Suppose that all the seven axioms (O), (I*), (C*), (RO), (SI), (RC) and (XC) of Hammond (1997) are satisfied throughout the domain $\Delta(Y^S; \mathbb{R}^\infty_+(\epsilon))$ of consequence lotteries with non-Archimedean objective probabilities ranging over $\mathbb{R}^\infty_+(\epsilon)$. Unless there is universal indifference over the whole domain, there exist*

- *a unique extended subjective probability measure $p(\cdot; \epsilon)$ that belongs to the space $\Delta(S, S; \mathbb{R}^\infty_+(\epsilon))$ of mappings $S \ni E \mapsto p(E; \epsilon) \in \mathbb{R}^\infty_+(\epsilon)$;*
- *a unique cardinal equivalence class of real-valued NMUFs $v : Y \to \mathbb{R}$*

such that the preference ordering $\succsim^S$ on $\Delta(Y^S; \mathbb{R}^\infty(\epsilon))$ is represented by the subjective expected utility function

$$\lambda^S \mapsto U^S(\lambda^S) \equiv \sum_{s \in S} p(s; \epsilon) \sum_{y \in Y} \lambda_s(y)\, v(y) \in \mathbb{R}^\infty(\epsilon) \tag{8}$$

on the domain $\Delta(Y^S; \mathbb{R}^\infty_+(\epsilon))$ of $\mathbb{R}^\infty_+(\epsilon)$-valued lotteries $\lambda^S \in \Delta(Y^S; \mathbb{R}^\infty_+(\epsilon))$ whose marginal distributions satisfy $\lambda_s \in \Delta(Y; \mathbb{R}^\infty_+(\epsilon))$ for all $s \in S$. Specifically,

$$\lambda^S \succsim^S \mu^S \Longleftrightarrow U^S(\lambda^S) \geq_L U^S(\mu^S)$$

3.7 Lexicographic Expected Utility

The subjective probability $p(s; \epsilon) \in \mathbb{R}^\infty_+(\epsilon)$ of every state $s \in S$ can be expressed as the power series $\sum_{k=0}^\infty p_k(s)\, \epsilon^k$. Thus, the SEU expression (8) can be re-written as the power series $U^S(\lambda^S) \equiv \sum_{k=0}^\infty u^S_k(\lambda^S)\, \epsilon^k$ whose coefficients of successive powers of ϵ are

$$u^S_k(\lambda^S) := \sum_{s \in S} p_k(s) \sum_{y \in Y} \lambda_s(y)\, v(y) \quad (k = 0, 1, 2, \ldots) \tag{9}$$

But then $\lambda^S \succsim^S \mu^S$, or equivalently $U^S(\lambda^S) \geq U^S(\mu^S)$, if and only if the two respective associated infinite hierarchies of coefficients $\langle u^S_k(\lambda^S)\rangle_{k=0}^\infty$ and $\langle u^S_k(\mu^S)\rangle_{k=0}^\infty$ in the power series satisfy

$$\langle u^S_k(\lambda^S)\rangle_{k=0}^\infty \geq_L \langle u^S_k(\mu^S)\rangle_{k=0}^\infty \tag{10}$$

w.r.t. the usual lexicographic total ordering $\geq_L$ on the space $\mathbb{R}^\infty$ of infinite sequences in $\mathbb{R}$. In this sense, the preference ordering $\succsim^S$ has a lexicographic expected utility representation.

4 Black Swans

4.1 Background

In 82 AD Juvenal (in Satires, VI, 165) had written *"rara avis in terris nigroque simillima cygno"* (a rare bird upon earth, and exceedingly like a black swan). That, however, was merely imaginative irony. Real black swans belonging to the biological species *Cygnus atratus* remained unknown to most of the world before 1697 when Willem de Vlamingh voyaged to what has since become Western Australia. There he became the first European to record seeing living black swans in their native habitat, which included the river he named "Swarte Swaene-Revier" (black swan river). This is now Swan River, which is the main waterway running through the capital city Perth.

Later John Stuart Mill, paraphrasing David Hume, wrote:

> No amount of observations of white swans can allow the inference that all swans are white, but the observation of a single black swan is sufficient to refute that conclusion.

In elementary philosophy, the existence of black swans has become a classical example of the limits to inferential reasoning.

Taleb's (2007) book provides many vivid examples of events, often related to finance or economics, which he sees as meeting his characterisation of a "Black Swan" event as an "outlier" with "an extreme impact" for which "human nature makes us concoct explanations after the event". The book was written before the recent crisis in global financial markets. Nevertheless, it does discuss several earlier ones like the stock market crash of October 1987 that are often plausibly blamed on faulty statistical models.

Indeed, at an early stage of his book, Taleb defines a "special case of 'gray' swans" which are rare but expected. More precisely, they have probability distributions described by "Mandelbrotian randomness", a particular class of fat-tailed probability distribution following a power law. These distributions put so much weight on outliers, or extreme values, of a random variable $v \in \mathbb{R}$ that, for large enough $k \in \mathbb{N}$, the expectation of the kth power of v, otherwise known as the kth moment of the distribution, becomes infinite. This is in stark contrast to the normal or Gaussian distribution, for which the tail of the distribution is so "thin" that all moments exist.

Yet the main issue with the random value of an asset, especially a derivative security, is typically not whether its distribution has fat or thin tails. Rather, for such assets there is typically a positive probability of losing everything. This potential loss cannot be captured by a Gaussian distribution, or by any "smooth" alternative such as a power law. But there is little really new here, since statisticians and financial economists, along with decision and game theorists, have long been coming to terms with probability distributions which do not correspond to a smooth density function.

4.2 Black Swan Events

Much more challenging than Taleb's "gray swans", however, are the true Black Swans which effectively break our existing scientific models. Indeed, the indisputable existence of the (black) swan species now called *Cygnus atratus* broke all previous biological models of the genus *Cygnus*. While Taleb does recognise that such events could occur, he regards them as "totally intractable", scientifically speaking. Nevertheless, biologists have formulated statistical models intended to forecast probabilistically the likely number of new species that one might expect to find in a poorly explored habitat. And of course economists have developed many models of economic growth with technical progress, which may be approximately treated as the accumulation of many small but typically favourable surprises. A notable example is Schumpeter's (1926, 1934) *The Theory of Economic Development* which sets out the view that, as entrepreneurs innovate, a capitalist market economy is subjected to repeated shocks that cannot be modelled in advance.

More generally, any practical model, especially in the social sciences, must have bounded scope and so must ignore some possibilities. As the statistician George Box wrote: "Essentially, all models are wrong, but some are useful." Should any unmodelled possibility such as a bank run or bank failure occur and have a noticeable impact, it will have to be recognised as an "aberrant" event which, by definition, lies outside the current model.

This is not to deny that any aberrant event could have appeared in an enriched version of the agent's model, if it had been imagined soon enough and then deemed worth modelling. But it was not. Instead, its occurrence demonstrates that the original model is broken and needs modifying accordingly. Such aberrant events lying outside the current model should be distinguished from events within the model which, like Taleb's "gray swans", have extremely low or even zero probability. By contrast, black swan events, unlike those described in Taleb's book, may not even be imagined *ex ante*. Thus, aberrance may be due to a failure of the imagination in constructing a decision model. This may be related to Shackle's (1953) concept of "surprise"—see also Hammond (2007). Indeed, there may be more phenomena in economics that can be explained by "asymmetric imagination" than by the widely used notion of asymmetric information. And not only in economics, but in culture, business, etc.

To summarize, sometimes models may change as their originators anticipate events that had to be excluded originally. To adapt the widely quoted saying by the statistician George Box: "Essentially, all useful models are incompletely specified." The excluded events would become aberrant if they were to occur before they could be included in a more accurate statistical model. Even so, their possible effects on the consequences of modelled current decisions can be allowed for, at least in principle, within a suitable EU decision model allowing an "enlivened" version of the usual decision tree. This is our next topic.

4.3 An Initial Simple Tree

Let Y be a fixed consequence domain. Consider a decision maker whose objective is to maximize the expected value of a von Neumann–Morgenstern utility function (NMUF) $v : Y \to \mathbb{R}$.

Consider an initial (dead) decision tree T:

- with an initial (decision) node n_0,
- at which the *agent* chooses a chance node n_1 in the set $N_1 := N_{+1}(n_0)$ of all nodes that immediately succeed n_0,
- at each of which *chance* determines an immediately succeeding terminal node n_2 in the set $N_2(n_1) := N_{+1}(n_1)$ of all nodes that immediately succeed n_1, using known transition *probabilities* $\pi(n_2|n_1)$ satisfying $\pi(\cdot|n_1) \in \Delta(N_2(n_1))$,
- each of which has a known *final consequence* $\gamma(n_2) \in Y$.

4.4 Initial Evaluation

In this initial simple tree there is a known consequence $\gamma(n_2) \in Y$, of reaching any terminal node n_2. The *initial evaluation* of reaching this node is evidently $w_2(n_2) = v(\gamma(n_2))$.

Working backwards, as usual in dynamic programming, the *conditional expected utility* of reaching any chance node $n_1 \in N_1$ is

$$w_1(n_1) = \mathbb{E}[w_2(n_2)|n_1] = \sum\nolimits_{n_2 \in N_2(n_1)} \pi(n_2|n_1)\, w_2(n_2) \tag{11}$$

Then an *optimal decision* $n_1^* \in N_1$ is any that maximizes $w_1(n_1)$ with respect to n_1, subject to $n_1 \in N_1$.

The above simple argument is a trivial application to an orthodox "unenlivened" decision model of the *optimality principle* of stochastic dynamic programming. That is, any current decision should be given a *continuation value* equal to the highest possible expected utility resulting from an appropriate plan for all subsequent decisions. Optimality requires the current decision to maximize the expectation of this continuation value.

4.5 Enriched Subtrees

One possible enrichment of the agent's decision model involves a new NMUF $v^+ : Y^+ \to \mathbb{R}$ defined on an enriched model consequence domain $Y^+ \supseteq Y$. But many other enrichments are also possible.

Before we discuss these, note first that the agent can hardly make an unmodelled decision. Accordingly, assume that a necessary and sufficient condition for being able to choose any $n_1 \in N_1$ is that node n_1 is included in the model. Hence the set $N_{+1}(n_0)$ remains fixed. So we assume that any enrichment of the tree takes place only after a particular chosen decision node $n_1^i \in N_{+1}(n_0)$ has already been reached.

What matters, however, is not just how the continuation subtree $T(n_1^i)$ after this particular node is enriched. Also relevant are the potential enrichments of the continuation subtrees $T(n_1)$ at all the other nodes $n_1 \in N_1 \setminus \{n_1^i\}$, since all these possible enrichments ultimately affect the relative expected values of moving to different nodes $n_1 \in N_1$.

Now, starting at each $n_1 \in N_1$, the original continuation subtree $T(n_1)$ had nodes $n_2 \in N_2(n_1)$. Instead there is now an *enriched* continuation subtree $T^+(n_1)$ with:

- an *expanded* set $N_{+1}^+(n_1) = N_2^+(n_1) \supseteq N_2(n_1)$ of immediately succeeding terminal nodes;
- *revised* transition probabilities $\pi^+(n_2^+|n_1)$ for all $n_2^+ \in N_2^+(n_1)$;
- *revised* consequences $\gamma^+(n_2^+) \in Y^+$ for all $n_2^+ \in N_2^+(n_1)$ with utilities $w_2^+(n_2^+) := v^+(\gamma^+(n_2^+))$.

Instead of (11), the *revised* expected utility of any decision at node n_0 to move to any node $n_1 \in N_1 = N_{+1}(n_0)$ is therefore

$$w_1^+(n_1) := \mathbb{E}^+[w_2^+(n_2^+)|n_1] := \sum\nolimits_{n_2^+ \in N_{+1}^+(n_1)} \pi^+(n_2^+|n_1)\, w_2^+(n_2^+) \tag{12}$$

4.6 Retrospective Evaluation in the Enlivened Tree

In this simple two-stage model, the *enriched tree* T^+ is the extension of T obtained by replacing each continuation subtree $T(n_1)$ ($n_1 \in N_1$) with its enrichment $T^+(n_1)$. We define the *enlivened tree* as the pair (T, T^+). Unlike botanical tree rings, this includes a complete record of how the tree has grown between:

1. the first period, when it was T;
2. the second period, when it has become T^+.

It is also a mathematical rather than a botanical growth process! For one thing, botanical trees may lose branches in windy conditions, whereas enlivened trees can only expand with time.

Analysed *ex post*, the appropriate decision at initial node n_0 would have been to maximize $w_1^+(n_1^i)$ with respect to $i \in I$. But *ex ante*, only the details of the original model can be used, by definition. What the agent can still do *ex ante*, however, is to recognize that the original evaluation function $w_1(n_1^i)$ may be revised to an as yet unknown and uncertain retrospective evaluation function $w_1^+(n_1^i)$ that ranges over a function space of possible evaluation functions. This is similar in spirit to the work of Koopmans (1964) and Kreps (1992) that allows uncertainty about future preferences—see also Dekel et al. (2001, 2007).

In other words, somewhat like Hansen and Sargent (2008, 2011), we can apply a robust decision analysis and choose the initial decision $i \in I$ in order to maximize $\mathbb{E}w_1^+(n_1^i)$ after allowing for uncertainty about the appropriate form of the function $i \mapsto w_1^+(n_1^i)$.

4.7 Cardinally Equivalent Evaluation Functions

Two evaluation functions $w_1, \tilde{w}_1 : N_1 \to \mathbb{R}$ are *cardinally equivalent*, with $w_1 \sim \tilde{w}_1$, just in case there exist:

- an additive constant $\alpha \in \mathbb{R}$
- a positive multiplicative constant $\rho \in \mathbb{R}_+$

such that $\tilde{w}_1(n_1^i) \equiv \alpha + \rho w_1(n_1^i)$.

The *value state space* Ω is defined as the set

- of all non-constant functions $n_1 \mapsto \omega(n_1)$ *normalized* to satisfy

$$\min_{n_1 \in N_1} \omega(n_1) = 0 \quad \text{and} \quad \max_{n_1 \in N_1} \omega(n_1) = 1$$

- together with the *normalized constant function* satisfying $\omega(n_1) = 0$ for all $n_1 \in N_{+1}(n_0)$, which represents complete indifference.

4.8 Uncertain Retrospective Evaluation

Enlivenment replaces the original evaluation function w_1 in T by an *uncertain* retrospective evaluation function w_1^+ derived in the tree T^+, which cannot even be modelled *ex ante*. Because the set N_1 is assumed to be finite, the function $w_1^+ : N_1 \to \mathbb{R}$ ranges over the space $\Omega \subseteq [0,1]^{N_1} \subset \mathbb{R}^{N_1}$—i.e., Ω is a subset of the unit hypercube in Euclidean space.

4.9 State-Dependent Consequence Domains

In this setting, applying standard subjective probability theory faces an obstacle. The relevant consequences are pairs $(n_1, \omega) \in N_1 \times \Omega$. So the consequence domain $N_1 \times \{\omega\}$ depends on the state $\omega \in \Omega$. This rules out Savage's *constant acts* $a : \Omega \to N_1$ with $a(\omega) = \bar{a}$ for all $\omega \in \Omega$.

In normative decision theory, Hammond (1998b, 1999) suggests a remedy for this kind of state-dependent consequence domain. It is to postulate the existence of an extended NMUF $U : N_1 \times \Omega \to \mathbb{R}$ whose expected value represents preferences $\succsim$ on $\Delta(N_1 \times \Omega)$, when one can choose, in addition to different nodes $n_1 \in N_1$, the probabilities of different states $\omega \in \Omega$.

Given any fixed state $\omega \in \Omega$, the expected values w.r.t. any $\nu \in \Delta(N_1)$ of the two functions $n_1 \mapsto U(n_1, \omega)$ and $n_1 \mapsto \omega(n_1)$ should represent preferences over corresponding lotteries $\nu \in \Delta(N_1)$ and $\nu \times \delta_\omega$. So the two functions $n_1 \mapsto U(n_1, \omega)$ and $n_1 \mapsto \omega(n_1)$ must be cardinally equivalent, for each fixed ω. That is, there must exist mappings $\omega \mapsto \alpha(\omega) \in \mathbb{R}$ and $\omega \mapsto \rho(\omega) \in \mathbb{R}_+$ such that $U(n_1, \omega) \equiv \alpha(\omega) + \rho(\omega)\, \omega(n_1)$.

4.10 *Subjective Expected Evaluation*

The agent's subjective expected utility objective in the enlivened tree (T, T^+) can (and should) use a *subjective probability measure* P over the Borel subsets of Ω. Then preferences over objective "roulette" lotteries $\nu \in \Delta(N_1)$ are ultimately represented by the *objectively expected* value $\mathbb{E}_\nu V$ of the *subjective expectation* function $N_1 \ni n_1 \mapsto V(n_1)$ defined by

$$V(n_1) := \int_\Omega U(n_1, \omega) P(d\omega) = \int_\Omega [\alpha(\omega) + \rho(\omega)\, \omega(n_1)] P(d\omega) \tag{13}$$

There is an obvious analogy here with Anscombe and Aumann (1963), who allow combinations of roulette and horse lotteries. An axiomatic justification, however, has yet to be developed, though it should be possible by combining the ideas of Myerson (1979), Fishburn (1982), and Hammond (1998b, 1999).

4.11 *Hubris Versus Enlivenment*

Tractable models are necessarily bounded in scope. Actions may have consequences that are not only unintended, but quite possibly unimagined, and certainly not included in whatever bounded model was used to analyse the agent's decision.

An agent's decision model, like any competent engineer's plan, will typically need to change as and when surprise events outside the model compel attention. Orthodox decision models ignore completely any possibility of model revision. In this sense, they are inherently *hubristic*.

4.12 Could There Be a Metamodel?

A decision model in discrete time amounts to a controlled stochastic process, or equivalently a decision tree that combines chance nodes with decision nodes where the decision is controlled by the decision-maker. Recognizing that the appropriate decision model is itself subject to uncertainty, is it possible, or even desirable, to construct a "metamodel" that embraces all possible decision models?

We will actually consider a simpler question: whether one can or should construct a metamodel in the form of a stochastic "metaprocess" defined on the space of all possible stochastic process models? The result would be a sequence of stochastic processes in which the state space is continually being enriched *unpredictably*.

Now, recall that the stochastic process model is based on Kolmogorov's extension theorem in probability theory. This result states that any "consistent" family of probability laws on finite Cartesian subproducts of an arbitrary collection of component measurable spaces can be extended to a probability law on the whole Cartesian product. The theorem, however, depends on significant topological assumptions such as the existence in each component measurable space of a *compact class* C of measurable sets—i.e., every sequence of sets in C whose finite intersections are non-empty has a non-empty infinite intersection—such that the probability of any measurable set must equal the supremum of the probabilities of all its subsets that lie in C.[1] It seems difficult to find a suitable topology on the class of all potentially relevant sequences of stochastic process models which allows an interesting probability measure to exist.

4.13 Should We Look for a Meta Stochastic Process?

El Aleph is a short story published by the distinguished Argentinian author Jorge Luis Borges in 1945. It begins with a quotation from Shakespeare's *Hamlet* Act II, Scene 2

> O God! I could be bounded in a nutshell,
> and count myself a King of infinite space ...

This could be regarded as Shakespeare's poetic description of a key requirement for a metamodel. Eventually we move to the heart of Borges' wonderful story[2]:

> He explained that an Aleph is one of the points in space that contains all other points. ... The Aleph's diameter was probably little more than an inch, but all space was there, actual and undiminished. Each thing (a mirror's face, let us say) was infinite things, since I distinctly saw it from every angle of the universe.

---

[1]See Neveu's (1965, p. 82) significant generalization of Kolmogorov's extension theorem, as described in Aliprantis and Border (1994, Sect. 14.6).

[2]The following brief extracts are from http://www.phinnweb.org/links/literature/borges/aleph.html, which reproduces the English translation on which Norman Thomas Di Giovanni collaborated with Borges himself.

Shortly thereafter the story takes a rather disturbing turn:

> I saw the Aleph from every point and angle, and in the earth the Aleph, and in the Aleph the earth; I saw my own face and my own bowels; I saw your face; and I felt dizzy and wept, for my eyes had seen that secret and conjectured object whose name is common to all men but which no man has looked upon—the unimaginable universe.
>
> I felt infinite wonder, infinite pity.

But eventually something like normality returns:

> Out on the street, going down the stairways inside Constitution Station, riding the subway, every one of the faces seemed familiar to me. I was afraid that not a single thing on earth would ever again surprise me; I was afraid I would never again be free of all I had seen. Happily, after a few sleepless nights, I was visited once more by oblivion.

A later postscript includes some explanation for Borges' choice of title:

> As is well known, the Aleph is the first letter of the Hebrew alphabet. Its use for the strange sphere in my story may not be accidental. For the Kabbala, the letter stands for the *En Soph*, the pure and boundless godhead; it is also said that it takes the shape of a man pointing to both heaven and earth, in order to show that the lower world is the map and mirror of the higher; for Cantor's *Mengenlehre* [set theory], it is the symbol of transfinite numbers, of which any part is as great as the whole.

Perhaps the moral of Borges' story is that in the end we should be relieved about how mathematically and conceptually intractable the problem of finding a stochastic metaprocess appears to be.

5 Concluding Remarks

A descriptive decision theory stands or falls by its capacity to explain what we observe. A prescriptive decision theory, on the other hand, stands or falls by its capacity to offer a normatively appealing approach to decision making. This work has set out alternative departures from standard prescriptive decision theory. These departures have been designed to deal separately with the three key phenomena of catastrophic risk, rare events, and true black swan events that transcend whatever decision model we may currently be using.

The work by Chichilnisky (1996, 2000, 2009, 2010) has set out heroically to deal with all these three phenomena within one integrated framework. In doing so, however, she follows Savage (1954) in relaxing the usual countable additivity property of probability measures, thus allowing probabilities that are only finitely additive. A conjecture to be settled by future research is that the same three phenomena could be accommodated within a different integrated framework which retains a countably additive probability measure. This framework would allow:

1. the kind of distinction between catastrophic and non-catastrophic consequences that was introduced in Sect. 2;
2. for rare events, non-Archimedean probabilities of the kind discussed in Sect. 3, but extended from a finite sample space S to a general measurable space $(S, \mathcal{S})$;
3. for true black swan events, enlivened trees of the kind sketched briefly in Sect. 4, with preferences represented by subjective expected utility based on extended probability measures over states of the world that correspond to possible retrospective evaluation functions defined for every modelled decision.

Note finally that rationality within bounded decision trees allows a *restricted* revealed preference hypothesis, applying only to options that receive serious consideration. But decision trees almost inevitably become *enlivened* in case the decision maker is forced to recognize the possibility of events which were excluded from earlier decision models. These unmodelled events are *truly unknown* "black swans", like the species *cygnus atratus* was to Europeans before Dutch explorers reached Western Australia. Such unmodelled events are *completely different* from the "highly improbable" but modelled events referred to as "grey swans" in Taleb (2007). Indeed, Taleb dismisses true black swans as completely intractable.

Acknowledgements Generous support during 2007–10 from a Marie Curie Chair funded by the European Commission under contract number MEXC-CT-2006-041121 is gratefully acknowledged. So is the opportunity to present this work at the AFOSR (Air Force Office of Scientific Research) workshop on Catastrophic Risk, organized by Graciela Chichilnisky, and held at SRI, Menlo Park, California, on May 31st and June 1st, 2012. Some of the material on black swans is adapted from the earlier informal online publication Hammond (2009).

References

Aliprantis, C. D., & Border, K. C. (1994). *Infinite dimensional analysis: A Hitchhiker's guide* (2nd ed. 1999). Berlin: Springer.

Anscombe, F., & Aumann, R. J. (1963). A definition of subjective probability. *Annals of Mathematical Statistics, 34*, 199–205.

Arrow, K. J. (1965). *Aspects of the theory of risk-bearing*. Helsinki: Yrjö Jahnssonin Säätiö.

Balasko, Y. (1978). Economic equilibrium and catastrophe theory: An introduction. *Econometrica, 46*, 557–569.

Chichilnisky, G. (1996). Updating von Neumann Morgenstern axioms for choice under uncertainty. In *Proceedings of a Conference on Catastrophic Risks*. Toronto: The Fields Institute for Mathematical Sciences.

Chichilnisky, G. (2000). An axiomatic approach to choice under uncertainty with catastrophic risks. *Resource and Energy Economics, 22*, 221–231.

Chichilnisky, G. (2009). The topology of fear. *Journal of Mathematical Economics, 45*, 807–816.

Chichilnisky, G. (2010). The foundations of statistics with black swans. *Mathematical Social Sciences, 59*, 184–192.

Dekel, E., Lipman, B. L., & Rustichini, A. (2001). Representing preferences with a unique subjective state space. *Econometrica, 69*, 891–934.

Dekel, E., Lipman, B. L., Rustichini, A., & Sarver, T. (2007). Representing preferences with a unique subjective state space: Corrigendum. *Econometrica, 75*, 591–600.

Drèze, J. H. (1962). L'utilité sociale d'une vie humaine. *Revue Française de Recherche Opérationnelle, 23*, 93–118.
Fishburn, P. C. (1982). *The foundations of expected utility*. Dordrecht: D. Reidel.
Halpern, J. Y. (2009). A nonstandard characterization of sequential equilibrium, perfect equilibrium, and proper equilibrium. *International Journal of Game Theory, 38*, 37–49.
Halpern, J. Y. (2010). Lexicographic probability, conditional probability, and nonstandard probability. *Games and Economic Behavior, 68*, 155–179.
Hammond, P. J. (1994). Elementary non-Archimedean representations of probability for decision theory and games. In P. Humphreys (Ed.), *Patrick Suppes: Scientific philosopher, vol. I: Probability and probabilistic causality* (pp. 25–59). Kluwer Academic Publishers. Ch. 2.
Hammond, P. J. (1997, 1999). Non-Archimedean subjective probabilities in decision theory and games. Stanford University Department of Economic Working Paper No. 97-038. *Mathematical Social Sciences, 38*, 139–156.
Hammond, P. J. (1998a). Objective expected utility: A consequentialist perspective. In S. Barberà, P.J. Hammond, & C. Seidl (Eds.) *Handbook of utility theory, vol. 1: Principles* (pp. 145–211). Boston: Kluwer Academic Publishers. Ch. 5.
Hammond, P. J. (1998b). Subjective expected utility. In S. Barberà, P.J. Hammond, & C. Seidl (Eds.), *Handbook of Utility Theory, Vol. 1: Principles* (pp. 213–271). Boston: Kluwer Academic Publishers. Ch. 6.
Hammond, P. J. (1999). Subjectively expected state-independent utility on state-dependent consequence domains. In M. J. Machina & B. Munier (Eds.), *Beliefs, interactions, and preferences in decision making* (pp. 7–21). Dordrecht: Kluwer Academic.
Hammond, P. J. (2007). Schumpeterian innovation in modelling decisions, games, and economic behaviour. *History of Economic Ideas, XV*, 179–195.
Hammond, P. J. (2009). Adapting to the entirely unpredictable: Black swans, fat tails, aberrant events, and hubristic models. http://www2.warwick.ac.uk/fac/soc/economics/research/centres/eri/bulletin/2009-10-1/hammond/.
Hansen, L. P., & Sargent, T. J. (2008). *Robustness*. Princeton: Princeton University Press.
Hansen, L. P., & Sargent, T. J. (2011). Wanting robustness in macroeconomics. In B.M. Friedman, & M. Woodford (Eds.), *Handbook of monetary economics* (Vol. 3B, pp. 1097–1157). Elsevier. Ch. 20.
Jensen, N. E. (1967). An introduction to Bernoullian utility theory, I: Utility functions. *Swedish Journal of Economics, 69*, 163–183.
Jones-Lee, M. W. (1974). The value of changes in the probability of death or injury. *Journal of Political Economy, 82*, 835–849.
Kolmogorov, A. N. (1933, 1956). *Grundbegriffe der Wahrscheinlichkeitsrechnung* (Berlin: Springer); translated as *Foundations of probability*. New York: Chelsea.
Koopmans, T. C. (1964). On flexibility of future preference. In M.W. Shelly, & G.L. Bryan (Eds.), *Human judgments and optimality* (pp. 243–254). New York: Wiley. Ch. 13.
Kreps, D. M. (1992). Static choice in the presence of unforeseen contingencies. In Dasgupta, P., Gale, D., Hart, O., & Maskin, E. (Eds.), *Economic analysis of markets and games: essays in honor of Frank Hahn* (pp. 258–281). Cambridge: M.I.T. Press.
Lightstone, A. H., & Robinson, A. (1975). *Nonarchimedean fields and asymptotic expansions*. Amsterdam: North-Holland.
McGonagall, W. (1880). The Tay Bridge disaster. http://www.mcgonagall-online.org.uk/gems/the-tay-bridge-disaster.
Myerson, R. B. (1979). An axiomatic derivation of subjective probability, utility, and evaluation functions. *Theory and Decision, 11*, 339–352.
Neveu, J. (1965). *Mathematical foundations of the calculus of probability*. San Francisco: Holden-Day.
Rényi, A. (1955). On a new axiomatic theory of probability. *Acta Mathematica Academiae Scientiarum Hungaricae, 6*, 285–335.

Rényi, A. (1956). On conditional probability spaces generated by a dimensionally ordered set of measures. *Theory of Probability and its Applications*, *1*, 61–71.
Robinson, A. (1973). Function theory on some nonarchimedean fields. *American Mathematical Monthly: Papers in the Foundations of Mathematics*, *80*, S87–S109.
Savage, L. J. (1954, 1972). *The foundations of statistics*. New York: Wiley; New York: Dover Publications.
Schumpeter, J. A. (1911, 1926). *Theorie der wirtschaftlichen Entwicklung; Eine Untersuchung über Unternehmergewinn, Kapital, Kredit, Zins und den Konjunkturzyklus* (2nd ed. 1926). München und Leipzig: Duncker und Humblot.
Schumpeter, J. A. (1934, 1961). *The theory of economic development: An inquiry into profits, capital, credit, interest, and the business cycle*. Translated from the German by Redvers Opie; with a new Introduction by John E. Elliott.
Selten, R. (1975). Re-examination of the perfectness concept for equilibrium points of extensive games. *International Journal of Game Theory*, *4*, 25–55.
Shackle, G. L. S. (1953). The logic of surprise. *Economica, New Series*, *20*, 112–117.
Taleb, N. N. (2007). *The black swan: The impact of the highly improbable*. New York: Random House.
Thom, R. (1973). *Stabilité struturelle et morphogéneèse (Paris: Interéditions); translated (1976) as structural stability and morphogenesis*. Reading, MA: W.A. Benjamin.
Villegas, C. (1964). On qualitative probability σ-algebras. *Annals of Mathematical Statistics*, *35*, 1787–1796.
von Neumann, J., & Morgenstern, O. (1944; 3rd edn. 1953). *Theory of games and economic behavior*. Princeton: Princeton University Press.
Zeeman, E. C. (1976). Catastrophe theory. *Scientific American*, 65–70 and 75–83.

Preference Representations for Catastrophic Risk Analysis

Richard E. Ericson and Jamie L. Kruse

1 Introduction

Catastrophic events, such as major hurricanes and earthquakes, tornados, floods and wild fires, have, over the past decade, become that focus of a growing body of economic literature.[1] These high (negative) impact, (very) low probability, events ("tail events," in Nordhaus 2012) appear to be imposing growing costs on society, raising policy issues that call for economic analysis. Such analysis must rest on an understanding of the human behavioral response to such events and the decisions taken prior to, and in anticipation of, their possible occurrence, in order to avert them and/or mitigate their impact. The need for such analysis is given some urgency by the perception that global climate change has rendered such events increasingly likely and volatile, and thus they have the potential to impose unprecedented and unpredictable damage when they occur. Indeed, at the global level, climate change itself may evolve into a massive catastrophe, threatening the ability of the earth to support modern civilization and standards of living, or even life itself.[2]

[1]This work is both reflected in, and stimulated by the 30 October 2006 Stern review on the economics of climate change. See references and discussion in, for example, Nordhaus (2007) and Weitzman (2007, 2009), and the reviews by Kunreuther et al. (2013), Heal and Millner (2014). Also see Parson (2007) review of Posner's influential book.
[2]Other massive catastrophes have also begun to receive some attention, e.g. the impact of a sufficiently large asteroid, capable of eliminating much, or most, life on earth. See Chichilnisky (2000).

R.E. Ericson (✉) · J.L. Kruse
Department of Economics, East Carolina University, Brewster A-427 Tenth Street,
Greenville, NC 27858, USA
e-mail: ericsonr@ecu.edu

J.L. Kruse
e-mail: KruseJ@ecu.edu

G. Chichilnisky and A. Rezai (eds.), *The Economics of the Global Environment*,
Studies in Economic Theory 29, DOI 10.1007/978-3-319-31943-8_3

1.1 The Problem

The economic analysis of such potential catastrophes faces several analytic challenges. Most fundamental is the true *uncertainty* surrounding both the likelihood, the potential magnitudes, and the timing of these rare events. While there is limited frequency data for some such events (e.g. 100-year floods, EF-5 tornadoes, Cat-5 Hurricanes), consequences of individual occurrences are often unprecedented (Hurricanes Katrina, Sandy), and others without precedent in recorded history (catastrophic asteroid hit, catastrophic climate change). Thus the distribution, the likelihood, of such events is truly unknown, and their occurrence may be arbitrarily far in the future, or 'the day after tomorrow'. This fundamental uncertainty compounds the better understood risks associated with any of the potential scenarios that might unfold in the face of these events, including scenarios in their absence.

As such events may only 'occur' well into the future, economic analysis should consider the interaction between uncertainty and the temporal dimension. Uncertainties regarding capabilities, technologies and resources, constraints and preferences increase as the horizon lengthens. And the kinds of decisions to be made, the available choices, will change over time, both exogenously and driven by the anticipation of the catastrophic events of interest. Further, there is a potential for 'learning' in all these dimensions and a 'demand' for the resolution of uncertainties, both of which will evolve as the uncertainties themselves evolve over time.

Finally, behind any economic analysis must stand a model of human behavior, a model of the objectives, preferences, and beliefs of those making decisions either in anticipation or under the impact of the catastrophic event, or that might influence its likelihood and/or impact. In most analysis this involves using an "expected utility" representation of suitably monotonic (non-satiated) preferences based on a clear understanding/knowledge of both the outcomes and consequences of decisions and their probabilities/likelihoods.[3] This models decision making in the face of well understood risks. While assuming such knowledge is a reasonable approximation in much policy analysis, it is questionable for extremely rare, unprecedented, and/or distant future events, particularly those of a catastrophic nature. Further, the standard model uses exponential discounting of (expected) future returns in evaluating the consequences of decisions, which effectively eliminates consideration of any events/consequences sufficiently far removed in time.[4]

The problem of modeling decisions, and their driving preferences, in the face of catastrophic risks, integrates all the other, logically separate, challenges. For any model of decision making must incorporate not only tastes and preferences with regard to outcomes, their uncertainties, and their timing, but also knowledge about constraints, possible actions, and consequences, and beliefs about the relevant

[3] Shaw and Woodward (2008) present arguments why this framework may not be adequate for the kind of issues we address.

[4] For example, a 3 % discount reduces \$1 million to \$493.10 in a quarter century. A 1 % discount requires 757 years to achieve the same degree of devaluation, while a 5 % discount requires only 14.85 years.

likelihoods of both major and minor events and changes in the constraints, technologies, threats and opportunities that may be faced. Here we explore only a part of that challenge, considering some analytic models of preference representation that we believe could be useful in analyzing human behavioral response to catastrophic events. We focus only on atemporal decision models, where future 'values' are appropriately discounted,[5] and the central question is: How do agents, at the time an *ex-ante* decision must be made, deal with fundamental, poorly defined uncertainties/risks? Thus we purposely ignore the growing body of literature exploring the impact of deep uncertainty in a dynamic framework,[6] where the appropriate discounting of the distant future and the updating of 'knowledge' about evolving states of the world become central, and focus on the *ex-ante* decision problem with all uncertainties summarized in a (generally unknown) distribution over future states.

1.2 Some Modeling Issues

There are a number of issues that should be kept in mind when developing an adequate model of decision making in the face of catastrophic risk. We highlight these issues not because we will or can deal with them adequately in this paper, but because they are important to interpreting our analysis, and will have to be dealt with in developing a fully adequate model of decisions in the face of such risks.

First, we consider 'catastrophic' risk to be more than the low-probability likelihood of extremely high losses. To be truly 'catastrophic' the extremely high losses/damage must be wide spread, impacting large numbers, and potentially causing a fundamental change in the socioeconomic system, in the 'way the world works'. In such a (rare) situation, not only agent expectations, but fundamental understandings about technologies, constraints, and how actions map into consequences, as well as the preferences of the agent, are apt to be different from the 'normal world'. And how all these will change is highly uncertain, indeed largely unknown and not considered, as agents make decisions *ex ante* with only the vaguest understanding of the circumstances with which they must deal in that rare/ unprecedented event. Thus catastrophic risk relates to a situation of fundamental uncertainty, where the likelihood of outcomes and consequences is truly unknown, and the decision makers'

[5]Here we abstract from the large and important question of exactly how future costs/benefits are assessed, a task for which the size and determination of both subjective (personal) and stochastic (market generated) discount factors are critical. Among the works addressing such issues are Weitzman (2009, 2010), Millner et al. (2013), and Millner (2013).

[6]A particularly important class of dynamic models dealing with fundamental uncertainty with regard to the evolving state of the world is that of "multiple-priors utility" admitting aversion to ambiguity (Chen and Epstein 2002). These have been applied to continuous time stochastic asset pricing and business cycle models with high frequency data, rather than to the analysis of uncertain discrete catastrophic events. Also see Ilut and Schneider (2014), Joeng et al. (2014), and Viale et al. (2014).

beliefs and 'tastes' for this fundamental uncertainty become critical components of any decision model.

Another issue that must be faced is the purpose of the modeling exercise, the objective of the analysis. A model of decisions in the face of catastrophic risk might be either 'descriptive' or 'prescriptive', i.e. either capture actual human decisions and behavior, including behavioral deviations from 'economic rationality', or provide a rational framework for making 'best' decisions or policies (social rationality). This is closely related to the 'subject' of the analysis: Who's preferences are being modeled? If we are studying merely extreme risks, with some observed, if low, frequency (measurable probability) of occurrences, then expected utility with appropriate risk and ambiguity preferences, and appropriate discounting, provides a useful decision tool; it reflects a 'social rationality', even if it does not capture the behavior of (some/many) individuals. For public policy analysis and decisions, we require intertemporally consistent, equitable, and probabilistically sophisticated, expectationally 'objective' benefit-cost analysis. Such analysis should incorporate fundamental uncertainty about what is know and what will be known about the 'risk' and its consequences, as well as the fundamental uncertainty about future 'technology' and consequent 'rates of return', even if the appropriate pure discount rate is zero.[7] We believe this is too much to assume for the analysis of individual behavior. Indeed, even for analysis of individual behavior in the face of moderate probability ('frequent' events), the model must consider individual time preference and 'behavioral' regularities, often violating standard axioms, such as: ignoring, or significantly overestimating ('fear factor'), very 'small' probabilities; underestimating, or accepting as 'certain', probabilities near one; hyperbolic or other non-exponential (time inconsistent) discounting; etc.

If a catastrophic event is of extremely small probability (rare event), the properties of the lower tail of distribution become significant for evaluation of 'measurable' events, as do (possibly non-measurable) 'tail events' for non-integral evaluation functionals. The standard analysis alluded to above is adequate for thin-tailed distributions (those dominated by Gaussian) over measurable events. However, uncertainty/ambiguity about even thin-tailed distributions can render the *decision-relevant distribution* 'heavy-tailed' (Weitzman 2009, 2010). Such distributions (power law; stable with $\alpha < 2$; regular variation at $-\infty$) lack finite moments of order above the first, and sometimes even the first moment, requiring direct evaluation of loss risk in the tail, related to the tail index of regular variation. And Chichilnisky (2000) has introduced catastrophes as 'non-measurable' events, requiring a non-integral evaluation functional, e.g. purely finitely additive functional, of 'events'.

Finally, any adequate model needs to be able to capture various agent (decision maker) responses to 'catastrophic risks'. This can range from ignoring to seriously exaggerating the dangers they pose, particularly when of very low, or unknown, (subjective) probability. 'Rational' behavior involves making choices ('acts') maximizing some value (minimizing some loss), given knowledge/beliefs/'fears' about

[7]This has been similarly noted in the economics of climate change literature, e.g. Kunreuther et al. (2013), Millner et al. (2013); Heal and Millner (2014).

likely consequences of those choices, including 'doing nothing' or 'waiting'. The valuations driving 'optimizing' choices are typically modeled in "preference representations" (utility functions) on a space of 'payoff relevant' outcomes, with 'beliefs' modeled by bounded, real-valued set functions (probabilities, capacities) on a space of payoff relevant 'events'. In this exercise, we believe that systematic deviations from modelled 'rational behavior' are best understood as a problem with the model—improper modelling of 'preferences' or 'beliefs'—requiring further development or new axiomatizations. If marginal utility is *unbounded* below ($x \downarrow$), the basic expected utility model suffers "tyranny of catastrophic risks"–outcomes with 'vanishingly small' probability, but an arbitrarily 'bad' result, dominate choice. The 'certainty equivalent' becomes the arbitrarily bad outcome implying an unbounded 'willingness to pay to avoid' (Buchholz and Schymura 2012).[8] On the other hand, if marginal utility is *bounded* below, the basic expected utility model eventually ignores risks, however catastrophic, of arbitrarily small probability ("black swans"). This renders expected (subjective) utility analysis somewhat impotent in the face of truly catastrophic risks of the sort described above.

1.3 What We Do

This note is a first step toward our exploration of these issues. Here we focus on the impact of perceptions of likelihood and the decision maker's 'taste for uncertainty'. We present an extremely simple (2 or 3-outcome) example of a world with catastrophic risk, incorporating only the minimal elements required for such an example, and develop the implications of several different models of preferences in the face of such risk. This model world allows us to explore some of the consequences of the axioms differentiating the various decision models, as well as exploring those models' implications for the behavior of agents when facing 'catastrophic' risks. We believe that this model world *may* provide a basis for future 'experimental' work.[9]

To capture agents' beliefs and 'taste' for fundamental uncertainty, we will use models with 'ambiguity', models initially developed to resolve the behavioral "paradoxes" of the (subjective) expected utility model, including the Ellsberg and Allais/Dreze paradoxes, and the kind of behavioral anomalies addressed in prospect theory.[10] These anomalies and paradoxes cast doubt on agents' use of mathematically consistent subjective probabilities in making decisions, particularly when

[8] This is also the implication of Weitzman (2009) "Dismal Theorem," where 'structural uncertainty' can drive the stochastic discount factor to infinity. See the thoughtful discussion with regard to long run climate change in Millner (2013).

[9] A recent survey of decision theory under ambiguity, similar to ours, can be found in Etner et al. (2012). They use a 2-state, 3-decision insurance example to illustrate a range of different decision models, but don't attempt to focus on catastrophic risks. A very cursory discussion, focused on climate change policy modelling, of 4 of the frameworks we study can be found in Heal and Millner (2014).

[10] A nice survey and introduction to the issues can be found in Gilboa (2009).

those probabilities are objectively unknown and/or near zero or one. We expect catastrophic risks to be particularly affected by these behavioral distortions, given their extreme rarity and frequently unprecedented nature when they do occur. Thus models in which the decision maker finds the likelihood/probabilities of events 'ambiguous' would seem particularly appropriate.

Below we present six such models, indicating their particular assumptions (axioms), and illustration their implications in our simple analytic example. The models we discuss are: (i) Choquet Expected Utility; (ii) Rank Dependent Expected Utility; (iii) Maximin Expected Utility; (iv) 'Smooth' Ambiguity Aversion (Second-Order Expected Utility); (v) Chichilnisky Model of "'Sensitivity' to 'Rare' Events"; and (vi) Variational Preferences.[11] The final model we believe has the potential to become an 'umbrella' model, encompassing all of the others, when properly specified for the atemporal decision situation modelled. These models share a common notation, summarized in the Appendix, where the full set of axioms used to derive the various representations is also presented.

2 Minimal Model of Catastrophic Risk

The idea of a 'catastrophe', rather than just an extremely bad outcome, must go beyond its rarity and unpredictability. It encompasses not only deep and wide spread (affecting a significant portion of the population) losses, but an aspect of fundamental change, an irreversibility of the resulting situation. Available resources and prior technologies can become sharply limited, standard production possibilities and means of interaction infeasible, requiring completely different ways of dealing with the new post-catastrophe world. And the resulting radical change in the 'choice set' suggests that preferences, including the taste for risk and uncertainty, are also apt to fundamentally change. While modeling in any detail such an unprecedented situation is perhaps an impossible task, it behoves us to incorporate whatever we can of the limited qualitative characteristics of such a world in our simple, reduced form representations. For any decisions in the face of the possibility, the risk, of such a catastrophic state must consider their consequences in that, as well as other, possible states.

With this in mind, we suggest that the following comprise a minimal set of essential components of any atemporal "Catastrophic Risk" model[12]:

[11]Models (iii), (iv), and (vi) are all multiple-prior models, albeit in a static setting. Chen and Epstein (2002) build their "recursive multiple-priors utility" on the static minmax expected utility model (iii). These models provide a foundation for dynamic models exploring the inpact of ambiguity aversion in finance and macroeconomics models. See the references in note 5 above, as well as Jahan-Parvar and Liu (2014), Ju and Miao (2012), Klibanoff et al. (2009), and Strzalecki (2013).

[12]Each component has a natural temporal dimension from which we abstract in this exercise. It will become important to consider that dimension in many applications.

- A set of 'states' in which consumption/production opportunities, and hence welfare, are severely limited;
  - Very low probability ('rare') and very low utility (high loss);
  - Severe restriction of production possibilities, limiting ability to recover or raise consumption;
  - Uncertain timing and incidence[13];
- Decisions/'Acts', each with impact/consequences across all 'states';
  - Consequences of ex-ante *decisions/acts* are dramatically different in catastrophes than in non-catastrophic states;
  - Learning and adjustment over time, as information about states evolves;
- Shift in the 'evaluation paradigm' in the event of catastrophe: "state-dependent utility"[14]
  - *ex-ante* uncertainty/'ambiguity' with regard to likelihood of 'catastrophic' states, appropriately updated in a dynamic analysis;
  - *ex-post* different ordering over consequences, and attitude toward risk, and handling of new information in 'normal' and 'catastrophic' states.

In this note we explore, for illustrative purposes, some of these aspects in an extremely simple 2- or 3-state example, aimed at revealing the implications of different approaches to modeling decision maker behavior in the face of uncertainty about the likelihood of the 'catastrophe'. Thus it attempts to capture, if only in highly reduced form, the differences in consequences and utilities between 'normal' and 'catastrophic' states. The illustrative examples, and the decision models discussed, will use a common notation within the Anscombe and Aumann (1963) framework.

2.1 Illustrative Examples

Two examples are used to illustrate the implications of differing formulations/models of decision making in the face of catastrophic risk. They allow us to clearly show how differing assumptions about agent behavior impact the evaluation of a decision/action to be taken. The first is in a fully reduced form, where preferences/utility only reflect the risk attitude toward the occurrence of a catastrophic state. Hence it illustrates how differing assumptions about agents' perceptions of the uncertainty

[13]Here collapsed into the 'valuations'.

[14]While it is typically not possible to identify unique state probabilities and utilities in the derivable additively separable representation, we believe it useful to choose a specific representation that captures the intuitively plausible impact of the state on ex-post (Bernoulli) utilities. See Karni (1985) for a thorough discussion.

affect their evaluations, the 'certainty equivalent' of the situation faced. The second allows (limited) exploration of how differing attitudes towards uncertainty, and differing risk attitudes in different states, might affect decisions.

2.1.1 2-Outcome Example

In the simplest example, outcomes, $x(s)$, give the agent's evaluation (expected utility) of the consequences of optimal decisions in state s, and her 'utility' function captures her attitude toward 'state risk'. Let $x(s_0) = 0.5$; $x(s_1) = 12$—the 'catastrophic' and 'normal' state outcomes, with probabilities $(0.005, 0.995)$ respectively, and $u(x) = 1 + \frac{x^{1-\eta}}{1-\eta}$ reflects her constant relative risk aversion, with $\eta = 2$. Then the expected value of this situation is $E(x) = 11.943$, and the standard (expected utility) model gives $Eu(x) = 0.907$, with a certainty equivalent of 10.776. This provides a benchmark for comparison with other approaches to evaluating 'state risk' uncertainty. This example is used to illustrate differences in evaluations in the first five representations.

2.1.2 3-Outcome Example

This example is only used with respect to the sixth representation. It allows the introduction of decisions (Savage 'acts') in the face of uncertainty about the true state. Let the underlying states be $S = \{0, 1\}$, and let there be potentially different outcomes in each of the states, $X = \{0.5, 3, 9\}$, $X_0 = \{0.5, 3\}$, $X_1 = \{3, 9\}$. Here the probabilities of outcomes depend on the decisions/acts of the agents, and those will depend on the beliefs of the agent about the probability of the 'bad' state, $s = 0$. In contrast to the 2-outcome example, there is no given probability for outcomes, and hence expected value is not well defined. Let p be the probability of the underlying 'bad' state. Then, for example, letting $f := (f_0, f_1)$ be the distribution over outcomes in the two states induced by the act $f \in \mathcal{F}$, the ex-ante probability of $x = 3$ is $p \cdot f_0(3) + (1-p) \cdot f_1(3)$.

To capture potential changes in preferences in the 'catastrophe' state (bad state, with worst outcome), we consider three Bernoulli utility functions, reflecting two different degrees of 'risk aversion'[15]:

$$u_0(x) = 1.15416 + \frac{x^{1-\gamma}}{1-\gamma} \text{ or } u_0(x) := v(x) = 0.5 + \frac{x^{1-\gamma}}{1-\gamma}, \; \gamma = 3; \; u_1(x) = \ln(x).$$

Note that the first u_0 and u_1 yield the same ex-post utility in the outcome ($x = 3$) that is common to both the catastrophe ($s = 0$), where it is 'best', and the normal state ($s = 1$), where it is worst. However, $v(3) < u_1(3)$, allowing illustration of the impact of distinctly lower welfare in the catastrophic state. Let $w_s(x)$ be the utility function

[15] Werner (2009) explores a new concept of risk for state contingent claims when states intrinsically matter to the agent, as in our example.

in state s. Then the (subjective) expected utility (SEU) model yields evaluation of act f,

$$Ew(f) = \sum_x p \cdot f_0(x)w_0(x) + (1-p) \cdot f_1(x)w_1(x),$$

with $w_0 = w_1 = w$ if utility is state independent.

3 Alternative Models

We look at six different models that capture uncertainty aversion, models which we feel show some promise for evaluation of situations with catastrophic risks. All but the fifth, Chichilnisky's "sensitivity to rare events," are built on the Anscombe and Aumann (1963) foundation, implicit in the axioms presented in the Appendix. And all are inspired by the failure of the Savage subjective utility framework to adequately capture behavior in the face of fundamental uncertainty, ambiguity with respect to underlying likelihoods of events.[16] The Ellsberg Paradox, in particular, has provided the challenge to which most of these models respond. And indeed, it is the true uncertainty, ignorance about likelihoods and consequences of truly catastrophic events that makes these models of decision making potentially important to their study.

3.1 'Behavioral Probabilities' and 'Ambiguity'

3.1.1 Choquet Expected Utility

We begin with the Schmeidler (1989) model of general non-additive valuation of risks using Choquet expected utility (CEU). This captures the idea that an agent's beliefs are not well specified, hence cannot be represented by a well specified probability. This representation derives from preferences satisfying axioms 1, 2a, 3, 4, and 6 in the Appendix. The representation allows uncertainty about likelihoods, and can reflect aversion to the ambiguity that uncertainty creates. Instead of a probability measure, the state space (S, Σ) is endowed with capacity,

$$\nu : \Sigma \longrightarrow [0,1]\,;\, \nu\left(\emptyset\right) = 0, \nu\left(S\right) = 1; \forall A, B \in \Sigma, A \subset B \Longrightarrow \nu\left(A\right) \le \nu\left(B\right).$$

Each potential act, f, is then evaluated using a Choquet integral,

$$I_\nu\left(f\right) = \int f d\nu \equiv \int_{-\infty}^{0} \left[\nu\left(f \ge t\right) - 1\right] dt + \int_{0}^{\infty} \nu\left(f \ge t\right) dt, \qquad (1)$$

$$f \succsim g \Longleftrightarrow I_\nu\left(f\right) \ge I_\nu\left(g\right)$$

[16]There is a good discussion in Gilboa (2009).

where, in general, $I_\nu(f+g) \neq I_\nu(f) + I_\nu(g)$ for differing acts f, g. A distaste for, an aversion to, ambiguity is captured in the *convexity* ('supermodularity') of the capacity ν:

$$\forall A, B \in \Sigma, \nu(A \cup B) + \nu(A \cap B) \geq \nu(A) + \nu(B).$$

An agent with such preferences uses a linear functional, $V(f) \equiv I_\nu(u \circ f) = \int u(f)\, d\nu$, to evaluate each act/decision that must be taken in the uncertain world. That is, $(f \succsim g) \Longleftrightarrow V(f) \geq V(g)$, and an optimal act is any $f^* \in \arg\max V(f) = \max_f I_\nu$ $(u \circ f)$.

This formulation can be used to capture other attitudes toward uncertainty/ambiguity, including pure SEU in situations without ambiguity. Indeed, if acts f, g are *co-monotonic*. i.e. $\exists p_\pi$, a probability vector for some permutation of states, π, such that

$$I_\nu(f) = \int_S f dp_\pi = \int_S g dp_\pi = I_\nu(g),$$

then $I_\nu(\cdot)$ becomes additive with respect to f and g; the comparison involves no ambiguity. Further, it can capture behavioral distortions of known probabilities, such as those revealed in the Allais paradox experiments and modeled in Prospect Theory.

3.1.2 Rank Dependent Expected Utility

A special case of non-additive 'probability' can be tractably analyzed using a second model, involving known risks, called Rank Dependent Expected Utility (Quiggin 1982). In this model, the finite set of outcomes, $x = (x_1, x_2, \ldots, x_N)$, is ordered from lowest (worst) to highest (best). Let $p = (p_1, p_2, \ldots, p_N)$ be the probabilities of each 'outcome', and $F(x_i)$ be the cumulative distribution function of x, evaluated at x_i. The agent is assumed to have a standard Bernoulli utility function over certain outcomes, $u(x)$, but to systematically distort the probabilities that she uses in evaluating prospects/decisions by using a probability weighting function that depends on the value/rank of the outcome. Such a function can capture the systematic deviations found in the experimental literature, such as overweighting or underweighting extreme (i.e. near zero or one) probabilities.

A representation of the preferences of such an agent is given by $V : \mathbb{R}^{2N} \longrightarrow \mathbb{R}$, the RDEU function:

$$V(\overline{x}, \overline{p}) = \sum_{i=1}^{N} u(x_i) h_i(\overline{p}), \tag{2}$$

where $h_i : \mathbb{R}^N \longrightarrow [0,1], i = 1, 2, \ldots, N; h_i(p) := q[F(x_i)] - q[F(x_{i-1})]$ weights outcome i as a function of the cumulative distribution of x, using a probability weighting function, $q : [0,1] \longrightarrow [0,1]$, $q(0) = 0$, $q' \geq 0$, that captures the agent's behavioral understanding of the probabilities. This allows non-linear (as revealed in behavioral

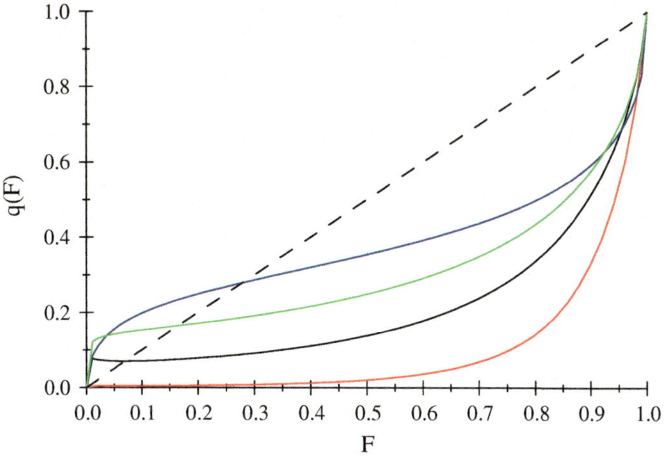

Fig. 1 Perception of outcome distribution

experiments) probabilities, while preserving first-order stochastic dominance. Quiggin's felicitous example is:

$$q(F) = \frac{F^{\gamma}}{\left(F^{\gamma} + (1-F)^{1-\gamma}\right)^{1/\gamma}};$$
$$\gamma \in (0, 1).$$

which 'overweights' extreme events [$h_i(\overline{p}) > p_i$, i near 1 or N], generalizing Tversky and Kahneman (1992). Thus it captures an empirical behavioral regularity at both the individual and group levels (experimental evidence; Gonzales and Wu 1999).[17] Note that a concave $q(F)$ implies classic risk aversion.

The distortion of probabilities that this implies is easy to see graphically. It is clear in the overall shape of Fig. 1, where the (dashed) diagonal shows the non-distorted cumulative probabilities. Different levels of distortion are indicated by color: $\gamma = 0.1$—red; $\gamma = 0.2$—black; $\gamma = 0.3$—green; $\gamma = 0.5$—blue. In Fig. 2, we magnify the graph in the vicinity of zero, near the probability of the worst possible outcome, showing that lower γ generates a greater increase in the exaggeration of the perceived probability of the worst event.

Analytic Example. The impact of preferences displaying rank dependent expected utility can be seen in the 2-outcome example. Such an agent dramatically overestimates the likelihood of the 'catastrophic' outcome, c, leading to a dramatically diminished valuation (certainty equivalent) of the prospect of facing that catastrophe. Letting $\gamma = 0.2$, $q(0.005) = 0.079453$, $q(0.995) = 0.93456$, and hence $RDE(x) =$

[17]Overweighting at the bottom may capture 'fear' of the worst outcome, as overweighting at the top may capture 'hope' for the best outcome.

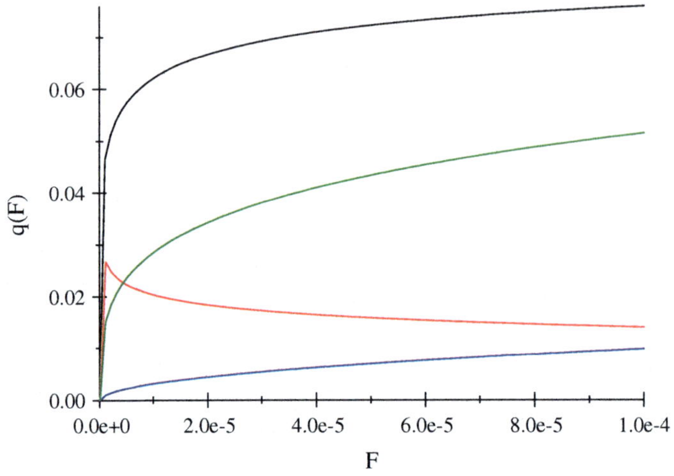

Fig. 2 'Rare event' probability perception

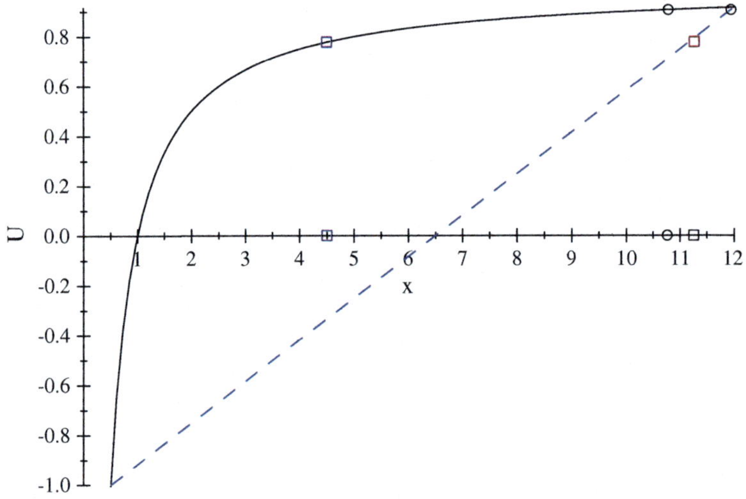

Fig. 3 Catastrophic risk: EU(x) & RDEU(x) certainty equivalent

11.254 and $RDEu(x) = 0.777$ with a u-certainty equivalent of 4.489. This is illustrated in Fig. 3, where 'circles' indicate expected utility and 'boxes' RDEU results. Note the non-additivity of the Quiggin probabilities, indicating extra weight at the extremes.

3.1.3 Maximin Expected Utility

Another promising model of preferences in the face of true uncertainty is the maxmin expected utility representation of Gilboa and Schmeidler (1989).[18] Rather than 'distorting' probabilities, this model captures an agent's ignorance/uncertainty about the likelihood of events in a set, $\Phi \subset \Delta(\Sigma)$, of potential/conceivable probability distributions over (S, Σ), with general S, and Φ assumed weak*-compact.[19] Preferences over outcomes are represented by $u(\cdot)$—strictly increasing, continuous, weakly concave—and the agent's evaluation of an 'act'/decision f, a real-valued, measurable, bounded function, is given by

$$V(f) \equiv \min_{\varphi \in \Phi} \left(\int_S u(f(s))\, d\varphi(s) \right). \tag{3}$$

This derives from preferences satisfying axioms 1, 2b, 3, 4, 5, and 6, where axiom 5, a 'convexity of preferences' or 'preference for hedging' assumption, gives the uncertainty aversion based on supermodularity/convexity of the capacity in CEU. Hence the optimal decision/act is,

$$f^* \in \arg\max V(f) = \max_f \left\{ \min_{\varphi \in \Phi} \left(\int_S u(f(s))\, d\varphi(s) \right) \right\},$$

the maximizing act against the minimizing distribution over outcomes.

This formulation provides a cognitive interpretation of Choquet Expected Utility, when $\Phi = core(v) = \{\varphi \in \Delta(\Sigma) |\ \varphi(A) \geq v(A),\ \forall A \subset \Sigma\}$, and $\Delta(\Sigma)$ is the space of all finitely additive probability measures on S. Then probabilities can be understood as based on past experience. However, Φ can be *more general*, containing $\varphi \notin core(v)$ for any capacity v.

This formulation of preferences clearly separates the agent's 'uncertainty', captured in Φ, and the agent's attitude/aversion towards that uncertainty, captured in $V(\cdot)$. Indeed, it displays a strong uncertainty aversion, an *unwillingness* to place any order, any likelihood, over the distributions in Φ, behaving as if the worst possible distribution there is actually true. The only limits to this pessimism in the face of ambiguity/uncertainty are in the size of the set Φ. Indeed, if Φ is a singleton, this criterion reduces to maximizing (subjective) expected utility—there is no ambiguity. On the other hand, if the agent is entirely ignorant of (or unwilling to contemplate) possible probability distributions, it reduces to the Wald (1945) Maximin Criterion:

$$\max_f \left\{ \min_s u(f(s)) \right\}.$$

---

[18]This is a canonical 'multiple-priors' model exploited in the financial literature and extended to dynamic models in Chen and Epstein (2002).

[19]This is the $\sigma(\Delta(\Sigma), B_0(\Sigma))$ topology: a net $\{p_\iota\}_{\iota \in \Upsilon}$ converges to p iff $p_\iota(A) \longrightarrow p(A),\ \forall A \in \Sigma$.

3.1.4 Smooth Ambiguity Aversion (Second Order Expected Utility)

This representation of preferences in the face of uncertainty/ambiguity extends the prior model by assuming a more sophisticated handling of the possible distributions of outcomes. Rather than assuming the worst, the decision maker places a (subjective) distribution over the set Φ, allowing her to weigh the likelihood of different distributions governing the impact of acts on (payoff relevant) outcomes. One particularly clear model is that of Klibanoff et al. (2005) which parametrizes the agent's uncertainty about a deeper (second-order) 'state', $\theta \in \Theta$, that determines the distribution of states, $\varphi_\theta \in \Delta(\Sigma)$, affecting the payoffs to acts.[20] The agent's subjective distribution ζ over Θ reflects her uncertainty about the distribution, φ_θ, governing 'events' in S. The attitude toward this 'state risk' is then modeled through introducing another strictly increasing function, $v : \mathbb{R} \longrightarrow \mathbb{R}$, which together with $u : X \longrightarrow \mathbb{R}$ captures the 'taste for ambiguity' of this agent. As usual, u reflects the agent's 'attitude toward risk'.

Based on the assumption that both first-order and second-order preferences are mutually consistent and have expected utility representations, the decision relevant preference representation over acts, f, becomes:

$$\begin{aligned} U(f) &= v^{-1}\mathbb{E}_\zeta v\left(u^{-1}\left(\mathbb{E}_{\varphi_\theta} u \circ f\right)\right) \equiv v^{-1}\mathbb{E}_\zeta \phi\left(\mathbb{E}_{\varphi_\theta} u \circ f\right) \\ &\equiv v^{-1}\left(\int_\Theta v\left(u^{-1}\left(\int_S u(f)\, d\varphi_\theta\right)\right) d\zeta(\theta)\right), \end{aligned} \tag{4}$$

where $u^{-1}\left(\int_S u(f)\, d\varphi_\theta\right)$ is the *certainty equivalent* of the gamble induced by decision f.

$$\begin{aligned} (f \succsim g) &\Longleftrightarrow U(f) \geq U(g) \\ &\Leftrightarrow v^{-1}\mathbb{E}_\zeta v\left(u^{-1}\left(\mathbb{E}_{\varphi_\theta} u \circ f\right)\right) \geq v^{-1}\mathbb{E}_\zeta v\left(u^{-1}\left(\mathbb{E}_{\varphi_\theta} u \circ g\right)\right). \end{aligned}$$

'Acts' are thus ranked by the 'certainty equivalent' (CE) of the induced (by ζ) distribution of the CEs of the 'lotteries' in each 'state'.[21] This formulation separates 'ambiguity' (beliefs) from 'attitude toward ambiguity' (tastes): the distribution ζ over Θ, $|\Theta| > 1$, captures ambiguity (beliefs); the composite function $\phi := v \circ u^{-1}$ captures *attitude* toward ambiguity – concavity $\Longleftrightarrow$ 'ambiguity aversion'. As a firmly (subjectively) held belief, ζ also reflects the pessimism/optimism of the agent, with a 'pessimistic' agent more heavily weighting the state(s), θ, with the greatest probability of disaster. Here, ambiguity aversion is an aversion to 'mean preserving spreads'

[20] This is derived assuming 'second order' preferences over 'acts' mapping the set of all probabilities over a sufficiently rich 'state space' directly into consequences, that is consistent with first-order preferences over Savage acts from 'states' to consequences, both satisfying the (subjective) expected utility hypothesis. Klibanoff et al. (2005), pp. 1856–9.

[21] $U(f)$ represents identical preferences to $V(f) \equiv \mathbb{E}_\zeta \phi\left(\mathbb{E}_{\varphi_\theta} u \circ f\right)$ (Klibanoff et al. 2005), as v is strictly monotonic. We find it more convenient to work with certainty equivalents in the fixed outcome space than with (subjective) utility values.

in the distribution of $\mathbb{E}_{\varphi_\theta} u \circ f$ induced by ζ and f. If, however, the composite function ϕ is linear, then we have 'ambiguity neutrality', implying the reducibility of the compound distribution, rendering the representation observationally equivalent to expected utility with the subjective prior ζ.

Analytic Example. The impact of these preferences can again be nicely illustrated in our 2-outcome example of catastrophic risk. Suppose that the agent's uncertainty about the state of the world is fully captured by 2 possible distributions: in θ_0 the probabilities are $(0.05, 0.95)$ while in state θ_1 they are $(0.005, 0.995)$ as above.[22] Hence in the worst case (θ_0), the catastrophic state is 10 times as likely as in our base case. Let the agent's underlying aversion to risk be reflected in the utility function, $u(x)$, as above, and let $v(r) := 1 + \frac{x^{1-\gamma}}{1-\gamma} = 1.0 - \frac{1}{3x^3}$ $(\gamma = 4)$, reflecting *ambiguity aversion;* $\phi(r) := v \circ u^{-1}(r) = \left(1 - \frac{1}{3\left(\frac{1}{r-1}\right)^3}\right) = 1 - \frac{1}{3}(r-1)^3$ is a clearly concave function. The two 'valuation' functions, $u(x)$ over outcomes and $v\left(u^{-1}(\cdot)\right)$ over certainty equivalents of risky prospects, are presented in Fig. 4, where the utilities are normalized to be equal in the best of all possible outcomes.[23]

The circles on $u(x)$ in Fig. 5 indicate the expected values and expected utilities of the gambles, and boxes give the certainty equivalents, in 'states' θ_0 and θ_1 respectively. The agent's uncertainty relates to which of these 'worlds' obtains, leading her to evaluate this ambiguity by 'weighting' these worlds with the distribution ζ.

Here the evaluation function, U, is given by:

$$U \equiv v^{-1}\left(\sum_{\theta} v\left(u^{-1}\left(p_c^\theta u(c) + p_n^\theta u(n)\right)\right) d\zeta(\theta)\right),$$

that is, it is the certainty equivalent of the ζ-weighting of the v-values of each u-certainty equivalent in each 'world', θ. Let the agent believe $\zeta = (0.4, 0.6)$, i.e. that there is a 40 % chance that the truly catastrophic world obtains. Then

$$\begin{aligned} U &= v^{-1}\left(0.4 \cdot v(5.5814) + 0.6 \cdot v(10.7623)\right) \\ &= v^{-1}(0.915\,932\,711) = 7.110299, \end{aligned}$$

indicated by circles on $v(x)$ [the upper curve] and the x-axis in Fig. 5. A more pessimistic view, $\zeta = (0.6, 0.4)$, yields $U = 6.42423$, as indicated in the graph with diamonds. Note that both these views are more optimistic than give by the Quiggin probability distortion in the RDEU model; see the + on $u(x)$ in Fig. 5.

---

[22]Note that this worst case places a lower probability (0.05) on the 'catastrophe' than does the RDEU distortion (0.795) above.

[23]In this normalization, if the certainty equivalent of a risk were the worst possible outcome, the 'utility' would be $v(0.5) = -1.75 < u(0.5) = -1$, a true catastrophe. The precise normalization of origin is, however, irrelevant to the decision; only the curvature has meaning.

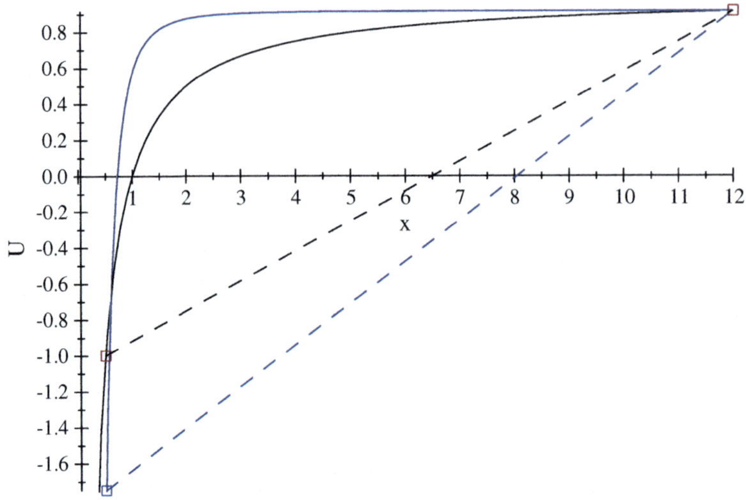

Fig. 4 $v(x)$ [*blue*] and $u(x)$ [*black*], equal at $x = 12$

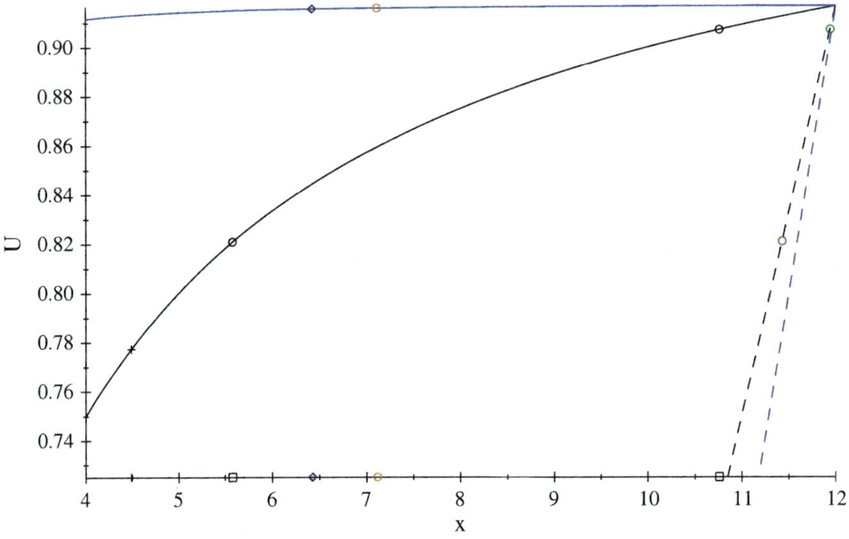

Fig. 5 SOEU evaluations with RDEU comparison

Indeed, this second order approach remains valid when first-order preferences are assumed to be RDEU (hence non-additive), so that the inner integral becomes a Choquet integral. If we look at the RDEU valuation in the θ_0 world, using $q(F)$ above, we find that the decision maker's weights assigned to the 2 outcomes sum to only 0.7413, as she assesses $q(0.05) = 0.0702$ and $q(0.95) = 0.6711$. Hence her rank-dependent 'expectation' of the outcome is 8.0883 and her RDEU valuation is 0.54498, as illustrated in Fig. 6.

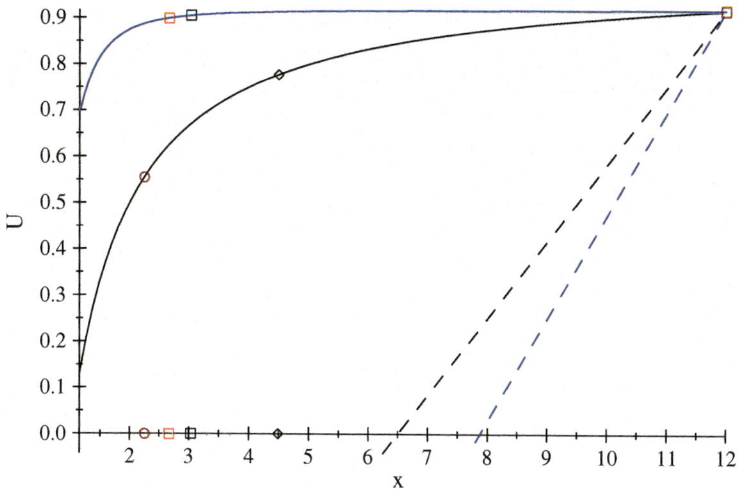

Fig. 6 RDEU CEs & SOEU evaluations

Here we also illustrate the SOEU when $\zeta = (0.4, 0.6)$—there's only a 40 % chance of the disastrous world in which there is a 5 % chance of a catastrophe, rather than just 0.5 %—with a 'box' on the 'second order' utility [upper curve], v:

$$U = v^{-1}\,(0.4 \cdot v\,(2.2471) + 0.6 \cdot v\,(10.7623)) = 3.036.$$

With the more pessimistic view, $\zeta = (0.6, 0.4)$, $U = 2.6589$ [lighter box]. Hence with Choquet/Quiggin expected utility over outcomes, the agent places much greater weight on the possibility of disaster, and a much lower evaluation of the world containing such a prospect, than does an agent with SOEU preferences and ζ beliefs.

3.2 *"Sensitivity to Rare Events: A Topology of Fear"*

This model takes a radically different approach to providing a formal, hence analytically useful, representation for preferences in the face of truly catastrophic risks. Rather than using the Anscombe-Aumann framework as above, this model begins from the von Neuman-Morgenstern axioms of weak preference, independence, monotonicity, continuity and boundedness of utility, and monotone continuity of beliefs (countably additive measure on the space/σ-field of events). Chichilnisky (2000, 2009, 2010a, b) argues that the resulting countably additive valuation functional, the expected utility representation, cannot, in principle, deal with catastrophic risk, as it necessarily ignores "rare events," the only events relevant for true catastrophe. Such a functional arises from the assumption of monotone continuity of preferences on the σ-field of events. Continuous bounded utility and monotone continuity imply the impact of any catastrophe, however large, becomes analytically negligible as its likelihood (Lebesgue measure of the event) diminishes to zero. Or, if marginal utility is unbounded, the potential catastrophic risk swamps all other considerations, however unlikely its occurrence may be Buchholz and Schymura (2012). Hence, she argues, to analyze truly catastrophic risks, we need to break free of the standard model, and develop new analytic foundations for that analysis.

To do so, Chichilnisky introduces axioms requiring "equal treatment" of both 'rare' and 'frequent' events, and dispenses with monotone continuity as contradicting sensitivity to rare events.[24] This requires a richer space of 'states' than other formulations—both unboundedness and cardinality no less than the continuum—one rich enough to capture all possible consequences, however unlikely, of any 'act'. To develop a basic representation, the simplest such space, $S = \mathbb{R}^1$, is used. It is endowed with Lebesgue measure $\mu(s)$, hence assuming that the underlying 'states' are 'equiprobable', with the likelihood/probability of an event depending on the 'number' of states in it. Functions $f, g : \mathbb{R} \longrightarrow \mathbb{R}$ are 'acts' generating "lotteries,", giving the 'utility' payoffs $u(f(s))$ and $u(g(s))$ to each 'state' s. The consequences, $u \circ f$, are assumed to be a.s.(λ) bounded functions. To insure that the impact of both frequent ($\mu(E) > 0$) and rare ($E \subset S$, $\mu(S) = 0$) events are captured, the space of 'acts', $\mathcal{F} : = L_\infty(\mathbb{R})$ [essentially bounded functions], is endowed with the topology generated by the esssup norm, $\|\cdot\|_\infty$, felicitously named the "topology of fear" (Chichilnisky 2009).

To generate a preference representation, Chichilnisky (2009, 809–810) applies a new axiom in the classic decision theory framework (Arrow 1971), "sensitivity to rare events." This gives 3 (summary) axioms for a ranking (preference representation):

Axiom 1. The ranking $W : L_\infty(\mathbb{R}) \rightarrow \mathbb{R}$ is linear and $\|\cdot\|_\infty$-continuous;

Axiom 2. The ranking $W : L_\infty(\mathbb{R}) \rightarrow \mathbb{R}$ is sensitive to rare events: the ordering of f and g can depend on consequences in events of arbitrarily small (Lebesgue) measure[25];

Axiom 3. The ranking $W : L_\infty(\mathbb{R}) \rightarrow \mathbb{R}$ is sensitive to frequent events: the ordering of f and g is dependent on payoffs in events of sufficiently large (Lebesgue) measure[26];

---

[24]The axiomatization sharply distinguishes Chichilnisky's model from those in which ambiguity and the decision maker's 'taste' for ambiguity are axiomatized as in the Appendix below.

[25]Formally, the negation of "insensitivity to rare events:" $\forall f, g$, $\exists \epsilon = \epsilon(f, g) > 0$, S.T. $W(f) > W(g) \Longleftrightarrow W(f') > W(g')$ $\forall f', g'$ satisfying $f' = f$, $g' = g$ a.e. on $A \subset \mathbb{R}$ when $\mu(A^c) < \epsilon$.

[26]Formally, the negation of "insensitivity to frequent events:" $\forall f, g$, $\exists \delta = \delta(f, g) \in (0, 1)$, S.T. $W(f) > W(g) \Longleftrightarrow W(f') > W(g')$ $\forall f', g'$ satisfying $f' = f$, $g' = g$ a.e. on $A \subset \mathbb{R}$ when $\mu(A^c) > \delta$.

Under these axioms, a preference representation for the evaluation of acts with potential extreme consequences is shown to exist.[27] The representation is a bounded linear functional on $L_\infty(\mathbb{R})$,

$$W(f) = \lambda \int_{\mathbb{R}} u(f(s))\, \phi_1(s)ds + (1-\lambda)\langle u \circ f, \phi_2 \rangle, \tag{5}$$

where $\lambda \in (0, 1]$, ϕ_1, $\phi_2 \in L_\infty^*$—continuous linear functionals on L_∞; $\phi_1 \in L_1(\mathbb{R})$, $\int_{\mathbb{R}} \phi_1(x)dx = 1$, and ϕ_2 is a purely finitely additive functional. The parameter λ is an essential aspect of the underlying preferences:

$$\begin{aligned}(f \succsim g) \Longleftrightarrow \lambda \int_{\mathbb{R}} u(f(s))\, \phi_1(s)ds + (1-\lambda)\langle u \circ f, \phi_2 \rangle \\ \geq \lambda \int_{\mathbb{R}} u(g(s))\, \phi_1(s)ds + (1-\lambda)\langle u \circ g, \phi_2 \rangle.\end{aligned}$$

λ reflects the decision maker's 'belief' in the reality of extremely rare events, while ϕ_2 provides her 'evaluation' of the consequences in such events. Both arise from a 'sensitivity' to rare events, with ϕ_2 placing a value on unmeasurable events within those of Lebesgue measure zero, and λ providing the 'utility weight' placed on the "normal" outcomes of the 'lottery' f. Notice that, in the absence of 'extreme outcomes', $W(f)$ reduces to classic expected utility. This could also occur if the agent refuses to consider the possibility of extreme events, or is "paralyzed by fear," hence setting $\lambda = 1$.

This model would seem to provide a very direct way to deal with true catastrophe, but its mathematical complexity, the non-constructive derivation, and general lack of explicit analytic form for the purely finitely additive (pfa) valuation functional make it hard to see how this representation can be applied, aside from elucidating some general principles.[28] Thus Chichilnisky has proposed a finite-state version of the representation, which however lacks the same firm mathematical foundation.

3.2.1 Finite-State Version with Extreme Event

Let $S, X \subset \mathbb{R}^S$, and $\mathcal{F}$ be as in the 'ambiguity' models above, and $s^* \in S$ be a 'rare' catastrophic state if its probability $\phi_{s^*} < \varepsilon$ for some arbitrarily small $\varepsilon > 0$. By analogy, we have the preference representation, $W : \mathbb{R}^S \longrightarrow \mathbb{R}$,

[27] The proof is non-constructive, using the axiom of choice.

[28] Some pfa functionals can be expressed as limits, as in the example Chichinilsky (2009, p. 814) gives. One satisfactory infinite dimensional application has been made in Figuieres and Tidball (2012) where limiting outcomes are evaluated. In the case of pure uncertainty, and it might be possible to isolate the valuation of 'rare' events along an ultrafilter net, but we do not see clearly how to do that.

$$W(f) = \lambda \cdot \langle \phi, u(f) \rangle + (1 - \lambda) \cdot \min_s u(f_s), \tag{6}$$

which puts extra 'weight' on the catastrophic outcome, $\min_s f(s)$. Here the first term is an 'expected utility' based on the 'subjective' probability vector, ϕ, and the utility vector, $u(f)$, resulting from the act, while the second puts independent weight on the worst that can happen. Maximizing $W(f)$ trades off maximization of expected utility against the minimization of catastrophic loss. Again, λ is a critical preference parameter, which might be derived from a constrained optimization formalizing that trade-off:

$$\max_{f \in F} \langle \phi, u(f) \rangle \;\; s.t. \;\; \min_s u(f_s) \geq \overline{u}, \tag{7}$$

or

$$\max_{f \in F} \min_{\mu \geq 0} \left[\langle \phi, u(f) \rangle + \mu \left(\min_s u(f_s) - \overline{u} \right) \right] \equiv \max_{f \in F} \min_{\mu \geq 0} L\left(u(f), \mu\right),$$

giving $\lambda = \frac{1}{1+\mu}$.

This formulation satisfies analogies of Axioms 2 and 3 above, but not Axiom 1: it is not continuous (Chichilnisky 2010), raising a question about its necessity and uniqueness. It also must assume (for relevance) that s^* at which $\min_s u(f_s)$ occurs (the 'catastrophe') is 'rare': $\varepsilon > \phi_{s^*} > 0$. However, it provides a wide range of valuations of potentially catastrophic lotteries, from standard EU ($\lambda = 1$) to the Wald maximin criterion ($\lambda = 0$), depending on the (subjective) value of λ. Finally, it is easier to apply than the rich model, while maintaining its intuition, and it appears to be justified by the axioms supporting the final model of preferences that we explore below, "variational preferences."

Analytic Example. The implications of this formulation can be easily seen in our 2-state example. There the probabilities and outcomes associated with the some act are given, so we parametrize the preference representation by λ:

$$\begin{aligned} W(f;\lambda) &= \lambda \cdot \langle \phi, u(f) \rangle + (1 - \lambda) \cdot \min_s u(f_s) \\ &= \lambda \left(0.005 u(c) + 0.995 u(n)\right) + (1 - \lambda)\, u(c)\,. \end{aligned}$$

Hence $W(f;\lambda) = 1.90708\lambda - 1.0$, $W(f;1) = 0.90708$, and $W(f;0) = -1.0$, showing the vast range of valuations that variation of λ generates. When $\lambda = 0.5$, $W = -0.04646$, and the certainty equivalent of this lottery becomes 0.9556, clearly reflecting "fear" of the 'bad' outcome. It is depicted in Fig. 7, where boxes show the expected utility, the certainty equivalent, and its utility. Indeed, the appropriate choice of λ can replicate the evaluation of any of our preference representations. For example, $\lambda = 0.92497 \Longrightarrow W = 0.764$, which is identical to the *RDEU* (with $\gamma = 0.2$) evaluation, reflecting just a little "fear"—see Figs. 3 and 7.

Clearly this representation can generate more extreme responses to extreme hazards than the other modifications of expected utility.

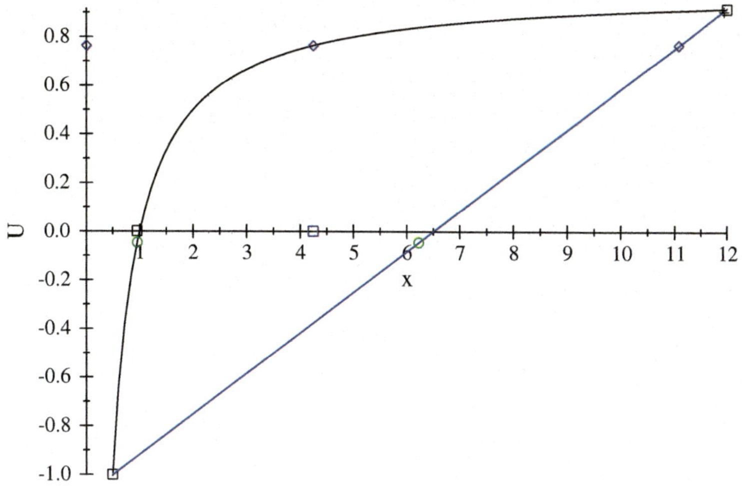

Fig. 7 'Topology of Fear' & RDEU(x) certainty equivalents

3.3 *Variational Preferences: An Umbrella Model?*

These preferences were introduced by Maccheroni et al. (2006) as a unifying framework for understanding the various models of decisions in the face of ambiguity based on a common behavioral (axiomatic) foundation. The models encompassed include those above, and the "multiplier preferences" model of Hansen and Sargent (2001), which is motivated by model uncertainty in control problems. These are a 'class' of preferences that can be specialized to cover most other forms of ambiguity respecting preferences, as discussed in Strzalecki (2011), as well as expected utility when the agent is 'ambiguity neutral'. In addition, variational preferences support the representation suggested for the finite state Chichilnisky model (6).

The common behavioral foundation is laid out in the notation and axioms in the Appendix. Under Axioms 1 through 6, preferences $\succsim$ defined on $\mathcal{F}$ are representable by

$$V(f) = \min_{p\in\Delta(\Sigma)} \left[\int u(f)dp + c(p) \right], \tag{8}$$

where $u : X \to \mathbb{R}$ is an affine utility; $c(p) : \Delta \to [0, 1]$ is an 'index' of ambiguity aversion, a convex lower-semicontinuous function with $\inf_{\Delta} c(p) = 0$. Over the space of countably additive probabilities, $\Delta^{\sigma}(\Sigma)$, $c(p) = \sup_{f\in\mathcal{F}} \left(u\left(x_f\right) - \int u(f)dp\right)$, where x_f is the certainty equivalent of act f, and, with Axiom 7, is unique. $c(p)$ is a "penalty" on less likely distributions, so lower $c(p)$ reflects higher 'ambiguity aversion'.

The other preference representations can be generated by specifying the form of the 'ambiguity index', $c(p)$. For example, "multiple priors" (maximin) preferences result from

$$c(p) = \delta_C(p) = \begin{cases} 0, & \text{if } p \in C \\ \infty, & \text{otherwise} \end{cases},$$

"multiplier preferences" from $c(p) = \theta R(p \parallel q)$ where $R(p \parallel q)$ is the relative entropy of p with respect to $q \in \Delta(\Sigma)$, a fixed, countably additive, non-atomic measure, and $c(p) = \theta G(p \parallel q)$ gives "mean-variance" preferences, with

$$V(f) = \int f dq - \frac{1}{2\theta} Var(f) = \min_{p \in \Delta(\Sigma)} \int f dp + \theta G(p \parallel q),$$

where $G(\cdot \parallel q) : \Delta \to [0, 1]$ is the relative Gini concentration index.[29] All other variational are ambiguity averse, as Axiom 5 indicates a "preference for hedging."[30]

Defining $\succsim_1$ as "more ambiguity averse" than $\succsim_2$ if, $\forall f \in \mathcal{F}$ and $x \in X$, $f \succsim_1 x \Longrightarrow f \succsim_2 x$, it is clear that $\{c_1 \leq c_2$ for affine equivalent $u_i, i = 1, 2\} \Leftrightarrow \{\succsim_1$ is more ambiguity averse than $\succsim_2\}$. "Ambiguity neutral" (subjective) expected utility, with prior q, arises from

$$c(p) = \delta_q(p) = \begin{cases} 0, & \text{if } p = q \\ \infty, & \text{otherwise} \end{cases};$$

this is the model with minimal ambiguity aversion. the case of maximal ambiguity aversion in this model is given by the Wald criterion: $c_m(p) = 0, \forall p \in \Delta$, so that

$$f \succsim g \Leftrightarrow \min_{p \in \Delta(\Sigma)} \int u(f) dp \geq \min_{p \in \Delta(\Sigma)} \int u(g) dp;$$
$$f \succsim g \Leftrightarrow \min_{s \in S} u(f(s)) \geq \min_{s \in S} u(g(s)).$$

Other cases of less extreme ambiguity aversion can be defined using a convex combination of these extremes. Let $\alpha \in (0, 1)$, and

$$V(f) = (1 - \alpha) \int u(f) dq + \alpha \min_{s \in S} u(f(s)).$$

Here, if $c(p) = \min_{p_1, p_2 \in \Delta} \{(1 - \alpha) c_q(p_2) + \alpha c_m(p_1) : (1 - \alpha) p_2 + \alpha p_1 = p\} = \delta_{(1-\alpha)q + \alpha\Delta}(p)$, then this is a special case of variational preferences that gives the same representation as the finite-state "topology of fear" preferences.

---

[29] Maccheroni et al. (2006), pp. 1449–50.

[30] There are several alternative definitions of 'ambiguity aversion' and 'ambiguity neutrality', and comparative ambiguity aversion, in the literature (Etner et al.; pp. 253–259). Their distinctions become more important in dynamic formulations where issues of probabilistic sophistication and consequentialism (Hammond 1988) become critical.

Analytic Example. This model can clearly replicate the outcomes of any of the above representations in our 2-outcome example; it can replicate any degree of 'aversion to uncertainty' in the face of a potentially catastrophic event. So we turn to the more elaborate (3-outcome) example with distinct outcomes in each of the two states, normal ($s = 1$) and catastrophic ($s = 0$), set up above (Sect. 2.1.2). Let there be 3 distinct acts mapping S into $\Delta(X)$, $X = \{0.5, 3, 9\}$:

$$\mathcal{F} = \left\{ \begin{array}{lll} f: & f_0 = (0.25, 0.75, 0) & f_1 = (0, 0.8, 0.2) \\ g: & g_0 = (0.5, 0.5, 0) & g_1 = (0, 0.6, 0.4) \\ h: & h_0 = (0.7, 0.3, 0) & h_1 = (0, 0.5, 0.5) \end{array} \right\}. \qquad (9)$$

These acts differ in mitigation effort with respect to the potential disaster, $x = 0.5$. Act f takes the possibility of disaster most seriously, placing the greatest effort on mitigation, diverting resources to that effort, and hence reducing the likelihood of the 'best' configuration of outcomes. Act h pays little attention to the possibility of a disastrous outcome, 'maximizing' the likelihood of the 'best' outcomes, as well as of the 'catastrophe'; it might be considered "business as usual" behavior. Finally, g cautiously compromises between these two approaches.

To illustrate the implications of variational preferences for choice in the face of 'catastrophe', we adapt a variational preferences representation with state dependent preferences:

$$V(a) = \min_{p \in [0,1]} \left\{ pE_0 \left\{ u_0(a_0) \right\} + (1-p) E_1 \left\{ u_1(a_1) \right\} + c(p) \right\},$$
$$a \in \{f, g, h\},$$

and let

$$c(p) = 5\,(p - 0.2)^2, p \in [0, 0.6]\,; c(p) = M, p \in (0.6, 1].$$

This assumes the agent believes that the 'most realistic' estimate the probability of the catastrophic state is $p = 0.2$, and that $p > 0.6$ is 'impossible'. We also assume that the agent is more risk averse in the 'catastrophic' state, and consider 2 cases: (i) Utility level independent of 'state' in the most likely configuration of outcomes, $x = 3$; (ii) Utility level is also "state dependent" even when different 'states' yield same outcomes ($x = 3$):

$$\text{(i)} \quad u_0(x) = 1.15416 + \frac{x^{1-\gamma}}{1-\gamma}; u_1(x) = \ln(x)\,; \gamma = 3; \quad u_0(3) = u_1(3);$$

$$\text{(ii)} \quad u_0(x) = 0.5 + \frac{x^{1-\gamma}}{1-\gamma}; u_1(x) = \ln(x)\,; \gamma = 3; \quad u_0(3) < u_1(3).$$

Note that the coefficient of relative risk aversion is 3 in state $s = 0$ and 1 in state $s = 1$. The consequences of the 3 acts, f, g, h, (diamond, box, and circle, respectively, in the graphs) in the catastrophic state, $s = 0$, are illustrated in Figs. 8 and 9 in terms

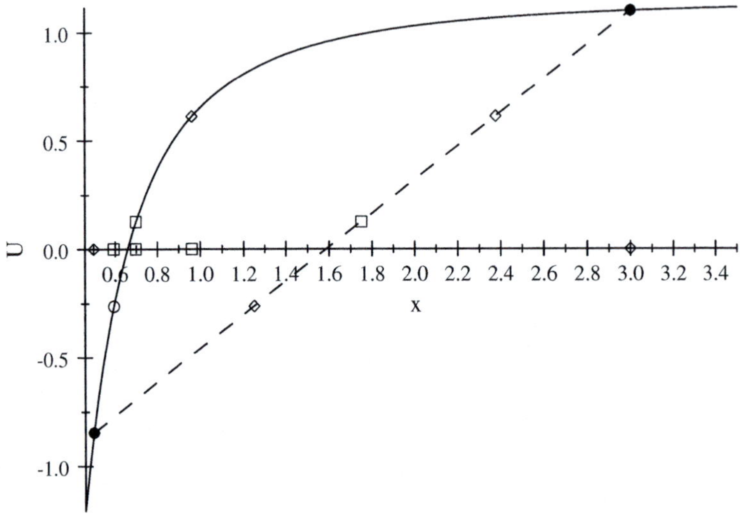

Fig. 8 'Catastrophe State' lotteries, $u_0(3) = u_1(3)$

of expected utilities (with outcomes) and certainty equivalents, with f yielding the highest value.

Note that the certainty equivalents remain the same despite the distinctly lower utility levels in case (ii), showing the irrelevance of *that* normalization *within a state*. But that irrelevance vanishes as soon as we study the full consequences, evaluating the outcomes of various acts across states. Case (i) is illustrated in Fig. 9, where the green lines connect the outcomes each of the acts f, g, h (least to highest spread) generates in each of the states.

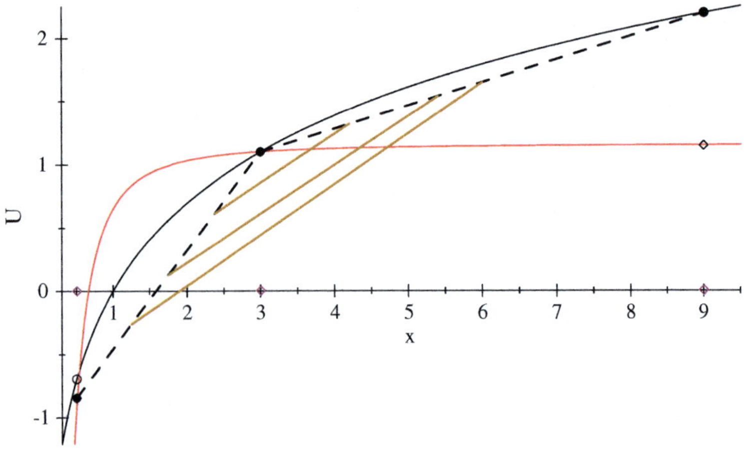

Fig. 9 $u(3)$ normalization, and the span of 'act' lotteries

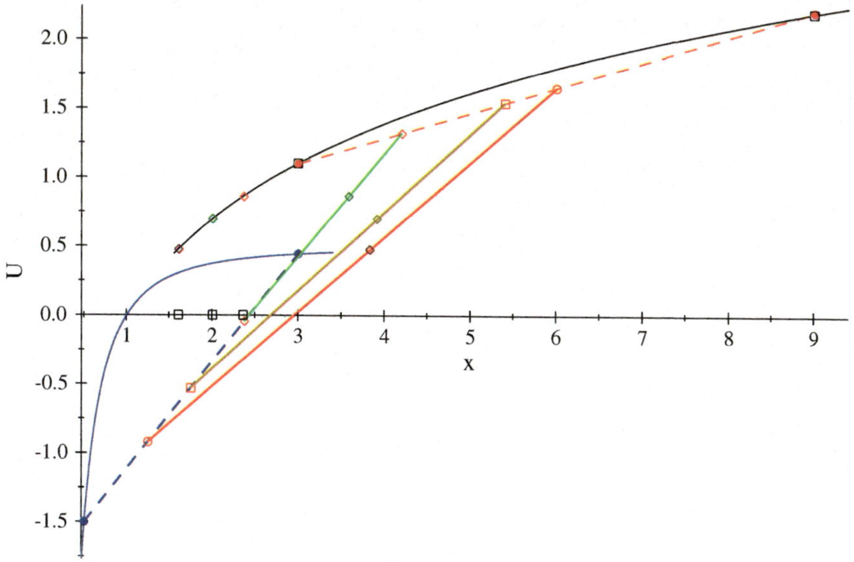

Fig. 10 $u_0(x) < u_1(x)$—expected values, CEs and utilities of f, g, h

When we take the loss of utility/wellbeing seriously in the catastrophic state, $u_0(x) < u_1(x)$ and more risk averse, the 'stakes' in the decisions/acts become much more serious, as can be seen in Fig. 10. Here the expected values of outcomes, as a function of p, of the acts f, g, and h are illustrated, where

$$\begin{aligned} V(f) &= 0.95385 \text{ at } p = 0.33600; \text{ Certainty Equivalent} : 2.3664; \\ V(g) &= 0.91151 \text{ at } p = 0.40658; \text{ Certainty Equivalent} : 2.00999; \\ V(h) &= 0.80615 \text{ at } p = 0.45646; \text{ Certainty Equivalent} : 1.611532, \end{aligned}$$

and $f \succ g \succ h$. Hence ambiguity aversion has led this agent to choose the most cautious, least remunerative act, f. With complete faith in her 'best estimate', reflecting absence of ambiguity, $p = 0.2$ [$\Leftrightarrow c(p) = 0$], and $h \succ g \succ f$, as $V(f) = 1.0463$, $V(g) = 1.1249$, and $V(h) = 1.1350$. This leads her to choose the least cautious, most remunerative, act, to pursue "business as usual." If the agent's preferences are described by the MEU (maximin) representation, then she believes $p = 0.6$, and $f \succ g \succ h$ as $V(f) = 1.3023$, $V(g) = 1.0986$, and $V(h) = 0.90917$.

It is straightforward to adjust this representation to incorporate RDEU probability distortions or "smooth ambiguity" preferences working with the certainty equivalents of the underlying 'act lotteries' and their 'ambiguity weightings'. More interesting is the impact of the finite version of the "topology of fear" model (Chichilnisky 2010b). As noted above, this arises from the representation (8),[31] $V(f) = \min_{p \in \Delta(\Sigma)} \int u(f)dp + c(p)$, with

[31] We set $\lambda = 1 - \alpha$ for notational consistency with that model.

$$c(p) = \min_{p_1,p_2 \in \Delta} \left\{ \lambda c_q(p_2) + (1-\lambda)\, c_m(p_1) \,:\, \lambda p_2 + (1-\lambda)\, p_1 = p \right\} = \delta_{\lambda q + (1-\lambda)\Delta}(p),$$

where

$$\delta_q(p) = \begin{cases} 0, & \text{if } p = q \\ \infty, & \text{otherwise} \end{cases},$$

giving for any act a,

$$\begin{aligned} V(a) &= \lambda \int u(a) dq + (1-\lambda) \min_{s \in S} u\left(a\left(s\right)\right). \qquad (10) \\ &= \left[q E_0 \left\{ u_0(a_0) \right\} + (1-q)\, E_1 \left\{ u_1(a_1) \right\} \right] + (1-\lambda)\, E_0 \left\{ u_0(a_0) \right\}. \end{aligned}$$

In our example, we need to specify the agent's (subjective) distribution over S, q, and the relevant expected utility in each state for each act. Using the 'best guess' from above, $q(s_0) = 0.2$, and the lottery generated by each act (9), we can calculate for each act the expected utility in each state, the impact of λ on the overall evaluation, $V(\cdot; \lambda)$, and the evaluation and certainty equivalent for $\lambda = 0.5$.

s	$Eu(f)$	$Eu(g)$	$Eu(h)$
0	−0.0417	−0.5278	−0.9167
1	1.3183	1.5381	1.6479
$V(\cdot;\lambda)$	$1.088\lambda - 0.0417$	$1.6527\lambda - 0.5278$	$2.0517\lambda - 0.9167$
$V(\cdot;.5)$	0.5023	0.29855	0.10915
Cert Equiv	1.6525	1.3479	1.1153

Clearly, if the agent puts full utility weight on the impact of the catastrophe ($\lambda = 0$), the most cautious act f is optimal, while if the agent places no special weight on the catastrophic outcome ($\lambda = 1$), the ordering is reversed, $h \succ g \succ f$, the least cautious being the best. That remains the ordering as long as $\lambda > 0.9747$ [the agent feels little fear], below which [$\lambda \in (0.90796, 0.9747)$] $g \succ h \succ f$. For λ below 0.8608, the most cautious strategy again becomes dominant: $f \succ g \succ h$. Thus only if the agent puts little weight (less than 0.14) on the catastrophic state will she opt for any but the most cautious act. Finally note that the certainty equivalents of these acts for $\lambda = 0.5$ are given in terms of the 'normal state' utility in the final row of the table above. In all cases they are substantially below those in our variational preferences example depicted in Figs. 9 and 10, indicating the extreme ambiguity aversion of preferences incorporating fear of a catastrophe.

4 Lessons and Conclusions

In this note we have reviewed and illustrated six different models of how decision makers might evaluate catastrophic risks, how their preferences deal with uncertain prospects. These models are grounded in a set of similar behavioral axioms, with differences in axioms identifying them. Each model shares a behavioral foundation based on a distaste for ('fear of') uncertainty/ambiguity, i.e. a lack of even probabilistic knowledge of, or firm beliefs about, the underlying events driving (potential) outcomes. And all yield a greater degree of caution, a desire to "hedge" against the unknown, than does the model of a well-informed agent with (subjective) expected utility preferences. Hence they each provide a cognitive explanation for the predictive shortcomings of the SEU model, providing a rational basis for observed human behavior in the face of true uncertainty.

In doing so, these models would appear better able to provide predictions of the human responses to potential catastrophe than the SEU model does. In an economic analysis of catastrophic risk, where the decisions, actions, and reactions of millions of (potentially impacted) individuals must be considered in policy formation, such models should be a critical tool. Whatever the policy considered, risk mitigation, management or reaction planning, an equilibrium analysis of individual behavioral responses which may further, undercut, or even negate policy measures, is essential. These models, and in particular that of "variational preferences," could be extremely helpful in that analysis.

Indeed, the cognitive structures that these models reflect may also be relevant to decision makers responsible for developing and implementing social policy with regard to catastrophic risks. On the other hand, it might be argued that relevant social preferences should differ from individual, captured in the models above, much as social discount rates should differ from those of individuals. Further, the analysis should become explicitly dynamic, as 'good' policy must cope with ever evolving information and changing circumstances, including changes in the population's ambiguity aversion, extending beyond the decision horizons of individuals.[32] Those are, however, arguments and analyses that go beyond the scope of the present paper. Even if the optimal rational social choice criterion is best reflected in minimally (or zero?) discounted SEU, where likelihoods (the 'state' distribution) are based on the best scientific evidence, each of these models admits SEU as an "ambiguity neutral" boundary case. Further, they each indicate the direction of potential bias that can arise from "uneasiness" about that scientific evidence/knowledge. With that in mind, we feel that the final model presented, that of "variational preferences," provides the most practical, analytically tractable model representation of ambiguity/uncertainty respecting preferences. Through its flexible specification of a convex 'ambiguity penalty' function, this model captures most of the other representations.

These representations with their clear axiomatic foundations lend themselves to experimental testing of which model better captures agent behavior in the face of

[32]In particular the evolution of stochastic discount factors and agents' demand for the resolution of ambiguity, dependent on intertemporal choice parameters in agents' preferences, would need to be modeled for proper policy analysis. There is some work in this direction in the finance literature, e.g. Chen and Epstein (2002), Hanany and Klibanoff (2009), and Ju and Miao (2012).

highly uncertain/unknown probability, high cost gambles.[33] It is now also important to develop specific models of situations involving great uncertainty and (potentially) vast costs to the decision maker, and to bring such models to data. One example might be the analysis of the mixed response in threatened populations to (mandatory) evacuation orders.[34] Finally, work is needed placing these models of individual behavior within a dynamic equilibrium framework within which policy decisions must be made. All these remain on our research agenda.

5 Appendix: Axiomatic Foundations of Representations

5.1 Notation

The following notation is common to all the models presented.

(S, Σ)	– states, with *algebra* of events;
$\Delta(\Sigma)$	– all finitely additive probability measures on (S, Σ)
X	– set of consequences, convex $\subset \mathbb{R}^n$
$\mathcal{F}$	– set of 'acts' $f : S \to X$, simple functions
$B_0(\Sigma)$	– real valued, measurable simple functions
$B(\Sigma)$	– supnorm closure of $B_0(\Sigma)$
$B_0(\Sigma, K)$	– simple maps into $K \subset \mathbb{R}$
$u : X \to \mathbb{R}$	– affine utility
$I : B_0(\Sigma, u(X)) \longrightarrow \mathbb{R}$	– "certainty equivalent" representing 'beliefs'

In the examples, we simplify to:

$S = \{0, 1\}$,	the 'catastrophic' and 'normal' states, respectively;
$X = X_0 \cup X_1 \subset \mathbb{R}$;	$X_0 \cap X_1 \neq \emptyset$;
$\Delta(X)$	a set of finitely additive measures on X;
$\mathcal{F} =$	$\{f : S \to \Delta(X) \mid f(0) \in \Delta X_0, f(1) \in \Delta X_1\}$
$u_s : X_s \to \mathbb{R}$,	a 'state-dependent' utility function.

5.2 Axioms

1. Weak Order: $\succsim$ complete, transitive on $\mathcal{F}$.
2. Weak Certainty Independence: $f, g \in \mathcal{F}, x, y \in X, \alpha \in (0, 1)$

[33]There is now a growing body of literature both experimentally and econometrically eliciting 'ambiguity aversion' attitudes in a financial environment. See, for example, Choi et al. (2007), Bonomo et al. (2011), Ju and Miao (2012), Ahn et al. (2014), and Joeng et al. (2014).

[34]See, for example, the studies of Dow and Cutter (2000) and Lindell et al. (2005).

$$\alpha f + (1-\alpha)x \gtrsim \alpha g + (1-\alpha)x \Longrightarrow \alpha f + (1-\alpha)y \gtrsim \alpha g + (1-\alpha)y.$$

(a) Independence: $f, g, h \in \mathcal{F}$, $\alpha \in (0,1)$

$$f \gtrsim g \Leftrightarrow \alpha f + (1-\alpha)h \gtrsim \alpha g + (1-\alpha)h.$$

(b) Comonotonicity Independence (Schmeidler 1989): $f, g, h \in \mathcal{F}$ pairwise *comonotonic*, $\alpha \in (0,1)$

$$f \succ g \Leftrightarrow \alpha f + (1-\alpha)h \succ \alpha g + (1-\alpha)h;$$

- f and g are *comonotonic* if $\neg\exists s, t \in S, f(s) \succ g(s) \wedge g(t) \succ f(t)$.

(c) Certainty Independence (CI: Gilboa-Schmeidler 2009):

$$f \gtrsim g \Leftrightarrow \alpha f + (1-\alpha)x \gtrsim \alpha g + (1-\alpha)x.$$

Lemma 1 $\gtrsim$ *satisfies CI* iff $f, g \in \mathcal{F}$, $x, y \in X$, $\alpha, \beta \in (0,1)$

$$\alpha f + (1-\alpha)x \gtrsim \alpha g + (1-\alpha)x \Longrightarrow \beta f + (1-\beta)y \gtrsim \beta g + (1-\beta)y.$$

3. Continuity: $\forall f, g, h \in \mathcal{F}$, $\{\alpha \in [0,1] \mid \alpha f + (1-\alpha)g \gtrsim h\}$ and $\{\alpha \in [0,1] \mid h \gtrsim \alpha f + (1-\alpha)g\}$ are closed.
4. Monotonicity [state independence condition]: $f, g \in \mathcal{F}, f(s) \gtrsim g(s), \forall s \Longrightarrow f \gtrsim g$.
5. Uncertainty/Ambiguity Aversion: $f, g \in \mathcal{F}$, $\alpha \in (0,1)$

$$f \sim g \Longrightarrow \alpha f + (1-\alpha)g \gtrsim f.$$

6. Nondegeneracy: $f \succ g$ for some $f, g \in \mathcal{F}(\Delta(Z))$.
7. Unboundedness: $\exists x \succ y \in X$ S.T. $\forall \alpha \in (0,1)$, $\exists z \in X$ satisfying either (i) $y \succ \alpha z + (1-\alpha)x$ or (ii) $\alpha z + (1-\alpha)y \succ x$.
8. Monotone Continuity: If, $x \in X\,[\pi \in \Delta(Z)]\ \{E_n\}_{n\geq 1} \in \Sigma$ with $E_1 \supseteq E_2 \supseteq \cdots$ and $\cap_{n\geq 1} E_n = \emptyset$, then $f \succ g \Longrightarrow \exists n_0 \geq 1$ S.T. $xE_{n_0}f \succ g$.

- Axioms 1, 3, 4 imply $\forall f \in \mathcal{F}$, $\exists x_f \in X$, a "certainty equivalent."
- Ambiguity Neutrality $\Longleftrightarrow$ SEU: $U(f) = \int u(f)dp$.

References

Ahn, D., Choi, S., Gale, D., & Kariv, S. (2014). Estimating ambiguity aversion in a portfolio choice model. *Quantitative Economics*, *5*(2), 195–223.

Anscombe, F. J., & Aumann, R. J. (1963). A definition of subjective probability. *Annals of Mathematical Statistics*, *34*, 199–205.

Arrow, K. J. (1971). *Essays in the Theory of Risk Bearing* (pp. 44–89). (Chicago: Markham Publishing, 1971). Essay 2.

Bonomo, M., Garcia, R., Meddahi, N., & Tedongap, R. (2011). Generalized disappointment aversion, long run volatility risk, and asset prices. *The Review of Financial Studies*, *24*(1), 82–122.

Buchholz, W., & Schymura, M. (2012). Expected utility theory and the tyranny of catastrophic risks. *Ecological Economics*, *77*(234–239), 2012.

Chen, Z., & Epstein, L. (2002). Ambiguity, risk, and asset returns in continuous time. *Econometrica*, *70*(4), 1403–1443.

Chichilnisky, G. (2000). An axiomatic approach to choice under uncertainty with catastrophic risks. *Resource and Energy Economics*, *22*(3), 221–231.

Chichilnisky, G. (2009). The topology of fear. *Journal of Mathematical Economics*, *45*(807–816), 2009.

Chichilnisky, G. (2010a). Subjective probability with black swans. *Journal of Probability and Statistics*.

Chichilnisky, G. (2010b). Catastrophic risks with finite or infinite states. *International Journal of Ecological Economics and Statistics*, *23*(11F), 3–18.

Chichilnisky, G., & Eisenberger, P. (2010). Asteroids: assessing the catastrophic risks. *Journal of Probability and Statistics*, Article ID 954750.

Choi, S., Fisman, R., Gale, D., & Kariv, S. (2007). Consistency and heterogeneity of individual behavior under uncertainty. *American Economic Review*, *97*(5), 1921–1938.

Dow, K., & Cutter, S. L. (2000). Public orders and personal opinions: Household strategies for hurricane risk reduction. *Global Environmental Change, Pert VB*, *2*(4), 143–155.

Etner, J., Jeleva, M., & Tallon, J.-M. (2012). Decision theory under ambiguity. *Journal of Economic Surveys*, *26*(2), 234–270.

Figuieres, C., & Tidball, M. (2012). Sustainable exploitation of a natural resource: A satisfying use of Chichilnisky's criterion. *Economic Theory*, *49*, 243–265.

Gilboa, I. (2009). *Theory of decision under uncertainty*. Cambridge: Cambridge University Press.

Gilboa, I., & Schmeidler, D. (1989). Maxmin expected utility with a non-unique prior. *Journal of Mathematical Economics*, *18*, 141–153.

Gonzales, R., & Wu, G. (1999). On the shape of the probability weighting function. *Cognitive Psychology*, *38*, 129–166.

Hammond, P. (1988). Consequentialist foundations for expected utility. *Theory and Decision*, *25*(1), 25–78.

Hanany, E., & Klibanoff, P. (2009). Updating ambiguity averse preferences. *B.E. Journal of Theoretical Economics*, *9*(1(Advances)), Article 37

Hansen, L., & Sargent, T. (2001). Robust control and model uncertainty. *American Economic Review*, *91*(1), 60–66.

Heal, G., & Millner, A. (2014). Uncertainty and decision makingin climate change economics. *Review of Environmental Economics and Policy*, *8*(1), 120–137.

Ilut, C., & Schneider, M. (2014). Ambiguous business cycles. *American Economic Review*, *104*(8), 2368–2399.

Jahan-Parvar, M., & Liu, H. (2014). Ambiguity aversion and asset prices in production economies. *Review of Financial Studies*, *27*(10), 2014.

Joeng, D., Kim, H., & Park, J. Y. (2014). Does ambiguity matter? Estimating asset price models with multiple-priors recursive utility. *Journal of Financial Economics*, *115*(forthcoming).

Ju, N., & Miao, J. (2012). Ambiguity, learning, and asset returns. *Econometrica*, *80*(2), 559–591.

Karni, E. (1985). *Decision making under uncertainty: The case of state-dependent preferences (Cambridge)*. Boston, MA: Harvard UP.

Klibanoff, P., Marinacci, M., & Mukerji, S. (2005). A smooth model of decision making under ambiguity. *Econometrica*, *73*(6), 1849–1892.

Klibanoff, P., Marinacci, M., & Mukerji, S. (2009). *Recursive smooth ambiguity preferences. JET*, *144*(3), 930–976.

Kunreuther, H., Heal, G., Allen, M., Edenhofer, O., Field, C. B., & Yohe, G. (2013). Risk managemant and climate change. *Nature Climate Change, 3*, 447–450.
Lindell, M., Lu, J. C., & Prater, C. S. (2005). Household decision making and evacuation in response to Hurricane Lili. *Natural Hazards Review, 6*(171–179), 2005.
Maccheroni, F., Marinacci, M., & Rustichini, A. (2006). Ambiguity aversion, robustness, and the variational representation of preferences. *Econometrica, 74*(6), 1447–1498.
Millner, A. (2013). On welfare frameworks and catastrophic climate risks. *Journal of Environmental Economics and Management, 65*(2), 310–32.
Millner, A., Dietz, S., & Heal, G. (2013). Scientific ambiguity and climate policy. *Environmental and Resource Economics, 55*(1), 21–46.
Nordhaus, W. (2007). The stern review on economics of climate change. *Journal of Economic Literature, 45*(3), 686–702.
Nordhaus, W. (2012). Economic policy in the face of severe tail events. *Journal of Public Economic Theory, 14*(2), 197–219.
Parson, E. A. (2007). The big one: Review of richard Posner's 'Catastrophic risk and response. *Journal of Economic Literature, 45*(1), 147–164.
Quiggin, J. (1982). A theory of anticipated utility. *Journal of Economic Behavior and Organization, 3*(225–243), 1982.
Schmeidler, D. (1989). Subjective probability and expected utility without additivity. *Econometrica, 57*(3), 571–588.
Shaw, W. D., & Woodward, R. T. (2008). Why environmental and resource economists should care about non-expected utility models. *Resource and Energy Economics, 30*(66–89), 2008.
Strzalecki, T. (2011). Axiomatic foundations of multiplier preferences. *Econometrica, 79*(1), 47–74.
Strzalecki, T. (2013). Temporal resolution of uncertainty and recursive models of ambiguity aversion. *Econometrica, 81*(3), 1039–1074.
Tversky, A., & Kahneman, D. (1992). Advances in prospect theory: Cumulative representation of uncertainty. *Journal of Risk and Uncertainty, 5*(297–323), 1992.
Viale, A. M., Garcia-Feijoo, L., & Giannetti, A. (2014). Safety first, learning under ambiguity, and the cross-section of stock returns. *The Review of Assett Pricing Studies, 4*(1), 118–159.
Wald, A. (1945). Statistical decision functions which minimize the maximum risk. *The Annals of Mathemetics, 46*(2), 265–280.
Weitzman, M. (2007). The stern review on economics of climate change. *Journal of Economic Literature, 45*(3), 703–724.
Weitzman, M. (2009). On modeling and interpreting the economics of catastrophic climate change. *Review of Economics and Statistics, 91*(1), 1–19.
Weitzman, M. (2010). Risk-adjusted gamma discounting. *Journal of Environmental Economics and Management, 60*(1), 1–13.
Werner, J. (2009). Risk and risk aversion when states of nature matter. *Economic Theory, 41*(2), 231–246.

Modeling Decisions Involving Ambiguous, Vague, or Rare Events

Louis Narens and Donald Saari

The Kolmogorov approach to probability theory, which defines probability as a normed σ-additive measure on a boolean algebra of events, has proved to be a fruitful foundation to understand issues from much of science. But there are exceptions where, for various reasons, a more flexible theory is needed. The purpose of doing so usually arises where there is a need to employ a more general form for the probability function, or to use a more general algebra of events. Both settings, for instance, occur in quantum mechanics.

This chapter describes a generalization for a normed finitely additive measure on a topology. The objectives of this extension are to present a new model of decision making that can incorporate well-documented features of human judgments of probability and to assess its "subjective rationality". Finally, the model's mathematical relationship to Chichilniski (2009) approach for catastrophic decision making is described.

The research for this chapter was supported by grants FA9550-08-1-0389 and FA9550-13-1-0012 from AFOSR. The chapter is based on a presentation given at the AFOSR Workshop on Catastrophic Risks, SRI Stanford, 2012.

L. Narens (✉)
Department of Cognitive Sciences, University of California, Irvine, USA
e-mail: lnarens@uci.edu

L. Narens
Institute for Mathematical Behavioral Science,
University of California, 2123 Social Science Plaza A, Irvine, CA, USA

D. Saari
Department of Mathematics, University of California, Irvine, USA

D. Saari
Department of Economics, University of California, Irvine, USA

G. Chichilnisky and A. Rezai (eds.), *The Economics of the Global Environment*,
Studies in Economic Theory 29, DOI 10.1007/978-3-319-31943-8_4

1 Topological Event Spaces

To start by reviewing some of the basic terms, a boolean algebra of events has the form,

$$\langle \mathcal{B}, \cup, \cap, -, X, \varnothing \rangle ,$$

where $\mathcal{B}$ is a collection of subsets of the nonempty set X closed under the set-theoretic operations of $\cup$, $\cap$, and $-$ and where X and $\varnothing$ are in $\mathcal{B}$. A topology has a similar form except that it is not required to be closed under the operation of set-theoretic complementation, $-$. Thus a topology has $\varnothing$, X, finite intersections, and arbitrary unions of subsets from $\mathcal{B}$ in $\mathcal{B}$. It is useful for applications to replace $-$ with a different complementation operator called "pseudo complementation".

To be specific, let $\mathcal{T}$ be a topology. By definition, for each A in $\mathcal{T}$ the *pseudo complement of* A, denoted as $\urcorner A$, is the largest element B of $\mathcal{T}$ such that $A \cap B = \varnothing$. By elementary properties of "topology", $\urcorner A$ always exists. With this operation, a *topological algebra of events* is defined to have the form

$$\langle \mathcal{T}, \cup, \cap, \urcorner, X, \varnothing \rangle ,$$

where X is the universe of $\mathcal{T}$. It is not difficult to show that, with this definition, a topological algebra of events where each open set is also a closed set is a σ-boolean algebra of events.

Boolean algebras of events correspond to the classical propositional calculus in logic, where "*c* implies *d*" in a classical presentation corresponds to an expression of the form $(-C) \cup D$ in a boolean algebra of events. A topological algebra of events corresponds to a different, well-known logic called the intuitionistic propositional calculus. Similar to the boolean algebra of events and classical logic, the operation $\cup$ corresponds to disjunction "or", and the operation $\cap$ corresponds to conjunction "and" in intuitionistic logic. But rather than $-$ corresponding to the "not" operation, for intuitionistic logic, $\urcorner$ corresponds to negation. Unlike boolean algebra of events, the operator corresponding to intuitionistic implication cannot be defined by a simple formula involving $\cup, \cap$, and $\urcorner$, although it has a purely topological definition. (For details involving topological algebras of events and their relationship to intuitionistic logic, see Narens 2015).

It is this difference in complementation operators that permits the logical structure of a topological algebra of events to differ from that of a boolean algebra of events. The following nine statements identify basic properties of $\urcorner$ for a topological event space. While the first eight remain valid by substituting $-$ for $\urcorner$, we call attention to Statement 9 because it becomes invalid under such a substitution. (The proofs for these statements can be found in Chap. 9 of Narens 2007.)

If $\langle \mathcal{T}, \subseteq, \cup, \cap, \urcorner, X, \varnothing \rangle$ is a topological algebra of events, then the following eight statements hold for all A and B in $\mathcal{X}$:

1. $\ulcorner X = \varnothing$ and $\ulcorner \varnothing = X$.
2. If $A \cap B = \varnothing$, then $B \subseteq \ulcorner A$.
3. $A \cap \ulcorner A = \varnothing$.
4. If $B \subseteq A$, then $\ulcorner A \subseteq \ulcorner B$.
5. $A \subseteq \ulcorner\ulcorner A$.
6. $\ulcorner A = \ulcorner\ulcorner\ulcorner A$.
7. $\ulcorner (A \cup B) = \ulcorner A \cap \ulcorner B$.
8. $\ulcorner A \cup \ulcorner B \subseteq \ulcorner (A \cap B)$.
9. There exists a topological algebra of events $\langle \mathcal{Y}, \subseteq, \cup, \cap, \ulcorner, Y, \varnothing \rangle$ such that the following three statements hold:
   - For some A in $\mathcal{Y}$, $A \cup \ulcorner A \neq Y$.
   - For some A in $\mathcal{Y}$, $\ulcorner\ulcorner A \neq A$.
   - For some A and B in $\mathcal{Y}$, $\ulcorner (A \cap B) \neq \ulcorner A \cup \ulcorner B$.

A rich and useful concept is the definition of a "probability function," which serves as a normed measure. In part, this concept is possible because the algebra inherent in a boolean algebra of events guarantees a sufficiently abundant subset of disjoint events. In contrast, the topological algebra of events need not enjoy this property of having a sufficiently generous subset of disjoint events. Closing this gap requires altering the concept of "probability function," which is needed to provide a decent theory of probability. The way to do so is to change the finite additivity clause in the definition of normed, finite measure to an expression that is logically equivalent for a boolean algebra of events:

$$\text{For all } A \text{ and } B \text{ in the topology, } \mathbb{P}(A \cup B) = \mathbb{P}(A) + \mathbb{P}(B) - \mathbb{P}(A \cap B)\,.$$

With this change of definition of "normed, finitely additive measure", $\mathbb{P}$ applied to a topological algebra of events $\mathcal{T}$ is called a *probability function* on $\mathcal{T}$.

2 Boundaries of Topological Events

Many applications of Kolmogorov probability theory begin with a topological event space from which a special boolean algebra of events is selected. Along with this algebra, a measure is chosen that assigns the probability of 0 to the boundary of an event. Thus the measure allows the boundaries of events to be ignored.

In applications of topological event theory, however, we may not wish to ignore these boundaries. Instead, the boundaries, although "small," could carry substantive interpretations that cannot be ignored. In other words, when assigning probabilities, it may be important to assign positive values to some boundaries, including their parts and even isolated boundary points.

As shown below, doing so allows for richer concepts to be developed in purely event terms that are not feasible when using a boolean algebra of events. In this chapter this is done for the specialized concepts of ambiguity and vagueness.

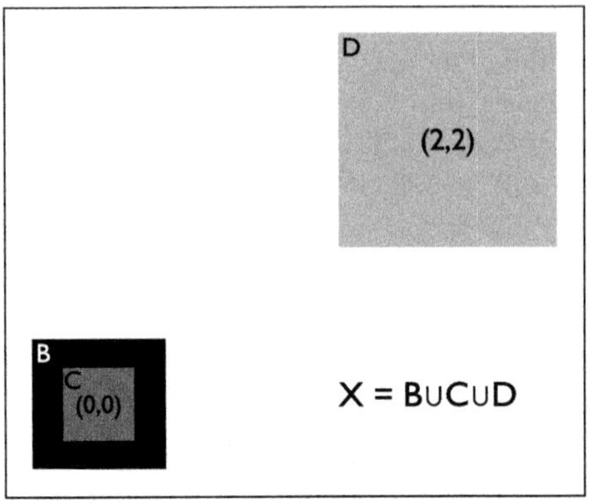

Fig. 1 $\mathcal{U} = \{X, B, C, D, \varnothing\}$, is a six element topological algebra of events with universal set X and an open set C with the "thick" boundary $B - C$

These notions are illustrated with Fig. 1, which describes a topology $\mathcal{T}$ consisting of 6 open sets; each set also is an open set in the Cartesian plane with the Euclidean topology:

- The (open) sets B, C, and D have the Euclidean areas of, respectively $\frac{1}{4}$, $\frac{1}{8}$, and $\frac{3}{4}$. Geometrically, sets B and C are centered at the point (0, 0), while D is centered at the point (2, 2).
- The other open sets are $X = B \cup D = B \cup C \cup D$ (which is the universe of $\mathcal{T}$), $C \cup D$, and $\varnothing$.

Note that $\neg B = \neg C = D$ and $B = \neg\neg B = \neg\neg C$. This last expression illustrates that it is possible to have $\neg\neg C \neq C$, which is condition 9 in the above list. Also note that the $\mathcal{T}$-topological boundary of C (i.e., the set of points a of X such that each element of $\mathcal{T}$ containing a intersects C and $X - C$) is $B - C$. Further note that although $X \neq C \cup D$, it is true that $\neg\neg(C \cup D) = X$.

For each event E in $\mathcal{T}$, if $\mathbb{P}(E)$ equals the area of E, then $\mathbb{P}$ is a probability function on the topological algebra of events $\mathcal{T}$. Because $\mathbb{P}(X) = 1$ and $\neg C = D$, it follows that

$$\mathbb{P}[C \cup (\neg C)] = \frac{1}{8} + \frac{3}{4} < 1 = \mathbb{P}(X). \tag{1}$$

Equation 1 is a probabilistic form of a well-known principle of intuitionistic logic that violates the *law of the excluded middle* coming from classical logic. That is, this example violates the condition $C \cup (\neg C) = X$.

It is clear what causes the inequality in Eq. 1; the boundary $B - C$ of $C \cup (\neg C)$ is ignored. This missing term has value for certain applications such as for human judgments of probability. For example, in Chap. 10 of Narens (2007), it is interpreted as potentially clear instances of C and cognitive non-instances of $\neg C$. Such instances are ignored in the cognitive calculation of the participant's subjective probability of $C \cup (\neg C)$.

Figure 1 illustrates a "thick" boundary, which is but one choice. Also the "measure method" for constructing the probability function $\mathbb{P}$ is only one way to construct probability functions for topologies. Of relevance for what follows is that a geometrically “thick” boundary is not needed in order to have a boundary, or part of a boundary, to behave as though it has a positive probability. Even individual points that are open sets can be assigned positive probabilities. In other words, a probability function on a topological algebra of events need not be produced in a usual mathematical way to derive a measure from a topology.

3 Application to Judgments of Probability

Many psychological experiments involving human judgments of uncertainty have the participants judge conditional probabilities that are of the forms $\mathbf{A}|\mathbf{Y}$ and $\mathbf{B}|\mathbf{Y}$, where $\mathbf{A}$, $\mathbf{B}$, and $\mathbf{Y}$ are descriptions, respectively, of the events A, B, and Y, with A and B being disjoint and $A \subseteq Y$ and $B \subseteq Y$. *Descriptions* are used here, because many experimental paradigms involve situations where different descriptions of the same event can lead to different results.

To provide an example, suppose the above event A is partitioned into the four events C, D, E, F with respective descriptions $\mathbf{C}$, $\mathbf{D}$, $\boldsymbol{E}$, $\mathbf{F}$. Suppose participants make judgments that are *sufficiently separated by time and design* so that the judgments do not influence one or another of

$$\mathbb{P}(\mathbf{A}|\mathbf{Y}),\ \mathbb{P}(\mathbf{B}|\mathbf{Y}),\ \mathbb{P}(\mathbf{C}|\mathbf{Y}),\ \mathbb{P}(\mathbf{D}|\mathbf{Y}),\ \mathbb{P}(\boldsymbol{E}|\mathbf{Y}),\ \mathbb{P}(\mathbf{F}|\mathbf{Y}).$$

The problem is that, rather than equality, many experimental studies show that

$$\mathbb{P}(\mathbf{C}|\mathbf{Y}) + \mathbb{P}(\mathbf{D}|\mathbf{Y}) + \mathbb{P}(\boldsymbol{E}|\mathbf{Y}) + \mathbb{P}(\mathbf{F}|\mathbf{Y}) > \mathbb{P}(\mathbf{A}|\mathbf{Y}), \tag{2}$$

with

$$\mathbb{P}(\mathbf{B}|\mathbf{Y}) + \mathbb{P}(\mathbf{C}|\mathbf{Y}) + \mathbb{P}(\mathbf{D}|\mathbf{Y}) + \mathbb{P}(\boldsymbol{E}|\mathbf{Y}) + \mathbb{P}(\mathbf{F}|\mathbf{Y}) \text{ being substantially} > 1\,. \tag{3}$$

As an example of Eqs. 2 and 3 consider the following experiment of Fox and Birke (2002):

(Jones Versus Clinton) 200 practicing attorneys were recruited (median reported experience: 17 years) at a national meeting of the American Bar Association (in November 1997). Of this group, 98 % reported that they knew at least “a little” about the sexual harassment allegation made by Paula Jones against President Clinton. At the time of that the survey, the case could have been disposed of by either A, which was an outcome other than a judicial verdict, or B, which was a judicial verdict. Furthermore, outcomes other than a judicial verdict can partition A into

(A1) settlement;
(A2) dismissal as a result of judicial action;
(A3) legislative grant of immunity to Clinton; and
(A4) withdrawal of the claim by Jones.

Table 1 Median judged probabilities for all events in study

(A) Other than a judicial verdict	0.75
(B) Judicial verdict	0.20
Binary partition total	0.95
(B) Judicial verdict	0.20
(A1) Settlement	0.85
(A2) Dismissal	0.25
(A3) Immunity	0.0
(A4) Withdrawal	0.19
Five fold partition total	1.49

Each attorney was randomly assigned to judge the probability of one of these six events. The results are given in Table 1.

The Jones versus Clinton example illustrates the core idea of the much investigated empirically based theory of probability judgments called *Support Theory,* which is due to Tversky and Koehler (1994) and modified by Rottenstreich and Tversky (1997). Chapter 10 of Narens (2007) employs algebras of topological events to provide a foundation for Support Theory and to model its empirical results.

This foundation is based on topological algebra of events that includes considerations about boundary points. The basic premise is that in making a judgment of probability, participants use cognitive heuristics like those proposed in various seminal articles of Kahneman and Tversky (e.g., Tversky and Kahneman 1974), and that these heuristics can be modeled through topological considerations.

For probability judgments the *availability heuristic* is particularly important. In this heuristic, the participant judges the probability of an event E in terms of the evidence for the occurrence on E and evidence for the non-occurrence of E. Namely, the judgments are based on the number and ease that instances of E are brought to mind by the event's description **E** as compared to the number and ease that non-instances of E are brought to mind by **not** E. In addition to availability, Chap. 10 of Narens (2007) models the "representativeness heuristic" by reducing it to the availability of properties of instances of an event. While it is beyond the scope of this current chapter to present a thorough discussion of Narens' foundation for Support Theory this brief description is intended to indicate the important kinds of cognitive instances of an event and their role in judgments based on the availability heuristic.

In general, the most important kind of cognitive instance of E is a *clear instance* base on **E**. These "instances" are the ones that come to mind; they are the ones that a participant views as definitely belonging to E when provided with the description **E**. The set of clear instances (to be denoted by CI) based on $\boldsymbol{E}$ is modeled as an open set $\mathsf{CI}_{E}(\mathrm{E})$ in a topology $\mathcal{T}$. The simpler notation "CI(E)" is employed when it is obvious that the set of clear instances are based on $\boldsymbol{E}$. A similar convention holds for the notation "CC(E)" that is presented next.

The *cognitive complement of the set of instances of* $\mathbf{E}$, $\mathsf{CC}(E)$, consists of *all instances that come to mind that are viewed by the participant as clearly not being clear instances of* $\mathbf{E}$. As with $\mathsf{CI}(E)$, it is assumed that $\mathsf{CC}(E)$ is an element of $\mathcal{T}$. It is also assumed that the structure of $\mathcal{T}$ is such that the pseudo complement of $\mathsf{CI}(E)$ with respect to $\mathcal{T}$, $\neg\, \mathsf{CI}(E)$, is the set of all instances i (in the domain X under consideration) such that *if* i *were presented to the participant as an instance described by* ***not E****, then she would consider it to be clearly not an instance of* ***not E***. (Note the counterfactual nature of the definition of $\neg\, \mathsf{CI}(E)$.) It is assumed that

$$\mathsf{CC}(E) \subseteq \neg\,\mathsf{CI}(\mathsf{E}).$$

This inclusion is a natural consequence of the meaning of "clear instance". Of interest, which is explored next, is that this expression need not be an equality.

Before considering other kinds of "instances", it is useful to describe what $\neg\neg\, \mathsf{CI}(E)$ corresponds to. It is assumed that $\mathcal{T}$ is such that $\neg\neg\, \mathsf{CI}(E)$ is the set of all instances i (in the domain X under consideration) such that *if* i *were presented to the participant as an instance described by* **not E**, *then she would consider it to be clearly not a clear instance of* ***not E***. In particular, each clear instance of **not E** is not in $\mathsf{CI}(E)$, and thus

$$\mathsf{CI}(E) \subseteq \neg\neg\,\mathsf{CI}(\mathsf{E}) \text{ and } \mathsf{CC}(E) \subseteq \neg\,\mathsf{CI}(E).$$

Because these expressions are of containment, but not necessarily of equality, the elements of $[\neg\neg\,\mathsf{CI}(E)] - \mathsf{CI}(E)]$ are of particular interest. These elements, which are called *potential clear instances of* $\mathbf{E}$, are possible clear instances of $\mathbf{E}$ that do not come to mind when judging the probability of E when presented the description $\mathbf{E}$; mathematically, this statement means that although these elements are related to $\mathsf{CI}(\mathsf{E})$, they are not in this set. Theoretically, they are responsible for empirical observations of Eq. 2 when the availability heuristic plays a primary role in probability estimations: A more specific description **F** of a subevent F of an event E is likely to bring to mind more clear instances of **F** than the subset of clear instances of **F** brought to mind when doing a probability estimation of E with a description $\mathbf{E}$ of E.

Indeed, it is the mathematical boundary structure, where even if a boundary point for a set is not in the set it still shares aspects of the set's structure, that provides an appropriate framework to describe two additional and important kinds of "cognitive instance"—ambiguity and vagueness. Element i is said to be a *weakly ambiguous instance of* $\mathbf{E}$ if and only if when making a probability judgment of E with $\mathbf{E}$, i is an element of the boundary of $\mathsf{CI}(E)$ and $\mathsf{CC}(E)$. Notice, i is not in either $\mathsf{CI}(E)$ or $\mathsf{CC}(E)$.

Similarly, i is said to be an *vague instance of* $\mathbf{E}$ if and only if when making a probability judgment of E with **E**, i is an element of the boundary of $\mathsf{CI}(E)$ but it is *not* an element of the boundary of $\mathsf{CC}(E)$.

A weakly ambiguous instance comes to mind in the judging of both **E** and **not E** and *the participant is aware of this*. Because of this awareness, it is neither a clear instance of **E** nor a clear instance of **not E**. A distinction is made between weakly ambiguous instances and another kind of ambiguous instance called "strongly ambiguous". Consider a situation where a participant judges $\mathbb{P}(\mathbf{E} \mid \mathbf{E} \text{ or } \mathbf{F})$ and later judges $\mathbb{P}(\mathbf{F} \mid E \text{ or } \mathbf{F})$, where the conjunction ***E* and F** describe the empty event. Then i is said to be *strongly ambiguous instance* of these judgments if and only if it is a clear instance of **E** when $(\mathbf{E} \mid \mathbf{E} \text{ or } \mathbf{F})$ is judged and it is a clear instance of **F** when $(\mathbf{F} \mid \mathbf{E} \text{ or } \mathbf{F})$ is judged.

Our reading of Tversky and Koehler (1994) suggests that Support Theory has participants ignoring weakly ambiguous and vague instances in their calculations of probability estimates. However, their calculations take into account strongly ambiguous instances, causing the sum,

$$\mathbb{P}(E \mid \mathbf{E} \text{ or } \mathbf{F}) + \mathbb{P}(\mathbf{F} \mid E \text{ or } \mathbf{F}),$$

to be an increase over what one would expect from standard probability theory, because of the strongly ambiguous instances that happen for both ***E*** and **F**.

In Support Theory, a participant's estimation $\mathbb{P}(A|A \cup B)$ of the conditional probability $A|A \cup B$, where $A \cap B = \varnothing$, is computed by the formula,

$$\mathbb{P}(A|A \cup B) = \frac{S(A)}{S(A) + S(B)}, \tag{4}$$

where S is a function with nonnegative real values. Tversky and Koehler (1994) calls S a *support function.* Equation 4 have been used by Luce (1959) and others to model choice situations where $\mathbb{P}$ is an observed probability function instead of a subjective estimation. Below, it is generalized slightly to model situations where a subject's probability estimations violate finite additivity.

The availability heuristic assumes that $S(A)$ and $S(B)$ are determined, respectively, by ease and number of instances of A and B come to mind when presented with appropriate instructions to the participant using descriptions **A** and **B**. Note that such instructions are asymmetric with respect to **A** and **B**: During this phase of the experiment, the participant is instructed to estimate the conditional probability of A given $A \cup B$, while no instruction (during this phase of the experiment) is given to estimate the conditional probability of B given $A \cup B$. The form of these instructions allows for asymmetric approaches for calculating $S(A)$ and $S(B)$.

In terms of the foundational concepts presented here, this asymmetry becomes more apparent. The reason is that the above foundation replaces Eq. 4 with

$$\mathbb{P}(A|A \cup B) = \frac{S[\mathsf{Cl}(A)]}{S[\mathsf{Cl}(A)] + S[\mathsf{CC}(A)]}. \tag{5}$$

An important difference is that Eq. 5 includes the possibility that the structure of $\mathcal{T}$ is such that $\mathsf{CC}(A) \neq \mathsf{Cl}(B)$. This provides the possibility for structurally asymmetric cognitive processing of A and B in the estimation of $\mathbb{P}(A|A \cup B)$, e.g., the clear instances A can be processed without consideration of the clear instances of B, but the processing of $\mathsf{CC}(A)$ requires also processing a relationship between A and B describing which clear instances come to mind that are instances B but not instances of A. It is assumed that subjects employ the processing described by Eq. 5. For some situations, this results in different predictions than the formula

$$\mathbb{P}(A|A \cup B) = \frac{S[\mathsf{Cl}(A)]}{S[\mathsf{Cl}(A)] + S[\mathsf{Cl}(B)]}.$$

Topological modeling of events is a promising alternative to boolean modeling for describing subjective probability estimations. The reason is that its internal "logic" matches better with the forms of cognitive processing entering into the estimations. This assertion becomes apparent when memory is involved. In particular, one of the more robust empirical findings in memory research is that, for the vast majority of times, recognition is easier than recall.

In fact, one of the simplest models of recall memory, which is called the *generation-recognition theory of recall,* relies on this fact. The model assumes that in response to a recall task of the "Name the wild African animals" type, the participant generates a set of animal names (the generation phase) and selects those that she believes are names of wild African animals (the recognition phase).

In contrast, the recognition task presents a list of animal names and asks to participant to select those that name wild African animals. This approach eliminates the need to generate possible names, which makes recognition generally an easier and more accurate task (in terms of percent correct) than recall.

Narens (2009) shows that the logical relationship of recognition and recall can be nicely modeled in a topological algebra of events by the operation of pseudo complementation, $\ulcorner$: To see how this is done, in the topological algebra $\mathcal{T}$, let E be a set of items that is recalled from a category $\boldsymbol{E}$ or a set of items that is recognizable as belonging to a category $\boldsymbol{E}$. For example, the universe of $\mathcal{T}$ can be the set of animals, $\boldsymbol{E}$ a description of the category of African animals, and E the items recalled with description $\boldsymbol{E}$. By definition, $\ulcorner E$ is the set of items of **not** $\boldsymbol{E}$ that is recognized. As **E** is a description of the category of African animals, it follows that $\ulcorner E$ is the set of animals that is recognized as being non-African. In turn, $\ulcorner\ulcorner E$ becomes the set of items of items of **not** $\boldsymbol{E}$ that is recognized, which coincides the set of items of $\boldsymbol{E}$ that is recognized. By properties of pseudo complementation,

$$E \subseteq (\ulcorner\ulcorner E).$$

When E is a set of recalled items of **E**, it is a subset of recognized items of **E**.

As these examples demonstrate, because various concepts derivable in topological algebras of events have structural correspondences with notions coming from cognitive psychology, a topological algebra of events can be an attractive alternative to a boolean algebras of events. Although Kolmogorov probability theory can be avoided for measuring uncertainty on a boolean algebra of events by using systems of weights on events instead of probabilities, such weightings do not have sufficient logical structure to provide a foundation for a subjective probability theory with a rich mathematical calculus for manipulating and calculating measurements of uncertainty. It is precisely having such a calculus that makes Kolmogorov probability theory so useful in applications. Narens (2007, 2015) show that topological algebras of events have rich probabilistic calculi.

4 Rationality

It is claimed by many that rational decision making under uncertainty requires that a particular model of decision making, the *Subjective Expected Utility Model*—or *SEU* for short—must hold. This model assumes that the decision maker has a utility function u over outcomes and lotteries and a finitely additive Kolmogorov probability function P over events such that for all lotteries

$$L = (a_1, A_1; \dots ; a_i, A_i; \dots ; a_n, A_n),$$

where a_i is a pure outcome, A_i is an event, and "a_i, A_i" stands for receiving a_i if A_i occurs, and

$$u(L) = \sum_{i=1}^{n} \frac{\mathsf{P}(A_i)}{\mathsf{P}(A_1) + \cdots + \mathsf{P}(A_n)} \cdot u(a_i). \tag{6}$$

Equation 6 is called the *SEU Model* for L.

Note that in Eq. 6,

$$\frac{\mathsf{P}(A_i)}{\mathsf{P}(A_1) + \cdots + \mathsf{P}(A_n)}$$

is the subjective conditional probability of A_i occurring given that $\bigcup_{i=1}^{n} A_i$ has occurred.

The basis for claiming the rationality of SEU rests on axiomatizations for which the individual axioms are argued to be rational, for example, the famous axiomatization of Savage (1954), or the axiomatization of a conditional form of SEU by Luce and Krantz (1971).

Humans tend to violate SEU in systematic ways. While economics generally consider these examples to be violations of rationality, some have argued that for human decision making, SEU is an inappropriate model of rationality. Instead, it is proposed that rationality should be evaluated in terms of a form of optimality that takes into account various constraints the decision maker encounters while making decisions. These include cognitive constraints like limitations of memory and the ability to make complicated mathematical calculations as well as inherent biological constraints such as the effects of emotion generated by the decision task on the final decision. Forms of rationality that take into account constraints like these are called *bounded rationality* (Simon 1957).

This section focuses on situations where the decision maker experiences different states while making decisions about lotteries, and it develops a notion of "rationality" for these situations. This form of rationality, which is called *cognitive rationality,* is illustrated in Fig. 3. It is distinguished from the rationality inherent in the SEU model called *objective rationality* and illustrated in Fig. 2.

Both Figs. 2 and 3 are concerned with a situation where lotteries from a set of lotteries $\mathcal{L}$ are presented to a participant. The elements of $\mathcal{L}$ are called the *objective lotteries*; they can be considered to be part of the everyday world.

For purposes of evaluating utilities, the participant needs to interpret them subjectively. From a mathematical perspective, objective rationality assumes there is an isomorphism between each objective lottery and a particular subjective representation that is used for calculating utilities. Namely, each item of an event in an objective lottery has a corresponding isomorphic item in the subjective representation.

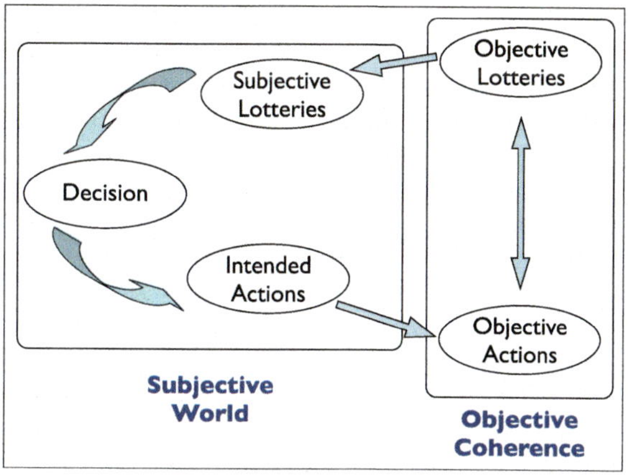

Fig. 2 Objective rationality

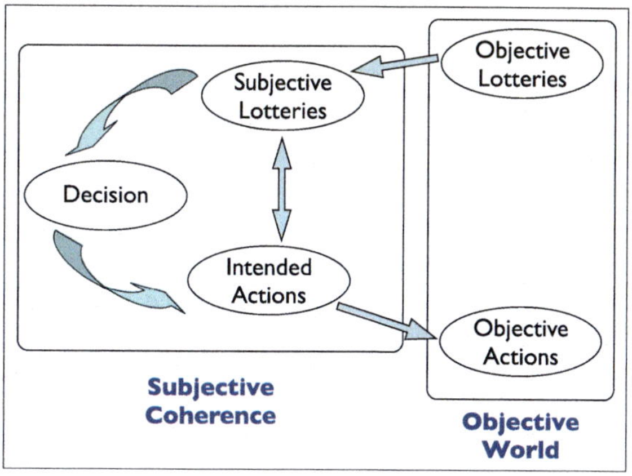

Fig. 3 Subjective rationality

While subjective rationality also assumes the existence of subjective representations of objective lotteries, the representations are not required to be isomorphisms of objective lotteries. They must, however, have the logical form of lotteries.

Both objective rationality and subjective rationality assume that each subjective lottery is an input to a decision process. The outcome of the decision process has two steps. The first yields intended actions, which then yield objective actions that take place in the everyday world. For this discussion, the intended actions can be assumed to produce preference orderings on the subjective lotteries, $\precsim_{\mathrm{obj}}$ for objective rationality, and $\precsim_{\mathrm{sub}}$ for subjective rationality. The intended actions are carried out in the everyday world producing preference orderings on objective lotteries, $\precsim'_{\mathrm{obj}}$ for objective rationality and $\precsim'_{\mathrm{sub}}$ for subjective rationality. Objective rationality assumes that its subjective lotteries with the ordering $\precsim_{\mathrm{obj}}$ is isomorphic to objective Lotteries with the ordering $\precsim'_{\mathrm{obj}}$. Subjective rationality does not make this assumption.

Notice how the principal difference between objective and subjective rationality is the kind of coherence that relates lotteries with preference orderings. Objective rationality assumes that $\precsim'_{\mathrm{obj}}$ is *objectively coherent* in that it demonstrates the following consistency with SEU: There is a utility function u_{obj} on the set of outcomes occurring in objective lotteries and a probability function $\mathsf{P}_{\mathrm{obj}}$ on the set of events occurring in objective lotteries such that

(i) each objective lottery satisfies the SEU Model (Eq. 6) with u_{obj} and $\mathsf{P}_{\mathrm{obj}}$, and
(ii) for all objective lotteries K and L,

$$K \precsim'_{\mathrm{obj}} L \text{ iff } u_{\mathrm{obj}}(K) \le u_{\mathrm{obj}}(L)\,.$$

Subjective reality assumes a similar kind of consistency for $\precsim_{\mathrm{sub}}$. Specifically, $\precsim_{\mathrm{sub}}$ is *subjectively consistent* if and only if there is a utility function u_{sub} on the set of outcomes occurring in objective lotteries and a probability function $\mathsf{P}_{\mathrm{sub}}$ on the set of events occurring in objective lotteries such that

(i) each objective lottery satisfies the SEU Model (Eq. 6) with u_{sub} and $\mathsf{P}_{\mathrm{sub}}$, and
(ii) for all objective lotteries K and L,

$$K \precsim_{\mathrm{sub}} L \text{ iff } u_{\mathrm{sub}}(K) \le u_{\mathrm{sub}}(L)\,.$$

The participant is assumed to enter into various states. Let S be the set of such states. It is important to understand how changes of state affect her subjective representations of lotteries and her subjective preference ordering. The following notation is useful for this. For each s in S and each objective lottery $L = (a_1, A_1; \ldots; a_i, A_i; \ldots; a_n, A_n)$, let

$$L^s = (a_1^s, A_1^s; \ldots; a_i^s, A_i^s; \ldots; a_n^s, A_n^s)$$

denote the participant's subjective representation of L when she is in state s. Various theories of subjective rationality can be formulated by postulating relationships among a_i, A_i, a_i^s, A_i^s, and a_i^t, A_i^t for states s and t. The following are the relationships postulated by a theory of Narens (2016) called *descriptive subjective expected utility* or *DSEU* for short, where

$$\mathcal{L}^S = \{L^s \mid s \in S\} = \text{ the set of objective lotteries}\,.$$

Suppose s and t are arbitrary states in S, $\mathcal{T}$ is a topological algebra of events, P is a probability function on $\mathcal{T}$, and u is a real valued function on the set of outcomes of objective lotteries. Then the following hold:

- Subjective rationality holds for $\mathcal{L}^S$.
- *Invariance of lotteries:* Each subjective lottery is a lottery with pure outcomes. (This holds automatically, because it is a consequence of subjective rationality. It is stated here to emphasize that for subjective rationality, the concept of "being a lottery" remains invariant under changes of state.)
- *Invariant utilities of outcomes:* $u(a^s) = u(a)$ for each outcome a of each objective lottery.
- *Invariance of disjointness of events across states:* For each event A of each objective lottery, A is in the topology $\mathcal{T}$ and

$$A^s \subseteq \ulcorner\ulcorner A \text{ and } A^t \subseteq \ulcorner\ulcorner A\,. \tag{7}$$

  Note that in the principle of invariance of disjointness of events across states, the topology $\mathcal{T}$ and the pseudo complementation operator $\ulcorner$ depends on the subject. (Also note that it implies that if C and D are distinct events occurring in some objective lottery, then $C^s \cap D^t = \varnothing$. This is part of the reason that this assumption is called "invariance of disjointness of events across states". It provides a much stronger constraint than invariance of lotteries. Also note that it provides a strong—but in applications a workable—constraint on the subjective representations of an objective event: They are related by Eq. 7.)
- *Subjective SEU with invariant probability and invariant utility across states:* For all objective lotteries $L = (a_1, A_1; \ldots; a_i, A_i; \ldots; a_n, A_n)$,

$$u(L^s) = \sum_{i=1}^{n} \frac{\mathsf{P}(A_i^s)}{\mathsf{P}(A_1^s) + \cdots + \mathsf{P}(A_n^s)} \cdot u(a_i)\,. \tag{8}$$

(Note that P and u do not depend on the state s.)

The idea behind the DSEU is to produce a model that satisfies much of the experimental literature designed to violate SEU while retaining much of the rationality expressed by SEU. Another approach in the economic literature for generalizing SEU replaces SEU's utility function with a family of utility functions, where the utility of an outcome can vary with state. This approach is reasonable for some situations, and DSEU can be easily modified to incorporate it as an additional feature.

However, there are many situations where it is unreasonable to think that the driving force for the failure of SEU is due to changes in utilities of outcomes. This appears to be likely for most situations involving emotional states, where changes in subjective probabilities appear to be a more reasonable choice.

5 Connections

An interesting feature of the above discussion is how the described method permits positive weights to be attached to boundary elements. Namely, part of the strength of this approach comes from the ability to assign added weight to important events that might otherwise be ignored.

A similar concern partly motivated the work of Chichilniski (2009), where she examined decision analysis in settings that include rare but catastrophic events. As she accurately points out, a weakness of standard expected utility approaches is that the small likelihood (the measure) of an horrific event could cause it to drop out of the decision analysis.

To see how this can happen, suppose an event has the extremely large negative utility of $-M$ where M has an arbitrary large value. But if the likelihood of this event is very rare, say M^{-10}, then in expected utility considerations the event becomes the unnoticeable $-M\frac{1}{M^{10}} = -\frac{1}{M^9}$. Stated in other words, with standard expected utility considerations, rare but crucially important events (such as earthquakes, attacks such as 9/11) might not receive sufficient consideration when it is part of a standard policy/decision analysis.

Resolutions for this kind of difficulty are immediate: The goal is to find ways to attach stronger, more commensurate attention to these concerns. This can be done through concepts involving the double negation operator $\neg\neg$. To review how this can be done, let the standard, everyday events be represented by E. In this setting, rare, possibly catastrophic events can be treated as being contained in the boundary of E: It can be shown that in many topologies

$$\text{boundary of } E \cap \text{boundary of } (\neg\neg\, E) - E$$

contain subsets of points that are natural candidates for representing rare, possibly catastrophic events. As described in the first section of this chapter, a difficulty with boundary events is that, with standard probabilistic approaches, they tend to be lost by being assigned a probability of zero. But similar to approaches described earlier, positive values can be attached to subsets contained in boundary of $E \cap$ boundary of $(\neg\neg\, E) - E$. However, unlike to the approaches described previously, such subsets in this case are not like events considered in these earlier approaches: They are not elements of the underlying topology.

Using this approach to boundaries, it becomes a direct exercise to convert the utility approach described in the previous sections into one that handles these kinds of subsets of boundaries. This is because the measure of a set supporting rare but important events that is, an event C contained in the subset of boundary points $[E \cap$ boundary of $(\neg\neg E) - E])$ can be assigned a weight commensurate with its actual importance, while retaining the measures of the non-rare events in E. This allows for an establishment of a coherent probability function without the use of inappropriate values coming from the mathematical structure of an adopted, but perhaps inappropriate decision method.

A new kind of interpretation needs to be given to the rare event C described just above. From the perspective of the decision method used for calculating E, C has very small but non-specifiable, non-infinitesimal chance of occurrence. Its non-specifiability puts it outside of the subsets determinable by the decision method with definite probabilities, whereas it is still described by **E**, and therefore is contained in $\neg\neg E$. A natural place for it is as a subset of the boundary of $E \cap$ the boundary of $(\neg\neg E) - E$. As such, C is not an open subset of E or an open subset of $\neg\neg E$, that is, $C \notin \mathcal{T}$. Events of $(\neg\neg E) - E$ are assigned probabilities by a new method. This gives rise to two probability functions, $\mathbb{P}_1$ by the old decision method for events in $\mathcal{T}$ and $\mathbb{P}_2$ by the new method for events contained in $(\neg\neg E) - E$ for each E in $\mathcal{T}$. Let

$$\mathcal{C} = \{C \mid C \notin \mathcal{T}, C \text{ is an event, and for some } E \text{ in } \mathcal{T}, [C \subseteq (\neg\neg E) - E]\}.$$

A probability function $\mathbb{P}$ is then defined on the boolean algebra generated $\mathcal{T} \cup \mathcal{C}$ having the following properties:

- On $\mathcal{T}$, $\mathbb{P} = \mathbb{P}_1$.
- On $\mathcal{C}$, $\mathbb{P} = \mathbb{P}_2$.
- On $\mathcal{T} \cup \mathcal{C}$, $\mathbb{P}$ is defined as the following weighted average: There exists $0 < \alpha < 1$ such that for all E in $\mathcal{T}$ and C in $\mathcal{C}$,

$$\mathbb{P}(E \cup C) = \alpha \mathbb{P}_1(E) + (1 - \alpha)\mathbb{P}_2(C).$$

Although Chichilniski adopted a different approach, it is interesting to note some of the similarities. She noted that if the utility function u is assumed to be in L^p, $p \geq 1$, (that is, the space of functions $f(x)$ where $\int |f(x)|^p\,dx$ is bounded), then the above same effect can occur causing an important rare event to be ignored. While the above example with the negative utility of $-M$ occurs on a set of measure M^{-10} will be picked up by placing the analysis in L^{11} (because now $\int |u(x)|^{11}\,dx$ includes the computation $|-M|^{11}(M^{-10}) = |M|$ where the $|M|$ value is noticed). But the same problems would be ignored in this space if the supporting measure is M^{-20} (because now the computation $|-M|^{11}(M^{-20}) = |M|^{-9}$ where the spike is ignored). In other words, a realistic issue is that, a priori, it is not known what would be the underlying measure of a serious rare event.

On the other hand, no matter how small the supporting measure, if it is positive, then this event will be picked up for functions in L^∞ (where the norm of a function can be viewed as being given by the supremum of $|f(x)|$ over sets of positive measure). With the above choice where u can have the negative utility of $-M$, no matter how small the supporting set for this value, if it has a positive measure, then the $|-M|$ value will dominate attention.

The next step is to find a way to determine the underlying measure *and* to find ways to assign positive values to small events. As a review to describe what is done, the well known Riesz representation theorem (e.g., Dunford and Schwartz 1957) states that a linear functional $\mathbb{L}(f)$ for $f \in L^p$ can be represented as $\mathbb{L}(f) = \int f(x)\, g(x)\, dx$ for a particular $g(x) \in L^q$ where $\frac{1}{p} + \frac{1}{q} = 1$. That is, the linear functional can be identified with an element in the dual space of L^p, which is L^q; the linear functional has the representation of a scalar product (which is an integral here).

While the $\frac{1}{p} + \frac{1}{q} = 1$ dual space representation holds only for finite $p, q > 1$, it suggests with Chichilniski's setting of $p = \infty$ (so $\frac{1}{p} = 0$ or $\frac{1}{q} = 1$) that the dual space (which defines the underlying measure) should involve L^1. It does; the dual space for L^∞ is the combination of L^1 with bounded, finitely additive signed measures that are absolutely continuous with the underlying measure. (See p. 296 of Dunford and Schwartz 1957). These finitely additive measures, which normally are difficult to use, are what provide the extra structure where positive weights can be assigned to objects of small size. In this way, the decision structure confronts and incorporates the rare but significant events into the decision analysis.

There is a certain similarity in how the two approaches elevate the importance of small but highly relevant sets; both incorporate a sense of the double dual, or double negation. In a degenerate topological space where every open set is also a closed set,

$$E = (\urcorner\urcorner\, E),$$

and the topology becomes a boolean algebra of events. That is, certain degenerate topological models would require a set to be equal to its second negation. A richer setting arises by adopting a modeling environment where the double negation introduces new sets through $E \subsetneq (\urcorner\urcorner\, E)$. As shown, the identity of these new sets vary with what is being modeled; they can range from the modeling of ambiguity or vagueness to providing a way to attach appropriate attention to rare but crucial events.

A similar mathematical effect occurs with expected utility theory with the duality operation. Here, the L^p spaces are reflexive in that the dual space for L^p is L^q and the dual space for L^q returns to L^p. That is, a fairly normal modeling environment is where a space equal to it second dual. But a richer setting arises by adopting a modeling environment where the second dual contains, but does not equal, the original space. This is the effect of assuming that the utility functions are in L^∞; the dual of this space, or the second dual of L^1, introduces the new finitely additive measures that can be used to handle rare but crucial events.

References

Chichilniski, G. (2009). The topology of fear. *Journal of Mathematical Economics*, *45*, 807–816.
Dunford, N., & Schwartz, J. (1957). *Linear operators*. New York: Interscience Publishers.
Fox, C. R., & Birke, R. (2002). Forecasting trial outcomes: Lawyers assign higher probabilities to scenarios that are described in greater detail. *Law and Human Behavior*, *26*, 159–173.
Luce, R. D. (1959). *Individual choice behavior*. New York: Wiley.
Luce, R. D., & Krantz, D. H. (1971). Conditional expected utility. *Econometrica*, *39*, 253–271.
Narens, L. (2007). *Theories of probability: An examination of logical and qualitative foundations*. London: World Scientific.
Narens, L. (2009). A foundation for support theory based on a non-boolean event space. *Journal of Mathematical Psychology*, *53*, 399–407.
Narens, L. (2015). *Probabilistic Lattices with Applications to Psychology*. Singapore: World Scientific.
Narens, L. (2016). Multimode utility theory. *Journal of Mathematical Psychology*, 1–17. http://dx.doi.org/10.1016/j.jmp.2016.02.003.
Rottenstreich, Y., & Tversky, A. (1997). Unpacking, repacking, and anchoring: Advances in support theory. *Psychological Review*, *104*, 203–231.
Savage, L. J. (1954). *The foundations of statistics*. New York: Wiley.
Simon, H. (1957). *Models of man*. New York: Wiley.
Tversky, A., & Kahneman, D. (1974). Judgment under uncertainty: Heuristics and biases. *Science*, *185*, 1124–1131.
Tversky, A., & Koehler, D. (1994). Support Theory: A nonextensional representation of subjective probability. *Psychological Review*, *101*, 547–567.

Modeling Uncertainty, Context, and Information Fusion via Lattice-Based Probability

Jun Zhang and Roman Ilin

1 Introduction

This chapter reviews and explores mathematical foundations for probabilistic inference, uncertainty representation, and fusion of disparate information sources. We will revisit probability measures defined on an event space that is modeled as a bounded distributive lattice—this includes as a special case Boolean lattice where each element has unique complementation and upon which standard probability theory has been axiomatized. Following the recent work of Narens (2009, 2011), we will invoke the relative pseudo-complementation operator on a distributive lattice, leading to Heyting algebra (as an extension of Boolean algebra) of event space that supports intuitionistic logic. We then consider basic probability assignment (b.p.a.) on finite distributive lattices, which are linked to lower probability (belief function) and upper probability (plausibility function) on such lattices. Making use of the fact that any topology on a set, that is, a system of subsets satisfying some requirements, forms a distributive lattice, pseudo-complementation can be addressed through the closure operation under such topology prescribed to an event space. Topology provides a rich semantics in terms of both the way subsets are categorized (open, closed, clopen) and the operations that characterize their properties (neighborhood, separation, etc.) and transformation (closure, interior, boundary, etc.). We therefore model contextual information for uncertainty as specifying a topology on the event space. The totality of all topologies (i.e., all contexts) themselves on an event space form

---

Draft: March 1, 2015.

J. Zhang (✉)
Department of Psychology and Department of Mathematics, University of Michigan, Ann Arbor, MI, USA
e-mail: junz@umich.edu

R. Ilin
AFRL Sensors Directorate, Wright-Patterson Air Force Base, OH, USA
e-mail: roman.ilin.1@us.af.mil

G. Chichilnisky and A. Rezai (eds.), *The Economics of the Global Environment*, Studies in Economic Theory 29, DOI 10.1007/978-3-319-31943-8_5

a bounded and hence complete lattice, ordered by coarse-grading, with the discrete topology (where each elementary event is treated as "clopen") as the top element, i.e., the finest/largest topology, and the indiscrete topology (consisting of only two events, the null-set and the full set) as the bottom element, i.e., the coarsest/smallest topology). This provides a setting for combining different b.p.a.'s, whose focal (i.e., with non-zero weight) assignments are stipulated to be only on open sets of a topology. Our lattice probability approach, identifying topology with context, deepens the upper/lower probability framework for dealing with uncertainty in two aspects: it provides a principled way for (i) defining "focal elements" (on open sets of the topology) while restraining b.p.a.'s in a given context to satisfy the lattice probability condition, and (ii) combining b.p.a.'s across different contexts through the lattice of topologies. Hence our framework provides a more fundamental mathematical framework compared with current theories (e.g., Dempster-Shafer belief function and Zadeh (1965) fuzzy probability).

1.1 Upper-Lower Probability Theory

A now-popular approach to uncertainty is through upper-lower probability theory, in which probabilistic assessment are given within an interval, meant to reflect tolerance to uncertainty. Starting from a basic probability assignment function $m()$, which assigns non-negative probability mass to (potentially all) subsets of the sample space Ω. The total probability is still required to be normalized to 1.0, but the assignment is not restricted just to its atomic elements (singleton subsets). The lower probability P_* (belief) and upper probability P^* (plausibility) are then defined as

$$P^*(A) = \sum_{X \cap A \neq \emptyset} m(X), \tag{1}$$

$$P_*(A) = \sum_{X \subseteq A} m(X), \tag{2}$$

with $0 \leq P_* \leq P^* \leq 1$. It can be shown that the lower probability P_* becomes a probability measure (and hence equals upper probability P^*) if and only if the basic probability assignment $m()$ is atomic (i.e., only to singletons). This is the case of Bayesian belief function. In general, a belief function is merely monotonic and does not satisfy the additive axiom of a probability measure (see below for more details). The belief function and the basic probability assignment are dual to each other, linked through the so-called Möbius transform:

$$m(A) = \sum_{X \subseteq A} (-1)^{|A|-|X|} P_*(X) \tag{3}$$

$$P_*(A) = \sum_{X \subseteq A} m(X). \tag{4}$$

Dempster-Shafer theory (see Yager and Liu 2008) provided a rule for evidence combination (more fashionably called "information fusion"), as well as a formula for conditioning. Though it has been extensively investigated in the past, its application to uncertainty reasoning in real systems has been limited due to (i) the need to hand-craft the basic probability assignment which is application-dependent; (ii) the lack of efficient computational algorithms to handle combinatoric explosion in the number of variables when computing belief functions defined on a power set.

The upper-lower probability theory provided an interval (with upper and lower bounds) representation of probability measure. This opens the door for representing ignorance and incomplete information. Researchers in Dempster-Shafer theory have been focused on basic probability assignment, and the evidence combination rules. They have rarely, if ever, invoked the theory of submodular functions (and Lovasz extension), which has been well developed in mathematics and recently begun to see wider applications in combinatoric optimization, machine learning, etc. It should be noted that in the past, submodular functions (variously called, capacity, Choquet integral, etc.) have been applied in economics and decision science, for instance, the so-called rank-dependent utility theory in particular. Through the technique of Lovasz extension, combinatoric optimization problems (in discrete variables) can be bypassed through applying convex programing to continuous variables in a vector space. This computation advance opens the door of applying upper-lower probability theory to uncertain reasoning and integrating disparate information in real systems.

The theory of belief functions has an alternative, equivalent approach owed to Fagin and Halpern (1991) who invoked inner and outer probability measures to deal with uncertainty. A non-measurable event, to which no probability measures can be assigned, is meant to be one that the agent does not have sufficient information to assign probability. Non-measurable events nevertheless are provided with an inner (outer) probability measure, which is the probability of the largest (smallest) measurable event contained in (containing) it and hence gives the lower (upper) bound of the degree of belief. The interval assigned by inner and outer measures thus characterizes the degree of uncertainty, akin to the interval provided by the belief-plausibility dichotomy. In fact, belief function and inner probability measure are equivalent. A further theoretical grounding of the upper-lower probability is the idea of rough sets (Pawlak 1982), who formally introduced the ideas of upper and lower sets based on a prescribed equivalence relationship on the set. However, this will not be pursued in the current chapter.

1.2 Non-Boolean Algebra with Pseudo-complementation

Standard probability theory is built upon Boolean algebra of an event space. Recall that given a ground set Ω, a probability measure $\Pr()$ is a function from the power-set $\mathbf{2}^{\Omega} \rightarrow [0, 1]$ that satisfies normalization condition:

$$\Pr(\emptyset) = 0, \quad \Pr(\Omega) = 1,$$

monotonicity condition

$$\Pr(A) \preceq \Pr(B) \text{ if } A \subseteq B,$$

and additivity condition

$$\Pr(A \cup B) = \Pr(A) + \Pr(B) - \Pr(A \cap B),$$

where A, B are any subsets of Ω. Traditionally, a probability measure is based on Boolean algebra over the event space, where an "event" is modeled as any subset of Ω. A collection of subsets (of a set) forms an algebra (of sets) if the unions and intersections of any finite members of the collection remain in that collection (in mathematical jargon, one says that the set-operations of union and intersection are "closed"). When the "union" and "intersection" operations in an algebra are replaced by the "sup" and "inf" operations with respect to the set-containment ordering, it becomes a lattice of sets. In a Boolean algebra, the collection is also required to include the (set-theoretic) complement (negation $\neg$) of each subset it contains. This leads to the Law of Excluded Middle, namely, the event (subset) A and its negation $A^{\neg}$ are not only mutually exclusive but complementary (in the sense that there is no third alternative). In intuitionistic logic, however, the Law of Excluded Middle is not enforced; this is done through relaxing the "unique complementation" requirement on the collection, but instead introducing a pseudo-complementation operation, defined in a way that the output is unique if it exists with respect to any collection. Shifting focus from unique complementation to pseudo-complementation turns the Boolean algebra into a non-Boolean one, thus providing a more general setting for studying propositional logic and for handling probabilistic inference. (There are other possible relaxations to Boolean algebra and Boolean logic, including quantum logic, which will not be discussed in this chapter.)

The study of non-Boolean algebra is closely associated with *lattice theory* the foundation of which, though traceable back to George Boole, was laid by Dedekind in a series of papers in the early 1900s. Lattice encodes algebraic behavior of entailment relation ("if-then" implications) and basic logical connectives (conjunction "and" and disjunction "or"), so it provides the appropriate framework of semantics of inference. However, it was not until 1930–40s when Birkhoff, von Neumann, Stone, Tarski, etc. fully brought out the power of lattice theory with algebraic rigor. McKinsey and Tarski (1944, 1946), in two ground-breaking papers, connected topology with modal logic. They linked the topological properties of the collection of open sets to the pseudo-complementation operation on distributive lattices of sets. By doing so, they showed that topological space provided rich semantics for intuitionistic logic. Their theory motivated modern extension of the so-called Heyting lattice for contemporary modal logic, which will not be further discussed here. Instead, we investigate the structure of probability measures, including belief functions (submodular functions) on distributive lattices and their implications for novel applications to information fusion and uncertainty management.

1.3 Why Lattice?

Lattice is an algebraic object that can be defined in two equivalent ways, (i) as a set with a non-strict partial order defined on it, and the set is closed with respect to the inf and sup operations induced from such non-strict partial order; (ii) as a set with two algebraic operations ("meet" and "join") defined on it satisfying basic axioms like commutativity and associativity, while the two operations must also be compatible by satisfying an "absorption" relation. The set is required to be closed with respect to the meet and join operations. The details will be reviewed below—here we emphasize the fact that a lattice has, *simultaneously*, order and algebraic structures. As an example, a collection of subsets of a set, under suitable constraints about the collection, can form a lattice (of sets), which behaves somewhat similarly (but with important differences) to the power-set; it provides the "right" amount of relaxation of Boolean algebra on the power-set. Two important classes of lattices, the distributive lattice and the (non-distributive) orthomodular lattice, turn out to be the mathematical tools in service of intuitionistic logic and quantum logic, respectively; they extend Boolean logic in different directions.

Recall that a *Boolean lattice (algebra)* $(B, \vee, \wedge, \neg, 0, 1)$ is a special kind of lattice with two binary operations join ($\vee$) and meet ($\wedge$) operations that distributes over one another, a bottom element 0 and a top element 1 such that $a \vee 0 = a$, $a \wedge 1 = a$ for all $a \in B$, and a unary operation $\neg$ "complementation" with unique output such that $a \vee \neg a = 1$ and $a \wedge \neg a = 0$, for all $a \in B$. A typical example of Boolean algebra is the lattice of power-set of a set, ordered by set inclusion; here, $\vee$ and $\wedge$ are set-theoretic $\cup$ and $\cap$, respectively. Boolean lattice is where classic probability theory with classic propositional logic is anchored upon. In fact, Stone (1936) proved an important result of Boolean algebra: a lattice is Boolean if and only if it is isomorphic to a field of sets.

One relaxation of Boolean lattices is the so-called "distributive lattice", namely, a lattice in which the operators $\wedge$ and $\vee$ still preserve their roles, but without the $\neg$ operation nor upper and/or lower bounds. An example of distributive lattice is the so-called *Brouwer lattice* $(B, \vee, \wedge, ', 0)$, which is bounded from below (the 0 element) and admits an additional unary operator $'$ called pseudo-complementation. More generally, the *Heyting algebra* $(H, \vee, \wedge, \rightarrow)$ is endowed with an additional binary operator $\rightarrow$, so-called relative pseudo-complementation, defined as follows: the relative pseudo-complement x of a with respect to b, denoted as $a \rightarrow b$, is the largest element x that meets a to b: i.e., $x \preceq (a \rightarrow b)$ iff $(a \wedge x) \preceq b$. The Brouwerian pseudo-complement $'$ is a special case with $a' \equiv a \rightarrow 0$, which satisfies $a \wedge a' = 0$. Yet $a = (a')'$ does not hold in general; nor does $a \vee a' = 1$ hold. So in a Brouwer lattice, $'$ stands in place of the complement $\neg$ operation as in Boolean algebra.

Pseudo-complementation operator may sound unnatural, but it embodies the "intuitionistic logic", which suspends the Law of Excluded Middle; it can be traced back to Brouwer's philosophy of mathematical foundation. It is a satisfying conclusion that every finite distributive lattice admits a relative pseudo-complementation operator. So distributive lattice provides a concise extension to Boolean lattice when one relaxes the Law of Excluded Middle. In the 1930s, Kolmogorov used Heyting algebra as a logic for describing mathematical constructions, while Gödel employed them as a basis for modal logics that are useful for understanding proof theory in mathematical logic. A recent interest of intuitionistic logic appeared in cognitive psychology, where a variant was employed as a basis for propositions that are neither verifiable nor refutable, and formed a basis for formulating a concept "incompleteness" or "ambiguity" that people presumably take into account in making probability judgments (Narens 2009, 2011). This chapter will follow this important move to provide an alternative to Bayesian probability theory based on distributive lattices.

2 Mathematical Background

2.1 An Introduction to Lattice Theory

2.1.1 Lattice as Poset

Lattice theory is a mature topic of mathematics, so here we follow standard introduction to this subject (e.g. Birkhoff 1933; Davey and Priestley 2002). Lattice is a kind of ordered set, that is, a set with a prescribed order structure. A *partially ordered set* (poset) $(X, \preceq)$ is a set X equipped with a binary relation $\preceq$ such that the binary relation is (i) reflexive; (ii) transitive; and (iii) antisymmetric. Reflexivity of $\preceq$ means that $x \preceq x$ always holds. Transitivity of $\preceq$ means that if $x \preceq y$, $y \preceq z$ then $x \preceq z$. Antisymmetricity of $\preceq$ means that x and y must be the same element whenever $x \preceq y$ and $y \preceq x$ hold at the same time. Note that, strictly speaking, the order $\preceq$ defined above should be called "non-strict partial order". The reflexivity requirement makes it more like the so-called "pre-order", which is a binary relation that only obeys (i) and (ii). On the other hand, if (i) is replaced by irreflexivity and (iii) is replaced by asymmetricity, then the binary relation is called *strict partial order*, usually denoted by $<$. In this case $\neg(x < x)$ (irreflexivity) holds and that if $x < y$ then it cannot be true that $y < x$ and verse versa (asymmetry). In the lattice theory, we focus on non-strict partial order $\preceq$.

In a poset X, it can happen that, between two arbitrary elements x, y, neither $x \preceq y$ nor $y \preceq x$ holds—we say x,y are *incomparable*. When all elements of a poset X are pairwise comparable, i.e., either $x \preceq y$ or $y \preceq x$, then the order is total, and the set is linearly ordered.

Let S be a subset of a poset X, $S \subseteq X$. If there is an element $x \in X$ such that $s \preceq x, \forall s \in S$, then x is said to be an *upper bound* of set S. An upper bound x is called a *least upper bound* (or "supremum") of S (denoted $x = \sup S$), if for any upper bound y of S, $x \preceq y$. Likewise, we define a *lower bound* of a set $S \subseteq X$ as any element x such that $x \preceq s, \forall s \in S$. The *greatest lower bound* (or "infimum") of S, denoted $\inf S$, is a lower bound x of S such that $y \preceq x$ for any other lower bound y of S. Note that, because of the anti-symmetric nature of $\preceq$, $\sup S$, if it exists, is unique. Likewise, $\inf S$ is unique if it exists.

Taking S in the above discussions to be a binary set $\{a, b\}$, we denote $a \vee b = \sup\{a, b\}$, $a \wedge b = \inf\{a, b\}$, where $\vee$ and $\wedge$ are called *join* and *meet*, respectively. A *lattice* is defined as a poset $(X, \preceq)$ in which both $\sup\{a, b\}$ and $\inf\{a, b\}$ exist for any pair a, b of elements of X. A *complete lattice* is a poset $(X, \preceq)$ of which every non-empty subset (not just the binary subsets as in a general lattice) has an infimum (greatest lower bound) and a supremum (least upper bound). If only either sup or inf is required to exist, then it is called a join or meet *semi-lattice*, respectively; semi-lattices are weaker concepts than lattices, of course.

Bounded lattice, complement, and pseudo-complement. A lattice $(X, \preceq)$ is called *bounded* if there is both a top and a bottom element, respectively denoted as $\mathtt{1}$ and $\mathtt{0}$, that is the upper bound and lower bound for all pairs of elements of X. A bounded lattice is, thus, always complete. In a bounded lattice, for any element a, its *complement* is defined as any element b such that $a \vee b = \mathtt{1}$ and $a \wedge b = \mathtt{0}$. In general, a lattice element may have more than one complement or none—this is very different from set-theoretic complementation, where the complement always exists and is unique. For instance, in a bounded lattice with linear order (i.e., a chain), $\mathtt{0}$ and $\mathtt{1}$ are the only elements that have complements. In a bounded lattice, for any two elements a, b, we can define *relative pseudo-complement* of a with respect to b as the largest element x that satisfy $a \wedge x \preceq b$. The above-mentioned chain has relative pseudo-complement for all pairs of its elements. Of course, on an arbitrary lattice, relative pseudo-complement may not exist for all pairs of elements. A bounded lattice in which relative pseudo-complement exists for all pairs of elements is called a Heyting lattice/algebra.

Joint-prime and meet-prime elements. In a lattice, we would like to distinguish certain elements that are more "primitive" than others, in the sense that they are not "generated" by joins and meets of other elements. Let $(X, \preceq)$ be a bounded lattice. We call an element $j \neq \mathtt{0}$ of X *join-prime* if $j \preceq a \vee b$ implies $j \preceq a$ or $j \preceq b$ for all $a, b \in X$. We use $J(X)$ denote the set of join-prime elements of X. Dually, we call an element $m \neq \mathtt{1}$ of X meet-prime if $a \wedge b \preceq m$ implies $a \preceq m$ or $b \preceq m$ for all $a, b \in X$. We use $M(X)$ to denote the set of meet-prime elements of X.

We call $U \subset X$ an *upset* of X if $x \in U$ and $x \preceq y$ imply $y \in U$. The set of all upsets of X is denoted $U(X)$, which forms a lattice itself, with set-containment $\subseteq$ as the (non-strict partial) order on $U(X)$. Dually, D is called a *downset* of X if $x \in A$ and $y \preceq x$ imply $y \in D$. The set of all downsets of X is denoted $D(X)$, which forms a lattice as well, with set-containment as induced order on $D(X)$. Furthermore, the mapping $x \mapsto U(x) = \{y \in X : x \preceq y\}$, when viewed as a map from X to $U(X)$, i.e., when viewing $U(x)$ as an element $U(X)$, preserves the order relation of elements in X. The same order-isomorphic property holds for the mapping $x \mapsto D(x) = \{y \in X, y \preceq x\}$.

The importance of $U(X)$ and $D(X)$ is that they provide a "good" model of the original set X—they are order-isomorphic with respect to $\preceq$, the order prescribed on X and used to construct $U(X)$, $D(X)$ in the first place. The discussions in the last paragraph can be summarized as the statement:

Lemma 2.1 *Let X be a set endowed with pre-order $\preceq$. For each $x, y \in X$, the following three conditions are equivalent:*

(i) $x \preceq y$;
(ii) $U(y) \subseteq U(x)$;
(iii) $D(x) \subseteq D(y)$.

More interestingly, while not all $x \in X$ are join-prime (or meet-prime) elements in X, $U(x)$ (or $D(x)$) is a join-prime (or meet-prime) element of $U(X)$ (or $D(X)$).

2.1.2 Lattice as Algebra

Note that $\wedge$ and $\vee$ are binary operations on a lattice: they map $X \times X \to X$. Both operations satisfy

(i) associativity: $a \wedge (b \wedge c) = (a \wedge b) \wedge c$ and $a \vee (b \vee c) = (a \vee b) \vee c$;
(ii) commutativity: $a \wedge b = b \wedge a$ and $a \vee b = b \vee a$;
(iii) absorption: $a \wedge (a \vee b) = a = a \vee (a \wedge b)$.

A special case of absorption is idempotency: $a \wedge a = a = a \vee a$, which can be obtained by replacing, in (iii), b with $a \wedge b$ or $a \vee b$. Viewed in another way, a lattice is a set X endowed with two binary operations $\wedge, \vee$ that satisfy (i)–(iii). In this case, letting $a \preceq b$ iff $a \wedge b = a$, or equivalently iff $a \vee b = b$, we turn a lattice as an algebra into a lattice as an ordered set. We use $L = (X, \preceq)$ to denote the lattice as a poset and $L = (X, \wedge, \vee)$ to denote the lattice as an algebra, but the reader should keep in mind this dualistic model of any lattice L.

Various forms of complementation Viewing lattice as an algebra allows the introduction of a variety of complementation operation on a lattice. Below, we investigate at least four such notions of "complementation" of a bounded lattice L, all as a unary map: $L \to L$.

(i) (Regular) Complement ($\neg$): $\neg a$ is any element x in L that satisfies $a \wedge x = \mathtt{0}$; $a \vee x = \mathtt{1}$. There maybe multiple such elements.
(ii) Orthocomplement ($\perp$): a special kind of complement, requiring *additionally* $(a^\perp)^\perp = a$ (involutive), and that $a \preceq b \longrightarrow b^\perp \preceq a^\perp$ (order-reversing). Hence, $\perp$ is an order-isomorphism: $a \preceq b$ if and only if $b^\perp \preceq a^\perp$.
(iii) De Morgan complement ($\sharp$): a bijective mapping $\sharp : L \to L$ such that for any $a, b \in L, \sharp(a \vee b) = (\sharp a) \wedge (\sharp b), \sharp(\sharp a) = a$, and $\sharp(\mathtt{1}) = \mathtt{0}$. In other words, $\sharp$ is "$\vee$-negation". Denote its inverse operation $(\sharp)^{-1} \equiv \flat$, the "$\wedge$-negation". Then it follows that $\flat(a \wedge b) = (\flat a) \vee (\flat b)$, and $\flat(\mathtt{0}) = \mathtt{1}$.
(iv) Pseudo-complement ($'$): weaker than regular complement, a' is the largest x (uniquely given) such that $a \wedge x = \mathtt{0}$ (without imposing the requirement of $a \vee x = \mathtt{1}$). Pseudo-complement is a special form of relative pseudo-complement, i.e., with respect to the element $\mathtt{0}$.

Note that while $\perp$ and the two De Morgan complements $\sharp$ and $\flat$ are involutive by definition (meaning that applying twice leads to identity mapping), $\neg$ and $'$ are not. De Morgan complement was introduced by Moisil (1935) and investigated by Monteiro (1980); a further requirement of $a \wedge \sharp(a) \preceq b \vee \sharp(b)$ for all a, b leads to the so-called Kleene algebra. These various "complementations" affect the existence and rules of probability measure defined on the corresponding lattices.

Compared with regular complement $\neg$, orthocomplement $\perp$ of a given element a selects out a special element out of many complements of a, with the additional property of being "orthogonal" to a—a binary relation $\perp$ (not necessarily symmetric) is said to describe "a orthogonal to b", $a \perp b$, iff $a \preceq b^{\perp}$. However, there can be more than one orthocomplementation operations definable on a complemented lattice. A lattice equipped with an orthocomplement operation is called ortholattice. If an ortholattice is uniquely complemented (i.e., if $\neg$ and hence $\perp$ is unique), then it is a Boolean lattice (algebra).

On the other hand, pseudo-complement $'$, if it exists, is always unique. A complete distributive lattice always admits a pseudo-complement for each element. If a distributive lattice is uniquely complemented (i.e., if $\neg$ is unique), then its complement $\neg$ must be the same as its pseudo-complement $'$—in this case the lattice is Boolean, which models the event space underlying classic probability measure.

As Narens (2014) pointed out, there are two ways to relax/generalize Boolean algebra/lattices to non-Boolean ones admitting appropriate notions of complementation. The first generalization is through distributive lattice which, when bounded, always admits pseudo-complementation. A distributive lattice endowed with pseudo-complementation is called a Brouwerian lattice (a subclass of Heyting algebra) which provides the setting for intuitionistic logic. The second generalization is through orthomodular lattice, a special kind of ortholattice (i.e., admitting "orthocomplement" operation) upon which the so-called "modularity law" is enforced on orthocomplement pairs. Modular lattice is a relaxation to distributive lattice (since all distributive lattices are modular lattices, but not vice versa) with imposing the modularity condition as instead of the more restrictive distributivity condition on all of its pairs. Ortholattice is, in general, non-modular, so orthomodular lattice is in general non-distributive, and provides the setting to model quantum logic. A lattice that is simultaneously orthocomplemented and distributive is a Boolean algebra. So in this sense, the above two approaches, namely, intuitionistic logic and quantum logic, provide two "independent and complementary" ways of relaxing Boolean lattice/algebra. Below, we focus on the first route of generalizing Boolean lattices, through distributive lattices.

2.2 *Distributive Lattice*

Viewing lattice as an algebra allows further classification of lattices. First, it can be shown that in any lattice $L = (X, \wedge, \vee)$,

$$(a \wedge b) \vee (a \wedge c) \preceq a \wedge (b \vee c), \tag{5}$$
$$a \vee (b \wedge c) \preceq (a \vee b) \wedge (a \vee c). \tag{6}$$

When equality in (5) holds, then we say in L "meet distributes over join"; when equality in (6) holds, then we say in L "join distributes over meet". It can be proven that these two equalities imply each other for any lattice, so if any one of them is satisfied, we call it *distributive lattice*. Another equivalent characterization of distributive lattice is that the following holds for any three elements a, b, c:

$$(a \wedge b) \vee (b \wedge c) \vee (c \wedge a) = (a \vee b) \wedge (b \vee c) \wedge (c \vee a).$$

Weaker than the notion of distributivity is the notion of modularity. First, in analogous to (5) and (6), the following holds in any lattices:

$$a \vee (b \wedge c) \preceq a \wedge (b \vee c), \quad \forall a, c \text{ such that } c \preceq a.$$

If the converse inequality holds in a certain lattice L, that is

$$a \wedge (b \vee c) \preceq a \vee (b \wedge c), \quad \forall a, c \text{ such that } c \preceq a,$$

then such a lattice L is called a *modular lattice*. Equivalently, if the equality

$$a \wedge (b \vee c) = a \vee (b \wedge c)$$

holds for all elements in L satisfying $c \preceq a$, then L is modular. A distributive lattice is always a modular lattice, but not vice versa (i.e., there exist non-distributive modular lattices).

Distributive lattices are also characterized by the absence of "pentagon" N_5 (two chains, with one and two elements each) and "crown" M_3 (three chains, each with one element) configurations as sublattices. As examples of distributive lattices, given any ground set X with pre-order on it, the set of all its upsets $(U(X), \cup, \cap)$ and the set of all its downsets $(D(X), \cup, \cap)$ both form distributive lattices (ordered by set-containment).

In a bounded distributive lattice, for each element a, one can define its *relative pseudo-complement* with respect to any other element b, denoted as $a \to b$ (or a^b):

$$x \preceq (a \to b) \text{ iff } (a \wedge x) \preceq b\ . \tag{7}$$

In a general lattice, an element a is said to be relative pseudo-complemented if $a \to b$ exists for all b. When $b = 0$, the relative pseudo-complement becomes pseudo-complement, and denoted $'$ as discussed before. A pseudo-complemented lattice that satisfies the relation $a' \vee (a')' = 1$ is called a Stone lattice. A pseudo-complemented lattice becomes a Boolean lattice iff $a \vee a' = 1, \forall a \in X$, that is, iff $a = (a')', \forall a \in X$.

When a distributive lattice is finite, it is relative pseudo-complemented for all of its elements; in particular, it is pseudo-complemented. On a Boolean lattice, pseudo-complement operation is identical to regular complement operation, and relative pseudo-complementation $a \rightarrow b$ is given by $\neg a \vee b$. Huntington's Theorem says the opposite is true as well: a lattice is Boolean iff it is pseudo-complemented and that the pseudo-complementation is also a complementation. Any element of a bounded distributive lattice can have at most one complement. So if a distributive lattice is complemented, then it must be uniquely complemented, and hence Boolean.

2.2.1 Brouwer and Heyting Algebra

We can define *Brouwer complementation operation* $'$ axiomatically as a unary operator satisfying the following properties (the use of the same symbol $'$ as we used for pseudo-complementation operation defined in terms of order is intentional, see below):

(i) $a \wedge a' = 0$;
(ii) $(a \vee b)' = a' \wedge b'$;
(iii) $a \preceq (a')'$, or equivalently, $a = a \wedge (a')'$;
(iv) $1' = 0$.

It can be deduced that $a' = ((a')')'$ and $0' = 1$. A lattice $L = (X, \wedge, \vee)$ with lower bound 0 and equipped with a unary operation $'$ is called a *Brouwer* algebra if it is closed with respect to the Brouwer complement) defined above. It turns out that a Brouwer algebra is necessarily a distributive lattice, and the Brouwer complementation defined above (when the lattice is viewed as a poset) is precisely the pseudo-complementation operation defined earlier when the (distributive lattice) is viewed as a poset. So Brouwer complement and pseudo-complement turns out to be equivalent, merely reflecting a difference in viewing the lattice as an algebra (former) versus a poset (latter).

De Morgan's laws under Brouwer algebra manifest as follows:

$$a' \wedge b' = (a \vee b)',$$
$$a' \vee b' \preceq (a \wedge b)',$$
$$(a \wedge b)' = ((a')' \wedge (b')')' = ((a' \vee b')')'.$$

Hence, it is useful to introduce the binary operator $\sqcup$ in a Brouwer algebra B:

$$a \sqcup b \equiv ((a \vee b)')'.$$

Consider the subset S of elements of B which satisfy $a = (a')'$. Then S is closed with respect to the operations $\wedge, \sqcup$, so $(S, \wedge, \sqcup)$ forms a Boolean algebra with respect to those two operations.

Recall that relative pseudo-complementation operation is defined in (7), when a lattice is viewed $L = (X, \preceq)$. There, $\wedge$ should be read as "greatest lower bound" using the language of ordered set, so relative pseudo-complement of a with respect to b is the largest element c such that the greatest lower bound, of this element c and a, should be less than b. When the lattice is viewed as an algebra $L = (X, \wedge, \vee)$, relative pseudo-complementation has been axiomatized by Monteiro (1980) as a binary operator satisfying:

(i) $a \to a = b \to b$;
(ii) $(a \to b) \wedge b = b$;
(iii) $(a \to b) \wedge a = a \wedge b$;
(iv) $a \to (b \wedge c) = (a \to b) \wedge (a \to c)$;
(v) $(a \vee b) \to c = (a \to b) \wedge (a \to c)$;
(vi) $0 \to a = 0$.

A distributive lattice $L = (X, \wedge, \vee)$ that admits the above operation can be called a Heyting algebra. The binary operation $\to$ is, unlike $\vee$ and $\wedge$, neither commutative nor associative. Brouwer complement $'$ is simply $a' \equiv a \to 0$.

2.2.2 Representation of Distributive Lattices

Representation of a lattice means to find a lattice isomorphism, typically using the lattice of sets as the target. For distributive lattices, the ring of set (closed under union and intersection operations) or field of set (closed under an additional set-complement operation) provide good candidates. It is well-known that:

(i) A lattice is distributive iff it is isomorphic to a ring of sets (Birkhoff 1933; Stone 1936);
(ii) A lattice is Boolean iff is isomorphic to a field of set (Stone 1936).

For sets with finite elements, the above results are intuitively understood. It is easy to envision that, in finite lattice, being Boolean is being isomorphic to the power-set of some finite set. For Boolean lattices with uncountable elements, much subtleties arise. Stone's characterization, for instance, involved topological considerations of compactness. The same complications apply to characterizing distributive lattices. In finite case, each distributive lattice $L = (X, \preceq)$ can be represented as the lattice of upsets (downsets) of some poset; that "some poset" is the dual poset $L^d = (X, \succeq)$ of the join-prime elements of the original lattice L. (Recall that join-prime elements means that they only have a single down-link in the Hasse diagram.) For distributive lattices possibly with infinitely many elements but without infinitely descending chains, each element a is the join of join-prime elements of X underneath a, that is: $a = \bigvee\{j \in J(X) : j \preceq a\}$. Such lattices X can be represented as the sublattice of upsets $U(\mathcal{F}(X))$ of the set of prime filters $\mathcal{F}(X)$ of X. So these characterization results become very technical. Priestley in 1990s found a characterization of bounded distributive lattices in terms of Priestley spaces, or equivalently pairwise

Stone spaces. A related characterization of Heyting algebra by the so-called Esakia space is also obtained using the framework of category theory.

2.2.3 Topology as Distributive Lattice

A topological space is a pair $(X, \mathcal{T})$, where X is a set and $\mathcal{T}$ is a collection of subsets of X, called *open sets*, containing $\emptyset$, X and closed under finite intersections and arbitrary unions. Immediately, one sees that this definition (as Hausdorff introduced) implies that any topology is a complete, distributive lattice of sets, in which set-containment is the order and $\vee$ and $\wedge$ are just $\cup$ and $\cap$; $\mathcal{T}$ is a sublattice of $P(X)$, the Boolean lattice of power-set of X. This is essentially Birkhoff (1933) characterization of distributive lattices.

Recall that a "closed set" of a topology is any set-theoretic complement of an open set of the topology. The collection $\mathcal{C}$ of the closed sets, containing $\emptyset$, X and closed under arbitrary intersection and finite union, also forms a complete distributive lattice and is a sublattice of $P(X)$.

In a topological space X, *open neighborhood* of a point x in X is defined as any open set containing x. We say x is in the interior of a subset $A \subseteq X$ if there is an open neighborhood U of x that is contained in A. The set $\text{Int}(A)$ contains all interior points of A. We say that x belongs to the closure of a subset $A \subseteq X$ if each open neighborhood U of x has nonempty intersection with A. The set $\text{Cl}(A)$ denotes the closure of A. It is easy to verify that the interior operator Int satisfies:

(i) $\text{Int}(X) = X$,
(ii) $\text{Int}(A) \subseteq A$,
(iii) $\text{Int}(A) = \text{Int}(\text{Int}(A))$,
(iv) $\text{Int}(A \cap B) = \text{Int}(A) \cap \text{Int}(B)$;

and the closure operator Cl satisfies

(i) $\text{Cl}(\emptyset) = \emptyset$,
(ii) $A \subseteq \text{Cl}(A)$,
(iii) $\text{Cl}(\text{Cl}(A)) = \text{Cl}(A)$,
(iv) $\text{Cl}(A \cup B) = \text{Cl}(A) \cup \text{Cl}(B)$.

The interior and closure operators are dual to each other: $\text{Int}(A) = X - \text{Cl}(X - A)$; $\text{Cl}(A) = X - \text{Int}(X - A)$. In fact, let any operator satisfy the four conditions of Cl above and call a subset $A \subseteq X$ "closed" if $A = \text{Cl}(A)$. Then $\mathcal{T} = \{A : X - A \text{ is closed}\}$ is a topology on X, and every topology on X arises this way.

A natural question arises: Is there a connection between the pseudo-complementation operator and closure operator (or the dually defined interior operator)? The answer was affirmatively provided by McKinsey and Tarski (1944, 1946): pseudo-complementation operation in a distributive lattice and interior operation in a topology are in one-to-one correspondence: $a \to b = \text{Int}((\neg a) \vee b)$.

2.3 *Probability and Belief Functions on Lattice*

We next review the known facts about the feasibility of introducing probability functions or belief functions on a lattice. The mathematical tool that plays a key role is Möbius transform on sets with partial order (Rota 1964).

2.3.1 Möbius Transform and Monotone Functions

Rota (1964) considered a poset $(X, \preceq)$ with a bottom element $\mathtt{0}$. For any function f on $(X, \preceq)$, the Möbius transform of f is a function $m : X \to R$ that is the solution of the equation:

$$f(x) = \sum_{y \preceq x} m(y).$$

The above equation always has a unique solution, given through the Möbius function $\mu : X \times X \to R$ by:

$$m(x) = \sum_{y \preceq x} \mu(y, x) f(y)$$

where μ is defined inductively by

$$\mu(x, y) = \begin{cases} 1, & \text{if } x = y \\ -\sum_{x \preceq t < y} \mu(x, t), & \text{if } x < y \\ 0, & \text{otherwise} \end{cases}.$$

Note that μ depends solely on X. One can also define the co-Möbius transform of f as:

$$g(x) = \sum_{y \succeq x} m(y).$$

A *capacity* on a set X is a function $f : 2^X \to [0, 1]$ such that (i) $f(\emptyset) = 0$, $f(X) = 1$ (normalization); and (ii) $A \subseteq B \subseteq X$ implies $f(A) \leq f(B)$ (monotonicity). Functions satisfying (ii) is called a *1-monotone* (or *strict monotone*) condition.

A function is said to be k-monotone[1] ($k \succeq 2$) if for *any* family of k subsets $A_1, \ldots, A_k$ of X, there holds:

$$f(\bigcup_{i=1}^{k} A_i) \geq \sum_{\emptyset \neq I \subseteq \{1,2,\ldots,k\}} (-1)^{|I|+1} f(\bigcap_{i \in I} A_i). \tag{8}$$

[1] Barthélemy (2000) used the term "weakly monotone".

In particular, the case for a 2-monotone function f is explicitly written as (condition of “convexity" or “supermodularity”):

$$f(A_1) + f(A_2) \leq f(A_1 \cap A_2) + f(A_1 \cup A_2) \tag{9}$$

for all subsets A_1, A_2 of X. Obviously, when f is k-monotone, it is k'-monotone for all $2 \leq k' \leq k$. If f is a k-monotone function and also satisfies $f(\emptyset) = 0$ and $f(\{x\}) \succeq 0$ for all $x \in X$, then f is 1-monotone—in this case, f then becomes a k-monotone capacity.

A function is said to be *totally monotone* if it is k-monotone for every $k \succeq 2$. It can be proved that when $|X| = n$, total monotonicity is equivalent to $(n-2)$-monotonicity for a capacity function.

For $k \succeq 2$, when equality in (8) holds, we say that the function f is a *k-valuation.* A *probability function* is both a capacity and a total valuation (i.e., k-valuation for every k). In fact, on a distributive lattice, the condition of 2-valuation (13) is sufficient for f to be a k-valuation for any k; the proof invokes the identity $0 = (1-1)^n$ with binomial expansion

$$\binom{m}{0} = \sum_{k=1}^{m} (-1)^{k-1} \binom{m}{k},$$

which expresses the inclusion-exclusion principle.

Characterization of capacity and probability functions. The characterization of a capacity function is through its Möbius transform. The following statement is known:

Lemma 2.2 (Chateauneuf and Jaffray 1989) *A set function f is a capacity if and only if its Möbius transform m satisfies $m(\emptyset) = 0$; $\sum_{A \subseteq X} m(A) = 1$ and that for all $A \subseteq X$,*

$$\sum_{\{x\} \subseteq A \subseteq X} m(A) \geq 0$$

for all $x \in X$. In particular, $m(\{x\}) \geq 0$ for all $x \in A$.

In their paper, Chateauneuf and Jaffray (1989) also discussed probability functions that dominate a given capacity, in the context of trans shipment problem.

Shafer (1976) showed that f is a k-monotone ($k \succeq 2$) capacity function if and only if its Möbius transform m satisfies

$$\sum_{C \subseteq B \subset X} m(B) \geq 0$$

for all subsets $C \subseteq X$ with $2 \leq |C| \leq k$. (For $k = 2$, the subsets with $|C| = k$ cannot be included.) Equivalently, the condition can be written as

$$\sum_{B \subseteq \bigcup_{i=1}^{k} A_i;\, B \notin \{A_1, \ldots, A_k\}} m(B) \geq 0$$

for any $A_1, \ldots, A_k \subseteq X$. In particular, 2-monotone functions are characterized by their Möbius transform satisfying

$$\sum_{B \subseteq (A_1 \cup A_2); B \nsubseteq A_1; B \nsubseteq A_2} m(B) \succeq 0$$

for all subsets A_1, A_2 of X; or by

$$\sum_{\{x_1, x_2\} \subseteq B \subseteq A} m(B) \succeq 0$$

for all subset A of X and all $x_1, x_2 \in X, x_1 \neq x_2$. Shafer (1976) showed that a totally monotone capacity is equivalent to its Möbius transform m being non-negative.

If any 1-monotone function is 2-monotone on a lattice L, then L must be linearly ordered. When the inequality sign of (9) is reversed:

$$f(A_1) + f(A_2) \geq f(A_1 \cap A_2) + f(A_1 \cup A_2)$$

the function f is called a *submodular* function. Submodular functions have the following so-called "diminishing marginal return" properties:

$$f(A \cup \{x\}) - f(A) \geq f(B \cup \{x\}) - f(B)$$

for all $A \subseteq B \subseteq X$ and $x \in X \backslash B$; and

$$f(A \cup \{x\}) + f(A \cup \{y\}) \geq f(A \cup \{x, y\}) + f(A)$$

for all $A \subseteq X$ and $x, y \in X \backslash A$. A 3-monotone function f satisfies:

$$f(A) + f(B) + f(C) + f(A \cap B \cap C) \leq f(A \cup B \cup C) + f(A \cap B) + f(B \cap C) + f(A \cap C).$$

Note that the concept of k-monotone function uses $\bigcap$-operation. As a counterpart, using $\bigcup$-operation instead leads to the so-called k-alternating function:

$$f(\bigcap_{i \in K} A_i) \preceq \sum_{I \subseteq K, I \neq \emptyset} (-1)^{|I|+1} f(\bigcup_{i \in I} A_i). \tag{10}$$

2.3.2 Valuation on a Lattice

Note that while the results in the last section are mostly dealing with real-valued functions on the power-set, now we study real-valued functions on a lattice. Valuation of a lattice is the assignment of a real-valued function on it. A valuation function f is called *strictly monotone* or simply *monotone* when $f(a) \leq f(b)$ iff $a \preceq b$.

Given any real-valued function f on $L = (X, \wedge, \vee)$, let us construct, for arbitrary $a, b \in X$,

$$\delta(a, b) \equiv f(a) + f(b) - 2f(a \wedge b).$$

It is easily seen that $\delta(a, a) = 0$; $\delta(a, b) = \delta(b, a)$. The triangular inequality

$$\delta(a, c) \leq \delta(a, b) + \delta(b, c)$$

amounts to the condition

$$f(a \wedge b) + f(b \wedge c) \leq f(b) + f(a \wedge c) \tag{11}$$

which, after taking $b = a \vee c$, leads to

$$f(a) + f(c) \leq f(a \vee c) + f(a \wedge c). \tag{12}$$

So $\delta(a, b)$ is a metric on L when f satisfies (12). Note that this is precisely the *2-monotone* condition except that f is defined on a lattice as opposed to be on the power-set as in (12).

When equality in (12) holds, that is,

$$f(a \wedge b) + f(a \vee b) = f(a) + f(b) \tag{13}$$

for $a, b \in X$, then f is called a *2-valuation*. In this case,

$$\delta(a, b) = f(a \vee b) - f(a \wedge b). \tag{14}$$

A lattice with monotone 2-valuation is called a *metric lattice*; in fact, one can show that the metric given by (14) also satisfies triangular inequality.

In analogous to 2-valuation, we call a function f on lattice L a *3-valuation* if

$$f(a) + f(b) + f(c) + f(a \wedge b \wedge c) = f(a \vee b \vee c) + f(a \wedge b) + f(b \wedge c) + f(a \wedge c)$$

for all lattice elements a, b, c. Clearly, 3-valuation implies 2-valuation, but not vice versa.

It is interesting to note that a metric lattice is always a modular lattice, which is weaker than a distributive lattice. A modular lattice of finite length is always metric.

The following is known Birkhoff (1967)—they link the properties of the lattice (modular or distributive) to the existence of strictly monotone valuations:

(i) L is modular if and only if it admits a strictly monotone 2-valuation;
(ii) L is distributive if and only if it is modular and every strictly monotone 2-valuation on L is a 3-valuation.

(iii) L is distributive if and only if it admits a strictly monotone 3-valuation;
(iv) L is distributive if and only if it is modular and every strictly monotone 2-valuation on L is a k-valuation for any $k > 2$.

In other words, the existence of a strict monotone 2-valuation characterizes modularity, while the existence of a strictly monotone k-valuation (any $k > 2$) characterizes distributivity.

2.3.3 Belief Function on a Lattice

A belief function has two equivalent definitions:

(i) A 1-monotone function whose Möbius transform m is non-negative;
(ii) A totally monotone capacity function.

The function m was called *basic probability assignment* (b.p.a.) in Dempster-Shafer Theory, and those subsets with non-zero probability assignment are called "focal" elements. The Möbius function on the elements of power-set is given as

$$\mu(A, B) = \begin{cases} (-1)^{B \setminus A} & \text{if } A \subseteq B \\ 0 & \text{otherwise} \end{cases}.$$

The theory of belief functions on general lattices was recently investigated by Barthélemy (2000), Grabisch (2008). An important conclusion is that any lattice admits a total monotone function:

Lemma 2.3 (Barthélemy 2000) *On any lattice L and for any function* $m : L \to [0, 1]$ *such that* $m(0) = 0$; $\sum_{x \in L} m(x) = 1$, *then the function* $f(x) = \sum_{y \preceq x} m(x)$ *is a totally monotone function and satisfies* $f(0) = 0$; $f(1) = 1$.

That is, for any mass assignment (b.p.a.), the corresponding inverse Möbius transform is a belief function. If two totally monotone functions on a lattice are identical, then their inducing b.p.a.'s must also be identical. Zhou (2013) showed that the converse is also true: the Möbius transform of any totally monotone capacity function on a lattice must be non-negative. In other words, for any capacity f on L, total monotonicity of f and non-negativity of its Möbius transform m are equivalent. Considering the smallest Boolean algebra (lattice) of which a given finite distributive lattice L is a sublattice, Zhou showed that a belief function on L is a probability function iff all focal elements (i.e., those with positive assignments of b.p.a.'s) are join-irreducible in L. So join-irreducible elements are akin to singletons in Boolean algebra.

To conclude, while a belief function can be defined on any lattice, a probability function (as a total-valuation, normalized and strictly monotone function) can only be defined on a distributive lattice.

3 Upper-Lower Probability Anchored on Topology

3.1 *Topologizing Dempster-Shafer Theory*

The Dempster-Shafer theory is constructed on the standard Boolean algebra of event space, with basic probability assignment (b.p.a.) on non-atomic element in general. Zadeh (1965) fuzzy set theory is also built upon Boolean algebra, with b.p.a. assigned to an ascending sequence of subsets. As a natural extension of event space structure for probability measures, belief functions on lattices are an interesting and natural topic of investigation. Indeed, recent studies (Barthélemy 2000; Grabisch 2008; Zhou 2013) show that any lattice admits a belief function (and hence an associated non-negative probability mass assignment), and any distributive lattice further admits a probability measure (that is, a 2-valuation that is a capacity). However, none of these research look at the role of pseudo-complementation operator in place of complementation operator of a Boolean lattice. Nor have they investigated the probability theory in a hierarchical setting, which is crucial for modeling multiple contexts and context change. As discussed earlier, pseudo-complementation amounts to specifying a topology on the ground set, which is important to provide semantics to a probability theory. Below, we initiate a new approach to information fusion by (i) investigating probability measures defined on a particular distributive lattice, namely, the topology of a set; and (ii) investigating the belief function defined on the full lattice of all topologies on the given set.

3.2 *Lattice of Topologies*

Given a set X, the set of all topologies on X form a bounded (and hence complete) lattice $\mathcal{L}_X$, ordered by "refinement", or relative coarseness (i.e., inclusion of collection of open sets), of the two topologies under comparison, see survey by Larson and Andima (1975). For two topologies τ and σ, their meet is set-wise intersection $\tau \cap \sigma$, $t \in (\tau \cap \sigma)$ if $t \in \tau$ and $t \in \sigma$; $\tau \cap \sigma$ contains the open sets common to the two topologies. Their join $\tau \cup \sigma$ is the topology *generated* by the intersections of all open sets $\{t_1 \cap t_2 | t_1 \in \tau; t_2 \in \sigma\}$. The top element of this lattice $\mathcal{L}_X$ is the discrete topology where each element is treated as a clopen set; this is the finest (largest) topology on X. The bottom element of $\mathcal{L}_X$ is the indiscrete topology, consisting of only $\emptyset, X$; this is the coarsest (smallest) topology. According to Larson and Andima (1975), there is no known formula for the number of topologies on a finite set, only that the number is between 2^n and $2^{n(n-1)}$ where $n = |X|$. For $n = 3, 4, 5, 6, 7$, the number of elements in the lattice is 29, 355, 6942, 209527, 9535241. Figure 1 gives the case of 29 topologies, organized as a lattice, for a three-element set $X = \{a, b, c\}$.

In general, the compact topologies (a topology that every cover has a finite cover) are closed downward in this lattice, since if a topology τ has fewer open sets than σ and σ is compact, then τ is compact. Similarly, the Hausdorff topologies (any

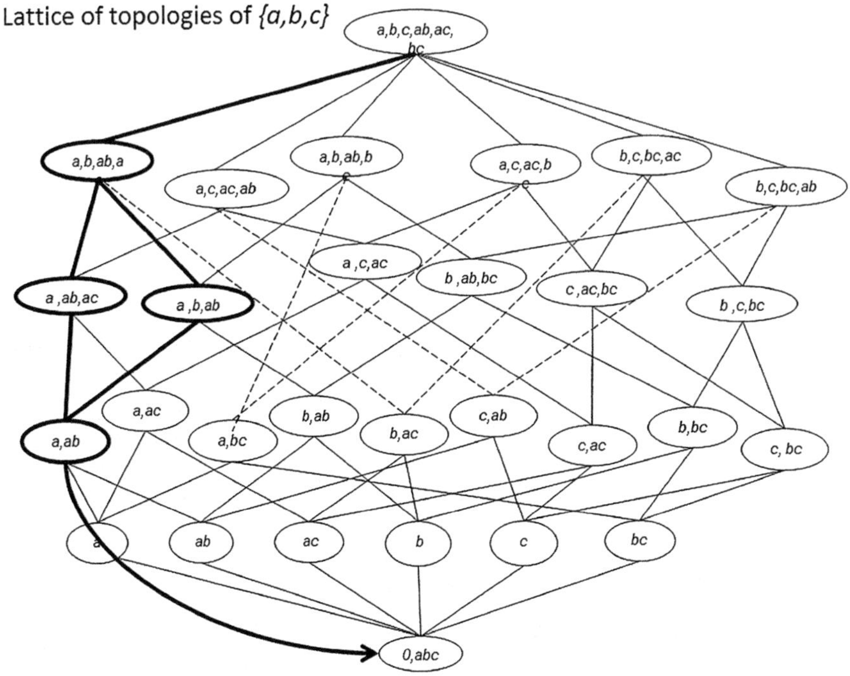

Fig. 1 The lattice of topologies for a 3-element set. Each node shows the elements of the corresponding topology with the empty set and the full set X omitted for best readability. The letter-string, say "bc", stands for the set $\{b, c\}$. The lattice from Fig. 3 is embedded in the complete lattice and shown by *thick lines*

two points are separated by disjoint open sets) are closed upward, since if τ is Hausdorff and contained in σ, then σ is Hausdorff. Thus, in this lattice of topology, the compact topologies inhabit the bottom of the lattice (where indiscrete topology lies as the extreme) and the Hausdorff topologies the top (where discrete topology lies as the extreme). These two types of topologies run into each other in the middle, known as compact Hausdorff topologies. They form an anti-chain in the lattice: no two compact Hausdorff topologies are comparable and they are all distinct.

The lattice of topologies $\mathcal{L}_X$, when the ground set X is denumerable, is also known to be complemented. Embedded in $\mathcal{L}_X$ is a sublattice of all T_1 topologies—a T_1 topology on a ground set X is one where each singleton subset $\{x\} \subset X$ is closed. The T_1 topology sublattice is, however, not complemented, not modular (and hence not distributive), but it is both upper and lower semi-modular.

Larson and Andima (1975) noted that, even for a finite set X, the lattice $\mathcal{L}_X$ is non-distributive, non-modular, neither upper nor lower semi-modular, and not self-dual. It has only trivial lattice homomorphisms. For any lattice l, there exists a ground set X such that l can be embedded into the lattice $\mathcal{L}_X$ of topologies on X. Moreover, Valent and Larson (1972) and Rosický (1975) showed that a finite lattice l is distributive if and only if it can be realized as an interval of T_1 topologies on a set X, that is, there are two T_1-topologies τ and σ on X such that the subinterval $[\tau, \sigma]$ of $\mathcal{L}_X$ is isomorphic to l. Knight et al. (1997) further strengthened its realizability from T_1 to Hausdorff.

Investigating the lattice of topologies along with each individual topologies allows us to construct a hierarchical scheme of upper-lower probabilities. This is done as follows. Referring to Fig. 1, which depicts all possible topologies on $X = \{a, b, c\}$. Each topology represents a distinct "context", and each will be assigned a probability measure. Figure 2 gives a few examples of the topologies, with graphic coding for open sets, closed sets, clopen sets, and sets that are neither open nor closed. On the lattice of topologies, we will prescribe a belief function (rather than a probability measure), from which we will construct the probability mass. This is the top-level of the hierarchical scheme, which models different contexts. Probability mass (b.p.a.) will, in general, not be assigned to singletons, i.e., any single topology, reflecting the fact that contexts are not "independent". In our hierarchical scheme, switching contexts amounts to switching topologies.

Though the lattice of topologies depicts all possible contexts, sometime we may restrict ourselves to a distributive sublattice for convenience. Figure 3 gives such a case. In this case, we may assign probability measure to such sublattice of topologies.

3.3 *A Hierarchical Scheme for Upper-Lower Probability*

As Zhou (2013) has recently shown, on a distributive lattice, basic probability assignment (b.p.a.'s) would only need to be given to joint-irreducible elements of a lattice for it to be consistent with the Bayesian framework. Any topology on a set fulfills the requirement of a distributive lattice L, where each open set is just an element of L, with set-inclusion $\subseteq$ identified as the order $\prec$ on L, and set-union $\cup$ and set-intersection $\cap$ are $\vee$ and $\wedge$ operations on L. The closure operation that comes with the given topology is related to the pseudo-complementation operation L. Following Narens (2009, 2011), we use this operation to model intuitionistic "negation" in propositional logic. Moreover, we simultaneously consider two or more topologies defined on the same ground set, each with its own b.p.a.'s. As the set of topologies form a lattice itself, and any lattice admits a belief function see Barthélemy (2000), we will endow a belief function on this "lattice of topology" $\mathcal{L}$ as specifying a higher-level (in a hierarchical Bayesian model) upper and lower probability of different contexts. This allows us to accomplish fusion of evidence (different b.p.a.'s) that is soundly rooted in topological semantics, and achieve a hierarchical inference structure that surpasses most current hierarchical schemes of probability inference.

At the bottom level of our hierarchical scheme, we will treat all open sets in a given topology of a ground set as focal elements for basic probability assignments. This is feasible because, for finite set at least, a topology amounts to a distributive lattice, and from Zhou (2013), Bayesian probability measure is possible as long as the b.p.a's are assigned to joint-irreducible elements of the lattice; these will be the "elementary events" in the topological event space.

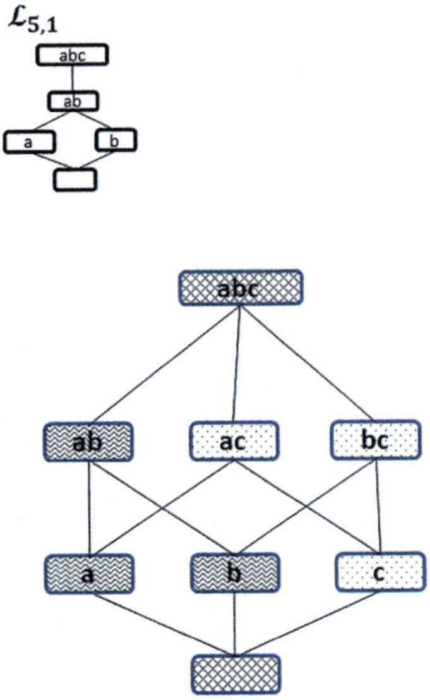

Set-theoretic complements of *Open* Sets (wave patterns) are the *Closed* Sets (dot patterns), listed below. Diamond fill are *Clopen* sets.

Complements

{}'={a,b,c}
{a}'={b,c}
{b}'={a,c}
{a,b}'={c}
{a,b,c}'={}

Topological Closures

Cl {}={}
Cl {a}={a,c}
Cl {b}={b,c}
Cl {c}={c}
Cl {a,b}={a,b,c}
Cl {a,c}={a,c}
Cl {b,c}={b,c}

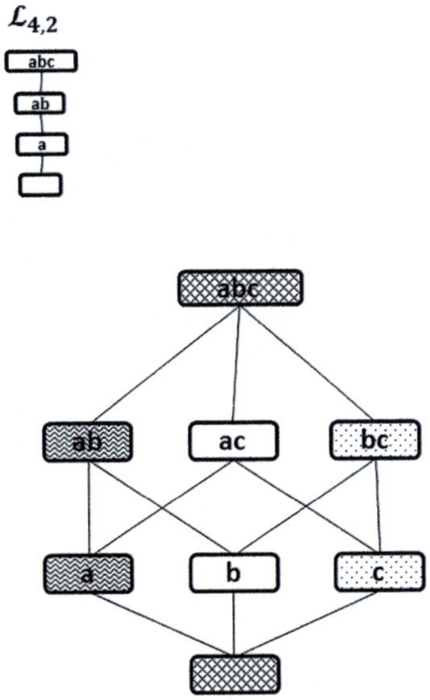

Set-theoretic complements of *Open* Sets (wave patterns) are the *Closed* Sets (dot patterns), listed below. Diamond fill are *Clopen* sets.

Complements

{}'={abc}
{a}'={bc}
{ab}'={c}
{abc}'={}

Topological Closures

Cl {}={}
Cl {a}={a,b,c}
Cl {b}={b,c}
Cl {c}={c}
Cl {a,b}={a,b,c}
Cl {a,c}={a,b,c}
Cl {b,c}={b,c}

Fig. 2 Some examples of topologies, where open sets (wave fill), closed sets (dot fill), clopen sets (diamond fill), and sets that are neither open nor closed (blank) are all graphically coded. Also indicated is the closure operation on each node

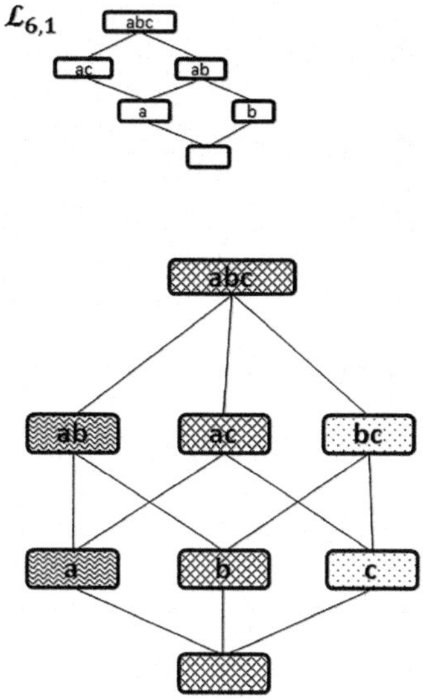

Set-theoretic complements of *Open* Sets (wave patterns) are the *Closed* Sets (dot patterns), listed below. Diamond fill are *Clopen* sets.

Complements

{}'={abc}
{a}'={bc}
{b}'={ac}
{ab}'={c}
{ac}'={b}
{abc}'={}

Topological Closures

Cl {}={}
Cl {a}={a,c}
Cl {b}={b}
Cl {c}={c}
Cl {a,b}={a,b,c}
Cl {a,c}={a,c}
Cl {b,c}={b,c}

Fig. 2 (continued)

A preliminary implementation of our hierarchical scheme of uncertainty reasoning and probabilistic inference based on topological event space is reported in Ilin and Zhang (2014). The general setting is sensor networks, where bodies of evidence for each sensor network needs to be "fused". There, we devised a flow-down algorithm for basic probability assignment only on join-irreducible elements of a distributive lattice (topological event space). We further invoke the lattice of topologies for representing different sensor networks, which are treated as different contexts for uncertainty reasoning.

To summarize, our scheme draws its source from two principled approaches to probabilistic inference and uncertainty management: the Dempster-Shafer theory for upper-lower probability constructed from basic probability assignments, and Narens' (2009, 2011), Narens and Saari (2015) approach to topological event space for lattice-based probability. Our idea, which is fueled by recent mathematical results on the existence of belief functions on a general lattice and probability measure on a distributive lattice, is to construct the upper-lower probability on topological event spaces by (i) stipulating a principled way for basic probability assignments to elements of a topology; and (ii) stipulating (even a sublattice of) the lattice of topologies on the same sample space for modeling switching between and integrating across contexts for b.p.a. assignments. Combining basic probability assignments in a topological event space to obtain upper-lower probabilities with the lattice of topologies to model

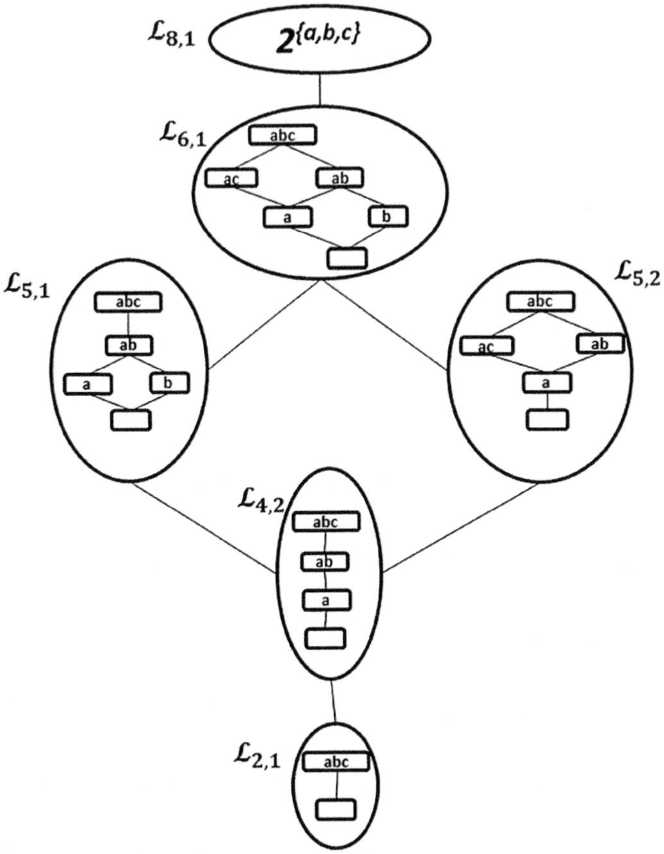

Fig. 3 A distributive sublattice of the lattice of topologies, which supports the assignment of probability measure to "singletons"

hierarchical inference structure has never, to our knowledge, been attempted. Ours can be viewed as a non-Bayesian hierarchical model. The advantage is its ability to model rich sets of contextual information using different topologies for the same underlying sample space and at the same time to reduce the size of the event space by not having to consider the full Boolean (combinatorial) structure.

4 Discussions

Our proposed scheme of upper-lower probability theory based on topological event space and lattice theory is complementary to several other recent developments in extending classical probability calculus.

4.1 Relation to Topological Characterization of "Rare Events"

Chichilnisky (2010) investigated a probability theory that is capable of dealing with unexpected contingencies ("black swan"). Specifically, she proposed a topological framework (Chichilnisky 2010) to deal with measure-zero ("rare") events which yet have catastrophic consequences. Chichilnisky's analysis centered on Villegas and Arrows "Axiom of Monotone Continuity", which invoked a topology that neglects rare events. By replacing it with sup-norm topology of the L_∞ space, she obtained a probability theory which contains a countably additive term as well as a purely finitely additive term.

Her careful analysis of measure-zero events is tightly related to the Ascending/Descending Chain Condition (ACC/DCC) in formulating probability measure. Recall that a partially ordered set (poset) is said to satisfy the ascending chain condition (ACC) if every strictly ascending sequence of elements $\{a_i, i \in \mathbb{N}\}$ eventually terminates. That is, given any sequence $a_1 \leq a_2 \leq a_3 \leq \cdots$, there exists a positive integer n such that $a_n = a_{n+1} = a_{n+2} = \cdots$. The descending chain condition (DCC) can be analogously defined. ACC and DCC are essentially finiteness properties satisfied by some algebraic structures, e.g., ideals in certain commutative rings.

It is interesting that an analysis of ascending and descending sequence also underlies the characterization of the properties of belief functions. Shafer (1979) defined the notion of "continuity" and "condensation" as regularity conditions for belief (lower-probability) functions to be defined on an infinite set. A belief function f is called continuous if

$$f(\bigcap_i A_i) = \lim_{i \to \infty} f(A_i)$$

for every decreasing sequence $A_1 \supseteq A_2 \supseteq \cdots$ of subsets of X. A similar definition for plausibility (upper-probability) functions g to be continuous is

$$g(\bigcup_i A_i) = \lim_{i \to \infty} g(A_i)$$

for every increasing sequence $A_1 \subseteq A_2 \subseteq \cdots$ of subsets of X. A belief function f is called condensable if

$$f(\bigcap \mathcal{A}) = \inf_{A \in \mathcal{A}} f(A)$$

where $\mathcal{A}$ is a down-net of X, that is, a collection of subsets such that if $A_1, A_2 \in \mathcal{A}$, then there exists $A_3 \in \mathcal{A}$ such that $A_3 \subseteq (A_1 \cap A_2)$. Similarly, a plausibility function g is called condensable if

$$g(\bigcup \mathcal{A}) = \sup_{A \in \mathcal{A}} g(A)$$

where $\mathcal{A}$ is a up-net of X, that is, a collection of subsets such that if $A_1, A_2 \in \mathcal{A}$, then there exists $A_3 \in \mathcal{A}$ such that $A_3 \supseteq (A_1 \cup A_2)$. The continuity and condensible conditions turn out to be sufficient and necessary conditions for representing a belief function by a $\cap$-homomorphism into the algebra of measure space (i.e., basic probability assignment on a multiplicative subclass of power-set), see Shafer (1979).

4.2 Relation to Quantum Logic and Quantum Probability

In recent years, quantum probability theory has been invoked in explaining certain phenomena in cognitive psychology, such as conjunction fallacy and order effect in human decision-making literature see Busemeyer and Bruza (2012). The computation model is, however, based on Hilbert space formulation of probability amplitude and quantum physical interpretation of probability measure. Narens (2014) proposed to interpret the cognitive phenomena using quantum logic rather than quantum physics. It is now accepted that the logic underlying quantum physical phenomenology is associated with orthomodular lattices, where orthocomplement is singled out as the "maximal" complement element that each element possesses, and where modularity requirement is imposed upon orthocomplemented pairs only.

Historically, von Neumann first attempted lattice-theoretic characterization of quantum measurements by resorting to orthocomplemented modular lattices. Modular lattice provides a good model for projective geometry. As discussed in Sect. 2.3.2, valuation of a lattice provides the tools for introducing the so-called dimension function on a lattice. In 1930s, von Neumann successfully gave lattice-theoretic treatment of dimension in complete complemented modular lattices. There, dimension is determined, up to a positive linear transformation, by the following two properties: It is conserved by perspective mappings ("perspectivities") and it is ordered by inclusion. According to Birkhoff, the deepest part of von Neumann's theory is the equivalence of perspectivity with "projectivity by decomposition", which gives rise to the transitivity of perspectivity as a corollary. The "dimension function" of the von Neumann algebra takes value not only in a discrete set {0,1,..., n}, but can also be a value in the unit interval [0, 1]. This is the "continuous geometry" setting. Kaplansky later (1955) showed that any orthocomplemented complete modular lattice is a continuous geometry.

It is rather curious that the existence of a strict monotone 2-valuation characterizes modularity, while the existence of a strictly monotone k-valuation (any $k > 2$) characterizes distributivity (see Birkhoff 1933). So for non-distributive modular lattices, a strict monotone 2-valuation does not satisfy 3-valuation condition, though any lattice admits totally monotone functions (which satisfy 2-monotone and 3-monotone conditions). So one wonders what prevents a 2-valuation from becoming a 3-valuation in a non-distributive modular lattice. Understanding this obstruction can lead to insights about distinct probability calculus in intuitionistic and quantum cases.

4.3 Closing Remarks

This chapter reviews some well-established mathematical theory of lattice and its connection to topology, as well as recent results about belief functions and probability measures defined on lattices. We then put forth the idea of a hierarchical scheme for modeling fusion of evidence based on constructing the lattice of topologies over a given sample space, where each topology encodes context for sensor measurement as specified by the basic probability assignment function. This approach provides a rigorous mathematical grounding for modeling uncertainty and information fusion based on upper and lower probabilities originally put forth by the Dempster-Shafer model.

References

Barthélemy, J.-P. (2000). Monotone functions on finite lattices: An ordinal approach to capacities, belief and necessity functions. *Preferences and decisions under incomplete knowledge* (pp. 195–208). Physica-Verlag HD.

Birkhoff, G. (1933). On the combination of subalgebras. *Proceedings of the Cambridge Philosophical Society, 29*, 441–464.

Birkhoff, G. (1967). *Lattice theory*. Providence, RI: American Mathematical Society.

Busemeyer, J., & Bruza, P. D. (2012). *Quantum models of cognition and decision*. Cambridge University Press.

Chateauneuf, A., & Jaffray, J.-Y. (1989). Some characterizations of lower probabilities and other monotone capacities through the use of Möbius inversion. *Mathematical Social Sciences, 17*, 263–283.

Chichilnisky, G. (2010). The foundations of statistics with Black Swans. *Mathematical Social Sciences, 59*, 184–192.

Davey, B., & Priestley, H. (2002). *Introduction to lattices and order*. Cambridge: University Press.

Fagin, R., & Halpern, J. Y. (1991). Uncertainty, belief, and probability. *Computational Intelligence, 7*, 160–173.

Grabisch, M. (2008). *Belief functions on lattice*. arXiv preprint arXiv:0811.3373v1.

Ilin, R., & Zhang, J. (2014). *2014*. Information fusion with uncertainty modeled on topological event spaces. In *Proceedings of IEEE Symposium on Foundation of Computation Intelligence (FOCI 2014)*

Kaplansky, I. (1955). Any orthocomplemented complete modular lattice is a continuous geometry. *Annals Mathematics, 61*, 524–541.

Knight, R. W., Gartside, P., & McIntyre, D. W. (1997). All finite distributive lattices occur as intervals between Hausdorff topologies. *Order, 14*, 259–265.

Larson, R. E., & Andima, S. J. (1975). The lattice of topologies: A survey. *Journal of Mathematics*, **5**(2).

McKinsey, J., & Tarski, A. (1944). The algebra of topology. *Annals of Mathematics, 45*, 141–191.

McKinsey, J., & Tarski, A. (1946). On closed elements in closure algebras. *Annals of Mathematics, 47*, 122–162.

Moisil, G. C. (1935). Recherches sur l'algebre de la logique. *Annales scientifiques de l'université de Jassy, 22*, 1–117.

Monteiro, A. (1980). Sur les algebres de Heyting symtriques. *Portugaliae Mathematica, 39*, 1–237.

Narens, L. (2011). Probabilistic lattices: Theory with an application to decision theory. In Dzhafarov et al. (Eds.) *Descriptive and normative approaches to human behavior* (pp. 161–202).

Narens, L., & Saari, D. (2015). Modeling decisions involving ambiguous, vague, and rare events. This volumn.
Narens, L. (2009). A foundation for support theory based on a non-Boolean event space. *Journal of Mathematical Psychology*, *53*, 399–407.
Narens, L. (2014). Alternative probability theories for cognitive psychology. *Topics in Cognitive Science*, *6*, 114–120.
Pawlak, Z. (1982). Rough sets. *International Journal of Computer and Information Sciences*, *11*, 341–356.
Rosický, J. (1975). Modular, distributive and simple intervals of the lattice of topologies. *Archivum Mathematicum*, *11*, 105–114.
Rota, G.-C. (1964). On the foundation of combinatorial theory. I. Theory of Möbius functions. *Probability Theory and Related Fields*, *2*, 340–368.
Shafer, G. (1976). *A Mathematical Theory of Evidence* (Vol. 1). Princeton: Princeton University Press.
Shafer, G. (1979). Allocations of probability. *The Annals of Probability*, *7*, 827–839.
Stone, M. H. (1936). The theory of representations for Boolean algebras. *Transactions of the American Mathematical Society*, *40*, 37–111.
Valent, R., & Larson, R. E. (1972). Basic intervals in the lattice of topologies. *Duke Mathematical Journal*, *39*, 401–411.
Yager, R. R., & Liu, L. (Eds.) (2008). *Classic works of the Dempster-Shafer theory of belief functions*. Springer.
Zadeh, L. A. (1965). Fuzzy sets. *Information and Control*, *8*, 338–353.
Zhou, C. (2013). Belief functions on distributive lattices. *Artificial Intelligence*, *201*, 1–31.

The Foundations of Uncertainty with Black Swans

Graciela Chichilnisky

1 Introduction

Black swans are rare events with major consequences, such as market crashes, natural hazards, global warming and major episodes of extinction. This article is about the foundations of probability when catastrophic events are at stake.It provides a new axiomatic foundation for subjective probability requiring sensitivity to rare and frequent events. The study culminates in Theorem 2, that proves existence and representation of a subjective probability satisfying three axioms. The last of these axioms requires that the subjective probability be sensitive to rare events, a property that is desirable but not respected by standard probabilities. The article discusses the connection between those axioms and the Axiom of Choice at the foundation of Mathematics. It defines a new type of subjective probabilities that coincide with standard distributions when the sample is populated by frequent events. Generally, however, they are a mixture of countable and finitely additive measures, assigning more weight to black swans than do normal distributions, and predicting more realistically the incidence of 'outliers,' 'power laws' and 'heavy tails'.

The article refines and extends the formulation of subjective probability as a decision characteristic of an agent choosing actions ("acts") in an uncertain world. It provides an argument, and formalization, that subjective probabilities must be additive functionals on $L_\infty(\mathcal{U})$, where $\mathcal{U}$ is a σ—field of "events" represented by their indicator (bounded, real valued) functions, that are neither countably additive nor finitely additive. The contribution is to provide an axiomatization showing that subjective

G. Chichilnisky (✉)
Columbia Consortium for Risk Management, Columbia University,
335 Riverside Drive, New York 10027, USA
e-mail: gc9@columbia.edu
URL: http://www.chichilnisky.com

G. Chichilnisky
Departments of Economics and Mathematical Statistics, Columbia University,
335 Riverside Drive, New York 10027, USA

G. Chichilnisky and A. Rezai (eds.), *The Economics of the Global Environment*,
Studies in Economic Theory 29, DOI 10.1007/978-3-319-31943-8_6

probabilities must lie in the full space L_∞^* rather than L_1 as the usual formalization (Arrow 1971) forcing countable additivity, implies. The new axioms incorporate both Savage (1972) axiomatization of finitely additive measures, as well as Villegas' (1964) and Arrow's (1971) that are based on countably additive measures, and extend both to deal more realistically with catastrophic events.

Savage (1972) axiomatized subjective probabilities as finitely additive measures representing the decision makers' beliefs, an approach that can ignore frequent events as shown in the Appendix. To overcome this, Villegas (1964) and Arrow (1971) introduced an additional continuity axiom (called 'monotone continuity') that yields countably additivity of the measures. However monotone continuity has unusual implications when the subject is confronted with rare events, for example it predicts that in exchange for a couple of cents, one should be willing to accept a small risk of death, a possibility that Arrow called 'outrageous' (1971, p. 48 and 49). This article defines a realistic solution: for some payoffs and in certain situations, one may be willing to accept a small risk of death—but not in others. This means that monotone continuity holds in some cases but not in others, a possibility that leads to the axiomatization proposed in this article and is consistent with the experimental observations reported in Chanel and Chichilnisky (2009a, b). The results are as follows. First we show that countably additive measures are insensitive to *black swans:* they assign negligible weight to rare events, no matter how important these may be, treating catastrophes as outliers, Chichilnisky (2009b, c). Finitely additive measures, on the other hand, may assign no weight to frequent events, which is equally troubling. Our new axiomatization balances the two approaches and extends both, requiring sensitivity to rare events as well as frequent events. We provide an existence theorem for probabilities that satisfy our axioms, and a characterization of all that do.

The results start from an axiomatic approach to choice under uncertainty and sustainable development introduced by the author Chichilnisky (1996, 2000, 2009a), and illuminate the classic issue of continuity that has always been at the core of 'subjective probability' axioms (Villegas 1964; Arrow 1971). To define continuity, we use a topology that tallies with the experimental evidence of how peple react to rare events that cause fear (Le Doux 1996; Chichilnisky 2009a), previously used by Debreu (1953) to formalize a market's Invisible Hand, and in Chichilnisky (2000, 2002) to axiomatize choice under uncertainty with rare events. The new results provided here show that a standard axiom of decision theory, Monotone Continuity, is equivalent to De Groot's Axiom SP_4 that lies at the foundation of classic likelyhood theory (Proposition 1) and that both of these axioms underestimate rare events no matter how catastrophic they may be. We introduce here a new Swan Axiom (Sect. 3) that negates them both, and show it is a combination of two axioms defined in Chichilnisky (2000, 2002) and prove that any subjective probability satisfying the Swan Axiom is neither countably additive nor finitely additive: it is both (Theorem 1). Theorem 2 provides a complete characterization of all subjective probabilities that satisfy linearity and the Swan Axiom, thus extending earlier results of Chichilnisky (2000, 2002, 2009a).

There are other approaches to subjective probability such as Choquet Expected Utility Model (CEU, Schmeidler 1989) and Prospect Theory (Kahneman and Tversky 1979 and 1992). They use a non-linear treatement of probabilities of likelyhoods (see e.g. Dreze 1987 or Bernstein 1996), while we retain linear probabilities. Both have a tendency to give higher weght to small probabilities, and are theoretical asnwers to experimental paradoxes (Allais 1953; Ellsberg 1961) refuting the *Independence Axiom* of the Subjective Expected Utility Utility (SEU) model. Our work focuses instead directly on the foundations of probability by taking the logical negation of the *Monotone Continuity Axiom.* It is striking that weakening or rejecting this axiom—respectively in decision theory and in probability theory—ends up in probability models that are more in tune with observed attitudes when facing catastrophic events. Presumably each approach has advantages and shortcomings. It seems that the approach offered here may be superior on four counts: (i) It retains linearity of probabilities, (ii) It identifies Monotone Continuity as the reason for underestimating catastrophic events, an axiom that depends on a technical definition of continuity and has no other compelling feature, (iii) it seems easier to explain and to grasp, and therefore (iv) it may be easier to use in applications.

2 The Mathematics of Uncertainty

Uncertainty is described by a set of distinctive and exhaustive possible **events** represented by a family of sets $\{U_\alpha\}, \alpha \in N$,whose union describes a universe $\mathcal{U} = \cup_\alpha U_\alpha$. An event $U \in \mathcal{U}$ is identified with its **characteristic function** $\phi_U : \mathcal{U} \to R$ where $\phi_U(x) = 1$ when $x \in U$ and $\phi_U(x) = 0$ when $x \notin U$, and therefore can be identified as a function $f(x) = \phi_U(x)$ for some U. The subjective probability of an event U is a real number $W(U)$ that measures how likely it is to occur according to the subject. Generally we assume that the proability of the universe is 1 and that of the empty set is zero $W(\emptyset) = 0$. In this article we make no difference between subjective probabilities and likelyhoods, using both terms intercheangeably. Classic axioms for subjective probability (respectively likelyhoods) are in Savage (1972) and De Groot (1970). The likelyhood of two disjoint events is the sum of their likelyhoods: $W(U_1 \cup U_2) = W(U_1) + W(U_2)$ when $U_1 \cap U_2 = \emptyset$. These properties correspond to the definition of a subjective probability or likelyhood as a *finite additive measure* on a family ($\sigma - algebra$) of measurable sets of $\mathcal{U}$, which is Savage (1972) definition of subjective probability. A *purely finitely additive probability* is one that is additive but not countably additive. Savage's subjective probabilities can be purely finitely additive or countably additive. In that sense they include all the probabilities in this article. However as seen below, this article excludes subjective probabilities that are either purely finitely additive, or countably additive, and therefore our characterization of a subjective probability is strictly finer than that Savage (1972), and different from the view of a measure as a countably additive set function. The following two axioms were introduced in Villegas (1964), see Arrow (1971) and De Groot (1970) for the purpose of obtaining countable additivity:

Monotone Continuity Axiom (MC), (Arrow 1971): for every two events f and g with $W(f) > W(g)$, and every **vanishing** sequence of events $\{E_\alpha\}_{=1,2,...}$ (defined as.$\forall\alpha$, $E_{\alpha+1} \subset E_\alpha$ and $\cap_{\alpha=1}^{\infty} E_\alpha = \emptyset$) there exists N such that altering arbitrarily the events f and g on the set E_i, where $i > N$, does not alter the subjective probability ranking of the events, namely $W(f') > W(g')$, where f' and g' are the altered events.

This axiom is equivalent to requiring that the probability of the sets along a vanishing sequence goes to zero. The decreasing sequence could consist of infinite intervals of the form (n, ∞) for $n = 1, 2, \ldots$ Monotone continuity therefore implies that the likelyhood of this sequence of events goes to zero, even though all its sets are unbounded. A similar example can be constructed with a decreasing sequence of bounded sets, $(-1/n, 1/n)$ for $n = 1, 2, \ldots$, which is also a vanishing sequence as it is a decreasing and their intersection is empty.

De Groot's Axiom SP_4 (De Groot 1970, Chap. 6, p. 71): if $A_1 \supset A_2 \supset \ldots$ is a decreasing sequence of events and B is some fixed event that is less likely than A_i for all i, then the probability of the intersection $\cap_i^\infty A_i$ is larger than that of B.

The following proposition establishes that these two axioms are one and the same; both imply countable additivity:

Proposition 1 *A relative likelihood satisfies the* ***Monotone Continuity*** *Axiom if and only if it satisfies Axiom SP_4. Each of the two axioms implies countable additivity.*

Proof Assume that De Groot's axiom SP_4 is satisfied. When the intersection of a decreasing sequence of events is empty $\cap_i A_i = \emptyset$ and the set B is less likely to occur than every set A_i, then the subset B must be as likely as the empty set, namely its probability must be zero. In other words, if B is more likely than the empty set, then regardless of how small is the set B, it is impossible for every set A_i to be as likely as B. Equivalently, the probability of the sets that are far away in the vanishing sequence must go to zero. Therefore SP_4 implies Monotone Continuity. Reciprocally, assume MC is satisfied. Consider a decreasing sequence of events A_i and define a new sequence by substracting from each set the intersection of the family, namely $A_1 - \cap_i^\infty A_i$, $A_2 - \cap_i^\infty A_i, \ldots$. Let B be a set that is more likely than the empty set but less likely than every A_i. Observe that the intersection of the new sequence is empty, $\cap_i A_i - \cap_i^\infty A_i = \emptyset$ and since $A_i \supset A_{i+1}$ the new sequence is, by definition, a vanishing sequence. Therefore by MC $\lim_i W(A_i - \cap_i^\infty A_i) = 0$. Since $W(B) > 0$, B must be more likely than $A_i - \cap_i^\infty A_i$ for some i onwards. Furthermore, $A_i = (A_i - \cap_i^\infty A_i) \cup (\cap_i^\infty A_i)$ and $(A_i - \cap_i^\infty A_i) \cap (\cap_i^\infty A_i) = \emptyset$, so that $W(A_i) > W(B)$ is equivalent to $W(A_i - \cap_i^\infty A_i) + W((\cap_i^\infty A_i) > W(B)$. Observe that $W(\cap_i^\infty A_i) < W(B)$ would contradict the inequality $W(A_i) = W(A_i - \cap_i^\infty A_i) + W((\cap_i^\infty A_i) > W(B)$, since as we saw above, by MC, $\lim_i W(A_i - \cap_i^\infty A_i) = 0$, and $W(A_i - \cap_i^\infty A_i) + W((\cap_i^\infty A_i)$ $> W(B)$. It follows that $W(\cap_i^\infty A_i) > W(B)$, which establishes De Groots's Axiom SP_4. Therefore Monotone Continuity is equivalent to De Groot's Axiom SP_4. A proof that each of the axioms implies countable additivity is in Villegas (1964); Arrow (1971) and De Groot (1970). ■

The next section shows that the two axioms, monotone continuity and SP_4 are biased against rare events no matter how important these may be.

3 The Value of Life

The best way to explain the role of *monotone continuity* is by means of an example provided by (Arrows 1971, p. 48 and 49). He explains that if a is an action that involves receiving one cent, b is another that involves receiving zero cents, and c is a third action involving receiving one cent and facing a small probability of death, then *Monotone Continuity* requires that the third action involving death and one cent should be preferred to the action with zero cents when the probability of death is small enough. Even Arrow says of his requirement 'this may sound outrageous at first blush ...' (Arrows 1971, p. 48 and 49). Outrageous or not, Monotone Continuity (MC) leads to neglect rare events with major consequences, like death. Death is a black swan.

To overcome the bias we introduce an axiom that is the logical negation of MC: this means that sometimes MC holds and others it does not. We call this the **Swan Axiom,** and it is stated formally below. To illustrate this, consider an experiment where subjects are offered a certain amount of money to choose a pill at random from a pile, which is known to contain one pill that causes death (Chanel and Chichilnisky 2009b). It was shown experimentally (Chanel and Chichilnisky 2009b) that in some cases people accept a sum of money and choose a pill provided the pile is large enough—namely when the probability of death is small enough—thus satisfying the monotone continuity axiom and determining the statistical value of their lives. But there are also cases where the subjects will not accept to choose any pill, no matter how large is the pile. Some people refuse the payment if it involves a small probability of death, no matter how small the probability may be (Chanel and Chichilnisky 2009a, b). This conflicts with the Monotone Continuity axiom, as explicitly presented by Arrow (1971).

Our Axiom provides a reasonable resolution to this dilemma that is realistic and consistent with the experimental evidence. It implies that there exist catastrophic outcomes such as the risk of death, so terrible that one is unwilling to face a small probability of death to obtain one cent versus half a cent, no matter how small the probability may be. According to our Axiom, no probability of death may be acceptable when one cent and half a cent are involved. Our Axiom also implies that in other cases there may be a small enough probability that the lottery involving death may be acceptable, or the payoff is large enough to justify the small risk. This is a possibility discussed by Arrow (1971). In other words: sometimes one is willing to take a risk with a small enough probability of a catastrophe, in other cases one is not. This is the content of our Axiom, which is formally stated below:

The **Swan Axiom**: There exist events f and g with $W(f) > W(g)$, and for every vanishing sequence of events $\{E_\alpha\}_{=1,2...}$ an $N > 0$ such that altering arbitrarily the events f and g on the set E_i, where $i > N$, does not alter the probability ranking of the events, namely $W(f') > W(g')$, where f' and g' are the altered events. For other events f and g with $W(f) > W(g)$, there exist vanishing sequence of events $\{E_\alpha\}_{=1,2...}$ where for every N, altering arbitrarily the events f and g on the set E_i, where $i > N$, does alter the probability ranking of the events, namely $W(f') < W(g')$, where f' and g' are the altered events.

Definition A probability W is said to be **biased against rare events** or **insensitive to rare events** when it neglects events that are small according to Villegas and Arrow; as stated in Arrow (1971, p. 48): " An event that is far out on a *vanishing sequence* is **'small'** by any reasonable standards" (Arrow 1971, p. 48). Formally, a probability is insensitive to rare events when given two events f and g and any vanishing sequence of events (E_j), $\exists\ N = N(f, g) > 0$, such that $W(f) > W(g) \Leftrightarrow W(f') > W(g')\ \forall f', g'$ satisfying $f' = f$ and $g' = g$ a.e. on $E_j^c \subset R$ when $j > N$, where E^c denotes the complement of the set E.

Proposition 2 *A subjective probability satisfies Monotone Continuity if and only if it is biased against rare events.*

Proof This is immediate from the definitions of both. ■

Corollary 1 *Countably additive probabilities are biased against rare events.*

Proof It follows from Propositions 1 and 2. ■

Proposition 3 *Purely finitely additive probabilties can be biased against frequent events.*

Proof See example in the Appendix. ■

Proposition 4 *A subjective probability that satisfies the Swan Axiom is neither biased against rare events, nor biased against frequent events.*

Proof This is immediate from the definition. ■

4 An Axiomatic Approach to Unbiased Statistics

This section proposes an axiomatic foundation for subjective probability that is unbiased against rare and frequent events. The axioms are as follows:

Axiom 1 Subjective probabilities are continuous and additive

Axiom 2 Subjective probabilities are unbiased against rare events

Axiom 3 Subjective probabilities are unbiased against frequent events.

Axioms 2 and 3 together are equivalent to the Swan Axiom defined in the previous section, which is required to avoid bias against rare and frequent events as shown in Sect. 3. Additivity is a natural condition and **continuity** captures the notion that 'nearby' events are thought as being similarly likely to occur; this property is important to ensure that 'sufficient statistics' exist. 'Nearby' has been defined by

Villegas (1964) and Arrow (1971) as follows: two events are **close** or **nearby** when they differ on a **small set** as defined in Arrow (1971), see previous section. We saw in Proposition 2 that the notion of continuity defined by Villegas and Arrow—namely monotone continuity—conflicts with the Swan Axiom. Indeed Proposition 2 shows that countably additive measures are biased against rare events. On the other hand, Proposition 3 and the Example in the Appendix show that purely finitely additive measures can be biased against frequent events. A natural question is whether after one eliminates both biases there is anything left. The following proposition addresses this issue:

Theorem 1 *A subjective probability that satisfies the Swan Axiom is neither finitely additive nor countably additive; it is a strict convex combination of both.*

Proof The result is immediate. The next Section develops the proofs and provides examples when the events are Borel sets in R or an interval (a, b). ■

Theorem 1 establishes that neither Savage's approach, nor Villegas' and Arrow's, satisfy the three axioms stated above. These three axioms require more than the additive subjective probabilties of Savage, since purely finitely additive probabilities are finitely additive and yet they must be excluded here. At the same time the axioms require less than the countably subjective additivity of Villegas and Arrow, since countably additive probabilities are biased against rare events. Theorem 1 above shows that a strict combination of both does the job.

Theorem 1 does not however prove the existence of likelihoods that satisfy all three axioms. What is missing is an appropriate definition of continuity that does not conflict with the Swan Axiom. The following Section shows that this can be achieved by identifying an event with its characteristic function, so that events are contained in the space of bounded real valued functions on the universe space $\mathcal{U}$, $L_\infty(\mathcal{U})$, and endowing this space with the sup norm. In this case the likelihood W: $L_\infty(\mathcal{U}) \rightarrow R$ is taken to be continuous with respect to the sup norm, a topology used in Debreu (1953).

5 Axiomatic Statistics on *R* or (a, b)

From now on events are taken to be the Borel sets of the real line R or the interval (a, b), a widely used case that make the results concrete and compare the results with the earlier axioms on choice under uncertainty of Chichilnisky (2000, 2002, 2009a). We use a concept of 'continuity' based on a topology that was used earlier in Debreu (1953) and in Chichilnisky (2000, 2002, 2009b, c): observable events are in the space of measurable and essentially bounded functions $L = L_\infty(R)$ with the sup norm $\| f \| = ess \sup_{x \in R} | f(x) |$. This is a sharper and more stringent definition of closeness than the one used by Villegas and Arrow, since an event can be small under the Villegas-Arrow definition but not under ours, see the Appendix. The difference as shown below determines sensitivity to rare events.

A subjective probabiliy satisfying the classic axioms in De Groot (1970) is called a **standard probability,** and is therefore countably additive. A classic result is that for any event $f \in L_\infty$ a standard probability has the form $W(f) = \int_R f(x).\phi(x)d\mu$, where $\phi \in L_1(R)$ is an integrable function in R.

The next step is to introduce the new axioms, show existence and characterize all the distributions that satisfy the axioms. We need more definitions. A subjective probability $W : L_\infty \to R$ is called **biased against rare events**, or **insensitive to rare events** when it neglects events that are small according to a probability measure μ on R that is absolutely continuous with respect to the Lebesgue measure. Formally, a probability is insensitive to rare events when given two events f and g $\exists\, \varepsilon = \varepsilon(f, g) > 0$, such that $W(f) > W(g) \Leftrightarrow W(f') > W(g')\ \forall f', g'$ satisfying $f' = f$ and $g' = g$ a.e. on $A \subset R$ and $\mu(A^c) < \varepsilon$. Here A^c denotes the complement of the set A. $W : L \to R$ is said to be **insensitive to frequent events** when given any two events f, g $\exists\, \varepsilon(f, g) > 0$ that $W(f) > W(g) \Leftrightarrow W(f') > W(g') \forall f', g'$ satisfying $f' = f$ and $g' = g$ a.e. on $A \subset R$ and $\mu(A^c) > 1 - \varepsilon$. W is called **sensitive** to rare (or frequent) events when it is **not insensitive** to rare (or frequent) events.

The following three axioms are identical to the axioms in last section, specialized to the case at hand:

Axiom 1 $W : L_\infty \to R$ is linear and continuous

Axiom 2 $W : L_\infty \to R$ is sensitive to frequent events

Axiom 3 $W : L_\infty \to R$ is sensitive to rare events

The first and the second axiom agree with classic theory and standard likelihoods satisfy them. The third axiom is new.

Lemma 1 *A standard probability satisfies Axioms 1 and 2, but it is biased against rare events and therefore does not satisfy Axiom 3.*

Proof Consider $W(f) = \int_R f(x)\phi(x)dx$, $\int_R \phi(x)dx = K < \infty$. Then

$$W(f) + W(g) = \int_R f(x)\phi(x)dx + \int_R g(x)\phi(x)dx = \int_R \mid f(x) + g(x) \mid \phi(x)dx = W(f + g),$$

and therefore W is linear. It is **continuous** with respect to the L_1 norm $\| f \|_1 = \int_R \mid f(x) \mid \phi(x)d\mu$ (1) because $\| f \|_\infty < \varepsilon$ implies

$$W(f) = \int_R f(x).\phi(x)dx \leq \int_R \mid f(x) \mid .\phi(x)dx \leq \varepsilon \int \phi(x)dx = \varepsilon K.$$

Since the sup norm is finer than the L_1 norm, continuity in L_1 implies continuity with respect to the sup norm (Dunford and Schwartz 1958). Thus a standard subjective probability satisfies Axiom 1. It is obvious that for every two events f, g, with $W(f) > W(g)$, the inequality is reversed namely $W(g') > W(f')$ when f' and g' are appropriate variations of f and g that differ from f and g on sets of sufficiently large

Lebesgue measure. Therefore Axiom 2 is satisfied. A standard subjective probability is however not sensitive to rare events, as shown in Chichilnisky (2000, 2002, 2009b, c), Chichilnisky and Wu (2006). ■

6 Existence and Representation

Theorem 2 *There exists a subjective probability* $W : L_\infty \to R$ *satisfying Axioms 1, 2, and 3. A probability satisfies Axioms 1, 2, and 3 if and only if there exist two continuous linear functions on* L_∞, *denoted* ϕ_1 *and* ϕ_2 *and a real number* λ, $0 < \lambda < 1$, *such that for any observable event* $f \in L_\infty$

$$W(f) = \lambda \int_{x\epsilon R} f(x)\phi_1(x)dx + (1-\lambda)\phi_2(f) \tag{1}$$

where $\phi_1 \in L_1(R, \mu)$ *defines a countably additive measure on* R *and* ϕ_2 *is a purely finitely additive measure.*

Proof This result follows from the representation Theorem in Chichilnisky (2000, 2009a). ■

Example 1 'Heavy' Tails.

The following illustrates the additional weight that the new axioms assign to rare events; in this example in the form of 'heavy tails'. The finitely additive measure ϕ_2 appearing in the second term in Eq. 1 can be illustrated as follows. On the subspace of events with limiting values at infinity, $L'_\infty = \{f \epsilon L_\infty : \lim_{x\to\infty}(x) < \infty\}$, define $\phi_2(f) = \lim_{x\to\infty} f(x)$ and extend this to a function on all of L_∞ using Hahn Banach's theorem. The difference between a standard probability and the likelihood defined in Eq. 1 is the second term ϕ_2, which focuses all the weight at infinity. This can be interpreted as a 'heavy tail' namely a part of the distribution that is not part of the standard density function ϕ_1 and gives more weight to the sets that contain *terminal* events namely sets of the form (x, ∞). ■

Corollary 2 *Absent rare events, a subjective probability that satisfies Axioms 1, 2, and 3 is consistent with classic axioms and yields a countably additive measure.*

Proof Axiom 3 is an empty requirement when there are no rare events while, as shown above, Axioms 1 and 2 are consistent with standard relative likelihood. ■

7 The Axiom of Choice

There is a connection between the axioms presented here and the Axiom of Choice in the foundation of mathematics (Godel 1940), which postulates that there exists a universal and consistent fashion to select an element from every set. The best way to describe the situation is by means of an example, see also Dunford and Schwartz (1958), Yosida (1952, 1974), Chichilnisky and Heal (1997) and Kadane and O'Hagan (1995).

Example 2 Illustration of a Purely Finitely Additive Measure

Consider a measure ρ that satisfies the following: for every interval $A \subset R$, $\rho(A) = 1$ if $A \supset \{x : x > a$, for some $a \in R\}$, and otherwise $\rho(A) = 0$. Then ρ is not countably additive, because the family of countably many disjoint sets $\{V_i\}_{i=0,1,...}$ defined as $V_i = (i, i+1] \bigcup (-i-1, -i]$, satisfy $V_i \bigcap V_i = \emptyset$ when $i \neq j$, and $\bigcup_{i=0}^{\infty} V_i = \bigcup_{i=0}^{\infty}(i, i+1] \bigcup (-i-1, -i] = R$, so that $\rho(\bigcup_{i=0}^{\infty} V_i) = 1$, while $\sum_{i=0}^{\infty} \rho(V_i) = 0$, which contradicts countable additivity. Since the contradiction arises from assuming that ρ is countably additive, ρ must be purely finitely additive. Observe that ρ assigns zero measure to any bounded set, and a positive measure only to unbounded sets that contain a 'terminal set' of the form

$$\{x \epsilon R : x > a \text{ for some } a \epsilon R\}.$$

One can define a function on L_∞ that represents this purely finitely additive measure ρ if we restrict our attention to the closed subspace L'_∞ of L_∞ consisting of those functions $f(x)$ in L_∞ that have a limit when $x \to \infty$, by the formula $\rho(f) = \lim_{x\to\infty} f(x)$, as in Example 1 of the previous section. The function $\rho(.)$ can be illustrated as a limit of a sequence of delta functions whose support increases without bound. The problem is now to extend the function ρ to another defined on the entire space L_∞. This could be achieved in various ways but as we will see, each of them requires the Axiom of Choice.

One can use Hahn-Banach's theorem (Dunford and Schwartz 1958) to extend the function ρ from the closed subspace $L'_\infty \subset L_\infty$ to the entire space L_∞ preserving its norm. However, in its general form Hahn-Banach's theorem requires the Axiom of Choice (Dunford and Schwartz 1958). Alternatively, one can extend the notion of a *limit* to encompass all functions in L_∞ including those with no standard limit. This can be achieved by using the notion of convergence along a *free ultrafilter* arising from compactifying the real line R as in Chichilnisky and Heal (1997). However the existence of a *free ultrafilter* also requires the Axiom of Choice.

This illustrates why the attempts to construct *purely finitely additive measures* that are representable as functions on L_∞, require the Axiom of Choice. Since our criteria require purely finitely additive measures, this provides a connection between the Axiom of Choice and our axioms for relative likelihood. It is somewhat surprising that the consideration of rare events that are neglected in standard statistical theory conjures up the Axiom of Choice, which is independent from the rest of mathematics (Godel 1940).

8 Appendix

Example 3 A Probability that is Biased Against Frequent Events

Consider $W(f) = \liminf_{x \epsilon R}(f(x))$. This is insensitive to frequent events of arbitrarily large Lebesgue measure (Dunford and Schwartz 1958) and therefore does not satisfy Axiom 2. In addition it is not linear, failing Axiom 1.

Example 4 **Two Approaches to 'Closeness'**

Consider the family $\{E_i\}$ where $E_i = [i, \infty), i = 1, 2, \ldots.$ This is a vanishing family because $\forall i \; E_i \supset E_{i+1}$ and $\bigcap_{i=1}^{\infty} E_i = \emptyset$. Consider now the events $f^i(t) = K$ when $t \in E_i$ and $f^i(t) = 0$ otherwise, and $g^i(t) = 2K$ when $t \in E_i$ and $g^i(t) = 0$ otherwise. Then for all i, $\sup_{E^i} \mid f^i(t) - g^i(t) \mid = K$. In the sup norm topology this implies that f^i and g^i are **not** 'close' to each other, as the difference $f^i - g^i$ does not converge to zero. No matter how far along we are along the vanishing sequence E^i the two events f^i, g^i differ by K. Yet since the events f^i, g^i differ from $f \equiv 0$ and $g \equiv 0$ respectively only in the set E_i, and $\{E_i\}$ is a vanishing sequence, for large enough i they are as 'close' as desired according to Villegas-Arrow's definition of 'nearby' events.

The dual space L_∞^*: countably additive and finitely additive measures

The space of continuous linear functions on L_∞ is the 'dual' of L_∞, and is denoted L_∞^*. It has been characterized e.g. in Yosida (1952, 1974). L_∞^* consists of the sum of two subspaces (*i*) L_1 functions g that define countably additive measures ν on R by the rule $\nu(A) = \int_A g(x)dx$ where $\int_R \mid g(x) \mid dx < \infty$ so that υ is *absolutely continuous* with respect to the Lebesgue measure, and (*ii*) a subspace consisting of purely finitely additive measure. A countable measure can be identified with an L_1 function, called its 'density,' but purely finitely additive measures cannot be identified by such functions.

Example 5 **A Finitely Additive Measure that is not Countably Additive**

See Example 2 in Sect. 7.

Acknowledgements This research was conducted at Columbia University's Program on Information and Resources and its Columbia Consortium for Risk Management (CCRM). We acknowledge support from Grant No 5222-72 of the US Air Force Office of Research directed by Professor Jun Zheng, Washington DC. The initial results (Chichilnisky (1996)) were presented at Stanford University's 1993 Seminar on Reconsideration of Values, organized by Professor Kenneth Arrow, at a 1996 Workshop on Catastrophic Risks organized at the Fields Institute for Mathematical Sciences of the University of Toronto, Canada, at the NBER Conference *Mathematical Economics: The Legacy of Gerard Debreu* at UC Berkeley, October 21, 2005, the Department of Economics of the University of Kansas National Bureau of Economic Research General Equilibrium Conference, September 2006, at the Departments of Statistics of the University of Oslo, Norway, Fall 2007, at a seminar organized by the late Professor Chris Heyde at the Department of Statistics of Columbia University, Fall 2007, and at seminars organized by Drs. Charles Figuieres and Mabel Tidball at LAMETA Universite de Montpellier, France December 19 and 20, 2008, and by Professor Alan Kirman at GREQAM Universite de Marseille, December 18 2008. We are grateful to the above institutions and individuals for supporting the research, and for helpful comments and suggestions. An anonymous referee provided insightful comments that improved this article.

References

Allais, M. (1953). Le Comportement de l'Homme Rationnel devant le Risque: Critique des Postulats et Axiomes de l'Ecole Americaine. *Econometrica*, *21*(4), 503–546.

Arrow, K. (1971). *Essays in the theory of risk bearing*. Amsterdam: North Holland.

Bernstein, P. L. (1996). *Against the gods: A remarkable story of risk*. New York: Wiley.

Chanel, O., & Chichilnisky, G. (2009a). The influence of fear in decisions: Experimental evidence. *Journal of Risk and Uncertainty*, *39*(3),

Chanel, O., & Chichilnisky, G. (2009b). *The value of life: Theory and experiments working paper, GREQE*. New York: Universite de Marseille and Columbia University

Chichilnisky, G. (1996). An axiomatic approach to sustainable development. *Soc Choice Welf*, *13*, 257–321.

Chichilnisky, G. (2000). An axiomatic approach to choice under uncertainty with catastrophic risks. *Resource Energy Economics*, *22*, 221–231.

Chichilnisky, G. (2002). Catastrophic risk, encyclopedia of environmetrics. In A.H. El-Shaarawi & W.W. Piegorsch (Eds.) (Vol. 1, pp. 274-279). Chichester: Wiley.

Chichilnisky, G. (2009a). The limits of econometrics: Non parametric estimation in Hilbert spaces. Journal of Economic Theory.

Chichilnisky, G. (2009b). Choice under uncertainty: The work and legacy of kenneth arrow encyclopedia of quantitative finance, ed Cont R. Chichester: Wiley (in press).

Chichilnisky, G. (2009c). The topology of fear. *Journal of Mathematical Economics*.

Chichilnisky, G., & Heal, G. M. (1997). Social choice with infinite populations. *Soc Choice Welf*, *14*, 303–319.

Chichilnisky, G., & Wu, H.-M. (2006). General equilibrium with endogenous uncertainty and default. *Journal of Mathematical Economics*, *42*, 499–524.

Debreu, G. (1953). Valuation equilibrium and pareto optimum. *Proceedings of the National Academy of Sciences*, *40*, 588–592.

De Groot, M. H. (1970, 2004). *Optimal statistical decisions*. Hoboken, New Jersey: Wiley.

Dreze, J. H. (1987). *Essays on economic decisions under uncertainty*. Cambridge: Cambridge University Press.

Dunford, N., & Schwartz, J. T. (1958). *Linear operators, Part I*. New York: Interscience.

Ellsberg, D. (1961). Risk, ambiguity, and the savage axioms. *Quartarly Journal of Economics*, *75*, 643–669.

Godel, K. (1940). *The consistency of the continuum hypothesis*. *Annals of Mathematical Studies 3*. Princeton: Princeton University Press.

Kadane, J. B., & O'Hagan, A. (1995). Using finitely additive probability: Uniform distributions of the natural numbers. *Journal of the American Statistical Association*, *90*, 525–631.

Kahneman, D., & Tversky, A. (1979). Prospect theory: An analysis of decisions under risk. *Econometrica*, *47*, 263–291.

Le Doux, J. (1996). *The emotional brain*. New York: Simon and Schuster.

Savage, L. (1972). *The foundations of statistics*. Wiley: New York (1054, revised edition).

Schmeidler, D. (1989). Subjective probability and expected utility without additivity. *Econometrica*, *47*, 263–291.

Tversky, A., & Kahneman, D. (1992). Advances in prospect theory: Cumulative representation of uncertainty. *Journal of Risk and Uncertainty*, *5*, 297–323.

Yosida, K. (1974). *Functional analysis* (4th ed.). New York, Heidelberg: Springer.

Yosida, K., & Hewitt, E. (1952). Finitely level independent measures. *Transactions of the American Mathematical Society*, *72*, 46–66.

Villegas, C. (1964). On quantitative probability σ–algebras. *Annals of Mathematical Statistics*, *35*, 1789–1800.

The Topology of Change Foundations of Probability with Black Swans
Dedicated to the Memory of Jerrold Marsden

Graciela Chichilnisky

1 Introduction

Classic probability theory treats rare events as 'outliers' and often disregards them. This is an unavoidable shortcoming of classic theory that has been known for some time and conflicts with observations about the distribution of rare events in natural and human systems, such as earthquakes and financial markets. It is now known that the shortcoming originates from the axioms created by Kolmogorov (1950) to provide a fundation for probability theory, Chichilnisky (2000, 2002). It turns out that the same phenomenon that underestimates rare events leads classic probability theory to underestimate the likelihood of change. In a situation of change, events that are rare become frequent and events that are frequent become rare. Therefore by ignoring rare events we tend to underestimate the possibility of change. In a slight abuse of language it could be said that classic probability theory leads us to 'ignore' change. The change we refer to includes rare events of great importance that should not be underestimated, for example *black swans* such as catastrophic climate change and major episodes of species extinction.

Sensitivity to change—or lack thereof–is a topological issue at its core. It measures how likelihoods change with changes in measurements or observations. If we are sensitive to change our responses change in harmony with the signals. To disregard change means that our response "needle" is either insensitive to, or at odds with, the signals. In mathematical terms this is all about continuity of the respose and as such it is defined and measured by topology. In a recent discovery it was found that an important continuity axiom of classic probability theory is responsible for the insensitivity to rare events. De Groot (1970) calls this axiom SP4, Arrow (1982) called it "monotone continuity" (Arrow (1971), and similar continuity axioms appear in Herstein and Milnor (1953), see Chichilnisky (2000, 2002)). The continuity that

G. Chichilnisky (✉)
Departments of Economics and Mathematical Statistics, Columbia University,
335 Riverside Drive, New York 10027, USA
e-mail: gc9@columbia.edu

G. Chichilnisky and A. Rezai (eds.), *The Economics of the Global Environment*,
Studies in Economic Theory 29, DOI 10.1007/978-3-319-31943-8_7

these axioms provide is coarse, and it was shown to be responsible for the insensitivity to rare events Chichilnisky (2009a, 2010). In that sense the clssic axioms lead to insensitivity about the likelihood of change. The fact is that a single continuity axiom explains why classic probabilty theory is insensitive to rare events and why it ignores change.

To overcome this limitation, new axioms for probability theory were created that balance the treatment of rare and frequent events, based on a more sensitive notion of continuity or a 'finer' topology—and new types of probability distributions have been identified as emerging from the new axioms (Chichilnisky 2000, 2002). In order to be sensitive to rare events, the new axioms have to use a different continuity criterion, a topology finer than that implicit in axiom SP4 or in the monotone continuity axiom, both of which involve averages. This new topology is about extremes not averages, and it is appropriately called "the topology of change" because it is more sensitive to the measurement of rare events that are often at stake in a stuation of change. This new topology is the *sup norm* topology on L_∞ that, while new in this area, has been used earlier by Debreu (1953) to formalize Adam Smith's theorem on the Invisible Hand, and was used in Chichilnisky (2000, 2002) to axiomatize choice under uncertainty. The sup norm provides a finer notion of continuity than "monotone continuity" and SP4. This sensitivity tallies with the experimental evidence of how people react to rare events (Le Doux (1996), Chichilnisky (2009a)). Using the topology of change, the new axioms of probability theory extend the classic foundations of probability, treating rare and frequent events in a more balanced fashion and providing a more balanced view on the likelihood of change.

This article provides new results in this framework, as follows. We introduce the Swan Axiom, a new axiom that is based on continuity in the topology of change. We show how the old and the new topologies differ, namely how continuity in the sense of monotone continuity and SP4 does not imply continuity in the topology of change. We also identify a new family of purely finitely additive measures that is continuous with respect to the "topology of change". Somewhat surprisingly, we show that the change in topology—from probability distributions that satisfy Monotone Continuity to those who satisfy the topology of change—does not necessarily give rise to discontinuity with respect to the Lebesgue measure on R, such as 'delta functions' or meaasures having or "atoms". Indeed the new results presented in this article show the opposite: each of the measures in the family we provide of purely finitely additive measures satifying the new axioms is in fact *absolutely continuous* with respect to the Lebesgue measure. Therefore the new notion of continuity that derives from the new axioms does not imply atoms nor assigns positive weights to sets of Lebesgue measure zero. These new results tally with the earlier characterization of probabilities measures satisfying the new axioms as combinations of purely finitely additive and countably additive measures, Chichilnisky (2000, 2002, 2009b). We contrasts the new measures with the those defined by Kolmogorov (1950), De Groot (1970), Arrow (1971), Dubins and Savage (1965), Savage (1972), Von Neumann and Morgernstern (1944), and Herstein and Milnor (1953). Finally we show that the new results rather than contradicting classic theory can be seen as an extension of it. The new theory of probabilty offered here is an extension of the old: the probability distributions implied by the new axioms coincide with classic countably additive distributions when the sample is populated only by frequent events, even though they are quite different in general. As already stated the new probability measures consist of a convex combination of countable and finitely additive measures with strictly positive elements of both which, in practical terms, assign more weight to black swans than do normal distributions, and predict more realistically the emergence of change and generally the incidence of 'outliers'.[1] When applied to decision theory under uncertainty, this givse rise to a new type of rationality that changes and updates Bayesian updating rules and also Von Neumann Morgernstern foundations of game theory Chichilnisky (1996, 2000, 2009a, 2010, 2011), appearing to coincide with the observations in Le Doux (1996) of how the brain makes decisions using both the amigdala and the cortex.

The article is organized as follows. First we show how the standard notion of continuity or topology that is used in clasic probability theory—"monotone continuity" as defned by Arrow (1971), and in Herstein and Milnor (1953), De Groot (1970)—implies *countably additive measures* that are by nature insensitive to rare events and hence to change: these probability measures assign a negligible weight to rare events, no matter how important these may be, treating such events as outliers, Chichilnisky (2009a, b). On the other hand the *purely finitely additive measures* defined by Dubins and Savage (1972) assign no weight to frequent events, which is equally troubling, as illustrated in the Appendix. Our new axiomatization for probability theory is shown to balance the two approaches and to extend both, requiring sensitivity to rare as well as to frequent events. This as we saw requires a notion of continuity that is defined with respect to a finer topology that is sensitive to rare as well as to frequent events, the topology of change. The results presented here highlight the classic issue of topology and continuity that have always been at the core of the axioms of probability theory (Villegas 1964; Arrow 1971).

2 The Mathematics of Uncertainty

Uncertainty is described by a distinctive and exhaustive set of **events** represented by sets $\{U_\alpha\}$ whose union describes a universe $\mathcal{U}$. An event is identified with its **characteristic function** $\phi_U : U \to R$[2]. The relative likelihood or probability of an event[3] is a real number $W(U)$ that measures how likely it is to occur. The proability of the universe is 1 and that of the empty set is zero. Classic axioms for subjective probability (respectively likelihoods) were introduced by Kolmogorov (1950), see Savage (1972) and De Groot (1970). The relative likelihood or probability of

[1]The theory presented here explains also Jump-Diffusion processes Chichilnisky (2012), the existence of 'heavy tails' in power law distributions, and the lumpiness of most of the physical systems that we observe and measure.

[2]$\phi_U(x) = 1$ when $x \in U$ and $\phi_U(x) = 0$ when $x \notin U$.

[3]In this article we make no difference between probabilities and relative likelihoods.

two disjoint events is the sum of their probabilities: $W(U_1 \cup U_2) = W(U_1) + W(U_2)$ when $U_1 \cap U_2 = \emptyset$. This corresponds to the definition of a probability as a *measure* on a family (σ–algebra) of measurable sets of $\mathcal{U}$.[4]

A measure is a continuous linear function that assigns to each event U a real number. The space of events can therefore be identified with the space of characteristic functions, which are measurable and essentially bounded functions. When $\mathcal{U} = R$, the characteristic functions are in $L_\infty(R)$, the space of Lebesgue measurable and essentially bounded real valued functions on R, which we endow with the *"topology of change"*, defined as the sup norm $f : R \to R$, namely $\| f \| = ess\sup_R | f(x) |$. Recall that the functions in L_∞ are defined a.e. with respect to the Lebesgue measure on R, and are each absolutely continuous with respect to the Lebesgue measure on R. Since measures are continuous real valued function on L_∞, they are by definition in the dual space of L_∞, denoted L_∞^*, namely in the space of all continuous real valued functions on L_∞ A measure μ therefore satisfies the usual conditions (1) $\mu(A \cup B) = \mu(a) + \mu(B)$ if A and B are disjoint, and $\mu(\emptyset) = 0$. A *countably additive measure* is an element of L_∞^* that satisfies also (2) $\mu(\sum A_i) = \sum_i \mu(A_i)$ $i = 1, \ldots \infty$, when the sets A_i are disjoint. A *purely finitely additive measure* is an element of L_∞^* that satisfies condition (1) but not condition (2); therefore for a purely finitely additive measure there are cases where the measure of an infinite sequence of disjoint sets is not the sum of the sequence of their measures. The *space of all purely finitely additive measures* is denoted *PA*.

It is well known that $L_\infty^* = L_1 + PA$ where L_1 is the space of integrable functions on R with respect to the Lebesgue measure; this is a classic representation theorem Yosida and Hewitt (1952). Indeed, each countably additive measure can be represented by an integrable continuous function on L_∞ (R) namely a function $g : R \to R$ in $L_1(R)$, where the representation takes the form $\mu(A) = \int_A g(x)dx$. This representation does not apply to purely finitely additive measures.[5] A *vanishing sequence of events* $\{E_\alpha\}_{=1,2,\ldots}$ is defined one satisfying.$\forall\alpha$, $E_{\alpha+1} \subset E_\alpha$ and $\cap_{\alpha=1}^{\infty} E_\alpha = \emptyset$ a.e.) The following two continuity axioms were introduced in Villegas (1964), see also Arrow (1971), Herstein and Milnor (1953) and De Groot (1970), in each case for the purpose of ensuring countable additivity:

Monotone Continuity Axiom (MC): For every **vanishing** sequence of events $\{E_\alpha\}_{=1,2\ldots}$ the probability $W(E_i) \to 0$ as $i \to \infty$.

In words, this axiom requires that the probability of the sets along a vanishing sequence goes to zero. For example consider the decreasing sequence made of infinite intervals of the form (n, ∞) for $n = 1, 2 \ldots$. This is a vanishing sequence. Monotone continuity implies that the likelihood of this sequence of events goes to zero even though all its sets are unbounded and essentially identical. A similar example can be constructed with a decreasing sequence of bounded sets, $(-1/n, 1/n)$ for

---

[4]This is Savage (1972) definition of probability.

[5]Savage's probabilities can be either purely finitely additive or countably additive. In that sense they include all the probabilities in this article. However this article will exclude probabilities that are either purely finitely additive, orthose that are countably additive, and therefore our characterization of a probability is strictly finer than that Savage (1972), and different from the view of a measure as a countably additive set function in De Groot (1970).

$n = 1, 2...$, which is also a vanishing sequence as it is a decreasing sequence and their intersection is a single point $\{0\}$: observe that the set consisting of a single point $\{0\}$ is almost everywhere (a.e.) equal to the empty set on R, and that the events in this section are always defined a.e. with respect to the Lebesgue measure of R.[6]

De Groot's Axiom SP_4[7]: If $A_1 \supset A_2 \supset ...$ is a decreasing sequence of events and B is some fixed event that is less likely than A_i for all i, then the probability or likelihood of the intersection $\cap_i^\infty A_i$ is larger than the probability or likelihood of the event B.

The following proposition establishes that the two axioms presented above, Monotone Continuity and SP_4, are equivalent and that both imply countable additivity:

Proposition 1 *A relative likelihood (or probability measure) satisfies the* ***Monotone Continuity*** *Axiom if and only if* it satisfies Axiom SP$_4$. *and each of the two axioms implies countable additivity of the corresponding relative likelihood.*

Proof Assume that Axiom SP_4 is satisfied. When the intersection of a decreasing (nested) vanishing sequence of events $\{A_i\}$ is empty namely $\cap_i A_i = \emptyset$ and the set B is less likely to occur than every set A_i, then the subset B must be as likely as the empty set, namely its probability must be zero. In other words, if B is more likely than the empty set, then regardless of how small is the set B, it is impossible for every set A_i to be as likely as B. Equivalently, the probability of the sets that are far away in the vanishing sequence $\{A_i\}$ must go to zero. Therefore SP_4 implies Monotone Continuity (MC). Reciprocally, assume MC is satisfied. Consider a decreasing sequence of events A_i and define a new sequence by substracting from each set the intersection of the family, namely $A_1 - \cap_i^\infty A_i$, $A_2 - \cap_i^\infty A_i$, ... Let B be a set that is more likely than the empty set but less likely than every A_i. Observe that the intersection of the new sequence is empty, $\cap_i A_i - \cap_i^\infty A_i = \emptyset$ and since $A_i \supset A_{i+1}$ the new sequence is, by definition, a vanishing sequence. Therefore by MC $\lim_i W(A_i - \cap_i^\infty A_i) = 0$. Since $W(B) > 0$, B must be more likely than $A_i - \cap_i^\infty A_i$ for some i onwards. Furthermore, $A_i = (A_i - \cap_i^\infty A_i) \cup (\cap_i^\infty A_i)$ and $(A_i - \cap_i^\infty A_i) \cap (\cap_i^\infty A_i) = \emptyset$, so that $W(A_i) > W(B)$ is equivalent to $W(A_i - \cap_i^\infty A_i) + W((\cap_i^\infty A_i) > W(B)$. Observe that $W(\cap_i^\infty A_i) < W(B)$ would contradict the inequality $W(A_i) = W(A_i - \cap_i^\infty A_i) + W((\cap_i^\infty A_i) > W(B)$, since as we saw above, by MC, $\lim_i W(A_i - \cap_i^\infty A_i) = 0$, and $W(A_i - \cap_i^\infty A_i) + W((\cap_i^\infty A_i) > W(B)$. It follows that $W(\cap_i^\infty A_i) > W(B)$, which establishes De Groots's Axiom SP_4. Therefore Monotone Continuity is equivalent to De Groot's Axiom SP_4. A proof that each of the two axioms implies countable additivity is in Villegas (1964), Arrow (1971) and De Groot (1970). ■

The next section shows that the two classic axioms, Monotone Continuity and SP_4, are biased against or neglect rare events, no matter how important these may be.

---

[6]An equivalent definition of Monotone Continuity is that for every two events E_1 and E_2 in $\{E_\alpha\}_{=1,2...}$ with $W(E_1) > W(E_2)$, there exists N such that altering arbitrarily the events E_1 and E_2 on a subset E^i, where $i > N$, does not alter the subjective probability ranking of the events, namely $W(E_1') > W(E_2')$ where E_1' and E_2' are the altered events.

[7]See De Groot (1970, Chap. 6, p. 71).

3 Rare Events and Change

The axioms presented in this article originated from Chichilnisky (1996, 2000, 2002), except for one new axiom—**the Swan Axiom**—that is introduced here and represents the essence of the new probability theory. Below we explain how the Swan Axiom relates to standard theory as its connection with Godel's incompleteness therem and the Axiom of Choice that are at the foundation of Mathematics.

To explain how the new theory intersects with standard probability or relative likelihood, we compare the results presented here with Savage (1972) axiomatization of probability measures as finitely additive measures, as well as with Villegas (1964) and Arrow (1971) classic work that is based on countably additive measures. Savage (1972) axiomatizes subjective probabilities as finitely additive measures representing the decision makers' beliefs, an approach that can ignore frequent events as shown in the Appendix. To overcome this, Villegas (1964) and Arrow (1971) introduced an additional continuity axiom (called 'Monotone Continuity') that ensures the countably additivity of the measures. However this requirement of monotone continuity has unusual implications when the subject is confronted with rare events. A practical example it discussed below: it predicts that in exchange for a couple of cents, one should be willing to accept a small risk of death, a possibility that Arrow himself described as 'outrageous' (1971, p. 48 and 49). The issue of course is the "smallness" of the risk and here is where topology enters. Monotone continuity has a low bar for smallness while the sup norm has a higerh bar as we shall wee below. This article defines a realistic solution, and it implies that for some very large payoffs and in certain special situations, one may be willing to accept a small risk of death—but not in others. This means that Monotone Continuity holds in some cases but not in others, a possibility that leads to the axiomatization proposed in this article, which is the logical negation f Monotoe Continuity—one that is consistent with recent experimental observations reported in Chanel and Chichilnisky (2009a, b).

The Experimental Value of Life
This section explains in what sense standard probability theory is biased against—or disregards—rare events. The next section defines new axioms for relative likelihood, and compares them with the classic axioms. In this section the definitions and results are given for a general measure space of events; the definitions are refined below when the events are Borel measurable sets in the real line R.

Definition A probability W is said to be **biased against rare events** or **insensitive to rare events** when it neglects events that are 'vanishing' according to the definition provided in Sect. 3 above. Formally, a probability is insensitive to rare events when given two events A and B and any vanishing sequence of events (E_j), $\exists N = N(f, g) > 0$, such that $W(A) > W(B) \Leftrightarrow W(A') > W(B')\ \forall A', B'$ satisfying $A' = A$ and $B' = B$ a.e. on $E_j^c \subset R$ when $j > N$.[8] As already discussed this implies a bias against the likelihood of change.

[8]Here E^c denotes the complement of the set E.

Proposition 2 *A probability satisfies Monotone Continuity if and only if it is biased against rare events and underestimates the likelihood of change.*

Proof Chichilnisky Chichilnisky (2000). ■

Corollary 1 *Countably additive probabilities are biased against rare events and underestimate change.*

Proof It follows from Propositions 1 and 2 and Chichilnisky (2000). ■

Proposition 3 *Purely finitely additive probabilties are biased against frequent events.*

Proof See Appendix. ■

The following example illustrates the role of Monotone Continuity and SP_4 in introducing a bias against rare events. The best way to explain the role of Monotone Continuity is by means of the example provided by Arrow (1971, p. 48 and 49). He explains that if a is an action that involves receiving one cent, b is another that involves receiving zero cents, and c is a third action involving receiving one cent and facing a small probability of death, then *Monotone Continuity* requires that the third action involving death and one cent should be preferred to the action with zero cents when the probability of death is small enough. Even Arrow says 'this may sound outrageous at first blush ...' Arrow (1971, p. 48 and 49). Outrageous or not, Monotone Continuity (MC) leads to neglect rare events that involve change with major consequences, like death. It can be said that death is a black swan: this is the content of Proposition 2 above.

4 New Axioms for Probability Theory: The Topology of Change

This section presents the new axiomatic foundation for probability theory that is neither biased against rare nor against frequent events (Chichilnisky 2000, 2002).

The new axioms for probability—or relative likelihoods—are as follows:

Axiom 1 Probabilities are additive and continuous in the topology of change

Axiom 2 Probabilities are unbiased against rare events

Axiom 3 Probabilities are unbiased against frequent events.

Additivity is a natural condition and the **continuity** captures the notion that 'nearby' events are thought as being similarly likely to occur; this property is important to ensure that 'sufficient statistics' exist and it is basaed on a finer topology than Monotone continuity—the sup norm of L_∞ that we called the "topology f change". However Axiom 1 defined continuity with respect to a finer topology Axioms 2 and 3 together are equivalent to the Swan Axiom defined in the previous section, which is required to avoid a bias against rare and frequent events as shown in Sect. 3. The concept of continuity requires some elaboration. Topology, provides the notion of what is meant by 'nearby'; different topologies define different notions of 'nearby' and therefore different notions of what is meant by 'continuity.' For example, 'nearby' was defined in Villegas (1964) and Arrow (1971) as follows: two events are **close** or **nearby** when they differ on a **small set**—thus reducing the problem to determine what is a small set. As stated in Arrow (1971, p. 48): "An event that is far out on a *vanishing sequence* "**is 'small' by any reasonable standards**" Arrow (1971, p. 48).

To overcome the bias against rare events, we introduce a new axiom that is the logical negation of MC: this means that sometimes MC holds and other times it does not. We call this the **Swan Axiom,** and stated it formally below:

Swan Axiom: There exist vanishing sequences of sets $\{U_i\}$—*namely*, $\forall i$, $U_{i+1} \subset U_i$ and $\cap U_i = \emptyset$—where the limit of the measures $\mu(U_i)$ as $i \to \infty$ is not zero.

Observe that in some cases the measures of the sets in a vanishing family may converge to zero and in other cases they do not. In words, this axiom is the logical negation of Monotone Continuity and can be equivalently described as follows: "There exist events A and B with $W(A) > W(B)$, and for every vanishing sequence of events $\{E_\alpha\}_{=1,2...}$ an $N > 0$ such that altering arbitrarily the events A and B on the set E^i, where $i > N$, alters the probability ranking of the events, namely $W(B') > W(A')$, where B' and A' are the altered events."

Proposition 4 *A probability that satisfies the Swan Axiom is neither biased against rare events, nor biased against frequent events.*

Proof This is immediate from the definition. ■

Example To illustrate how this axiom works in practice consider an experiment where the subjects are offered a certain amount of money to choose a pill at random from a pile that contains one pill that causes death Chanel and Chichilnisky (2009a), Chanel and Chichilnisky (2009b). Experimentally, it is observed that in some cases people accept a sum of money and choose a pill provided the pile is large enough—namely when the probability of death is small enough—thus satisfying the monotone continuity axiom and in the process determining a statistical value for their lives. But there are also cases where the subjects will not accept to choose any pill, no matter how large is the pile. Some people refuse the payment if it involves a small probability of death, no matter how small the probability may be Chanel and Chichilnisky (2009a), Chanel and Chichilnisky (2009b). This conflicts with the Monotone Continuity axiom, as explicitly presented by Arrow (1971). Our Axiom provides a reasonable resolution to this dilemma that is realistic and consistent with the experimental evidence. It implies that there exist catastrophic outcomes such as the risk of death, so terrible that one is unwilling to face a small probability of death to obtain one cent versus half a cent, no matter how small the probability may be. According to our Swan Axiom, no probability of death may be acceptable when one cent and half a cent are involved. Our Axiom also implies that in other cases there may be a small enough probability that the lottery involving death may be acceptable, or that the payoff is large enough to justify the small risk. This is a possibility discussed by Arrow (1971), where he explains that for large payoffs (for example, one billion US dollars, one may be willing to accept a small probability of death. In other words: sometimes one is willing to take a risk of death with a small enough probability of a catastrophe, and in other cases one is not. This is the content of the Swan Axiom.

We saw in Proposition 2 that the notion of continuity defined by Villegas and Arrow—namely Monotone Continuity—conflicts with the Swan Axiom and neglects rare events. Indeed Proposition 2 shows that countably additive measures are biased against rare events. On the other hand, Proposition 3 and the Example in the Appendix show that purely finitely additive measures can also be biased, in this case against frequent events. A natural question is whether it is possible to eliminate simultaneously both biases. The following proposition addresses this issue:

Theorem 1 *A probability that satisfies the Swan Axiom is neither biased against frequent nor against rare events. The resulting measures are neither purely finitely additive nor countably additive. They are a strict convex combinations of both.*

Proof The next section contains a proof of Theorem 1 and provides examples when the events are Borel sets in R or within an interval $(a, b) \subset R$. ■

Theorem 1 establishes that neither Savage's approach, nor Villegas' and Arrow's approaches, satisfy the three new axioms stated above. These three axioms require more than the additive probabilties of Savage, since purely finitely additive probabilities are finitely additive and yet they must be excluded here; at the same time the axioms require less than the countably additivity of Villegas and Arrow, since countably additive probabilities are biased against rare events. Theorem 1 above shows that a strict combination of both does the job.

Theorem 1 shows how the Swan Axiom resolves the bias problem against frequent and rare events, but it does not by itself prove the existence of likelihoods that satisfy all three axioms. What is missing is an appropriate definition of 'nearby', namely of topology and continuity, that does not conflict with the Swan Axiom. The following shows that this can be achieved.

We now specialize the space of measurable sets so they are Borel measurable subsets of the real line R, and consider the Lebbesgue measure on R. In this context a probability or likelihood function $W: L_\infty \to R$ is called **biased against rare events**, or **insensitive to rare events** when it neglects events that are small according to a probability measure μ on R that is absolutely continuous with respect to the Lebesgue measure. Formally:

Definition A probability is insensitive to rare events when given two events f and g $\exists\, \varepsilon = \varepsilon(f, g) > 0$, such that $W(f) > W(g) \Leftrightarrow W(f') > W(g')\ \forall f', g'$ satisfying $f' = f$ and $g' = g$ a.e. on $A \subset R$ and $\mu(A^c) < \varepsilon$. Here A^c denotes the complement of the set A.

Definition A probability or likelihood function $W : L \to R$ is said to be **insensitive to frequent events** when given any two events $f, g\ \exists\, \varepsilon(f, g) > 0$ that $W(f) > W(g) \Leftrightarrow W(f') > W(g')\ \forall f', g'$ satisfying $f' = f$ and $g' = g$ a.e. on $A \subset R$ and $\mu(A^c) > 1 - \varepsilon$.

Definition W is called **sensitive** to rare (or frequent) events when it is **not insensitive** to rare (or frequent) events.

Below we identify an event with its characteristic function, so that events are contained in the space of bounded real valued functions on the universe space $\mathcal{U}$, $L_\infty(\mathcal{R})$, and endow this space with the sup norm rather than with the notion of smallness and continuity defined by Arrow and Villegas as described above. In this case the probability or likelihood $W: L_\infty(\mathcal{U}) \to R$ is taken to be continuous with respect to the sup norm. Events are elements of the Borel measurable sets of the real line R or an interval (a, b), they are identified with the characteristic functions, denoted f, g etc., and 'continuity' is based on a topology used earlier in Debreu (1953) and in Chichilnisky (2000, 2002, 2009a, b), the sup norm $\| f \| = ess \sup_{x \in R} | f(x) |$. This is a sharper and more stringent definition of closeness than the one used by Villegas and Arrow, since an event can be small under the Villegas-Arrow definition but not under ours, see the Appendix for examples. The difference in the use of topologies as shown below achieves sensitivity to rare events. To simply notation, a probabiliy that satisfies the classic axioms in De Groot (1970) is from now on called a **standard probability,** and is therefore countably additive. As already mentioned, a classic representation result is that for any event $f \in L_\infty$ a standard (countably additive) probability has the form $W(f) = \int_R f(x).\phi(x)d\mu$, where $\phi \in L_1(R)$ is an integrable function in R.

The next step is to show existence and characterize all the likelihoods or probability distributions that satisfy the 3 new axioms. The following three axioms are identical to the axioms above, specialized to the case at hand, Borel sets of R, and measures in L_∞ with the topology defined by the *sup norm* on $L_\infty(R)$, which we called "the topology of change".

Axiom 4 $W : L_\infty \to R$ is linear and continuous with the sup normor "topology of change"

Axiom 5 $W : L_\infty \to R$ is sensitive to frequent events

Axiom 6 $W : L_\infty \to R$ is sensitive to rare events

The first and the second axiom agree with classic theory and standard likelihoods satisfy them. The third axiom is new.

Lemma 1 *A standard probability satisfies Axioms 1 and 2, but it is biased against rare events and therefore does not satisfy Axiom 3.*

Proof Consider $W(f) = \int_R f(x)\phi(x)dx$, $\int_R \phi(x)dx = K < \infty$. Then

$$W(f) + W(g) = \int_R f(x)\phi(x)dx + \int_R g(x)\phi(x)dx = \int_R f(x) + g(x).\phi(x)dx = W(f + g),$$

and therefore W is linear. It is **continuous** with respect to the L_1 norm $\| f \|_1 = \int_R | f(x) | \phi(x) d\mu$ because $\| f \|_\infty < \varepsilon$ implies

$$W(f) = \int_R f(x).\phi(x)dx \leq \int_R | f(x) | .\phi(x)dx \leq \varepsilon \int \phi(x)dx = \varepsilon K.$$

Since the sup norm is finer than the L_1 norm, continuity in L_1 implies continuity with respect to the sup norm (Dunford and Schwartz 1958). Thus a standard probability satisfies Axiom 1. It is obvious that for every two events f, g, with $W(f) > W(g)$, the inequality is reversed namely $W(g') > W(f')$ when f' and g' are appropriate variations of f and g that differ from f and g on sets of sufficiently large Lebesgue measure. Therefore Axiom 2 is satisfied. A standard probability is however not sensitive to rare events, as shown in Chichilnisky (2000, 2002, 2009a, b); Chichilnisky and Wu (2006). ■

5 Existence and Representation Theorems

Theorem 2 *There exists a probability distribution or likelihood function* $W : L_\infty \to R$ *satisfying the new Axioms 1, 2, and 3. A probability distribution satisfies Axioms 1, 2 and 3 if and only if there exist two continuous linear functions on* L_∞, *denoted* ϕ_1 *and* ϕ_2, *and a real number* λ, $0 < \lambda < 1$, *such that for any observable event* $f \in L_\infty$

$$W(f) = \lambda \int_{x \epsilon R} f(x)\phi_1(x)dx + (1 - \lambda)\phi_2(f) \tag{1}$$

where $\phi_1 \in L_1(R, \mu)$ *defines a countably additive measure on* R *and where* ϕ_2 *is a purely finitely additive measure.*

Proof This result follows from the representation Theorem in Chichilnisky (2000, 2002). ■

Corollary 3 *Absent rare events, a probability that satisfies Axioms 1, 2, and 3 is consistent with classic axioms and yields a countably additive measure.*

Proof Axiom 3 is an empty requirement when there are no rare events while, as shown above, Axioms 1 and 2 are consistent with standard relative likelihood. ■

6 Heavy Tails and Families of Purely Finitely Additive Measures

This section presents new results adding to the introduction of the Swan Axiom defined in Sect. 4 above: the different ntions of continuity, how heay tailsoriginate from the new axioms and defines a family of purely finitely additive measures that are each absolutely continuous with respect to the Lebesgue measure on R.

A main difference introduced by the new axioms is the use of a finer topology—the "topology of change", which is the sup norm on L_∞, in Axiom 1 that defines the continuity properties of probability distributions. In the classic axioms a probabiloty dstribution is contiuous if it satisfies Monotone Continuity or equivaletly SP4. Here the continuity required in Axiom 1 is with respect to the topology of change, which is a finer topology. The following example explains the difference that this makes on the concept of continuity of probability distibutions:

6.1 *Contrasting Monotone Continuity and the Topology of Change*

Two different topologies define two different approaches to 'continuity' as show in the following. Consider the family $\{E^i\}$ where$E^i = [i, \infty)$, $i = 1, 2, \ldots$. This is a vanishing family because $\forall i\ E^i \supset E^{i+1}$ and $\bigcap_{i=1}^{\infty} E^i = \emptyset$. Consider now the events $f^i(t) = K > 0$ when $t \in E^i$ and$f^i(t) = 0$ otherwise, and $g^i(t) = 2K$ when $t \in E^i$ and $g^i(t) = 0$ otherwise. Then for all i, $\sup_{E^i} \mid f^i(t) - g^i(t) \mid = K$. In the sup norm topology this implies thatf^i and g^i are **not** 'close' to each other, as the difference$f^i - g^i$ does not converge to zero. No matter how far along we are along the vanishing sequence E^i the two eventsf^i, g^i differ by at least the number K. Yet since the events f^i, g^i differ from$f \equiv 0$ and $g \equiv 0$ respectively only in the set E^i, and $\{E^i\}$ is a vanishing sequence, for large enough i they are as 'close' as desired according to Villegas-Arrow's definition of 'nearby' events.

6.2 *Heavy Tails*

The following illustrates the additional weight that the new axioms assign to rare events; in this example in the form of 'heavy tails' (e.g. Chichilnisky (2000). The finitely additive measure ϕ_2 appearing in the second term in (1) can be illustrated as follows. On the subspace of events with limiting values at infinity, $L'_\infty = \{f \epsilon L_\infty :$ $\lim_{x \to \infty}(x) < \infty\}$, define $\phi_2(f) = \lim_{x \to \infty} f(x)$ and extend this to a function on all of L_∞ using Hahn Banach's theorem. The difference between a standard probability and the likelihood defined in (1) is the second term ϕ_2, which focuses all the weight at infinity. This can be interpreted as a 'heavy tail' namely a part of the distribution that is not part of the standard density function ϕ_1 and gives more weight to the sets that contain *terminal* events namely sets of the form (x, ∞). ■

6.3 *The Family PA of Purely Finitely Additve Measures on R*

This section provides a new family of purely fnitely additive measures and studies ts properties.

Definition An open neighborhood of a real number $x \in R$ has the standard meaning under the usual topology of the line R. An 'open neighborhood of ∞' is defined to be a set of the form $\{x \in R : x. > r \text{ for } r \in R\}$. As already stated, the word "essentially" means a.e. with respect to the Lebesgue measure on R that has been used to define the space L_∞.

We now define a property on measures in the space L^*_∞ :

Definition (*Property (P)*) A measure in L^*_∞ is said to satisfy Property (P) at x if it assigns measure zero to any set that is essentially contained in the complement of an open neighborhood of x. A measure in L^*_∞ is said to satisfy Property (P) at ∞, if it assigns measure 0 to any measurable set that is essentially contained in the complement of an open neighborhood of ∞ as defined above. A measure is said to satisfy Property (P) if it satisfies Property (P) either at ∞ or at any $x \in R$.

Lemma 2 *A measure satisfying property* (P) *is always purely finitely additive.*

Proof Consider first the case where the measure has property (P) at ∞. Define a countable family of disjoint sets $F = \{A_1, A_2 \dots\}$ recursively as follows: $A_1 = \{x : -1 < x < 1\}$ and for all n, $A_n = \{x : -n < x < n\} - A_{n-1}$. Observe that each set A_n has measure zero, since by assumption μ satisfies property (P), and that each of the sets in the family F is bounded. The sets in the family F are also disjoint by construction. If μ was countably additive, then we should have $\mu(\cup F) = \mu(\cup_{n=1}^{\infty} A_n) = \sum_{n=1}^{\infty} \mu(A_n) = 0$. Yet the measure of the union of the countable family F is not 0, because $\cup F = R$, the entire real line, so that $\mu(\cup F) = 1$. Therefore μ fails to be countably additive on the countable and disjoint family F. Since by definition μ is a measure, and it fails to be countably additive, it must be a purely finitely additive measure.

A similar argument can be given for the case where the measure has property (P) at a finite number $x \in R$. Define now $F = \{A_n\}_{n=1,2,\dots}$ recursively as follows: $A_1 = [x - 1, x + 1]^C$ where the superindex C denotes the complement of a set, and for all $n \geq 1$, $A_n = [x - 1/n, x + 1/n]^C - A_{n-1}$. Observe that each set in the family F has measure 0. The union of the family is not the whole space as before - since the point $\{x\}$ is not in the union; yet the entire space minus $\{x\}$ should have the same measure than the space as a whole, because by definition a measure is a continuous linear function on L_∞, the space of measurable and essentially bounded functions with the Lebesgue measure on R, which means that the measure must provide the same value to functions in L_∞ that are essentially equal, in the sense of differing only in a set of Lebesgue measure 0. The characteristic functions of two measurable sets differing in a set of measure zero, must therefore be are assigned the same value by a measure, so the union of the family F must be assigned the same measure as the entire space, namely $\mu(\cup F)= 1$. Therefore the measure μ fails to be countably additive, and since it is a measure it must be purely finitely additive. ■

Observe that in Lemma 1 the same argument applies for a measure that has property (P) at x for a finite $x \in R$, or one that has property (P) at $\{\infty\}$. The "test" family F is defined similarly in both cases, where for a finite x $A_1 = \{x : -\epsilon < x < \varepsilon\}$, and $A_n = \{x : -n < x < n\} - A_{n-1}$. The only difference in the argument arises from the fact that, for a finite $\{x\}$, the union of the family $\cup F$ is not all of R, but rather $R - \{x\}$. But this is essentially the same as R in the Lebesgue measure used to define $L\infty$.

Lemma 3 *Using Hahn-Banach's theorem it is possible to define purely finitely additive measures on R.*

Proof Lemma 1 started from assuming the existence of a measure in L_∞^* that satisfies property (P) at ∞. Using Hahn Banach's theorem—we now define the desired measure, namely a continuous linear function h from L_∞ to R, and show that it satisfies (P) at ∞. Therefore by Lemma 1, the function h is a purely finitely additive measure, as we wished to prove.

Consider the subspace CL_∞ of all functions f in L_∞ that are continuous and have an essential limit at ∞. CL_∞ is a closed linear subspace of the Banach space $L\infty$. On the subspace CL_∞ define the function $h(f) = ess \lim_{x\to\infty} f(x)$. By construction the function h is well defined on CL_∞; this function is continuous, linear and has norm 1. The function h can therefore be extended by using Hahn-Banach's theorem to all of L_∞, as a continuous, linear function that preserves the norm of h. Since h has norm 1 the extension is not the zero function. Call this extension h as well; by contruction, $h \in L_\infty^*$. Therefore by definition, the extended function h defines a measure. Now observe that $h : L_\infty \to R$ satisfies Property (P) since when applied to characteristics functions of bounded sets, it assigns to them measure zero. A similar argument can be replicated to show the existence of purely finite measures that satisfy property (P) at any $x \in R$. ■

We have mentioned that it is not possible to contruct a purely finitely additive measure on R the same way as one constructs a countably aditive measure on R. This is not surprising since the Hanh-Banach Theorem that is used to define a purely finitely additive measure in Lemma 2 is itself not constructible. The next and last section show the connection between the new axioms for probability (or relative likelihoods) presented here and the Axioms of Choice and Godel (1940) work.

7 The Axiom of Choice and Godel's Incompleteness Theorem

There is a connection between Axioms 1, 2 and 3 presented here and the Axiom of Choice that is at the foundation of mathematics (Godel 1940). The Axiom of Choice postulates that there exists a universal and consistent fashion to select an element from every set.

The best way to describe the situation is by means of an example, see also Dunford and Schwartz (1958), Yosida and Hewitt (1952), Yosida (1974), Chichilnisky and Heal (1997) and Kadane and O'Hagan (1995).

Example Representing a Purely Finitely Additive Measure

Define a measure ρ as follows: for every Borel measurable set $A \subset R$, $\rho(A) = 1$ if $A \supset \{x : x > a$, for some $a \in R\}$, and otherwise $\rho(A) = 0$. Then ρ is not countably additive, because the family of countably many disjoint sets $\{V_i\}_{i=0,1,\ldots}$ defined as $V_i = (i, i+1] \bigcup (-i-1, -i]$, satisfy $V_i \bigcap V_i = \emptyset$ when $i \neq j$, and $\bigcup_{i=0}^{\infty} V_i = \bigcup_{i=0}^{\infty} (i, i+1] \bigcup (-i-1, -i] = R$, so that $\rho(\bigcup_{i=0}^{\infty} V_i) = 1$, while $\sum_{i=0}^{\infty} \rho(V_i) = 0$, which contradicts countable additivity. Since the contradiction arises from assuming that ρ is countably additive, ρ must be purely finitely additive. Observe that ρ assigns zero measure to any bounded set, and a positive measure only to unbounded sets that contain a 'terminal set' of the form

$$\{x \epsilon R : x > a \text{ for some } a \epsilon R\}.$$

One can define a function on L_∞ that represents this purely finitely additive measure ρ if we restrict our attention to the closed subspace L'_∞ of L_∞ consisting of those functions $f(x)$ in L_∞ that have a limit when $x \to \infty$, by the formula $\rho(f) = \lim_{x \to \infty} f(x)$, as in Example 1 of the previous section. The function $\rho(.)$ can be seen as a limit of a sequence of delta functions whose support increases without bound. The problem is now to extend the function ρ to another defined on the entire space L_∞. This could be achieved in various ways but as we will see, each of them requires the Axiom of Choice.

One can use Hahn-Banach's theorem (Dunford Schwartz 1958) to extend the function ρ from the closed subspace $L'_\infty \subset L_\infty$ to the entire space L_∞ preserving its norm. However, in its general form Hahn-Banach's theorem requires the Axiom of Choice (Dunford Schwartz 1958). Alternatively, one can extend the notion of a *limit* to encompass all functions in L_∞ including those with no standard limit. This can be achieved by using the notion of convergence along a *free ultrafilter* arising from compactifying the real line R as in Heal (1997). However the existence of a *free ultrafilter* also requires the Axiom of Choice.

This illustrates why the attempts to construct *purely finitely additive measures* that are representable as functions on L_∞, require the Axiom of Choice. Since our criteria require purely finitely additive measures, this provides a connection between the Axiom of Choice and our axioms for relative likelihood. It is somewhat surprising that the consideration of rare events that are neglected in standard statistical theory conjures up the Axiom of Choice, which is independent from the rest of mathematics (Godel 1940).

8 Appendix

Example A Probability that is Biased Against Frequent Events

Consider $W(f) = \liminf_{x \epsilon R}(f(x))$. This is insensitive to frequent events of arbitrarily large Lebesgue measure (Dunford and Schwartz 1958) and therefore does not satisfy Axiom 2. In addition it is not linear, failing Axiom 1.

Example The dual space L^*_∞ consists of countably additive and finitely additive measures

The space of continuous linear functions on L_∞ is the 'dual' of L_∞, and is denoted L^*_∞. It has been characterized e.g. in Yosida and Hewitt (1952), Yosida (1974). L^*_∞ consists of the sum of two subspaces (i) L_1 functions g that define countably additive measures ν on R by the rule $\nu(A) = \int_A g(x)dx$ where $\int_R \mid g(x) \mid dx < \infty$ so that υ is *absolutely continuous* with respect to the Lebesgue measure, and (ii) a subspace consisting of purely finitely additive measure. A countable measure can be identified with an L_1 function, called its 'density,' but purely finitely additive measures cannot be identified by such functions.

Example A Finitely Additive Measure that is not Countably Additive

See Example in Sect. 7.

Acknowledgements This article is an expression of gratitude to the memory of Jerry Marsden, a great mathematician and a wonderful man. As his first PhD student in pure Mathematics when he was a professor at the Mathematics Department of UC Berkeley, the author is indebted to Jerry Marsden for counseling and support in obtaining the first of her two PhDs at UC Berkeley, in pure Mathematics. The second PhD in Economics at UC Berkeley was obtained by the author with the counseling of the Nobel Laureate economist, Gerard Debreu. Jerry Marsden was critical to encourage the growth of the research in this article on new and more realistic axiomatic foundations of probability theory; he invited the author to organize a Workshop on Catastrophic Risks at the Fields Institute in 1996 where this research was introduced, and strongly encouraged since 1996 the continuation and growth of this research.

The author is Director, Columbia Consortium for Risk Management (CCRM) Columbia University, and Professor of Economics and of Mathematical Statistics, Columbia University, New York 10027, 335 Riverside Drive, NY NY 10025, tel. 212 678 1148, email: chichilnisky@columbia.edu; website: www.chichilnisky.com. We acknowledge support from Grant No 5222-72 of the US Air Force Office of Research directed by Professor Jun Zheng, Washington DC from 2009 to 2012. Initial results on Sustainable Development were presented at Stanford University's 1993 Seminar on Reconsideration of Values, organized by Kenneth Arrow, at the National Bureau of Economic Research Conference *Mathematical Economics: The Legacy of Gerard Debreu* at UC Berkeley, October 21, 2005, the Department of Economics of the University of Kansas National Bureau of Economic Research General Equilibrium Conference, September 2006, at the Departments of Statistics of the University of Oslo, Norway, Fall 2007, at a seminar organized by the former Professor Christopher Heyde at the Department of Statistics of Columbia University, Fall 2007, at seminars organized by Drs. Charles Figuieres and Mabel Tidball at LAMETA Universite de Montpellier, France December 19 and 20, 2008, and by Professor Alan Kirman at GREQAM Universite de Marseille, December 18 2008. In December 8 2012, the work presented here and its applications were presented at an invited Plenary Key Note Presentation by the author to the Annual Meetings of the Canadian Mathematical Society Montreal Canada, December 8 2012. The work presented in this article is also the subject of a forthoming Plenary Key Note Presentation to the Annual Meeting of the Australian Mathematical Society in Sidney, Australia, December 18, 2013. We are grateful to the above institutions and individuals for supporting the research, and for helpful comments and suggestions.

References

Arrow, K. (1971). *Essays in the theory of risk bearing*. Amsterdam: North Holland.

Chanel, O., & Chichilnisky, G. (2009a). The influence of fear in decisions: Experimental evidence. *Journal of Risk and Uncertainty*, *39*(3),

Chanel, O. & Chichilnisky, G. (2009b). *The Value of Life: Theory and Experiments Working Paper, GREQE*. New York: Universite de Marseille and Columbia University.

Chichilnisky, G. (1996). An axiomatic approach to sustainable development. *Social Choice and Welfare*, *13*, 257–321.

Chichilnisky, G. (2000). An axiomatic approach to choice under uncertainty with catastrophic risks. *Resource Energy Economics*, *22*, 221–231.

Chichilnisky, G. (2002). Catastrophic Risk. In A. H. El-Shaarawi & W. W. Piegorsch (Eds.), *Encyclopedia of environmetrics* (Vol. 1, pp. 274–279). Chichester: Wiley.

Chichilnisky, G. (2009a). The limits of econometrics: Non parametric estimation in Hilbert spaces. *Economic Theory* 2009.

Chichilnisky, G. (2009b). The topology of fear. *Journal of Mathematical Economics* 2009.

Chichilnisky, G. (2010). The foundations of statistics with black swans. *Mathematical Social Sciences*, *59*(2), 184–192.

Chichilnisky, G. (2011). Sustainable markets with short sales. *Economic Theory*, *49*(2), 293–307.

Chichilnisky, G. (2012). Economic theory and the global environment. *Economic Theory*, *59*(2), 217–225.

Chichilnisky, G., & Heal, G. M. (1997). Social choice with infinite populations. *Social Choice and Welfare*, *14*, 303–319.

Chichilnisky, G., & Wu, H.-M. (2006). General equilibrium with endogenous uncertainty and default. *Journal of Mathematical Economics*, *42*, 499–524.

De Groot, M. H. (1970 and 2004). *Optimal statistical decisions*. Hoboken, New Jersey: Wiley.

Debreu, G. (1953). Valuation equilibrium and pareto optimum. *Proceedings of the National Academy of Sciences*, *40*, 588–592.

Dubins, L. (1975). Finitely additive conditional probabilities, conglomerability and disintegration. *The Annals of Probability*, *3*, 89–99.

Dubins, L., & Savage, L. (1965). *How to gamble if you must*. New York: McGraw Hill.

Dunford, N., & Schwartz, J. T. (1958). *Linear operators, Part I*. New York: Interscience.

Godel, K. (1940). The consistency of the continuum hypothesis. *Annals of mathematical studies 3*. Princeton NJ, USA: Princeton University Press.

Herstein, I. N., & Milnor, J. (1953). An axiomatic approach to measurable utility. *Econometrica*, *21*, 291–297.

Kadane, J. B., & O'Hagan, A. (1995). Using finitely additive probability: uniform distributions of the natural numbers. *Journal of the American Statistical Association*, *90*, 525–631.

Kolmogorov, A.N. (1933/1950) *Foundations of the theory of probability*. New York: Chelsea Publishing Co.

Le Doux, J. (1996). *The Emotional Brain*. New York: Simon and Schuster.

Purves, R. W., & Sudderth., (1976). Some finitely additive probability. *The Annals of Probability*, *4*, 259–276.

Savage, L. J. (1972). *The foundations of statistics* New York: Wiley. (1054, revised edition).

Villegas, C. (1964). On quantitative probability σ–algebras. *Annals of Mathematical Statistics, 35*, 1789–1800.
Von Neumann, J. (1932/1955). *Mathematical foundations of quantuum theory*. New Jersey: Princeton University Press.
Von Neumann, J., & Morgernstern, O. (1944). *Theory of games and economic behavior*. New Jersey: Princeton University Press.
Yosida, K. (1974). *Functional analysis* (4th ed.). New York, Heidelberg: Springer.
Yosida, K., & Hewitt, E. (1952). Finitely level independent measures. *Transactions of the American Mathematical Society*, *72*, 46–66.

Sustainable Markets with Short Sales

Graciela Chichilnisky

1 Introduction

Sustainable development requires satisfying the basic needs of the present without sacrificing the needs of future generations. This seems to set up a confrontation between market objectives, which are typically short term, and the requirements of

Received: 20 April 2011/Accepted: 23 April 2011.

This research was conducted at the Columbia Consortium for Risk Management (CCRM) directed by the author at Columbia University in New York, and its Program on Information and Resources (PIR). We gratefully acknowledge support from Grant No 5222-72 of the US Air Force Office of Research (US AFOSR) and its officer Professor Jun Zheng, Arlington VA. The initial results on sustainable preferences were presented as an invited address at Stanford University's 1993 Seminar on Reconsideration of Values, organized by Kenneth Arrow, and more recently in 2000 at the Fields Institute for Mathematical Sciences in Toronto Canada, at the NBER Conference *Mathematical Economics: The Legacy of Gerard Debreu* at UC Berkeley, October 21, 2005, the Department of Economics of the University of Kansas National Bureau of Economic Research General Equilibrium Conference, September 2006, the Department of Statistics of the University of Oslo, Norway, Fall 2007, the Department of Statistics of Columbia University, Fall 2007, the US AFOSR in Arlington VA, January 2010, and the Colloqium at the Mathematics Department and the Institute for Behavioral Sciences of the University of California at Irvine, May 18, 2010. We thank the above institutions and individuals for supporting the research as well as Eduardo Jose Chichilnisky, Salk Institute, La Jolla, California, Economic Theory editor Nicholas Yannelis and two anonymous reviewers, for valuable comments and suggestions.

Reprinted chapter: Originally published in Chichilnisky - Sustainable Markets with Short Sales - DOI:10.1007/s00199-011-0626-6

G. Chichilnisky (✉)
Departments of Economics and Mathematical Statistics, Columbia University,
335 Riverside Drive, New York, NY 10027, USA
e-mail: chichilnisky@columbia.edu; chichilnisky1@gmail.com
URL: http://www.chichilnisky.com

G. Chichilnisky
Columbia Consortium for Risk Management (CCRM), New York, NY, USA

G. Chichilnisky and A. Rezai (eds.), *The Economics of the Global Environment*,
Studies in Economic Theory 29, DOI 10.1007/978-3-319-31943-8_8

sustainable development. Behind this confrontation is a standard feature of classic markets: the concept of *impatience*, an axiom that was introduced in T. Koopmans's seminal work on economics over time and is a requirement of the Arrow Debreu theory of markets. A dollar tomorrow is worth less than a dollar today and this, in its simplest form, introduces an economic bias against the future (Chichilnisky 1996a, b; Chichilnisky and Heal 1998). The problem is quite general. Even in markets with infinite horizons the existence of a solutions seems to require some form of *impatience*. For example the "cone condition" of Chichilnisky and Kalman (1980), also known as "properness" and frequently to proved existence of market equilibrium with infinite horizons, still requires a form of impatience, see e.g. Yannelis and Zame (1986) and Chichilnisky (1993).[1]

Markets and sustainability seem opposed to each other. Is this a correct view of markets? A natural question is whether a society committed to sustainability must discard markets. This would be a major change, since markets are a widespread form of organization.

The article argues that it is possible to overcome a market's bias against the future. This requires defining new types of markets where traders have 'sustainable preferences', as introduced in Chichilnisky (1996a, b), and a corresponding new definition of market equilibrium. Because markets follow the priorities of the traders, when traders have sustainable preferences markets become sustainable. This, in a nutshell, explains the results of the article.

Whether markets clash with sustainability hinges therefore on whether traders have *sustainable preferences*. These are a new type of preferences that overcome the impatience axiom: they are based on new axioms that require equal treatment of the present and the future, Chichilnisky (1996a, b). These new axioms reflect an increasing body of empirical evidence about how humans value the long term future (Chichilnisky 1996a, b). Based on the concept of sustainable preferences, this article defines *sustainable markets* that differ from standard Arrow-Debreu markets in two ways: traders (i) they have *sustainable preferences* and (ii) engage in short trades. Markets with sustainable preferences overcome impatience because they are neither dictatorial for the present nor for the future; they are in fact sensitive to the needs of both as required for sustainable development. In sustainable markets, market prices take a new role: they represent the value of instantaneous consumption as well as the value of future consumption. This approach resolves the conflict between a market's short-term objectives and the goals of sustainability, without eliminating market organization. We require instead a different type of market than the one we had until now, namely *sustainable markets*, which are not based on impatience. Sustainable preferences are linear as time preferences, the same property satisfied by standard present discounted utility functions.

[1]Chichilnisky (1993) showed that the original "cone condition" is the same as the later condition named "properness".

To reconcile the opposing needs of the present and the future we establish the existence of equilibrium in sustainable markets. We show that when markets have sustainable preferences, a single condition—*limited arbitrage*—is necessary and sufficient for the existence of Pareto efficient market equilibrium, without bounds on short sales. This ensures the logical consistency of *sustainable markets.* To achieve this, we have to overcome three interlinked technical issues (i) continuity of preferences and prices, (ii) compactness of trading sets and efficient allocations, both of which (i) and (ii) are used to prove existence of solutions, and (iii) appropriate supporting prices for efficient allocations. With infinite horizons, compactness requires weaker topologies that can imply a form of impatience (Yannelis and Zame 1986). Furthermore supporting prices that are continuous with the sup norm can lead to paradoxical results (Chichilnisky and Kalman 1980). To resolve the continuity-compactness dilemma and avoid paradoxes we rely on the properties of sustainable preferences that are sensitive to the present and the future, and use the notion of "limited arbitrage" introduced in Chichilnisky (1991, 1995, 1994a, 1996c, 1998). Taken together, sustainable preferences and the notion of limited arbitrage, overcome the problems of impatience and unlimited short sales by limiting somewhat the diversity of the traders (Chichilnisky 1994b), bounding the "gains from trade" that they can achieve trading with each other, while creating sensitivity to the present and the future and ensuring existence of solutions. In Proposition 2, we show that limited arbitrage is equivalent to bounded gains from trade, and it ensures the compactness of Pareto utility allocations, as needed for the existence of solutions, based on earlier results by Chichilnisky (1991, 1995, 1998) and Chichilnisky and Heal (1984, 1998). To complete the existence result, Theorem 1 proves that, in a sustainable market, limited arbitrage is equivalent to the compactness of the set of Pareto efficient allocations, and Theorem 2 establishes that it is necessary and sufficient for the existence of sustainable as well as efficient competitive market equilibrium. Section 4 discusses the role of prices in a sustainable market economy. These assign economic value both to instantaneous and long-run consumption, providing a connection with the axiom of choice that is at the foundation of mathematics. It ought to be clarified that a sustainable market equilibrium is shown to exist, but it is a somewhat different concept than what is normally defined as a market equilibrium, since in sustainable markets market prices take a new role: they represent the value of instantaneous consumption as well as the value of long run future consumption—a concept that differs from standard market equilibrium prices. Indeed, when sustainable constrains are taken into consideration, new concepts and features of economics arise as demonstrated throughout this entire special volume on the economics of the global environment, see Asheim et al. (2016), Burniaux and Martins (2016), Chipman and Tian (2016), Dutta and Radner (2016), Figuieres and Tidball (2016), Karp and Zhang (2016), Lauwers (2016), Lecocq and Hourcade (2016) and Rezai et al. (2016).

2 Sustainable Markets

This section defines *sustainable markets*.

2.1 Definitions

A competitive market has $H \geq 2$ traders and $N \geq 2$ commodities that are traded over time $t \in R_+$. The consumption of commodities yields utility $u(x(t))$ at each period of time t, where $x(t) \in R^N$, and $u(x) : R^N \to R_+$ is a concave increasing real valued function that represents instantaneous utility in period t. Following the classic work of Debreu (1953), and Chichilnisky (1996a, b, 2009a, b) one can view consumption paths over time $\left(x(t)_{t \in R+}\right)$ as elements of $L_\infty(R^N)$. Similarly utility paths over time $f(t) = u(x(t))$ are elements of the linear space $L_\infty(R)$, where L_∞ is the space of all essentially bounded measurable real valued functions on R with the sup *norm* $\|f\| = ess\,\sup_{t \in R} |f(t)|$. In this context a *preference over time* $U : L_\infty \to R$ is a real valued *linear* function ranking utility paths $u(x(t))$, while $U : L_\infty\left(R^N\right) \to R$ denotes the ranking of consumption paths $(x(t))$, which is based on a concave instantaneous utility $u : R^N \to R$, and is generally *non linear*. We say that the preference over time U is a *dictatorship of the present* when it disregards all utility beyond a period T, namely $U(f) > U(g) \Leftrightarrow U(f') > U(g')$ for any a.e. modification of f and g that occurs beyond T, i.e. when $f'(t) = f(t)$ and $g'(t) = g(t)$ for all $t > T$. A ranking is a *dictatorship of the future* when it disregards utility modifications in the present: formally, for any two paths f, g, there exists a period $T \in R : U(f) > U(g) \Leftrightarrow U(f') > U(g')$ for any a.e. modification of f, g that occurs prior to period T, i e. whenever $f'(t) = f(t)$ and $g'(t) = g(t)$ for all $t < T$. The logical negation of these two dictatorship properties defines *non-dictatorship* of the present and *non-dictatorship* of the future.

2.2 Axioms for Sustainable Preferences

A *sustainable preference* U is an increasing ranking that as a time preference satisfies three axioms.[2]

Axiom 1 $U : L_\infty \to R$ is continuous and linear[3]

Axiom 2 $U : L_\infty \to R$ is a non-dictatorship of the future

Axiom 3 $U : L_\infty \to R$ is a non-dictatorship of the present

[2]The axioms for sustainable preferences were introduced in Chichilnisky (1996a, b), and similar axioms were introduced for the foundations of preferences under uncertainty, for NP econometrics (Chichilnisky 2009a, b), for relative likelihoods and the foundations of probability and statistics (Chichilnisky 2010a, b).

[3]The time preference U ranks paths over time $u(t) \in L_\infty$ and Axiom 1 requires U to be continuous and linear. Observe that since the instantaneous utility function $u : R^N \to R$ is concave and need not be linear, the ranking of consumption paths need not be linear as a function of consumption, $x(t)$.

These axioms were introduced in Chichilnisky (1996a, b). The first two are consistent with T. Koopman's classic axioms of choice over time and are satisfied by present discounted utility

$$U(f) = U(x(t)) = \int_{R+} u(x(t))e^{-\delta t}dt, \qquad \delta > 1$$

where $f \in L_\infty$ represents a time path $u(x(t))$, δ is a time 'discount factor'. Observe that the present discounted utility $U(f)$ defined above is linear on utility paths $u(t)$, and thus satisfies Axiom 1, but it may not be linear in consumption x. Sustainable preferences that satisfy Axioms 1, 2 and 3 are also linear on utility paths but may not be linear on consumption. The third axiom however is not satisfied by present discounted utilities (Chichilnisky 1996a, b). **Sustainable preferences** have been characterized in a representation theorem established in Chichilnisky (1996a, b, 2009a, b, 2010a, b) to be of the form

$$U(f) = \lambda U_1(f) + (1 - \lambda)U_2(f) \tag{1}$$

where $U_1(.)$ is a function in L_1 and $U_2(.)$ is in $L^*_\infty - L_1$,[4] $0 < \lambda < 1$, both $U_1(f)$ and $U_2(f)$ are increasing and non-zero, and specifically:

$$U(f) = U(x(t)) = \lambda \int_{R+} u(x)\phi(x)dt + (1 - \lambda)\chi(u(x))$$

where $U_1(f) = \lambda \int_{R_+} u(x)\phi(x)dt$, $U_2(f) = (1 - \lambda)\chi(u(x), 0 < \lambda < 1, \phi \in L_1$, e.g. $\phi(t) = e^{-\delta t}$, and $\chi \in L^*_\infty - L_1$ is a purely finitely additive measure on R (for a proof see Chichilnisky 1996a, b, 2009a, b, 2010a, b).

2.3 Sustainable Markets

Definition A sustainable market is an Arrow-Debreu market where traders have infinite horizons, no bounds on short sales and sustainable preferences over time.

A sustainable market economy can be represented as $E = \{X, \Omega_h, U_h : X \to R, h = 1, \dots H\}$. It has $H \geq 2$ traders indexed by $h = 1, 2, \dots, H$, $N \geq 2$ commodities that are traded over time $t \in R_+$; the *consumption space* or *trading space* is the Banach space $X = L_\infty$ with the sup norm $\|.\|_{\sup}$ (Debreu 1953; Chichilnisky 1996a, b, 2009a, b); this assumption implies no bounds on short sales. $\Omega_h \in X$ represents trader $h's$ property rights, $\Omega = \sum_h \Omega_h$ represents society's total resources over time; and traders' preferences over time $U_h : L_\infty \to R_+$ are based as above on concave instantaneous utility $u_h : R^N \to R_+$ and define sustainable time preferences.

[4] L^*_∞ is the dual space of L_∞, the space of all continuous, linear, real valued functions on L_∞.

Traders may have zero endowments of some goods, and endowments could be negative or positive; since the trading space is $X = L_\infty$ short selling is allowed. We consider general preferences where the normalized gradients to indifference surfaces define either an open or a closed map on every indifference surface, namely (i) indifference surfaces contain no half-lines, for example strictly convex preferences, or (ii) the normalized gradients to any closed set of indifferent vectors define a closed set, for example linear preferences. In this article for simplicity we identify case *(ii)* with linear preferences. The assumptions and the results of the paper are ordinal, and $U_h(0) = 0$ and $\sup_{x \in X} U_h(x) = \infty$. Preferences are increasing so that $U_h(x(t)) > U_h(y(t))$ when for all t, $x(t) \geq y(t)$ and for a set of positive Lebesgue measure, $x(t) > y(t)$. In addition we assume the traders's preferences are uniformly non-satiated, which means that they can be represented by a utility U with a bounded rate of increase: for smooth preferences, which are Frechet differentiable, $\exists \varepsilon, K > 0 : \forall x \in X, K > \|DU(x)\| > \varepsilon$. If a utility function is uniformly non-satiated, its indifference surfaces are within a uniform distance from each other: $\forall r, s \in R, \exists N(r, s) \in R$ such that $f \in U^{-1}(r) \Rightarrow \exists y \in U^{-1}(s)$ with $\|f - g\| \leq N(r, s)$, see Chichilnisky and Heal (1998).

Assumption 1 Each trader has a sustainable time preference, satisfying Axioms 1, 2 and 3, which is represented by an increasing, uniformly non-satiated function of consumption paths over time $U : L_\infty \to R^+$ based on a concave instantaneous utility $u : R^N \to R$ such that $U(0) = 0$ and $\sup_{f \in X} U(f) = \infty$.

Prices are real valued linear functions on X that are continuous with the sup norm (Debreu 1953). The space of *feasible allocations* over time is $\{(f_1(t), \ldots, f_H(t)) \in X^H : \sum_{h=1}^{H} f_h(t) = \sum_{h=1}^{H} \Omega_h = \Omega\}$ To simplify notation when it is clear we obviate the time variable t. A utility vector $U = (U_1(f_1) \ldots U_H(f_H))$ is *feasible* if the allocation $(f_1, \ldots, f_H)$ is feasible.

The set of *individually rational feasible allocations* is the set of utility allocations $\{U_1(f_1) \ldots U_H(f_H)\}$ that are feasible and preferred to the initial endowments, $\forall h, U_h(f_h) \geq U_h(\Omega_h)$. A utility vector $U = (U_1(f_1), \ldots, U_H(f_H))$—which need not be feasible—is *efficient or undominated* if there is no allocation $G = (g_1, .., g_H)$ such that $\forall h, U_h(g_h) \geq U_h(f_h)$ and $U_k(g_k) > U_k(f_k)$ for some k, and there exists a sequence of feasible allocations $(f_1^j, \ldots, f_H^j)_{j=1,2,\ldots}$ such that $G = \lim_{j \to \infty} (f_1^j, \ldots, f_H^j)_{j=1,2,\ldots}$. A *feasible efficient allocation* is a feasible allocation that is also efficient.

The *Pareto Frontier* $P(E) \subset R_+^H$ is the set of individually rational and efficient feasible utility vectors. A *competitive equilibrium of the economy E* consists of a price vector $p^* \in X_+^*$ and an allocation $(f_1^*, \ldots, f_H^*) \in X^H$ such that f_h^* optimizes U_h over the budget set $B_h(p^*) = \{f \in X : \langle p^*, f \rangle = \langle p^*, \Omega_h \rangle\}$ and clears the markets $\sum_{h=1}^{H} f_h^* - \Omega_h = 0$. A feasible allocation $(f_1, \ldots, f_H)$ is a *quasiequilibrium* when there is a price $p \neq 0$ with $\forall h, \langle p, \Omega_h \rangle = \langle p, f_h \rangle$, and $\langle p, g \rangle \geq \langle p, f_h \rangle$ for any g implies $U_h(g) \geq U_h(f_h)$. A quasi-equilibrium is a *competitive equilibrium* when $U_h(g) > U_h(f_h) \Rightarrow \langle p, g \rangle > \langle p, f_h \rangle$.

The following concept of a **global cone** contains global information about a trader since, in ordinal terms, the sequences in this cone achieve utility values that eventually exceed those of all trades. The global cone was introduced in Chichilnisky (1991, 1995, 1994a, b, 1996c, d), see also Chichilnisky and Heal (1998).

Definition The cone A_h consists of all sequences of net trades $\{f^j\}$ in X along which the hth trader's utility increases and exceeds that of any other vector in the space; it can be based on rays of directions in X along which the hth trader's utility exceeds eventually all utility values:

$$A_h(\Omega_h) = \{\{f^j\} : \forall g \in X, \exists j : U_h(f^j) > U_h(g)\}$$

Definition The *global cone* $G_h(\Omega_h)$ is the set of all sequences of net trades in X along which the *hth* trader's utility never ceases to increase; it can be based on rays of directions with ever increasing utility:

$$G_h(\Omega_h) = \{\{f^j\} :\sim \exists Max_j U_h(f^j)\}.$$

We assume that $G_h(\Omega_h)$ has a simple structure, which was established in different forms in Chichilnisky (1991, 1995, 1994b, 1998), Chichilnisky and Heal (1998): when preferences have no half-lines in their indifferences, case (i), then $G_h(\Omega_h)$ is the closure of $A_h(\Omega_h)$ and in case (ii) when preferences have half-lines in their indifference surfaces, for example linear preferences, then $G_h(\Omega_h) = A_h(\Omega_h)$.

Definition The *Market Cone* $D_h(\Omega_h)$ is

$$D_h(\Omega_h) = \{p \in X : \forall\{g\} \in G_h(\Omega_h), \exists i : \langle g^j, p\rangle > 0 \text{ for } j > i\}$$

This is the set of all prices assigning eventually strictly positive value to net trades in the global cone. We assume the results of the following proposition, which was established in different forms in Chichilnisky (1991, 1995, 1994a, b, 1996c, d, 1998), Chichilnisky and Heal (1998), and is used in proving the connection between limited arbitrage and the existence of a sustainable market equilibrium:

Proposition 1 *If a utility* $U : X \to R$ *is uniformly non-satiated, then*

(A) $A(\Omega) \neq \emptyset$, and the cones $G(\Omega)$ and $D(\Omega)$, are all convex and uniform across vectors Ω in X.[5] For general preferences $G(\Omega)$ and $D(\Omega)$ may not be uniform, Chichilnisky (1998), Chichilnisky and Heal (1998).
(B) In *case* (i), preferences have no half lines in their indifferences, $G_h = \overline{A_h}$; with linear preferences case (ii) $G_h = A_h$.

[5]The cones $C(\Omega) = \{\{f\} \subset X : \lim_{j\to\infty} f^j = U(j^{j_0}) \text{ for some } j_0\}$ are also convex and uniform across vectors Ω.

2.4 Limited Arbitrage and Gains from Trade with Short Sales

This section defines limited arbitrage and provides an intuitive interpretation in terms of gains from trade. The following definitions and results are used in establishing the existence of a competitive equilibrium, and are based on Chichilnisky (1991, 1995, 1994a, b, 1996c, d), Chichilnisky and Heal (1998).

Definition *Gains from trade* are defined as

$$\mathcal{G}(E) = \sup\{\sum_{h=1}^{H}(U_h(f_h) - U_h(\Omega_h))\} \text{ where } \forall h,\ f_h = (f_h^j) \text{ satisfies}$$

$$\sum_{h=1}^{H}(f_h - \Omega_h) = 0 \text{ and } U_h(f_h^{j+1}) > U_h(f_h^j) > U_h(\Omega_h) \geq 0.$$

Definition The economy E satisfies *limited arbitrage* when

$$\bigcap_{h=1}^{H} D_h \neq \emptyset \tag{2}$$

Geometrically, *Limited Arbitrage* (2) bounds arbitrage opportunities in the economy by limiting the utility that can be achieved by the traders when trading with each other. Under the assumptions, Proposition 2 applies in case (i) and (ii): either indifference surfaces contain no half lines (e.g. strictly convex preferences) of (ii) linear preferences.

Proposition 2 *Limited arbitrage implies bounded gains from trade, namely* $\mathcal{G}(E) < \infty$.

Proof The proof relies on limited arbitrage, and follows the proofs of similar propositions in Chichilnisky (1991, 1995, 1998), Chichilnisky and Heal (1998) adapted to markets with sustainable preferences. Along the way we also highlight properties of sustainable preferences that are useful for understanding the structure of sustainable preferences, and the existence of a competitive equilibrium in sustainable markets.

Assume E has limited arbitrage and without loss of generality that $\forall h$, $\Omega_h = 0$. For every h, let $U_h = U_{1h} + U_{2h}$ where U_{1h} and U_{2h} are the two (non-zero) parts of the sustainable preference U_h that exist according to the representation of sustainable preferences provided in (1) Sect. 2.2, see Chichilnisky (1996a, b). If gains from trade $\mathcal{G}(E)$ were not bounded there would be a sequence of feasible, individually rational allocations of increasing utility $\{g^j\} = \{g_1^j, \ldots, g_H^j\}_{j=1,2\ldots}$ satisfying (i) $\forall j, \sum_{h=1}^{H} g_h^j = 0$, (ii) $\forall h, j\ U_h(g_h^{j+1}) > U_h(g_h^j)$ and (iii) for some k, $\lim_{j\to\infty}(U_k(g_k^j)) = \infty$, which implies that $\lim_{j\to\infty} \|g_k^j\|_\infty = \infty$. Define the set of traders K by $k \in K \Longleftrightarrow \lim_{j\to\infty} U_k(g_k^j) = \infty$ so that in particular $\lim_j \|g_k^j\| = \infty$; then by assumption $K \neq \emptyset$. We show that limited arbitrage contradicts (i), (ii) and (iii) so that gains from trade

$\mathcal{G}(E)$ cannot be unbounded with limited arbitrage. By definition of limited arbitrage (2) for $j > j_0$ there exists a p and a j_0 such that $\sum_{h \in K} \langle p, g_h^j \rangle > 0$ for $j > j_0$, because (ii), (iii) imply that $\forall\, h$, $\{g_h^j\}$ is in $G_h(0)$. However by (i) $\forall j$, $\sum_{h=1}^{H} g_h^j = 0$ so that $\forall p > 0$, $\sum_{h \in J} \langle p, g_h^j \rangle = 0$, a contradiction. The contraction arises from assuming that $\mathcal{G}(E)$ is not bounded. Therefore limited arbitrage implies bounded gains from trade, as we wanted to show.

Next we derive properties of general sustainable preferences, as stated above. Observe that, under limited arbitrage, when the sequence of purely finitely additive utilities $\{U_{2h}(g_h^j)\}$ in (2) grows without bound as $j \to \infty$, so does the countably additive sequence $\{U_{1h}(g_h^j)\}_{j\to\infty}$ in (2). Assume, to the contrary, that $\{U_{2h}(g_h^j)\}$ grows without bound but $\{U_{1h}(g_h^j)\}$ is bounded. Since as we saw above gains from trade $\sum_h U_h(g_h^j)$ are bounded under limited arbitrage, for each h, $\{U_h(g_h^j)\}_j$ is bounded. However for each j, $U_{2h}^j = U_h^j - U_{1h}^j$ and the right hand side is bounded by assumption, because U_h^j is bounded and we just assumed U_{1h}^j to be bounded as well. Therefore the sequence $\{U_{2h}(g_h^j)\}_j$ must be bounded, which is a contradiction. Therefore, under the conditions, when the sequence of purely finitely additive utilities $\{U_{2h}(g_h^j)\}_{j\to\infty}$ grows without bound so does the countably additive sequence of utilities $\{U_{1h}(g_h^j)\}$. For each h consider the sequence of normalized vectors $(\frac{g_h^j}{\|g_h^j\|})$, denoted also $\{g_h^j\}$. We now show that the sequence of normalized vectors $\{g_h^j\}$ has a weak* limit, and that its weak* limit is not zero. First observe that the normalized sequence $\{g_h^j\}$ is contained in the unit sphere of L_∞, which is weak* compact by Alaoglu's theorem. Consider a subsequence with a weak* limit; we show that this weak* limit is not zero. Since $X = L_\infty$ and utilities are continuous and sustainable, the preferred sets have non-empty interiors and by the properties of sustainable preferences presented in Sect. 2.2 there exist two non-zero prices $p_1 \in L_1$ and $p_2 \in L_\infty^* - L_1$,[6] such that p_1 supports the preferred set of U_{1h} at $\{0\}$, denoted U_{1h}^0, and p_2 supports the preferred set of U_{2h} at $\{0\}$ denoted U_{2h}^0, so that $p = p_1 + p_2$ supports the preferred set of U_h at $\{0\}$, U_h^0. We saw that, under limited arbitrage and with sustainable preferences, for any h, $\lim_j U_{2h}(g_h^j) = \infty$ implies that $\lim_{j\to\infty} U_{1h}(g_h^j) = \infty$, namely when purely finitely additive utility values grow without bound the corresponding countably additive parts do too. This implies in turn that $\forall h$, when the limiting utility values $\lim_j U_h(g_h^j) = \lim_j (U_{1h}(g_h^j) + U_{2h}(g_h^j)) = \infty$, then $\lim_{j\to\infty} U_{1h}(g_h^j) = \infty$, since $\lim_j U_h(g_h^j) = \infty$ and $U_h^j(g_h^j) = U_{1h}(g_h^j) + U_{2h}(g_h^j)$ implies that either $\lim_{j\to\infty} U_{1h}(g_h^j) = \infty$ as we wish to prove, or else $\lim_{j\to\infty} U_{2h}(g_h^j) = \infty$ which in turn implies that $\lim_{j\to\infty} U_{1h}(g_h^j) = \infty$ as seen above. Thus in all cases $\lim_j U_h(g_h^j) \to \infty$ implies $\lim_{j\to\infty} U_{1h}(g_h^j) = \infty$ as we wished to prove. Next observe that for all h there exist a subsequence denoted also $\{g_h^j\}$, j_0 and $r > 0$, such

[6] This follows from the initial results of Chichilnisky (1996a, b). $L_\infty^* - L_1$ denotes the complement of L_1 in L_∞^*, namely the space of purely finitely additive measures, see also the Appendix.

that $\langle p_1, g_h^j \rangle \geq r$ when $j > j_0$. Otherwise, $\lim_j \langle p_1, g_h^j \rangle = 0$ and in particular for any $t > 0$, and $\varepsilon > 0$, $\exists j : U_{1h}(g_h^j) > t$ and $(p_1, g_h^j) < \varepsilon$. But $\forall y$ satisfying $\langle p_1, y \rangle < 0$, $U_{1h}(y) < U_{1h}^0$ because p_1 supports U_{1h}^0. Therefore by continuity $\lim_j U_{1h}(g_h^j) \leq 0$, a contradiction since we showed that $\lim_{j\to\infty} U_{1h}(g_h^j) = \infty$. Therefore

$$\forall j > j_0, \qquad \left\langle p_1, g_h^j \right\rangle \geq r > 0, \tag{3}$$

which implies that the sequence $\{g_h^j\}$ is weak* bounded away from zero, by definition, since $p_1 \in L_1$. Therefore we have shown that the weak* compact sequence $\{g_h^j\}$ contains a subsequence, denoted also $\{g_h^j\}$, with a weak* limit denoted g_h, which is not zero because of (3). Consider now the cone C defined by all strictly positive convex combinations of the vectors g_h for all h. Either C is strictly contained in a half-space, or it defines a subspace of X. Since by construction $\sum_{h=1}^{H} g_h = 0$, C cannot be strictly contained in a half space. Therefore C defines a subspace. In particular there is $H' \subset H$, k, and $\forall h \in H'$, $\lambda_h > 0$, such that $(*)\ -g_k = \sum_{h=1}^{H'} \lambda_h g_h$. □

Corollary 1 *Limited arbitrage is necessary and sufficient for bounded gains from trade in* case *(ii).*

Proof Consider preferences in *case* (ii), which include linear preferences. Since $G_h = A_h$ in this case as shown in Proposition 1, the set of traders K defined by $k \in K \Longleftrightarrow \lim_{j\to\infty} U_k(g_k^j) = \infty$ equals H. In this case bounded gains from trade imply there can be no sequence $\{g_h^j\}$ satisfying (i), (ii) and (iii) in Proposition 2, so the reciprocal of the statement of this proposition is immediate. Thus limited arbitrage is necessary and sufficient for bounded gains from trade when preferences are in *case* (ii). □

3 Existence of Sustainable Market Equilibrium with Short Sales

As already mentioned the markets considered in this article allow unbounded short sales, since the trading domain X is the entire space. Yet this Section shows that under *limited arbitrage* traders only wish to engage in bounded trades with each other, and implies that the set of efficient trades is compact:

Theorem 1 *Limited arbitrage is necessary and sufficient for the compactness of a non-empty Pareto Frontier $P(E)$.*

Proof This follows Chichilnisky (1991, 1995), Chichilnisky and Heal (1998) and Propositions 1 and 2. Assume Limited Arbitrage. Observe that the Pareto Frontier is in euclidean space$P(E) \subset R^H_+$. Proposition 2 showed that with limited arbitrage, $P(E)$ is always bounded. To show compactness it suffices to show that $P(E)$ is closed with limited arbitrage. Without loss of generality, consider a sequence of allocations $\{g^j_h\}_{j=1,2,\ldots}$ satisfying $\forall j, \sum_{h=1}^H g^j_h = 0$, so that $\lim_{j\to\infty} \sum_{h=1}^H g^j_h = 0$, and the corresponding utility values $U_1(g^j_1), \ldots, U_H(g^j_H) \in R^H_+$. Assume that the utility values converge either to ∞ or to a utility allocation $V = \left(V_1, \ldots, V_H\right) \in R^H_+$ that is undominated by the utility allocation of any other feasible allocation; V may or not be a utility allocation corresponding to a feasible allocation. When limited arbitrage is satisfied, we show that V is the utility allocation corrsponding to a feasible allocation. It suffices to consider the case where the sequence of feasible utility allocations $\{U_1(g^j_1), \ldots, U_H(g^j_H)\}$, and therefore the corresponding allocations $\{g^j_h\}_{j=1,2,\ldots}$, $h = 1, \ldots, H$ are unbounded. Observe that, as shown in the proof of Proposition 2, the countably additive parts of the utilities $\{U_{11}(g^j_1), \ldots, U_{1H}(g^j_H)\}_{j\to\infty}$ are also unbounded in this case; the the normalized sequence $\{\frac{g^j_h}{\|g^j_h\|}\}_j = 1, 2, \ldots$, denoted also $\{g^j_h\}_{j=1,2,\ldots}$ is weak* precompact by Alaoglu's theorem and as shown in the proof of Proposition 2 it has a weak* convergent subsequence, denoted also $\{g^j_h\}_{j=1,2,\ldots}$ with a non zero weak* limit $g_h = \lim_{j\to\infty}\{g^j_h\}_{j=1,2,\ldots}$. If $\forall h, g^j_h \notin G_h$ then eventually the utility values of the traders attain their limit for all h, the utility vector V is achieved by a feasible allocation and the proof is complete. It remains to consider the case when for some trader k, $g^j_k \in G_k$; without loss assume that $\forall h, g^j_k \in G_k$. As in Proposition 2, consider the open convex cone C of strictly positive linear combinations of the (non zero) vectors $g_h, h = 1, 2, \ldots, H$, $C = \{w = \sum_h \mu_h g_h$ where $\forall h, \mu_h > 0\}$. Either (*a*) C is contained strictly in a half-space of X or else (*b*) C is a subspace of X. By construction $\forall j, \sum_h g^j_h = 0$, which eliminates case (*a*). Therefore case (*b*) must hold, in particular, there exists k, $g_k \in K$ and $\forall h, \lambda_h \geq 0$ satisfying

$$-g_k = \sum_{h=1}^{H} \lambda_h g_h$$

However limited arbitrage implies that $\exists p \in \cap_h D_h$ so that $\forall h, \langle p, g_h\rangle \geq 0$, which contradicts $-g_k = \sum_{h=1}^H \lambda_h g_h$. The contradiction arises from assuming that the Pareto frontier is not closed under limited arbitrage, therefore $P(E)$ must be closed. Limited arbitrage implies therefore a closed non-empty Pareto Frontier $P(E) \subset R^H$ which, from Proposition 2, is also bounded and hence compact. This establishes sufficiency. The reciprocal is established as follows. Failure of limited arbitrage means as seen above that for any $(U_1(g_1), \ldots, U(g_H)) \in P(E)$, there exists $(v_1, \ldots, v_H)$ satisfying $\forall h, \sum_{h=1}^H v_h = 0$ and $U_h(g_h + v_h) > U_h(g_h)$, a contradiction. Therefore limited arbitrage is necessary for a compact non-empty $P(E)$. □

Corollary 2 *Limited arbitrage implies that the Pareto frontier P(E) is homeomorphic to a simplex.*

This follows from Theorem 1, and the convexity of preferences, Arrow and Hahn (1971), Lemma 3, Chap. 5, p. 81.

Theorem 2 *Consider a sustainable market economy* $E = \{X, U_h, \Omega_h, h = 1, \ldots, H\}$ *where* $H \geq 2$, $X = L_\infty$, *and* $\forall h$, *trader h has a sustainable preference* U_h. *Then the economy E has a sustainable market equilibrium if and only if it satisfies limited arbitrage, and the sustainable equilibrium is Pareto efficient.*

Proof Necessity first. Without loss assume that $\forall h, \Omega_h = 0 \in X$. Let p^* be a price equilibrium and let $f^* = (f_1^*, \ldots f_H^*)$ be the corresponding equilibrium allocation. If limited arbitrage fails, $\exists h$ and $\{g^j\} \in G_h$ such that $\langle p^*, g^j \rangle \leq 0$ for some $j > j_0$ namely g^j is affordable at prices p^*. Recall that G_h is the same for every allocation by Proposition 1. It follows that $\exists j_0 > 0$ such that for $j > j_0$, $U_h(f_h^* + g^j) > U_h(f_h^*)$ which, together with the affordability of g^j, contradicts the fact that f^* is an equilibrium allocation. Limited arbitrage is therefore necessary for existence of an equilibrium.

For sufficiency, Theorem 1 established that the Pareto frontier is homeomorphic to a simplex when limited arbitrage is satisfied. The standard Negishi fixed point argument on the Pareto frontier $P(E)$ in utility space R^H establishes therefore the existence of a pseudoequilibrium, see Negishi (1960) and Chichilnisky and Heal (1984). To complete the proof, observe that $\forall h = 1, 2, \ldots H$ there exists always an allocation in X of strictly lower value than the pseudo equilibrium f_h^* at the price p^*. Therefore by Lemma 3, Chap. 4, p. 81 of Arrow and Hahn (1971) the quasi equilibrium $\langle p^*, g^* \rangle$ is also a competitive equilibrium, completing the proof of existence. Pareto efficiency follows from the fact that the equilibrium is in the Pareto frontier $P(E)$. □

4 Market Value and the Axiom of Choice

The existence of a sustainable equilibrium ensures the logic consistency of the concept of sustainable markets. The main condition required for existence is limited arbitrage, a condition that has been used to prove existence in the literature (Chichilnisky 1991, 1995, 1994a, b, 1996c, d; Chichilnisky and Heal 1998), applied in this case to markets where the traders have sustainable preferences. The notion of an equilibrium price in a sustainable market requires however further discussion. An equilibrium price is defined here—as is usual—as a continuous linear function on commodities or trades, and this price establishes the economic value of commodities at a market equilibrium. The space of prices is here L_∞^*, the dual of the space of commodities L_∞ that has been characterized (see the Appendix) as consisting of the linear sum of two subspaces, one subspace consisting of prices in L_1 that have a ready interpretation, and the second subspace consisting of finitely additive measures on R that require further explanation. Since preferences are sustainable, a price equilibrium will have the same form as a sustainable preference as characterized in Sect. 2.2 and in Chichilnisky (1996a, b), namely a convex combination of a path of prices through time that is an element of L_1 and of a purely finitely additive measure, for example a measure that focuses its weight on unbounded sets in R. In this context, therefore, a market price may assign two types of economic values: (i) an instantaneous value to commodities through time, and in addition (ii) a value to the long run future. The second term (ii) may seem unusual in standard markets, but it seems entirely appropriate for a sustainable market equilibrium. It modifies the conventional notion of prices just as needed to value the long run, as seems required for sustainable market solutions.

The finitely additive part of the price that assigns value to the long run future establishes a connection between sustainable markets and the Axiom of Choice in the foundation of mathematics, which postulates that there exists a universal and consistent fashion to select an element from every set; see Dunford and Schwartz (1958), Yosida (1974), Yosida and Hewitt (1952) (20, 21), Chichilnisky and Heal (1997), Kadane and O'Hagan (1995), Purves and Sudderth (1976), Dubins (1975) and Dubins and Savage (1965). It is possible to illustrate—but not in general construct—a purely finitely additive measure on R, or on any finite open interval (a, b) of R, see examples in Chichilnisky (2010a, b). This issue of constructibility is not unique to sustainable markets: it is an issue shared by the proof of the second fundamental theorem of welfare economics, see Debreu (1953), which requires Hahn—Banach Theorem and therefore the Axiom of Choice. The proof of existence of such purely finitely additive functions can be achieved in various ways but each requires the Axiom of Choice or a related result. To illustrate the problem consider the function $\phi(g(t)) = \lim_{t\to\infty} g(t)$ that is defined only on a closed strict subspace L'_∞ of L_∞ consisting of functions that have a limit at infinity. This function is continuous and linear on L'_∞. One can use Hahn—Banach's theorem to extend this function ϕ from the closed subspace $L'_\infty \subset L_\infty$ to the entire space L_∞ preserving the norm. Since the extension is not in L_1 it defines a purely finitely additive measure, as shown in the Appendix. However, in a general form Hahn—Banach's theorem requires the Axiom of Choice, which has been shown to be independent from the rest of the axioms of Mathematics (Godel 1940). Alternatively, one can extend the notion of a *limit* to encompass all functions in L_∞ including those with no standard limit. This can be achieved by defining convergence along a *free ultrafilter* arising from Stone-Cech compactification of the real line R as in Chichilnisky and Heal (1997). However the existence of a *free ultrafilter* requires once again the Axiom of Choice (Godel 1940). This illustrates why the actual construction of a *purely finitely additive measure* requires the Axiom of Choice. Since sustainable markets have prices that include purely finitely additive measures, this provides a connection between the Axiom of Choice and sustainable markets. It appears that the consideration of sustainable goals about consumption in the long run future conjures up the Axiom of Choice that is independent from the rest of mathematics (Godel 1940).

Appendix

Example A preference that is insensitive to the present

Consider $W(f) = \liminf_{x \epsilon R}(f(x))$. This utility is insensitive to the present and therefore does not satisfy Axiom 2. In addition this map is not linear, failing Axiom 1.

The dual space L_∞^*: countably additive and purely finitely additive measures

See Yosida (1974), Yosida and Hewitt (1952), Dunford and Schwartz (1958). A measure η is called *finitely additive* when for any family of pairwise disjoint measurable sets $\{A_i\}_{i=1,\ldots N}$ $\eta(\bigcup_{i=1}^{N} A_i) = \sum_{i=1}^{N} \eta(A_i)$. The measure η is called *countably additive* when for any countable family of pairwise disjoint measurable sets $\{A_i\}_{i=1,\ldots\infty}$ $\eta(\bigcup_{i=1}^{\infty} A_i) = \sum_{i=1}^{\infty} \eta(A_i)$. The space of continuous linear functions on L_∞ is the 'dual' of L_∞,and is denoted L_∞^*. This space has been characterized e.g. in Yosida and Hewitt (1952), Yosida (1974). $L_\infty^* = L_1 + (L_\infty^1 - L_1)$: it consists of L_1 functions g that define countably additive measures ν on R by the rule $\nu(A) = \int_A g(x)dx$ where $\int_R \mid g(x) \mid dx < \infty$ so that υ is *absolutely continuous* with respect to the Lebesgue measure, plus measures that are not countably additive, also called purely finitely additive measures, forming a subspace denoted $L_\infty^1 - L_1$. While a countable measure can be identified with an L_1 function, namely its so called 'density', purely finitely additive measures cannot be identified by such functions.

References

Arrow, K. J., & Hahn, F. (1971). *General competitive analysis*. San Francisco: Holden Day.

Asheim, G. B., Mitra, T., & Tungodden, B. (2016) sustainable resoursive social welfare functions. *Economic Theory* (Special Issue on the Global Environment, this issue).

Burniaux, J. -M., & Martins, J.O. (2016). Carbon leakages: A general equilibrium view. *Economic Theory* (Special Issue on the Global Environment, this issue).

Chichilnisky, G. (1991, 1995). Limited arbitrage is necessary and sufficient for the existence of competitive equilibrium with or without short sales, Discussion Paper No. 650, Columbia University Department of Economics, December 1991. *Economic Theory*, *95*, 79–108.

Chichilnisky, G. (1993). The cone condition, properness and extremely desirable commodities. *Economic Theory*, *3*, 177–182.

Chichilnisky, G. (1994a). Limited arbitrage is necessary and sufficient for the existence of a competitive equilibrium and the core, and it limits voting cycles. *Economic Letters*, *46*, 321–331.

Chichilnisky, G. (1994b). Social diversity, arbitrage and gains from trade: A unified perspective on resource allocation. *American Economic Review*, *84*, 427–434.

Chichilnisky, G. (1996a). An axiomatic approach to sustainable development. *Social Choice and Welfare*, *13*(2), 231–257.

Chichilnisky, G. (1996b). What is sustainable development. *Resource Energy Economics*, *73*(4), 467–491.

Chichilnisky, G. (1996c). Markets and games: A simple equivalence among the core. *Equilibrium and Limited Arbitrage, Metroeconomica*, *47*, 266–280.

Chichilnisky, G. (1996d). Limited arbitrage is necessary and sufficient for the non emptiness of the core. *Economic Letters*, *52*, 177–180.
Chichilnisky, G. (1998). A unified perspective on resource allocation: Limited arbitrage is necessary and sufficient for the existence of a competitive equilibrium, the core and social choice. In G. Chichilnisky (Ed.), *Topology and Markets* (p. 2002). Toronto, Canada: Fields Institute for Research in Mathematical Sciences and the American Mathematical Society.
Chichilnisky, G. (2009a). The limits of econometrics: Non parametric estimation in Hilbert spaces. *Econometric Theory*, *25*(04), 1070–1086.
Chichilnisky, G. (2009b). The topology of fear. *Journal of Mathematical Economics*, *45*, 807–816.
Chichilnisky, G. (2010a). The foundations of statistics with black swans. *Mathematical Social Sciences*, *59*(2), 184–192.
Chichilnisky, G. (2010b) The foundations of probability with black swans. *Journal of Probability and Statistics*, *2010*, 1–11. Article ID 838240.
Chichilnisky, G., & Heal, G.M. (1993). Existence of a competitive equilibrium in sobolev spaces without bounds on short sales. *Journal of Economic Theory*, *59*, 364–384. Initially circulated as Chichilnisky, G., & Heal, G. M. Existence of a Competitive Equilibrium in L_p and Sobolev Spaces IMA Preprint series #79, Institute for Mathematics and its Applications, University of Minnessota, Minneapolis, Minnesota, June 1984.
Chichilnisky, G., & Heal, G. M. (1997). Social choice with infinite populations. *Social Choice and Welfare*, *14*, 303–319.
Chichilnisky, G., & Heal, G. M. (1998). A unified treatment of finite and infinite economies: Limited arbitrage is necessary and sufficient for the existence of equilibrium and the core. *Economic Theory*, *12*, 163–176.
Chichilnisky, G., & Kalman, P. (1980). An application of functional analysis to models of efficient allocation of economic resources. *Journal of Optimization Theory and Applications*, *30*, 19–32.
Chipman, J., & Tian, G. (2016). Detrimental externalities pollution rights and the "Coase Theorem". *Economic Theory* (Special Issue on the Global Environment, this issue).
Debreu, G. (1953). Valuation equilibrium and pareto optimum. *Proceedings of the National Academy of Sciences*, *40*, 588–592.
Dubins, L. (1975). Finitely additive conditional probabilities, conglomerability and disintegration. *The Annals of Probability*, *3*, 89–99.
Dubins, L., & Savage, L. (1965). *Inequalities for stochastic processes: How to gamble if you must*. New York: McGraw Hill.
Dunford, N., & Schwartz, J. T. (1958). *Linear operators, Part I*. New York: Interscience.
Dutta, P., & Radner, R. (2016). Capital growth in a global warming model. *Economic Theory* (Special Issue on the Global Environment, this issue).
Figuieres, C., & Tidball, M. (2016) Sustainable exploitation of a natural resource: A satisfying use of chichilnisky's criterion. *Economic Theory* (Special Issue on the Global Environment, this issue).
Godel, K. (1940). The consistency of the continuum hypothesis. *Annals of Mathematical Studies*, *3*. (Princeton University Press, Princeton)
Kadane, J. B., & O'Hagan, A. (1995). Using finitely additive probability: uniform distributions of the natural numbers. *Journal of American Statistical Association*, *90*, 525–631.
Karp, L., & Zhang, J. (2016). Taxes v. quantities for a stock pollutant with endogenous abatement costs and assymetric information. *Economic Theory* (Special Issue on the Global Environment, this issue).
Lauwers, L. (2016) Intergenerational equity, efficiency and constructibility. *Economic Theory* (Special Issue on the Global Environment, this issue).
Lecocq, F., & Hourcade, J. C. (2016). Unspoken ethical issues in the climate affair: Insights from a theoretical analysis of the negotiations. *Economic Theory* (Special Issue on the Global Environment, this issue).
Negishi, T. (1960). Welfare economics and existence of an equilibrium for a competitive economy. *Metroeconomica*, *12*, 92–97.

Purves, R. W., & Sudderth, W. D. (1976). Some finitely additive probability. *The Annals of Probability*, *4*, 259–276.
Rezai, A., Foley, D. K., & Taylor, L. (2016). Global warming and economic externalities. *Economic Theory* (Special Issue on the Global Environment, this issue).
Yosida, K. (1974). *Functional analysis* (4th ed.). New York: Springer.
Yosida, K., & Hewitt, E. (1952). Finitely level independent measures. *Transactions of the American Mathematical Society*, *72*, 46–66.
Yannelis, N., & Zame, W. (1986). Equilibria in banach lattices without ordered preferences. *Journal of Mathematical Economics*, *15*, 85–110.

Part II
Ethical and Welfare Considerations

Sustainable Recursive Social Welfare Functions

Geir B. Asheim, Tapan Mitra and Bertil Tungodden

1 Introduction

How should we treat future generations? From a normative point of view, what are the present generation's obligations towards the future? What ethical criterion for intergenerational justice should be adopted if one seeks to respect the interests of

We are grateful for helpful discussions with Wolfgang Buchholz, and many constructive comments by anonymous referees, Larry Blume, Walter Bossert, Graciela Chichilnisky, John Hartwick, Aanund Hylland, Larry Karp, Luc Lauwers, Mohamed Mabrouk, Marco Mariotti, Tomoichi Shinotsuka, Yves Sprumont, Kotaro Suzumura, Peter Wakker and seminar participants at Cornell University, Queen's University, Universität Heidelberg, Universität Osnabrück, Université catholique de Louvain and Université de Montréal. Asheim thanks Cornell University and University of California at Santa Barbara for hospitality. This paper is part of the research activities at the center of Equality, Social Organization, and Performance (ESOP) at the Department of Economics at the University of Oslo. ESOP is supported by the Research Council of Norway.

Reprinted with kind permission from the Authors: Originally published in *Economic Theory*, Volume 49, Number 2, February 2012 DOI 10.1007/s00199-010-0573-7.

G.B. Asheim (✉)
Department of Economics, University of Oslo, Blindern, P.O. Box 1095, 0317 Oslo, Norway
e-mail: g.b.asheim@econ.uio.no

T. Mitra
Department of Economics, Cornell University, 448 Uris Hall, Ithaca, NY 14853, USA
e-mail: tm19@cornell.edu

B. Tungodden
Department of Economics, Norwegian School of Economics and Business Administration, Helleveien 30, 5045 Bergen, Norway
e-mail: Bertil.Tungodden@nhh.no

B. Tungodden
Chr. Michelsen Institute, P.O. Box 6033, 5892 Bergen, Norway

G. Chichilnisky and A. Rezai (eds.), *The Economics of the Global Environment*, Studies in Economic Theory 29, DOI 10.1007/978-3-319-31943-8_9

future generations? Answering such questions is essential when faced with the task of managing the global environment, e.g., in the context of climate change.[1]

These questions can be approached and answered in at least two ways:

1. Through an axiomatic analysis one can investigate on what ethical conditions various criteria for intergenerational justice are based, and then proceed to evaluate the normative appeal of these conditions.
2. By considering different technological environments, one can explore the consequences of various criteria for intergenerational justice, and compare the properties of the intergenerational well-being streams that are generated.

It is consistent with Rawls (1971) *reflective equilibrium* to do both: criteria for intergenerational justice should be judged both by the ethical conditions on which they build and by their consequences in specific technological environments. In particular, we may question the appropriateness of a criterion for intergenerational justice if it produces unacceptable outcomes in relevant technological environments. This view has been supported by many scholars, including Koopmans (1967), Dasgupta and Heal (1979, p. 311), and Atkinson (2001, p. 206).

When evaluating long-term policies, economists usually suggest to maximize the sum of discounted utilities. On the one hand, such *discounted utilitarianism* has been given a solid axiomatic foundation by Koopmans (1960).[2] On the other hand, this criterion has ethically questionable implications when applied to economic models with resource constraints. This is demonstrated by Dasgupta and Heal (1974) in the so-called Dasgupta-Heal-Solow (DHS) model of capital accumulation and resource depletion (Dasgupta and Heal 1974, 1979; Solow 1974), where discounted utilitarianism for any positive discount rate undermines the well-being of generations in far future, even if sustainable streams with non-decreasing well-being are feasible.

In this paper we revisit Koopmans (1960) framework, with numerical representability, sensitivity and stationarity as its key features. In Sect. 2 we consider conditions that are sufficient to numerically represent the social welfare relation by means of a recursive social welfare function satisfying sensitivity, stationarity and a condition requiring that the evaluation of two streams with the same present well-being *not* depend on what that level of well-being is, thereby echoing the analysis of Koopmans (1960, Sects. 3–7). In this framework we introduce an equity condition we call "Hammond Equity for the Future", capturing the following ethical intuition: A sacrifice by the present generation leading to a uniform gain for all future gen-

[1]A separate set of important questions relates to how to implement policies that are designed to respect the interests of future generations and to assess their effectiveness; see e.g., in the context of climate change, Burniaux and Martins (2016), Dutta and Radner (2016), Karp and Zhang (2016) and Ostrom (2016). In this context it is also of interest to investigate the validity of the 'Coase theorem', as done by Chipman and Tian (2016). Moreover, as pointed out by Lecocq and Hourcade (2016), optimal policies may require estimates of future *intra*generational distribution. Finally, as illustrated by Rezai et al. (2016), in some cases, such policies may benefit all generations, and thus do not represent a question of intergenerational justice.

[2]For an alternative set of axioms leading to discounted utilitarianism, see Lauwers (1997).

erations cannot lead to a less desirable stream of well-being if the present remains better-off than the future even after the sacrifice.[3]

In Sect. 3 we point out that "Hammond Equity for the Future" is weak, as it is implied by all the standard consequentialist equity conditions suggested in the literature. We show that adding this condition leads to a class of sustainable recursive social welfare functions, where the well-being of the present generation is taken into account if and only if the future is better-off. Furthermore, we establish general existence by means of an algorithmic construction. Finally, we show that any member of this class of sustainable recursive social welfare functions satisfies the key axioms of Chichilnisky's (1996) "sustainable preferences", namely "No Dictatorship of the Present" and "No Dictatorship of the Future".[4]

In Sect. 4 we offer results that identify which of the conditions used by Koopmans (1960) to axiomatize discounted utilitarianism is particularly questionable from an ethical perspective. The condition in question, referred to as "Independent Present" by us and listed as Postulate 3′a by Koopmans (1960, Sect. 14), requires that the evaluation of two streams which differ during only the first two periods *not* depend on what the common continuation stream is. It is only by means of "Independent Present" that Koopmans (1960, Sect. 14) moves beyond the recursive form to arrive at discounted utilitarianism, since this condition allows for additively separable representations when combined with stationarity and the requirement that the evaluation of two streams with the same present well-being not depend on what that level of well-being is (Debreu 1960; Gorman 1968a; Koopmans 1986a).

We suggest in Sect. 4 that "Independent Present"—which in the words of Heal (2005) is "restrictive" and "surely not innocent"—may not be supported by ethical intuition, as it is not obvious that the resolution of a conflict between the first two generations should be independent of how their well-being compares to the well-being of later generations. In our formal analysis we single out "Independent Present" as the culprit by showing that the addition of this condition contradicts both "Hammond Equity for the Future" and the Chichilnisky's (1996) conditions.

In Sect. 5 we apply sustainable recursive social welfare functions for studying optimal harvesting of a renewable resource that yields amenities. In a companion paper (Asheim and Mitra 2010) it is demonstrated how such functions can be used to solve the distributional conflicts in the DHS model. In both settings, our new criterion yields consequences that differ from those of discounted utilitarianism.

Koopmans (1960) has often been interpreted as presenting the definitive case for discounted utilitarianism. In Sect. 6 we discuss how our results contribute to a weakening of this impression, by exploring other avenues within the general setting of his approach. We also investigate the scope for our new equity condition "Hammond Equity for the Future" outside the Koopmans (1960) framework by *not* imposing that the social welfare relation is numerically representable.

All lemmas and proofs are relegated to an appendix.

[3]Our condition is inspired from Hammond (1976) Equity condition, but—as we will see—it is weaker and has not only an egalitarian justification.

[4]See Chichilnisky (2016) for an analysis of markets where traders have "sustainable preferences".

2 Formal Setting and Basic Result

Let $\mathbb{R}$ denote the set of real numbers and $\mathbb{Z}_+$ the set of non-negative integers. Denote by ${}_0\mathbf{x} = (x_0, x_1, \dots, x_t, \dots,)$ an infinite stream, where $x_t \in Y$ is a one-dimensional indicator of the well-being of generation t, and $Y \subseteq \mathbb{R}$ is a non-degenerate interval of admissible well-beings.[5] We will consider the set $\mathbf{X}$ of infinite streams bounded in well-being (see Koopmans 1986b, p. 89); i.e., $\mathbf{X}$ is given by

$$\mathbf{X} = \{{}_0\mathbf{x} \in \mathbb{R}^{\mathbb{Z}_+} \mid [\inf_t x_t, \sup_t x_t] \subseteq Y\} .$$

By setting $Y = [0, 1]$, this includes the important special case where $\mathbf{X} = [0, 1]^{\mathbb{Z}_+}$. However, the formulation allows for cases where Y is not compact.

Denote by ${}_0\mathbf{x}_{T-1} = (x_0, x_1, \dots, x_{T-1})$ and ${}_T\mathbf{x} = (x_T, x_{T+1}, \dots, x_{T+t}, \dots,)$ the T-head and the T-tail of ${}_0\mathbf{x}$. Write ${}_{\text{con}}z = (z, z, \dots)$ for the stream of a constant level of well-being equal to $z \in Y$. Throughout this paper we assume that the indicator of well-being is at least ordinally measurable and level comparable across generations; Blackorby et al. (1984) call this "level-plus comparability".

For all ${}_0\mathbf{x}$, ${}_0\mathbf{y} \in \mathbf{X}$, we write ${}_0\mathbf{x} \geq {}_0\mathbf{y}$ if and only if $x_t \geq y_t$ for all $t \in \mathbb{Z}_+$, ${}_0\mathbf{x} > {}_0\mathbf{y}$ if and only if ${}_0\mathbf{x} \geq {}_0\mathbf{y}$ and ${}_0\mathbf{x} \neq {}_0\mathbf{y}$, and ${}_0\mathbf{x} \gg {}_0\mathbf{y}$ if and only if $x_t > y_t$ for all $t \in \mathbb{Z}_+$.

A *social welfare relation* (SWR) is a binary relation $\succsim$ on $\mathbf{X}$, where for all ${}_0\mathbf{x}$, ${}_0\mathbf{y} \in \mathbf{X}$, ${}_0\mathbf{x} \succsim {}_0\mathbf{y}$ stands for $({}_0\mathbf{x}, {}_0\mathbf{y}) \in \succsim$ and entails that ${}_0\mathbf{x}$ is deemed socially at least as good as ${}_0\mathbf{y}$. Denote by $\sim$ and $\succ$ the symmetric and asymmetric parts of $\succsim$; i.e., ${}_0\mathbf{x} \sim {}_0\mathbf{y}$ is equivalent to ${}_0\mathbf{x} \succsim {}_0\mathbf{y}$ and ${}_0\mathbf{y} \succsim {}_0\mathbf{x}$ and entails that ${}_0\mathbf{x}$ is deemed socially indifferent to ${}_0\mathbf{y}$, while ${}_0\mathbf{x} \succ {}_0\mathbf{y}$ is equivalent to ${}_0\mathbf{x} \succsim {}_0\mathbf{y}$ and ${}_0\mathbf{y} \not\succsim {}_0\mathbf{x}$ and entails that ${}_0\mathbf{x}$ is deemed socially preferred to ${}_0\mathbf{y}$.

All comparisons are made at time 0. We abuse notation slightly by writing, for T, $T' \geq 0$, ${}_T\mathbf{x}$ and ${}_{T'}\mathbf{y}$ when referring to ${}_0\mathbf{x}'$ and ${}_0\mathbf{y}'$ where, for all t, $x'_t = x_{T+t}$ and $y'_t = y_{T'+t}$. This notational convention allows us to write ${}_T\mathbf{x}$, ${}_{T'}\mathbf{y} \in \mathbf{X}$ and ${}_T\mathbf{x} \succsim {}_{T'}\mathbf{y}$. It is used throughout the paper; e.g., in the definition of condition **IF**, in the statement of Lemma 2, and in the proofs of Proposition 2 and Lemma 3.

A *social welfare function* (SWF) representing $\succsim$ is a mapping $W : \mathbf{X} \to \mathbb{R}$ with the property that for all ${}_0\mathbf{x}$, ${}_0\mathbf{y} \in \mathbf{X}$, $W({}_0\mathbf{x}) \geq W({}_0\mathbf{y})$ if and only if ${}_0\mathbf{x} \succsim {}_0\mathbf{y}$. A mapping $W : \mathbf{X} \to \mathbb{R}$ is *monotone* if ${}_0\mathbf{x} \geq {}_0\mathbf{y}$ implies $W({}_0\mathbf{x}) \geq W({}_0\mathbf{y})$.

In the present section we impose conditions on the SWR sufficient to obtain a numerical representation in terms of an SWF with a recursive structure (see Proposition 2 below), similar to but not identical to Koopmans (1960, Sects. 3–7).

To obtain a numerical representation, we impose two conditions.

Condition O (*Order*) $\succsim$ is complete (for all ${}_0\mathbf{x}$, ${}_0\mathbf{y} \in \mathbf{X}$, ${}_0\mathbf{x} \succsim {}_0\mathbf{y}$ or ${}_0\mathbf{y} \succsim {}_0\mathbf{x}$) and transitive (for all ${}_0\mathbf{x}$, ${}_0\mathbf{y}$, ${}_0\mathbf{z} \in \mathbf{X}$, ${}_0\mathbf{x} \succsim {}_0\mathbf{y}$ and ${}_0\mathbf{y} \succsim {}_0\mathbf{z}$ imply ${}_0\mathbf{x} \succsim {}_0\mathbf{z}$).

---

[5]A more general framework is, as used by Koopmans (1960), to assume that the well-being of generation t depends on an n-dimensional vector $\mathbf{x}_t$ that takes on values in a connected set $\mathbf{Y}$. However, by representing the well-being of generation t by a scalar x_t, we can focus on intergenerational issues. In doing so, we follow, e.g., Diamond (1965), Svensson (1980), Chichilnisky's (1996), Basu and Mitra (2003) and Bossert et al. (2007).

Condition RC (*Restricted Continuity*) For all ${}_0\mathbf{x}, {}_0\mathbf{y} \in \mathbf{X}$, if ${}_0\mathbf{x}$ satisfies $x_t = z$ for all $t \geq 1$, and the sequence of streams $\langle {}_0\mathbf{x}^n \rangle_{n\in\mathbb{N}}$ satisfies $\lim_{n\to\infty} \sup_t |x_t^n - x_t| = 0$ with, for each $n \in \mathbb{N}$, ${}_0\mathbf{x}^n \in \mathbf{X}$ and ${}_0\mathbf{x}^n \not\prec {}_0\mathbf{y}$ (resp. ${}_0\mathbf{x}^n \not\succ {}_0\mathbf{y}$), then ${}_0\mathbf{x} \not\prec {}_0\mathbf{y}$ (resp. ${}_0\mathbf{x} \not\succ {}_0\mathbf{y}$).

Condition **RC** is weaker than ordinary supnorm continuity as, under condition **RC**, the stream ${}_0\mathbf{x}$ to which the sequence $\langle {}_0\mathbf{x}^n \rangle_{n\in\mathbb{N}}$ converges is restricted to having a constant level of well-being from period 1 on.

Condition C (*Continuity*) For all ${}_0\mathbf{x}, {}_0\mathbf{y} \in \mathbf{X}$, if the sequence of streams $\langle {}_0\mathbf{x}^n \rangle_{n\in\mathbb{N}}$ satisfies $\lim_{n\to\infty} \sup_t |x_t^n - x_t| = 0$ with, for each $n \in \mathbb{N}$, ${}_0\mathbf{x}^n \in \mathbf{X}$ and ${}_0\mathbf{x}^n \not\prec {}_0\mathbf{y}$ (resp. ${}_0\mathbf{x}^n \not\succ {}_0\mathbf{y}$), then ${}_0\mathbf{x} \not\prec {}_0\mathbf{y}$ (resp. ${}_0\mathbf{x} \not\succ {}_0\mathbf{y}$).

Condition **C** is entailed by Koopmans (1960) Postulate 1. As the analysis of Sect. 3 shows, the weaker continuity condition **RC** enables us to show existence of sustainable recursive social welfare functions.

The central condition in Koopmans (1960) analysis is the stationarity postulate (Postulate 4). Combined with Koopmans' Postulate 3b (the condition requiring that the evaluation of two streams with the same present well-being not depend on what that level of well-being is), the stationarity postulate is equivalent to the following independence condition (where we borrow the name that Fleurbaey and Michel (2003) use for a slightly stronger version of this condition).

Condition IF (*Independent Future*) For all ${}_0\mathbf{x}, {}_0\mathbf{y} \in \mathbf{X}$ with $x_0 = y_0$, ${}_0\mathbf{x} \succsim {}_0\mathbf{y}$ if and only if ${}_1\mathbf{x} \succsim {}_1\mathbf{y}$.

Condition **IF** means that an evaluation concerning only generations from the next period on can be made as if the present time (time 0) was actually at time 1; i.e., as if generations $\{0, 1, \dots\}$ would have taken the place of generations $\{1, 2, \dots\}$. If we extended our framework to also include comparisons at future times, then **IF** would imply time consistency as long as the SWR is time invariant.

With the well-being of each generation t expressed by a one-dimensional indicator x_t, it is uncontroversial to ensure through the following condition that a higher value of x_t cannot lead to a socially less preferred stream.

Condition M (*Monotonicity*) For all ${}_0\mathbf{x}, {}_0\mathbf{y} \in \mathbf{X}$, if ${}_0\mathbf{x} > {}_0\mathbf{y}$, then ${}_0\mathbf{y} \not\succ {}_0\mathbf{x}$.

Combined with the completeness part of condition **O**, it follows from condition **M** that, for all ${}_0\mathbf{x}, {}_0\mathbf{y} \in \mathbf{X}$, if ${}_0\mathbf{x} \geq {}_0\mathbf{y}$, then ${}_0\mathbf{x} \succsim {}_0\mathbf{y}$. Condition **M** is obviously implied by the "Strong Pareto" condition.

Condition SP (*Strong Pareto*) For all ${}_0\mathbf{x}, {}_0\mathbf{y} \in \mathbf{X}$, if ${}_0\mathbf{x} > {}_0\mathbf{y}$, then ${}_0\mathbf{x} \succ {}_0\mathbf{y}$.

With condition **M** we need not impose Koopmans (1960) extreme streams postulate (Postulate 5) and can consider the set of infinite streams bounded in well-being.

As the fifth and final condition of our basic representation result (Proposition 2), we impose the following efficiency condition.

Condition RD (*Restricted Dominance*) For all $x, z \in Y$, if $x < z$, then $(x, {}_{\text{con}}z) \prec {}_{\text{con}}z$.

To evaluate the implications of **RD**, consider the following three conditions.

Condition WS (*Weak Sensitivity*) There exist ${}_0\mathbf{x}, {}_0\mathbf{y}, {}_0\mathbf{z} \in \mathbf{X}$ such that $(x_0, {}_1\mathbf{z}) \succ (y_0, {}_1\mathbf{z})$.

Condition DF (*Dictatorship of the Future*) For all ${}_0\mathbf{x}, {}_0\mathbf{y} \in \mathbf{X}$ such that ${}_0\mathbf{x} \succ {}_0\mathbf{y}$, there exist $\underline{y}, \bar{y} \in Y$, with $\underline{y} \leq x_t, y_t \leq \bar{y}$ for all $t \in \mathbb{Z}_+$, and $T' \in \mathbb{Z}_+$ such that, for every ${}_0\mathbf{z}, {}_0\mathbf{v} \in [\underline{y}, \bar{y}]^{\mathbb{Z}_+}$, $({}_0\mathbf{z}_{T-1}, {}_T\mathbf{x}) \succ ({}_0\mathbf{v}_{T-1}, {}_T\mathbf{y})$ for all $T > T'$.

Condition NDF (*No Dictatorship of the Future*) Condition **DF** does not hold.
Condition **SP** implies condition **RD**, which in turn implies condition **WS**. Condition **WS** coincides with Koopmans (1960) Postulate 2. Condition **NDF** generalizes one of Chichilnisky's (1996) two main axioms to our setting where we consider the set of infinite streams bounded in well-being.

Proposition 1 *Assume that the* SWR $\succsim$ *satisfies conditions **O** and **IF**. Then **WS** is equivalent to **NDF**.*

As already noted at the end of the introduction, the proof of this and later results are provided in an appendix.

Since **RD** strengthens **WS**, it follows from Proposition 1 that **RD** ensures "No Dictatorship of the Future", provided that the SWR satisfies conditions **O** and **IF**. To appreciate why we cannot replace **RD** with an even stronger efficiency condition, we refer to the analysis of Sect. 3 and the impossibility result of Proposition 4.

To state Proposition 2, we introduce the following notation:

$$\begin{aligned} \mathcal{U} &:= \{U : Y \to \mathbb{R} \mid U \text{ is continuous and non-decreasing;} \\ &\qquad\qquad U(Y) \text{ is not a singleton}\} \\ \mathcal{U}_I &:= \{U : Y \to \mathbb{R} \mid U \text{ is continuous and increasing}\} \\ \mathcal{V}(U) &:= \{V : U(Y)^2 \to \mathbb{R} \mid V \text{ satisfies (V.0), (V.1), (V.2), and (V.3)}\}, \end{aligned}$$

where, for all $U \in \mathcal{U}$, $U(Y)$ denotes the range of U, and the properties of the *aggregator function* V, (V.0)–(V.3), are as follows:

(V.0) $V(u, w)$ is continuous in (u, w) on $U(Y)^2$.
(V.1) $V(u, w)$ is non-decreasing in u for given w.
(V.2) $V(u, w)$ is increasing in w for given u.
(V.3) $V(u, w) < w$ for $u < w$, and $V(u, w) = w$ for $u = w$.

Proposition 2 *The following two statements are equivalent.*

(1) The SWR $\succsim$ *satisfies conditions **O**, **RC**, **IF**, **M**, and **RD**.*
(2) There exists a monotone SWF $W : \mathbf{X} \to \mathbb{R}$ *representing* $\succsim$ *and satisfying, for some* $U \in \mathcal{U}_I$ *and* $V \in \mathcal{V}(U)$, $W({}_0\mathbf{x}) = V(U(x_0), W({}_1\mathbf{x}))$ *for all* ${}_0\mathbf{x} \in \mathbf{X}$ *and* $W({}_{\text{con}}z) = U(z)$ *for all* $z \in Y$.

For a given representation W (with associated utility function U) of an SWR satisfying conditions **O**, **RC**, **IF**, **M**, and **RD**, we refer to $U(x_t)$ as the *utility* of generation t and $W({}_0\mathbf{x})$ as the *welfare* derived from the infinite stream ${}_0\mathbf{x}$.

3 Hammond Equity for the Future

Discounted utilitarianism satisfies conditions **O**, **RC**, **IF**, **M**, and **RD**. Hence, these conditions do not by themselves prevent "Dictatorship of the Present", in the terminology of Chichilnisky's (1996).

Condition DP (*Dictatorship of the Present*) For all ${}_0\mathbf{x}, {}_0\mathbf{y} \in \mathbf{X}$ such that ${}_0\mathbf{x} \succ {}_0\mathbf{y}$, there exist $\underline{y}, \bar{y} \in Y$, with $\underline{y} \le x_t, y_t \le \bar{y}$ for all $t \in \mathbb{Z}_+$, and $T' \in \mathbb{Z}_+$ such that, for any ${}_0\mathbf{z}, {}_0\mathbf{v} \in [\underline{y}, \bar{y}]^{\mathbb{Z}_+}$, $({}_0\mathbf{x}_{T-1}, {}_T\mathbf{z}) \succ ({}_0\mathbf{y}_{T-1}, {}_T\mathbf{v})$ for all $T > T'$.

Condition NDP (*No Dictatorship of the Present*) Condition **DP** does not hold.

Condition **NDP** generalizes the other of Chichilnisky's (1996)'s two main axioms to our setting where we consider the set of infinite streams bounded in well-being.

We impose a weak new equity condition that ensures **NDP**. Combined with **RC**, this condition entails that the interest of the present are taken into account only if the present is worse-off than the future. Consider a stream $(x, {}_{\text{con}}z)$ having the property that well-being is constant from the second period on. For such a stream we may unequivocally say that, if $x < z$, then the present is worse- off than the future. Likewise, if $x > z$, then the present is better-off than the future.

Condition HEF (*Hammond Equity for the Future*) For all $x, y, z, v \in Y$, if $x > y > v > z$, then $(x, {}_{\text{con}}z) \not\succ (y, {}_{\text{con}}v)$.[6]

For streams where well-being is constant from the second period on, condition **HEF** captures the idea of giving priority to an infinite number of future generations in the choice between alternatives where the future is worse-off compared to the present in both alternatives. If the present is better-off than the future and a sacrifice now leads to a uniform gain for all future generations, then such a transfer from the present to the future cannot lead to a less desirable stream, as long as the present remains better-off than the future.

To appreciate the weakness of condition **HEF**, consider weak versions of the standard "Hammond Equity" condition (Hammond 1976) and Lauwers (1998) non-substitution condition.

Condition WHE (*Weak Hammond Equity*) For all ${}_0\mathbf{x}, {}_0\mathbf{y} \in \mathbf{X}$, if ${}_0\mathbf{x}$ and ${}_0\mathbf{y}$ satisfy that there exists a pair τ', τ'' such that $x_{\tau'} > y_{\tau'} > y_{\tau''} > x_{\tau''}$ and $x_t = y_t$ for all $t \neq \tau', \tau''$, then ${}_0\mathbf{x} \not\succ {}_0\mathbf{y}$.[7]

Condition WNS (*Weak Non-Substitution*) For all $x, y, z, v \in Y$, if $v > z$, then $(x, {}_{\text{con}}z) \not\succ (y, {}_{\text{con}}v)$.

By assuming, in addition, that well-beings are at least cardinally measurable and fully comparable, we may also consider weak versions of the Lorenz Domination and Pigou-Dalton principles. Such equity conditions have been used in the setting of infinite streams by, e.g., Birchenhall and Grout (1979), Asheim (1991), Fleurbaey and Michel (2001), and Hara et al. (2008).

Condition WLD (*Weak Lorenz Domination*) For all ${}_0\mathbf{x}, {}_0\mathbf{y} \in \mathbf{X}$, if ${}_0\mathbf{x}$ and ${}_0\mathbf{y}$ are such that ${}_0\mathbf{y}_{T-1}$ weakly Lorenz dominates ${}_0\mathbf{x}_{T-1}$ and ${}_T\mathbf{x} = {}_T\mathbf{y}$ for some $T > 1$, then ${}_0\mathbf{x} \not\succ {}_0\mathbf{y}$.[8]

---

[6]Condition **HEF** was introduced in a predecessor to this paper (Asheim and Tungodden 2004b) and has been analyzed by Banerjee (2006), Asheim et al. (2007), Asheim and Mitra (2010), and Alcantud and García-Sanz (2010).

[7]Under completeness, condition **WHE** corresponds to the standard "Hammond Equity" condition, where the premise implies ${}_0\mathbf{x} \precsim {}_0\mathbf{y}$.

[8]For any $T > 0$, ${}_0\mathbf{y}_{T-1}$ *weakly Lorenz dominates* ${}_0\mathbf{x}_{T-1}$ if and only if (i) $\sum_{\tau=0}^{T-1} y_\tau = \sum_{\tau=0}^{T-1} x_\tau$ and (ii) if φ and ψ are permutations on $\{0, \dots, T-1\}$ such that $y_{\varphi(1)} \le \cdots \le y_{\varphi(T-1)}$ and $x_{\psi(1)} \le \cdots \le x_{\psi(T-1)}$, then $\sum_{\tau=0}^{t} y_{\varphi(\tau)} \ge \sum_{\tau=0}^{t} x_{\psi(\tau)}$ for every $t = 0, \dots, T-1$.

Condition WPD (*Weak Pigou-Dalton*) For all ${}_0\mathbf{x}, {}_0\mathbf{y} \in \mathbf{X}$, if ${}_0\mathbf{x}$ and ${}_0\mathbf{y}$ are such that there exist a positive number ϵ and a pair τ', τ'' satisfying $x_{\tau'} - \epsilon = y_{\tau'} \geq y_{\tau''} = x_{\tau''} + \epsilon$ and $x_t = y_t$ for all $t \neq \tau', \tau''$, then ${}_0\mathbf{x} \not\prec {}_0\mathbf{y}$.

While it is clear that condition **HEF** is implied by **WNS**, it is perhaps less obvious that, under **O** and **M**, **HEF** is at least as weak as *each* of **WHE**, **WPD**, and **WLD**.

Proposition 3 *Assume that the* SWR $\succsim$ *satisfies conditions* ***O*** *and* ***M****. Then each of* ***WHE, WPD****, and* ***WLD*** *implies* ***HEF****.*

Note that condition **HEF** involves a comparison between a sacrifice by a single generation and a uniform gain for each member of an infinite set of generations that are worse-off. Hence, contrary to the standard "Hammond Equity" condition, if well-beings are made (at least) cardinally measurable and fully comparable, then the transfer from the better-off present to the worse-off future specified in condition **HEF** increases the sum of well-beings for a sufficiently large number T of generations. This entails that condition **HEF** is implied by both **WPD** and **WLD**, independently of what specific cardinal scale of well-beings is imposed (provided that conditions **O** and **M** are satisfied). Hence, "Hammond Equity for the Future" can be endorsed from both an egalitarian and utilitarian point of view. In particular, condition **HEF** is weaker and more compelling than the standard "Hammond Equity" condition.

However, in line with the Diamond-Yaari impossibility result (Diamond 1965) on the inconsistency of equity and efficiency conditions under continuity,[9] the equity condition **HEF** is in conflict with the following weak efficiency condition under **RC**.

Condition RS (*Restricted Sensitivity*) There exist $x, z \in Y$ with $x > z$ such that $(x, {}_{\text{con}}z) \succ {}_{\text{con}}z$.

Condition **SP** implies condition **RS**, which in turn implies condition **WS**.

Proposition 4 *There is no* SWR $\succsim$ *satisfying conditions* ***RC, RS****, and* ***HEF****.*

Impossibility results arising from **HEF** are further explored in Asheim et al. (2007). Here we concentrate on SWRs that satisfy **HEF**. We note that it follows from Proposition 4 that **RD** is the strongest efficiency condition compatible with **HEF** under **RC**, when comparing streams $(x, {}_{\text{con}}z)$ where well-being is constant from the second period on with constant streams ${}_{\text{con}}z$.

The following result establishes that "Dictatorship of the Present" is indeed ruled out by adding condition **HEF** to conditions **O**, **RC**, **IF**, and **M**.

Proposition 5 *Assume that the* SWR $\succsim$ *satisfies conditions* ***O, RC, IF****, and* ***M****. Then* ***HEF*** *implies* ***NDP****.*

How does the basic representation result of Proposition 2 change if we also impose condition **HEF** on an SWR $\succsim$ satisfying conditions **O**, **RC**, **IF**, **M**, and **RD**? To investigate this question, introduce the following notation:

---

[9]The Diamond-Yaari impossibility result states that the equity condition of "Weak Anonymity" (deeming two streams socially indifferent if one is obtained from the other through a finite permutation of well-beings) is inconsistent with the efficiency condition **SP** given **C**. See also Basu and Mitra (2003) and Fleurbaey and Michel (2003).

$$\mathcal{V}_S(U) := \{V : U(Y)^2 \to \mathbb{R} \mid V \text{ satisfies (V.0), (V.1), (V.2), and (V.3}')\},$$

where (V.3′) is given as follows:

(V.3′) $V(u, w) < w$ for $u < w$, and $V(u, w) = w$ for $u \geq w$.

Note that, for each $U \in \mathcal{U}$, $\mathcal{V}_S(U) \subseteq \mathcal{V}(U)$.

Proposition 6 *The following two statements are equivalent.*

(1) The SWR $\succsim$ *satisfies conditions* ***O****,* ***RC****,* ***IF****,* ***M****,* ***RD****, and* ***HEF****.*

(2) There exists a monotone SWF $W : \mathbf{X} \to \mathbb{R}$ *representing* $\succsim$ *and satisfying, for some* $U \in \mathcal{U}_I$ *and* $V \in \mathcal{V}_S(U)$*,* $W({}_0\mathbf{x}) = V(U(x_0), W({}_1\mathbf{x}))$ *for all* ${}_0\mathbf{x} \in \mathbf{X}$ *and* $W({}_{\text{con}}z) = U(z)$ *for all* $z \in Y$*.*

We refer to a mapping satisfying the property presented in statement (2) of Proposition 6 as a *sustainable recursive* SWF. Proposition 6 does not address the question whether there exists a sustainable recursive SWF for any $U \in \mathcal{U}_I$ and $V \in \mathcal{V}_S(U)$. This question of existence is resolved through the following proposition, which also characterizes the asymptotic properties of such social welfare functions.

Proposition 7 *For all* $U \in \mathcal{U}_I$ *and* $V \in \mathcal{V}_S(U)$*, there exists a monotone mapping* $W : \mathbf{X} \to \mathbb{R}$ *satisfying* $W({}_0\mathbf{x}) = V(U(x_0), W({}_1\mathbf{x}))$ *for all* ${}_0\mathbf{x} \in \mathbf{X}$ *and* $W({}_{con}z) = U(z)$ *for all* $z \in Y$*. Any such mapping* W *satisfies, for each* ${}_0\mathbf{x} \in \mathbf{X}$*,*

$$\lim_{T\to\infty} W({}_T\mathbf{x}) = \liminf_{t\to\infty} U(x_t).$$

By combining Propositions 6 and 7 we obtain our first main result.

Theorem 1 *There exists a class of* SWRs $\succsim$ *satisfying conditions* ***O****,* ***RC****,* ***IF****,* ***M****,* ***RD****, and* ***HEF****.*

The proof of the existence part of Proposition 7 is based on an algorithmic construction. For any ${}_0\mathbf{x} \in \mathbf{X}$ and each $T \in \mathbb{Z}_+$, consider the following finite sequence:

$$\left.\begin{array}{l} w(T,T) = \liminf_{t\to\infty} U(x_t) \\ w(T-1,T) = V(U(x_{T-1}), w(T,T)) \\ \dots \\ w(0,T) = V(U(x_0), w(1,T)) \end{array}\right\} \tag{1}$$

Define the mapping $W_\sigma : \mathbf{X} \to \mathbb{R}$ by

$$W_\sigma({}_0\mathbf{x}) := \lim_{T\to\infty} w(0,T). \tag{W}$$

In the proof of Proposition 7 we show that W_σ is a sustainable recursive SWF.

It is an open question whether W_σ is the *unique* sustainable recursive SWF given $U \in \mathcal{U}_I$ and $V \in \mathcal{V}_S(U)$. As reported in the following proposition, we can show uniqueness if the aggregator function satisfies a condition introduced by Koopmans et al. (1964, p. 88): $V \in \mathcal{V}(U)$ satisfies the property of *weak time perspective* if there exists a continuous increasing transformation $g : \mathbb{R} \to \mathbb{R}$ such that $g(w) - g(V(u, w))$ is a non-decreasing function of w for given u.

Proposition 8 *Let $U \in \mathcal{U}_I$ and $V \in \mathcal{V}_S(U)$. If V satisfies the property of weak time perspective, then there exists a unique monotone mapping $W : \mathbf{X} \to \mathbb{R}$ satisfying $W({}_0\mathbf{x}) = V(U(x_0), W({}_1\mathbf{x}))$ for all ${}_0\mathbf{x} \in \mathbf{X}$ and $W({}_{con}z) = U(z)$ for all $z \in Y$. This mapping, W_σ, is defined by (W).*

We have not been able to establish that the property of weak time perspective follows from the conditions we have imposed. However, it is satisfied in special cases; e.g., with V given by

$$V(u, w) = \begin{cases} (1-\delta)u + \delta w & \text{if } u < w \\ w & \text{if } u \geq w, \end{cases} \tag{2}$$

where $\delta \in (0, 1)$.[10] We can also show that the set of supnorm continuous sustainable recursive SWFs contains at most W_σ. However, even though W_σ is continuous in the weak sense implied by condition **RC**, it need not be supnorm continuous.

Once we drop one of the conditions **RC**, **IF**, and **RD**, and combine the remaining two conditions with **O**, **M**, and **HEF**, new possibilities open up. It is clear that:

- The mapping $W : \mathbf{X} \to \mathbb{R}$ defined by $W({}_0\mathbf{x}) := \liminf_{t\to\infty} U(x_t)$ for some $U \in \mathcal{U}_I$ represents an SWR satisfying **O**, **RC**, **IF**, **M**, and **HEF**, but not **RD**.
- The maximin SWR satisfies **O**, **RC**, **M**, **RD**, and **HEF**, but not **IF**.
- Leximin and undiscounted utilitarian SWRs for infinite streams satisfy **O**, **IF**, **M**, **RD**, and **HEF**, but not **RC** (cf. Proposition 13).

It follows from Propositions 1, 5, and 6 that any sustainable recursive SWF represents an SWR satisfying **NDF** and **NDP**. Chichilnisky's (1996, Definition 6) defines "sustainable preferences" by imposing **NDF** and **NDP** as well as numerical representability and **SP**. When showing existence in her Theorem 1, she considers SWRs violating condition **IF**. Hence, through showing general existence for our sustainable recursive SWF, we demonstrate that **NDF** and **NDP** can be combined with (a) numerical representability, (b) condition **IF** which implies stationarity, and (c) sensitivity to present well-being—and thus be imposed within the Koopmans (1960) framework—provided that **SP** is replaced by weaker dominance conditions.[11]

---

[10] Sustainable recursive SWFs with aggregator function given by (2) are analyzed in the companion paper (Asheim and Mitra 2010). Note that an SWR $\succsim$ represented by such a sustainable recursive SWF satisfies the following restricted form of the **IP** condition introduced in the next section: For all ${}_0\mathbf{x}, {}_0\mathbf{y}, {}_0\mathbf{z}, {}_0\mathbf{v} \in \mathbf{X}$ such that $(x_0, x_1, {}_2\mathbf{z})$, $(y_0, y_1, {}_2\mathbf{z})$, $(x_0, x_1, {}_2\mathbf{v})$, $(y_0, y_1, {}_2\mathbf{v})$ are non-decreasing, $(x_0, x_1, {}_2\mathbf{z}) \succsim (y_0, y_1, {}_2\mathbf{z})$ if and only if $(x_0, x_1, {}_2\mathbf{v}) \succsim (y_0, y_1, {}_2\mathbf{v})$.

[11] Mitra (2008) shows by means of an example that "sustainable preferences" can be combined with **IF** in the case where $Y = [0, 1]$ if we are willing to give up **RC**. See also Lauwers (2016) where the constructibility of Chichilnisky's (1996) criterion is investigated.

4 Independent Present

The following condition is invoked as Postulate 3′a in Koopmans (1960)' characterization of discounted utilitarianism.

Condition IP (*Independent Present*) For all ${}_0\mathbf{x}, {}_0\mathbf{y}, {}_0\mathbf{z}, {}_0\mathbf{v} \in \mathbf{X}$, $(x_0, x_1, {}_2\mathbf{z}) \succsim (y_0, y_1, {}_2\mathbf{z})$ if and only if $(x_0, x_1, {}_2\mathbf{v}) \succsim (y_0, y_1, {}_2\mathbf{v})$.

Condition **IP** requires that the evaluation of two streams differing only in the first two periods *not* depend on what the common continuation stream is. We suggest in this section that this condition may not be compelling, both through appeal to ethical intuition, and through formal results.

We suggest that it might be supported by ethical intuition to accept that the stream $(1, 4, 5, 5, 5, \ldots)$ is socially better than $(2, 2, 5, 5, 5, \ldots)$, while not accepting that $(1, 4, 2, 2, 2, \ldots)$ is socially better than $(2, 2, 2, 2, 2, \ldots)$. It is not obvious that we should treat the conflict between the worst-off and the second worst-off generation presented by the first comparison in the same manner as we treat the conflict between the worst-off and the best-off generation put forward by the second comparison.

Turn now to the formal results. Koopmans (1960) characterizes discounted utilitarianism by means of conditions **IF**, **WS**, and **IP**. However, it turns out that conditions **IF**, **WS**, and **IP** contradict **HEF** under **RC** and **M**. Furthermore, this conclusion is tight, in the sense that an SWR exists if any one of these conditions is dropped. This is our second main result.

Theorem 2 *There is no* SWR $\succsim$ *satisfying conditions* ***RC****,* ***IF****,* ***M****,* ***WS****,* ***HEF****, and* ***IP****. If one of the conditions* ***RC****,* ***IF****,* ***M****,* ***WS****,* ***HEF****, and* ***IP*** *is dropped, then there exists an* SWR $\succsim$ *satisfying the remaining five conditions as well as condition* ***O****.*

In the following proposition, we reproduce Koopmans (1960) characterization of discounted utilitarianism within the formal setting of this paper.[12]

Proposition 9 *The following two statements are equivalent.*

(1) The SWR $\succsim$ *satisfies conditions* ***O****,* ***RC****,* ***IF****,* ***M****,* ***WS****, and* ***IP****.*
(2) There exists a monotone SWF $W : \mathbf{X} \to \mathbb{R}$ *representing* $\succsim$ *and satisfying, for some* $U \in \mathcal{U}$ *and* $\delta \in (0, 1)$, $W({}_0\mathbf{x}) = (1 - \delta)U(x_0) + \delta W({}_1\mathbf{x})$ *for all* ${}_0\mathbf{x} \in \mathbf{X}$.

Strengthening ***WS*** *to* ***RD*** *in statement (1) is equivalent to replacing* $\mathcal{U}$ *by* $\mathcal{U}_I$ *in statement (2).*

This proposition follows from standard results for additively separable representations (Debreu 1960; Gorman 1968a; Koopmans 1986a), by exploiting the overlap of periods that conditions **IF** and **IP** give rise to (cf. Lemma 3).

Furthermore, we note that the discounted utilitarian SWF exists and is unique.

---

[12] See Bleichrodt et al. (2008) for a simplified characterization of discounted utilitarianism on an extended domain, as well as an overview of related literature.

Proposition 10 *For all $U \in \mathcal{U}$ and $\delta \in (0,1)$, there exists a unique monotone mapping $W : \mathbf{X} \to \mathbb{R}$ satisfying $W({}_0\mathbf{x}) = (1-\delta)U(x_0) + \delta W({}_1\mathbf{x})$ for all ${}_0\mathbf{x} \in \mathbf{X}$. This mapping, W_δ, is defined by, for each ${}_0\mathbf{x} \in \mathbf{X}$,*

$$W_\delta({}_0\mathbf{x}) = (1-\delta)\sum_{t=0}^{\infty} \delta^t U(x_t)\,.$$

Propositions 9 and 10 have the following implication.

Proposition 11 *There is no* SWR $\succsim$ *satisfying conditions **O**, **RC**, **IF**, **M**, **IP**, **NDP**, and **NDF**.*

To summarize, it follows from Theorem 2 and Propositions 1 and 11 that, within a Koopmans framework where **O**, **RC**, **IF**, **M**, and **WS** are imposed, condition **IP** contradicts both **HEF** and **NDP**. Hence, in such a framework, **IP** is in conflict with consequentialist equity conditions that respect the interests of future generations.

5 Applying Sustainable Recursive SWFs

We apply sustainable recursive SWFs for studying optimal harvesting of a renewable resource where, following Krautkraemer (1985), well-being may be derived directly from the resource stock. Using discounted utilitarianism in this setting reduces the resource stock below the green golden-rule (defined below) and leads to resource deterioration for sufficiently high discounting (Heal 1998).

Maximizing sustainable recursive SWFs leads to very different conclusions, as reported in Proposition 12. Before stating this result, we introduce the model.

The law of motion governing the bio-mass of the renewable resource, k, is given by a standard increasing, concave stock-recruitment function, f, and therefore the production framework is formally the same as the standard neoclassical aggregate model of economic growth. The function $f : \mathbb{R}_+ \to \mathbb{R}_+$ is assumed to satisfy:

(i) $f(0) = 0$,
(ii) f is continuous, increasing and strictly concave on $\mathbb{R}_+$
(iii) $\lim_{k\to 0} \frac{f(k)}{k} > 1$ and $\lim_{k\to\infty} \frac{f(k)}{k} < 1$.

It can be shown that there exists a unique number $\bar{k} > 0$ such that $f(\bar{k}) = \bar{k}$ and $f(k) > k$ for $k \in (0, \bar{k})$.

A feasible path from $k \in [0, \bar{k}]$ is a sequence of *resource stocks* ${}_0\mathbf{k}$ satisfying:

$$k_0 = k\,, \quad 0 \le k_{t+1} \le f(k_t) \;\; \text{for } t > 0\,.$$

It follows from the definition of $\bar{k}$ that $k_t \in [0, \bar{k}]$ for $t > 0$. Hence, $\bar{k}$ is the maximal attainable resource stock if one starts from an initial stock in $[0, \bar{k}]$. Associated with a feasible path ${}_0\mathbf{k}$ from $k \in [0, \bar{k}]$ is a *consumption* stream ${}_0\mathbf{c}$, defined by

$$c_t = f(k_t) - k_{t+1} \quad \text{for } t \geq 0\,.$$

Well-being, x, depends on consumption and resource amenities through a function $x : [0, \bar{k}]^2 \to \mathbb{R}$, which is assumed to satisfy:

$$\begin{aligned} &\text{(i) } x \text{ is continuous and quasi-concave on } [0,\bar{k}]^2, \\ &\text{(ii) } x \text{ is non-decreasing in } (c,k), \text{ and increasing in } c \text{ (when } k > 0). \end{aligned} \tag{3}$$

The set of of admissible well-beings is given by $Y := [x(0,0), x(\bar{k},\bar{k})]$. Associated with a feasible path ${}_0\mathbf{k}$ from $k \in [0,\bar{k}]$ is a well-being stream ${}_0\mathbf{x}$, defined by

$$x_t = x(f(k_t) - k_{t+1}, k_t) \quad \text{for } t \geq 0\,.$$

For any $k \in [0,\bar{k}]$, the set of well-being streams associated with feasible resource paths from k is contained in $\mathbf{X} = Y^{\mathbb{Z}_+}$.

It follows from the continuity and strict concavity of f and the continuity and quasi-concavity of x, combined with property (3)(ii), that there exists a unique number $k^* \in [0,\bar{k}]$ such that $x(f(k^*) - k^*, k^*) \geq x(f(k) - k, k)$ for all $k \in [0,\bar{k}]$. Since, for any $k \in (0,\bar{k})$, $x(f(k) - k, k) > x(f(0) - 0, 0) = x(0,0)$, we have that $k^* > 0$. Clearly, an additional assumption can be imposed to ensure the existence of $k \in (0,\bar{k})$ such that $x(f(k) - k, k) > x(f(\bar{k}) - \bar{k}, \bar{k}) = x(0,\bar{k})$, so that $k^* < \bar{k}$. The subsequent analysis holds with (and without) any such assumption.

We write $c^* := f(k^*) - k^*$ and $x^* := x(c^*, k^*)$. By keeping the resource stock constant at k^*, a maximum sustainable well-being equal to x^* is attained; this corresponds to the *green golden-rule* (Chichilnisky et al. 1995). The following result shows that if $k \in [k^*,\bar{k}]$ and a sustainable recursive SWF is maximized, then welfare corresponds to the green golden-rule, and the resource stock never falls below the green golden-rule level.

Proposition 12 *Assume that an economy maximizes a sustainable recursive* SWF $W : \mathbf{X} \to \mathbb{R}$ *on the set of well-being streams associated with feasible resource paths from* $k \in [k^*,\bar{k}]$. *Then an optimum exists, and for any optimal resource path* ${}_0\hat{\mathbf{k}}$, *with associated well-being stream* ${}_0\hat{\mathbf{x}}$,

$$W({}_t\hat{\mathbf{x}}) = W({}_{\text{con}}x^*), \;\; \hat{x}_t \geq x^*\,, \;\; \textit{and} \;\; \hat{k}_t \geq k^* \; \textit{for } t \geq 0\,.$$

Hence, in contrast to the existence problem encountered when Chichilnisky's (1996) "sustainable preferences" are applied to such a setting (see Figuieres and Tidball 2016 where this problem motivates an interesting analysis), optima exist when sustainable recursive SWFs are used to evaluate streams (at least, for $k \in [k^*,\bar{k}]$). Moreover, in contrast to the outcome under discounted utilitarianism, sustainable recursive SWFs sustain well-being at or above its maximum sustainable level, by sustaining the resource stock at or above the green golden-rule level.

In a companion paper (Asheim and Mitra 2010) it is demonstrated how sustainable recursive SWFs can be used to resolve in an appealing way the interesting distributional conflicts that arise in the DHS model of capital accumulation and resource depletion. In particular, applying sustainable recursive SWFs in this setting leads to growth and development at first when capital is productive, while protecting the generations in the distant future from the grave consequences of discounting when the vanishing resource stock undermines capital productivity.

6 Concluding Remarks

Koopmans (1960) has often been interpreted as presenting the definitive case for discounted utilitarianism. In Sects. 2 and 3 we have sought to weaken this impression by exploring other avenues within the general setting of his approach. In particular, by not imposing condition **IP**, used by Koopmans (1960) to characterize discounted utilitarianism, we have been able to combine our new equity condition **HEF** with the essential features of the Koopmans (1960) framework: (a) numerical representability, (b) sensitivity to the interests of the present generation, and (c) condition **IF** which includes Koopmans (1960)' stationarity postulate. This leads to a non-empty class of sustainable recursive social welfare functions. We have argued that condition **HEF** is weak, as it is implied by all the standard consequentialist equity conditions suggested in the literature, yet strong enough to ensure that the Chichilnisky's (1996) conditions are satisfied. As we have discussed in Sect. 5, sustainable recursive social welfare functions are applicable and yield consequences that differ from those of discounted utilitarianism.

In this final section we note that even wider possibilities open up if we are willing to give up numerical representability by not imposing **RC**. In particular, we are then able to combine the equity condition **HEF** and the independence condition **IP** with our basic conditions **O** and **IF**, while strengthening our efficiency conditions **M** and **RD** to condition **SP**.

Proposition 13 *There exists an* SWR $\succsim$ *satisfying conditions **O**, **IF**, **SP**, **HEF**, and **IP**.*

The proof of this proposition employs the leximin and undiscounted utilitarian SWRs for infinite streams that have been axiomatized in recent contributions (see, Asheim and Tungodden 2004a; Basu and Mitra 2007; Bossert et al. 2007).

We end by making the observation that continuity is not simply a "technical" condition without ethical content. In a setting where **RC** (or a stronger continuity condition like **C**) is combined with **RS** (or a stronger efficiency condition like **SP**), it follows from Proposition 4 that condition **HEF** is not satisfied. Hence, on this basis one may claim that, in combination with a sufficiently strong efficiency condition, continuity rules out SWFs that protect the interests of future generations by implying that the equity condition **HEF** does not hold. In the main analysis of this paper we have avoided the trade-off between continuity and numerical representability on the one hand, and the ability to impose the equity condition **HEF** on the other hand, by weakening the efficiency condition in an appropriate way.

Appendix: Proofs

Proof (of Proposition 1) *Part I:* **WS** implies **NDF**. Assume that the SWR $\succsim$ satisfies conditions **O** and **WS**. By **WS**, there exist ${}_0\mathbf{x}$, ${}_0\mathbf{y} \in \mathbf{X}$ with ${}_1\mathbf{x} = {}_1\mathbf{y}$ such that ${}_0\mathbf{x} \succ {}_0\mathbf{y}$. Let ${}_0\mathbf{z}$, ${}_0\mathbf{v} \in \mathbf{X}$ be given by ${}_0\mathbf{z} = {}_0\mathbf{v} = {}_0\mathbf{x}$. We have that, for any $\underline{y}, \bar{y} \in Y$ satisfying $\underline{y} \leq x_t, y_t \leq \bar{y}$ for all $t \in \mathbb{Z}_+$, ${}_0\mathbf{z}$, ${}_0\mathbf{v} \in [\underline{y}, \bar{y}]^{\mathbb{Z}_+}$. Still, for all $T > 0$, $({}_0\mathbf{z}_{T-1}, {}_T\mathbf{x}) = {}_0\mathbf{x} = ({}_0\mathbf{x}_{T-1}, {}_T\mathbf{y}) = ({}_0\mathbf{v}_{T-1}, {}_T\mathbf{y})$, implying by **O** that $({}_0\mathbf{z}_{T-1}, {}_T\mathbf{x}) \sim ({}_0\mathbf{v}_{T-1}, {}_T\mathbf{y})$. This contradicts **DF**.

Part II: **NDF** *implies* **WS**. Assume that the SWR $\succsim$ satisfies conditions **O** and **IF**. Suppose that **WS** does not hold, i.e., for all ${}_0\mathbf{x}'$, ${}_0\mathbf{y}' \in \mathbf{X}$ with ${}_1\mathbf{x}' = {}_1\mathbf{y}'$, ${}_0\mathbf{x}' \sim {}_0\mathbf{y}'$.

Case (i): There exist ${}_0\mathbf{x}$, ${}_0\mathbf{y} \in \mathbf{X}$ *such that* ${}_0\mathbf{x} \succ {}_0\mathbf{y}$. Suppose ${}_0\mathbf{x}$, ${}_0\mathbf{y} \in \mathbf{X}$ are such that ${}_0\mathbf{x} \succ {}_0\mathbf{y}$. Let ${}_0\mathbf{z}$, ${}_0\mathbf{v}$ be arbitrary streams in $\mathbf{X}$. We have that ${}_{T-1}\mathbf{x} \sim (z_{T-1}, {}_T\mathbf{x})$ for all $T > 0$ since **WS** does not hold. By **IF** and the above argument,

$$ {}_{T-2}\mathbf{x} = (x_{T-2}, {}_{T-1}\mathbf{x}) \sim (x_{T-2}, z_{T-1}, {}_T\mathbf{x}) \sim ({}_{T-2}\mathbf{z}_{T-1}, {}_T\mathbf{x}) . $$

By invoking **O** and applying **IF** and the above argument repeatedly, it follows that ${}_0\mathbf{x} \sim ({}_0\mathbf{z}_{T-1}, {}_T\mathbf{x})$ for all $T > 0$. Likewise, ${}_0\mathbf{y} \sim ({}_0\mathbf{v}_{T-1}, {}_T\mathbf{y})$ for all $T > 0$. By **O**, $({}_0\mathbf{z}_{T-1}, {}_T\mathbf{x}) \succ ({}_0\mathbf{v}_{T-1}, {}_T\mathbf{y})$ for all $T > 0$. This establishes **DF**, implying that **NDF** does not hold.

Case (ii): There do not exist ${}_0\mathbf{x}$, ${}_0\mathbf{y} \in \mathbf{X}$ *such that* ${}_0\mathbf{x} \succ {}_0\mathbf{y}$. Then **DF** is true trivially, implying that **NDF** does not hold in this case either. □

The following lemma is useful for proving Proposition 2 and subsequent results.

Lemma 1 *Assume that the* SWR $\succsim$ *satisfies conditions* ***O****,* ***RC****,* ***M****. Then, for all* ${}_0\mathbf{x} \in \mathbf{X}$*, there exists* $z \in Y$ *such that* ${}_{\text{con}}z \sim {}_0\mathbf{x}$*. If condition* ***RD*** *is added, then* z *is unique.*

Proof Assume that the SWR $\succsim$ satisfies conditions **O**, **RC**, and **M**. By **O**, **M**, and the definition of $\mathbf{X}$, there exists $z \in Y$ such that $\inf\{v \in Y \mid {}_{\text{con}}v \succsim {}_0\mathbf{x}\} \leq z \leq \sup\{v \in Y \mid {}_{\text{con}}v \precsim {}_0\mathbf{x}\}$. By **O** and **RC**, ${}_{\text{con}}z \sim {}_0\mathbf{x}$.

If condition **RD** is added, then by **O**, **M**, and **RD** we have that

$$ {}_{\text{con}}v = (v, {}_{\text{con}}v) \precsim (v, {}_{\text{con}}z) \prec {}_{\text{con}}z \quad \text{if } v < z , \tag{4} $$

so that $\inf\{v \in Y \mid {}_{\text{con}}v \succsim {}_0\mathbf{x}\} = \sup\{v \in Y \mid {}_{\text{con}}v \precsim {}_0\mathbf{x}\}$ and z is unique. □

Proof (of Proposition 2) *Part I: (1) implies (2).* Assume that the SWR $\succsim$ satisfies conditions **O**, **RC**, **IF**, **M**, and **RD**. In view of Lemma 1, determine $W : \mathbf{X} \to Y$ by, for all ${}_0\mathbf{x} \in \mathbf{X}$, $W({}_0\mathbf{x}) = z$ where ${}_{\text{con}}z \sim {}_0\mathbf{x}$. By **O** and (4), $W({}_0\mathbf{x}) \geq W({}_0\mathbf{y})$ if and only if ${}_0\mathbf{x} \succsim {}_0\mathbf{y}$. By **M**, W is monotone.

Let $U \in \mathcal{U}_I$ be given by $U(x) = x$ for all $x \in Y$, implying that $U(Y) = Y$. Hence, by construction of W, $W({}_{\text{con}}z) = z = U(z)$ for all $z \in Y$. It follows from **IF** that, for given $x_0 \in Y$, there exists an increasing transformation $V(U(x_0), \cdot) : Y \to Y$ such that, for all ${}_1\mathbf{x} \in \mathbf{X}$, $W(x_0, {}_1\mathbf{x}) = V(U(x_0), W({}_1\mathbf{x}))$. This determines $V : Y \times Y \to Y$, where $V(u, w)$ is increasing in w for given u, establishing that V satisfies (V.2). By

M, $V(u, w)$ is non-decreasing in u for given w, establishing that V satisfies (V.1). Since $(x,_{\text{con}} z) \not\prec {}_{\text{con}}v$ (resp. $(x, {}_{\text{con}}z) \not\succ {}_{\text{con}}v$) if and only if

$$V(x, z) = V(U(x), W({}_{\text{con}}z)) = W(x, {}_{\text{con}}z) \geq v \quad (\text{resp. } \leq v),$$

RC implies that V satisfies (V.0). Finally, since

$$V(z, z) = V(U(z), W({}_{\text{con}}z)) = W({}_{\text{con}}z) = z$$
$$V(x, z) = V(U(x), W({}_{\text{con}}z)) = W(x, {}_{\text{con}}z) < W({}_{\text{con}}z) = z \text{ if } x < z\,,$$

by invoking **RD**, it follows that V satisfies (V.3). Hence, $V \in \mathcal{V}(U)$.

Part II: (2) implies (1). Assume that the monotone mapping $W : \mathbf{X} \to \mathbb{R}$ is an SWF and satisfies, for some $U \in \mathcal{U}_I$ and $V \in \mathcal{V}(U)$, $W({}_0\mathbf{x}) = V(U(x_0), W({}_1\mathbf{x}))$ for all ${}_0\mathbf{x} \in \mathbf{X}$ and $W({}_{\text{con}}z) = U(z)$ for all $z \in Y$. Since the SWR $\succsim$ is represented by the SWF W, it follows that $\succsim$ satisfies **O**. Moreover, $\succsim$ satisfies **M** since W is monotone, $\succsim$ satisfies **IF** since V satisfies (V.2), and $\succsim$ satisfies **RD** since $U \in \mathcal{U}_I$ and V satisfies (V.3). The following argument shows that $\succsim$ satisfies **RC**.

Let ${}_0\mathbf{x}, {}_0\mathbf{y} \in \mathbf{X}$, and let $x_t = z$ for all $t \geq 1$. Let ${}_0\mathbf{x}^n \in \mathbf{X}$ for $n \in \mathbb{N}$, with the property that $\lim_{n\to\infty} \sup_t |x_t^n - x_t| = 0$ and, for each $n \in \mathbb{N}$, ${}_0\mathbf{x}^n \not\prec {}_0\mathbf{y}$. We have to show that ${}_0\mathbf{x} \not\prec {}_0\mathbf{y}$, or equivalently, $W({}_0\mathbf{x}) \geq W({}_0\mathbf{y})$. Define $\epsilon_0(n)$ and $\epsilon(n)$ for $n \in \mathbb{N}$ by, for each $n \in \mathbb{N}$, $\epsilon_0(n) := \max\{0, x_0^n - x_0\}$ and $\epsilon(n) := \max\{0, \sup_{t\geq 1}(x_t^n - x_t)\}$, so that $\lim_{n\to\infty} \epsilon_0(n) = 0$ and $\lim_{n\to\infty} \epsilon(n) = 0$. For each $n \in \mathbb{N}$,

$$\begin{aligned} V(U(x_0 + \epsilon_0(n)), U(z + \epsilon(n))) &= V(U(x_0 + \epsilon_0(n)), W({}_{\text{con}}(z + \epsilon(n)))) \\ &= W(x_0 + \epsilon_0(n), {}_{\text{con}}(z + \epsilon(n))) \\ &\geq W({}_0\mathbf{x}^n) \geq W({}_0\mathbf{y}) \end{aligned}$$

since W is monotone and represents $\succsim$, and ${}_0\mathbf{x}^n \not\prec {}_0\mathbf{y}$. This implies that

$$W({}_0\mathbf{x}) = V(U(x_0), W({}_{\text{con}}z)) = V(U(x_0), U(z)) \geq W({}_0\mathbf{y})$$

since U and V are continuous and $\lim_{n\to\infty} \epsilon(n) = 0$. The same kind of argument can be used to show that ${}_0\mathbf{x} \not\succ {}_0\mathbf{y}$ if, for each $n \in \mathbb{N}$, ${}_0\mathbf{x}^n \not\succ {}_0\mathbf{y}$. □

Proof (of Proposition 3) Assume $x > y > v > z$. We must show under **O** and **M** that each of **WHE**, **WLD**, and **WPD** implies $(x, {}_{\text{con}}z) \not\succ (y, {}_{\text{con}}v)$.

Since $x > y > v > z$, there exist an integer T and utilities $x', z' \in Y$ satisfying $y > x' \geq v > z' > z$ and $x - x' = T(z' - z)$.

By **O** (completeness) and **WHE**, $(x', z', {}_{\text{con}}z) \succsim (x, {}_{\text{con}}z)$, and by **M**, $(y, {}_{\text{con}}v) \succsim (x', z', {}_{\text{con}}z)$. By **O** (transitivity), $(y, {}_{\text{con}}v) \succsim (x, {}_{\text{con}}z)$.

Consider next **WLD** and **WPD**. Let ${}_0\mathbf{x}^0 = (x, {}_{\text{con}}z)$, and define ${}_0\mathbf{x}^n$ for $n \in \{1, \ldots, T\}$ inductively as follows:

$$x_t^n = x_t^{n-1} - (z' - z) \qquad \text{for } t = 0$$
$$x_t^n = z' \qquad \text{for } t = n$$
$$x_t^n = x_t^{n-1} \qquad \text{for } t \neq 0, n\,.$$

By **O** (completeness) and **WLD**, ${}_0\mathbf{x}^T \succsim {}_0\mathbf{x}^0$, and by **M**, $(y, {}_{\text{con}}v) \succsim {}_0\mathbf{x}^T$. By **O** (transitivity), $(y, {}_{\text{con}}v) \succsim (x, {}_{\text{con}}z)$ since ${}_0\mathbf{x}^0 = (x, {}_{\text{con}}z)$.

By **O** (completeness) and **WPD**, ${}_0\mathbf{x}^n \succsim {}_0\mathbf{x}^{n-1}$ for $n \in \{1, \dots, T\}$, and by **M**, $(y, {}_{\text{con}}v) \succsim {}_0\mathbf{x}^T$. By **O** (transitivity), $(y, {}_{\text{con}}v) \succsim (x, {}_{\text{con}}z)$ since ${}_0\mathbf{x}^0 = (x, {}_{\text{con}}z)$. □

Proof (of Proposition 4) This follows from Asheim et al. (2007, Proposition 2). □

Proof (of Proposition 5) Assume that the SWR $\succsim$ satisfies conditions **O**, **RC**, **M**, **IF**, and **HEF**. Let ${}_0\mathbf{x}, {}_0\mathbf{y} \in \mathbf{X}$ satisfy ${}_0\mathbf{x} \succ {}_0\mathbf{y}$, and let $\underline{y}, \bar{y} \in Y$ satisfy $\underline{y} \le x_t, y_t \le \bar{y}$ for all $t \in \mathbb{Z}_+$. For any $T > 0$ with $x_{T-1} > \underline{y}$, Proposition 4 implies that $(x_{T-1}, {}_{\text{con}}\underline{y}) \succ {}_{\text{con}}\underline{y}$ would contradict **RC** and **HEF**. Since $x_{T-1} \ge \underline{y}$, it follows from **O** and **M** that $(x_{T-1}, {}_{\text{con}}\underline{y}) \sim {}_{\text{con}}\underline{y}$ for all $T > 0$. By **IF** and the above argument,

$$({}_{T-2}\mathbf{x}_{T-1}, {}_{\text{con}}\underline{y}) = (x_{T-2}, x_{T-1}, {}_{\text{con}}\underline{y}) \sim (x_{T-2}, {}_{\text{con}}\underline{y}) \sim {}_{\text{con}}\underline{y}$$

for all $T > 1$. By invoking **O** and applying **IF** and the above argument repeatedly, $({}_0\mathbf{x}_{T-1}, {}_{\text{con}}\underline{y}) \sim {}_{\text{con}}\underline{y}$ for all $T > 0$. Likewise, $({}_0\mathbf{y}_{T-1}, {}_{\text{con}}\underline{y}) \sim {}_{\text{con}}\underline{y}$ for all $T > 0$.

Let ${}_0\mathbf{z}, {}_0\mathbf{v} \in [\underline{y}, \bar{y}]^{\mathbb{Z}_+}$ be given by ${}_0\mathbf{z} = {}_0\mathbf{v} = {}_{\text{con}}\underline{y}$. Since $({}_0\mathbf{x}_{T-1}, {}_{\text{con}}\underline{y}) \sim {}_{\text{con}}\underline{y} \sim ({}_0\mathbf{y}_{T-1}, {}_{\text{con}}\underline{y})$ for all $T > 0$, we have by **O** that $({}_0\mathbf{x}_{T-1}, {}_T\mathbf{z}) \sim ({}_0\mathbf{y}_{T-1}, {}_T\mathbf{v})$ for all $T > 0$. This contradicts **DP**. □

The following result is useful for the proof of Proposition 6.

Lemma 2 *Assume that the* SWR $\succsim$ *satisfies conditions* ***O****,* ***RC****,* ***IF****,* ***M****,* ***RD****, and* ***HEF****. Then, for all* ${}_0\mathbf{x} \in \mathbf{X}$ *and* $T \in \mathbb{Z}_+$, ${}_T\mathbf{x} \precsim {}_{T+1}\mathbf{x}$.

Proof Assume that the SWR $\succsim$ satisfies conditions **O**, **RC**, **IF**, **M**, **RD**, and **HEF**. By the interpretation of ${}_T\mathbf{x}$, it is sufficient to show that ${}_0\mathbf{x} \precsim {}_1\mathbf{x}$. Suppose on the contrary that ${}_0\mathbf{x} \succ {}_1\mathbf{x}$. By Lemma 1, there exist $z^0, z^1 \in Y$ such that ${}_{\text{con}}z^0 \sim {}_0\mathbf{x}$ and ${}_{\text{con}}z^1 \sim {}_1\mathbf{x}$, where, by **O**, **M**, and ${}_0\mathbf{x} \succ {}_1\mathbf{x}$, it follows that $z^0 > z^1$. Furthermore, since ${}_1\mathbf{x} \sim {}_{\text{con}}z^1$, it follows by **IF** that $(x_0, {}_1\mathbf{x}) \sim (x_0, {}_{\text{con}}z^1)$. Hence, ${}_0\mathbf{x} \sim (x_0, {}_{\text{con}}z^1)$.

If $x_0 \le z^0$, then,

$$\begin{aligned} {}_0\mathbf{x} &\sim (x_0, {}_{\text{con}}z^1) \prec (x_0, {}_{\text{con}}z^0) && \text{by (4) and condition } \textbf{IF} \text{ since} z^1 < z^0 \\ &\precsim (z^0, {}_{\text{con}}z^0) = {}_{\text{con}}z^0 \sim {}_0\mathbf{x} && \text{by conditions } \textbf{O} \text{ and } \textbf{M} \text{ since } x_0 \le z^0. \end{aligned}$$

This contradicts condition **O**, ruling out this case. If $x_0 > z^0$, then, by selecting some $v \in (z^1, z^0)$,

$$\begin{aligned} {}_0\mathbf{x} &\sim (x_0, {}_{\text{con}}z^1) \precsim (z^0, {}_{\text{con}}v) && \text{by conditions } \textbf{O} \text{ and } \textbf{HEF} \\ & && \text{since } x_0 > z^0 > v > z^1 \\ &\prec (z^0, {}_{\text{con}}z^0) \sim {}_0\mathbf{x} && \text{by (4) and condition } \textbf{IF} \text{ since } v < z^0. \end{aligned}$$

This contradicts condition **O**, ruling out also this case. □

Proof (of Proposition 6) *Part I: (1) implies (2).* Assume that the SWR $\succsim$ satisfies conditions **O**, **RC**, **IF**, **M**, **RD**, and **HEF**. By Proposition 2, the SWR $\succsim$ is represented by a monotone SWF $W : \mathbf{X} \to \mathbb{R}$ satisfying, for some $U \in \mathcal{U}_I$ and $V \in \mathcal{V}(U)$, $W({}_0\mathbf{x}) = V(U(x_0), W({}_1\mathbf{x}))$ for all ${}_0\mathbf{x} \in \mathbf{X}$ and $W({}_{\text{con}}z) = U(z)$ for all $z \in Y$. It remains to be shown that $V(u, w) = w$ for $u > w$, implying that V satisfies (V.3′) and, thus, $V \in \mathcal{V}_S(U)$.

Since $V(u, w)$ is non-decreasing in u for given $w \in U(Y)$ and $V(u, w) = w$ for $u = w$, suppose that $V(u, w) > w$ for some u, $w \in U(Y)$ with $u > w$. Since $U \in \mathcal{U}_I$, the properties of W imply that there exist $x, z \in Y$ with $x > z$ such that

$$\begin{aligned} W(x, {}_{\text{con}}z) &= V(U(x), W({}_{\text{con}}z)) = V(U(x), U(z)) \\ &= V(u, w) > w = U(z) = W({}_{\text{con}}z) . \end{aligned}$$

Since the SWR $\succsim$ is represented by the SWF W, it follows that $(x, {}_{\text{con}}z) \succ {}_{\text{con}}z$. This contradicts Lemma 2.

Part II: (2) implies (1). Assume that the monotone mapping $W : \mathbf{X} \to \mathbb{R}$ is an SWF and satisfies, for some $U \in \mathcal{U}_I$ and $V \in \mathcal{V}_S(U)$, $W({}_0\mathbf{x}) = V(U(x_0), W({}_1\mathbf{x}))$ for all ${}_0\mathbf{x} \in \mathbf{X}$ and $W({}_{\text{con}}z) = U(z)$ for all $z \in Y$. By Proposition 2, it remains to be shown that the SWR $\succsim$, represented by the SWF W, satisfies **HEF**. We now provide this argument.

Let $x, y, z, v \in Y$ satisfy $x > y > v > z$. We have to show that $(x, {}_{\text{con}}z) \not\succ (y, {}_{\text{con}}v)$, or equivalently, $W(x, {}_{\text{con}}z) \leq W(y, {}_{\text{con}}v)$. By the properties of W,

$$\begin{aligned} W(x, {}_{\text{con}}z) &= V(U(x), W({}_{\text{con}}z)) = V(U(x), U(z)) = U(z) \\ &< U(v) = V(U(y), U(v)) = V(U(y), W({}_{\text{con}}v)) = W(y, {}_{\text{con}}v) , \end{aligned}$$

since $x > y > v > z$, $U \in \mathcal{U}_I$, and $V \in \mathcal{V}_S(U)$ □

Proof (of Proposition 7) Fix $U \in \mathcal{U}_I$ and $V \in \mathcal{V}_S(U)$. The proof has two parts.

Part I: $\lim_{T\to\infty} W({}_T\mathbf{x}) = \liminf_{t\to\infty} U(x_t)$. Assume that the monotone mapping $W : \mathbf{X} \to \mathbb{R}$ satisfies $W({}_0\mathbf{x}) = V(U(x_0), W({}_1\mathbf{x}))$ for all ${}_0\mathbf{x} \in \mathbf{X}$ and $W({}_{\text{con}}z) = U(z)$ for all $z \in Y$. Hence, by Proposition 6, the SWF W represents an SWR $\succsim$ satisfying **O**, **RC**, **M**, **RD**, **IF**, and **HEF**. By Lemma 1, for all ${}_0\mathbf{x} \in \mathbf{X}$, there exists $z \in Y$ such that ${}_{\text{con}}z \sim {}_0\mathbf{x}$. By Lemma 2, $W({}_t\mathbf{x})$ is non-decreasing in t.

Step 1: $\lim_{t\to\infty} W({}_t\mathbf{x})$ *exists.* Suppose $W({}_\tau\mathbf{x}) > \limsup_{t\to\infty} U(x_t)$ for some $\tau \in \mathbb{Z}_+$. By the premise and the fact that $U \in \mathcal{U}_I$, there exists $z \in Y$ satisfying

$$W({}_\tau\mathbf{x}) \geq U(z) > \limsup_{t\to\infty} U(x_t)$$

and $T \geq \tau$ such that $z > v := \sup_{t\geq T} x_t$. By **RD**, **O**, and **M**, ${}_{\text{con}}z \succ (v, {}_{\text{con}}z) \succsim {}_T\mathbf{x}$, and hence, by **O**, ${}_{\text{con}}z \succ {}_T\mathbf{x}$. However, since $W({}_t\mathbf{x})$ is non-decreasing in t, $W({}_T\mathbf{x}) \geq W({}_\tau\mathbf{x}) \geq U(z)$. This contradicts that W is an SWF. Hence, $W({}_t\mathbf{x})$ is bounded above by $\limsup_{t\to\infty} U(x_t)$, and the result follows since $W({}_t\mathbf{x})$ is non-decreasing in t.

Step 2: $\lim_{t\to\infty} W({}_t\mathbf{x}) \geq \liminf_{t\to\infty} U(x_t)$. Suppose

$$\lim_{t\to\infty} W({}_t\mathbf{x}) < \liminf_{t\to\infty} U(x_t)\,.$$

By the premise and the fact that $U \in \mathcal{U}_I$, there exists $z \in Y$ satisfying

$$\lim_{t\to\infty} W({}_t\mathbf{x}) \ \leq\ U(z)\ <\ \liminf_{t\to\infty} U(x_t)$$

and $T \geq 0$ such that $z < v := \inf_{t\geq T} x_t$. By **O**, **M**, and **RD**, ${}_{\text{con}}z \precsim (z, {}_{\text{con}}v) \prec {}_{\text{con}}v \precsim {}_T\mathbf{x}$, and hence, by **O**, ${}_{\text{con}}z \prec {}_T\mathbf{x}$. However, since $W({}_t\mathbf{x})$ is non-decreasing in t, $W({}_T\mathbf{x}) \leq \lim_{t\to\infty} W({}_t\mathbf{x}) \leq U(z)$. This contradicts that W is an SWF.

Step 3: $\lim_{t\to\infty} W({}_t\mathbf{x}) \leq \liminf_{t\to\infty} U(x_t)$. Suppose

$$\lim_{t\to\infty} W({}_t\mathbf{x}) > \liminf_{t\to\infty} U(x_t)\,.$$

By Lemma 1, there exists, for all $t \in \mathbb{Z}_+$, $z^t \in Y$ such that ${}_{\text{con}}z^t \sim {}_t\mathbf{x}$. Since $U \in \mathcal{U}_I$, $z \in Y$ defined by $z := \lim_{t\to\infty} z^t$ satisfies $U(z) = \lim_{t\to\infty} W({}_t\mathbf{x})$. By the premise and the fact that $U \in \mathcal{U}_I$, there exists $x \in Y$ satisfying

$$\liminf_{t\to\infty} U(x_t)\ <\ U(x)\ <\ U(z)$$

and a subsequence $(x_{t_\tau}, z^{t_\tau})_{\tau\in\mathbb{Z}_+}$ such that, for all $\tau \in \mathbb{Z}_+$, $x_{t_\tau} \leq x < z^{t_\tau}$. Then

$${}_{\text{con}}z^{t_\tau} \sim {}_{t_\tau}\mathbf{x} = (x_{t_\tau}, {}_{t_\tau+1}\mathbf{x}) \precsim (x, {}_{\text{con}}z^{t_{\tau+1}}) \precsim (x, {}_{\text{con}}z)\,,$$

since z^t is non-decreasing in t. By **O**, **RC**, and the definition of z, ${}_{\text{con}}z \precsim (x, {}_{\text{con}}z)$. Since $x < z$, this contradicts **RD**.

Part II: Existence. Let ${}_0\mathbf{x} \in \mathbf{X}$. This implies that there exist $\underline{y}, \bar{y} \in Y$ such that, for all $t \in \mathbb{Z}_+$, $\underline{y} \leq x_t \leq \bar{y}$. For each $T \in \mathbb{Z}_+$, consider $\{w(t,T)\}_{t=0}^T$ determined by (1).

Step 1: $w(t,T)$ is non-increasing in T for given $t \leq T$. Given $T \in \mathbb{Z}_+$,

$$\begin{aligned} w(T, T+1) &= V(U(x_T), w(T+1, T+1)) \\ &\leq w(T+1, T+1) = \liminf_{t\to\infty} U(x_t) = w(T,T) \end{aligned}$$

by (1) and (V.3′). Thus, applying (V.2), we have

$$\begin{aligned} w(T-1, T+1) &= V(U(x_{T-1}), w(T, T+1)) \\ &\leq V(U(x_{T-1}), w(T,T)) = w(T-1, T)\,. \end{aligned}$$

Using (V.2) repeatedly, we obtain

$$w(t, T+1) \leq w(t,T) \ \text{ for all } t \in \{0, ..., T-1\}\,,$$

which establishes that $w(t,T)$ is non-increasing in T for given $t \leq T$.

Step 2: $w(t,T)$ is bounded below by $U(\underline{y})$. By (1), (V.1), (V.2), and (V.3′), $w(T,T) = \liminf_{t\to\infty} U(x_t) \geq U(\underline{y})$, and for all $t \in \{0, ..., T-1\}$,

$$w(t+1,T) \geq U(\underline{y}) \Rightarrow w(t,T) = V(U(x_t), w(t+1,T)) \geq V(U(\underline{y}), U(\underline{y})) = U(\underline{y}).$$

Hence, it follows by induction that $w(t,T)$ is bounded below by $U(\underline{y})$.

Step 3: Definition and properties of W_σ. By steps 1 and 2, $\lim_{T\to\infty} w(t,T)$ exists for all $t \in \mathbb{Z}_+$. Define the mapping $W_\sigma : \mathbf{X} \to \mathbb{R}$ by (W). We have that W_σ is monotone by (1), (V.1), and (V.2). As $w(0,T) = V(U(x_0),\ w(1,T))$ and V satisfies (V.0), we have that $W_\sigma({}_0\mathbf{x}) = V(U(x_0), W_\sigma({}_1\mathbf{x}))$. Finally, if ${}_0\mathbf{x} = {}_{\text{con}}z$ for some $z \in Y$, then it follows from (1) and (V.3′) that $w(t,T) = U(z)$ for all $T \in \mathbb{Z}_+$ and $t \in \{0, ..., T\}$, implying that $W_\sigma({}_0\mathbf{x}) = U(z)$. □

Proof (of Proposition 8) Suppose there exists a monotone mapping $W : \mathbf{X} \to \mathbb{R}$ satisfying $W({}_0\mathbf{y}) = V(U(y_0), W({}_1\mathbf{y}))$ for all ${}_0\mathbf{y} \in \mathbf{X}$ and $W({}_{\text{con}}z) = U(z)$ for all $z \in Y$ such that $W({}_0\mathbf{x}) \neq W_\sigma({}_0\mathbf{x})$. Since V satisfies the property of weak time perspective, there is a continuous increasing transformation $g : \mathbb{R} \to \mathbb{R}$ such that $|g(W({}_0\mathbf{x})) - g(W_\sigma({}_0\mathbf{x}))| = \epsilon > 0$, and furthermore, $|g(W({}_t\mathbf{x})) - g(W_\sigma({}_t\mathbf{x}))| = |g(V(U(x_t), W({}_{t+1}\mathbf{x}))) - g(V(U(x_t), W_\sigma({}_{t+1}\mathbf{x})))| \leq |g(W({}_{t+1}\mathbf{x})) - g(W_\sigma({}_{t+1}\mathbf{x}))|$ for all $t \in \mathbb{Z}_+$. It now follows, by induction, that

$$|g(W({}_T\mathbf{x})) - g(W_\sigma({}_T\mathbf{x}))| \geq \epsilon > 0$$

for all $T \in \mathbb{Z}_+$. However this contradicts that, for all $T \in \mathbb{Z}_+$,

$$\lim_{T\to\infty} W({}_T\mathbf{x}) = \liminf_{t\to\infty} U(x_t) = \lim_{T\to\infty} W_\sigma(\mathbf{x})$$

by Proposition 7, since g is a continuous increasing transformation. □

For the proofs of the results of Sect. 4, the following notation is useful, where ${}_0\mathbf{z} = (z_0, {}_1\mathbf{z}) = (z_0, z_1, {}_2\mathbf{z}) \in \mathbf{X}$ is a fixed but arbitrary reference stream:

$x_0 \succsim_0^{\mathbf{z}} y_0$	means	$(x_0, {}_1\mathbf{z}) \succsim (y_0, {}_1\mathbf{z})$
${}_1\mathbf{x}\ {}_1\!\succsim^{\mathbf{z}}\ {}_1\mathbf{y}$	means	$(z_0, {}_1\mathbf{x}) \succsim (z_0, {}_1\mathbf{y})$
$(x_0, x_1)\ {}_0\!\succsim_1^{\mathbf{z}} (y_0, y_1)$	means	$(x_0, x_1, {}_2\mathbf{z}) \succsim (y_0, y_1, {}_2\mathbf{z})$
${}_2\mathbf{x}\ {}_2\!\succsim^{\mathbf{z}}\ {}_2\mathbf{y}$	means	$(z_0, z_1, {}_2\mathbf{x}) \succsim (z_0, z_1, {}_2\mathbf{y})$
$x_1 \succsim_1^{\mathbf{z}} y_1$	means	$(z_0, x_1, {}_2\mathbf{z}) \succsim (z_0, y_1, {}_2\mathbf{z})$.

Say that $\succsim_0^{\mathbf{z}}$ is *independent of* ${}_0\mathbf{z}$ if, for all ${}_0\mathbf{x}, {}_0\mathbf{y}, {}_0\mathbf{z}, {}_0\mathbf{v} \in \mathbf{X}$, $x_0 \succsim_0^{\mathbf{z}} y_0$ if and only if $x_0 \succsim_0^{\mathbf{v}} y_0$, and likewise for ${}_1\!\succsim^{\mathbf{z}}$, ${}_0\!\succsim_1^{\mathbf{z}}$, ${}_2\!\succsim^{\mathbf{z}}$, and $\succsim_1^{\mathbf{z}}$. In this notation and terminology, condition **IF** implies that ${}_1\!\succsim^{\mathbf{z}}$ is independent of ${}_0\mathbf{z}$, while condition **IP** states that ${}_0\!\succsim_1^{\mathbf{z}}$ is independent of ${}_0\mathbf{z}$. The following result due to Gorman (1968b) indicates that imposing condition **IP** is consequential.

Lemma 3 *Assume that the* SWR $\succsim$ *satisfies conditions **IF** and **IP**. Then* $\succsim_0^{\mathbf{z}}$, ${}_1\!\succsim^{\mathbf{z}}$, ${}_0\!\succsim_1^{\mathbf{z}}$, ${}_2\!\succsim^{\mathbf{z}}$, *and* $\succsim_1^{\mathbf{z}}$ *are independent of* ${}_0\mathbf{z}$.

Proof Assume that the SWR $\succsim$ satisfies conditions **IF** and **IP**. By repeated application of **IF**, ${}_1\succsim^{\mathbf{z}}$ and ${}_2\succsim^{\mathbf{z}}$ are independent of ${}_0\mathbf{z}$, while **IP** states that ${}_0\succsim_1^{\mathbf{z}}$ is independent of ${}_0\mathbf{z}$. By **IF**, $(x_1, {}_2\mathbf{z}) \succsim (y_1, {}_2\mathbf{z})$ is equivalent to $(z_0, x_1, {}_2\mathbf{z}) \succsim (z_0, y_1, {}_2\mathbf{z})$, which, by **IP**, is equivalent to $(z_0, x_1, {}_2\mathbf{v}) \succsim (z_0, y_1, {}_2\mathbf{v})$, which in turn, by **IF**, is equivalent to $(x_1, {}_2\mathbf{v}) \succsim (y_1, {}_2\mathbf{v})$, which finally, by **IF**, is equivalent to $(v_0, x_1, {}_2\mathbf{v}) \succsim (v_0, y_1, {}_2\mathbf{v})$, where ${}_0\mathbf{v} \in \mathbf{X}$ is some arbitrary stream. Hence, $\succsim_0^z$ and $\succsim_1^z$ are independent of ${}_0\mathbf{z}$. □

Proof (of Theorem 2) *Part I:* This part is proved in three steps.

Step 1: By Lemma 3, **IF** and **IP** imply that $\succsim_0^{\mathbf{z}}$ is independent of ${}_0\mathbf{z}$.

Step 2: By condition **WS**, there exist ${}_0\mathbf{x}, {}_0\mathbf{y}, {}_0\mathbf{z} \in \mathbf{X}$ such that $x_0 \succ_0^{\mathbf{z}} y_0$. This rules out that $x_0 = y_0$, and by **M**, $x_0 < y_0$ would lead to a contradiction. Hence, $x_0 > y_0$. Since $\succsim_0^{\mathbf{z}}$ is independent of ${}_0\mathbf{z}$, this implies **RS**.

Step 3: By Proposition 4, there is no SWR $\succsim$ satisfying **RC**, **RS**, and **HEF**.

Part II: To establish this part, consider dropping a single condition.

Dropping **IP**. Existence follows from Theorem 1 since **RD** implies **WS**.

Dropping **HEF**. Existence follows from Propositions 9 and 10.

Dropping **WS**. All the remaining conditions are satisfied by the SWF $\succsim$ being represented by the mapping $W : \mathbf{X} \to \mathbb{R}$ defined by $W({}_0\mathbf{x}) := \liminf_{t\to\infty} x_t$.

Dropping **M**. All the remaining conditions are satisfied by the SWF $\succsim$ being represented by the mapping $W : \mathbf{X} \to \mathbb{R}$ defined by $W({}_0\mathbf{x}) := -x_0 + \liminf_{t\to\infty} x_t$.

Dropping **IF**. All the remaining conditions are satisfied by the SWF $\succsim$ being represented by the mapping $W : \mathbf{X} \to \mathbb{R}$ defined by $W({}_0\mathbf{x}) := \min\{x_0, x_1\}$.

Dropping **RC**. Existence follows from Proposition 13 since **SP** implies **M** and **WS**. □

Proof (of Proposition 9) The proof is based on standard results for additively separable representations (Debreu 1960; Gorman 1968a; Koopmans 1986a), and is available at http://folk.uio.no/gasheim/srswfs2.pdf □

Proof (of Proposition 10) Available at http://folk.uio.no/gasheim/srswfs2.pdf □

Proof (of Proposition 11) Assume that the SWR $\succsim$ satisfies conditions **O**, **RC**, **IF**, **M**, **IP**, and **NDF**. By Proposition 1, **O**, **IF**, and **NDF** imply **WS**. Hence, by Propositions 9 and 10, the SWR $\succsim$ is represented by $W_\delta : \mathbf{X} \to \mathbb{R}$ defined by, for each ${}_0\mathbf{x} \in \mathbf{X}$,

$$W_\delta({}_0\mathbf{x}) = (1-\delta)\sum\nolimits_{t=0}^{\infty} \delta^t U(x_t),$$

for some $U \in \mathcal{U}$ and $\delta \in (0, 1)$. This implies **DP**, thus contradicting **NDP**. □

The proof of Proposition 12 needs some preliminaries. A sustainable recursive SWF $W : \mathbf{X} \to \mathbb{R}$ is given, with $W({}_0x) = V(U(x_0), W({}_1x))$ for all ${}_0x \in \mathbf{X}$ and $W({}_{con}z) = U(z)$ for all $z \in Y$. A utility stream ${}_0\mathbf{u}$ is associated with a feasible path ${}_0\mathbf{k}$ from $k \in [0, \bar{k}]$ if $u_t = U(x(f(k_t) - k_{t+1}, k_t))$ for $t \geq 0$. Write $u^* \equiv U(x^*) = U(x(c^*, k^*)) = U(x(f(k^*) - k^*, k^*))$.

Write $S := \{(c, k) \in [0, \bar{k}] \mid x(c, k) = x^*\}$. Since $S \neq \emptyset$, we can define:

$$I = \{k \in (0, \bar{k}] \mid \text{there is some } c \geq 0 \text{ satisfying } (c, k) \in S\}$$

Note that $k^* \in I$. Let $k \in I$; then there is $c \geq 0$ such that $x(c, k) = x^*$. Now, let $k' \in I$ satisfy $k' > k$. Then $x(c, k') \geq x(c, k) = x^*$, while $x(0, k') \leq x(0, \bar{k}) = x(f(\bar{k}) - \bar{k}, \bar{k}) \leq x^*$. Thus, by continuity of x, there is some $c' \geq 0$, such that $x(c', k') = x^*$. This shows that I is a sub-interval of $(0, \bar{k}]$, containing $[k^*, \bar{k}]$.

Define, for each $k \in I$, the set $\phi(k) = \{c \geq 0 \mid (c, k) \in S\}$. By definition of I, $\phi(k)$ is non-empty for each $k \in I$. Since $k \in I$ implies $k > 0$, $\phi(k)$ is a singleton by property (3)(ii) of the function x. Thus, ϕ is a function from I to $\mathbb{R}_+$, and by definition, $x(\phi(k), k) = x^*$ for all $k \in I$, so $c^* = \phi(k^*)$. By property (3)(ii), ϕ is non-increasing on I.

Lemma 4 *For every $k \in [k^*, \bar{k}]$, there exists a feasible resource path, ${}_0\hat{\mathbf{k}}$, from k where the associated well-being stream, ${}_0\hat{\mathbf{x}}$, satisfies $W({}_t\hat{\mathbf{x}}) = W({}_{\text{con}}x^*)$ for $t \geq 0$.*

Proof Let $k \in [k^*, \bar{k}]$, and consider the resource path ${}_0\hat{\mathbf{k}}$ defined by

$$k_0 = k\,, \quad k_{t+1} = f(k_t) - \phi(k_t) \text{ for } t > 0\,.$$

Note that, if $k_t \in [k^*, \bar{k}]$, then

$$\bar{k} \geq f(\bar{k}) \geq f(k_t) - \phi(k_t) \geq f(k_t) - \phi(k^*) = f(k_t) - [f(k^*) - k^*] \geq k^*\,.$$

Hence, $k_{t+1} \in [k^*, \bar{k}]$ and, by induction, $k_t \in [k^*, \bar{k}]$ for $t \geq 0$. This shows that ${}_0\hat{\mathbf{k}}$ is feasible from $k \in [k^*, \bar{k}]$. By the definition of $\phi(\cdot)$, $x_t = x(\phi(k_t), k_t) = x^*$ for $t \geq 0$. □

Lemma 5 *Let ${}_0\hat{\mathbf{k}}$ be a feasible resource path from $k \in [0, \bar{k}]$ with associated utility stream, ${}_0\mathbf{u}$. Given any $\varepsilon > 0$, there is some $T \geq 0$ such that $u_T < u^* + \varepsilon$.*

Proof Suppose, on the contrary, there is some $\varepsilon > 0$, such that $u_t \geq u^* + \varepsilon$ for all $t \geq 0$. By continuity of U, there is $\delta > 0$, such that whenever $x \in Y$ and $|x - x^*| < \delta$, we have $|U(x) - U(x^*)| < \varepsilon$. Thus, we must have $|x_t - x^*| \geq \delta$ for all $t \geq 0$. Further, since U is an increasing function, we must have $x_t \geq x^* + \delta$ for all $t \geq 0$. This implies:

$$x(f(k_t) - k_{t+1}, k_t) = x(c_t, k_t) > x^* \text{ for all } t \geq 0\,.$$

Since $x(f(k_t) - k_t, k_t) \leq x^*$, property (3)(ii) implies that $k_{t+1} < k_t$ for all $t \geq 0$. Thus, ${}_0\mathbf{k}$ must converge to some $\kappa \in [0, \bar{k}]$. The continuity of f and x then imply that $x(f(\kappa) - \kappa, \kappa) \geq x^* + \delta$, and this contradicts the definition of x^*. □

Lemma 6 *Let ${}_0\hat{\mathbf{k}}$ be a feasible resource path from $k \in [0, \bar{k}]$ with associated well-being stream, ${}_0\hat{\mathbf{x}}$. Then, we have $W({}_0\hat{\mathbf{x}}) \leq W({}_{\text{con}}x^*)$.*

Proof Suppose, by way of contradiction, that there exist $k \in [0, \bar{k}]$ and a feasible resource path, ${}_0\hat{\mathbf{k}}$, from k where the associated well-being stream, ${}_0\hat{\mathbf{x}}$, satisfies $W({}_0\hat{\mathbf{x}}) > W({}_{\text{con}}x^*) = U(x^*) = u^*$. Denote by ${}_0\hat{\mathbf{u}}$ the associated utility stream (i.e.,

$\hat{u}_t = U(\hat{x}_t)$ for $t \geq 0$). Since $W({}_t\hat{\mathbf{x}})$ is non-decreasing in t, and is bounded above by $U(x(\bar{k}, \bar{k}))$ (by the properties of a sustainable recursive SWF), it converges to some $\omega \leq U(x(\bar{k}, \bar{k}))$. Hence, $\omega \geq W({}_0\hat{\mathbf{x}}) > u^*$. Since the aggregator function V satisfies (V.3′), we must have $V(u^*, \omega) < V(\omega, \omega) = \omega$. Using the continuity of V, we can find $\varepsilon > 0$ such that

$$V(u^* + \varepsilon, \omega) < V(\omega, \omega) = \omega\,. \tag{5}$$

Write $\theta := \omega - V(u^* + \varepsilon, \omega)$. By (5), $\theta > 0$.

Choose $T \in Z_+$ large enough so that for all $t \geq T$, $W({}_t\hat{\mathbf{x}}) \geq \omega - (\theta/2)$. By Lemma 5, $\hat{u}_t < u^* + \varepsilon$ for some $t \geq T$. Let τ be the first period ($\geq T$) for which $\hat{u}_t < u^* + \varepsilon$. Then:

$$\begin{aligned} \omega - \tfrac{\theta}{2} \; &\leq \; W({}_\tau\hat{\mathbf{x}}) \; = \; V(u_\tau, W({}_{\tau+1}\hat{\mathbf{x}})) \\ &\leq \; V(u^* + \varepsilon, W({}_{\tau+1}\hat{\mathbf{x}})) \; \leq \; V(u^* + \varepsilon, \omega) \; = \; \omega - \theta \; < \; \omega - \tfrac{\theta}{2}\,, \end{aligned}$$

which is a contradiction. □

Lemma 7 *If a feasible resource path,* ${}_0\hat{\mathbf{k}}$, *from* $k \in [0, \bar{k}]$ *has an associated well-being stream,* ${}_0\hat{\mathbf{x}}$, *which satisfies* $W({}_0\hat{\mathbf{x}}) = W({}_{\text{con}}x^*)$, *then (i)* $\hat{x}_t \geq x^*$ *for all* $t \geq 0$*; and (ii)* $\hat{k}_t \geq k^*$ *for all* $t \geq 0$.

Proof Assume that a feasible resource path, ${}_0\hat{\mathbf{k}}$, from $k \in [0, \bar{k}]$ has an associated well-being stream, ${}_0\hat{\mathbf{x}}$, which satisfies $W({}_0\hat{\mathbf{x}}) = W({}_{\text{con}}x^*)$. Since $W({}_t\hat{\mathbf{x}})$ is non-decreasing in t, it follows from Lemma 6 that $W({}_t\hat{\mathbf{x}}) = W({}_{\text{con}}x^*) = u^*$ for all $t \geq 0$.

To establish (i), suppose, by way of contradiction, that $\hat{x}_t < x^*$ for some $t \geq 0$. Then, since $U(\hat{x}_t) < U(x^*) = u^*$, (V.3′) implies:

$$u^* \; = \; W({}_t\hat{\mathbf{x}}) \; = \; V(U(\hat{x}_t), u^*) \; < \; V(u^*, u^*) \; = \; u^*,$$

which is a contradiction.

To establish (ii), suppose, on the contrary, that $\hat{k}_\tau < k^*$ for some $\tau \geq 0$. Then, by the fact that $x(f(k) - k, k) < x^*$ if $k \neq k^*$, we have $x(f(\hat{k}_\tau) - \hat{k}_\tau, \hat{k}_\tau) < x^*$, while:

$$x^* \leq \hat{x}_\tau = x(\hat{c}_\tau, \hat{k}_\tau) = x(f(\hat{k}_\tau) - \hat{k}_{\tau+1}, \hat{k}_\tau)\,.$$

So, $\hat{k}_{\tau+1} < \hat{k}_\tau < k^*$, and repeating this step, $\hat{k}_{t+1} < \hat{k}_t$ for all $t \geq \tau$. Thus, ${}_0\hat{k}$ must converge to some $\kappa \in [0, \bar{k}]$, with $\kappa \leq \hat{k}_\tau < k^*$. The continuity of f and x then imply that:

$$x(f(\kappa) - \kappa, \kappa) \geq x^*$$

using (i). But, this contradicts that $x(f(k) - k, k) < x^*$ if $k \neq k^*$. □

Proof (of Proposition 12) Lemmas 4 and 6 establish existence of an optimum and that any optimal well-being stream satisfies $W({}_t\hat{\mathbf{x}}) = W({}_{\text{con}}x^*)$ for $t \geq 0$. Lemma 7 shows that any optimal resource path ${}_0\hat{\mathbf{k}}$, with associated well-being stream ${}_0\hat{\mathbf{x}}$, satisfies $\hat{x}_t \geq x^*$ and $\hat{k}_t \geq k^*$ for $t \geq 0$. □

Proof (of Proposition 13) Asheim and Tungodden (2004a), Basu and Mitra (2007), and Bossert et al. (2007) define different incomplete leximin and undiscounted utilitarian SWRs, each of which is given an axiomatic characterization. Denote by $\succsim$ one such incomplete SWR. It can be verified that $\succsim$ is reflexive, transitive and satisfies **IF**, **SP**, **HEF** (with $(x, {}_{\text{con}}z) \precsim (y, {}_{\text{con}}v)$ if $x > y > v > z$), and **IP**. Completeness (and thereby condition **O**) can be satisfied by invoking Arrow's (1951) version of Szpilrajn's (1930) extension theorem (see also Svensson 1980).

Since $\succsim$ satisfies conditions **SP** and **HEF** (with $(x, {}_{\text{con}}z) \precsim (y, {}_{\text{con}}v)$ if $x > y > v > z$), so will any completion. Since, for all ${}_0\mathbf{x}, {}_0\mathbf{y}, {}_0\mathbf{z} \in \mathbf{X}$, $(x_0, x_1)\ {}_0\succsim_1^{\mathbf{z}}\ (y_0, y_1)$ or $(x_0, x_1)\ {}_0\precsim_1^{\mathbf{z}}\ (y_0, y_1)$, and $\succsim$ satisfies **IP**, so will any completion. However, special care must be taken to ensure that the completion satisfies **IF**.

Consider $\mathbf{X}_0^2 = \{({}_0\mathbf{x}, {}_0\mathbf{y}) \in \mathbf{X}^2 \mid x_0 \neq y_0\}$, and invoke Arrow's (1951) version of Szpilrajn's (1930) extension theorem to complete $\succsim$ on this subset of $\mathbf{X}^2$. For any $({}_0\mathbf{x}, {}_0\mathbf{y}) \in \mathbf{X}$ with ${}_0\mathbf{x} \neq {}_0\mathbf{y}$, let ${}_0\mathbf{x}$ be at least as good as ${}_0\mathbf{y}$ if and only if ${}_T\mathbf{x}$ is at least as good as ${}_T\mathbf{y}$ according to the completion of $\succsim$ on $\mathbf{X}_0^2$, where $T := \min\{t \mid x_t \neq y_t\}$. Since $\succsim$ satisfies **IF**, this construction constitutes a complete SWR satisfying **IF**. □

References

Alcantud, J. C. R., & García-Sanz, M. D. (2010). Paretian evaluation of infinite utility streams: An egalitarian criterion. *Economics Letters*, *106*, 209–211.

Arrow, K. J. (1951). *Social choice and individual values*. New York: Wiley.

Asheim, G. B. (1991). Unjust intergenerational allocations. *Journal of Economic Theory*, *54*, 350–371.

Asheim, G. B., & Mitra, T. (2010). Sustainability and discounted utilitarianism in models of economic growth. *Mathematical Social Sciences*, *59*, 148–169.

Asheim, G. B., & Tungodden, B. (2004a). Resolving distributional conflicts between generations. *Economic Theory*, *24*, 221–230.

Asheim, G.B., & Tungodden, B. (2004b). Do Koopmans' postulates lead to discounted utilitarianism? Discussion Paper 32/04, Norwegian School of Economics and Business Administration.

Asheim, G. B., Mitra, T., & Tungodden, B. (2007). A new equity condition for infinite utility streams and the possibility of being Paretian. In J. Roemer & K. Suzumura (Eds.), *Intergenerational equity and sustainability* (pp. 55–68). Basingstoke: Palgrave Macmillan.

Atkinson, A. B. (2001). The strange disappearance of welfare economics. *Kyklos*, *54*, 193–206.

Banerjee, K. (2006). On the equity-efficiency trade off in aggregating infinite utility streams. *Economic Letters*, *93*, 63–67.

Basu, K., & Mitra, T. (2003). Aggregating infinite utility streams with intergenerational equity: The impossibility of being Paretian. *Econometrica*, *32*, 1557–1563.

Basu, K., & Mitra, T. (2007). Utilitarianism for infinite utility streams: A new welfare criterion and its axiomatic characterization. *Journal of Economic Theory*, *133*, 350–373.

Birchenhall, C. R., & Grout, P. (1979). On equal plans with an infinite horizon. *Journal of Economic Theory*, *21*, 249–264.

Blackorby, C., Donaldson, D., & Weymark, J. A. (1984). Social choice with interpersonal utility comparisons: A diagrammatic introduction. *International Economic Review*, *25*, 327–356.

Bleichrodt, H., Rohde, K. I. M., & Wakker, P. P. (2008). Koopmans' constant discounting for intertemporal choice: A simplification and a generalization. *Journal of Mathematical Psychology*, *52*, 341–347.

Bossert, W., Sprumont, Y., & Suzumura, K. (2007). Ordering infinite utility streams. *Journal of Economic Theory, 135*, 579–589.

Burniaux, J.-M., & Martins, J. O. (2016). Carbon leakages: A general equilibrium view. *Economic Theory* (reprinted in this volume).

Chichilnisky, G. (1996). An axiomatic approach to sustainable development. *Social Choice and Welfare, 13*, 231–257.

Chichilnisky, G. (2016). Sustainable markets with short sales. *Economic Theory* (reprinted in this volume).

Chichilnisky, G., Heal, G. M., & Beltratti, A. (1995). The green golden rule. *Economic Letters, 49*, 175–179.

Chipman, J. S., Tian, G. (2016). Detrimental externalities, pollution rights, and the "Coase Theorem". *Economic Theory* (reprinted in this volume).

Dasgupta, P. S., & Heal, G. M. (1974). The optimal depletion of exhaustible resources. In *Review of Economic Studies*(Symposium) (pp. 3–28).

Dasgupta, P. S., & Heal, G. M. (1979). *Economic theory and exhaustible resources*. Cambridge: Cambridge University Press.

Debreu, G. (1960). Topological methods in cardinal utility theory. In K. J. Arrow, S. Karlin, & P. Suppes (Eds.), *Mathematical methods in the social sciences* (pp. 16–26). Stanford: Stanford University Press.

Diamond, P. (1965). The evaluation of infinite utility streams. *Econometrica, 33*, 170–177.

Dutta, P. K., & Radner, R. (2016). Capital growth in a global warming model: Will China and India sign a climate treaty? *Economic Theory* (reprinted in this volume).

Figuieres, C., & Tidball, M. (2016). Sustainable exploitation of a natural resource: A satisfying use of Chichilnisky's criterion. *Economic Theory* (reprinted in this volume).

Fleurbaey, M., & Michel, P. (2001). Transfer principles and inequality aversion, with application to optimal growth. *Mathematical Social Sciences, 42*, 1–11.

Fleurbaey, M., & Michel, P. (2003). Intertemporal equity and the extension of the Ramsey criterion. *Journal of Mathematical Economics, 39*, 777–802.

Gorman, W. M. (1968a). Conditions for additive separability. *Econometrica, 36*, 605–609.

Gorman, W. M. (1968b). The structure of utility functions. *Review of Economic Studies, 35*, 367–390.

Hammond, P. J. (1976). Equity, Arrow's conditions and Rawls' difference principle. *Econometrica, 44*, 793–804.

Hara, C., Shinotsuka, T., Suzumura, K., & Xu, Y. (2008). Continuity and egalitarianism in the evaluation of infinite utility streams. *Social Choice and Welfare, 31*, 179–191.

Heal, G. M. (1998). *Valuing the future: Economic theory and sustainability*. New York: Columbia University Press.

Heal, G. M. (2005). Intergenerational welfare economics and the environment. In K.-G. Mäler & J. Vincent (Eds.), *Handbook of environmental economics*. Amsterdam: North-Holland.

Karp, C., & Zhang, J. (2016) Taxes versus quantities for a stock pollutant with endogenous abatement costs and asymmetric information. *Economic Theory* (reprinted in this volume).

Koopmans, T. C. (1960). Stationary ordinal utility and impatience. *Econometrica, 28*, 287–309.

Koopmans, T. C. (1967). Intertemporal distribution and optimal aggregate economic growth. In *Ten economic studies in the tradition of irving fisher*. New York: Wiley.

Koopmans, T. C. (1986a). Representation of preference orderings with independent components of consumption. In C.B. McGuire, & R. Radner, (Eds.), *Decision and organization: A volume in honor of Jacob Marschak* (2nd ed., pp. 57–78). Previously published in 1972. Minneapolis: University of Minnesota Press.

Koopmans, T. C. (1986b). Representation of preference orderings over time. In C.B. McGuire, & R. Radner (Eds.), *Decision and organization: A volume in honor of Jacob Marschak* (2nd ed., pp. 79–100). Previously published in 1972. Minneapolis: University of Minnesota Press.

Koopmans, T. C., Diamond, P. E., & Williamson, R. E. (1964). Stationary utility and time perspective. *Econometrica, 32*, 82–100.

Krautkraemer, J. A. (1985). Optimal growth, resource amenities and the preservation of natural environments. *Review of Economic Studies*, *52*, 153–170.
Lauwers, L. (1997). Rawlsian equity and generalised utilitarianism with an infinite population. *Economic Theory*, *9*, 143–150.
Lauwers, L. (1998). Intertemporal objective functions: Strong Pareto versus anonymity. *Mathematical Social Sciences*, *35*, 37–55.
Lauwers, L. (2016). Intergenerational equity, efficiency, and constructibility. *Economic Theory* (reprinted in this volume).
Lecocq, F., & Hourcade, J.-C. (2016). Insights from a theoretical analysis of negotiation mandates: Unspoken ethical issues in the climate affair. *Economic Theory* (reprinted in this volume).
Mitra, T. (2008). *Stationary paretian SWF: An example*. Mimeo: Cornell University.
Ostrom, E. (2016). Nested externalities and polycentric institutions: must we wait for global solutions to climate change before taking actions at other scales? *Economic Theory* (reprinted in this volume).
Rawls, J. (1971). *A theory of justice*. Cambridge: Harvard University Press.
Rezai, A., Foley, D. K., & Taylor, L. (2016). Global warming and economic externalities. *Economic Theory* (reprinted in this volume).
Solow, R. M. (1974). Intergenerational equity and exhaustible resources. In *Review of Economic Studies* (Symposium) (pp. 29–45).
Svensson, L.-G. (1980). Equity among generations. *Econometrica*, *48*, 1251–1256.
Szpilrajn, E. (1930). Sur l'extension du l'ordre partiel. *Fundamenta Mathematicae*, *16*, 386–389.

Intergenerational Equity, Efficiency, and Constructibility

Luc Lauwers

1 Introduction

Global environmental issues are long term issues. Traditional discounting is unable to take the long run into account. In contrast, overtaking and Chichilnisky criteria do select long run strategies. The next example recalls these observations. Then, we further discuss (i) the incompleteness of the overtaking criterion, and (ii) one of the Chichilnisky axioms (non-dictatorship of the present). In particular, we indicate a route to decrease the incompleteness of overtaking, and we show that non-dictatorship of the present involves non-constructible mathematics. We provide a balanced interpretation of these results.

Example An economy uses trees as a necessary input to production or consumption. The dynamics of tree reproduction are as follows. If n out of $2n$ subsequent generations cut the forest at a maximal rate, the species become extinct after the

I thank Norbert Brunner, Graciela Chichilnisky, Koen Decancq, Marc Fleurbaey, Johan Quaegebeur, and Luc Van Liedekerke for helpful conversations. I am grateful for the many constructive comments by the referee. A first version was presented at the workshop "Intergenerational equity in climate negotiations, overlapping generations and social welfare" organized by Claude d'Aspremont and Thierry Bréchet (CORE, April 27–28, 2006). This text presents research results of the Belgian Programme of Interuniversity Poles of Attraction initiated under the Science Policy Programmes of the Prime Minister's Office, Belgium. The scientific responsibility is assumed by its authors.

Originally published in Economic Theory, Volume 49, Number 2, February 2012 DOI 10.1007/s00199-010-0573-7.

L. Lauwers (✉)
Center for Economic Studies, University of Leuven—KU Leuven,
Naamsestraat 69—Bus 3565, 3000 Leuven, Belgium
e-mail: Luc.Lauwers@econ.kuleuven.be; luc.lauwers@kuleuven.be

L. Lauwers
Departments of Economics, University of Leuven—KU Leuven,
Naamsestraat 69—Bus 3565, 3000 Leuven, Belgium

G. Chichilnisky and A. Rezai (eds.), *The Economics of the Global Environment*,
Studies in Economic Theory 29, DOI 10.1007/978-3-319-31943-8_10

$2n$th generation, in which case there is zero utility at every period from then on. This strategy results in utility streams of the form $u^n = (0, \ldots, 0; 1, \ldots, 1; 0, 0, \ldots)$ with the first (resp. last) 1 at the $n + 1$ (resp. $2n$)'th place, in which generations $n + 1, \ldots, 2n$ cut at a full capacity and exhaust the forest. When the consumption of the forest is delayed and n becomes larger, the forest slightly expands and more generations can benefit. Alternatively, generations can invest in the forest and only cut at an equilibrium rate which allows the forest to survive. This strategy results in the utility stream $u^\infty = (0.20, 0.20, \ldots, 0.20, \ldots)$ in which each generation reaches the same utility level.

We evaluate the different policies by means of the normalized[1] discounting rule

$$u = (u_1, u_2, \ldots, u_t, \ldots) \longmapsto D_\beta(u) = (1 - \beta)(u_1 + \beta\, u_2 + \cdots + \beta^{t-1} u_t + \cdots).$$

We obtain $D_\beta(u^n) = \beta^n - \beta^{2n}$ and $D_\beta(u^\infty) = 0.20$. For each β in the open interval (0, 1), there exists an n^* such that $\beta^{n^*} - \beta^{2n^*} = 0.25$. Optimization with respect to a discounting rule leads to the elimination of this forest.[2] If we judge the long term future important, then we should use the right tools to evaluate a long run policy.

The literature on intergenerational equity provides such tools.[3] Let me focus on two criteria. *First*, the overtaking criterion considers an infinite stream u better than v if for some T in $\mathbb{N}_0$, the undiscounted sum $u_1 + u_2 + \cdots + u_t$ is larger than $v_1 + v_2 + \cdots + v_t$ as soon $t \geq T$. According to this criterium, the sustainable stream u^∞ is better than u^n for each n. *Second*, Chichilnisky (1996) proposes a convex sum "$C_{\lambda,\beta} = \lambda D_\beta + (1 - \lambda) Lim$" of the discounting rule and a value that captures the limiting behavior of the utility stream. Since $\lim_{t\to\infty} u^n = 0$ and $\lim_{t\to\infty} u^\infty = 0.2$, we obtain

$$C_{\lambda,\beta}(u^\infty) = 0.2\,\lambda + 0.2(1 - \lambda) = 0.20, \quad \text{and} \quad C_{\lambda,\beta}(u^{n^*}) = 0.25\,\lambda.$$

As soon as the weight λ of the discounting rule is less then 0.80, the Chichilnisky criterion $C_{\lambda,\beta}$ ranks the sustainable stream u^∞ at the top.

Both criteria have their merits and shortcomings. *First*, the overtaking criterion combines equity or finite anonymity and Pareto but fails completeness. This is an inevitable consequence of the Lauwers (2010a), Zame (2007) impossibility result: the existence of a complete, equitable, and Paretian criterion relies on non-constructive mathematics (such as the Axiom of Choice). Recent contributions in this track of literature concentrate on constructible, equitable, and Paretian criteria. For example, the imposition of an anonymity demand stronger than finite

[1] An evaluation F of infinite utility streams is said to be normalized if $F(r, r, \ldots, r, \ldots) = r$ for each r in $\mathbb{R}$. Due to this normalization the discounted sum is premultiplied with $(1 - \beta)$.

[2] This conclusion extends to, for example, the widely used Dasgupta-Heal-Solow growth model.

[3] Asheim (2010) provides an excellent survey.

anonymity decreases incompleteness[4]; e.g. Lauwers (1997), Fleurbaey and Michel (2003), Mitra and Basu (2007), Banerjee (2006), Asheim et al. (2016), Asheim and Banerjee (2010), and Kamaga and Kojima (2009a, b). In this note, I explore the boundaries of combining different equity principles and different Pareto principles without imposing completeness. I show that a maximal equity principle, compatible with Pareto, is a non-constructible object.

Second, Chichilnisky (1996, 2009b) translates the requirement of equal treatment for the present and the future into two new axioms for sustainable development. These two axioms (non dictatorship of the present and non dictatorship of the future)[5] in combination of independence, characterize the class $C_{\lambda,\beta}$ of sustainable social welfare functions. Unfortunately, in many economic models of growth there does not exist a utility stream that is optimal under this criterion. Recent contributions in this track of literature concentrate on this issue of applicability; e.g. Heal (1998), Li and Löfgren (2000), and Figuières and Tidball (2011). Furthermore, Alvarez-Cuadrado and Van Long (2009) defend the axioms of non dictatorship, propose a Bentham-Rawls criterion, and show the existence of optimal paths. Similarly, Asheim et al. (2011) introduce the concept of a sustainable recursive social welfare function, of which Asheim and Mitra's (2010) sustainable discounted utilitarianism is special case. Their criteria also satisfy the two Chichilnisky axioms and are applicable.[6] Besides this problem of applicability, there is a problem of non-constructibility. The "distinct future"-part in the Chichilnisky criterion is an integral against a purely finitely additive measure. Such a measure is a non-constructible object.[7]

The results that 'maximal anonymity' and 'finitely additive measures' involve non-constructive mathematics should be interpreted with care. The use of non-constructive mathematics within economic theory is well known. For example, Debreu's proof of the second welfare theorem (each Pareto allocation can be realized as the market equilibrium of some economy) uses—similar to the proof of existence of Chichilnisky criteria—non-constructive mathematics (see Chichilnisky 2009a, 2011). In the practical context of optimal growth, however, an explicit description

[4]Consider the streams $w = (1, -1, -1, 1; 1, -1, -1, 1; \dots; 1, -1, -1, 1; \dots)$ and $z = (0, 0, \dots, 0, \dots)$. The overtaking criterion is unable to rank w and z. An utilitarian overtaking criterion that satisfies fixed step anonymity considers w and z equally good.

[5]A welfare function displays 'dictatorship of the present' if it is insensitive for changes that affect the distinct future. A welfare function displays 'dictatorship of the future' if it is insensitive for changes that do not affect the limiting behavior criterion.

[6]Burniaux and Martins (2011), Chipman and Tian (2011), Dutta and Radner (2011), Karp and Zhang (2011), and Ostrom (2011) tackle the question of how to implement policies that respect the interests of future generations and to assess their effectiveness in the context of global externalities with long-lasting effects. Lecocq and Hourcade (2011) argue that optimal policies may require estimates of future intragenerational distributions. Rezai et al. (2011) show that, in some cases, such policies may benefit all generations.

[7]Purely finitely additive measures are typically obtained via non-constructive mathematics (Hahn-Banach's theorem or ultrafilters, cf. Chichilnisky 2009a, 2011). This observation can be strengthened: it is impossible to create a purely finitely additive measure on $\mathbb{N}_0$ without recurse to non-constructive methods.

of the ordering describing social preferences would generally be needed in order to compute the optimal path (Fleurbaey and Michel 2003, p. 794).

In view of this, I propose a positive interpretation. *First*, the result on maximal anonymity should be seen as an additional defense of the 'fixed step anonymity' axiom. Fixed step anonymity is stronger than finite anonymity, decreases incomparability, and is based upon the constructible group of fixed step permutations. *Second*, the Chichilnisky criterion can be made constructible by restricting the domain to, for example, those infinite utility streams which exhibit a well defined and finite limiting behavior (Chichilnisky 2009a). In this restricted domain, the limiting value is well defined (without recurse to non-constructive mathematics) and captures the long run value. Also, one can weaken the independence axiom and allow for alternatives to capture the very long run behavior. For example, the maps lim inf and lim sup do not involve non-constructive mathematics, violate independence,[8] and fit in the Chichilnisky approach.

The next section collects preliminaries. Section 2.1 recalls the notions of a social welfare relation and the basic notations. Section 2.2 characterizes Pareto-compatible anonymity demands (Mitra and Basu 2007). Surprisingly, an anonymity demand is compatible with strong Pareto if and only if it is compatible with weak Pareto. Section 2.3 recalls the Axiom of Choice and the notion of ultrafilter. Section 3 develops the main result: a maximal anonymity condition involves an ultrafilter on the lattice of partitions. As a consequence, the group of fixed step permutations is not maximal. Section 4 concentrates on the non-constructibility of purely finitely additive measures.

2 Preliminaries

2.1 Social Welfare Relations

Let $\mathbb{N}_0 = \{1, 2, 3, \ldots\}$ denote the set of positive integers and $\mathbb{R}$ the set of real numbers. Let the interval $Y \subseteq \mathbb{R}$ be the set of all possible utility levels. We assume that Y contains 0 and 1. The set $X = Y^{\mathbb{N}_0}$ collects all possible utility streams and is called the domain. An infinite utility stream x is a vector in X. Each x in X can be viewed as a map from $\mathbb{N}_0$ to Y, associating with each t in $\mathbb{N}_0$ the element x_t in Y. Vector inequalities are denoted $\leq$, $<$, and $\ll$. For each x in X, $\liminf(x)$ is the infimum (and $\limsup(x)$ is the supremum) of the set of accumulation points of x.

A social welfare relation (SWR) is a reflexive and transitive binary relation in the domain X. The symmetric and the asymmetric component of the SWR $\precsim$ are denoted by $\sim$ and $\prec$. The SWR $\precsim$ is complete if for each x and y in X we have that either $x \precsim y$ or $y \precsim x$. The SWR $\precsim_1$ is a subrelation to a SWR $\precsim_2$ if for each x and y in X we have (*i*) $x \precsim_1 y$ implies $x \precsim_2 y$ and (*ii*) $x \prec_1 y$ implies $x \prec_2 y$.

[8] The map lim inf (see Sect. 2.1) is not additive: $\liminf(1, 0, 1, 0, \ldots) = \liminf(0, 1, 0, 1, \ldots) = 0$; while $\liminf[(1, 0, 1, 0, \ldots) + (0, 1, 0, 1, \ldots)] = \liminf(1, 1, 1, 1, \ldots) = 1$.

A permutation π on $\mathbb{N}_0$ is a one-to-one map from $\mathbb{N}_0$ to $\mathbb{N}_0$. For each x in X, the composite map $x \circ \pi$ is a map from $\mathbb{N}_0$ to Y and can be written as the infinite utility stream

$$x \circ \pi = \left(x_{\pi(1)}, x_{\pi(2)}, \ldots, x_{\pi(t)}, \ldots\right).$$

Let $\mathrm{Sym}(\mathbb{N}_0)$ collect all permutations on $\mathbb{N}_0$. The set $\mathrm{Sym}(\mathbb{N}_0)$ when equipped with the composition operation becomes a group. The next definition collects a monotonicity demand, two infinite versions of the Pareto axiom, and one concept related to permutations.

Definition • A SWR $\precsim$ is *monotonic* if for each x and y in X we have that $x \leq y$ implies $x \precsim y$.

- A SWR $\precsim$ satisfies the *Pareto* axiom if $\precsim$ is monotonic and for each x and y in X we have that $x < y$ implies $x \prec y$.
- A SWR $\precsim$ satisfies the *weak Pareto* axiom if $\precsim$ is monotonic and for each x and y in X, we have that $x \ll y$ implies $x \prec y$.
- Let Q be a group of permutations. A SWR $\precsim$ satisfies *Q-anonymity* if for each π in Q and for each x in X we have $x \sim x \circ \pi$.

The Pareto axiom, also known as the strong Pareto axiom, postulates sensitivity in each coordinate. The SWRs represented by the maps lim inf and lim sup combine completeness, $\mathrm{Sym}(\mathbb{N}_0)$-anonymity, and monotonicity, and violate weak Pareto. Indeed, the infinite sequences $z = (0, 0, \ldots, 0, \ldots)$ and $y = (1, 1/2, \ldots, 1/k, \ldots)$ have one single accumulation point ($\lim z = \lim y = 0$), have the same lim inf- and lim sup-values, and satisfy $z \ll y$. Chambers (2009) characterizes both SWRs.

With respect to anonymity, we only consider groups of permutations that include the group of finite permutations. Hereby, the permutation π is said to be finite if there exists a T in $\mathbb{N}_0$ such that $\pi(t) = t$ for each $t \geq T$. Let Q_{fn} collect all finite permutations. A SWR is said to be finite anonymous if it satisfies Q_{fn}-anonymity. The overtaking criterion satisfies finite anonymity, the Chichilnisky criterion violates finite anonymity. Furthermore, a permutation π is said to be fixed step if there exists a natural number n, such that $\pi(\{1, 2, \ldots, kn\}) = \{1, 2, \ldots, kn\}$ for each k in $\mathbb{N}_0$. Let Q_{fs} collect all fixed step permutations. Observe the inclusion $Q_{\mathrm{fn}} \subset Q_{\mathrm{fs}}$. Finally, for each group Q of permutations we define the SWR $\precsim_Q$ as follows: for each x and y in X, we have

$$x \precsim_Q y \qquad \text{if and only if} \qquad \text{there is a } \pi \text{ in } Q \text{ such that } x \circ \pi \leq y.$$

This relation is Q-anonymous, reflexive (the identity permutation belongs to the group Q), and transitive (the group Q is closed under composition).

2.2 Pareto-Compatible Permutations

Anonymity axioms are based upon groups of permutations. In this subsection we characterize permutations that are compatible with Pareto.

Let π be a permutation on the set $\mathbb{N}_0$. The vector $(k, \pi(k), \pi^2(k), \pi^3(k), \dots)$ is said to be the cycle generated by π on k in $\mathbb{N}_0$. Each permutation can be written as a succession of cycles on disjoint sets (Hall 1976, Chap. 5). For example, the permutation

$$\pi_1 = (1, 2)(3, 4)(5, 6) \cdots (2n - 1, 2n) \cdots$$

switches the odd and even numbers: for each n in $\mathbb{N}_0$ the number $2n - 1$ is mapped upon $2n$ and $2n$ is mapped upon $2n - 1$. The final element in a cycle is mapped upon the first element in that cycle. The permutation

$$\pi_2 = (1)(2, 3)(4, 5) \cdots (2n, 2n + 1) \cdots$$

keeps the number 1 fixed and then switches the even and odd numbers. A permutation on $\mathbb{N}_0$ might generate a cycle of infinite length. The permutation

$$\pi_3 = (\dots, 9, 7, 5, 3, 1, 2, 4, 6, 8, \dots)$$

maps 1 upon 2. Furthermore, π_3 maps an even number upon its even successor and an odd number upon its odd predecessor, as such $\pi_3(123) = 121$ and $\pi_3(100) = 102$. We keep the references π_1, π_2, and π_3 throughout this note. The decomposition of a permutation into pairwise disjoint cycles is unique, except for the order in which the cycles are written, also within each cycle the numbers are allowed to be permuted cyclically. As such, the permutations $(1, 2)(3)(4, 5, 6, 7)$ and $(3)(1, 2)(5, 6, 7, 4)$ coincide.

A permutation representable by an infinite sequence of finite cycles is said to be *cyclic*. The permutations π_1 and π_2 are cyclic, π_3 is not cyclic. Finite permutations and fixed step permutations are cyclic. The set of all cyclic permutations is denoted by $\mathcal{P}$. The set $\mathcal{P}$ is not a group: the composition $\pi_1 \circ \pi_2$ of two cyclic permutations results in the permutation π_3. The next lemma highlights the main motivation to study cyclic permutations. The lemma already appeared in Mitra and Basu (2007, Lemma 1). Their proof uses coordinatewise convergent sequences of infinite utility streams. The proof below only uses 0–1-utility streams and therefore strengthens their result.

Lemma 1 *A permutation π is cyclic if and only if there is no x in X satisfying $x < x \circ \pi$.*

Proof The only-if-part is straightforward. If the permutation π is cyclic, then it can be decomposed as an infinite juxta position of permutations on finite sets. Each permutation on a finite set is unable to conflict with the Pareto principle.
The if-part (if there is no conflict with Pareto, then the permutation is cyclic) is done by contraposition. Hence, consider a permutation π with an infinite cycle at m in $\mathbb{N}_0$:

$$(\dots, \pi^{-4}(m), \pi^{-3}(m), \pi^{-2}(m), \pi^{-1}(m), m, \pi^1(m), \pi^2(m), \pi^3(m), \pi^4(m), \dots).$$

Relabel this cycle (let 1 denote m) to obtain the cycle π_3 and consider the following table:

$$\begin{aligned} \pi_3 &= (\ldots, 9, 7, 5, 3, 1, 2, 4, 6, 8, \ldots), \\ x &= (\ldots, 0, 0, 0, 0, 0, 1, 1, 1, 1, \ldots), \\ y = x \circ \pi_3 &= (\ldots, 0, 0, 0, 0, 1, 1, 1, 1, 1, \ldots). \end{aligned}$$

The first line in this table is a cycle of infinite length. The second line presents an infinitely long utility stream in X. This utility stream is made up of two sequences, a sequence of 'ones' is attached to the even positions ($x_{2n} = 1$) and a sequence of zeros is attached to the odd positions ($x_{2n-1} = 0$). The final line presents the permuted utility stream $y = x \circ \pi_3$ (recall that $y_i = x_{\pi(i)}$). The utility stream y dominates x (indeed, $x_1 < y_1$). □

The infinite cycle π_3 generates a second domination result. There exists an x in X such that $x \ll (x \circ \pi_3)$:

$$\begin{aligned} \pi_3 &= (\ldots, 9, 7, 5, 3, \quad 1, \quad 2, \quad 4, \quad 6, \quad 8, \quad \ldots), \\ x &= (\ldots, \tfrac{1}{9}, \tfrac{1}{7}, \tfrac{1}{5}, \tfrac{1}{3}, \quad 1, \; 2-\tfrac{1}{2}, 2-\tfrac{1}{4}, 2-\tfrac{1}{6}, \; 2-\tfrac{1}{8}, \; \ldots), \\ x \circ \pi_3 &= (\ldots, \tfrac{1}{7}, \tfrac{1}{5}, \tfrac{1}{3}, 1, 2-\tfrac{1}{2}, 2-\tfrac{1}{4}, 2-\tfrac{1}{6}, 2-\tfrac{1}{8}, 2-\tfrac{1}{10}, \ldots). \end{aligned}$$

Lemma 1, thus, holds when Pareto is weakened to weak Pareto. We summarize. Let Q be a group of permutations. Then,

Q – anonymity and Pareto are compatible	$\Longleftrightarrow$	Q – anonymity and weak Pareto are compatible	$\Longleftrightarrow$	the group Q only contains cyclic permutations.

Within the class of transitive and reflexive relations, there is no trade-off between the Pareto axioms and anonymity. Finally, if Q is a group of cyclic permutations, then (*i*) the relation $\precsim_Q$ extends the Suppes-Sen grading principle, (*ii*) is the smallest (for inclusion) SWR that satisfies Q-anonymity and Pareto (Banerjee 2006), and (*iii*) satisfies

- $x \sim_Q y$ if and only if there exists a π in Q such that $x \circ \pi = y$, and
- $x \prec_Q y$ if and only if there exists a π in Q such that $x \circ \pi < y$.

We only verify the first item. Suppose both $x \precsim_Q y$ and $y \precsim_Q x$ hold. Then there exist two permutations π and σ in Q such that $x \circ \pi \leq y$ and $y \circ \sigma \leq x$. Therefore,

$$x \circ \pi \circ \sigma \leq y \circ \sigma \leq x.$$

Since Q is a group, the permutation $\pi \circ \sigma$ is cyclic. The inequalities become equalities, hence $y \circ \sigma = x$, and $x \sim_Q y$.

2.3 The Axiom of Choice, Ultrafilters

The Axiom of Choice (AC) postulates for each nonempty family $\mathcal{D}$ of nonempty sets the existence of a function f such that $f(S) \in S$ for each set S in the family $\mathcal{D}$. The function f is referred to as a choice function. AC does not provide an explicit way to construct such a choice function and provoked considerable criticism in the aftermath of Zermelo's formulation in 1904.[9] AC implies a number of paradoxes such as the decomposition of a sphere into a sphere of smaller size and the existence of a nonmeasurable set of real numbers. The nonconstructive character of AC is further revealed by Dianonescu (1975) who showed that AC implies the law of the excluded middle.[10] Constructive mathematics rejects the law of the excluded middle and hence rejects AC.

This note appeals to the non-constructible object "free ultrafilter". We will use free ultrafilters (i) in the definition of maximal groups of cyclic permutations (Sect. 3), and (ii) in the definition of the "distinct future"-part of the Chichilnisky criterion (Sect. 4). The definition of a filter and an ultrafilter is as follows.

Let S be a set. A filter on S is a nonempty family $\mathcal{F}$ of subsets of S that satisfies

- $\varnothing$ is not in $\mathcal{F}$,
- if A and B are in $\mathcal{F}$, then $A \cap B$ is in $\mathcal{F}$ (intersection property),
- if A is in $\mathcal{F}$ and $A \subseteq B$, then B is in $\mathcal{F}$.

If, in addition,

- for each $A \subseteq S$, either $A \in \mathcal{F}$ or $S - A \in \mathcal{F}$,

then $\mathcal{F}$ is an ultrafilter. An ultrafilter is a filter that is maximal for inclusion. The family of all subsets of S that contain a given element s of S is an ultrafilter on S and is said to be principal. An ultrafilter $\mathcal{F}$ that is not principal is said to be free and satisfies $\cap_{\mathcal{F}} A = \varnothing$. AC (reformulated as Zorn's lemma) implies the existence of free ultrafilters on infinite sets. The non-constructiveness of free ultrafilters is well known (Jech 1973).

3 Maximal Pareto-Compatible Anonymity Conditions

From Sect. 2.2 we know that Pareto-compatible anonymity axioms are based upon groups of 'cyclic' permutations. This section introduces partition groups of cyclic permutations. We show that a maximal (for inclusion) partition group of cyclic

---

[9] AC is (i) consistent and (ii) independent: (i) AC can be added to the Zermelo-Fraenkel axioms of set theory (ZF) without yielding a contradiction, and (ii) AC is not a theorem of ZF (Fraenkel et al. 1973).

[10] The law of the excluded middle states the truth of 'P or not-P' for each proposition P and can be used to claim the existence of certain objects without any hint to its construction. For example, the real number $c = \sqrt{2}^{\sqrt{2}}$ *either* is rational (in which case one sets $a = b = \sqrt{2}$) *or* is not rational (in which case one sets $a = c$ and $b = \sqrt{2}$). Conclude the *existence* of irrational numbers a and b for which a^b is rational.

permutations involves an ultrafilter on the lattice of partitions and is therefore a non constructible object.

The notion of a filter on sets extends to a filter on a lattice of partitions. Let us recall the definitions (Halbeisen and Löwe 2001). A partition of $\mathbb{N}_0$ is a family of pairwise disjoint nonempty sets such that their union coincides with $\mathbb{N}_0$. If A and B are two partitions of $\mathbb{N}_0$, we say that A is coarser than B (or that B is finer than A) and we write $A \sqsubseteq B$ if each piece in A is a union of pieces of B. The coarsest partition of $\mathbb{N}_0$ (everything in one piece) is denoted by $0 = \{\mathbb{N}_0\}$, the finest partition (all pieces of which are singletons) by 1. Each partition is in between 0 and 1.

Let Ω_0 collect those partitions of $\mathbb{N}_0$ that consist out of infinitely many finite pieces. Partitions containing one (or more) infinite piece(s) are not distinguished, they are denoted by 0. We endow the class $\Omega = \Omega_0 \cup \{0\}$ with two operations $\cup$ and $\cap$. The partition $A \cup B$ is the coarsest partition in Ω that refines A and B, and the partition $A \cap B$ is the finest partition in Ω that is coarser than A and B. In case the partition $A \cap B$ contains an infinite piece, we put $A \cap B$ equal to 0. The couple $(\Omega, \sqsubseteq)$ is a lattice.

A filter on the lattice $(\Omega, \sqsubseteq)$ is a collection $\mathcal{F}$ of members of Ω that satisfies

- 0 is not in $\mathcal{F}$,
- if both A and B are in $\mathcal{F}$, then $A \cap B$ is in $\mathcal{F}$,
- if B is in $\mathcal{F}$ and $B \sqsubseteq A$ (with A in Ω), then A is in $\mathcal{F}$.

A family $\mathcal{B} \subseteq \Omega$ is said to be a *filter base* if (*i*) $0 \notin \mathcal{B}$, and (*ii*) for each A_1 and A_2 in $\mathcal{B}$, there is a B in $\mathcal{B}$ such that $B \sqsubseteq A_1 \cap A_2$. In case $\mathcal{B}$ is a filter base, then the family

$$\mathcal{B}^+ = \{ A \in \Omega \mid \text{there is a } B \text{ in } \mathcal{B} \text{ such that } B \sqsubseteq A \}$$

is a filter on the lattice $(\Omega, \sqsubseteq)$. The filter $\mathcal{B}^+$ coincides with the intersection of all filters that include $\mathcal{B}$. A filter that is maximal for inclusion is said to be an ultrafilter. Each ultrafilter $\mathcal{F}$ on $(\Omega, \sqsubseteq)$ is free, i.e. $\bigcap \{ A \mid A \in \mathcal{F} \} = 0$.

We recall two facts on ultrafilters (Facts 2.1–2 in Halbeisen and Löwe 2001, p. 321).

- A family $\mathcal{F}$ is an ultrafilter on $(\Omega, \sqsubseteq)$ if and only if for each A in Ω either $A \in \mathcal{F}$ or there is a B in $\mathcal{F}$ such that $A \cap B = 0$ (the 'either-or' being exclusive).
- If $\mathcal{B}$ is a family of elements of Ω with the finite intersection property (for each finite subfamily $\{ A_1, A_2, \dots, A_n \} \subseteq \mathcal{B}$ we have $A_1 \cap A_2 \cap \dots \cap A_n \neq 0$), then there is an ultrafilter $\mathcal{F}$ on $(\Omega, \sqsubseteq)$ with $\mathcal{B} \subseteq \mathcal{F}$.

The second fact is implied by AC (reformulated as Zorn's lemma). The notion "ultrafilter on a lattice" generalizes the notion "free ultrafilter on a set". The next example clarifies this statement.

Example Each infinite subset $S = \{n_1, n_2, \dots, n_k, \dots\}$, with $n_1 < n_2 < \cdots < n_k < \cdots$, of $\mathbb{N}_0$ induces a partition

$$V_S = \{ \{1, 2, \dots, n_1\}, \{n_1 + 1, n_1 + 2, \dots, n_2\}, \dots, \{n_k + 1, n_k + 2, \dots, n_{k+1}\}, \dots \}.$$

Now, let $\mathcal{F}_{\mathbb{N}_0}$ be a free filter on the set $\mathbb{N}_0$. Then, the family of all partitions generated by elements in $\mathcal{F}_{\mathbb{N}_0}$, denoted by $\mathcal{F}_\Omega = \{ V_S \in \Omega \mid S \in \mathcal{F}_{\mathbb{N}_0} \}$, is a filter on the lattice $(\Omega, \sqsubseteq)$. Moreover, $\mathcal{F}_\Omega$ is an ultrafilter on the lattice $(\Omega, \sqsubseteq)$ if and only if $\mathcal{F}_{\mathbb{N}_0}$ is a free ultrafilter on the set $\mathbb{N}_0$.

The link towards cyclic permutations is as follows. Each permutation partitions the set $\mathbb{N}_0$: present the permutation as a juxta position of cycles and replace the brackets (and) by { and }. Each cyclic permutation partitions the set $\mathbb{N}_0$ into an infinite sequence of finite sets. For example, the partition induced by the permutation π_1 is equal to

$$\text{Part}(\pi_1) = \Big\{ \{1,2\}, \{3,4\}, \ldots, \{2n-1, 2n\}, \ldots \Big\}.$$

Consider the partition $A = \{ N_1, N_2, \ldots, N_k, \ldots \}$ in Ω_0. We will refer to

$$\text{Sym}(A) = \text{Sym}(N_1) \times \text{Sym}(N_2) \times \cdots \times \text{Sym}(N_k) \times \cdots,$$

with $\text{Sym}(N_k)$ the group of all permutations on the finite set N_k, as the symmetric group of the partition A. The group $\text{Sym}(A)$ stabilizes the partition A, i.e. this group collects all the permutations with an induced partition that is equal to or finer than A. We shorten $\text{Sym}(\text{Part}(\pi))$ to $\text{Sym}(\pi)$. A group $\mathcal{Q}$ of permutations that includes $\text{Sym}(\pi)$ for each π in $\mathcal{Q}$ is said to be a *partition group*.

An anonymity condition based upon a partition group $\mathcal{Q}$ of cyclic permutations is Pareto-compatible. In order to enlarge the group $\mathcal{Q}$ (towards a maximal subgroup of cyclic permutations), we add a cyclic permutation to the group $\mathcal{Q}$, and we consider the group generated by $\mathcal{Q}$ and this additional permutation. The next lemma investigates the effect of enlarging a partition group, its proof is in the Appendix.

Lemma 2 *Let σ_A and σ_B be two cyclic permutations on $\mathbb{N}_0$. Then,* $\text{Sym}(\sigma_B)$ *contains a permutation ρ such that $\rho \circ \sigma_A$ generates the partition* $\text{Part}(\sigma_A) \cap \text{Part}(\sigma_B)$.

Adding a cyclic permutation to a partition group smuggles in permutations with courser partitions. As such, one runs the risk of ending up with a non-cyclic permutation. E.g. the addition of π_1 to the partition group $\text{Sym}(\pi_2)$ generates the non-cyclic permutation π_3.

We continue with some further notation. Let $\mathcal{B}$ be a family of partitions in Ω. Consider the set

$$\{ \pi \mid \text{there is a } B \text{ in } \mathcal{B} \text{ such that } B \sqsubseteq \text{Part}(\pi) \}$$

of all permutations that stabilize an element of $\mathcal{B}$. Denote by $\mathcal{Q}_\mathcal{B}$ the smallest partition group that includes this set of stabilizers. If $\mathcal{B}$ is a filter base, then $\mathcal{Q}_\mathcal{B}$ and $\mathcal{Q}_{\mathcal{B}+}$ coincide. For example, let $\mathcal{FS}$ collect the partitions

$$\{ \{1,2,\ldots,n\}, \{n+1, n+2, \ldots, 2n\}, \ldots, \{kn+1, kn+2, \ldots, (k+1)n\}, \ldots \}$$

with $n = 1, 2, \ldots$. The family $\mathcal{FS}$ is a filter base and the partition group $\mathcal{Q}_{\mathcal{FS}}$ coincides with the group $\mathcal{Q}_{\text{fs}}$ of fixed step permutations. Proposition 1 characterizes maximal partition groups.

Proposition 1 *Let $\mathcal{B}$ be a family of partitions in Ω. Then, $\mathcal{Q}_{\mathcal{B}}$ is a maximal group of cyclic permutations if and only if $\mathcal{B}^+$ is an ultrafilter.*

Proof The if-part. Let $\mathcal{B}^+$ be a filter. Then, $0 \notin \mathcal{B}$, and $\mathcal{Q}_{\mathcal{B}}$ only contains cyclic permutations. If π and ρ belong to $\mathcal{Q}_{\mathcal{B}}$, then $\text{Part}(\pi) \cap \text{Part}(\rho)$ belongs to $\mathcal{B}^+$. Hence, $\mathcal{Q}_{\mathcal{B}}$ is closed for composition. Next, observe that the partition induced by a permutation coincides with the partition induced by its inverse permutation. Therefore, $\mathcal{Q}_{\mathcal{B}}$ is a (partition) group of cyclic permutations.

Now, suppose that $\mathcal{B}^+$ is an ultrafilter. We have to show that $\mathcal{Q}_{\mathcal{B}}$ is maximal. Therefore, assume that the cyclic permutation π is not in $\mathcal{Q}_{\mathcal{B}}$. The induced partition $A = \text{Part}(\pi)$ does not belong to the ultrafilter $\mathcal{B}^+$. Hence, there is a B in $\mathcal{B}^+$ such that $A \cap B = 0$. Lemma 2 implies the existence of a permutation in $\text{Sym}(B)$ such that the composition with π induces the partition 0. This composed permutation has an infinite cycle. Therefore, the permutation π cannot be added to $\mathcal{Q}_{\mathcal{F}}$ to generate a larger group of cyclic permutations.

The only-if-part. Let $\mathcal{Q}_{\mathcal{B}}$ be a maximal subgroup of cyclic permutations. We have to show that $\mathcal{B}^+$ is an ultrafilter. Since only cyclic permutations are involved, $0 \notin \mathcal{B}$. Next, assume that the partition A is not in $\mathcal{B}^+$. We have to show the existence of a partition B in $\mathcal{B}^+$ with $A \cap B = 0$. A permutation π that induces A does not belong to $\mathcal{Q}_{\mathcal{B}}$. Since the group $\mathcal{Q}_{\mathcal{B}}$ is maximal, there is a σ in $\mathcal{Q}_{\mathcal{B}}$ such that $\pi \circ \sigma$ is not cyclic. Conclude that $A \cap \text{Part}(\sigma) \sqsubseteq \text{Part}(\pi \circ \sigma) = 0$ with $\text{Part}(\sigma)$ in $\mathcal{B}^+$. □

Mitra and Basu (2007) formulate the question whether the group $\mathcal{Q}_{\text{fs}}$ of fixed step permutations is a maximal (for inclusion) group of cyclic permutations. Proposition 1 answers this question in the negative. The filter generated by the family $\mathcal{FS}$ is not an ultrafilter and the partition group $\mathcal{Q}_{\mathcal{FS}} = \mathcal{Q}_{\text{fs}}$ is not maximal.

There is one further concern. We should check whether larger partition groups (and stronger anonymity demands) reduce the incompleteness of the social welfare relation. Let the partition group $\mathcal{G}'$ be larger than the partition group $\mathcal{G}$. Then, the relation $\precsim_{\mathcal{G}}$ is a subrelation to $\precsim_{\mathcal{G}'}$. The next proposition studies the indifference sets of these relations and uses the concept of permissible permutations. The definition is as follows. Let $\precsim$ be a SWR in X. The set of *permissible partitions* is defined as

$$\Pi(\precsim) \;=\; \{\, A \in \Omega \mid \text{for each } \pi \text{ in } \text{Sym}(A) \text{ and for each } x \text{ in } X \text{ we have } \pi(x) \sim x \,\}.$$

If the SWR $\precsim_1$ is a subrelation to the Paretian SWR $\precsim_2$, then $\Pi(\precsim_1) \subseteq \Pi(\precsim_2)$. If, in addition $\Pi(\precsim_1)$ is a strict subset of $\Pi(\precsim_2)$, then $\precsim_1$ is a strict subrelation to $\precsim_2$ (i.e. $\precsim_2$ is less incomplete than $\precsim_1$). Proposition 2 investigates the link between partition groups and permissible partitions. We use $\precsim_{\mathcal{B}}$ as a shorthand for the social welfare relation $\precsim_{\mathcal{Q}_{\mathcal{B}}}$.

Proposition 2 *Let the family $\mathcal{B}$ of partitions in Ω be a filter base. Then, the relation $\precsim_{\mathcal{B}}$ is reflexive, transitive, Paretian, and $\mathcal{B}$-anonymous. Furthermore, the set $\Pi(\precsim_{\mathcal{B}})$ of permissible partitions coincides with the filter $\mathcal{B}^+$.*

Proof The conditions imposed upon $\mathcal{B}$ turn $\mathcal{Q}_{\mathcal{B}}$ into a partition group of cyclic permutations. This group $\mathcal{Q}_{\mathcal{B}}$ coincides with $\mathcal{Q}_{\mathcal{B}^+}$. Mitra and Basu (2007, Proposition 3) show that for each group $\mathcal{G}$ of cyclic permutations, the relation $\precsim_{\mathcal{G}}$ is reflexive, transitive, Paretian, and $\mathcal{G}$-anonymous. Apply their result for $\mathcal{G} = \mathcal{Q}_{\mathcal{B}}$ and conclude that $\precsim_{\mathcal{B}}$ satisfies the properties as listed.
Let us now verify that $\Pi(\precsim_{\mathcal{B}})$ coincides with $\mathcal{B}^+$. The inclusion $\mathcal{B}^+ \subseteq \Pi(\precsim_{\mathcal{B}})$ is immediate. In case $\mathcal{B}^+$ is an ultrafilter also the reverse inclusion holds (otherwise, there exists a cyclic permutation π outside the group $\mathcal{Q}_{\mathcal{B}}$ that keeps the indifference relation; as $\mathcal{Q}_{\mathcal{B}}$ is maximal $\mathcal{Q}_{\mathcal{B}} \cup \{\pi\}$ generates noncyclic permutations and a contradiction is obtained).
There remains one single statement to be proved: the inclusion $\Pi(\precsim_{\mathcal{B}}) \subseteq \mathcal{B}^+$ under the assumption that $\mathcal{B}^+$ is not an ultrafilter. We show this inclusion by contradiction and assume $A \notin \mathcal{B}^+$. There exists an ultrafilter $\mathcal{F}$ that extends $\mathcal{B}$ and does not contain A (in the family $\mathcal{A}$ of all filters which do not contain A each chain has a maximal element, so by Zorn's lemma $\mathcal{A}$ has a maximal element that appears to be an ultrafilter; cf. Ax 1968, Sect. 11a). The relation $\precsim_{\mathcal{B}}$ is a subrelation to $\precsim_{\mathcal{F}}$, and $A \notin \Pi(\precsim_{\mathcal{F}})$. Hence, A does not belong to $\Pi(\precsim_{\mathcal{B}})$. □

Propositions 1 and 2 justify the statements claimed in the introduction. Anonymity demands are formulated in terms of groups of cyclic permutations. We focussed on partition groups of cyclic permutations. Enlarging the partition group, strengthens the anonymity demand, and decreases the incomparability. The strongest Pareto-compatible anonymity demand based upon a partition group of cyclic permutations, involves an ultrafilter on the lattice of partitions and is therefore a non-constructible object.

4 Measures on $\mathbb{N}_0$, the Chichilnisky Criterion

A finitely additive measure μ on $\mathbb{N}_0$ assigns to each subset of $\mathbb{N}_0$ a nonnegative real number and assigns to the union of two disjoint sets the sum of their numbers. The measure μ is said to be countably additive if the measure of a countable union of pairwise disjoint sets is equal to the sum of the measures of those sets. The finitely additive measure ν is dominated by μ (and we write $\nu \leq \mu$) is for each subset S of $\mathbb{N}_0$, we have $\nu(S) \leq \mu(S)$. The finitely additive measure μ is said to be purely finitely additive if the inequalities $0 \leq \nu \leq \mu$ with ν countably additive imply that $\nu = 0$. From Yosida and Hewitt (1952) and Rao (1958) we know that each finitely additive measure uniquely decomposes as the sum of a countably additive and a purely additive measure. This decomposition result is at the heart of the Chichilnisky criterion: the discounting rule (non dictatorship of the future) takes the role of the countably additive measure and the "distinct future"-part (non dictatorship of the present) is a purely finitely additive measure.

Typically, a purely finitely additive measure is obtained by means of Hahn-Banach's theorem or by means of a free ultrafilter (e.g. Chichilnisky 2009a, b). We only describe the second route.

A free ultrafilter $\mathcal{F}$ on $\mathbb{N}_0$ defines a limit on X. Consider a sequence x in X and all of its limit points. Each limit point is the limit of a subsequence. There is only one limit point with a converging subsequence $x_{i_1}, x_{i_2}, \ldots, x_{i_t}, \ldots$ for which the set $\{i_1, i_2, \ldots, i_t, \ldots\}$ of indices belongs to $\mathcal{F}$. Define $\lim_{\mathcal{F}}(x) = \lim_{t\to\infty} x_{i_t}$. Due to the intersection property of $\mathcal{F}$, we have $\lim_{\mathcal{F}}(x+y) = \lim_{\mathcal{F}}(x) + \lim_{\mathcal{F}}(y)$ for each x and y in X. The ultrafilter-based-limit $\lim_{\mathcal{F}}$ defines a finitely additive measure:

$$\mu_{\mathcal{F}}(S) = \lim_{\mathcal{F}} s_t \qquad \text{with } s_t = \frac{\#(S \cap \{1,2,\ldots,t\})}{t},$$

and S a subset of $\mathbb{N}_0$. If the sequence $s_1, s_2, \ldots, s_t, \ldots$ has only one accumulation point, then $\mu_{\mathcal{F}}(S)$ coincides with 'the' limit of this sequence and is known as the natural density of S. For example, the set of even numbers has a natural density equal to 0.5; the set of all multiples of 20 has a natural density equal to 0.05. Unfortunately, not every subset of $\mathbb{N}_0$ has a natural density. For example, the set

$$S_1 = \{1, 10, 11, \ldots, 19, 100, 101, \ldots, 199, 1000, 1001, \ldots\}$$

of all natural numbers having their first digit equal to 1 has no natural density. The measure $\mu_{\mathcal{F}}(S_1)$ depends upon the particular (non-constructible) ultrafilter $\mathcal{F}$ and can take any value between 1/9 and 5/9.[11]

Both routes to obtain purely finitely additive measures (Hahn-Banach's theorem and a free ultrafilter) rely upon AC. As a consequence, both ways to obtain a purely finitely additive measure involve non-constructive methods. Obviously, one cannot conclude from this that purely finitely additive measures are non-constructible objects. The knowledge that non-constructive methods can be used to obtain a purely finitely additive measure, does not answer the question whether a purely finitely additive measure can be obtained without recurse to non-constructive methods.

The question whether or not a purely finitely additive measure on $\mathbb{N}_0$ is a constructible object is tackled by Lauwers (2010b). Not surprisingly, the answer is negative: the existence of a purely finitely measure relies upon AC.

Proposition 3 (Lauwers 2010b). *The existence of a purely finitely additive measure on $\mathbb{N}_0$ entails the existence of a non-Ramsey set (from Mathias (1977) we know that a non-Ramsey set is a non-constructible object).*

As already mentioned in the introduction, the normative question on how to evaluate policies that involve the distant future should by no means be answered through the Axiom of Choice. Only constructible and well defined criteria can take part in the discussions. Although the maps lim inf and lim sup violate additivity

[11] In this example, $\liminf(s_t)$ is the limit of the sequence $1/9, 11/99, 111/999, \ldots$ and is equal to 1/9; $\limsup(s_t)$ is the limit of the sequence $1, 11/19, 111/199, \ldots$ and is equal to 5/9.

(cf. footnote 8), with respect to the "distinct future"-part in the Chichilnisky criterion, the maps lim inf and lim sup provide a constructible alternative. A convex combination of a discounting rule, lim inf, and lim sup remains in the spirit of the Chichilnisky criteria. Alternatively, the Chichilnisky criterion is constructible when applied to a restricted domain, e.g. the domain of infinite paths which have a well defined and finite limiting behavior. In such a restricted domain, the usual limit of a path is defined, captures the distinct future value, and does not depend upon non-constructive methods.

5 Appendix, Proof of Lemma 2

Lemma 2 connects the partition induced by the product of two cyclic permutations to the intersection of the partitions induced by the permutations. In general, the relation $\mathrm{Part}(\sigma_1) \cap \mathrm{Part}(\sigma_2) \sqsubseteq \mathrm{Part}(\sigma_1 \circ \sigma_2)$ holds. For example, consider the following cyclic permutations:

$$\sigma_1 = (1)(2,3,5,6,7,4)(8,11,13,14,15,12,10,9)\underline{(16,19,21,22,23,20,18,17)}\cdots,$$
$$\sigma_2 = (1,2,3)(4,8,10,11,7,5)(6)(9)(12,16,18,19,15,13)(14)(17)\underline{(20,24,26,27,23,21)(22)(25)}\cdots.$$

The representation continues by repeating the underlined cycles taking into account a shift of +8. Here, $\mathrm{Part}(\sigma_1) \cap \mathrm{Part}(\sigma_2) = \mathbb{N}_0$ while both compositions $\sigma_2 \circ \sigma_1$ and $\sigma_1 \circ \sigma_2$ are cyclic:

$$\sigma_2 \circ \sigma_1 = \pi_1 = (1,2)(3,4)(5,6)(7,8)\cdots, \text{and}$$
$$\sigma_1 \circ \sigma_2 = (1,3)(2,5)(4,11)(6,7)(8,9)(10,13)(12,19)(14,15)(16,17)(18,21)\cdots.$$

Lemma 2 *Let σ_A and σ_B be two cyclic permutations on $\mathbb{N}_0$. Then,* $\mathrm{Sym}(\sigma_B)$ *contains a permutation ρ such that $\rho \circ \sigma_A$ generates the partition* $\mathrm{Part}(\sigma_A) \cap \mathrm{Part}(\sigma_B)$.

Proof Denote $A = \mathrm{Part}(\sigma_A)$ and $B = \mathrm{Part}(\sigma_B)$. We prove the lemma in case $C = A \cap B$ consists out of an infinite number of finite sets. In case the partition C contains an infinite piece, the same ideas apply.

Without loss (otherwise re-enumerate $\mathbb{N}_0$), we assume the existence of an increasing sequence $n_1, n_2, \ldots, n_k, \ldots$ in $\mathbb{N}_0$ such that the partition C can be written as

$$C = \Big\{ \underbrace{[1,n_1]}_{S}, [n_1+1, n_2], \ldots, [n_k+1, n_{k+1}], \ldots \Big\}.$$

Both A and B are finer than C. We focus on one of the pieces in C, say $S = [1, n_1]$. Again, without loss, we assume that the restriction of σ_A to S is as follows

$$\sigma_A|_S = (1,2,\ldots,k_1)(k_1+1,k_1+2,\ldots,k_2)\cdots(k_{m-1}+1,k_{m-1}+2,\ldots,n_1).$$

Denote the partition classes by $S_1 = [1,k_1], S_2 = [k_1+1,k_2], \ldots, S_m = [k_{m-1}+1,n_1]$.

We construct a permutation ρ in $\mathrm{Sym}(B|_S)$ by induction. The partition $A \cap B$—when restricted to S—is equal to S. Hence, there exists a couple (ℓ_1, ℓ^1) in $S_1 \times (S - S_1)$ both belonging to one piece of B. Put $\rho(\ell_1) = (\ell^1)$. Let ℓ^1 belong to $S^1 = S_i$. Move on to the set $S_2 = S_1 \cup S^1$. Again, there exists a couple (ℓ_2, ℓ^2) in $S_2 \times (S - S_2)$ that both belong to one piece of B. Put $\rho(\ell_2) = \ell^2$. This procedure ends after m steps. Put the permutation ρ equal to $(\ell_1, \ell^1)(\ell_2, \ell^2) \cdots (\ell_m, \ell^m)$, elements of S that are not listed remain fixed.

The permutation $\rho \circ \sigma_A$ generates the cycle S in one piece. Repeat the whole construction for the other pieces in C and paste together the corresponding permutations to obtain the result. □

References

Alvarez-Cuadrado, F., & Van Long, N. (2009). A mixed Bentham-Rawls criterion for intergenerational equity: Theory and implications. *Journal of Environmental Economics and Management, 58*, 154–168.

Asheim, G. B. (2010). Intergenerational equity. *Annual Review of Economics, 2*, 197–222.

Asheim, G. B., d'Aspremont, C., & Banerjee, K. (2016). Generalized time-invariant overtaking. *Journal of Mathematical Economics*,. doi:10.1016/j.jmateco.2010.03.005.

Asheim, G. B., & Banerjee, K. (2010). Fixed-step anonymous overtaking and catching up. *International Journal of Economic Theory, 6*, 149–165.

Asheim, G. B., & Mitra, T. (2010). Sustainability and discounted utilitarianism in models of economic growth. *Mathematical Social Sciences, 59*, 148–169.

Asheim, G. B., Mitra, T., & Tungodden, B. (2011). Sustainable recursive social welfare functions. *Economic Theory* (reprinted in this volume). doi:10.1007/s00199-010-0573-7.

Ax, J. (1968). The elementary theory of finite fields. *Annals of Mathematics, 88*, 239–271.

Banerjee, K. (2006). On the extension of the utilitarian and Suppes-Sen social welfare relations to infinite utility streams. *Social Choice and Welfare, 27*, 327–339.

Burniaux, J.-M., & Martins, J. O. (2011). Carbon leakages: A general equilibrium view. *Economic Theory* (reprinted in this volume). doi:10.1007/s00199-010-0598-y.

Chambers, C. P. (2009). Intergenerational equity: sup, inf, lim sup, and lim inf. *Social Choice and Welfare, 32*, 243–252.

Chichilnisky, G. (1996). An axiomatic approach to sustainable development. *Social Choice and Welfare, 13*, 231–257.

Chichilnisky, G. (2009a). The topology of fear. *Journal of Mathematical Economics, 45*, 807–816.

Chichilnisky, G. (2009b). Avoiding extinction: Equal treatment of the present and the future. *The Open-Access, Open-Assessment E-Journal, 3*, 2009–2032.

Chichilnisky, G. (2011). Sustainable markets with short sales. *Economic Theory* (reprinted in this volume).

Chipman, J. S., & Tian, G. (2011). Detrimental externalities, pollution rights, and the "Coase theorem". *Economic Theory* (reprinted in this volume). doi:10.1007/s00199-011-0602-1.

Dianonescu, R. (1975). Axiom of choice and complementation. *Proceedings of the American Mathematical Society, 51*, 176–178.

Dutta, P. K., & Radner, R. (2011). Capital growth in a global warming model: will China and India sign a climate treaty? *Economic Theory* (reprinted in this volume). doi:10.1007/s00199-010-0590-6.

Figuières, C., & Tidball, M. (2011). Sustainable exploitation of a natural resource: a satisfying use of Chichilnisky's criterion. *Economic Theory* (reprinted in this volume). doi:10.1007/s00199-010-0579-1.

Fleurbaey, M., & Michel, P. (2003). Intertemporal equity and the extension of the Ramsey criterion. *Journal of Mathematical Economics*, *39*, 777–802.
Fraenkel, A. A., Bar-Hillel, Y., & Levy, A. (1973). *Foundation of Set Theory*. Amsterdam: North Holland.
Halbeisen, L., & Löwe, B. (2001). Ultrafilter spaces on the semilattice of partitions. *Topology and its Applications*, *115*, 317–332.
Hall, M, Jr. (1976). *The theory of groups*. New York: Chelsea Publishing Company.
Heal, G. M. (1998). *Valuing the future: Economic theory and sustainability*. New York: Columbia University Press.
Jech, T. J. (1973). *The axiom of choice*. Amsterdam: North-Holland.
Kamaga, K., & Kojima, T. (2009a). Q-anonymous social welfare relations on infinite utility streams. *Social Choice and Welfare*, *33*, 405–413.
Kamaga, K., & Kojima, T. (2009b). *On the leximin and utilitarian criteria with extended anonymity*. Tokyo, Japan: Waseda University.
Karp, L., & Zhang, J. (2011). Taxes versus quantities for a stock pollutant with endogenous abatement costs and asymmetric information. *Economic Theory* (reprinted in this volume). doi:10.1007/s00199-010-0561-y.
Lauwers, L. (1997). Infinite utility: Insisting on strong monotonicity. *Australasian Journal of Philosophy*, *75*, 222–233.
Lauwers., L. (2010a). Ordering infinite utility streams comes at the cost of a non-Ramsey set. *Journal of Mathematical Economics*, *46*, 32–37.
Lauwers, L. (2010b). *Purely finitely additive measures are non-constructible objects*. K.U.Leuven: DPS 1010 Center for Economic Studies.
Lecocq, F., & Hourcade, J.-C. (2011). Unspoken ethical issues in the climate affair, insights from a theoretical analysis of negotiation mandates. *Economic Theory* (reprinted in this volume). doi:10.1007/s00199-010-0589-z.
Li, C.-Z., & Löfgren, K.-G. (2000). Renewable resources and economic sustainability: A dynamic analysis with heterogeneous time preferences. *Journal of Environmental Economics and Management*, *40*, 236–250.
Mathias, A. R. D. (1977). Happy families. *Annals of Pure and Applied Logic*, *12*, 59–111.
Mitra, T., & Basu, K. (2007). On the existence of Paretian social welfare relations for infinite utility streams with extended anonymity. In J. Roemer & K. Suzumura (Eds.), *Intergenerational equity and sustainability* (pp. 85–99). London: Palgrave.
Ostrom, E. (2011). Nested externalities and polycentric institutions: Must we wait for global solutions to climate change before taking actions at other scales? *Economic Theory* (reprinted in this volume). doi:10.1007/s00199-010-0558-6.
Rao, R. R. (1958). A note on finitely additive measures. *Indian Journal of Statistics*, *19*, 27–28.
Rezai, A., Foley, D. K., & Taylor, L. (2011). Global warming and economic externalities. *Economic Theory* (reprinted in this volume). doi:10.1007/s00199-010-0592-4.
Yosida, K., & Hewitt, E. (1952). Finitely additive measures. *Transactions of the American Mathematical Society*, *72*, 46–66.
Zame, W. (2007). Can intergenerational equity be operationalized? *Theoretical Economics*, *2*, 187–202.

Sustainable Exploitation of a Natural Resource: A Satisfying Use of Chichilnisky's Criterion

Charles Figuières and Mabel Tidball

1 Introduction

Global environmental problems, such as biodiversity loss or climate change, present us with at least two major sets of issues. The first is "what decisions Humanity should take in its own interest?" The second set of questions is largely complementary to the first: "what are the difficulties of collective action and how might they be overcome?". By and large, the papers collected in this special issue of Economic Theory strive to find answers to either or both of these challenges. Our contribution, like that of Lauwers (2010) and Asheim et al. (2010), focuses only on the first question. Links to the second question will be discussed in the Sect. 6, in connection with the insighs offered by the other contributions of this special issue.

One salient aspect of the literature on *intertemporal social choice* is the concern for intergenerational ethics in the exploitation of natural resources over time.[1] In this field of research, there is a fascinating and long lasting challenge, starting with the seminal contribution of Pigou (1920), later followed by Ramsey (1928), Fisher (1930) and many others, to examine whether intertemporal *social welfare criteria* (SWC hereafter) could be designed so as to satisfy both a concern for efficiency and for justice. The former is captured precisely by the Pareto Principle, either weak

[1]For other aspects, see Pezzey (1992).

The authors would like to thank Hubert Stahn, Antoine Soubeyran, their collegues at LAMETA (in particular those of the PERENE team), the participants at the ISDG worshop in Segovia, September 2006, Gilles Rotillon, Katheline Schubert, Michel De Lara, Vincent Martinet and Luc Doyen, and Graciela Chichilnisky for interesting and helpful comments. Many thanks also to an anonymous referee for his/her constructive suggestions.

Originally published in Economic Theory, Volume 49, Number 2, February 2012 DOI 10.1007/s00199-010-0573-7.

C. Figuières (✉) · M. Tidball
INRA, UMR LAMETA, 2 Place Viala, 34060 Montpellier, France
e-mail: figuiere@supagro.inra.fr

G. Chichilnisky and A. Rezai (eds.), *The Economics of the Global Environment*, Studies in Economic Theory 29, DOI 10.1007/978-3-319-31943-8_11

(WP) or strong (SP),[2] whereas the later is generally embodied in some sort of equity condition. A bulk of knowledge has accumulated, oscillating between optimistic and pessimistic answers.

The pessimistic side obtains when equity is thought of as equal treatment of all generations, identified in the literature with the so-called *weak anonymity axiom* (WA), and when a representation of the SWC by a function is imposed. As carefully described by Fleurbaey and Michel (2003), efforts to construct a continuous, complete and transitive SWC that satisfies both efficiency, even in the sense of WP, and WA easily end up in a stalemate (see Koopmans 1960; Diamond 1965; Basu and Mitra 2003). Using an axiomatic approach Koopmans (1960) tried to avoid incompleteness. Furthermore, he sought an order that respected other reasonable properties (continuity, a strong Pareto principle, separability and stationarity) which in principle are not linked to a form of impatience. However, and this is the surprising nature of Koopmans' analysis, the logical implication of these properties is ... the discounted criterion, a clear violation of the requirement of anonymity. Asheim et al. (2010) were able to identify the particular assumption among those made by Koopmans (1960) which leads to the discounted criterion. The culprit is a separability condition, listed as Postulate 3a' by Koopmans (1960) and referred to as *independent present*[3] by Asheim et al. (2010).

The optimistic side is rather tiny. Svensson (1980) shows that the difficulty to combine SP and WA with continuity properties critically depends on the chosen topology. Impossibility to obtain a SWC represented by a real valued function can be shown if one insists on continuity in the sup metric (Diamond 1965); but lowering the demand to the existence of an ordering, compatibility obtains with a stronger topology (Svensson 1980, Theorem 3, Sect. 3). Unfortunately, Svensson's proof rests on a non-constructive method that leaves us with an existence result but no explicit SWC. Fleurbaey and Michel (2003) (see their conclusion on p. 794) make the conjecture that no explicit complete continuous ordering will ever be found that satisfies efficiency and WA. A recent result by Lauwers (2010) confirms this conjecture. See also Zame (2007) who attacks this issue with a measure-theoretic approach. From a practical point of view, this Lawers–Zame conclusion is equivalent to an inexistence result.

In this perspective, Chichilnisky's contribution (1996) appears very useful. She sticks to the sup metric and SP, but she rests on a weaker view of equity, summarized in two axioms: no dictatorship of the present (NDP) and no-dictatorship of the future (NDF). Roughly speaking, her first axiom requires the SWC to be sensitive to the welfare of the very long run generations, whereas the second axiom requires the SWC to be sensitive to the welfare of both the present and the finitely distant generations. Adding a boundedness condition on per period utilities and a linearity

[2]According to WP, an exploitation path should be deemed better than another if under the former all generations are strictly better-off. According to SP, an exploitation path should be deemed better than another if under the former all generations are better-off, with one generation at least being strictly better-off.

[3]This axioms requires that the evaluation of two streams of utilities which differ during only the first two periods not depend on what the common continuation stream is.

requirement, a continuous SWC (in the sup norm topology) emerges; Chichilnisky's SWC (CSWC) takes the form of a convex combination of the traditional discounted utilitarist SWC and of the limit value of utility over time.[4] Clearly, with such a SWC the finitely distant generations are not favored compared to the present one. Put differently, an outside observer under the *veil of ignorance,* who would assign a diffuse prior probability of living at any date (formally giving an equal probability to living at any time), would prefer a SWC that respects the axiom of WA, rather than NDP and NDF; those two last axioms do capture a concern for intergenerational equity, but in a weaker sense compared to WA. A closely related approach is Li and Lofgren (2000) who, motivated by CSWC, propose a criterion with a discount factor that declines towards zero: the dictatorhsips of the present and the future is avoided and there exists maximal paths.[5] This criterion is unbounded, which brings technical issues, and it is not stationary hence it naturally raises the issue of time consistency.

However, if the aim is to offer a SWC at the level of applicability of discounted utilitarianism, optimism should be temperate: the optimal path to Li and Lofgren's criterion is to be approximated by sequences, and for renewable resources it can be shown that no particular path achieves the upper bound of Chichilnisky SWC.[6] Should this inexistence of an optimal path be viewed as a serious flaw (Asheim 1996)? A positive answer would certainly be too harsh; as explained in the course of the paper, one can pick up an exploitation path that gives CSWC a value arbitrarily close to its upper bound. In a nutshell, Chichilnisky arrives at a continuous and explicit SWC at the cost of relaxing somehow the equity requirement, and with some arbitrarily small loss of efficiency when the resource is renewable.

Decisions about the management of natural resources are to be made now and tomorrow; and the mere fact that CSWC is an explicit complete continuous crite-

---

[4]When NDP, NDF, Pareto, linearity and completeness are required simultaneously, this is the only such criterion, or more precisely family of criterions. But other different possibilities appear if some of these conditions are relaxed. One may investigate, as Lauwers (2010), what maximal anonymity properties can be consistent with Pareto. On another hand, dropping Strong Pareto can end up to the recursive social welfare functions proposed by Asheim et al. (2010). Lastly, dropping linearity, Alvarez-Cuadrado and Ngo Van Long (2009) propose a weighted average of the maximin and the discounted utilitarian ordering, which they call MBR criterion (a shortcut for Mixed Bentham-Rawls).

[5]Li and Lofgren (2000) propose a foundation for the criterion with a decreasing discount factor, i.e. with individual variation in time preferences. They consider a society that consists of two individuals, an utilitarian and a conservationist. The utilitarian wants to maximize the discounted utilitarian criterion with constant discount. The conservationist wants to maximize the discounted utilitarian criterion with constant discount when this discount tends to zero. The society wants to maximize a convex combination of these criteria and end up with utilitarian criterion with a declining discount factor. The authors prove that the optimal solution exists and can be approximated by sequences. They characterize the steady state, that is the golden rule path, and prove that both, social and conservationist optimal solution, converge to the golden rule path. In particular the social optimal consumption is between the optimal utilitarian and optimal conservationist consumptions.

[6]Though in the case of an exhaustible resource, an optimal solution generally exists; see Chichilnisky (1997), Sect. 5.B, or Heal (1998), Chap. 6. With renewable resources, an optimal solution also exists generally if and only if the discount factor decreases to zero as time tends to infinity (Chichilnisky 1997, Theorem 3).

rion that combines successfully WP and a minimal concern for intergenerational equity, justifies some further efforts to deal with its weaknesses. We shall focus in this paper on the inexistence issue of CSWC. Could sensible principles guide us to single out a unique path among the many possible admissible ones? Can we avoid a loss of efficiency? More generally, what can we hope for in this research direction? In particular, do we achieve the degree of applicability of discounted utilitarianism? This paper contributes an exploration of those questions in the context of natural resources.

The next section presents a discrete-time natural resource framework. Section 3 explains our suggestion to cope with the inexistence issue, namely to restrict the space of controls in a specific way. It also presents the conditions on the fundamentals of the economy under which a "restricted" optimal exploitation policy exists (Theorem 1). Section 4 applies this methodology to a parametric example and derives further results. Section 5 offers a discussion of the ethical properties of our solution and compares it with another recent way out to the existence problem recently offered by Chichilnisky (2009). Under some conditions it turns out that it implies no loss of efficiency but it has redistributive consequences in favor of intermediate generations. Section 6 concludes with a discussion on policy implications. Proofs are relegated to the Appendix.

2 A Natural Resource Framework

Consider a simple framework for the management of a natural resource, whose value at date t is denoted $x_t \in [0, +\infty[$; this resource evolves according to the recurrence equation:

$$x_{t+1} = G\left(x_t - c_t\right), \quad x_0 \text{ given}, \tag{1}$$

where $c_t \in C_t := [0, x_t]$ denotes the human extraction at date t, and $G(.)$ is a transition function.

Throughout the paper, attention is restricted to converging sequences only, i.e. c_t and x_t both converge to a finite limit as t tends to infinity. The set C of *admissible extraction paths* is made of converging sequences $\{c_t\}_{t=0}^{\infty}$, $c_t \leq x_t$, $\lim_{t\to\infty} c_t < \infty$, that generate sequences in the admissible set $\mathcal{X}$ of converging paths for the state of the resource $x_0, x_1, x_2, \ldots, \lim_{t\to\infty} x_t < \infty$, via the dynamic Eq. (1).

Each generation is endowed with a utility function:

$$U(c_t, x_t) \tag{2}$$

defined over the current consumption and, possibly, over the current level of the stock.

The following criterion belongs to the family of SWC proposed by Chichilnisky (1996) to rank admissible sequences of harvests:

$$J^s = \theta \sum_{t=0}^{\infty} \beta^t U(c_t, x_t) + (1-\theta) \lim_{t\to\infty} U(c_t, x_t) \,, \tag{3}$$

where $\beta \in]0,1[$ is a discount factor; the arbitrary parameter $\theta \in]0,1[$ constructs CSWC as a convex combination of the discounted utilitarian criterion and the utility of the infinitely distant generation. It is worth recalling that this particular criterion belongs to the unique family of SWC that meets a number of seductive axioms for sustainability; in particular Criterion J^s is neither a *dictatorship of the present*, nor it is a *dictatorship of the future* generations (see Chichilnisky 1996). This criterion is also *sensitive* in the sense that it increases with an increase of the utility of any generation (so it satisfies both SP and WP). The following assumption ensures this criterion is well defined (Chichilnisky 1996, pp. 290–291):

Assumption 1 The utility function $U(c,x)$ is continuous and its values are bounded from below and from above, i.e. $\underline{U} \leq U(c,x) \leq \overline{U}$ for all admissible pairs (c,x).

Remark 1 Note that if all the admissible pairs (c,x) belong to a compact set, then every continuous utility function is bounded. That is, the above assumption does not overly restrict the class of admissible utility functions. More on this later.

Remark 2 This assumption is not required only to guarantee CSWC is well-defined; it is also required in this paper, along with the assumption $\beta \in]0,1[$ and assumptions on the transition function, to use a contraction argument for proving the existence of a solution associated to the discounted utilitarian problem. And it will also play a key role in the main result of this paper.

We shall call Problem $\mathcal{P}$ the quest for the *most sustainable* admissible exploitation path $\left\{c_t\right\}_{t=0}^{\infty} \in C$, i.e. the path that renders criterion J^s as high as possible, and we adopt the notation $\left\{\hat{c}_t, \hat{x}_t\right\}_{t=0}^{\infty}$ for a solution to Problem $\mathcal{P}$, with $\hat{x} = \lim_{t\to\infty} \hat{x}_t$.

It has been shown (see Beltratti et al. 1994, pp. 334–335; Heal 1998, pp. 95–98) that one can approximate the independant maximization of both terms in the maximand and thereby its upper bound. To achieve this, candidate solutions would start with a consumption slightly lower than the utilitarian consumption, to be arbitrarily close to the optimal path for the discounted utilitarian problem until the date where the stock reaches the level of the green golden rule. At that date, it is then optimal to change the consumption pattern so as to remain at the green golden rule. Generally there is no solution to Problem $\mathcal{P}$ (Beltratti et al. 1994; Heal 1998, Chap. 7, Proposition 20). Indeed, ever increasing the initial and subsequent consumptions (while keeping them lower than the utilitarian ones) would increase the first part of the maximand without detracting to the second part. This can be seen as a lack of compactness in admissible consumptions, that could approach but not reach exactly upper bound levels (for a development around the lack of compactness problem, see Chichilnisky 2009). Equivalently, ever postponing the switching date makes it possible to increase the first part of CSWC while maintaining the value of the second part, which is the explanation for this inexistence result. Whatever the way of seeing the problem, near optimal controls are not stationnary: at some date there is a change of regime.

This inexistence result may leave the reader with the feeling that nothing can be done to adopt a catch policy that is more cautious with future generations. Upon reflection, this is of course a flawed conclusion. Obviously, one can easily obtain a "near optimal" trajectory for Chichilnisky's criterion by choosing arbitrarily a switching date. But, is the choice of a non optimal policy necessarily arbitrary? Or can it be guided by sensible principles? In the next section, we offer an answer to those questions.

3 A Restricted Optimal Solution

The essence of the inexistence problem stems from the impossibility to achieve the upper bound for the criterion J^s. As we have seen in the previous section, the possibility of non stationarity of controls creates the problem. This suggests to restrict the space of admissible controls to stationary controls. Also, since by varying θ Problem $\mathcal{P}$ becomes arbitrarily close to either the discounted utilitarian problem or the green golden rule problem, that is two problems for which optimal and stationary solutions generally exist, it would be welcome that the restricted class encompasses both programs. Finally, as explained in Heal (1998) (Chap. 6, Proposition 17) and Chichilnisky (2009) (Theorem 3), a path which is optimal with respect to CSWC must satisfy the necessary conditions for the maximization of the discounted utilitarian problem. Intuitively, restricted controls should depart from those conditions as little as possible, and in a direction that is best for the second term $\lim_{t\to\infty} U(c_t, x_t)$. A restricted class of controls that meets those requirements then suggests itself: the space of convex combinations between the optimal discounted utilitarian program and stationary programs leading to the green golden rule.

This section identifies conditions on the fundamentals of the economy (the transition function and the utility function) under which there exists an optimal program in the set of convex combinations between the optimal discounted utilitarian policy and a stationary program leading to the GGR (Theorem 1). Those conditions can be listed as follows.

Assumption 2 The transition function $G(.)$ is compact-valued, continuous, strictly increasing and concave.

Assumption 3 The utility function is non decreasing in both arguments c and x. Besides, for any $0 \le \sigma \le 1$ and any admissible pairs $\left(x_t, x_{t+1}\right)$ and $\left(x'_t, x'_{t+1}\right)$ such that

$$x_{t+1} \in \left[0, G\left(x_t\right)\right] \quad \text{and} \quad x'_{t+1} \in \left[0, G\left(x'_t\right)\right] \tag{4}$$

the utility function satisfies

$$U\left[\sigma x_t + (1-\sigma)x'_t - G^{-1}\left(\sigma x_{t+1} + (1-\sigma)x'_{t+1}\right), \sigma x_{t+1} + (1-\sigma)x'_{t+1}\right]$$
$$\geq \sigma U\left[x_t - G^{-1}\left(x_{t+1}\right), x_{t+1}\right] + (1-\sigma)U\left[x'_t - G^{-1}\left(x'_{t+1}\right), x'_{t+1}\right],$$

with a strict inequality if $x_t \neq x'_t$.

Remark 3 Under the last assumptions, the function

$$F\left(x_t, x_{t+1}\right) \equiv U\left[x_t - G^{-1}\left(x_{t+1}\right), x_t\right] \tag{5}$$

is strictly concave, a standard assumption in growth models.

Proposition 1 *Under Assumptions 1, 2 and 3 there exists a unique and continuous policy function* $c_t^{DU} = \phi_{DU}\left(x_t\right)$ *for the discounted utilitarian problem.*

Proof Under Assumptions 1 and 2, existence follows from Theorems 4.5 and 4.6 in Stokey and Lucas (1989), Chap. 4. Under Assumptions 1, 2 and 3, unicity and continuity is stated in Theorem 4.8, same chapter. □

Assumption 4 The utility function $U(.,.)$ along with the transition function $G(.)$ satisfy the Inada-like condition

$$U_c(0,0)\left(1 - \frac{1}{G'(0)}\right) + U_x(0,0) > 0.$$

This assumption rules out $c_t = 0$ as a possibility for the golden rule consumption.

Assumption 5 The transition function satisfies $G(0) = 0$, $G'(0) > 1$ and $\lim_{x \to +\infty} G(x) < x$.

Proposition 2 *Under Assumptions 1, 4 and 5, there exists a unique interior solution to the green golden rule problem and a linear (hence continuous) policy function* $c_t^{GGR} = \phi_{GGR}\left(x_t\right)$ *such that the economy converges towards the green golden rule.*

Proof Appendix A. □

When the above particular solutions exist, then

Lemma 1 *Any convex combination of* $\phi_{DU}\left(x_t\right)$ *and* $\phi_{GGR}\left(x_t\right)$, $c_t^{\gamma} = \gamma\phi_{DU}\left(x_t\right) + (1-\gamma)\phi_{GGR}\left(x_t\right)$, $\gamma \in [0,1]$, *is also an admissible path.*

Proof Appendix B. □

It can also be established that

Lemma 2 *The solution to*

$$x_{t+1}^{\gamma} = G\left(x_t^{\gamma}, \gamma\phi_{DU}\left(x_t^{\gamma}\right) + (1-\gamma)\phi_{GGR}\left(x_t^{\gamma}\right)\right) = \overline{G}_t(\gamma),$$

x_0 *given, is continuous in* γ.

Proof Appendix C. □

Theorem 1 *Under Assumptions 1, 2, 3, 4 and 5, there exists a convex combination of the discounted utilitarianism and of the green golden rule, among the set of such combinations, that maximizes CSWC.*

Proof Appendix D. □

Four last remarks are in order.

First, Assumptions 1, 2, 3, and 4 are sufficient but not necessary. The example given in the next section uses a logarithmic utility function which does not meet by itself Assumption 1 for instance; still, the methodology works well for this example.

Second, the existence result provided in Theorem 1 allows a very simple and general approach to find restricted optimal solutions in concrete examples: (i) compute the discounted utilitarian optimum and the green golden rule, (ii) find out the optimal convex combination between the two. Numerical procedures to carry on the second step can be very simple. For instance, one can try a finite number of values for γ, chosen on a pre-specified grid.

Third, the restricted optimal approach suggested here could presumably be applied in substantially more general contexts than the one offered in this section, for instance to an economy with multiple stock variables including private capital. Though we won't carry on further investigations, it seems that, provided boundedness, continuity and strict concavity requirements are met, many generalizations are conceivable.

The last remark is about the property of time consistency. Would the restricted optimal path remain optimal if it were to be reconsidered later in the future? Answering this question is usually done by checking that the policy of interest obey *Bellman equation*. In its original form, it is not adapted to guarantee this property in the present context. Because the objective is not a simple addition of (discounted) flow payoffs, the optimization problem here does not belong to the class of problems for which this functional equation has been derived. But an adapted formulation of Bellman Principle of Optimality can be identified. Consider a sequence ${}_t\gamma = (\gamma, \gamma, \ldots)$ of constant optimal convex combinations optimally computed at date t. Remember that:

$$\begin{aligned} c_t^\gamma &= \gamma \phi_{DU}\left(x_t\right) + (1-\gamma)\,\phi_{GGR}\left(x_t\right),\ \gamma \in [0,1], \\ x_{t+1}^\gamma &= G\left(x_t^\gamma, \gamma\phi_{DU}\left(x_t^\gamma\right) + (1-\gamma)\,\phi_{GGR}\left(x_t^\gamma\right)\right) = \overline{G}_t\left(\gamma\right), \end{aligned}$$

and define:

$$W\left(x_t,\ {}_t\gamma\right) \equiv U(c_t^\gamma, x_t) + \sum_{s=t+1}^{\infty} \beta^{s-t} U(c_s^\gamma, x_s^\gamma), \tag{6}$$

as the value of the discounted utilitarian criterion, starting at date t with a stock x_t, under the restricted optimal solution. Next denote:

$$V\left(x_t\right) \equiv \theta \left[U(c_t^\gamma, x_t) + \sum_{s=t+1}^{\infty} \beta^{s-t} U(c_s^\gamma, x_s^\gamma)\right] + (1-\theta) \lim_{s\to\infty} U(c_s^\gamma, x_s^\gamma), \tag{7}$$

the value function obtained by plugging the solution into CSWC, from date t onwards. Clearly, one can also write:

$$V\left(x_t\right) = \theta \left[U(c_t^\gamma, x_t) + \beta W\left(x_{t+1}^\gamma,\ {}_{t+1}\gamma\right)\right] + (1-\theta) \lim_{s\to\infty} U(c_s^\gamma, x_s^\gamma). \tag{8}$$

If γ is time consistent then

$$\begin{aligned} \gamma = \arg\max_{\varphi} \theta \left[U(c_t^\varphi, x_t) + \beta W\left(x_{t+1}^\varphi,\ {}_{t+1}\gamma\right)\right] \\ + (1-\theta) \lim_{s\to\infty} U(c_s^\gamma, x_s^\gamma), \quad \forall t, \end{aligned} \tag{9}$$

and

$$\begin{aligned} V\left(x_t\right) = \max_{\varphi} \theta \left[U(c_t^\varphi, x_t) + \beta W\left(x_{t+1}^\varphi,\ {}_{t+1}\gamma\right)\right] \\ + (1-\theta) \lim_{s\to\infty} U(c_s^\gamma, x_s^\gamma), \quad \forall t. \end{aligned} \tag{10}$$

This adapted principle of optimality makes use of two value functions, instead of only one. Equation (9) provides a test that our constant restricted solution should pass if it is time consistent. Alternatively, the functional Eq. (10) provides a recursive method to compute a time consistent sequence of optimal convex combinations, that may or may not be stationary. Presumably, a stationary optimal convex combination does systematically satisfy (9), though it is verified in the illustration provided in the next section.

4 An Illustration

Let the dynamics be:

$$x_{t+1} = \left(x_t - c_t\right)^\alpha, \qquad 0 < \alpha < 1, \quad x_0 \text{ given}, \tag{11}$$

This formulation captures the case of a renewable resource ($\alpha < 1$) but discards the possibility of an exhaustible resource (that would occur with $\alpha = 1$).

Let each generation be endowed with the utility function:

$$U(c_t, x_t) = \ln c_t + \pi \ln x_t \,, \quad \pi \geq 0, \tag{12}$$

Note that this cardinal representation of preferences verifies Assumption 4. Therefore zero consumption (and $x = 0$) cannot be a solution to the green golden rule problem; also, it can be checked that, provided that the initial stock is strictly positive, consumption and the stock variable is never zero along the optimal discounted utilitarian program. And none of those two trajectories tends to infinity. As a result, within our restricted set of controls, i.e. convex combinations between those two particular programs, the boundedness condition (1) is satisfied, even though it is not a property of the logarithmic function itself.

Solving the associated discounted utilitarian problem,

$$\max_{\{c_t\}_{t=0}^{\infty}} \sum_{t=0}^{\infty} \beta^t \left[\ln c_t + \pi \ln x_t\right], \quad 0 < \beta < 1, \tag{13}$$

subject to (11), the solution is a linear feedback

$$c_t = \frac{1 - \alpha\beta}{1 + \pi\alpha\beta} x_t \,. \tag{14}$$

Also, the associated golden rule problem consists in finding the pair (x, c) that solves the program:

$$\max_{x \geq 0, c \in [0,x]} \{\ln c + \pi \ln x, \quad x = (x - c)^{\alpha}\}.$$

Again the solution takes a linear feedback form

$$x^{GGR} = \left[\frac{\alpha(1+\pi)}{1+\pi\alpha}\right]^{\frac{\alpha}{1-\alpha}}, \quad c^{GGR} = \frac{1-\alpha}{1+\pi\alpha} x.$$

Remember that there exists no path that maximizes Chichilnisky's criterion. However, from the previous section we know that a restricted optimal solution can be looked for within the set of convex combinations between the discounted utilitarian solution and the green golden rule. Since those two particular outcomes are linear feedbacks, so does any convex combination of them. In the sequel, we look for an optimal solution within the set of linear feedback rules and we check afterwards that the optimal linear rule is indeed a convex combination of the discounted utilitarian solution and of the linear program leading to the green golden rule.

Consider the policy function

$$c_t = c\left(x_t\right) = \mu x_t\,. \tag{15}$$

Using this decision rule into the dynamic Eq. (1), the value of the stock at any date can be expressed as

$$x_t = (1-\mu)^{\alpha+\alpha^2+\cdots+\alpha^t} x_0^{\alpha^t} \tag{16}$$

and it becomes possible to express the resulting value for CSWC as a function of parameter μ. Indeed the two components of this criterion can be fully characterized as functions of parameter μ. The discounted utilitarian criterion is

$$\begin{aligned} J^{du} &= \sum_{t=0}^{\infty} \beta^t \left[ln\left(c_t\left(x_t\right)\right) + \pi \ln x_t\right] = \sum_{t=0}^{\infty} \beta^t \left[ln\mu + (1+\pi) ln\, x_t\right], \\ &= \sum_{t=0}^{\infty} \beta^t [ln\, \mu + (1+\pi)(\alpha+\alpha^2+\cdots+\alpha^t) ln\, (1-\mu) + \alpha^t ln\, x_0]. \end{aligned}$$

Then, as

$$\begin{aligned} \sum_{t=0}^{\infty}\sum_{s=1}^{t} \alpha^s \beta^t &= \sum_{t=0}^{\infty} \beta^t \alpha \sum_{s=1}^{t} \alpha^{s-1} = \alpha \sum_{t=0}^{\infty} \frac{1-\alpha^t}{1-\alpha}\beta^t \\ &= \frac{\alpha\beta}{(1-\beta)(1-\alpha\beta)}, \end{aligned}$$

the value of the discounted utilitarian objective as a function of μ turns out to be

$$\begin{aligned} J^{du}(\mu) = \sum_{t=0}^{\infty} \beta^t \left[ln\left(c_t\left(x_t\right)\right) + \pi \ln x_t\right] &= \frac{ln\, \mu}{1-\beta} \\ &+ \frac{\alpha\beta(1+\pi)}{(1-\beta)(1-\alpha\beta)} ln(1-\mu) + \text{constant}. \end{aligned} \tag{17}$$

As for the second part of CSWC, it worth noting from (15) and (16) that

$$\lim_{t\to\infty} x_t = (1-\mu)^{\frac{\alpha}{1-\alpha}} \tag{18}$$

Then the problem $\mathcal{P}$ in the set of linear feedback controls is equivalent to find $\mu \in]0,1]$ that maximizes

$$\begin{aligned} J^s(\mu) = &\theta \left[\frac{ln\, \mu}{1-\beta} + \frac{\alpha\beta(1+\pi)}{(1-\beta)(1-\alpha\beta)} ln(1-\mu)\right] \\ &+ (1-\theta)\left[ln\, \mu + \frac{\alpha}{1-\alpha}(1+\pi)\, ln(1-\mu)\right]. \end{aligned} \tag{19}$$

The solution for this optimization problem is interior; it solves the first order condition:

$$\theta\left[\frac{1}{(1-\beta)\,\mu}-\frac{\alpha\beta(1+\pi)}{(1-\beta)(1-\alpha\beta)(1-\mu)}\right]+(1-\theta)\left[\frac{1}{\mu}-\frac{\alpha\,(1+\pi)}{(1-\alpha)\,(1-\mu)}\right]=0, \tag{20}$$

or explicitly:

$$\mu=\frac{(1-\alpha)\,(\alpha\beta-1)\,(\theta\beta-\beta+1)}{(1-\theta)\,[\alpha\beta\,(1-\beta)+\beta]+\theta\alpha-1+\pi\left[\alpha\beta\,(1+\alpha-2\theta)-\alpha\,(1-\theta)\left(1+\alpha\beta^{2}\right)\right]} \tag{21}$$

Proposition 3 *The optimal linear stationary decision rule is a convex combination of the discounted utilitarian rule and the stationary linear rule leading to the green golden rule*

Proof Appendix E. □

The restriction ex ante to stationary controls definitely gives a restricted optimal answer to our problem. This comes as no surprise, for the inexistence issue is entirely due to the possibility for the controls to depend on time; once this dependence is ruled out, the problem disappears. Put differently, restricting attention to stationary controls creates a bridge over the two components of CSWC, and we shall see now that this bridge makes operational a trade-off between the conflicting interests of the generations.

From expression (21):

Proposition 4 *The higher (resp. lower) θ, the closer the restricted solution to the discounted utilitarian control (resp. green golden rule).*

Proof From Proposition (3), $\mu=\gamma\frac{1-\alpha\beta}{1+\pi\alpha\beta}+(1-\gamma)\frac{1-\alpha}{1+\pi\alpha}$ for some γ. At the end of Appendix E, it is shown that γ is an increasing function of θ. □

One feature of the unrestricted Problem P is the possibility to separate out the maximization of the first part of Chichilnisky's criterion from the maximization of its second part. As a consequence any candidate trajectory for optimization by no means depends on the parameter θ. In other words, increasing the weight of, say, future generations in the criterion would have no consequence on the solution of the problem, if it exists. On the contrary, with a restricted optimal solution, such an increase leads to a higher steady state, making operational the trade off between the two terms of CSWC.

A few other intuititive properties are worth noting.

Remark 4 When $\theta\to 1$, in the limit, i.e. for the discounted utilitarian problem, the feedback solution is stationary. It is

$$c_t=\frac{1-\alpha\beta}{1+\pi\alpha\beta}\,x_t\,. \tag{22}$$

Remark 5 When $\theta \to 1$ and $\beta \to 1$, in the limit the classical utilitarianism problem obtains, with solution

$$c_t = \frac{1-\alpha}{1+\pi\alpha} x_t. \tag{23}$$

One can check that this is indeed the optimal extraction path according to Ramsey's criterion:

$$\sum_{t=0}^{\infty} \left(U\left(c_t, x_t\right) - \hat{U} \right)$$

where

$$\hat{U} = \max\left\{ \ln c + \pi \ln x, \quad x = (x-c)^{\alpha} \right\} \tag{24}$$

is the maximum level of utility, or bliss point.

Remark 6 Conversely, what happens when $\theta \to 0$? In the limit, the golden rule problem emerges, which consists in finding the pair (x, c) that achieves the bliss point. It is

$$x^{GGR} = \left[\frac{\alpha(1+\pi)}{1+\pi\alpha} \right]^{\frac{\alpha}{1-\alpha}}, \quad c^{GGR} = \frac{1-\alpha}{1+\pi\alpha} x. \tag{25}$$

It is worth noting the linear feedback form of the consumption. Note also, there is a unique stationary program that navigates the economy to the GGR: here it is exactly the solution to the classical utilitarian problem above.

Remark 7 The restricted optimal rule (21) does not depend on the initial condition, hence it is time consistent in the illustration of this section. This property does not carry over to the general framework considered in the paper. Looking at the maximand (19) one observes that it does not depend on x_0, hence the optimal feedback cannot be configured by this parameter. More generally, one can expect this property to hold whenever the maximand can be made separable in μ and the state variable, as in the log example. For other economies, the restricted solution will not be time consistent.

5 Restricted Solution, Intermediate Generations and Shadow Discount Factor

Note that, so far, we have addressed the problem from a mathematical point of view. The problem is due to a lack of compactness and to restore existence one can, more or less arbitrarily, restrict the admissible controls so as to optimize on a compact set. However, there are many such restrictions one could conceive in order to obtain to compact set. How are we to choose among them? On what ground? Chichilnisky (2009) has offered recently yet another approach to the existence question. Her idea is to focus on a particular constraint for the stock of the resource at infinity. Interestingly enough, when controls are continuous and with a uniform bound on their variations, her approach not only compactifies the problem, but it also identifies one value for parameter θ of particular interest in the context of natural resource: that value is related to the marginal value[7] of the resource at the point of extinction.

We feel that Chichilnisky (2009)'s solution can be seen as a useful indication of where ethical thoughs may start. Admissible solutions should be restricted to avoid extinction at any date and a possibility to achieve this is either to impose a long run constraint in the discounted utilitarian problem or to impose an upper bound $\overline{\theta}$ to parameter θ when using her criterion: the two possibilities are equivalent (Chichilnisky 2009, Theorem 3). But there are many possible solutions that avoid extinction (imposing a long run stock larger than the one of extinction, or equivalently imposing $\theta < \overline{\theta}$ will do). For those cases, there is room for an autonomous ethical thought. Keeping this in mind, two important properties emerge from the illustration of the previous section.

1. The restricted approach benefits some intermediate generations; intuitively this is so because the care given to the very long run generation induces some concern for some intermediate generations, for with stationary exploitation rules natural resources cannot be lead to a situation that benefits the former without a transition that somehow benefits also the later; the sharpest way to grasp this property in the logarithmic example is to consider the situation where the discount factor is very low while parameter θ is close to zero. Because of a huge preference for the present, the discounted utilitarian rule is to consume from the start as much as possible, leaving little for future generations. And an almost optimal solution under Chichilnisky's criterion will be arbitrarily close to this path, accumulating slowly and switch to the golden rule in the very far future. On the other hand, with a sustainable preferences that favors heavily the future (low θ), the restricted solution will be arbitrarily close to the plan leading to the green golden rule, with a much lower consumption for the current generation, hence a larger stock for the subsequent generations to exploit.
2. The restricted optimal rule could obtain from the discounted utilitarian program with a different, actually lower, preference for the present (that is to say, a higher discount factor). Indeed, on the one hand observe from expression (22) that the discounted utilitarian rule is an decreasing function of β. And as this parameter increases toward the unitary value, the utilitarian discounted rule tends to the feedback leading to the green golden rule. On the other hand, from Proposition 3, we know that the restricted optimal feedback for Chichilisky's criterion lies between the discounted utilitarian feedback and the green golden rule feedback. Those two observations together imply that there exists another discount factor, call it β^c, that yields exactly the restricted feedback. To put it differently, the restricted solution to Chichilnisky's problem maximizes a fictitious discounted utilitarian criterion where the "shadow" discount factor is larger than

[7]"Marginal" here is to be understood in the sense of the Frechet derivative of the discounted utilitarian criterion.

the original discount factor, $\beta^c > \beta$ (preference for the present is lower). Consequently, the restricted solution does not produce a loss of efficiency but it clearly has a redistributive consequence: it curves the pattern of exploitation of the natural resource in favor of future generations, as one may expect.
Does this redistributive property carry over more general frameworks? We will conclude this paper with a brief investigation of this question. Assume the following properties hold:

Assumption 6 The utilitarian discounted rule is a continuous function of the discount factor when this parameter is close to unity.

With an abuse of notation let us emphasize this property by writing the discounted utilitarian rule as a function of the discount factor $\phi_{DU}(\beta)$.

Assumption 7 When discounting vanishes, the discounted utilitarian rule tends to the GGR rule, $\lim_{\beta \to 1} \phi_{DU}(\beta) = \phi_{GGR}$.

Assumption 8 When β is close to one, there exists a unique shadow discount factor that replicates the restricted solution, i.e.

$$\phi_{DU}(\beta^c) = \gamma \phi_{DU}(\beta) + (1 - \gamma)\phi_{GGR} \tag{26}$$

All those assumptions are satisfied in the logarithmic example, even when β is not close to one. Given the above assumptions, can we ascertain that the restricted solution is akin to an implicit discounted utilitarian solution featuring a lower preference for the present?

Expression (26) gives β^c as an implicit function of β. It states that when $\beta \to 1$, then $\phi_{DU}(\beta^c) \to \phi_{GGR}$; hence

$$\lim_{\beta \to 1} \beta^c(\beta) = 1 \tag{27}$$

Also, applying the implicit functions theorem to expression (26) one has

$$(\beta^c(\beta))' = \gamma \frac{\phi'_{DU}(\beta)}{\phi'_{DU}(\beta^c)}, \tag{28}$$

from which one can deduce

$$\lim_{\beta \to 1} (\beta^c(\beta))' = \gamma < 1. \tag{29}$$

To summarize, for values of β that tend to one, the shadow discount factor tends to one and $\beta^c(\beta) > \beta$. This proves:

Proposition 5 *Under Assumptions 6, 7, 8 the restricted solution to Chichilnisky's problem is consistent with a discounted utilitarian path under a lower discounting of future generations.*

One may wonder whether the above property restore a dictatorhsip role for the present? Our answer is finely-shaded, being both yes and no. *Stricto sensus*, *Non Dictatorship of the present* requires that the criterion is sensitive to the infinetely distant generations' well-being (Chichilnisky 1996, pp. 237 and 241), and clearly this is still the case under our approach, since we make no change to Chichilnisky's criterion. But one has to go beyond the formal definition of no dictatorship of the present. Leaving this deontologic point of view, sensitivity for the future is desirable insofar that, in cost-advantage or in optimization exercises, it prevents Society to totally disregard the future. One can suspect a dictatorship of the present because the restricted approach may prevent sensitivity to the future to translate into a social choice that cares about the long run. Under this consequentialist point of view, one may indeed fear a dictatorship of the present. Let us list the arguments in favor and against this argument.

- In favor, as stated in Proposition 5, under some conditions there exists a relationship between the restricted solution to CSWC and the solution to a fictitious discounted utilitarian criterion, the later reflecting a dictatorship of the present however large the discount factor may be.
- Against, there are a couple of reasons. Firstly the connection between the two problems is of a local nature, when β is close to one. It is possible that the shadow discount factor is exactly one, since $\beta^c > \beta$, in which case the associated criterion becomes the pure utilitarian one, that treats anonymously all generations. That fictitious criterion is unbounded hence it may or may not have an optimal solution. If there is one, it is not a dictatorship of the present. If there is not, and remembering that there exists a restricted solution, the relationship between the two problems collapses. Secondly, it is clear from the properties stated in Propositions 3 and 4 (existence of an operational trade-off between the present and future) that the restricted solution is sensitive to the weight given to the second term of CSWC. In particular the steady state stock of the resource can be made arbitrarily close to the reen golden rule if the weight given to the future is large enough.

This equivalence with a "shadow" discounted utilitarian problem is presumably an important property, given the recurrent debates about the proper discount rate to be used in cost-benefit analysis. It is often argued, for environmental issues in particular, that authorities should use discount rates lower than the rate prevailing on markets. The last proposition provides a further argument supporting this view: discount rates should be cut down if one is to be consistent with sustainable preferences.

6 Conclusions

Chichilnisky criterion for sustainability has the merit to be the unique explicit, complete, linear and continuous social welfare criterion that combines successfully the requirement of efficiency with an operational notion of intergenerational equity (*no dictatorship of the present* and *no dictatorship of the future*). But it has one important drawback: when applied in the context of renewable resources with a fixed discount factor, there exists no exploitation path that maximizes this criterion.

Could something be done to cope with this weakness? The present article offers a positive answer. Firstly we have recalled the reason of the inexistence problem. Second, elaborating on this information, we have identified a restricted class of controls in which there exists an optimal exploitation path for Chichilnisky's criterion. This class turned out to be the set of convex combinations between the optimal discounted utilitarian program and the golden rule program. Importantly, it has been shown in Theorem 1 that the idea of looking for an optimal path in the set of such convex combinations can be applied in a general framework, under rather weak sufficient conditions on the fundamentals of the problem. Also, it has been shown in a example that the larger the weight given to infinitely distant generations, the closer the restricted optimal path to the green golden rule. Finally, some ethical properties of this approach have been discussed. In some cases, it turns out that the restricted solution implies no loss of efficiency and benefits intermediate and infinitely distant generations.

The next natural step is to ask how a desirable path of the kind studied in our paper might emerge, not as the point of view of some sort of benevolent planner which is developed for the needs of a normative analysis but, switching to a more descriptive standpoint, as the result of interactions between a number of decentralized decision-making units? Of course, when it comes to implementation many conceptual and practical challenges appear, some of which are tackled by the contributions in this special issue of Economic Theory.

How to foster cooperation among decentralized decision units? The lack of cooperation is likely to result in under-investment in mitigation of Green House Gas emissions, even though Pareto improvments are possible (Rezai et al. 2010). But threats to fall back to the business-as-usual scenario might not be sufficient to sustain a Pareto optimal path as a subgame perfect non cooperative equilibrium. However, foreign aid can help. If the slower growth economies—like the United States and Western Europe—are willing to make transfers to China and India then the latter can be incentivized to cut emissions (Dutta and Radner 2010). Another proactive approach to ensure cooperation is through correctly designed protocols of negociations (Lecocq and Hourcade 2010) or, in a similar spirit, using a coasian solution (Chipman et al. (2010) , identify a necessary and sufficient condition for this last possibility). It is often argued that for global problems like climate change, solutions must be found at global levels. The effectiveness of partial or unilateral action to curb carbon emissions has been dismissed because of possible "carbon leakages", this referring to the rise of emissions in non-participating countries. Burniaux and Oliviera Martins (2010) offers a general equilibrium (GE) exploration of the key mechanisms and factors underlying the size of carbon leakages. However, Ostrom (2010) put forwards several reasons to focus also on lower scales of decisions. She underlines the merits of a polycentric approach to the problem of climate change in order to gain the benefits at multiple scales. Whatever the level under scrutiny, the relative merits of different usual regulations, taxes or quotas, is affected by the asymmetry of information between the regulator and firms (Karp and Zhang 2010).

Knowledge about the modern challenges posed by the environment is accumulating rapidly, and the contributions of this special issue is somehow a photo of this movement. On that picture, one observes that papers concerned by implementation in a dynamic context almost invariably use the discounted utilitarian criterion as a normative benchmark. The normative compass points to efficiency, but it is biased in favor of the present. One can guess that a promising direction of research might be to better articulate and integrate the results obtained via the social choice approach (such as Chichilnisky 1996; Lauwers 2010; Asheim et al. 2010 or ours) with results on the implementation front. Chichilnisly (2010) is a clear effort in this direction. Markets aggregate information revealed by stakeholders. If their preferences feature impatience, then markets and prices will be biased against long term goals. In particular, one may expect interest rates to be excessively high. Chichilnisky shows that *limited arbitrage* is a necessary and sufficient condition for *sustainable markets,* where the invisible hand delivers sustainable/equitable and efficient outcomes.

Appendix

Appendix A: Existence of a Stationary Program Leading the Green Golden Rule

The green golden rule problem reads as

$$\max_{(c,x)} \; U\,(c,x) \;, \tag{30}$$

subject to

$$x = G(x-c)\;, 0 \leq c \leq x\,. \tag{31}$$

It is worth starting the analysis of this problem with an investigation of the possible steady states, noted generically (c^s, x^s), before looking for those among them giving the highest utility. Fix $c = 0$. $G(.)$ is concave, $G'(0) > 1$ and $\lim_{x \to +\infty} G(x) < x$, therefore there are only two steady states, $x^s = 0$ which is unstable and another one, x^{sup}, which is globally asymptotically stable: indeed, starting from a positive initial stock lower (resp. larger) than the steady state, the resource monotonically increases (resp. decreases) while consumption is constant (zero actually), therefore the utility function is non-decreasing (resp. non-increasing) along the trajectory and it can be used to construct a Lyapunov function to prove the global asymptotic stability of x^{sup}. Now for any non-zero stationary consumption, $0 < c < x_t, \forall t$, applying the implicit function theorem to relation (31), one can check that the larger the stationary consumption, the lower the stable stationary stock:

$$\left.\frac{dx}{dc}\right|_{(c,x^{\text{sup}})} = \frac{G'}{G'-1} < 0\,, \tag{32}$$

where the inequality obtains since at any positive steady state, function $G(.)$ crosses the 45° line from above, hence $0 < G' < 1$. To summarize, as the stationary consumption increases away from zero, the locus of steady states $(c, x^{\sup})$ is characterized by lower stationary stocks. Then, in order to find the best steady state, note that the relevant space of stationary consumptions and stocks is compact:

$$x \in \left[0, x^{\sup}\right], \; c \in \left[0, x^{\sup}\right] \tag{33}$$

Since the transition function is bijective, we can write:

$$c = x - G^{-1}(x) \leq x \;\text{ for }\; x \geq 0. \tag{34}$$

Then rewrite the problem as follows:

$$\max_{x \in [0, x^{\sup}]} \; U\left[x - G^{-1}(x), x\right] . \tag{35}$$

There necessarily exists a solution since the function to be optimized is continuous and defined over a compact set. The solution cannot be zero since, by Assumption 4, the above function is strictly increasing at $x = 0$. Let us note $0 < c^* < x^*$ this solution ($c^* = x^*$ is not possible because $G(0) = 0$ would imply $c^* = x^* = 0$ for the second steady state as well).

It remains to show the existence of a stationary plan leading to the GGR. Clearly a possible, though not unique, such plan is the linear one:

$$c_t = \frac{c^*}{x^*} x_t . \tag{36}$$

Plugging back this plan into the dynamics, one has:

$$x_t = G\left[(1 - \frac{c^*}{x^*}) x_t\right] = g\left(x_t\right). \tag{37}$$

Function $g(.)$ is similar to function $G(.)$ upon a positive linear transformation of its arguments (since $c^* < x^*$): its dynamic properties are the same, the economy converges towards the golden rule (c^*, x^*). □

Appendix B: Admissibility of Convex Combinations

To state that $\left\{c_t^{\gamma}\right\}_{t=0}^{\infty}$ is admissible it must be proven that: (i) $c_t^{\gamma} \leq x_t$, $\forall t$ and (ii) $\lim_{t \to \infty} c_t^{\gamma} = c^{\gamma} < \infty$. Part (i) is true since

$$c_t^{\gamma} \leq \max\left\{\phi_{DU}\left(x_t\right), \phi_{GGR}\left(x_t\right)\right\} \leq x_t. \tag{38}$$

As for part (ii), note that

$$\lim_{t\to\infty} c_t^\gamma = \gamma \lim_{t\to\infty} \phi_{DU}\left(x_t\right) + (1-\gamma)\lim_{t\to\infty} \phi_{GGR}\left(x_t\right)$$
$$= \gamma x^{DU} + (1-\gamma)x^{GR} < \infty.$$

□

Appendix C: Continuity of Convex Combinations

This property can be established recursively. Note first that

$$\gamma\phi_{DU}\left(x_0\right) + (1-\gamma)\,\phi_{GGR}\left(x_0\right) \tag{39}$$

varies continuously with γ. Therefore, x_1^γ is also a continuous function of γ because $G(.)$ is a continuous function. Now it is easy to see that if this property holds for x_s^γ, $\forall s \le t$ for some t, it must hold for x_{t+1}^γ as well by Assumption 2, which completes the proof. □

Appendix D: Existence of a Restricted Optimal Program for CSWC

Under Assumptions 1, 2 and 3, each term in the series

$$J_T = \sum_{t=0}^{T} \beta^t U(c_t^\gamma, x_t^\gamma) \tag{40}$$

is continuous in γ. Let $U^{\text{sup}} = \max\left\{\left|\underline{U}\right|, \left|\overline{U}\right|\right\}$. Under the boundedness condition in Assumption 1

$$J_\infty^s = \theta \sum_{t=0}^{\infty} \left\|\beta^t U(c_t^\gamma, x_t^\gamma)\right\|_\infty + (1-\theta)\lim_{t\to\infty} U(c_t^\gamma, x_t^\gamma)$$
$$\le \theta U^{\text{sup}} \sum_{t=0}^{\infty} \beta^t + (1-\theta)\,U^{\text{sup}} = \frac{\theta U^{\text{sup}}}{1-\beta} + (1-\theta)\,U^{\text{sup}} < \infty\,,$$

where $\|.\|_\infty$ denotes the sup norm. By definition:

$$J^s = \theta \sum_{t=0}^{\infty} \beta^t U(c_t^\gamma, x_t^\gamma) + (1-\theta)\lim_{t\to\infty} U(c_t^\gamma, x_t^\gamma)$$
$$\le J_\infty^s = \theta \sum_{t=0}^{\infty} \left\|\beta^t U(c_t^\gamma, x_t^\gamma)\right\|_\infty + (1-\theta)\lim_{t\to\infty} U(c_t^\gamma, x_t^\gamma) < \infty.$$

So the series

$$J_T^s = \theta \sum_{t=0}^{T} \beta^t U(c_t^\gamma, x_t^\gamma) + (1-\theta)\,U(c_T^\gamma, x_T^\gamma) \tag{41}$$

converges normally towards the function J^s, which implies that it also converges uniformly, that is

$$\forall \epsilon > 0, \exists N \in N, \forall \gamma \in [0,1], \;\; \forall T > N, \; \left|J_T^s - J^s\right| < \epsilon. \tag{42}$$

This last property ensures that J^s is a continuous function of γ. The demonstration is then completed, since the continuous mapping of a compact set is itself a compact set, with a maximal point. □

Appendix E: Proof of Proposition 3

The problem is to show that the coefficient of the optimal linear feedback,

$$\mu = \frac{(1-\alpha)(\alpha\beta - 1)(\theta\beta - \beta + 1)}{(1-\theta)[\alpha\beta(1-\beta)+\beta] + \theta\alpha - 1 + \pi\left[\alpha\beta(1+\alpha-2\theta) - \alpha(1-\theta)\left(1+\alpha\beta^2\right)\right]}, \tag{43}$$

is actually a convex combination

$$\gamma\frac{1-\alpha\beta}{1+\pi\alpha\beta} + (1-\gamma)\frac{1-\alpha}{1+\pi\alpha}, \;\; \gamma \in [0,1] \;, \tag{44}$$

of the coefficients of the discounted utilitarian optimal feedback and of the green golden rule.

Solving the equation

$$\gamma\frac{1-\alpha\beta}{1+\pi\alpha\beta} + (1-\gamma)\frac{1-\alpha}{1+\pi\alpha} = \frac{(1-\alpha)(\alpha\beta - 1)(\theta\beta - \beta + 1)}{(1-\theta)[\alpha\beta(1-\beta)+\beta] + \theta\alpha - 1 + \pi\left[\alpha\beta(1+\alpha-2\theta) - \alpha(1-\theta)\left(1+\alpha\beta^2\right)\right]}$$

for γ, one finds

$$\gamma = \frac{(\alpha - 1)\theta(\pi\alpha\beta + 1)}{D}, \tag{45}$$

where

$$D = \begin{array}{c} \beta - \pi\alpha + \alpha\beta(1 + \pi - \beta + \pi\alpha - \pi\alpha\beta) - 1 \\ +\theta\left(\alpha - \beta + \pi\alpha - \alpha\beta - 2\pi\alpha\beta + \alpha\beta^2 + \pi\alpha^2\beta^2\right) \end{array}. \tag{46}$$

Let us investigate the properties of the above expression, viewed as a function $\gamma(\theta)$. Note first that

$$\gamma(0) = 0\,, \quad \gamma(1) = 1\,. \tag{47}$$

Also,

$$\gamma'(\theta) = \frac{N}{D^2} > 0\,. \tag{48}$$

where

$$N = (\pi\alpha\beta + 1)(\pi\alpha + 1)(1 - \alpha\beta)(1 - \beta)(1 - \alpha) > 0 . \tag{49}$$

This means that $\gamma(\theta) \in [0, 1]$, hence the result. □

References

Alvarez-Cuadrado, F., & Ngo Van Long, N. (2009). A mixed Bentham-Rawls criterion for intergenerational equity: Theory and implications. *Journal of Environmental Economics and Management, 58*, 154–168.

Asheim, G. (1996). Ethical preferences in the presence of resource constraints. *Nordic Journal of Political Economy, 23*, 55–67.

Asheim, G., Mitra, T., & Tungodden, B. (2010). Sustainable recursive social welfare functions. *Economic Theory*.

Basu, K., & Mitra, T. (2003). Aggregating infinite utility streams with intergenerational equity: The impossibility of being Paretian. *Econometrica, 71*, 1557–1563.

Beltratti, A., Chichilnisky, G., & Heal, G. (1994). The environment and the long run: A comparison of different criteria. *Richerche Economiche, 48*, 319–340.

Burniaux, J.-M., & Oliviera Martins, J. (2010). Carbon leakages: A general equilibrium view. *Economic Theory*.

Chichilnisky, G. (1996). An axiomatic approach to sustainable development. *Social Choice and Welfare, 13*, 231–57.

Chichilnisky, G. (1997). What is sustainable development? *Land Economics, 73*(4), 467–91.

Chichilnisky, G. (2009). Avoiding extinction: Equal treatment of the present and the future. *The Open-Access, Open-Assessment E-Journal, 3*, 2009–2032.

Chichilnisly, G. (2010). Sustainable markets with short sales. *Economic Theory*.

Chipman, J. S., & Tian, G. (2010). Detrimental Externalities, Pollution Rights, and the "CoaseTheorem". *Economic Theory*.

Diamond, P. A. (1965). The evaluation of infinite utility stream. *Econometrica, 33*, 170–177.

Dutta, P. K., & Radner, R. (2010). Capital growth in a global warming model: Will China and India sign a climate treaty? *Economic Theory*.

Fleurbaey, M., & Michel, P. (2003). Intertemporal equity and the extension of the Ramsey criterion. *Journal of Mathematical Economics, 39*, 777–802.

Fisher, I. (1930). *"The theory of interest: As determined by impatience to spend income and opportunity to invest it", 1954 Reprint*. New York: Kelley and Millman.

Heal, G. (1998). *Valuing the future: Economic theory and sustainability*. New York: Columbia University Press.

Karp, L., & Zhang, J. (2010). Taxes versus quantities for a stock pollutant with endogenous abatement costs and asymmetric information. *Economic Theory*.

Koopmans, T. (1960). Stationary ordinal utility and impatience. *Econometrica, 28*, 287–309.

Lauwers, L. (2010). Intergenerational equity, efficiency, and constructibility. *Economic Theory*.

Lecocq, F., & Hourcade, J. C. (2010). Unspoken ethical issues in the climate affair: Insights from a theoretical analysis of negotiation mandates. *Economic Theory*.

Li, C.-Z., & Lofgren, K.-G. (2000). *Journal of Environmental Economics and Management, 40*, 236–250.

Ostrom, E. (2010). Nested externalities and polycentric institutions: Must we wait for global solutions to climate change before taking actions at other scales? *Economic Theory*.

Pezzey, J. (1992). Sustainable development concepts. *World Bank Environment*. Paper Number 2.

Pigou, A. C. (1920). *The economics of welfare*. London: Macmillan.

Ramsey, F. (1928). A mathematical theory of savings. *Economic Journal, 38*, 543–559.

Rezai, A., Foley, D. K., & Taylor, L. (2010). Global warming and economic externalities. *Economic Theory*.
Stokey, N. L., & Lucas, R. E. (with E.C. Prescott) (1989). *Recursive methods in economic dynamics*. Harvard University Press
Svensson, L. G. (1980). Equity among generations. *Econometrica*, *48*, 1251–1256.
Zame, W. (2007). Can utilitarianism be operationalized? *Theoretical Economics*, *2*, 187–202.

The Axiomatic Approach to the Ranking of Infinite Streams

Luc Lauwers

1 Introduction

Global environmental issues—like biodiversity conservation or climate change—are in reality long term issues that are not properly taken into account with traditional models that incorporate the impatience axiom manifested in fixed discount factors and in the use of present discounted utility criteria.

This social impatience conflicts with the utilitarian tradition of moral philosophy where it is recommended to treat different generations equally. As (Sidgwick 1907, p. 414) writes: "the time at which a man exists cannot affect the value of his happiness from a universal point of view; and [...] the interests of posterity must concern a utilitarian as much as those of his contemporaries." Ramsey (1928), in one of the first formal studies on the evaluation of social welfare in an intertemporal framework, strongly endorses this view. Despite this position, he nevertheless introduces a rate of discount in some of the investigations. Simply because undiscounted utilitarianism provides no unique answer in case the maximum total welfare is infinite and this infinite value is achieved by many feasible paths. Later on, the same doubt on the sustainability of the equal treatment principle occurs when Koopmans (1960) characterizes the discounted utilitarian rule on the basis of five appealing axioms. Section 3 recalls this result of Koopmans.

It is only recently (Lauwers 2010a; Zame 2007) that the deep cause of the conflict as experienced by Ramsey and Koopmans was exposed: it is not possible to define in a "*constructive*" way a complete ranking on the set of infinite consumption paths

I thank Graciela Chichilnisky and Armon Rezai for inviting me to write this chapter. I am also grateful for the comments from a referee, Tom Potoms, and Luc Van Liedekerke.
24 juli 2014.

L. Lauwers (✉)
Center for Economic Studies, Katholieke Universiteit Leuven,
Naamsestraat 69, 3000 Leuven, Belgium
e-mail: Luc.Lauwers@econ.kuleuven.be; luc.lauwers@kuleuven.be

G. Chichilnisky and A. Rezai (eds.), *The Economics of the Global Environment*,
Studies in Economic Theory 29, DOI 10.1007/978-3-319-31943-8_12

that combines anonymity (an axiom that captures equal treatment) and Pareto (or sensitivity).[1] Since the question of how to evaluate policies that involve the distant future is normative, it should by no means be answered through "*non-constructive*" mathematics such as the Axiom of Choice or Szpilrajn's Lemma.[2] Only constructible and well defined criteria can take part in the discussions. The inevitable implication, then, is that at most two of the three requirements—completeness, anonymity, and Pareto—are compatible. Either one has to drop the requirement of completeness, or one has to weaken the requirement of anonymity and/or Pareto. Anyway, one cannot but accept the above incompatibility and temper the ambition to find a representable ordering (as encapsulated in Koopmans' postulates). Incompleteness of the criterion should be interpreted as a consequence of an unbiased position in the discussion. In order to leave as many options open, the primitive should be a partial ordering. Section 4 starts with the introduction of Diamond's axiom of equal treatment and closes with the statement and the interpretation of the Lauwers-Zame impossibility theorem.

Sections 5 and 6 show two different routes to extend in a constructive way a sequence of finite dimensional orderings towards a partial ordering on the set of infinite streams. If the finite dimensional orderings all satisfy anonymity and Pareto, then the resulting infinite dimensional partial ordering inherits both these properties.

Section 7 drops the Pareto principle, strengthens anonymity, imposes monotonicity and completeness, and discusses rules in the Rawlsian spirit such as the infimum rule and the limit inferior. Also, the rank-discounted utilitarian rule (Zuber and Asheim 2012) is discussed. Although each of these rules violates the weakest form of Pareto, they may be extremely useful in a two step procedure as proposed by Ferejohn and Page (1978, p. 274)[3]:

> Our result suggests that the search for a fair rate of discount is a vain one. Instead of searching for the right number, i.e. 'the' social rate of discount, we must look to broader principles of social choice to incorporate ideas of intertemporal equity. Once found, these principles might be used as side conditions in a discounting procedure to rule out gross inequities that can arise with discounting, even with a low discount rate

A strongly anonymous welfare function might indeed be used as a first step. As a strongly anonymous welfare function typically has thick levels sets, a second step can further investigate the set of optimal paths obtained in the first step.

Sections 8 and 9 return to social welfare functions. At the center of the sustainable discounted utilitarian rule (Asheim and Mitra 2010) is the axiom of Hammond equity for the future, according to which a sustained improvement at the cost of the

[1] In his review on intergenerational equity, Asheim (2010, Sect. 3.2) coins this result as the Lauwers-Zame impossibility theorem.

[2] I want to mention already here that the combination of continuity with respect to the sup-topology and representability does not guarantee that the ranking rule is constructible. The other way around, the representation of a non-constructible ordering, is in itself a non-constructible object.

[3] Ferejohn and Page (1978) have shown that unrestricted domain, Pareto, Arrovian independence, and stationarity results in dictatorship of the first generation. This result has been strengthened by Packell (1980).

present generation is considered an improvement only in case this first generation is ex post still the better off. They succeed in modifying the axioms of Koopmans towards a characterization. Finally, Chichilnisky (1996, 1997) introduces the axioms of non-dictatorship of the present and non-dictatorship of the future. She proposes a convex combination of the discounted utilitarian rule and a map that captures the limiting behavior of an infinite stream. The discounted rule prevents the future from dictatorship, the limit-part prevents the present from dictatorship. She coins those social welfare functions as sustainable preferences.

This overview is organized as follows. Section 2 introduces notation and describes the problem. The next sections unfold as described above. As the focus is on the axiomatic approach, we recall and discuss different appealing properties or axioms. Thomson (2001, p. 349) motivates as follows:

> The objective of the axiomatic program is to give as detailed as possible a description of the implications of properties of interest, singly or in combination, and in particular to trace out the boundary that separates combinations of properties that are compatible from combinations of properties that are not.

Applied to the ranking of infinite streams, one investigates on what ethical conditions various ranking criteria are based and proceeds to evaluate the normative appeal of these conditions. Although not included in this overview, an alternative approach that confronts the criteria with different technological environments and compares the properties of the intergenerational well-being streams that are generated, is undoubtedly a necessary route in the debate.[4]

2 Notation

We consider a model with successive generations, each generation living exactly one period. Time is discrete and starts with period 1. Let $\mathbb{N} = \{1, 2, \ldots, t, \ldots\}$ be the set of natural numbers, $\mathbb{R}$ the set of real numbers, and let Y be a subset of $\mathbb{R}$. Let X be the infinite cartesian product $Y^{\mathbb{N}}$. A sequence $x = (x_1, x_2, \ldots, x_t, \ldots)$ in X is said to be an infinite stream of generational well-being, for each t in $\mathbb{N}$ the real number x_t indicates the average well-being of generation t. The indicator of well-being is assumed to be at least ordinally measurable and level comparable across generations. Also, for each generation, the distribution of resources among the individuals of a same generation is neglected. Furthermore, the population size is assumed to be given and constant over time.

For each n in $\mathbb{N}$ and for each x in X, we write $x = (x_{-n}, x_{+n})$ with $x_{-n} = (x_1, x_2, \ldots, x_n)$ and $x_{+n} = (x_{n+1}, x_{n+2}, \ldots)$. Note that $x_{-1} = (x_1)$. For each x and y in X, we write

[4]Section 3 provides a simple example to indicate that paths optimal with respect to a sustainable-equitable approach might differ substantially from optima generated by the discounted utilitarian rule.

- $x \geq y$ if $x_t \geq y_t$ for each t in $\mathbb{N}$,
- $x > y$ if $x \geq y$ and $x \neq y$,
- $x \gg y$ if $x_t > y_t$ for each t in $\mathbb{N}$,
- $d_s(x, y) = \sup_{t \text{ in } \mathbb{N}} |x_t - y_t|$.

The different inequalities will be used to formulate axioms of monotonicity, the distance function d_s generates the sup-topology and is used to formulate an axiom of continuity.

The object we look for is a partial social welfare ordering, denoted by $\succsim$, on the set X of infinite streams. That is, $\succsim$ is a relation in X, and $x \succsim y$ means that the infinite stream x is at least as good as y. The relation $\succsim$ is assumed to be (*i*) transitive, for each x, y, and z in X, we have that $x \succsim y$ and $y \succsim z$ implies $x \succsim z$, and (*ii*) reflexive, for each x in X, we have that $x \succsim x$. The symmetric and the asymmetric parts of $\succsim$ are denoted by $\sim$ and $\succ$. The partial ordering $\succsim_e$ is said to extend the partial ordering $\succsim$, or $\succsim$ is said to be a subrelation to $\succsim_e$, in case, for each x and y in X, $x \succ y$ implies $x \succ_e y$ and $x \sim y$ implies $x \sim_e y$.

The term social welfare order refers to a partial social welfare ordering $\succsim$ that is complete, i.e. for each x and y in X, we have $x \succsim y$ or $y \succsim x$. The term social welfare function refers to a map $f : X \to \mathbb{R}$ that represents some social welfare order $\succsim$: for each x and y in X, we have $f(x) \geq f(y)$ if and only if $x \succsim y$. Recall that in view of the Lauwers-Zame impossibility result, completeness (or representability) is not a neutral requirement as it already excludes the combination of the axioms of anonymity and Pareto.

3 Discounted Utilitarianism

Koopmans (1972b)[5] considers a social welfare order $\succsim$ on the set of all bounded consumption streams, i.e. infinite streams for which the supremum and the infimum are both finite,[6] and investigates the next five postulates.

Continuity. The relation $\succsim$ is continuous with respect to the sup-topology, i.e. the topology generated by the distance function d_s.

Sensitivity. There exist infinite streams x and $y = (y_1, x_{+1})$ such that $x \succ y$.

The next axiom appeals to the following rankings generated by $\succsim$ and some fixed reference stream $z = (z_1, z_2, z_3, \ldots, z_t, \ldots)$. The orderings $\succsim_{z_{+1}}$ on Y, $\succsim_{z_{+2}}$ on Y^2, and $\succsim_{z_{-1}}$ on X are defined by

[5] See also Koopmans (1960, 1965, 1972a) and Koopmans et al. (1964).

[6] For simplicity, $Y = \mathbb{R}$, x_t is the consumption of generation t, and the set ℓ_∞ of bounded streams takes the role of X.

- for each x and y in Y, we write $x \succsim_{z_{+1}} y$ if $(x, z_{+1}) \succsim (y, z_{+1})$,
- for each (x_1, x_2) and (y_1, y_2) in Y^2,
  we write $(x_1, x_2) \succsim_{z_{+2}} (y_1, y_2)$ if $(x_1, x_2, z_{+2}) \succsim (y_1, y_2, z_{+2})$, and
- for each x and y in X, we write $x_{+1} \succsim_{z_{-1}} y_{+1}$ if $(z_1, x_{+1}) \succsim (z_1, y_{+1})$.

Independence. The three orderings, $\succsim_{z_{+1}}$ on Y, $\succsim_{z_{+2}}$ on Y^2, and $\succsim_{z_{-1}}$ on X, do not depend on the reference stream z.

Stationarity$_0$. There exists an x_1^* in Y such that

$$(x_1^*, x_2, x_3, \ldots, x_t, \ldots) \succsim (x_1^*, y_2, y_3, \ldots, y_t, \ldots)$$

if and only if

$$(x_2, x_3, \ldots, x_t, \ldots) \succsim (y_2, y_3, \ldots, y_t, \ldots).$$

Monotonicity.[7] Let x and y in Y satisfy $x_t \geq y_t$ for each t in $\mathbb{N}$. Then, $x \succsim y$.

We briefly discuss these five axioms. Koopmans (1972a) motivates the continuity axiom: a small change in a prospect cannot drastically change the position of that prospect in the ranking of all other prospects. Continuity in combination with a monotonicity axiom that imposes $x \succ y$ in case $x > y$ and $x \succsim y$ in case $x \geq y$ implies that the ranking $\succsim$ is representable by a real valued function (Diamond 1965; Lauwers 1997a).[8] The usefulness of a representation by a continuous function lies primarily in the availability of stronger mathematical techniques.[9] Sensitivity excludes the ordering $\succsim$ from being trivial in the sense that all infinite streams are equally good. This axiom also prohibits dictatorship of the future (see Sect. 9): the ranking of infinite streams is not solely based upon the limiting behavior of the infinite streams. Independence removes all complementarity between the well-being of different (subsequent) generations, cannot be regarded as a realistic assumption, and should be looked upon as a way to facilitate the investigations (Koopmans 1972b, p. 83).[10] Independence implies that the particular value x^* in the axiom of stationarity$_0$ can be replaced with any value in Y. In case the axiom of independence is imposed upon the ranking $\succsim$, stationarity$_0$ strengthens to the next axiom.

Stationarity. For each x_1^* in Y we have

$$(x_1^*, x_2, x_3, \ldots, x_t, \ldots) \succsim (x_1^*, y_2, y_3, \ldots, y_t, \ldots)$$

[7]Koopmans considers infinite streams of vectors instead of scalars. The axiom of monotonicity is a one-dimensional version of Koopmans' axiom.

[8]Diamond (1965) follows Debreu (1954) to prove this result.

[9]This motivation, however, is wrong in case the continuous social welfare function represents a non-constructible order.

[10]See also Fleurbaey and Michel (2003, Sect. 3.4).

if and only if

$$(x_2, x_3, \ldots, x_t, \ldots) \succsim (y_2, y_3, \ldots, y_t, \ldots).$$

Stationarity compares infinite streams with a common first period value in the same way as the infinite streams that are obtained by deleting these first period values and advancing the timing of all subsequent values by one period. Repeated use of stationarity results in comparing two streams (x_{-t}, x_{+t}) and (x_{-t}, y_{+t}) with the common head x_{-t} in the same way as the infinite tails x_{+t} and y_{+t}. In other words, if the first t generations are not affected the ranking is made as if the present time (date 1) actually was in time t. The passage of time has no effect on preferences. Stationarity also implies that the ranking of two infinite streams is not altered if both streams are postponed by one unit of time and identical values are plugged in at time period 1. Monotonicity demands that if each generation is at least as good off in x than in y, the infinite stream x should be considered at least as good as y.

These five postulates characterize the discounted utilitarian rule.

Theorem 1 (Koopmans 1972b)[11] *Let the social welfare ordering $\succsim$ on the set of bounded streams satisfy continuity with respect to the sup-topology, sensitivity, independence, stationarity$_0$, and monotonicity. Then, the ordering $\succsim$ is represented by a continuous function*

$$D : (x_1, x_2, \ldots, x_t, \ldots) \longmapsto (1-\alpha) \sum_{t=1}^{\infty} \alpha^{t-1} u(x_t),$$

with u nowhere decreasing and continuous and with α in the open interval $(0, 1)$.[12]

A first step towards this result investigates the representation of $\succsim$ restricted to the subset of infinite streams with a fixed tail, say z_{+t}. Next, the domain of infinite streams with a constant tail is considered. The partial results are then generalized towards the full domain of bounded streams. Discounted utilitarianism satisfies a recursive relation:

$$D(x_1, x_2, \ldots, x_t, \ldots) = (1-\alpha)\, u(x_1) + \alpha\, D(x_2, x_3, \ldots, x_{t+1}, \ldots),$$

for each infinite stream x. The map D attaches the weight $1-\alpha$ to the utility allocated to the present period and the complementary weight α to the aggregated utility of all future periods.

Although (Koopmans 1965, Sect. 6) holds an ethical preference for neutrality, he provides an argument for the discounted utilitarian rule:

---

[11]Different axiomatizations are obtained by Lauwers (1997c), Bleichrodt et al. (2008), and Asheim et al. (2012).

[12]The factor $(1-\alpha)$ in the definition of D ensures that $D(x, x, \ldots, x, \ldots) = u(x)$. Hence, the weights with which the $u(x_t)$ are multiplied add up to 1.

> We admit to an ethical preference for neutrality as between the welfare of different generations. ... A previous investigation has shown that there does not exist a utility function of all consumptions paths, which at the same time exhibits timing neutrality and satisfies other reasonable postulates which all utility functions used sofar have agreed with.

The mathematical conflict between neutrality and the five postulates overrules the ethical preference for neutrality. In his 1975 Nobel Memorial Lecture, Koopmans (1977) returns to the issue of a positive discount factor:

> Thus, the impatience expressed by a positive discount rate merely denies to uncounted distant generations a permanently higher level of consumption because that would necessitate a substantially smaller present consumption. Perhaps a pity, but not a sin.

Arrow (1999) considers a world in which an investment by the first generation generates a perpetual stream of benefits. The undiscounted total gain exceeds the finite loss to the first generation, and the optimal path would almost sacrifice the first generation. He concludes "that the strong ethical requirement that all generations be treated alike, itself reasonable, contradicts a very strong intuition that it is not morally acceptable to demand excessively high saving rates of any one generation, or even every generation." This argument, however, assumes a non-decreasing path and leaves open the case where a period of economic growth is followed by a period of economic regression. Then, the 'small sin' might take problematic proportions.

The set of five axioms leads to the 'class' of discounted utilitarian rules. The axioms, however, do not pass any judgment about the value of α and the particular form of u. Let us illustrate the effect of a change in α. The next table (Fleurbaey and Zuber 2012, Table 1) considers α equal to 0.9862 and 0.9737 and shows the minimum return a one dollar investment for the future should have in order to be considered better than consuming it now.

	$\alpha = 0.9862$	$\alpha = 0.9737$	Ratio
Time period 50	2.00	3.79	1.89
Time period 100	4.02	14.36	3.57
Time period 200	16.13	206.11	12.78
Time period 1000	1091327.24	371914916666.52	340791.38

The minimal return sufficient to defend a one dollar investment today

The value $r = (1/\alpha) - 1$ has the interpretation of a discount rate, $\alpha = 0.9862$ (resp. 0.9737) corresponds to $r = 1.4\,\%$ (resp. 2.7 %). The final column has the following alternative interpretation. When the discount rate r jumps from 1.4 to 2.7 % the minimal return that is sufficient to defend a one dollar investment at year 1 at the benefit of year 200 becomes 12.78 times larger. The huge ratios in the final column show the impact of a change in the discount rate.

We close this section with a simple example illustrating the shortsightedness of the time-discounted utilitarian approach (Lauwers 2012). Consider an economy in which trees are a necessary input to production or consumption. The dynamics of tree reproduction are as follows. If n out of $2n$ subsequent generations cut the forest at a maximal rate, the species become extinct after the $2n$'th generation, in which case there is zero utility at every period from then on. Assume this strategy results in utility streams of the form $u^n = (0.1, 0.1, \dots, 0.1; 1, 1, \dots, 1; 0, 0, \dots)$ with the first (resp. last) 1 at the $n+1$ (resp. $2n$)'th place, in which generations $n+1, \dots, 2n$ cut at a full capacity and exhaust the forest. When the consumption of the forest is delayed and n becomes larger, the forest slightly expands and more generations can benefit. Alternatively, generations can invest in the forest and only cut at an equilibrium rate which allows the forest to survive. This strategy results in the utility stream $u^\infty = (0.25, 0.25, \dots, 0.25, \dots)$ in which each generation reaches the same utility level. Optimization with respect to a discounted utilitarian rule leads to the elimination of the forest.[13] If the long term future is considered important, the constant stream u^∞ should be ranked strictly above a stream where within a finite horizon the forest is consumed.

The need for alternatives to the discounted utilitarian rule should be clear. It is a natural step in the axiomatic approach to confront the rule characterized through a set of axioms with the consequences it generates in specific environments. In case a set of desirable axioms leads to undesirable consequences, there is always the invitation to reconsider the axioms. This iterative process between moral principles and their examination in particular models is supported by, for example, Atkinson (2001). Dasgupta and Heal (1979) state

> ... it is legitimate to revise or criticize ethical norms in the light of their implications.

4 Finite Anonymity

Diamond (1965) continues the axiomatic approach initiated by Koopmans. He considers infinite utility streams: x_t is the one period utility level associated with consumption in period t and Y is the closed interval $[0, 1]$. In order to investigate whether or not impatience is unavoidable when ranking infinite streams, he introduces the next axiom. Let $\succsim$ be a partial order on the set X.

Finite anonymity. For each $x = (x_1, x_2, \dots, x_t, \dots)$ in X and for each t in $\mathbb{N}$, we have

$$x \sim x^t = (x_t, x_2, \dots, x_{t-1}, x_1, x_{t+1}, \dots).$$

The infinite stream x^t is obtained from x by switching the coordinates x_1 and x_t. A finitely anonymous evaluation is indifferent between two infinite streams that are equal up to a finite number switches in the coordinates. A ranking $\succsim$ which treats all generations equally should satisfies this condition. Diamond keeps the axiom of continuity with respect to the sup-topology and imposes the following strengthening of the axiom of sensitivity.

---

[13]The map $x \mapsto (1-\beta)(x_1 + \beta x_2 + \cdots + \beta^{t-1} x_t + \cdots)$ obtains a maximal value, equal to 0.3025, in one of the streams of type t^n; while the stream u^∞ obtains a lower value of 0.25.

Strong Pareto. For each x and y in X, we have $x \succ y$ as soon $x > y$.

The imposition of strong Pareto requires that the welfare ordering judges a utility stream superior to another as soon at least one period obtains a higher utility while all other periods obtain at least the same utility. This ability to detect an improvement in a single period makes strong Pareto a very demanding axiom in the study of infinite streams. In order to make the axioms of strong Pareto and finite anonymity meaningful instruments, a common zero level is presupposed. A finite context is sufficient to explain this statement. The combination of strong Pareto and finite anonymity is unable to compare the couples $(3, 8)$ and $(5, 4)$. However, after a shift in the 'zero' levels a decision might be made: the combination of strong Pareto and anonymity ranks $(3, 8) + (1, -1)$ above $(5, 4) + (1, -1)$ since $(4, 7)$ dominates $(3, 6)$. Furthermore, recall that the combination of strong Pareto and continuity with respect to the sup-topology entails that the ranking $\succsim$ is representable by a continuous real valued map.

The following theorem, a result that Diamond attributes to Yaari, reveals a fundamental conflict.

Theorem 2 (Diamond 1965) *There does not exist a social welfare order* $\succsim$ *on the set* X *that satisfies continuity, strong Pareto, and finite anonymity.*

This impossibility result has been the starting point of an extensive literature on its robustness. The axiom of continuity with respect to the sup-topology is, in contrast to strong Pareto and finite anonymity, a rather technical condition. Furthermore, as different distance functions generate different notions of continuity, the axiom of continuity is manipulable. For example, with respect to the discrete topology on X, each ordering $\succsim$ becomes continuous and continuity becomes an empty concept. Therefore, axioms of continuity in the infinite dimensional framework are considered as controversial. Svensson (1980) shows the existence of a complete and transitive relation that combines strong Pareto, finite anonymity, and a very weak continuity requirement. Basu and Mitra (2003) insist on representability by a real valued function, drop continuity, and obtain again an impossibility result. Fleurbaey and Michel (2003) consider the following Pareto axiom.

Weak Pareto. For each x and y in X, we have (*i*) $x \succsim y$ as soon $x \geq y$ and (*ii*) $x \succ y$ as soon $x \gg y$.

Weak Pareto strengthens monotonicity: a utility stream is considered at least as good as another as soon each period obtains at least the same utility. Furthermore, according to weak Pareto a utility stream is superior to another as soon each period obtains a higher utility. A ranking that satisfies strong Pareto also satisfies weak Pareto. Fleurbaey and Michel (2003) strengthen Diamond's theorem: a social welfare ordering cannot simultaneously satisfy continuity with respect to the sup-topology, finite anonymity, and weak Pareto.

A deep result in this track of research was conjectured by Fleurbaey and Michel (2003) and confirmed by Lauwers (2010a) and Zame (2007). We present the discrete version with $Y = \{0, 1\}$. A discrete version of the Pareto principle is needed.

Intermediate Pareto. For each x and y in X, we have (i) $x \succsim y$ as soon $x \geq y$ and (ii) $x \succ y$ as soon $x_i > y_i$ for infinitely many i in $\mathbb{N}$.

Theorem 3 (Lauwers 2010a) *A partial order on the set* $X = \{0, 1\}^{\mathbb{N}}$ *of infinite utility streams made up of zeros and ones that satisfies intermediate Pareto and finite anonymity either is incomplete, or is a non-constructive object (and hence has no explicit description).*

This theorem considers only two levels of utility. Then, intermediate Pareto is the weakest version of Pareto that is non-trivial. In this limited framework a partial order is unable to combine (in a constructive way) completeness, intermediate Pareto, and finite anonymity.

With respect to the change from a continuous towards a discrete setting, one might argue that there exists a smallest unit (or quantum) of utility and that a discrete level set Y is natural. In such a framework, intermediate Pareto seems appropriate. If the set Y of utility levels is equal to $\mathbb{N}$, then the map $X \to \mathbb{R} : x \mapsto \min_t x$ defines a complete, finitely anonymous, and weakly Paretian order on X that violates intermediate Pareto. If, however, the set Y is the unit interval $[0, 1]$, the minimum of an infinite stream is not well defined while the map $x \mapsto \inf_t x$ violates weak Pareto. Indeed, the streams

$$x = (1, 1/2, 1/3, \ldots, 1/n, \ldots) \text{ and } y = (0, 0, \ldots, 0, \ldots)$$

dominate each other ($x \gg y$) while they have the same infimum. This kind of situations do not occur in a discrete setting.[14] Intermediate Pareto and minimum should be seen as the discrete analogue to weak Pareto and infimum.

The above theorem appeals to the concept of constructibility. In order to explain this, consider Brouwer's fixed theorem: each continuous function from a convex compact subset of a Euclidean space to itself has a fixed point. A well known proof of this result is based on algebraic topology and is not constructive: the proof shows the existence of a fixed point but does not specify where it is located. Only later on constructive proofs, algorithms to detect the location of a fixed point, were provided. Theorem 3 is in the same spirit. Svensson (1980) already showed the *existence* of a complete and transitive relation that combines strong Pareto and finite anonymity. Svensson's proof, however, uses the Axiom of Choice, which is within the axiomatic setup of set theory in mathematics a non-constructive axiom. Theorem 3 shows the impossibility to provide a constructive proof of Svensson's result. The use of a non-constructive axiom (in the spirit of the Axiom of Choice) cannot be avoided to obtain Svensson's result. As a consequence, Svensson's existence proof contributes almost nothing to the discussion on how exactly infinite streams should be ordered.

---

[14]Dubey (2011) and Dubey and Mitra (2011) investigate the role of the set Y of possible utility levels and refine the results of Lauwers (2010a) and Zame (2007).

5 Pareto Dominating Tails

The framework with an infinite number of periods generates incompatibilities that are easy to reconcile in a finite work. This section explains how to construct, starting from an infinite sequence of finite dimensional partial orderings, a partial ordering on X that combines finite anonymity and strong Pareto. The following idea is at the basis of the construction. If the infinite stream x Pareto dominates y, all generations unanimously (and independently) agree to rank x above y. If, for some T, the infinite tail x_{+T} Pareto dominates y_{+T}, then from $T+1$ onwards each generation agrees to rank x above y. The problem, in this case, reduces to check whether the anonymous aggregative decision of the finite horizon society $\{1, 2, \dots, T\}$ agrees with the unanimous decision of generations $T+1, T+2, \dots, T+k, \dots$.

Before we explain the construction, let us list four well documented ranking relations on the finite dimensional Euclidean space $\mathbb{R}^n$: the Suppes-Sen grading principle $\succsim_n^S$, the utilitarian ordering $\succsim_n^U$, the leximin ordering $\succsim_n^L$, and the generalized Lorenz partial ordering $\succsim_n^G$.[15]

We need some extra notation. For each n-tuple a in $\mathbb{R}^n$ let a^+ be a rearrangement of a that satisfies $a_{[1]} \le a_{[2]} \le \cdots \le a_{[n]}$. Let $\geq_n^L$ denote the lexicographic ordering on the set of non-decreasing n-tuples: $a^+ \geq_n^L b^+$ if $(a_{[1]}, a_{[2]}, \cdots, a_{[k-1]}) = (b_{[1]}, b_{[2]}, \dots, b_{[k-1]})$ and $a_{[k]} \ge b_{[k]}$ for some $k = 1, 2, \dots, n$.

For each a and b in $\mathbb{R}^n$ we have

$$\begin{array}{lll} a \succsim_n^S b & \text{if} & a^+ \ge b^+, \\ a \succsim_n^U b & \text{if} & a_1 + a_2 + \cdots + a_n \ge b_1 + b_2 + \cdots + b_n, \\ a \succsim_n^L b & \text{if} & a^+ \geq_n^L b^+, \\ a \succsim_n^G b & \text{if} & a_{[1]} + a_{[2]} + \cdots + a_{[k]} \ge b_{[1]} + b_{[2]} + \cdots + b_{[k]},\ k = 1, 2, \dots, n, \end{array}$$

All four ranking rules combine (finite) anonymity and strong Pareto. As a matter of fact, the Suppes-Sen grading principle $\succsim_n^S$ is a subrelation to each partial ordering on $\mathbb{R}^n$ that satisfies anonymity and strong Pareto extends. Next, the rankings $\succsim_n^U$ and $\succsim_n^L$ are both complete. The utilitarian rule orders vectors according to the sum of the utilities. The Suppes-Sen grading principle, the leximin rule, and the generalized Lorenz criterion make decisions after rewriting the n-tuples in increasing order. The leximin rule judges the n-tuple with the highest lowest utility level as being better; if these lowest levels are the same for the two n-tuples, then the ranking is based on the second lowest utilities; and so forth. The generalized Lorenz criterion is incomplete, e.g. it is unable to compare the vectors $(0, 3)$ and $(1, 1)$. All these finite dimensional relations are easy to extend towards the framework of infinite streams. We will explain this by means of an arbitrary sequence of orderings.

[15] We refer to Suppes (1966), Sen (1971), Hammond (1976), d'Aspremont and Gevers (1977), and Shorrocks (1983). Bossert and Weymark (2004) provide an excellent overview.

Consider a sequence $\succsim_1, \succsim_2, \ldots, \succsim_n, \ldots$ of (partial) rankings the subscript of which reflects the dimension or length of the vectors it compares, i.e. for each n, the relation $\succsim_n$ is defined on $\mathbb{R}^n$. A first method to construct a relation $\succsim_\infty$ on the set X is as follows:

$$x \succsim_\infty y \quad \text{if there exists a } T \text{ in } \mathbb{N} \text{ such that } x_{-T} \succsim_T y_{-T} \text{ and } x_{+T} \geq y_{+T}.$$

Two infinite streams x and y for which the infinite tail x_{+T} dominates y_{+T} are comparable if, according to $\succsim_T$, the head x_{-T} is not worse than y_{-T}. In order to decide whether the infinite stream x is at least as good as y it is necessary that (*i*) from some generation $T + 1$ onwards each individual generation t prefers x to y on the basis that the level x_t is at least as good as y_t and (*ii*) the finite horizon society $\{1, 2, \ldots, T\}$ considers x_{-T} at least as good as y_{-T} on the basis of the relation $\succsim_T$. In other words, a finite society judges on the basis of the finite social welfare relation and all future generations unanimously concur this judgement. If the relation $\succsim_\infty$ is able to rank two infinite streams, then Pareto dominance applies to their tails. The relation $\succsim_\infty$ is a partial ranking. The following transfer of properties from $\succsim_t$ towards $\succsim_\infty$ is obvious: in case each $\succsim_t$ satisfies anonymity or strong Pareto, then the infinite version $\succsim_\infty$ also meets the axiom.

To illustrate the relations $\succsim_\infty^U$, $\succsim_\infty^L$, $\succsim_\infty^S$, and $\succsim_\infty^G$ we consider the streams

$$x = (\underbrace{0.3, 0.3, \ldots, 0.3}_{3001 \text{ times}}, 0.5, 0.5, \ldots, 0.5, \ldots),$$
$$y = (\underbrace{0.2, 0.2, \ldots, 0.2}_{2000 \text{ times}}, 0.5, 0.5, \ldots, 0.5, \ldots),$$
$$z = (\underbrace{0.1, 0.1, \ldots, 0.1}_{1000 \text{ times}}, \underbrace{0.3, 0.3, \ldots, 0.3}_{1000 \text{ times}}, 0.5, 0.5, \ldots, 0.5, \ldots).$$

The infinite horizon utilitarian rule $\succsim_\infty^U$ considers y and z equally good, and strictly better than x. The infinite horizon leximin rule $\succsim_\infty^L$ ranks x strictly above y, and y strictly above z. The rule $\succsim_\infty^G$ ranks y strictly above z and is unable to compare x and y, and x and z. The infinite Suppes-Sen grading principle $\succsim_\infty^S$ is a subrelation to each ranking criterion that satisfies finite anonymity and strong Pareto. Characterizations were obtained by means of the following axioms.

Partial translation scale invariance (Basu and Mitra 2007). For each x and y in X and for each α in $\mathbb{R}^{\mathbb{N}}$, if $x \succsim y$, $x_{+T} = y_{+T}$, and $x + \alpha$ and $y + \alpha$ belong to X, then $x + \alpha \succsim y + \alpha$.

The axiom requires that preferences are invariant to changes in the origins of the utility indices used in the various periods and should be interpreted as an infinite version of unit interpersonal comparability (Sen 1977; d'Aspremont and Gevers 1977).

Hammond equity (Hammond 1976; Asheim and Tungodden 2004). For each x and y in X, for each i and j in $\mathbb{N}$, if $x_t = y_t$ for each t different from i and j, and $y_j > x_j > x_i > y_i$, then $x \succsim y$.

Start from an infinite stream y. Bring a better-off generation j and a worse-off generation i closer to each other. The resulting infinite stream x is not worse than the original stream y.

Strict transfer principle (Hara et al. 2008; Bossert et al. 2007). For each x and y in X, for each i and j in $\mathbb{N}$, if $x_t = y_t$ for each t different from i and j, $y_j > x_j \geq x_i > y_i$, and $x_i + x_j = y_i + y_j$, then $x \succsim y$.

Start from an infinite stream y. Execute a Pigou-Dalton transfer, i.e. a transfer of a positive amount from a better-off to a worse-off generation so that the relative ranking of the two agents does not change. The resulting infinite stream x is not worse than the original stream y.

The next theorem characterizes the four partial orderings.

Theorem 4 *A partial ordering $\succsim$ on the set X satisfies strong Pareto and finite anonymity if and only if $\succsim^S_\infty$ is a subrelation to $\succsim$. A partial ordering $\succsim$ on the set X satisfies strong Pareto, finite anonymity, and partial translation scale invariance (resp. Hammond equity, or the strict transfer principle) if and only if $\succsim^U_\infty$ (resp. $\succsim^L_\infty$, or $\succsim^G_\infty$) is a subrelation to $\succsim$.*

For the characterization of the infinite horizon Suppes-Sen grading principle $\succsim^S_\infty$ we refer to Banerjee (2006b) and Mitra and Basu (2007). Basu and Mitra (2007) introduce and characterize the infinite horizon utilitarian rule $\succsim^U_\infty$. The transfer-sensitive infinite horizon rule $\succsim^G_\infty$ and the infinite horizon leximin rule $\succsim^L_\infty$ are introduced and characterized by Bossert et al. (2007).

6 Decisive Sets of Horizons

Again we start from an infinite sequence $\succsim_1, \succsim_2, \dots, \succsim_n, \dots$ of finite dimensional criteria. Let x and y be two infinite streams. For each t let the finite horizon society $\{1, 2, \dots, t\}$ decide, on the basis of $\succsim_t$, whether or not x_{-t} should be considered at least as good as y_{-t}. Consider the set

$$N(x, y) = \{t \text{ in } \mathbb{N} \mid x_{-t} \succsim_t y_{-t}\}.$$

The set $N(x, y)$ collects all finite time horizons t for which the truncated vector x_{-t} is at least as good as y_{-t}. In this section, the problem of ranking x not below y is reduced to the question of whether or not the set $N(x, y)$ is large enough. In case $N(x, y)$ is equal to $\mathbb{N}$ up to a finite set, one can argue that the infinite stream x should not be ranked below y. Indeed, for each t larger than some T, the aggregative judgement (on the basis of $\succsim_t$) of the finite horizon society $\{1, 2, \dots, t\}$ considers x not worse than y.

Let $\mathcal{F}$ denote the collection of all subsets of $\mathbb{N}$ that are equal to $\mathbb{N}$ up to a finite set. Then, we can define the following relation on the set X of infinite streams:

$$x \succsim_{\mathcal{F}} y \qquad \text{if} \qquad N(x,y) \in \mathcal{F}.$$

An element of the collection $\mathcal{F}$ is a subset of $\mathbb{N}$ and can be interpreted as a decisive set. If the set $N(x,y)$ belongs to $\mathcal{F}$, then $N(x,y)$ is decisive and the stream x is judged to be at least as good as y. If, in addition, also $N(y,x)$ belongs to $\mathcal{F}$, then x and y are considered equally good. If, however, $N(y,x)$ does not belong to $\mathcal{F}$, then x is strictly preferred to y.[16]

Let us list the relevant properties of $\mathcal{F}$. The empty set does not belong to $\mathcal{F}$, the empty set is never decisive. The collection $\mathcal{F}$ is closed for intersection, i.e. for each A and B in $\mathcal{F}$, the intersection $A \cap B$ also belongs to $\mathcal{F}$. As a consequence, the relation $\succsim_{\mathcal{F}}$ is transitive. Furthermore, the collection $\mathcal{F}$ is closed for supersets, i.e. if A belongs to $\mathcal{F}$, then a superset $B \supset A$ also belongs to $\mathcal{F}$. A superset of a decisive coalition is in its turn decisive. These three properties turn the collection $\mathcal{F}$ into a filter.

To illustrate the approach, start from the sequence $\succsim_n^U$ of utilitarian orderings and consider the infinite streams

$$u = (0.2, 0, 0.1, 0, 0.1, 0, 0.1, 0, \ldots) \qquad \text{and} \qquad v = (0, 0.1, 0, 0.1, 0, 0.1, 0, 0.1, \ldots).$$

The odd-indexed generations prefer u to v, while the even-indexed generations prefer v to u.[17] Hence, the tails do not dominate each other and the almost-unanimity-approach, discussed in the previous section and denoted by $\succsim_\infty^U$, is unable to rank these streams. On the other hand, the set $N(u,v)$ coincides with $\mathbb{N}$ and $N(v,u)$ is empty, each finite horizon society has an aggregative strict preference for u. We obtain

$$N(u,v) \in \mathcal{F} \text{ and } N(v,u) \notin \mathcal{F}, \qquad \text{hence} \qquad u \succ_{\mathcal{F}}^U v.$$

Whatever the horizon t, the finite horizon society $\{1, 2, \ldots, t\}$ when equipped with the utilitarian rule $\succsim_t^U$ considers—as a group—the stream u better than v.

The relation $\succ_{\mathcal{F}}^U$ is known as the utilitarian overtaking criterion (Atsumi 1965; von Weizsäcker 1965). Fleurbaey and Michel (2003) study different versions and extensions of this overtaking criterion and introduce the method of decisive horizons. We refer to Brock (1970), Asheim and Tungodden (2004), Basu and Mitra (2007), and Asheim (2010) for the axiomatizations of the utilitarian and the leximin overtaking rule and their catching-up versions. Starting from the axiomatizations in Theorem 4, an additional consistency demand is sufficient.

Let us return to the sequence $\succsim_1, \succsim_2, \ldots, \succsim_n, \ldots$ and the filter $\mathcal{F}$. It is obvious that the relation $\succsim_{\mathcal{F}}$ is not complete, even in case each $\succsim_n$ is complete. In order to reduce

[16]Note the similarity with the decisive sets in Arrow's impossibility theorem. See also Fleurbaey and Michel (2003).

[17]Basu and Mitra (2007) discuss this example.

the incompleteness of $\succsim_{\mathcal{F}}$ it is sufficient to enlarge the collection $\mathcal{F}$ of decisive sets. The more decisive sets, the more pairs of streams can be ranked.

A strengthening of the finite anonymity axiom can be used to extend the partial order $\succsim_{\mathcal{F}}$. Let us introduce some further notation. A permutation is a bijective map $\pi : \mathbb{N} \to \mathbb{N}$. For each infinite stream x, the composition $x \circ \pi$ is a map from $\mathbb{N}$ to Y and can be written as

$$x \circ \pi \;=\; (x_{\pi(1)}, x_{\pi(2)}, \ldots, x_{\pi(t)}, \ldots).$$

A permutation is said to be finite if for some T in $\mathbb{N}$ we have $\pi(t) = t$ for each $t \geq T$. A permutation is said to be fixed step if there exists a natural number k, such that $\pi(\{1, 2, \ldots, kn\}) = \{1, 2, \ldots, kn\}$ for each n in $\mathbb{N}$. The permutation σ that switches the numbers $2j - 1$ and $2j$ for each j is a fixed step permutation with $k = 2$. Applied to the stream $y = (0, 0.1, 0, 0.1, 0, 0.1, 0, 0.1, \ldots)$ we obtain

$$y \circ \sigma \;=\; (0.1, 0, 0.1, 0, 0.1, 0, 0.1, 0, \ldots).$$

Note that a finite permutation is fixed step. We now formulate an anonymity axiom that is stronger than finite anonymity.

Fixed step anonymity (Lauwers 1997b). Let x be an infinite stream and let π be a fixed step permutation. Then, $x \circ \pi$ and x are equally good.

The imposition of fixed step anonymity forces indifference between the utility stream.

$$v = (0, 0.1, 0, 0.1, 0, 0.1, 0, 0.1, \ldots) \qquad \text{and} \qquad z = v \circ \sigma = (0.1, 0, 0.1, 0, 0.1, 0, 0.1, 0, \ldots).$$

A preference for z above v indeed reveals some form of impatience. Fixed step anonymity is compatible with strong Pareto (Lauwers 1997b). Moreover, the combination of fixed step anonymity and Pareto dominance ranks rank $u = (0.2, 0, 0.1, 0, 0.1, 0, 0.1, 0, \ldots)$ above v.

In contrast to finite anonymity, however, the imposition of fixed step anonymity conflicts with the combination of strong Pareto and stationarity. Indeed, a fixed step anonymous rule considers the streams

$$a = (1, 0, 1, 0, \ldots, 1, 0, \ldots),\; b = (0, 1, 1, 0, \ldots, 1, 0, \ldots),\; \text{and } c = (0, 1, 0, 1, \ldots, 0, 1, \ldots)$$

equally valuable. Stream b is equal to a up to the switch in the first two coordinates, and stream c is obtained from a after switching the odd and even coordinates. Stationarity implies indifference between b_{+1} and c_{+1}. Since c_{+1} coincides with a, we have indifference between c_{+1} and c. Because of strong Pareto, the stream b_{+1} is strictly preferred to c. Thus, the incompatibility is established.[18]

[18]Demichelis et al. (2010) study axioms of anonymity in combination with strong Pareto and stationarity.

Let $\mathcal{G}$ be the collection of sets that are up to a finite set equal to $k\mathbb{N} = \{k, 2k, \dots, nk, \dots\}$ for some k. The collection $\mathcal{G}$ is a filter. If the finite dimensional relations $\succsim_n$ are anonymous, then the relation $\succsim_{\mathcal{G}}$ is fixed step anonymous and extends the relation $\succsim_{\mathcal{F}}$. Lauwers (1997b), Fleurbaey and Michel (2003), Kamaga and Kojima (2009, 2010), Asheim (2010), and Sakai (2010) discuss and axiomatize fixed step anonymous criteria.

In order to extend the partial ordering $\succsim_{\mathcal{G}}$ to a complete ordering, the relations $\succsim_n$ need to be complete and the collection $\mathcal{U}$ of decisive sets should meet the following requirement:

$$\text{for each } A \subset \mathbb{N}, \text{either } A \in \mathcal{U} \text{ or } \mathbb{N} - A \in \mathcal{U},$$

the either-or being exclusive. Indeed, if the collection $\mathcal{U}$ satisfies this demand, then for an arbitrary pair x and y, either the set $N(x, y)$, or $N(y, x)$, or both belong to $\mathcal{U}$ and the infinite streams x and y are comparable. The following definition and lemma summarizes.

Definition (*ultrafilter*) A collection $\mathcal{U}$ of subsets of $\mathbb{N}$ is said to be an ultrafilter, if

- the empty set $\emptyset$ does not belong to $\mathcal{U}$,
- for each A and B in $\mathcal{U}$, the intersection $A \cap B$ belongs to $\mathcal{U}$,
- for each $A \subset \mathbb{N}$ either $A \in \mathcal{U}$ or its complement $\mathbb{N} - A \in \mathcal{U}$.

An ultrafilter $\mathcal{U}$ that includes $\mathcal{F}$ is said to be free and satisfies $\cap_{\mathcal{U}} A = \emptyset$. A finite subset of $\mathbb{N}$ does not belong to a free ultrafilter. The existence of a free ultrafilter follows from the Axiom of Choice (formulated as Zorn's Lemma). Free ultrafilters are non-constructible objects. As a consequence, the next lemma (Fleurbaey and Michel 2003; Lauwers 2010a) provides a way to obtain non-constructive existence results. In contrast, the filter $\mathcal{G}$ and the relation is $\succsim_{\mathcal{G}}$ are well defined.

Lemma *Let $\succsim_t$ be a relation on Y^t for each t in $\mathbb{N}$ and let $\mathcal{U}$ be a free ultrafilter on $\mathbb{N}$. Define the relation $\succsim_{\mathcal{U}}$ on $X = Y^{\mathbb{N}}$ by*

$$x \succsim_{\mathcal{U}} y \qquad \text{if} \qquad N(x, y) \in \mathcal{U}.$$

If each relation $\succsim_t$ satisfies the axiom of transitivity, completeness, finite anonymity, strong Pareto, Hammond equity, or the strict transfer principle, then the relation $\succsim_{\mathcal{U}}$ satisfies the same axioms.

While the relation $\succsim_{\mathcal{U}}$ is not relevant from a practical point of view, the subrelation $\succsim_{\mathcal{G}}$ will be incomplete, but is still defined in a constructive way. On the other hand, the above lemma cannot be used to conclude that a certain set of axioms necessarily leads to a ranking rule that is non-constructible. The proof of Theorem 3, for example, appeals to the concept of non-Ramsey sets and shows how the existence of an ordering that combines intermediate Pareto and finite anonymity implies the existence of such a non-constructible object.[19]

7 Strong Anonymity

The principles of finite anonymity and fixed step anonymity are concepts of procedural equity: the action of a permutation upon an infinite stream does not change the distribution of the levels in the infinite stream. In contrast equity principles as Hammond's equity and the Pigou-Dalton transfer principle are called consequentialist equity concepts: they judge the effect of a change in the distribution in the levels. This section discusses the strongest form of procedural equity, labeled strong anonymity.

Strong anonymity. For each infinite stream x in X and each permutation $\pi : \mathbb{N} \to \mathbb{N}$, we have $x \sim x \circ \pi$.

This axiom conflicts with weak Pareto (Fleurbaey and Michel 2003). Let us recall their example:

$$\begin{array}{ll} t & (\dots, 9, 7, 5, 3, \quad 1, \quad 2, \quad 4, \quad 6, \quad 8, \quad \dots), \\ x = & (\dots, \frac{1}{9}, \frac{1}{7}, \frac{1}{5}, \frac{1}{3}, \quad 1, \quad 2 - \frac{1}{2}, 2 - \frac{1}{4}, 2 - \frac{1}{6}, \; 2 - \frac{1}{8}, \; \dots), \\ y = & (\dots, \frac{1}{7}, \frac{1}{5}, \frac{1}{3}, 1, 2 - \frac{1}{2}, 2 - \frac{1}{4}, 2 - \frac{1}{6}, 2 - \frac{1}{8}, 2 - \frac{1}{10}, \dots). \end{array}$$

The first line lists the moments in time. The even indexed moments are written in increasing order, the odd indexed moments are written in decreasing order (or reading from right to left, in increasing order). The second line presents the infinite stream x. The even indexed values increase an have a limit equal to 2. The odd indexed values decrease (as time moves forward) and have a limit equal to 0. The third line lists the very same values in the same order as in the previous line. Each value, however, is shifted one place to the left. The result is that the stream y is just a permutation of the stream x that strongly dominates x, i.e. for each t we have $y_t > x_t$. Conclude that strong anonymity conflicts with weak Pareto.

Consider next the discrete setting. The next two streams shelter an infinite number of zeroes and an infinite number of ones:

$$z_1 = (\underbrace{1, 1, \dots, 1}_{99 \text{ times}}, 0; \dots; \underbrace{1, 1, \dots, 1}_{99 \text{ times}}, 0; \dots) \sim z_{99} = (1, \underbrace{0, 0, \dots, 0}_{99 \text{ times}}; \dots; 1, \underbrace{0, 0, \dots, 0}_{99 \text{ times}}; \dots).$$

[19] A non-Ramsey set is a subset $\mathcal{N}$ of the collection $\mathbb{N}_\infty$ of all infinite subsets of $\mathbb{N}$ such that for each element J in $\mathcal{N}$ the collection of infinite subsets of J intersects both $\mathcal{N}$ and its complement $\mathbb{N}_\infty - \mathcal{N}$. The technique developed in Lauwers (2010a) to define non-Ramsey sets has been used by Dubey and Mitra (2013) to show that a complete ranking that combines strong Pareto and Hammond equity (or the strict transfer principle) is non-constructible. See also Dubey (2011), Dubey and Mitra (2011, 2012), and Banerjee and Dubey (2013).

There is a bijective map (permutation) that transforms z_1 into z_{99}. Hence, the infinite stream with 99 % of the generations at level 1 is equally good as the stream with 1 % at level 1. In the discrete setting, with $Y = \{0, 1\}$ strong anonymity conflicts with intermediate Pareto (Van Liedekerke and Lauwers 1997).

Note the difference between the two examples. In the first example, the distribution of the different levels remains untouched after being permuted (because all the different periods are equipped with different levels). In contrast, in the second example the volume—and hence the distribution—of zeros changes from 1 to 99 % when permuting z_1 into z_{99}.

The most familiar rules that satisfies strong anonymity are the Rawlsian infimum-rule and the limit inferior-rule (Rawls 1999). Both these rules satisfy strong anonymity, monotonicity (if $x \geq y$, then $x \succsim y$), continuity with respect to the sup-topology, and Hammond equity. Characterizations are obtained by Lauwers (1997c) and Chambers (2009).[20]

We close this section with the rank-discounted utilitarian rule. This rule, introduced by Zuber and Asheim (2012), also satisfies strong anonymity. First, we discuss two sets, X^+ and $\bar{X}$, of infinite utility streams for which the axioms of strong Pareto and strong anonymity are not in conflict. The domain X^+ collects the nowhere decreasing streams, i.e. streams x with $x_t \leq x_{t+1}$ for each t in $\mathbb{N}$. For each stream x in X^+, it is impossible to permute x into y such that y strongly Pareto dominates x. The set $\bar{X}$ is defined as the set of infinite streams that can be rewritten as a nowhere decreasing stream. The set $\bar{X}$ is closed under all permutations and within the set $\bar{X}$ the axioms of strong anonymity and strong Pareto are compatible. In addition, each infinite stream in $\bar{X}$ has a well defined lowest level, second lowest level, and so forth. Hence, for each infinite stream x in $\bar{X}$ a corresponding non-decreasing stream $x_{[\,]} = (x_{[1]}, x_{[2]}, \ldots, x_{[t]}, \ldots)$ is defined. Furthermore, the limit $\ell(x)$ is well defined for each x in $\bar{X}$. Indeed, the increasing stream $x_{[\,]}$ associated to x has a unique point of accumulation. The infinite stream x in $\bar{X}$ either remains below $\ell(x)$ (for each t, $x_t < \ell(x)$), or reaches $\ell(x)$ after a finite number of time moments (for each t big enough, $x_t = \ell(x)$).

Zuber and Asheim (2012) define the rank-discounted utilitarian rule

$$R : \bar{X} \longrightarrow \mathbb{R} : x \longmapsto R(x) = (1 - \beta) \sum \beta^{r-1} u(x_{[r]}),$$

with $0 < \beta < 1$ and u a continuous and increasing map. The factor $(1 - \beta)$ normalizes the total sum of the weights to 1. In contrast to the time-discounted utilitarian rule, the weight $(1 - \beta)\beta^{r-1}$ corresponds to the rank a particular value $x_t = x_{[r]}$ obtains after rewriting x in increasing order. The highest weight $(1 - \beta)$ is attached to the moment t for which x_t is the lowest level, the second highest weight is attached to the moment with the second lowest level, and so forth. The lower the value x_t, the higher the weight attached to generation t.

---

[20]Doyen and Martinet (2012) apply the maximin rule in a general dynamic economic model.

The axiomatization of Koopmans' rule can be used to characterize the rank-discounted utilitarian rule R when restricted to the domain X^+ of increasing streams. Roughly, the axioms of Koopmans are not imposed upon the whole collection X of infinite streams. Restricted axioms are imposed to order the set X^+. Strong anonymity then extends the rule R to the set $\bar{X}$. When applied to the set $\bar{X}$, the criterion R attaches weights to a generation on the basis of the rank this generation obtains. Koopmans' rule uses discounting according to the position in time, in contrast, the welfare function R uses discounting according to the rank after rewriting the stream in increasing order.

Finally, Zuber and Asheim (2012) introduce an extended rank-discounted utilitarian social welfare function on the set X of all infinite streams:

$$R : X \longrightarrow \mathbb{R} : x \longmapsto R(x) = u(\ell(x)) + (1-\beta) \sum_{r=1}^{|L(x)|} \beta^{r-1}\Big(u(x_{[r]}) - u(\ell(x))\Big),$$

with $0 < \beta < 1$, $\ell(x)$ the limit inferior of x, $L(x) \subset \mathbb{N}$ the set of indices t for which $x_t < \ell(x)$, and u a continuous and increasing real valued function. The length of the discounted sum in $R(x)$ either is finite (if $|L(x)| < +\infty$) or infinite (if $|L(x)| = +\infty$). In words, take an infinite stream x, let $\ell(x) = \liminf(x)$ be the limit inferior of x, let $L(x)$ collect all generations t for which $x_t < \ell(x)$, and apply the rank-discounted utilitarian rule upon the stream x restricted to $L(x)$ which is, if necessary, supplemented with infinitely many values $\ell(x)$.

For example, the welfare function R attaches value zero to the above infinite streams z_1 and z_{99}. The imposition of strong anonymity (on the set X) implies a cost: the welfare function R does not satisfy weak Pareto. The infinite stream $(1, 1/2, 1/3, \ldots, 1/n, \ldots)$ is equally good as the zero stream. In conclusion, when applied to the set X of all infinite streams, the criterion R refines the 'Rawlsian' limit inferior as it only pays attention to those generations that obtain a level below or equal to this limit inferior.[21]

8 Sustainable Discounted Utilitarianism

The infinite horizon utilitarian rule $\succsim_\infty^U$ imposes

$$x = (x_1, z+\varepsilon, z+\varepsilon, \ldots, z+\varepsilon, \ldots) \succ y = (y_1, z, z, \ldots, z, \ldots),$$

for each x_1, y_1, z, and $\varepsilon > 0$. Whatever the sacrifice $y_1 - x_1$ of the first generation, the infinitely many ε's bridge the gap and overtake the infinite stream y. The axiom of Hammond equity for the future modifies the above ranking as follows.

[21] Asheim and Zuber (2013) study the behavior of the rank-discounted utilitarian rule as β goes to zero and show the convergence of R towards a strongly anonymous leximin relation.

Hammond equity for the future (Asheim and Tungodden 2004). For each x_1, y_1, z, and $\varepsilon > 0$, we have

$$x = (x_1, z+\varepsilon, z+\varepsilon, \ldots, z+\varepsilon, \ldots) \succ y = (y_1, z, z, \ldots, z, \ldots),$$

as soon $y_1 > x_1 > z + \varepsilon$.

Similar to Hammond's equity axiom, bringing the levels y_1 and z closer to each other (towards x_1 and $z+\varepsilon$) results in a better stream. In contrast to Hammond's axiom, where only two generations are involved, now each generation is involved. The transfer from the better-off first generation leads to a sustained increase in the level of all subsequent generations while the first generation remains the better off.[22] The early generations are not condemned to starvation in order to maximize the welfare for later generations.

The imposition of Hammond equity of the future comes at the cost of weakening strong Pareto and the axiom of independence. For example, an improvement for the first generation is not taken into account in case this first generation is better off than the future generations. Asheim (2010), Asheim et al. (2012) introduce and characterize the following class of sustainable discounted utilitarian rules:

$$S : X \longrightarrow \mathbb{R} : x \longmapsto \begin{cases} (1-\beta)U(x_1) + \beta S(x_{+1}) & \text{if } U(x_1) \leq S(x_{+1}), \\ S(x_{+1}) & \text{if } U(x_1) > S(x_{+1}), \end{cases}$$

with β in the open interval (0, 1) and $U(x) = S(x, x, \ldots, x, \ldots)$. Observe the similarity with the time-discounted utilitarian rule,

$$D(x) = (1-\alpha)\, u(x_1) + \alpha\, D(x_{+1}).$$

Both social welfare functions are defined in a recursive way. The sustainable discounted utilitarian rule gives zero weight to those generations that are better off then their future generations. As a consequence the sustainable discounted utilitarian rule violates weak Pareto: again, the infinite stream $(1, 1/2, 1/3, \ldots, 1/n, \ldots)$ and the zero stream are considered equally good. On the other hand, when restricted to the domain of non-decreasing streams, the rule coincides with both the time-discounted and the rank-discounted utilitarian rule.

The sustainable discounted utilitarian rule satisfies continuity with respect to the sup-topology, monotonicity, stationarity, and the following weakening of the independence axiom.

Separable future. For each x and y in X and for each t in $\mathbb{N}$,

$$\text{if } x \succsim (x_{-t}, y_{+t}), \text{ then } (y_{-t}, x_{+t}) \succsim y.$$

[22] Also Banerjee (2006a), Asheim et al. (2007), Alcantud and García-Sanz (2010), Dubey and Mitra (2013) consider Hammond equity for the future.

In words, if two infinite streams have the same head up to time T, then the ranking of these streams does not depend upon this common head. Asheim (2010) provides the following example to motivate the rejection of the other part of Koopmans' independence axiom. Consider the streams

$$a = (0, 0.75, 1, 1, \dots, 1, \dots) \text{ and } b = (0.25, 0.25, 1, 1, \dots, 1, \dots).$$

Assume that a is preferred to b. Then, it is not obvious to rank the modifications

$$a^* = (0, 0.75, 0.25, 0.25, 0.25, \dots, 0.25, \dots) \text{ and } b^* = (0.25, 0.25, 0.25, 0.25, \dots, 0.25, \dots)$$

in the same way. The common tails might influence our look upon the conflict between the couple $(0, 0.75)$ and $(0.25, 0.25)$. The value 0.75 is the second worst-off in a, while it is the best-off in a^*. The sustainable discounted utilitarian rule S treats the same value but at different ranks in a different way.

9 Chichilnisky's Sustainable Preference

Chichilnisky (1996, 1997) introduces the axioms of non-dictatorship and characterizes a whole class of social welfare orderings that satisfies these axioms in combination with completeness, continuity, and representability. We recall these axioms.

Dictatorship of the present. For each x, y, v, and w in X, if $x \succ y$, then there exists a T in $\mathbb{N}$ such that

$$(x_{-(T+k)}, v_{+(T+k)}) \succ (y_{-(T+k)}, w_{+(T+k)}) \text{ for each } k \text{ in } \mathbb{N}.$$

A rule that satisfies this axiom ranks two infinite streams on the basis of their heads (i.e. the truncated streams). Time-discounted utilitarianism is the prime example that satisfies this axiom. The time-discounted rule just puts the (very) long run offside.

Dictatorship of the future. For each x, y, v, and w in X, if $x \succ y$, then there exists a T in $\mathbb{N}$ such that

$$(v_{-(T+k)}, x_{+(T+k)}) \succ (w_{-(T+k)}, y_{+(T+k)}) \text{ for each } k \text{ in } \mathbb{N}.$$

A rule that satisfies this axiom ranks two infinite streams on the basis of their tails. The map $x \mapsto \liminf x$ is an example. This map looks for the infimum of the set of accumulation points, and is not sensitive for changes in the head of the infinite stream. As a consequence, each rule that satisfies strong Pareto violates dictatorship of the future.

The axioms of non-dictatorship impose that the axioms of dictatorship do not hold. Many rules meet the axioms of non-dictatorship. The Pareto dominating tail rules $\succsim_\infty^S$, $\succsim_\infty^U$, $\succsim_\infty^L$, and $\succsim_\infty^U$; the fixed step anonymous $\mathcal{G}$-rules $\succsim_\mathcal{G}^S$, $\succsim_\mathcal{G}^U$, $\succsim_\mathcal{G}^L$, and $\succsim_\mathcal{G}^U$ all satisfy non-dictatorship of the future (as they all satisfy strong Pareto) and non-dictatorship of the present (the criteria are sensitive for shifts in the tails). None of these criteria, however, is complete.

Theorem 5 (Chichilnisky 1996) *Let the ordering $\succsim$ on the set of bounded streams satisfy continuity with respect to the sup-topology, independence, non-dictatorship of the present, and non-dictatorship of the future. Then, the ordering $\succsim$ is represented by a continuous function*

$$C : x \longmapsto \sum_{t=1}^{\infty} \lambda_t x_t \ + \ \varphi(x),$$

where the real numbers λ_t are all positive and add up to a finite number, and φ a purely finitely additive measure.

Chichilnisky uses the term sustainable preferences for the social welfare functions characterized by the previous theorem. The sustainable preference C decomposes into two parts. The first part, $x \mapsto \sum \lambda_t x_t$, is countably additive, satisfies strong Pareto, and captures the short run. The second part, $x \mapsto \varphi(x)$, is purely finitely additive and captures the long run. This decomposition follows from the representation of a finitely additive measure on the set $\mathbb{N}$, i.e. a map that (*i*) assigns to each subset of $\mathbb{N}$ a nonnegative number, and (*ii*) assigns to the union of two disjoint sets the sum of their numbers.[23]

A purely finitely additive map is typically obtained by means of Hahn-Banach's theorem or by means of a free ultrafilter (e.g. it selects the unique accumulation point the converging subsequence of which is indexed by a set that belongs to the free ultrafilter). It has been shown that such a purely finitely additive map, although it is continuous with respect to the sup-topology and finitely additive, is a non-constructible object (Lauwers 2009, 2010b). Obviously, the representability of a non-constructible relation does not change the non-constructible nature of the relation. Here, the map φ provides us an example of a continuous and additive representation of a non-constructible relation. Representability by means of a continuous, monotonic, and additive map does not imply constructibility.[24]

There are, however, at least two ways to circumvent this problem. One can restrict the domain to, for example, those infinite utility streams which exhibit a well defined —without recurse to non-constructive mathematics—and finite limiting behavior (Chichilnisky 2009). Alternatively, one can replace the map φ with, for example, the map lim inf which looks for the infimum of the set of accumulation points.[25]

[23]We refer to Yosida and Hewitt (1952), Rao (1958), and Peressini (1967).

[24]Dubey and Mitra (2013) provide an example of a non-constructible relation on X that satisfies the Pigou-Dalton transfer principle while its representation has been established by Sakamoto (2012).

[25]The map lim inf violates additivity: let $x = (1, 0, 1, 0, \ldots)$ and $y = (0, 1, 0, 1, \ldots)$, then $\liminf(x) = \liminf(y) = 0$ while $\liminf(x + y) = 1$. The map lim inf, however, still fits in the Chichilnisky approach.

Finally, we observe the similarities between the results of Chichilnisky and Koopmans. Both social welfare functions satisfy continuity with respect to the sup-topology, strong Pareto, and independence. Stationarity is violated by Chichilnisky's criterion but satisfied by discounted utilitarianism. Non-dictatorship of the present is satisfied by Chichilnisky's criterion but violated by the discounted utilitarian rule.

References

Alcantud, J. C. R., & García-Sanz, M. D. (2010). Paretian evaluation of infinite utility streams: An egalitarian criterion. *Economics Letters*, *106*, 209–211.

Arrow, K. J. (1999). Discounting, morality, and gaming. In J. P. Portney & J. P. Weyant (Eds.), *Discounting and intergenerational equity*. Washington DC: Resources for the Future.

Asheim, G. B. (2010) Intergenerational equity. In K. J. Arrow & T. F. Breshanan (Eds.), Book series, *Annual review of economics* (Vol. 2, pp. 197–222).

Asheim, G. B., & Banerjee, K. (2010). Fixed-step anonymous overtaking and catching-up. *International Journal of Economic Theory*, *6*, 149–165.

Asheim, G. B., & Mitra, T. (2010). Sustainability and discounted utilitarianism in models of economic growth. *Mathematical Social Sciences*, *59*, 148–169.

Asheim, G. B., Mitra, T., & Tungodden, B. (2007). A new equity condition for infinite utility streams and the possibility of being Paretian. In J. Roemer & K. Suzumura (Eds.), *Intergenerational equity and sustainability*. London: Palgrave.

Asheim, G. B., Mitra, T., & Tungodden, B. (2012). Sustainable recursive social welfare functions. *Economic Theory*, *49*, 267–292.

Asheim, G. B., & Tungodden, B. (2004). Resolving distributional conflicts between generations. *Economic Theory*, *24*, 221–230.

Asheim, G. B., & Zuber, S. (2013). A complete and strongly anonymous leximin relation on infinite streams. *Social Choice and Welfare*, *41*, 819–834.

Atkinson, A. B. (2001). The strange disappearance of welfare economics. *Kyklos*, *54*, 193–206.

Atsumi, H. (1965). Neoclassical growth and the efficient program of capital accumulation. *Review of Economic Studies*, *32*, 127–136.

Banerjee, K. (2006a). On the equity-efficiency trade off in aggregating infinite utility streams. *Economic Letters*, *93*, 63–67.

Banerjee, K. (2006b). On the extension of the utilitarian and Suppes-Sen social welfare relations to infinite utility streams. *Social Choice and Welfare*, *27*, 327–339.

Banerjee, K., & Dubey, R. S. (2013). Do all constructive strongly monotonic inter-temporal orders exhibit impatience. Working paper: Florida Atlantic University, Montclair State University, USA.

Basu, K., & Mitra, T. (2003). Aggregating infinite utility streams with intergenerational equity: The impossibility of being Paretian. *Econometrica*, *71*, 1557–1563.

Basu, K., & Mitra, T. (2007). Utilitarianism for infinite utility streams: A new welfare criterion and its axiomatic characterization. *Journal of Economic Theory*, *133*, 350–373.

Bleichrodt, H., Rohde, K. I. M., & Wakker, P. P. (2008). Koopmans' constant discounting for intertemporal choice: A simplification and a generalization. *Journal of Mathematical Psychology*, *52*, 341–347.

Bossert, W., Sprumont, Y., & Suzumura, K. (2007). Ordering infinite utility streams. *Journal of Economic Theory*, *135*, 579–589.

Bossert, W., & Weymark, J. A. (2004). Utility in social choice. In S. Barbera, P. J. Hammond, & C. H. Seidl (Eds.), *Handbook of utility theory* (Vol. 2). Amsterdam: Kluwer Academic Publishers.

Brock, W. A. (1970). An axiomatic basis for the Ramsey-Weizsäcker overtaking criterion. *Econometrica*, *38*, 927–929.

Chambers, C. P. (2009). Intergenerational equity: Sup, inf, lim sup, and lim inf. *Social Choice and Welfare, 32*, 243–252.
Chichilnisky, G. (1996). An axiomatic approach to sustainable development. *Social Choice and Welfare, 13*, 231–257.
Chichilnisky, G. (1997). What is sustainable development? *Land Economics, 73*, 467–491.
Chichilnisky, G. (2009). Avoiding extinction: Equal treatment of the present and the future. Economics, The Open-Access, Open-Assessment E-Journal 3, Article number 200932. http://www.economics-ejournal.org/economics/journalarticles/2009-32.
Dasgupta, P. S., & Heal, G. M. (1979). *Economic theory and exhaustible resources*. Cambridge UK: Cambridge University Press.
d'Aspremont, C., & Gevers, L. (1977). Equity and the informational basis of collective choice. *Review of Economic Studies, 46*, 199–210.
Debreu, G. (1954). Representation of a preference ordering by a numerical function. In R. M. Thrall, C. H. Coombs, & R. L. Davis (Eds.), *Chapter XI in Decision processes*. New York.
Demichelis, S., Mitra, T., & Sorger, G. (2010). Intergenerational equity and stationarity. Working Paper No 1003: Department of Economics, University of Vienna.
Diamond, P. A. (1965). The evaluation of infinite utility streams. *Econometrica, 33*, 170–177.
Doyen, L., & Martinet, V. (2012). Maximin, viability and sustainability. *Journal of Economic Dynamics and Control, 36*, 1414–1430.
Dubey, R. S. (2011). Fleurbaey-Michel conjecture on equitable weak Paretian social welfare order. *Journal of Mathematical Economics, 47*, 434–439.
Dubey, R. S., & Mitra, T. (2011). On equitable social welfare functions satisfying the weak Pareto axiom: A complete characterization. *International Journal of Economic Theory, 7*, 231–250.
Dubey, R. S., Mitra, T. (2012). On monotone social welfare orders satisfying the strong equity axiom: Construction and representation. Working paper: Montclair State University, Cornell University Ithaca, USA.
Dubey, R. S., Mitra, T. (2013). On representation and construction of equitable social welfare orders on infinite utility streams. Working paper: Montclair State University, Cornell University Ithaca, USA.
Ferejohn, J., & Page, T. (1978). On the foundations of intertemporal choice. *American Journal of Agricultural Economics, 60*, 269–275.
Fleurbaey, M., & Michel, P. (2003). Intertemporal equity and the extension of the Ramsey criterion. *Journal of Mathematical Economics, 39*, 777–802.
Fleurbaey, M., Zuber, S. (2012). Climate policies deserve a negative discount rate. Fondation Maison des sciences de l'homme, WP-2012-19.
Hammond, P. J. (1976). Equity, Arrow's conditions, and Rawls' difference principle. *Econometrica, 44*, 793–804.
Hara, C., Shinotsuka, T., Suzumura, K., & Xu, Y. (2008). Continuity and egalitarianism in the evaluation of infinite utility streams. *Social Choice and Welfare, 31*, 179–191.
Kamaga, K., & Kojima, T. (2009). Q-anonymous social welfare relations on infinite utility streams. *Social Choice and Welfare, 33*, 405–413.
Kamaga, K., & Kojima, T. (2010). On the leximin and utilitarian overtaking criteria with extended anonymity. *Social Choice and Welfare, 35*, 377–392.
Koopmans, T. C. (1960). Stationary ordinal utility and impatience. *Econometrica, 28*, 287–309.
Koopmans, T. C. (1965). On the concept of optimal economic growth. *Pontificiae Academiae Scientiarum Scripta Varia, 28*, 225–300.
Koopmans, T. C. (1972a). Representation of preference orderings with independent components of consumption. In C. B. McGuire & R. Radner (Eds.), *Decision and organization*. Amsterdam: North-Holland.
Koopmans, T. C. (1972b). Representation of preference orderings over time. In C. B. McGuire & R. Radner (Eds.), *Decision and organization*. Amsterdam: North-Holland.
Koopmans, T. C. (1977). Concepts of optimality and their uses. *American Economic Review, 67*(3), 261–274.

Koopmans, T. C., Diamond, P. A., & Williamson, R. E. (1964). Stationary utility and time perspective. *Econometrica*, *32*, 82–100.

Lauwers, L. (1997a). Continuity and equity with infinite horizons. *Social Choice and Welfare*, *14*, 345–356.

Lauwers, L. (1997b). Infinite utility, insisting on strong monotonicity. *Australasian Journal of Philosophy*, *75*, 222–233.

Lauwers, L. (1997c). Rawlsian equity and generalized utilitarianism with an infinite population. *Economic Theory*, *9*, 143–150.

Lauwers, L. (2009). A report on Graciela Chichilnisky: Avoiding extinction equal treatment of the present and the future. Economics, The Open-Access, Open-Assessment E-Journal. http://www.economics-ejournal.org/economics/discussionpapers/2009-8#comments.

Lauwers, L. (2010a). Ordering infinite utility streams comes at the cost of a non-Ramsey set. *Journal of Mathematical Economics*, *46*, 32–37.

Lauwers, L. (2010b). Center for Economic Studies, DPS: Purely finitely additive measures are nonconstructible objects. KU Leuven.

Lauwers, L. (2012). Intergenerational equity, efficiency, and constructibility. *Economic Theory*, *49*, 227–242.

Mitra, T., & Basu, K. (2007). On the existence of Paretian social welfare relations for infinite utility streams with extended anonymity. In J. Roemer & K. Suzumura (Eds.), *Intergenerational equity and sustainability*. London: Palgrave.

Packell, E. (1980). Impossibility results in the axiomatic theory of intertemporal choice. *Public Choice*, *35*, 219–227.

Peressini, A. (1967). *Ordered topological vector spaces*. New York: Harper and Row.

Ramsey, F. P. (1928). A mathematical theory of savings. *The Economic Journal*, *38*, 543–559.

Rao, R. R. (1958). A note on finitely additive measures. *The Indian Journal of Statistics*, *19*, 27–28.

Rawls, J. (1999). *A theory of justice* (revised ed.). Cambridge: The Belknap Press of the Harvard University Press.

Sakai, T. (2010). Intergenerational equity and an explicit construction of welfare criteria. *Social Choice and Welfare*, *35*, 393–414.

Sakamoto, N. (2012). Impossibilities of Paretian social welfare functions for infinite utility streams with distributive equity. *Hitotsubashi Journal of Economics*, *53*, 60–64.

Sen, A. K. (1971). *Collective choice and social welfare*. Edinburgh: Oliver and Boyd.

Sen, A. K. (1977). On weights and measures: Informational constraints in social welfare analysis. *Econometrica*, *45*, 1539–1572.

Shorrocks, A. F. (1983). Ranking income disctributions. *Economica*, *50*, 3–17.

Sidgwick, H. (1907). *The methods of ethics*. London: Macmillan.

Suppes, P. (1966). Some formal models of grading principles. *Synthese*, *6*, 284–306.

Svensson, L.-G. (1980). Equity among generations. *Econometrica*, *48*, 1251–1256.

Szpilrajn, E. (1930). Sur l'extension du l'ordre partiel. *Fundamenta Mathematicae*, *16*, 386–389.

Thomson, W. (2001). On the axiomatic method and its recent applications to game theory and resource allocation. *Social Choice and Welfare*, *18*(2), 327–386.

Van Liedekerke, L., & Lauwers, L. (1997). Sacrificing the patrol: Utilitarianism, future generations and infinity. *Economics and Philosophy*, *13*, 159–174.

von Weizsäcker, C. C. (1965). Existence of optimal programs of accumulation for an infinite time horizon. *Review of Economic Studies*, *32*, 85–104.

Yosida, K., & Hewitt, E. (1952). Finitely additive measures. *Transactions of the American Mathematical Society*, *72*, 46–66.

Zame, W. R. (2007). Can intergenerational equity be operationalised. *Theoretical Economics*, *2*, 187–202.

Zuber, S., & Asheim, G. B. (2012). Justifying social discounting: The rank-discounted utilitarian approach. *Journal of Economic Theory*, *147*, 1572–1601.

Part III
The Environment in a Global Context

Nested Externalities and Polycentric Institutions: Must We Wait for Global Solutions to Climate Change Before Taking Actions at Other Scales?

Elinor Ostrom

1 Introduction

The Fourth Assessment Report of the Intergovernmental Panel on Climate Change (2007) and the Stern Report (Stern 2007) both stress the need to recognize the impact of human actions on the global environment. Even though there is now a relatively high level of agreement among scientists about the danger that humans are facing related to the uncorrected negative externalities of greenhouse gas emissions (Rezai et al. 2010), little agreement exists about what should and could be done (Dutta and Radner 2010; Schelling 2007). Further, agreement among citizens concerning the seriousness of global warming is falling. In the March 2010 Gallup Poll on the Environment, 48 % of those surveyed responded that the seriousness of global warming was generally exaggerated—a 13 % increase as contrasted with poll results in 2008 (Newport 2010).

Thanks to Dan Cole Eric Coleman Maria Claudia Lopez Jimmy Walker an anonymous reviewer and the editor of this journal for very useful comments on the earlier draft of this paper and to Patty Lezotte for her excellent editing. Sections of this paper draw on a report that I submitted to the World Bank entitled "A Polycentric Approach for Coping with Climate Change." Financial support from the National Science Foundation and the MacArthur Foundation is gratefully acknowledged.

Originally published in Economic Theory, Volume 49, Number 2, February 2012 DOI 10.1007/s00199-010-0573-7.

E. Ostrom (✉)
Workshop in Political Theory and Policy Analysis, Indiana University,
Bloomington, IN 47408, USA
e-mail: ostrom@indiana.edu

E. Ostrom
Center for the Study of Institutional Diversity, Arizona State University,
Tempe, AZ 85287, USA

G. Chichilnisky and A. Rezai (eds.), *The Economics of the Global Environment*,
Studies in Economic Theory 29, DOI 10.1007/978-3-319-31943-8_13

The Kyoto Protocol to the United Nations Framework Convention on Climate Change is an international environmental treaty created and signed at the Conference of the Parties of the UNFCC in Kyoto in 1997. More than 180 countries have now ratified the protocol, but the United States has not. Considerable disagreements exist even among the major states that have signed regarding how large a reduction in emissions should be imposed (Matthews and Caldeira 2008). Major debates exist over a number of issues related to achieving efficient and equitable mechanisms at a global level. One issue relates to who is responsible for the current and immediate future levels of CO_2 in the atmosphere (Botsen et al. 2008; Dellink et al. 2009; den Elzen et al. 2005; Lauwers 2010; Lecocq and Hourcade 2010). In other words, who should bear the primary burden of paying for solutions? (Chichilnisky and Heal 1994, 2000; Baer et al. 2000; Posner and Sunstein 2008). Other debates address whether taxes or quotas are the best instrument for achieving abatement (Karp and Zhang 2010). Similar scholarly concerns have also been raised regarding claims that Payments for Ecosystem Services (PES) can increase carbon sequestration while at the same time enhancing species conservation on the same landscape (Nelson et al. 2008).

Given the failure to reach agreement at the international level on efficient, fair, and enforceable reductions of greenhouse gas emissions, continuing to wait without investing in efforts at multiple scales may defeat the possibilities of significant abatements and mitigations in enough time to prevent tragic disasters. We need to make a scholarly investment in a more appropriate theory of global change that offers a better explanation of micro-level incentives and outcomes (Chipman and Tian 2010; Chichilnisky 2010; Asheim et al. 2010) as well as being a foundation for more effective public policies. This paper represents an effort to posit a theory of nested externalities at multiple scales to provide a better foundation for analyzing the multiple scales involved in reducing the threat of climate change. Another goal is to balance the arguments made in the policy literature that a global solution is the *only* way to cope with climate change. "Global solutions" negotiated at a global level—if not backed up by a variety of efforts at national, regional, and local levels—are not guaranteed to work effectively.

The problem of averting massive climate change is a global "public good" (Chichilnisky and Heal 2000; Sandler 2004). Millions of actors affect the global atmosphere and they all benefit from reduced greenhouse gas emissions. The problem is they benefit whether or not they pay any of the costs since beneficiaries cannot be excluded. Trying to solve the problem of providing a public good is a classic collective-action dilemma (Cole 2008). It is probably the largest dilemma the world has ever knowingly faced. Many analysts call for an institutional solution at the global level (see Stavins 1997; C. Miller 2004; Wiener 2007).

Given the widespread presumption that any collective-action problem that has global effects must be "solved" entirely at the global level, several theoretical questions need to be addressed as analysts undertake the next round of research on climate change. They include:

1. How may polycentric institutions improve exclusive reliance on a global approach to cope with global climate issues?
2. Are multiple, nested externalities produced by decisions made at less than a global scale?
3. What types of actions are being taken at less-than-global scale to reduce greenhouse gas emissions?
4. Are large-scale governments usually better equipped to cope with collective-action problems that have outcomes that are large scale themselves?
5. If multiple governments and other organizations work to reduce energy consumption and greenhouse gas emissions, does that only produce leakage, chaotic systems, and potentially counterproductive processes?

Each of these questions will be theoretically and empirically addressed below.

2 A Polycentric Approach

Let us briefly review the origin of the term "polycentricity." During the 1950s, massive academic criticism was leveled at metropolitan areas across the United States and Europe due to the large number of small-, medium-, and large-scale governmental units operating at the same time. Scholars thought this was chaotic. Vincent Ostrom et al, Charles Tiebout, and Robert Warren wrote a classic article in (1961) entitled "The Organization of Government in Metropolitan Areas: A Theoretical Inquiry." The authors reasoned that a simple dichotomy between "the" market and "the" government was not a good scientific approach to the study of public economies. Further, "the" market is not a single unit. It is composed of many small-, medium-, and large-scale firms. The expected efficiency of a market disappears if it were consolidated into a monopoly. There is no reason to presume that a monopoly government is more efficient than a system of governmental units at multiple scales.

Economic theory teaches us about the dangers of allocating all capabilities to a single unit even though one cannot apply *all* lessons derived from the analysis of market economies to the public sector (Williamson 1975, 1985, 2000). Ostrom (1999: 57) referred to a polycentric system as "one where many elements are capable of making mutual adjustments for ordering their relationships with one another within a general system of rules where each element acts with independence of other elements" (see also Ostrom 2008a, b; McGinnis 1999a, b, 2000). A polycentric system exists when multiple public and private organizations at multiple scales jointly affect collective benefits and costs. The early theoretical work on polycentricity stimulated intensive research on the governance of one of the major public goods for urban areas—that of providing public safety (Ostrom et al. 1978)—and is a foundation for the theory presented herein.

Readers of this article may ask: What is the relevance of polycentric systems for the analysis of *global* public goods? The initial relevance of the polycentric approach is the parallel between the earlier theoretical presumption that *only* the largest scale was relevant for the provision and production of public goods for metropolitan areas, and the contemporary presumption that *only* the global scale is relevant for policies related to climate change. Extensive empirical research found that while large-scale units were an essential part of effective governance of metropolitan areas, small- and medium-scale units were also *necessary* components (Parks and Ostrom 1999). An important lesson is that relying entirely on international efforts to solve global climate problems needs to be rethought.

3 Do Nested, Positive Externalities Exist at Multiple Levels from Reducing Emissions?

Greenhouse gas emissions are the result of many actions taken at multiple scales. The positive externalities of reduced greenhouse gas emissions are also distributed across scales—from the household to the globe. Nested externalities occur when actions taken within one decision-making unit simultaneously generate costs or benefits for other units organized at different scales.

Decisions within a household as to what form of transportation to use for various purposes, what car to purchase, what investments to make regarding power consumption within their home, all have small effects on the global atmosphere and relatively larger effects at a smaller scale. Better health is enhanced by members of a household who bike to work rather than driving. Family expenditures allocated to heating and electricity may be reduced when investments have been made in better construction of a building, reconstruction of existing buildings, investment in solar panels, and many other investments in equipment that families as well as private firms can make that pay off in the long run. Similar decisions within firms are also important as buildings used by government offices, businesses, and as private homes account for "more than 70 % of the electricity used and almost 40 % of greenhouse gas emissions in the United States" (Fuller et al. 2009).

No change at a small scale can be expected without shared knowledge about the costs and benefits of actions and shifts in preference functions to take into account previously unrecognized benefits for self as well as others. As the scientific community has achieved a higher level of agreement about human impacts on the global atmosphere, knowledge of the effects of individual and family actions is becoming more available.[1] In local discussions and meetings, information is generated about the prevailing unrecognized costs of individual and family activities. Discussions within the family and with neighbors in a community about actions that can be

[1]Many Web pages are now available for households and businesses to learn about new ways of saving energy. See, for example, the stories about ways to save energy in homes on the Environmental Defense Fund Web pages at http://www.fightglobalwarming.com/page.cfm?tagID=262 (accessed 20 February 2009).

taken to reduce greenhouse gas emissions are also important factors leading to the potential for change (see, for example, Miller 2009). Even without major taxes imposed on energy at a national level, however, families who decide to invest in better insulation, more efficient furnaces and other appliances, to join a carpool whenever feasible, and other energy-reducing actions, can save funds over the long run as well as reducing emissions. They may face high up-front investments to achieve some of these benefits, but the important point is that positive benefits can be achieved that offset costs at a household or neighborhood level.

Jurisdictions that have established power networks that enable households to invest in solar power to be used for household energy production, and when not needed is contributed to the network, can also potentially reduce local energy costs by working out complex network arrangements as well as reducing greenhouse gas emissions. In Japan, for example, the Ministry of Trade and Industry issued "A New Purchase System for Solar Power-Generated Electricity" that requires electric utilities to purchase solar power electricity that exceeds the needs of households. The national government also subsidizes households that install solar energy. As a result, sales of solar panels rose by 21 % during 2009—the highest level since 1981 (Sato 2010). Investments in better waste disposal facilities and to reduce pollution levels also generate local benefits as well as helping on global emissions. Given that many of the actions generating greenhouse gas emissions are taken at multiple scales, activities to reduce emissions can also be organized at multiple scales ranging from households to the globe (Kates and Wilbanks 2003).

4 What Efforts to Reduce Greenhouse Gas Emissions Now Occur at Less Than a Global Scale?

It is not possible to list the large number of projects going on across the world at multiple scales. What I will do is focus on some of the projects that have been organized at a local level as part of the Clean Development Mechanism of the Kyoto Protocol, at the level of a state government in the United States, at a regional level, and discuss some of the efforts in Europe to substantially reduce emissions. Schreurs (2008) and Hoffman and Eidelman (2009) have identified a large number of experiments at multiple levels that reflect action by diverse governance arrangements to take climate change seriously and take actions to reduce the threat.

4.1 Local-Level Projects and Alliances to Reduce Local-Level Externalities

One of the most successful efforts made by local governments across the United States, and supported by the U.S. Clean Air Act, has been to reduce the level of fine-particulate air pollution (which in some cases has also reduced greenhouse gas emissions as well). Pope et al. (2009) have completed a major study of the level of the impact on life expectancy of particulate matter in the air sampled over the period from 1979 to 2000 for 51 metropolitan areas (including more than 200 counties). Metropolitan areas across the nation have reduced air pollution levels by one-third. They also found that increased life expectancy during this period was associated with reductions in fine-particulate air pollution after controlling for socioeconomic, demographic, and other variables associated with life expectancy. Given their statistical analysis, the average life expectation that could be attributed to reduction in air pollutants was one-third of a year.

"Buildings use 40 % of the primary energy supplied in the United States, and more than 70 % of all generated electricity, primarily for heating, cooling, and lighting" (Gershenfeld et al. 2010: 1086). Dietz et al. (2009) have identified seventeen actions that can be taken within a home or a business facility that can cumulatively have a major impact on carbon emissions. Thus, retrofitting buildings to add insulation, solar photovoltaics, and more efficient heating systems is another important strategy that can be taken at a local level and may actually generate a long-term savings to the firm or family that takes such actions in energy costs as well as reducing greenhouse gas emissions.

The up-front costs of such efforts are frequently daunting, even when the private investment will reduce private costs over the long run. By a public ballot approved by 81 % of the voters, Berkeley, California, has adopted a general policy to reduce emissions substantially over time. Berkeley FIRST (Financing Initiative for Renewable and Solar Technology) is designed to reduce the barrier of up-front costs. To participate in the program, a commercial or residential property owner asks a contractor for an estimate of the costs of new solar energy equipment and improvements to the energy efficiency of the building. The estimate is submitted to the city for review and to ensure that the owner has a clear title.

> After the municipality approves the application, the work is completed, a lien is placed on the property, and a check is issued to the property owner. A special tax is added to future property bills. If the property is sold before the end of the 20-year repayment period, the new owner pays the remaining special taxes as part of their property's annual tax bill. The interest component of the special tax payments will be tax deductible, similar to a home equity line or home mortgage (Pope et al. 2009: 25).

The demand for long-term and reasonable public loans has been high and Berkeley plans to increase the funds available to support this program over time.

Some local utilities in the United States are now also actively finding ways of reducing energy consumption by developing local monitoring systems that are then reported on the bills that customers receive. The Sacramento Municipal Utility District, for example, has tried various techniques including rebates for energy-saving appliances, but recently found a more effective technique.

> Last April (2008), it began sending out statements to 35,000 randomly selected customers, rating them on their energy use compared with that of neighbors in 100 homes of similar

> size that used the same heating fuel. The customers were also compared with 20 neighbors who were especially efficient in saving energy.
>
> Customers who score high earned two smiley faces on their statements. "Good" conservation got a single smiley face (Kaufman 2009).

The utility company conducted an initial assessment of this new strategy after using it for 6 months. The assessment found "that customers who received the personalized report reduced energy use by 2 % more than those who got standard statements" (Kaufman 2009). Using various forms of competition among households and groups, and feedback as to who is doing the best of reducing energy use, is a strategy for reducing emissions that is increasingly being adopted by college campuses, small cities, and utility firms around the country. University efforts to stimulate competition among campus dormitories to see who can reduce electricity consumption are proving to be effective (Peterson et al. 2007). Contemporary psychological studies have found that framing problems related to resource use in a social context do affect actions (Schultz et al. 2007; Mumford 2007).

Methods for developing reliable city-scale greenhouse gas inventories have been developed and tested (Ramaswami et al. 2008; Hillman and Ramaswami 1902). These are being used by many of the large number of cities across in multiple countries that have pledged to reduce GHG emissions consistent with the Kyoto Protocol. In the United States alone, the mayors of 1,026 cities have now joined the U.S. Conference of Mayors' Climate Protection Agreement to reduce GHG emissions of at least 5 % relative to 1990 levels (U.S. Mayors' Climate Protection Agreement 2010).

Multiple cities have started to initiate a variety of "green" initiatives that are prominently displayed on their home pages on the Web. The city of Toronto, for example, has established an "environmental portal" that announces more than a dozen current city policies, related publications, and meetings that are focused on climate change.[2] The city has supported a number of renewable energy projects including major investments averaging around $100,000 each for building rooftop gardens, solar photovoltaic panels on houses, and solar water-heating systems. The city also funds smaller projects to support neighborhood efforts to enhance the forested areas of local parks, local gardens, and for organizations at the local level that are working with communities to hold planning meetings to discuss better bicycle paths and other activities that can be undertaken at a small, neighborhood scale.

Large city mayors are also banding together to discuss actions to reduce carbon emissions that can be taken locally but if taken jointly, can have a much bigger effect. In October 2005, eighteen large cities sent representatives to London to examine actions that could be taken at a municipal level to reexamine various urban policies that could be revised including their own purchasing policies and ways of encouraging more investment in climate-friendly technologies in their cities. The C40 Large Cities Climate Summit occurred in May 2007 for the exchange of

---

[2]http://www.toronto.ca/environment/index.htm (accessed 9 February 2009).

information about many policies adopted to reduce emissions and the announcement of a $5 billion global Energy Efficiency Building Retrofit Program by the Clinton Climate Initiative.[3]

4.2 State-Level Projects in the United States

California is not only the twelfth largest emitter of greenhouse gases in the world—comparable to Australia's emissions—but it is now one of the leading governments to adopt policies related to climate change (Engel 2006). For example, in 2006, the California legislature passed legislation called the Global Warming Solutions Act, aimed at reducing greenhouse gas emissions by 25 % by 2020 by requiring drastic reductions from major industries including oil and gas refineries and utility plants.[4] The California Air Resources Board is charged with developing a market-based cap-and-trade program to implement the policy (Goulder 2007). This program is essentially a local version of the carbon market developed in the Kyoto Protocol. This is another example of how state-level policies can be designed to carry out policies originally formulated for a global level. The California policy reflects both its exposure to dramatic sea-level rises, if emission levels are not reduced, as well as a spur to the U.S. government to begin adopting policies at a national level.

The Colorado legislature passed State House Bill 08–1350, which was signed into law in 2008, to enable local governments to adopt policies similar to the Berkeley FIRST described above. The legislation allows municipalities in Colorado to finance approved building improvements and enables property owners to pay off capital investments made to decrease their use of fossil fuels for heating and electricity through a repayment over 20 years. In July of 2007, Governor Charlie Crist brought together government, business, and scientific leaders from across the state of Florida to discuss what actions could be taken by Florida to address climate change issues. At the conclusion of the meeting, several executive orders were signed to set out targets for reducing greenhouse gas emissions in Florida and to change the building code to require increased energy efficiency in new construction.[5]

4.3 Regional Efforts

Efforts are also being made among the states to develop joint policies. The Regional Greenhouse Gas Initiative (RGGI), joined by ten states located in the northeast and

---

[3]http://www.c40cities.org/ (accessed 1 February 2009).

[4]Global Warming Solutions Act of 2006, Calif. Assembly Bill 32.

[5]http://www.dep.state.fl.us/climatechange/ (accessed 27 June 2008).

mid-Atlantic regions of the United States, plans to cap CO_2 from the power sector by 10 % by 2018.[6] Further, RGGI is one of the first market-based efforts in the United States aimed at reducing greenhouse gas emissions by auctioning emission allowances and investing the proceeds in various forms of clean energy technologies and to green jobs in each of the states.

4.4 European Efforts

In Europe, various interventions tend to combine local, national, and European levels. The EU Emissions Trading Scheme (EU-ETS) was developed so as to reduce the economic costs of meeting its Kyoto target of 8 % CO_2 reduction by 2012. The EU-ETS is a major manifestation of the carbon market envisioned in the Kyoto Protocol. Around 10,000 large industrial plants in the power generation, iron and steel, glass, brick, and pottery industries in Europe are included, but not the transport sector. Operators of these facilities receive emission allowances that are good for a 1-year period. If they are not fully used by the assigned operator (after verification), the unused portion may be sold to other facilities that have not yet met their assigned target. The official data issued by the European Environmental Agency (EEA) in (2006) show that the EU members that had signed the Kyoto Agreement were able to achieve a 2 % cut in CO_2 emissions in 2005 compared to 1990 levels. CO_2 emissions are projected to decline further by 2010 compared to 2004 levels (EEA 2006: Sections 8 and 9). Thus, the decentralized impact of markets—resulting from the price of carbon that is itself now reflecting the externalities of climate change—helps to break up a global policy of the Kyoto treaty into individual actions by businesses and consumers.

5 Are Large-Scale Governments Usually Better Able to Cope with Collective Action?

While the presumption is made in many policy discussions that global solutions are necessary for coping with the problems of climate change because of the inadequacy of local and regional efforts, few of these analyses examine the problems that large-scale units themselves face in developing effective policies related to resources. Before making a commitment that the global level is the *only* scale in which to address climate change, one should at least reflect on past efforts to adopt uniform policies by very large entities intended to correct for problems of collective action.

[6] http://rggi.org/home/ (accessed 7 February 2009).

Contemporary assignments of regional, national, or international governments with the *exclusive* responsibility for providing local public goods and common-pool resources remove authority from local officials and citizens to solve local problems that differ from one location to the next. Doug Wilson, Research Director for the Institute for Fisheries Management and Coastal Community Development in Denmark, has recently reflected on the evolution of fisheries policies in the European Union.

> The Common Fisheries Policy (CFP) as it is called is an 'exclusive competence' of the European Union (EU) meaning that all decisions are taken at the level of the Union …
>
> The CFP is not only politically important within the overall effort to build a new kind of polity in Europe; it is also failing to do a very good job of maintaining sustainable fish stocks. Fisheries scientists tell us that, in 2003, 22 % of the fish caught from stocks managed by the CFP were taken from stocks that were smaller than they should have been for sustainable fishing. Neither scientists, fishers, government agencies, nor marine conservation groups are happy with the CFP, and there are myriad attempts to reform it. The reforms include better policy, better data gathering, a reduction in perverse subsidies to the fishing industry and, finally 30 years after most other fisheries management agencies had moved beyond top-down management, some serious attempts at stakeholder involvement (Wilson 2006: 7).

Other policies related to fisheries adopted by large-scale units have also exhibited major problems.[7] Exclusive Economic Zones (EEZs) were created in 1982 that extend 200 nautical miles along the borders between the ocean and coastal states and extended full sovereign powers to these states to manage these fisheries so that they are not overexploited (United Nations 1982). Instead of reducing overharvesting, however, many national governments subsidized expansions of fishing fleets that increased the demand on coastal fisheries and placed more in danger of overexploitation (Walters 1986). The models of fishery dynamics used by national governments tended to be relatively crude and led to inaccurate assessment of fishery stocks (Wilson 2002).[8]

Problems have also been noted regarding the way the Clean Development Mechanism (CDM) authorized by the Kyoto Protocol is being implemented in some settings. Several CDM processes are involved. One CDM process is supposed to substitute carbon-emitting energy-production processes with "green energy production." This process works approximately in this fashion: (1) a developing country decides to forego the construction of a power plant emitting substantial

---

[7]See Clark (2006) for a review of policies that have been adopted by national governments related to fisheries that initially led to perverse outcomes—some of which were eventually reversed.

[8]The Department of Fisheries and Oceans in Canada, for example, developed a model of stock regeneration for northern cod that scientists later found to be flawed. Local cod fishers in Newfoundland raised serious questions in the late 1980s and predicted a near-term collapse; the Canadian government refused to listen and assured doubters that their model was correct. In 1992, however, the cod stock collapsed and the Canadian government declared a moratorium on all fishing in Canadian waters, which has generated very substantial costs for local fishing villages dependent upon that stock that they had earlier managed relatively effectively (Finlayson 1994; Finlayson and McCay 1998).

greenhouse gases, (2) it plans to build a wind farm that is more "carbon friendly," and (3) the country applies for credit in the form of Certified Emissions Reductions (CERs) to sell to industrialized nations wishing to buy CERs as authorized by the Kyoto Protocol (Lohmann 2008). The income from selling the CERs can then, in principle, be allocated to the construction of the more expensive wind farm.

One problem with this highly complicated and flexible system is that it can be gamed (Sovacool and Brown 2009). Only 300 of the thousands of CDM projects that are underway have received accreditation by the UN. As it turns out, a large proportion of the CERs relate to triflouromethane, HFC-23, a greenhouse gas that is not associated with transportation or the production of power, but rather is used as a refrigerant—and a highly profitable greenhouse gas to claim to have "averted." As Sovacool and Brown (2009) conclude, the CDM has unfortunately made HFC-23 abatement too profitable.

> The sale of carbon credits generated from CERS for HFC-23 has become far more valuable than its production in the first place. Manufacturers of HFC-23, responding to market demand for CERs, started producing it just to offset it. Researchers at Stanford University have calculated that, at a result, payments to refrigerant manufacturers and carbon market investors to governments and compliance buyers for HFC-23 credits has exceeded €4.7 *billion* when the costs of merely abating HFC-23 would have been about €100 *million*—a major distortion of the market (Sovacool and Brown 2009: 14; citing Wara 2007 and Wara and Victor 2008).

Since the Bali round of negotiations held in December 2007, efforts to reduce emissions from deforestation and degradation (REDD) have been added to the portfolio of activities authorized under the Kyoto Protocol. Forest ecosystems do store an immense quantity of carbon, and the scientific foundation for adopting REDD is quite strong. Designing REDD projects so that new projects do not just lead to further leakage is a substantial problem. Ensuring that the rights of indigenous peoples are, at least, protected and ideally, enhanced as a result of support of their management of forest ecologies, is a goal that is widely shared by social activists at multiple scales. Accomplishing this goal while expanding the amount of forested land in developing counties would be economically efficient but a difficult challenge.[9] Currently there is considerable debate about this program and too few projects have been adopted to make a serious evaluation of the possibilities and threats (see Angelsen 2009; O'Sullivan 2008; Streck et al. 2008; Corbera and Brown 2008).

The discussion of problematic policies of large-scale governmental units related to climate change and other environmental policies is not meant to challenge the need for global policies related to climate change. The intent is to balance the major

[9]John Vidal (2008), in an article in *The Guardian* (17 October 2008), stressed that recognizing forest community rights would be a more cost-effective mechanism for reducing emissions than paying organizations to plant trees. "A study by Jeffrey Hatcher, an analyst with Rights and Resources in Washington, found that it costs about $3.50 (€2) per hectare to recognize forest people's land. The costs of protecting forests under REDD have been estimated at about €2000 per hectare.".

attention that has been given in the policy literature to the need for global solutions as the *only* strategy for coping with climate change. Extensive research on institutions related to environmental policies has repeatedly shown that creative, effective, and efficient policies, as well as disasters, have been implemented at *all* scales. Dealing with the complexity of environmental problems can lead to "negative learning" by scientists and policymakers at all scales (Oppenheimer et al. 2008). Reliance on a single "solution" may be more of a problem than a solution (Pritchett and Woolcock 2003).

It is important that we recognize that devising policies related to complex environmental processes is a grand challenge and reliance on one scale and one model alone to solve these problems is naïve. On the other hand, climate mitigation policies must eventually involve all of the countries of the world. Countries that are low emitters today, such as those in Africa and Latin America, are likely to increase their contributions significantly in the future. Further, as discussed below, those countries that are not included in agreements can undermine the efforts of those that are through "leakage" and behaving generally as free-riders. The efforts of many organizations at less-than-global scale can help reduce remissions to some extent, and they can also spur their own governments to take necessary national and international efforts.

6 Are There Too Many Actors Working on Climate Change?

One criticism leveled at current efforts to reduce greenhouse gas emissions is that the system is chaotic. Unquestionably, many problems characterize the current efforts. Many of these do relate to the lack of effective policies at an international level. Further, some of the projects that are overtly aimed at reducing greenhouse gas emissions may well be ineffective, too costly, and rewarding actors who are not genuinely interested in reducing the threat of climate change, but are rather looking for opportunities to gain funds and search for minimal ways of meeting project announcements.

Thus, it is important that we examine some of the key problems that have been identified as plaguing efforts to control greenhouse gas emissions. Recognition of problems is essential to start serious efforts to find methods to reduce them. The problems raised most frequently relate to leakage, inconsistent policies, free-riding, and inadequate certification.

Leakage is one of the problems frequently identified with subnational projects aimed at reducing carbon emissions (Burniaux and Martins 2010). Two types of leakage can occur from policies adopted at less-than-global scale: location and market leakage (Ebeling 2008: 49–51). Leakage between locations occurs when an activity that would have occurred in X location is shifted to Y location because of a climate change project that occurs in X location (Sovacool and Brown 2009). The EU's efforts to reduce emissions from industrial producers may, in some cases, simply shift the emissions that would have been produced by a European chemical firm to another location in a developing country where the costs of production may be lower. Carbon is still emitted, however, in the production of chemicals plus the carbon emitted in transportation of the chemicals to European locations (Chomitz 2002). Market leakage refers to the changes in the price structure that may occur by restrictions placed on harvesting from forests. Such restrictions reduce the volume of timber and other forest products generated in one area. This stimulates an increase in the prices of these products. If everything goes well, higher prices encourage the intensification of agricultural and forest production in other areas and it does not stimulate more deforestation. "In a less favorable scenario, particularly when land-use regulations are poorly enforced, higher prices provide an additional incentive to clear forests for timber or agriculture elsewhere, thereby reducing the net benefits of the climate mitigation project" (Ebeling 2008: 50).

Whenever actions taken by some individuals or organizations benefit a larger group, a risk exists that some participants will free-ride on the efforts of others and not contribute at all or not contribute an appropriate share. At the current time, there are many governmental and private entities at multiple scales that are increasing their greenhouse gas emissions substantially—especially in the developing world—without adopting any policies to reduce emissions. This is a major problem. Current debates over who caused the human threat and thus who should pay the most in the future are legitimate debates. At the same time, they may also cover a free-riding strategy by at least some of those involved.

For policies adopted at any scale that provide diverse rewards for projects that reduce greenhouse gas emissions, a need exists for skilled personnel to certify that the project does indeed reduce ambient CO_2 by some specified amount over a defined time period. A very active new industry of "global consultants" has emerged. While many consultants do have good scientific training, the greatly increased need for certification has generated opportunities for at least some contractors who lack appropriate skills to earn money in the new "certification game." Sovacool and Brown (2009: 14) report on one study that evaluated 93 randomly chosen CDM projects and "found that in a majority of cases the consultants hired to validate CERs did not possess the requisite knowledge needed to approve projects, were overworked, did not follow instructions, and spent only a few hours evaluating each case."

Problems do exist in the design and administration of projects at multiple scales trying to deal with climate change. There is a lot to learn, however, from these efforts. It is essential that we recognize: (1) the complexity of causes of climate change; (2) the challenge of acquiring knowledge about causes and effects in a world that is changing rapidly; (3) the wide diversity of policies that can lead to reduced emissions but might also enable opportunistic efforts to obtain a flow of funds by appearing to reduce emissions while not having a real impact, or worse, effectively increasing rather than decreasing emissions; (4) the opportunities that major sources of funding open up for policy experiments if funds are also allocated to monitoring and evaluation of the benefits and costs of the experiment; and (5) that all policies adopted at any scale can generate errors, but without trial and error, learning cannot occur.

Acknowledging the complexity of the problem, as well as the relatively recent agreement among scientists about the human causes of climate change, leads to recognition that just waiting for effective policies to be established at the global level is unreasonable. Rather than only a global effort, it would be better to self-consciously adopt a polycentric approach to the problem of climate change in order to gain the benefits at multiple scales as well as to encourage experimentation and learning from diverse policies adopted by multiple scales. Less-than-global efforts may also spur essential efforts at a global level.

Further, the extensive empirical research on collective action discussed above has repeatedly identified a necessary central core of trust and reciprocity among those involved to be associated with successful levels of collective action. If the *only* policy adopted related to climate change was at the global scale, it is particularly difficult to increase the trust that citizens and firms need to have that other citizens and firms located halfway around the globe as well as nearby are taking similar actions. Effective monitoring is needed both to catch offenders as well as assuring those who cooperate with costly policies that they are not suckers. One of the core findings from recent research on the sustainability of forests in a dozen countries around the world is the importance of users having a strong commitment to collective action to protect their forests. As a result, in the forests where users themselves contribute to monitoring efforts, their forests are in better condition (Gibson et al. 2005; Hayes and Ostrom 2005; Ostrom and Nagendra 2006; Coleman 2009; Chhatre and Agrawal 2008). In these settings, users are able to engage in sustainable exploitation of natural resources (Figuières 2010 and Tidball 2010). Citizens living in a community that has adopted policies to restrain the emissions of greenhouse gases interact in a variety of local settings where they can directly question each other if inconsistent behavior is observed. When most of their friends, neighbors, and coworkers appear to be following rules to reduce their carbon emissions, each citizen gains trust that they are not foolish for complying themselves. This is another complementary aspect of adopting policies at local levels that are consistent with the goals of policies at regional, national, and global levels.

7 Conclusion

Given that the recognition of the danger of climate change among citizens and public officials is still relatively recent, and that major debates about potential solutions are continuing, one cannot expect a global solution to be constructed in the near future. Building a global regime is a necessity (Barrett 2007), but building a polycentric system starts the process of reducing greenhouses gas emissions and acts as a spur to national and international regimes to get their act together!

Recognizing the potential of building more effective ways of reducing energy use at multiple scales is thus an important step forward. Further, an important strategy for reducing CO_2 in the atmosphere is developing more effective policies for protecting ecosystem services—particularly those related to carbon sequestration. Developing effective and adaptive programs, however, requires selecting appropriate areas, developing plans for leaving some areas untouched, and for making major investments in the flora and fauna as well as the technological infrastructure of other areas (Michel 2009). This requires substantial investment in scientific modeling (Nelson et al. 2009) and use of geographic information systems combined with in-depth knowledge of the biophysical settings to map ecological systems over time (Daily et al. 2009). The models, however, need to be developed at multiple scales so that relevant decision-making units can address what policies can be adopted to improve carbon sequestration that fits the ecology at that particular scale.

Building a strong commitment to finding ways of reducing individual emissions is an important element for coping with climate change. Building such a commitment can be more effectively undertaken in small- to medium-scale governance units that are linked together through information networks and monitoring at all levels. Global policies are indeed necessary but they are not sufficient.

References

Angelsen, A. (Ed.). (2009). *Realising REDD+: National strategy and policy options*. Bogor, Indonesia: CIFOR.

Asheim, G. B., Mitra, T., & Tungodden, B. (2010). Sustainable recursive social welfare functions. *Economic Theory*.

Baer, P., Harte, J., Haya, B., et al. (2000). Equity and greenhouse gas responsibility. *Science, 289*, 2287.

Barrett, S. (2007). *Why cooperate? The incentive to supply global public goods*. Oxford: Oxford University Press.

Botsen, W. J. W., Gowdy, J. M., & Van den Bergh, J. C. J. M. (2008). Cumulative CO_2 emissions: Shifting international responsibilities for climate debt. *Climate Policy, 8*, 569–576.

Burniaux, J.-M., & Martins, J. O. (2010). Carbon leakages: A general equilibrium view. *Economic Theory*.

Chhatre, A., & Agrawal, A. (2008). Forest commons and local enforcement. *PNAS, 105*, 13286–13291.

Chichilnisky, G. (2010). Sustainable Markets with short sales. *Economic Theory*.

Chichilnisky, G., & Heal, G. (1994). Who should abate carbon emissions?: An international viewpoint. *Economics Letters, 44*, 443–449.

Chichilnisky, G., & Heal, G. (2000). *Environmental markets: Equity and efficiency*. New York: Columbia University Press.

Chipman, J.S., Tian, G.: Detrimental externalities, pollution rights, and the 'Coase theorem'. *Economic Theory* (2010).

Chomitz, K. M. (2002). Baseline, leakage and measurement issues: How do forestry and energy projects compare? *Climate Policy, 2*, 35–49.

Clark, C. W. (2006). *The worldwide crisis in fisheries: Economic models and human behavior*. Cambridge: Cambridge University Press.

Cole, D. H. (2008). Climate change and collective action. *Current Legal Problems, 61*, 229–264.

Coleman, E. (2009). Institutional factors affecting ecological outcomes in forest management. *Journal of Policy Analysis and Management, 28*, 122–146.

Corbera, E., & Brown, K. (2008). Building institutions to trade ecosystem services: Marketing forest carbon in Mexico. *World Development, 36*, 1956–1979.

Daily, G. C., Polasky, S., Goldstein, J., et al. (2009). Ecosystem services in decision making: Time to deliver. *Frontiers in Ecology and the Environment, 7*, 21–28.

Dellink, R., Dekker, T., den Elzen, M., et al. (2009). *Sharing the burden of adaptation financing*. Amsterdam: Institute for Environmental Studies Newsletter.

den Elzen, M. G. J., Schaeffer, M., & Lucas, P. L. (2005). Differentiating future commitments on the basis of countries' relative historical responsibility for climate change: Uncertainties in the 'Brazilian Proposal' in the context of a policy implementation. *Climate Change, 71*, 277–301.
Dietz, T., Gardner, G. T., Gilligan, J., Stern, P. C., & Vandenbergh, M. P. (2009). Household actions can provide a behavioral wedge to rapidly reduce US carbon emissions. *PNAS, 106*, 18452–18456.
Dutta, P. K., & Radner, R. (2010). Capital growth in a global warming model: Will China and India sign a climate treaty? *Economic Theory*.
Ebeling, J. (2008). Risks and criticisms of forestry-based climate change mitigation and carbon trading. In C. Streck, et al. (Eds.), *Climate change and forests: Emerging policy and market opportunities* (pp. 43–58). London: Chatham House.
Engel, K. H. (2006). State and local climate change initiatives: What is motivating state and local governments to address a global problem and what does this say about federalism and environmental law? Discussion Paper 06–36. Tucson: University of Arizona, James E. Rogers College of Law, Arizona Legal Studies (2006).
European Environmental Agency (EEA). (2006). Greenhouse Gas Emission Trends and Projections in Europe 2006. Luxembourg: Office for Official Publications of the European Communities. EEA Report No. 9, ISSN 1725–9177.
Figuières, C., & Tidball, M. (2010). Sustainable exploitation of a natural resource: A satisfying use of Chichilnisky's criterion. *Economic Theory*.
Finlayson, A. C. (1994). Fishing for truth: A sociological analysis of northern cod stock assessment from 1977–1990. St. Johns: Memorial University of New Foundland, New Foundland Institution of Social and Economic Research.
Finlayson, A. C., & McCay, B. J. (1998). Crossing the threshold of ecosystem resilience: The commercial extension of northern cod. In F. Berkes, & C. Folke (Eds.), *Linking social and ecological systems: Management practices and social mechanisms for building resilience* (pp. 311–338). Cambridge: Cambridge University Press.
Fuller, M. C., Portis, S. C., & Kammen, D. M. (2009). Toward a low-carbon economy: Municipal financing for energy efficiency and solar power. *Environment, 51*, 22–32.
Gershenfeld, N., Samouhos, S., & Nordman, B. (2010). Intelligent infrastructure for energy efficiency. *Science, 327*, 1086–1087.
Gibson, C., Williams, J., & Ostrom, E. (2005). Local enforcement and better forests. *World Development, 33*, 273–284.
Goulder, L. H. (2009). California's bold new climate policy. *The Economists' Voice, 4*. Retrieved February 20, 2009, from http://www.bepress.com/ev/vol4/iss3/art5/.
Hayes, T., & Ostrom, E. (2005). Conserving the world's forests: Are protected areas the only way? *Indiana Law Review, 38*, 595–617.
Hillman, T., & Ramaswami, A. (2010). Greenhouse gas emission footprints and energy use benchmarks for eight U.S. cities. *Environmental Science & Technology, 44*, 1902–1910.
Hoffman, M. J., & Eidelman, G. (2009). Experimenting with climate governance. Paper presented at the Amsterdam Human Dimensions of Environmental Change Meeting.
Intergovernmental Panel on Climate Change. (2007). Climate Change 2007: Impacts, Adaptation and Vulnerability. Contribution of Working Group II to the Fourth Assessment Report. Cambridge University Press.
Karp, L., & Zhang, J. (2010). Taxes versus quantities for a stock pollutant with endogenous abatement costs and asymmetric information. *Economic Theory*.
Kates, R. W., & Wilbanks, T. J. (2003). Making the global local: Responding to climate change concerns from the ground up. *Environment, 45*, 12–23.
Kaufman, L. (2009). A desire to keep up with neighbors is spurring conservation. *New York Times*. Retrieved January 30, 2009, from http://www.nytimes.com/2009/01/31/science/earth/31compete.html.
Lauwers, L. (2010). Intergenerational equity, efficiency, and constructability. *Economic Theory*.
Lecocq, F., & Hourcade, J.-C. (2010). Unspoken ethical issues in the climate affair: Insights from a theoretical analysis of negotiation mandates. *Economic Theory*.

Lohmann, L. (2008). Carbon trading, climate justice and the production of ignorance: Ten examples. *Development, 51*, 359–365.
Matthews, H. D., & Caldeira, K. (2008). Stabilizing climate requires near zero emissions. *Geophysical Research Letters, 35*, 1–5.
McGinnis, M. (Ed.). (1999a). *Polycentric governance and development: Readings from the workshop in political theory and policy analysis*. Ann Arbor: University of Michigan Press.
McGinnis, M. (Ed.). (1999b). *Polycentricity and local public economies: Readings from the workshop in political theory and policy analysis*. Ann Arbor: University of Michigan Press.
McGinnis, M. (Ed.). (2000). *Polycentric games and institutions: Readings from the workshop in political theory and policy analysis*. Ann Arbor: University of Michigan Press.
Michel, D. (2009). Foxes, hedgehogs, and greenhouse governance: Knowledge, uncertainty, and international policy-making in a warming world. *Applied Energy, 86*, 258–264.
Miller, C. A. (2004). Climate science and the making of a global political order. In S. Jasanoff (Ed.), *States of Knowledge: The coproduction of science and social order* (pp. 46–66). New York: Routledge.
Miller, P. (2009). Saving energy: It starts at home. *Natl. Geogr., 215*, 60–81.
Mumford, G. (2007). Psychology's ability to curb energy use. *Monitor on Psychology, 38*, 20–21.
Nelson, E., Polasky, S., Lewis, D. J., et al. (2008). Efficiency of incentives to jointly increase carbon sequestration and species conservation on a landscape. *PNAS, 105*, 9471–9476.
Nelson, E., Mendoza, G., Regetz, J., et al. (2009). Modeling multiple ecosystem services, biodiversity conservation, commodity production, and tradeoffs at landscape scales. *Frontiers in Ecology and the Environment, 7*, 4–11.
Newport, F. (2010). Americans' global warming concerns continue to drop. Retrieved March 15, 2010, from http://www.gallup.com/poll/126560/americans-global-warming-concerns-continue-drop.aspx.
O'Sullivan, R. (2008). Reducing emissions from deforestation in developing countries. In C. Streck, R. O'Sullivan, T. Janson-Smith, & R. Tarasofsky (Eds.), *Climate change and forests: Emerging policy and market opportunities* (pp. 179–190). London: Chatham House.
Oppenheimer, M., O'Neill, B. C., & Webster, M. (2008). Negative learning. *Climate Change, 89*, 155–172.
Ostrom, V. (1999). Polycentricity—Part 1. In M. McGinnis (Ed.), *Polycentricity and local public economies: Readings from the workshop in political theory and policy analysis* (pp. 52–74). Ann Arbor: University of Michigan Press.
Ostrom, V. (2008a). *The intellectual crisis in american public administration* (3rd ed.). Tuscaloosa: University of Alabama Press.
Ostrom, V. (2008b). *The political theory of a compound republic: Designing the American experiment* (3rd ed.). Lanham, MD: Lexington Books.
Ostrom, E., & Nagendra, H. (2006). Insights on linking forests, trees, and people from the air, on the ground, and in the laboratory. *PNAS, 103*, 19224–19231.
Ostrom, V., Tiebout, C. M., & Warren, R. (1961). The organization of government in metropolitan areas: A theoretical inquiry. *The American Political Science Review, 55*, 831–842.
Ostrom, E., Parks, R. B., & Whitaker, G. P. (1978). *Patterns of metropolitan policing*. Cambridge, MA: Ballinger.
Parks, R. B., & Ostrom, E. (1999). Complex models of urban service systems. In M. McGinnis (Ed.), *Polycentricity and local public economies: Readings from the workshop in political theory and policy analysis* (pp. 355–380). Ann Arbor: University of Michigan Press.
Peterson, J. E., Shunturov, V., Janda, K., Platt, G., & Weinberger, K. (2007). Dormitory residents reduce electricity consumption when exposed to real-time visual feedback and incentives. *International Journal of Sustainability in Higher Education, 8*, 16–33.
Pope, C. A., Ezzati, M., & Dockery, D. W. (2009). Fine-particulate air pollution and life expectancy in the United States. *The New England Journal of Medicine, 360*, 376–386.
Posner, E. A., & Sunstein, C. (2008). Justice and climate change. Discussion Paper 08–04. Cambridge, MA: Harvard University Project on International Climate Agreements.

Pritchett, L., & Woolcock, M. (2003). Solutions when the solution is the problem: Arraying the disarray in development. *World Development, 35*, 435–461.
Ramaswami, A., Hillman, T., Janson, B., Reiner, M., & Thomas, G. (2008). A demand-centered, hybrid life-cycle methodology for city-scale greenhouse gas inventories. *Environment Science & Technology, 42*, 6455–6461.
Rezai, A., Foley, D. K., & Taylor, L. (2010). Global warming and economic externalities. *Economic Theory*.
Sandler, T. (2004). *Global collective action*. Cambridge: Cambridge University Press.
Sato, S. (2010). *Japan's solar panel sales rise to record on subsidy*. Retrieved April 21, 2010, from http://www.bloomberg.com/apps/news?pid=20601072&sid=avbxkfyJQgGQ.
Schelling, T. C. (2009). Climate change: The uncertainties, the certainties, and what they imply about action. *The Economists' Voice*, ***4***. Retrieved February 20, 2009, from http://www.bepress.com/ev/vol4/iss3/art3/.
Schreurs, M. A. (2008). From the bottom up: Local and subnational climate change politics. *Journal of Environment & Development, 17*, 343–355.
Schultz, P., Nolan, J., Cialdini, R., et al. (2007). The constructive, destructive, and reconstructive power of social norms. *Psychological Science, 18*, 429–434.
Sovacool, B. K., & Brown, M. A. (2009). Addressing climate change: Global and local? In F. P. Sioshansi (Ed.), *Generating electricity in a carbon constrained world*. New York: Elsevier.
Stavins, R. (1997). *Policy instruments for climate change: How can national governments address a global problem?* University of Chicago Legal Forum, vol. 1997: Rethinking Environmental Protection for the 21st Century (pp. 293–329).
Stern, N. (2007). *The economics of climate change: The stern review*. Cambridge: Cambridge University Press (2007).
Streck, C., O'Sullivan, R., Janson-Smith, T., & Tarosofsky, R. (2008). *Climate change and forests: Emerging policy and market opportunities*. London: Chatham House.
United Nations. (1982). *Final act of the third conference on the law of the seas*. Montenegro Bay, Jamaica: United Nations (1982).
U.S. Conference of Mayors' Climate Protection Agreement. (2010). Retrieved May 1, 2010, from http://www.usmayors.org/climateprotection/revised/.
Vidal, J. (2008). *Pay indigenous people to protect rainforests, conservation groups urge. The Guardian*. Retrieved October 23, 2008, from http://www.guardian.co.uk/environment/2008/oct/17/forests-endangeredhabitats.
Walters, C. J. (1986). *Adaptive management of renewable resources*. New York: Macmillan.
Wara, M. (2007). Is the global carbon market working? *Nature, 445*, 595–596.
Wara, M., & Victor, D. G. (2008). A realistic policy on international carbon offsets. Working Paper no. 74. Stanford: Stanford University Program on Energy and Sustainable Development.
Wiener, J. B. (2007). Think globally, act globally: The limits of local climate policies. *The University of Pennsylvania Law Review, 155*, 1961–1979.
Williamson, O. E. (1975). *Markets and hierarchies: Analysis and antitrust implications*. New York: Free Press.
Williamson, O. E. (1985). *The economic institutions of capitalism: Firms, markets, relational contracting*. New York: Free Press.
Williamson, O. E. (2000). The new institutional economics: Take stock, looking ahead. *The Journal of Economic Literature, 38*, 593–613.
Wilson, J. (2002). Scientific uncertainty, complex systems, and the design of common-pool institutions. In E. Ostrom, T. Dietz, N. Dolšak, P. Stern, S. Stonich, E. Weber (Eds.), *National Research Council, Committee on the Human Dimensions of Global Change. The Drama of the Commons* (pp. 327–359). Washington, DC: National Academy Press (2002).
Wilson, D. (2006). Predictions about European commons. *The Common Property Resource Digest, 77*, 7–8.

Capital Growth in a Global Warming Model: Will China and India Sign a Climate Treaty?

Prajit K. Dutta and Roy Radner

1 Introduction

This paper addresses the following question: (when) will the fast growing economies of the East, China and India for example, agree to caps on their greenhouse gas emissions? This introductory section contextualizes the question and then provides a summary of the answers contained in this paper.

1.1 The East Versus West Debate

Global climate change (CC) has emerged as the most important environmental issue of our times and, arguably, the one with the most critical long-run import. The observed rise in temperatures and variability of climate—the hot summers in Europe and the United States, the increased frequency of storms and hurricanes including Katerina, the melting of the polar ice-caps and glaciers on Asian mountain-tops threatening to dry the rivers that water that continent, the rise in sea-levels—have all placed the problem center-stage. Since the climate change problem involves a classic "commons"—that irrespective of the source of greenhouse gas emissions it is the common stock of it that affects the global climate—it can only be solved by an international effort at reaching agreement. For such an agreement to get carried out,

Originally published in Economic Theory, Volume 49, Number 2, February 2012 DOI 10.1007/s00199-010-0573-7.

P.K. Dutta (✉)
Department of Economics, Columbia University, New York, NY 10025, USA
e-mail: pkd1@columbia.edu

R. Radner
Stern School of Business, KMC 8-87, New York University, 44 W. Fourth Street, New York, NY 10012, USA
e-mail: rradner@stern.nyu.edu

G. Chichilnisky and A. Rezai (eds.), *The Economics of the Global Environment*, Studies in Economic Theory 29, DOI 10.1007/978-3-319-31943-8_14

however, it has to align the incentives of the signatory nations so that countries will, in fact, carry out their promises. At the same time, to meaningfully contain emissions an agreement has to be signed by all the major emitting countries, both developed and developing, and they have to commit to possibly deep cuts in emissions now and in the future. In other words for an agreement to be effective it has to balance two competing forces—large enough cuts that make a difference to the climate that are yet "small enough" that countries will not cheat on their promises.

And herein lies the rub. Since emissions are tied to economic activity, countries that are growing the fastest, such as China and India, are reluctant to sign onto emission cuts that they fear will compromise their growth. They point, moreover, to the "legacy effect"—that the vast majority of existing greenhouse gas stock was accumulated in the last 100 years due to the industrialization of the West—and the per capita numbers—that per person their citizens contribute a fraction of the per capita emissions from the United States and the European Union. They argue, therefore, that they should not be asked to clean up a problem not of their making. On the other hand, leaving these countries out of a climate change treaty is simply not going to solve the problem since their growth path of emissions is high, China's total emissions are already on par with the United States and unless the emissions of the developing world are reduced they will rapidly out-strip those of today's developed economies and make it impossible to sove the climate change problem.

Put another way, finding a solution to the US/Europe versus the China/India standoff is perhaps the most critical step in arriving at a meaningful climate change treaty. This paper is a modest attempt at analyzing that problem, critiquing a solution that has been suggested and offering an alternative that we believe is attractive.

Before getting to all that though, here are some facts on current greenhouse gas emissions related to the arguments above (details on sources and years may be found in the footnotes):

1. In the last 100 years, 63 % of the cumulative emissions of greenhouse gases have come from the developed economies. Of that, the US has accounted for 25 % and Western Europe for 21 %. China and India, home to 40 % of the world's population, have contributed, respectively, 7 and 2 % of the last 100 years of cumulative emissions.[1]
2. Of 2004 emissions, the United States accounted for 22 % of the total, China for 18 % and the European Union for 15 %. (And since then, China has surpassed the US in total emissions.) The next set of countries—each roughly at 5 %—included Japan, India and Russia.[2]

---

[1]The data is drawn from the World Resource Institute web-site and credits two studies published in 2000—one by Houghton and Hackler and the other by Marland et al. For details see http://earthtrends.wri.org/features/view_feature.php?fid=3&theme=3.

[2]The data, corresponding to emissions in 2004, was collected in 2007 by the CDIAC (Carbon Dixide Information Analysis Center) of the US Department of Energy for the United Nations. The data considers only carbon dioxide emissions from the burning of fossil fuels. See http://en.wikipedia.org/wiki/List_of_countries_by_carbon_dioxide_emissions.

3. Whilst total greenhouse gas emissions are currently lower in the developing world than in the developed economies, the rapid growth in the economies and populations of the former is expected to reverse that by 2015. According to some estimates, in the next 20 years, emissions in the developing economies will double while growing about 20 % in the developed economies.[3]

Given all this, the question is—what will induce China and India to sign a treaty that limits their emission growth, a treaty that they will then comply with? One possible answer is that they will perceive that the costs of climate change are so high for their economies that they have no option but to sign. These costs include the rise in sea-level along their coast-lines, the drying up of the mighty rivers that feed their agricultural plains, the possible migration into their countries from neighbors such as Bangladesh who are severely affected etc. The problem though is that these climate change induced costs still seem remote in time whereas the economic cost of abandoning a high economic growth path is immediate.

In a recent well-advertized (July 19, 2009) incident, the US Secretary of State, Hillary Clinton, was lectured to by Jairam Ramesh, India's Environment and Forestry Minister who declared "We are simply not in a position to take over legally binding emission reduction targets". As the New York Times went on to observe "Both countries (China and India) say their economic growth should not be constrained when the West never faced such restrictions during its industrialization." Indeed Secretary Clinton hastened to add that "No one wants to, in any way, stall or undermine economic growth that is necessary to lift millions more people out of poverty. The United States does not, and will not, do anything that would limit India's economic progress."[4]

In a parallel diplomatic incident (reported July 15, 2009), the US Commerce and Energy Secretaries Steven Chu and Gary Locke—themselves of Chinese ethnicity—warned the Chinese leadership on a recent visit to the country—"If China's emissions of global warming gases keep growing at the pace of the last 30 years, the country will emit more such gases in the next three decades than the United States has in its entire history" (Chu) and "50 years from now, we do not want the world to lay the blame for environmental catastrophe at the feet of China" (Locke).[5]

1.2 A Discussion of the Model and the Main Results

The present paper is part of an ongoing research project in which we have addressed certain elements of the global warming problem from a strategic and economic per-

[3]These numbers are drawn from the US EPA (Environmental Protection Agency) web-site that quotes an article published in the Energy Journal. For details see http://www.epa.gov/climatechange/emissions/globalghg.html.

[4]All this and more at http://www.nytimes.com/2009/07/20/world/asia/20diplo.html?scp=5&sq=Hillary%20Clinton%20climate%20change%20India%20visit&st=cse .

[5]All this and more at http://www.nytimes.com/2009/07/16/world/asia/16warming.html.

spective. For other studies in the current project, see Dutta and Radner (2004, 2006, 2009).

By now the basic mechanism of the greenhouse effect is well-known. The build-up of greenhouse gases—primarily CO_2—during the course of industrialization of Western economies traps heat in a manner analogous to a greenhouse. Currently, the burning of fossil fuels accounts for most of the carbon emissions produced by humans and almost all of the burning of fossil fuels is done for the purpose of producing energy. Carbon emissions can be reduced in three different ways. Over time technology changes and typically this leads to a progressive "decarbonization"of energy production. For example this has coincided with the movement from coal to oil and natural gas. Another source of decarbonization is increased efficiency in the utilization of energy, coming from improvements in the design of electric generation and transmission systems, electric motors, combustion engines, heating and cooling systems, etc. A third source of decarbonization is a lowering of emissions through reduced utilization of energy.

The costs of climate change are subject to considerable uncertainty and debate. Roughly speaking, the costs are themselves the results of two primary effects: (1) a rise in the sea level, and (2) climate changes. The rise in the sea level, caused by melting of glacial ice, and to some extent by the thermal expansion of sea water, would damage, and even eliminate, many coastlines. Climate changes are more complex. Parts of the world, such as Sub-Saharan Africa, would probably become more arid and less productive agriculturally. Other effects would include increased energy requirements for air-conditioning, curtailed water supplies, damage to human health, increased hurricane and fire damage, costly increased immigration, etc.

The efforts to avoid CC will, of course, be costly as well. Immediate costs would be incurred if economies were forced to substitute more expensive but less carbon intensive technologies for producing energy. Cutbacks in energy use would also be costly in terms of lower levels of output of goods and services, including "amenities"such as household cooling. What is particularly significant here is the role of capital accumulation. Capital and energy are, presumably, complementary inputs in the production process. Hence, the cost imposed on a country, when energy usage is curtailed, will depend on the size of its capital stock. Constant technology, the cost of energy curtailment is therefore going to be higher when capital stocks are larger—or equivalently, the long-term costs will be higher when capital grows at a faster rate. And that, of course, is part of the objection of China and India to emission cuts, that their fast growth (of capital) will imply that they have the most to lose from a climate change treaty and its attendant emission cuts.

As mentioned above, this paper is part of a project exmining climate treaties. Our approach in the project is unique in that we are the only ones to have analyzed a fully dynamic and fully game-theoretic model. By fully dynamic we mean a model in which actions in the current period have effects that persist into the future. Such intertemporal linkages are vital to the CC problem because the prominent greenhouse gas, CO_2 has a half-life of a hundred years. A game-theoretic approach is required because on the international scale of this problem there is no court that can enforce contracts and there are ndeed a few big "players". (Recall from fact 2 above that six "countries", taking the Europen Union as a single decision-making entity, produce over 70 % of the current emissions.)[6] The players in our game are countries, and it is assumed that each country has the authority and political will to control its own rate of emission of greenhouse gases. In the model, each country can control its emissions essentially by controlling its level of economic activity.[7] What we look for is a treaty that countries will sign and then comply with. In game-theoretic terminology what we look for are (subgame perfect) equilibria of a dynamic game of climate change.

In our model each country emits greenhouse gases and gets a short-term benefit from doing so. The size of that benefit depends on country-specific welfare parameters and on the size of its capital stock. This capital stock grows exogenously and geometrically and hence the size of the short-term benefit itself changes over time along with the size of capital stock. The cost of CC depends on the global common—the stock of greenhouse gases that have been built up over time. We make one important simplification—that the marginal cost of CC is independent of the size of this stock. The reasoning behind this simplification is discussed at length in Dutta and Radner (2009) but it suffices to mention that our model lends itself to calibration and hence deduction of numerical magnitudes in closed form which a non-linear model would not allow.

We start in Sect. 2 with quick review of the initial results from Dutta and Radner (2009), a model in which capital stock is fixed through time.[8] In that paper, the basic result shows that there is a simple Markov Perfect equilibrium, termed the "Business as Usual" (BAU) equilibrium. This equilibrium exhibits a tragedy of the common in that it leads to emissions that exceed those under any Global Pareto Optimal (GPO) solution. It is further shown that there are better equilibria than the BAU including a class of equilibria whose norm behavior on emissions is sustained by the threat of reverting to the BAU. If countries are sufficiently patient GPO emissions can be sustained as an equilibrium norm as well. These results parallel the well-known results from Repeated Games using trigger strategies.[9]

In Sect. 3 we introduce exogenous capital accumulation into the model. Again there is a BAU equilibrium—termed a Generalized Business as Usual Equilibrium (GBAU) in this more general model. And it involves over-emission relative to the Generalized Global Pareto Optima (GGPO). The one difference though is that the size of the emissions, in both the GBAU as well as the GGPO, depends on the size of capital stock (on account of the fact that capital and energy/emissions are complementary inputs in the benefit function.) In particular, we show that the tragedy is

[6]Models that are fully dynamic but not strategic include Nordhaus and Boyer (2000) and Nordhaus and Yang (1996). Models that are fully strategic but not dynamic include Barrett (2003) and Finus (2001). Also see the fuller bibliographic discussion in Sect. 6.

[7]One other determinant of economic activity, beyond capital and energy, is labor but that is assumed to remain fixed.

[8]Please note that the short-term benefit function is taken to be a Cobb-Douglas function here but is a more general concave function in Dutta and Radner (2009).

[9]Though, as noted above, the model is dynamic with intertemporal linkages rather than the static model that a repeated game studies.

worse under capital accumulation—in that it worsens over time as capital grows—when evaluated in terms of the differences btween GGPO and GBAU emission levels.

That makes the search for better equilibria more pressing. Our first port of call is to find analogs of the trigger strategy equilibria that we analyzed in the model without capital. And here we discover the first surprise—by way of a negative result. The fastest growing country, i.e., the one with the fastest rate of capital accumulation, will never sign a treaty that requires it to emit at GGPO levels forever. The reasoning is related to the fact—established in Sect. 3—that both GGPO and GBAU emissions in each country grow at its rate of effective growth of capital.[10] What that means is that the short-term cost to the fastest growing country in adopting the GGPO emission norm rather than the GBAU emission rate also grows at that rate. In order to bear this cost there has to be, of course, a compensatory future gain from following the GGPO path. In the standard trigger strategy logic that gain arises from the lower emissions of the other countries following the GGPO path. Since the cost is growing the benefit needs to grow as well and at the same rate. However, the gain—by similar logic—grows at the rate of capital accumulation of the other countries. And, hence grows more slowly. Hence no matter what the initial conditions, at some point the gain is simply not big enough to offset the loss in own utility even though the gain persists over the infinite future.[11] Section 4 concludes by showing that the same logic applies to any uniform cut in GBAU emissions; sanctions that slow growing countries can muster are simply not potent enough to dissuade the fastest growing economies from their preferred emissions.

Although the sanctions route is not promising—as the Indian Minister seemed also to intimate—there is a "carrot" that works better than the "stick". And that carrot has to do with foreign aid (that is conditional on emissions). The foreign aid that we examine is made up of transfers made from the slower growing economies—like, presumably, the US and the European Union—to the fast growing economies, like China and India.[12] The aid is "budget-balanced" in that in every period the total donation equals the total received. The aid is also conditional in that aid continues just as long as the emission norm—such as the GGPO emission path—is observed but is cut off forever after in the event of a deviation. The starting intuition is that slow- growing countries might be willing to share the benefits that they get from the fast growing countries' lower trajectory of emissions. Using the analogy above, the slower growing countries benefit grows at the same rate as the fast growing countries' rate of capital accumulation. If this benefit is transferred over (in part) to a fast

[10]The effective growth rate is precisely defined in Sect. 4. It coincides with the actual growth rate of capital when there is constant returns to scale.

[11]The logic will, of course, be detailed in Sect. 4. But one quick way to see it is to take the extreme case where the other countries'capital does not grow at all. Then the future gain to the fastest growing country, to all other countries following the GGPO emission rather than the GBAU emission, is some finite amount. However, its short-term cost is proportional to is capital stock. As capital stock grows infinitely large, at some period, the short-term cost overwhelms.

[12]An example of such a conditional transfer—or foreign assistance—policy is the World Bank's Climate Investment Fund (CIF).

growing country it might be willing to suffer the loss in its own welfare due to following GGPO emissions. So a "bribe"—aka foreign aid—might work where a threat does not.

In Sect. 5 we prove three results. First we show that there exists a policy of (zero-sum) foreign aid transfers such that the "bribe" of conditional foreign aid—transfers made if and only if the GGPO emissions policy is followed—sustains GGPO emissions as an equilibrium outcome (at a high enough discount factor). Second, inclusive of foreign aid, both recipients as well as donors are better off than under the GBAU. Third, there is a continuum of such emission policies all of which involve uniform emission cuts to the GBAU which can be sustained as equilibria. And, again inclusive of foreign aid, both recipients as well as donors are better off than under the GBAU. These results stand in sharp contrast to the results in Sect. 4 which showed that the threat of sanctions is not effective.[13]

The paper concludes in Sect. 6 with some observations on how the model should be elaborated and generalized to make it more realistic and a brief discussion of other parts of this research project.

2 A Simple Climate Change Game

In this section we present the model and first results of the simplified "climate change game" studied in detail in Dutta and Radner (2009).

2.1 Benchmark Model

There are I countries. The *emission* of (a scalar index of) greenhouse gases during period t by country i is denoted by $a_i(t)$. [Time is discrete, with $t = 0, 1, 2, \ldots$, ad inf., and the $a_i(t)$ are nonegative.] Let $A(t)$ denote the global (total) emission during period t;

$$A(t) = \sum_{i=1}^{I} a_i(t). \tag{1}$$

The total (global) stock of greenhouse gases (GHGs) at the beginning of period t is denoted by $g(t) + g_0$, where g_0 is what the "normal"steady-state stock of GHGs

[13] A referee has suggested that the solution offered in this paper—the benefits of foreign aid in ameliorating climate change—is being realized in practice in current climate agreements, and has been implemented through the UNFCCC rules. The referre points out that the solution proposed theoretically in this paper agrees with the actual structure of the Kyoto Protocol carbon market—which is now international law since 2005, and trading in the European Union Emissions Trading System—that allows such foreign aid transfers through the structure of the UNFCCC Clean Development Mechanism, a mechanism that has already transferred over \$26 billion to nations such as China and India to create similar incentives for clean development projects.

would be if there were negligible emissions from human sources (e.g., the level of GHGs in the year 1800). We might call $g(t)$ the *excess GHG*, but we shall usually suppress the word "excess." The law of motion for the GHG is

$$g(t+1) = A(t) + \sigma g(t), \tag{2}$$

where σ is a given parameter ($0 < \sigma < 1$). We may interpret $(1 - \sigma)$ as the fraction of the beginning-of-period stock of GHG that is dissipated from the atmosphere during the period. The "surviving"stock, $\sigma g(t)$, is augmented by the quantity of global emissions, $A(t)$, during the same period. (Note: A realistic model of GHG dynamics would be more complicated; see (Thomson 1997) but the one above has been fairly widely used.)

Suppose that the utility of country i in period t is

$$v_i(t) = [a_i(t)]^{\beta_i} - c_i g(t). \tag{3}$$

The function $[a_i(t)]^{\beta_i}$ represents, for example, what country i's gross national product would be at different levels of its own emissions, holding the global level of GHG constant.[14] This function reflects the costs and benefits of producing and using energy from alternative sources, including fossil fuels. The parameter $c_i > 0$ represents the marginal cost to the country of increasing the global stock of GHG. Of course, it is not the stock of GHG itself that is costly, but the associated climatic conditions. In a more general model, the cost would be nonlinear. The total payoff (utility) for country i is

$$v_i = \sum_{t=0}^{\infty} \delta^t v_i(t) dt. \tag{4}$$

For the sake of simplicity, we have taken the discount factor, δ, to be the same for all countries. (Note: It has implicitly been assumed here that each country's population is constant in time. The case of changing populations can be examined without much additional difficulty; we do so in Dutta and Radner 2006.)

A *strategy* for a country determines for each period the country's emission level as a function of the entire past history of the system, including the past actions of all the countries. A *stationary strategy* for country i is a function that maps the current state, g, into a current action, a_i. As usual, a *Nash Equilibrium* is a profile of strategies such that no individual country can increase its payoff by *unilaterally* changing its strategy. A *Markov Perfect Equilibrium (MPE)* is a Nash Equilibrium in which every country's strategy is stationary. A *Subgame Perfect Equilibrium (SPE)* is a profile of strategies, not necessarily stationary, that constitutes a Nash Equilibrium after every history.

---

[14]In Dutta and Radner (2009) we consider a more general form of felicity function that includes the Cobb-Douglas form $[a_i(t)]^{\beta_i}$.

2.2 The GPO

Let $x = (x_i)$ be a vector of positive numbers, one for each country. A GPO corresponding to x is a profile of strategies that maximizes the weighted sum of country payoffs,

$$v = \sum_i x_i v_i, \tag{5}$$

which we shall call the *global welfare*. Without loss of generality, we may take the weights, x_i, to sum to I.

Theorem 1 *Let $\hat{V}(g)$ be the maximum attainable global welfare starting with an initial GHG stock equal to g; then there are a set of constant emissions $\hat{a}_i$ determined by*

$$\hat{a}_i = \left(\frac{\beta_i x_i}{\delta w}\right)^{\frac{1}{1-\beta_i}}$$

where $w = \frac{1}{1-\delta\sigma}\sum_i x_i c_i$, that constitute the GPO emissions. Writing $\hat{A} = \sum_i \hat{a}_i$ for the total emissions, the lifetime GPO payoffs are

$$\hat{V}(g) = u - wg, \tag{6}$$

$$u = \frac{1}{1-\delta}\left[\sum_i x_i \hat{a}_i^{\beta_i} - \delta w \hat{A}\right].$$

Proof The proof uses a standard dynamic programming argument. Let $a = (a_i)$. It is sufficient to show that the value function, $\hat{V}$, given above satisfies the functional equation:

$$\hat{V}(g) = \max_a \left\{\sum_j x_j \left[\hat{a}_j^{\beta j} - c_j g\right] + \delta \hat{V}\left[\sum_j a_j + \sigma g\right]\right\}. \tag{7}$$

The first-order condition for a maximum is that, for each i,

$$x_j \beta_j \hat{a}_j^{\beta j - 1} + \delta \hat{V}'\left[\sum_j a_j + \sigma g\right] = 0.$$

But $\hat{V}' = -w$, so the optimal emission is independent of g, and is given by (7). The values of u and w are now determined by the equation:

$$\hat{V}(g) = \sum_j x_j \left[\hat{a}_j^{\beta j} - c_j g\right] + \delta \hat{V}\left[\sum_j a_j + \sigma g\right],$$

which must be satisfied for all values of g. □

2.3 The BAU Equilibrium

The next proposition describes a Markov Perfect equilibrium, which we call the *BAU* equilibrium. This MPE has the unusual feature that the equilibrium emission rate of each country is constant in time, and it is the unique MPE with this property.

Theorem 2 (BAU) *Let g be the initial stock of GHG. For each country i, let $\overline{a}_i$ be determined by*

$$\overline{a}_i = \left(\frac{\beta_i}{\delta w_i}\right)^{\frac{1}{1-\beta_i}}$$

where $w_i = \frac{c_i}{1-\delta\sigma}$, and let its strategy be to use a constant emission equal to $\overline{a}_i$ in each period; then this strategy profile is a MPE, and, writing $\overline{A} = \sum_j \overline{a}_j$ for the aggregate emissions, country i's corresponding payoff is

$$\overline{V}_i(g) = u_i - w_i g, \tag{8}$$
$$u_i = \frac{1}{1-\delta}\left[\overline{a}_i^{\beta_i} - \delta w_i \overline{A}\right].$$

Proof The proof uses an argument similar to that of Theorem 1. If the emissions of all countries other than i are constant, say a_j for country j, then country i faces a standard dynamic programming problem. It is sufficient to show that the value function $\overline{V}_i$ satisfies the functional equation,

$$\overline{V}_i = \max_{a_i}\left\{a_i^{\beta_i} - c_i g + \delta \overline{V}_i\left(\sum_j a_j + \sigma g\right)\right\}.$$

The argument now proceeds as in the proof of Theorem 1. □

If the cost of the stock of GHG were nonlinear, then one would expect the GPO and BAU emissions to vary with the stock, and in fact one would expect higher stocks to lead to lower emissions. In the next section we will see that, once we introduce capital stock, emissions will no longer be constant in time.

2.4 Comparison of the GPOs and the BAU

The preceding results enable us to compare the emissions in the GPOs with those in the BAU equilibrium:

$$\text{GPO}: \quad \beta_i \hat{a}_i^{\beta_i - 1} = \frac{\delta \sum_j x_j c_j}{x_i(1-\delta\sigma)}, \tag{9}$$

$$\text{BAU}: \quad \beta_i \overline{a}_i^{\beta_i - 1} = \frac{\delta c_i}{1-\delta\sigma}.$$

From

$$x_i c_i < \sum_j x_j c_j,$$

it follows that

$$\frac{\delta c_i}{1-\delta\sigma} < \frac{\delta \sum_j x_j c_j}{x_i(1-\delta\sigma)}.$$

Since $a_i^{\beta_i}$ is concave, it follows that

$$\overline{a}_i > \hat{a}_i. \tag{10}$$

Note that this inequality holds for all vectors of strictly positive weights (x_i).[15] It follows from these results that there is an open set of strictly positive weights (x_i) such that the corresponding GPO is strictly Pareto superior to the BAU. We are therefore led to search for (non-Markovian) Nash equilibria of the dynamic game that sustain a GPO, or at least are superior to the BAU.

2.5 BAU Sanctions

In Dutta and Radner (2009) we further characterize equilibria in this game that are sustained by the threat of reverting to the BAU equilibrium. We report here, without proof, two of the main results. First, for all discount factors, the third-best solution qualitatively mirrors the BAU and GPO solutions; there is a constant emission level $\widetilde{a}_i$ that country i emits, independently of the stock of GHGs. Second, if discount factors are high enough, then, in fact, the GPO emission levels are themselves the third-best solution.

Let $x = (x_i)$ be a vector of positive numbers, one for each country. A *Third-Best Optimum (TBO)* corresponding to x is a profile of "norm" strategies that maximizes the weighted sum of country payoffs:

$$v = \sum_i x_i v_i, \tag{11}$$

[15]We conjecture that this inequality would hold in a variety of models. It certainly does in the concave model of Dutta and Radner (2009). Indeed, one can show in a quite general model that a GPO cannot be a BAU, or even that, starting from a GPO, each country will want to increase its emissions unilaterally by a small amount.

subject to BAU reversion, i.e., subject to the constraint—detailed below—that should any country i not follow the norm, all countries would switch to BAU emissions vector $\overline{a} = (\overline{a}_1, \overline{a}_2, \ldots \overline{a}_I)$ forever from the following period on. As before, and without loss of generality, we may take the weights, x_i, to sum to I. The following result characterizes the TBO:

Proposition 1 *There exists a vector $\tilde{a}$ of constant emission levels $\tilde{a}_i$ such that on the equilibrium path country i's TBO strategy is to use a constant emission equal to $\tilde{a}_i$ in all periods, where $\tilde{a}_i$ satisfies the incentive constraint:*

$$\textit{for every } i, \quad \widetilde{a}_i^{\beta_i} - \delta w_i\left(\widetilde{a}_i + \delta \sum_{j \neq i} \widetilde{a}_j\right) \geq \overline{a}_i^{\beta_i} - \delta w_i\left[\overline{a}_i + \delta \sum_{j \neq i} \overline{a}_j\right].$$

It is immediate that the BAU emission policy is sustainable by the threat of BAU reversion—of course!—since the inequality is trivially satisfied when $\widetilde{a}_i = \overline{a}_i$. What is also not very difficult to show is that the GPO emission policy also becomes sustainable at a high enough δ. Formally, we have:

Proposition 2 *(a) The welfare that is achievable under the threat of BAU emissions is at least as high as $\overline{u}$.*

(b) Suppose that the GPO solution under equal country weighting, ($x_i = x_j$ for all i,j) Pareto-dominates the BAU solution for all high $\delta \geq \delta'$ Then, there is a cut-off discount factor $\widetilde{\delta} \in (\delta', 1)$ such that, above it, the GPO emission policy is sustainable as an equibrium norm.

3 A Generalized Model with Exogenous Capital Accumulation

We now generalize the model of Sect. 2.1 to include the possibility that the capital stock in each country changes exogenously over time. For simplicity, we assume that each capital stock evolves geometrically, although models with other stock dynamics would also be tractable.

3.1 The Model

Let $K_i(t)$ denote the size of capital stock of country i at the beginning of period t, and let $K(t)$ be the vector with coordinates $K_i(t)$. The *state of the system at the beginning of period t* is now the pair $[g(t), K(t)]$.

Corresponding to (3) of Sect. 2.1, the utility of country i in period t is

$$v_i(t) = \left[a_i(t)\right]^{\beta_i} \left[K_i(t)\right]^{\gamma_i} - c_i g(t) \tag{12}$$

where the coefficients β_i and γ_i are positive fractions. (Of course one special situation is the CRS case $\beta_i + \gamma_i = 1$.) It is convenient, though not necessary, to think of the utility function, $\left[a_i(t)\right]^{\beta_i} \left[K_i(t)\right]^{\gamma_i}$, as the nation's GDP function. Note that emissions are an "input" into the GDP "production" function because there is a one to one link between emissions and energy usage in the economy—and energy usage is an actual input into the production function. In the Cobb-Douglas form assumed here, emissions/energy and capital stock are complementary inputs in that the marginal product of one input increases in the level of the other input. Again, the total payoff (utility) for country i is given by the sum of discounted one-period utilities, as in (4) of Sect. 2.

A Markov strategy for country i is a function that maps the current state, (g, K) into a current action, a_i. As in (2) of Sect. 2.1, the level of greenhouse gas evolves according to the linear difference equation

$$g(t+1) = A(t) + \sigma g(t), \tag{13}$$

where $A(t) = \sum_{i=1}^{I} a_i(t)$. We assume that the capital stock in country i evolves according to the geometric growth equation

$$K_i(t+1) = \theta_i K_i(t), \tag{14}$$

where the parameter θ_i satisfies $\theta_i > 1$. Thus the capital stock in country i becomes unboundedly large. To preserve boundedness of solutions, we shall require that discounted growth is bounded, i.e., that $\delta\theta_i^{\frac{\gamma_i}{1-\beta_i}} < 1$ for all countries i. Note that in the CRS case—$\gamma_i = 1 - \beta_i$—this condition reduces to the more familiar one that $\delta\theta_i < 1, \forall i$.

3.2 GBAU Equilibrium

Reversing the order followed in Sects. 2.2 and 2.3, we first derive the analog of the Markov Perfect Equilibrium that was called there "business-as usual"(BAU); hereinafter, Generalized "business-as usual"equilibrium (GBAU).

Theorem 3 (GBAU Equilibrium) *Let g be the initial stock of greenhouse gas, and let K be the vector of initial capital stocks. For each i, let country i use the Markovian strategy* $\overline{a}_i = \overline{\alpha}_i(K_i)$ *determined by*

$$\beta_i \overline{a}_i^{\beta_i - 1} K_i^{\gamma_i} = \delta w_i, \tag{15}$$

where $w_i = \frac{c_i}{1-\delta\sigma}$. *Then this strategy profile is a MPE, and country i's corresponding payoff is*

$$\overline{V}_i(g, K) = \overline{u}_i(K) - w_i g, \tag{16}$$

where the function $\overline{u}_i(K)$ *is separable in being the sum of two functions,* $\overline{u}_i(K) = \overline{u}_i^i(K_i) + \sum_{j \neq i}^{I} \overline{u}_i^j(K_j)$, *and, furthermore,* $\overline{u}_i^i(K_i) = \Phi_i^i K_i^{\frac{\gamma_i}{1-\beta_i}}$ *and* $\overline{u}_i^j(K_j) = \Phi_i^j K_j^{\frac{\gamma_j}{1-\beta_j}}$ *both of which are continuous in their arguments and solve the functional equations*

$$\overline{u}_i^i(K_i) = \overline{a}_i^{\beta_i} K_i^{\gamma_i} + \delta[\overline{u}_i^i(K_i') - w_i \overline{a}_i(K_i)], \tag{17}$$

$$\overline{u}_i^j(K_j) = -\delta w_i \overline{a}_j(K_j) + \delta \overline{u}_i^j(K_j'), \tag{18}$$

$$K_i' \equiv \theta_i K_i.$$

Proof That the value associated with the strategies given by (15) is continuous and separable of the form given in (17) is established by way of a bootstrapping argument and the Bellman equation. Presuming that the value function is of that form, we write the Bellman equation as:

$$\overline{u}_i^i(K_i) + \sum_{j \neq i}^{I} \overline{u}_i^j(K_j) = \max_{a_i} \left[a_i^{\beta_i} K_i^{\gamma_i} - \delta w_i a_i\right] + \delta \overline{u}_i^i(K_i')$$
$$+ \delta \sum_{j \neq i}^{I} \left[-w_i \overline{a}_j(K_j) + \overline{u}_i^j(K_j')\right]. \tag{19}$$

It is seen that the Bellman equation preserves both properties, continuity and separability. Substituting the maximizing emission values

$$a_i = \left[\frac{\beta_i K_i^{\gamma_i}}{\delta w_i}\right]^{\frac{1}{1-\beta_i}}, a_j = \left[\frac{\beta_j K_j^{\gamma_j}}{\delta w_j}\right]^{\frac{1}{1-\beta_j}}$$

and recognizing the separable nature of the equation we get that the above reduces to

$$\overline{u}_i^i(K_i) = f_i(\beta_i) K_i^{\frac{\gamma_i}{1-\beta_i}} + \delta \overline{u}_i^i(K_i')$$

where $f_i(\beta_i) = \left[\frac{\beta_i}{\delta w_i}\right]^{\frac{\beta_i}{1-\beta_i}} \left[1 - \beta_i\right]$ and

$$\overline{u}^j_i(K_j) = \delta \left[-g_j(\beta_j) K_j^{\frac{\gamma_j}{1-\beta_j}} + \overline{u}^j_i(K'_j) \right] ;j \neq i$$

where $g_j(\beta_j) = w_i \left[\frac{\beta_j}{\delta w_j}\right]^{\frac{1}{1-\beta_j}}$. Writing

$$\Phi^i_i = \frac{f_i(\beta_i)}{1 - \delta\theta_i^{\frac{\gamma_i}{1-\beta_i}}}, \quad \Phi^j_i = \frac{-\delta g_j(\beta_j)}{1 - \delta\theta_j^{\frac{\gamma_j}{1-\beta_j}}}$$

it further follows that

$$\overline{u}^i_i(K_i) = \Phi^i_i K_i^{\frac{\gamma_i}{1-\beta_i}},$$

$$\overline{u}^j_i(K_j) = \Phi^j_i K_j^{\frac{\gamma_j}{1-\beta_j}}.$$

Standard arguments then show that the space of continuous, power functions is a complete metric space. The Bellman equation is a contraction and hence it has a fixed point, i.e., the value function. Finally, the characterization of the GBAU emissions follows immediately from the maximization above. The theorem is proved. □

Remark 1. In the GBAU-equilibrium strategy of country i, the current emission depends only on the country's own current capital stock. Own value u^i_i is also affected only by own capital stock K_i.

2. For *any* profile of stationary strategies with property that a country's current action depends only on its own current capital stock, the value function of country i has the separable form given by (17), with w_i given above.
3. It should be clear that if the growth rates of capital stock are not equal, i.e., if $\theta_i \neq \theta_j$ then the country with the highest growth rate will eventually come to dominate in terms of utility. This happens both because its own utility $\overline{u}^i_i$ grows at the fastest rate and also because the disutility it imposes on others through its own emissions, $\overline{u}^i_j$, $j \neq i$, grows at the fastest rate as well.

3.3 GGPO

We define a GPO as in Sect. 2.2. The following theorem, which corresponds to Theorem 1, characterizes the GGPO for a given set of welfare weights, (x_j). The proof is omitted, since the method is similar to that used in the previous theorem.

Theorem 4 (GGPO) *Given strictly positive welfare weights* (x_i)*, let* $\hat{V}(g, K)$ *be the maximum attainable global welfare starting with an initial GHG stock equal to g and capital stocks K; then, after writing* $w = \sum_i x_i w_i$*,*

$$\hat{V}(g, K) = \widehat{u}(K) - wg, \tag{20}$$

where

$$\widehat{u}(K) = \sum_i x_i \widehat{u}_i(K_i) \tag{21}$$

and the $\widehat{u}_i$ *are the solution of the functional equation*

$$x_i \widehat{u}_i(K_i) = x_i \widehat{a}_i^{\beta_i} K_i^{\gamma_i} + \delta[x_i u_i(K_i') - w\widehat{\alpha}_i(K_i)].$$

Country i's GGPO emission $\widehat{\alpha}_i(K_i)$ *is the stationary strategy determined by*

$$x_i \beta_i \widehat{a}_i^{\beta_i - 1} K_i^{\gamma_i} = \delta w. \tag{22}$$

3.4 Comparison of BAU and GPO Emission Rates

Comparing the GBAU and GGPO strategies, we have:

$$GBAU: \quad \beta_i \overline{a}_i^{\beta_i - 1} K_i^{\gamma_i} = \delta w_i, \tag{23}$$

$$GGPO: \quad \beta_i \widehat{a}_i^{\beta_i - 1} K_i^{\gamma_i} = \delta \left(\frac{1}{x_i}\right) \sum_j x_j w_j. \tag{24}$$

Therefore, since $\left(\frac{1}{x_i}\right) \sum_j x_j w_j > w_i$, for all K, i, and vectors (x_i),

$$\overline{\alpha}_i(K_i) > \widehat{\alpha}_i(K_i), \tag{25}$$

i.e., *the BAU emission rates will exceed the GPO emission rates.*

Indeed, for future usage, it will be useful to note the exact relationship between the two emission levels:

$$\frac{\widehat{\alpha}_i(K_i)}{\overline{\alpha}_i(K_i)} = \left[\frac{w_i x_i}{w}\right]^{\frac{1}{1-\beta_i}}.$$

Note in particular that the ratio of emission levels is actually *independent* of the size of capital stock even though each emission is a function of that variable. Put another way, the GGPO emission level $\widehat{\alpha}_i(K_i)$ is a constant fraction of the GBAU emission level $\overline{\alpha}_i(K_i)$ and the size of the fraction is independent of the capital stock. Put yet another way, the GGPO is achieved by a simple across the board cut in emissions from the GBAU level. All cuts of this form we will call *uniform cuts*.

Definition A uniform cut in emissions is achieved by a (capital-dependent) emission policy $\widetilde{a}_i(.)$ where, for all capital stock K_i, emissions are a constant fraction, say λ_i, of the GBAU emission level, i.e.,

$$\widetilde{a}_i(K_i) = \lambda_i \overline{\alpha}_i(K_i).$$

The reader will notice that the Kyoto agreement attempted to bring about just such a uniform cut in emissions. In the next two sections we shall investigate the sustainability of such uniform emission cuts.

3.5 *Effects of Capital Stock on Emission Levels*

As we saw in the previous subsection, there is a tragedy of the common with capital stocks, exactly as there was without. The question though is: does the presence of capital exacerbate the tragedy, possibly because capital and energy are complementary inputs in the production function? As we shall now see, the answer is that the tragedy does indeed get worse when we consider absolute levels of emissions but not when we consider percentages (or ratios of emission levels). Note that the GBAU as well as the GGPO emission level for country i only depends on its own capital stock.

Theorem 5 (Capital Effect on Tragedy of the Common)

(i) Absolute Levels—Consider the absolute difference in emission levels $\overline{a}_i - \widehat{a}_i$. That increases at the rate $K_i^{\frac{\gamma_i}{1-\beta_i}}$.

(ii) Percentages—Consider the ratio difference in emission levels $\frac{\overline{a}_i}{\widehat{a}_i}$. That is independent of the size of capital stock.

The proof follows immediately from the characterizations provided in the previous subsections. Indeed, the second part was explicitly derived in the immediate prequel.

4 Uniform Emission Cuts Under BAU Sanctions

In this subsection, we characterize the emission policies that are sustainable under the threat of BAU reversion in the model with capital. The answers are largely negative. We start by asking whether the GGPO policy can be so sustained. The answer, it will turn out, is in general *no*. The GGPO is an example of a broader class of emission policies that involve uniform cuts from GBAU emission levels. So the next question that we then ask is whether *any* uniform cut emissions policy is sustainable as part of an equilibrium norm. And the answer, again, is *no*.

The reason why the GGPO cannot be sustained as a SPE, by threatening to revert to the GBAU emission, is critically linked to the growth of capital. To understand the intuition, suppose for a moment that there are two countries and suppose, furthermore, that production is subject to CRS, i.e., that $\frac{\gamma_i}{1-\beta_i} = 1$. Finally, without loss of generality, let us suppose that the growth rate of capital is higher in country 1 i.e., that $\theta_1 > \theta_2$.

From the discussion in the previous section it follows that emissions in each country grows at rate K_i. Of course, under the GGPO emissions are lower; they are a (fixed) fraction of emissions under the GBAU. Imagine that an agrement is proposed in which the two countries are to cut their GBAU emissions to the fractions that the GGPO represents. The clear "loss" for country i in doing so is the loss every period t in own utility, $[a_i(t)]^{\beta_i} [K_i(t)]^{1-\beta_i} - \delta w_i a_i(t)$, where by loss we mean the difference between own utility under the GBAU and that under the GGPO.[16] By definition, this loss is proportional to K_i since the emissions $a_i(t)$ are proportional to K_i. The "gain" though for country i is that the other country is also going to reduce its emissions and hence the damage inflicted by the other country—$-\delta w_i a_j(t)$ – is lower if the GGPO agrement is adhered to. How much lower though and does it offset the loss in own utility? Well, the gain is—by similar logic as above—of order K_j. So country 1, in our two country example, gives up own utility which loss grows at rate θ_1 in return for a gain in damages as imposed by country 2's emissions. Yet that gain only grows at rate θ_2. It is clear that no matter what the initial conditions are, at some point the gain is simply not going to be big enough to offset the loss in own utility. Put another way, at that point in the future, the agreement will break down. Knowing that—or given the constraints of subgame perfection—such an agrement will never get written in period 0. In the proof of the theorem below, it will be seen that the logic generalizes when there are many countries and when there is not CRS in the production function.

Recall from the last section that the difference in greenhouse gas growth rates are proportional to $K_i^{\frac{\gamma_i}{1-\beta_i}}$ which is effectively proportional to $\theta_i^{\frac{\gamma_i}{1-\beta_i}}$. Based on that, let us call $\theta_i^{\frac{\gamma_i}{1-\beta_i}}$ the effective growth rate of capital stock. We shall say that there is a *unique maximal effective growth* rate if, without loss of generality

$$\theta_1^{\frac{\gamma_1}{1-\beta_1}} > \theta_i^{\frac{\gamma_i}{1-\beta_i}}, \ \forall i \neq 1$$

[16]That this is the utility consequence to country i from emission $a_i(t)$ is easily seen by noting that that level of emission causes first, an immediate "GDP" payoff $[a_i(t)]^{\beta_i} [K_i(t)]^{1-\beta_i}$ (where we have used the CRS simplification). However, next period there is $\sigma c_i a_i(t)$ of GHG damage, the period after that $\sigma^2 c_i a_i(t)$, two periods after that $\sigma^3 c_i a_i(t)$, all of which discounted back at rate δ yields a present discounted cost of $\delta w_i a_i(t)$.

In the sequel we shall prove two results. The first says that the GGPO emission level cannot be sustained by the threat of reverting to the GBAU emission. Then we go on to show the more general result that no emission policy that involves a uniform reduction from the GBAU is sustainable. (This is a more general result since—as we have seen in the previous section—the GGPO does in fact involve a uniform reduction in emissions from the GBAU.) Indeed that is the result we prove.

Theorem 6 *Suppose that there is a unique maximal effective growth rate. Then, no matter what the discount factor, and no matter what the initial levels of capital stock are, the GGPO cannot be sustained as part of a SPE by the threat of reverting to the GBAU. In particular for country* 1, *with the maximal effective growth rate, there will be a date, say* T_1, *such that it will deviate from the GGPO agrement in every period after* T_1.

Theorem 7 *Suppose that there is a unique maximal growth rate. Then, no matter what the discount factor, and no matter what the initial levels of capital stock are, no emission policy involving uniform cuts from the GBAU can be sustained as part of a SPE by the threat of reverting to the GBAU. In particular for country* 1, *with the maximal effective growth rate, there will be a date, say* T_1, *such that it will deviate from the GGPO agrement in every period after* T_1.

Proof of Theorem 7 Recall the GBAU emission policy $\overline{\alpha}_i(K_i) = \left(\frac{\beta_i}{\delta w_i}\right)^{\frac{1}{1-\beta_i}} K_i^{\frac{\gamma i}{1-\beta_i}}$. By extension, for an emission policy that involves a uniform cut in the GBAU emissions we have $\vec{a}_i(K_i) = \lambda_i \overline{\alpha}_i(K_i)$, where λ_i is any fraction. Consider the life-time payoff to any such emission policy for country i. By the decomposition given by (19), which we repeat here in slightly modified form for easy access, we have

$$\vec{u}_i^i(K_i) = \lambda_i \overline{a}_i^{\beta_i} K_i^{\gamma_i} - \delta w_i \lambda_i \overline{a}_i + \delta \vec{u}_i^i(K_i')$$

and

$$\sum_{j \neq i}^{I} \vec{u}_i^j(K_j) = \delta \sum_{j \neq i}^{I} \left[-w_i \lambda_j \overline{a}_j(K_j) + \vec{u}_i^j(K_j') \right].$$

The first equation above yields by simple substitution

$$\vec{u}_i^i(K_i) = \left[\frac{\beta_i}{\delta w_i}\right]^{\frac{\beta_i}{1-\beta_i}} \left[\lambda_i^{\beta_i} - \beta_i \lambda_i\right] K_i^{\frac{\gamma i}{1-\beta_i}} + \delta \vec{u}_i^i(K_i').$$

Note that the immediate own-payoff—the first term in the expression above—is maximized at the GBAU emission, i.e., is maximized when $\lambda_i = 1$. Substituting a conjectured solution $\vec{u}_i^i(K_i) = \vec{\Phi}_i^i K_i^{\frac{\gamma i}{1-\beta_i}}$ and using the fact that $K_i' = \theta_i K_i$ we can see right away that

$$\vec{u}_i^i(K_i) = \frac{\left[\frac{\beta_i}{\delta w_i}\right]^{\frac{\beta_i}{1-\beta_i}} \left[\lambda_i^{\beta_i} - \beta_i \lambda_i\right]}{1 - \delta\theta_i^{\frac{\gamma_i}{1-\beta_i}}} K_i^{\frac{\gamma_i}{1-\beta_i}}.$$

By similar logic

$$\vec{u}_i^j(K_j) = \frac{-\delta w_i \lambda_j}{1 - \delta\theta_j^{\frac{\gamma_j}{1-\beta_j}}} \left(\frac{\beta_j}{\delta w_j}\right)^{\frac{1}{1-\beta_j}} K_j^{\frac{\gamma_j}{1-\beta_j}}.$$

Again, it is clear that the greatest damage is inflicted in the GBAU case, i.e., when $\lambda_j = 1$.

Armed with the lifetime payoffs, we now turn to the sustainability of any uniform cut policy. We shall show that such a policy cannot be sustained by simply showing that for country 1, the one with the highest effective growth rate of capital, the lifetime payoff under GBAU must eventually exceed the lifetime payoff from the uniform cut policy. Say it exceeds by time $T + 1$. In particular therefore, at time T, country 1 has no further incentive to continue with the cuts since—by construction—next period onwards the payoffs are higher by switching to the GBAU policy. That switch can be affected by deviating in the current period when own payoffs are in any case going to be higher from the deviation.[17] Using the results above, the difference between GBAU and uniform emissions payoffs is given by

$$A_1 K_1^{\frac{\gamma_1}{1-\beta_1}} - \delta \sum_{j \neq 1}^{I} B_{1j} K_j^{\frac{\gamma_j}{1-\beta_j}}$$

where

$$A_1 = \frac{\left[\frac{\beta_1}{\delta w_1}\right]^{\frac{\beta_1}{1-\beta_1}}}{1 - \delta\theta_1^{\frac{\gamma_1}{1-\beta_1}}} \left[1 - \lambda_1^{\beta_1} - \beta_1(1 - \lambda_1)\right] > 0$$

and

$$B_{1j} = \frac{-\delta w_i}{1 - \delta\theta_j^{\frac{\gamma_j}{1-\beta_j}}} \left(\frac{\beta_j}{\delta w_j}\right) (\lambda_j - 1) > 0.$$

[17] As with all Nash equilibrium logic, other countries—whom country 1 is best responding to at time T—will be presumed to be carrying on with the cuts in that period. Hence the T period payoff consequence from the others' actions is identical for country 1 whether it deviates or not.

Since $K_1^{\frac{\gamma_1}{1-\beta_1}}$ is arbitrarily bigger than $K_j^{\frac{\gamma_j}{1-\beta_j}}$ by some time period, say $T+1$, it follows that the expression must be strictly ositive from that period onwards. The theorem is proved. □

5 Foreign Aid

The main point of the previous section is that capital growth makes it difficult to sustain equilibria better than the GBAU. It certainly makes it impossible to sustain the most natural ones that involve a uniform cut in BAU emissions. The reason is straightforward enough as we saw above. Countries where capital accumulation is fastest have an ever increasing "potential loss" from agreeing to emission cuts—they would prefer the BAU emissions and lose by reducing emissions to, say, GPO levels. The fact that capital is complementary to emissions means that larger and larger amounts of capital amplify this loss in own-welfare. The only way then that such a country would agree to emissions reductions is if it is "made good" on this loss. One way the loss can be made good is by the threat of other countries raising their own emissions in the event that the fast-growing country does not cut its own emissions. That is the way in which a reversion to GBAU levels works. However, as we saw in the previous section, the threat is not strong enough since it is, in turn, tied to the rate of capital expansion in those countries. And if country 1 is the fastest growing country then the threat of being affected by the slower growth of country 2's GBAU emissions is simply not enough of a threat.

In this section we show that foreign aid, however, works. The starting point is that another way in which a fast growing country can be "made good" on the loss from not pursuing BAU emissions is that other countries might be willing to share the benefits that they get from this country's lower trajectory of emissions. Using the analogy above, country 2—the slower growing one—benefits from country 1's reduced emissions and it will be seen that this benefit grows at the same rate as country 1's capital growth (since 1's emissions are indeed linked to its rate of capital expansion). Now if this benefit is transferred over in part to country 1 it might be willing to suffer the loss in its own welfare due to following GGPO emissions. So a "bribe"—aka foreign aid—might work where a threat does not.

Even if such a bribe works to induce the fast-growing country to limit its emissions, one may wonder whether the bribe will be given. Put differently, would the foreign aid donor, inclusive of aid, be better off relative to the GBAU? Put yet differently, can foreign aid be Pareto improving?

In this section we prove three results. First we show that there exists a policy of (zero-sum) foreign aid transfers such that the "bribe" of conditional foreign aid—transfers made if and only if the GGPO emissions policy is followed—sustains GGPO emissions as an equilibrium outcome (at a high enough discount factor). Second, inclusive of foreign aid, both recipients as well as donors are better off than under the GBAU. Third, there is a continuum of such emission policies all of which involve uniform emission cuts to the GBAU which can be sustained as equilibria. And, again inclusive of foreign aid, both recipients as well as donors are better off than under the GBAU. These results stand in sharp contrast to the results in the previous section which showed that the threat of sanctions is not effective.

5.1 Foreign Aid: A Definition

First a defiinition regarding foreign aid. We shall consider a "clearing-house" mechanism of providing foreign aid rather than bilateral aid between countries. (Though we believe that the results and the intuition can be carried forward to the bilateral case as well.) Imagine that there is an international aid agency, much like the World Bank, which makes a transfer Υ_i to country i. We will adopt the usual convention that $\Upsilon_i > 0$ implies that country i is a recipient of aid whilst $\Upsilon_i < 0$ implies that it is a donor.

Definition A *feasible foreign aid policy* (related to climate change) is a sequence of time and emission-dependent aid levels $\{\Upsilon_{it}\}$ with the requirement that in every period the transfers aggregate to zero, i.e.,

$$\sum_i \Upsilon_{it} = 0, \forall t.$$

Furthermore, the transfers are made to country i in period t only if the appropriate emissions are recorded for country i in that period.[18]

5.2 Sustainability of the GGPO Emission Policy Under Foreign Aid

In this subsection we show that there is a feasible foreign aid policy such that the equally wighted GGPO emission level can be sustained as part of a SPE by sufficiently patient countries.[19] In the next subsection we will then show that indeed there is a continuum of such emission reduction policies that are also sustainable—though possibly at different discount factors.

[18]The closest institutional mechanism to climate change related foreign aid is the aid that is disbursed by the World Bank via its Climate Investment Fund (CIF). In the CIF though, there is no requirement that the transfers should aggregate to zero. Clearly having that as an additional requirement only makes our task of showing the beneficial effects of aid more difficult. Equally clearly, some kind of budget-balance, but possibly over a long horizon, will be required of any such policy. We choose to work with the most stringent budget balance policy.

[19]By the equally weighted GGPO emission level what we mean is that we consider the GGPO where each country is given equal weight. In terms of the notation of Sect. 3, the weight $x_i = \frac{1}{I}$, for all i.

Definition The *Aid Induced GGPO strategy* that we consider is the following:

Norm: Start at period 0, given capital stock K_{i0}, by following GGPO emission level $\widehat{\alpha}_i(K_{i0})$ and transferring Υ_{i0} upon observing it. Follow thereafter in every period t with GGPO emission level $\widehat{\alpha}_i(K_{it})$ and transferring Υ_{it} provided these emissions have been followed in the past.

Punishment: In the event that there has been a unilateral deviation in period t country i did not emit at GGPO levels or withheld promised foreign aid, switch for all countries j to the GBAU emissions $\overline{a}_j(K_{jt+1})$ from period $t+1$ onwards with no foreign aid from period t onwards.[20]

Recall that $\theta_i^{\frac{\gamma_i}{1-\beta_i}}$ is the effective growth rate of capital stock in country i (θ_i being the actual growth rate, γ_i the coefficient for capital in the production function and β_i the emissions coefficient).[21] Recall too that, without loss of generality, we have adopted the convention that this growth rate is highest in country 1, i.e.,

$$\theta_1^{\frac{\gamma_1}{1-\beta_1}} \geq \theta_i^{\frac{\gamma_i}{1-\beta_i}}, \ \forall i$$

Note that—unlike in the previous section—the above is a weak inequality, i.e., that country 1 need not have the uniquely maximum effective growth rate. Recall too that for the problem to be bounded we have imposed the restriction that $\delta\theta_1^{\frac{\gamma_1}{1-\beta_1}} < 1$. Call any such discount factor *feasible*.

Theorem 8 *There is a cut-off discount factor $\widehat{\delta} < \theta_1^{\frac{\gamma_1}{\beta_1-1}}$ and a feasible foreign aid policy with the property that the Aid Induced GGPO strategy defined above is a SPE for all feasible discount factors above $\widehat{\delta}$. Furthermore, for every country i, donor as well as recipient, life-time payoffs inclusive of foreign aid strictly Pareto dominates the GBAU lifetime payoffs.*

Proof Evidently the proposed aid induced GGPO strategy is an equilibrium if no country i has a profitable deviation against it at any time τ, i.e., if for all i and all τ it is the case that

$$\sum_{t=\tau}^{\infty} \delta^{t-\tau} \left(\widehat{\alpha}_{it}^{\beta_i} K_{it}^{\gamma_i} - \delta w_i \widehat{\alpha}_{it} - \delta w_i \sum_{j \neq i} \widehat{\alpha}_{jt} + \Upsilon_{it} \right)$$
$$\geq \max_{a_i} [a_i^{\beta_i} K_{i\tau}^{\gamma_i} - \delta w_i a_i] - \delta w_i \sum_{j \neq i} \widehat{\alpha}_{j\tau}$$
$$+ \sum_{t=\tau+1}^{\infty} \delta^{t-\tau-1} \left(\overline{a}_{it}^{\beta_i} K_{it}^{\gamma_i} - \delta w_i \overline{a}_{it} - \delta w_i \sum_{j \neq i} \overline{a}_{jt} \right).$$

---

[20]As always, given Nash equilibrium logic, one can ignore multiple simultaneous deviations.

[21]In the CRS case the effective and actual growth rates of capital coincide.

It is immediate that the best deviation, the solution to the maximimization above, is attained at the GBAU emission associated with capital stock $K_{i\tau}$, what we have denoted $\overline{a}_{i\tau}$. Substituting that—and doing a bit of re-arranging—we can show that the above is equivalent to the holding of the following *Individual Incentive Constraints* (IIC):

$$\sum_{t=\tau}^{\infty} \delta^{t-\tau} \left(\widehat{\alpha}_{it}^{\beta_i} K_{it}^{\gamma_i} - \delta w_i \widehat{\alpha}_{it} - \delta w_i \sum_{j \neq i} \widehat{\alpha}_{jt} + \Upsilon_{it} \right)$$
$$\geq \sum_{t=\tau}^{\infty} \delta^{t-\tau} \left(\overline{a}_{it}^{\beta_i} K_{it}^{\gamma_i} - \delta w_i \overline{a}_{it} - \delta w_i \sum_{j \neq i} \overline{a}_{jt} \right)$$
$$+ \delta w_i \sum_{j \neq i} (\overline{a}_{j\tau} - \widehat{\alpha}_{j\tau}), \textit{for all } i, \tau. \tag{26}$$

The proof will rely on the following "Aggregation Lemma" which essentially says that we can replace the I incentive constraints, one for each country, with a single incentive constraint that sums up—across countries—both sides of the I individual constraints. The proof that this single incentive constraint is all that is required to be checked, is in the Appendix. The intuition for it is that the simplifying force of foreign aid is just this—if there is sufficient slack in the incentives of some countries then they can transfer some of that slack via foreign aid to those countries whose incentives are not being met. Is there enough slack to make those transfers, i.e., to make up the shortfall? Yes, if the total slack is more than the total shortfall.

Aggregation Lemma *The IIC above, as given by* (26), *hold if and only if the following Aggregate Incentive Constraints (AIC) hold*

$$\sum_{i=1}^{I} \sum_{t=\tau}^{\infty} \delta^{t-\tau} \left(\widehat{\alpha}_{it}^{\beta_i} K_{it}^{\gamma_i} - \delta w_i \widehat{\alpha}_{it} - \delta w_i \sum_{j \neq i} \widehat{\alpha}_{jt} \right)$$
$$\geq \sum_{i=1}^{I} \sum_{t=\tau}^{\infty} \delta^{t-\tau} \left(\overline{a}_{it}^{\beta_i} K_{it}^{\gamma_i} - \delta w_i \overline{a}_{it} - \delta w_i \sum_{j \neq i} \overline{a}_{jt} \right)$$
$$+ \delta \sum_{i=1}^{I} w_i \sum_{j \neq i} (\overline{a}_{j\tau} - \widehat{\alpha}_{j\tau}), \textit{for all } \tau. \tag{27}$$

Proof In Appendix. □

Continuing with the proof of the theorem, we shall now show that the AIC holds, i.e., that (27) holds (at every τ). To conserve on notation—and because the cases are qualitatively identical—we shall focus in the immediate sequel on the case $\tau = 0$, i.e., we will show that

$$\sum_{i=1}^{I}\sum_{t=0}^{\infty}\delta^t\left(\widehat{\alpha}_{it}^{\beta_i}K_{it}^{\gamma_i}-\delta w_i\widehat{\alpha}_{it}-\delta w_i\sum_{j\neq i}\widehat{\alpha}_{jt}\right)$$
$$\geq\sum_{i=1}^{I}\left[\sum_{t=0}^{\infty}\delta^t\left(\overline{a}_{it}^{\beta_i}K_{it}^{\gamma_i}-\delta w_i\overline{a}_{it}-\delta w_i\sum_{j\neq i}\overline{a}_{jt}\right)\right.$$
$$\left.+\,\delta w_i\sum_{j\neq i}(\overline{a}_{j0}-\widehat{\alpha}_{j0})\right]. \tag{28}$$

Interchanging the order of summation in (28) we get that the requirement is

$$\sum_{t=0}^{\infty}\sum_{i=1}^{I}\delta^t\left(\widehat{\alpha}_{it}^{\beta_i}K_{it}^{\gamma_i}-\delta w_i\widehat{\alpha}_{it}-\delta w_i\sum_{j\neq i}\widehat{\alpha}_{jt}\right)$$
$$\geq\sum_{t=0}^{\infty}\sum_{i=1}^{I}\delta^t\left(\overline{a}_{it}^{\beta_i}K_{it}^{\gamma_i}-\delta w_i\overline{a}_{it}-\delta w_i\sum_{j\neq i}\overline{a}_{jt}\right)$$
$$+\,\delta\sum_{i=1}^{I}w_i\sum_{j\neq i}(\overline{a}_{j0}-\widehat{\alpha}_{j0}). \tag{29}$$

Clearly (29) can be re-written as

$$\sum_{t=0}^{\infty}\delta^t\sum_{i=1}^{I}\left(\widehat{\alpha}_{it}^{\beta_i}K_{it}^{\gamma_i}-\delta\widehat{\alpha}_{it}\sum_{j}w_j\right)$$
$$\geq\sum_{t=0}^{\infty}\delta^t\sum_{i=1}^{I}\left(\overline{a}_{it}^{\beta_i}K_{it}^{\gamma_i}-\delta\overline{a}_{it}\sum_{j}w_j\right)$$
$$+\,\delta\sum_{i=1}^{I}w_i\sum_{j\neq i}(\overline{a}_{j0}-\widehat{\alpha}_{j0}) \tag{30}$$

which is, of course, equivalent to

$$\sum_{i=1}^{I}\sum_{t=0}^{\infty}\delta^t\frac{1}{I}\left[\left(\widehat{\alpha}_{it}^{\beta_i}K_{it}^{\gamma_i}-\delta\widehat{\alpha}_{it}\sum_{j}w_j\right)-\left(\overline{a}_{it}^{\beta_i}K_{it}^{\gamma_i}-\delta\overline{a}_{it}\sum_{j}w_j\right)\right]$$
$$\geq\delta\frac{1}{I}\sum_{i=1}^{I}w_i\sum_{j\neq i}(\overline{a}_{j0}-\widehat{\alpha}_{j0}). \tag{31}$$

Term by term, for every i, the LHS of (31) is precisely the difference between the GGPO lifetime payoffs (when using the GGPO welfare function with equal weights for all countries) and the lifetime payoffs under the same welfare function but under

GBAU emissions. From the construction of the GGPO each term, for each country, is strictly positive. We will now show that more is true. That in fact the LHS blows up to infinity as $\delta \uparrow \theta_1^{\frac{\gamma_1}{\beta_1 - 1}}$. To see this, note that from the characterization of the GBAU and GGPO emission levels in Sect. 3 it follows that the difference in country 1's payoffs is

$$\sum_{t=0}^{\infty} \delta^t \frac{1}{I} \left[\left(\widehat{\alpha}_{1t}^{\beta_1} K_{1t}^{\gamma_1} - \delta \widehat{\alpha}_{1t} \sum_j w_j \right) - \left(\overline{a}_{1t}^{\beta_1} K_{1t}^{\gamma_1} - \delta \overline{a}_{1t} \sum_j w_j \right) \right] = \frac{(\widehat{\Phi}_1 - \overline{\Phi}_1) K_{10}^{\frac{\gamma_1}{1-\beta_1}}}{1 - \delta \theta_1^{\frac{\gamma_1}{1-\beta_1}}}$$

where $\widehat{\Phi}_1 - \overline{\Phi}_1 = \frac{1}{I} \left(\left[\frac{\beta_i}{\delta \sum_j w_j} \right]^{\frac{\beta_i}{1-\beta_i}} (1 - \beta_i) - \left[\frac{\beta_i}{\delta w_i} \right]^{\frac{\beta_i}{1-\beta_i}} (1 - \beta_i \frac{\sum_j w_j}{w_i} \right)$ which by construction is positive. Re-writing the incentive constraint we have the requirement that

$$\sum_{i=1}^{I} \frac{(\widehat{\Phi}_i - \overline{\Phi}_i) K_{i0}^{\frac{\gamma_i}{1-\beta_i}}}{1 - \delta \theta_i^{\frac{\gamma_i}{1-\beta_i}}} \geq \delta w \sum_{i=1}^{I} (\overline{a}_{i0} - \widehat{\alpha}_{i0})$$

where $w = \frac{1}{I} \sum_{i=1}^{I} w_i$. Note that $(\overline{a}_{i0} - \widehat{\alpha}_{i0})$ is also proportional to $K_{i0}^{\frac{\gamma_i}{1-\beta_i}}$, say is equal to $\lambda_i K_{i0}^{\frac{\gamma_i}{1-\beta_i}}$. Hence we need to show that

$$\sum_{i=1}^{I} \left[\frac{(\widehat{\Phi}_i - \overline{\Phi}_i)}{1 - \delta \theta_i^{\frac{\gamma_i}{1-\beta_i}}} - \lambda_i \right] K_{i0}^{\frac{\gamma_i}{1-\beta_i}} \geq 0$$

The coefficients on the LHS stay bounded away from zero even as $\delta \uparrow \theta_1^{\frac{\gamma_1}{\beta_1 - 1}}$. Naturally it follows that the LHS of the above inequality blows up and hence is strictly positive above a feasible cut-off discount factor.

To simplify notation we had taken the starting point of the deviation to be $\tau = 0$. What if the deviation happens at $\tau > 0$? It is easily seen the arguments repeat with no change other than notation. Given the observation that $K_{it}^{\frac{\gamma_i}{1-\beta_i}}$ is equal to $\theta_i^{\frac{\gamma_i t}{1-\beta_i}} K_{i0}$ means that the positive terms—such as the incentive slack for country 1—only get disproportionately larger than th negative terms. Hence the inequality holds at every time period if it holds at period 0.

Evidently, there is no profitable deviation in which a country withholds aid after lowering its emissions. Since in that case it gets the GBAU emissions from the next period onwards and loses out on aid as well. If there is going to be a deviation it might as well be on emissions as well. Which we have shown to be unprofitable. Finally, the GBAU punishment regime, once started, does get carried out, i.e., the punishment is credible.

We have so far shown that the Aid Induced GGPO strategy is a subgame perfect equilibrium. To see that—inclusive of aid—it implies a Pareto improvement vis-a-vis the GBAU one need only look at (26). The theorem is proved. □

5.3 *Sustainability of Other Emission Reduction Policies*

In this subsection we examine the sustainability of other emission reduction policies. In particular, we will consider any policy that involves uniform reductions from the GGPO but is at least as high an emission level as the equally weighted GGPO. Modifying the definition given above we reproduce it here for easy access:

Definition A uniform cut in emissions is achieved by a (capital-dependent) emission policy $\widetilde{a}_i(.)$ where, for all capital stock K_i, emissions are a convex combination, with weight say λ_i, of the GBAU and equally weighted GGPO emission level, i.e.,

$$\widetilde{a}_i(K_i;\lambda_i) = \lambda_i \overline{\alpha}_i(K_i) + (1-\lambda_i)\widehat{a}_i(K_i).$$

We shall consider—as in the previous subsection—an aid induced emissions policy with the obvious difference that the Norm emission policy will be given by $\widetilde{a}_i(K_i;\lambda_i)$ rather than the GGPO emissions. The punishment—as above—will be the witholding of aid coupled with a reversal to the GBAU emissions.

Theorem 9 *There is a cut-off discount factor $\delta(\lambda) < \theta_1^{\frac{\gamma_1}{\beta_1-1}}$ and a feasible foreign aid policy with the property that the Aid Induced emission reduction strategy $\widetilde{a}_i(.;\lambda_i)$ is a SPE for all feasible discount factors above $\delta(\lambda)$. Furthermore, for every country i, donor as well as recipient, life-time payoffs inclusive of foreign aid strictly Pareto dominates the GBAU lifetime payoffs.*

Proof The proof is identical to that for the proof of the immediately preceding theorem, with the obvious changes of notation. Note first that (26) is the IIC with the norm emission policy being $\widetilde{a}_i(.)$ rather than the GGPO emission policy considered above. The Aggregation Lemma applies without any change because it clearly made no use of the specific emission policy. Hence, after making the same substitutions and interchanges as we made in the previous proof we get

$$\sum_{i=1}^{I}\sum_{t=0}^{\infty}\delta^t\frac{1}{I}\left[\left(\widetilde{a}_{it}^{\beta_i}K_{it}^{\gamma_i}-\delta\widetilde{a}_{it}\sum_j w_j\right)-\left(\overline{a}_{it}^{\beta_i}K_{it}^{\gamma_i}-\delta\overline{a}_{it}\sum_j w_j\right)\right]$$
$$\geq \delta\frac{1}{I}\sum_{i=1}^{I}w_i\sum_{j\neq i}(\overline{a}_{j0}-\widetilde{a}_{j0}). \tag{32}$$

The function $a_{it}^{\beta_i}K_{it}^{\gamma_i}-\delta a_{it}\sum_j w_j$ is strictly concave and at every K_{it} it reaches a maximum at $\widehat{a}_{it}$. Hence it follows that the bracketed terms are all strictly positive. More is true since $\widetilde{a}_i(K_i;\lambda_i)$ is proportional to the effective capital—$K_{i0}^{\frac{\gamma_i}{1-\beta_i}}$ since the component emissions—the GGPO and GBAU emissions—are. Using that fact and aggregating payoffs in the same way that we did above we get that the AIC above holds iff

$$\sum_{i=1}^{I}\left[\frac{(\widetilde{\Phi}_i-\overline{\Phi}_i)}{1-\delta\theta_i^{\frac{\gamma_i}{1-\beta_i}}}-\lambda_i\right]K_{i0}^{\frac{\gamma_i}{1-\beta_i}}\geq 0$$

where $\widetilde{\Phi}_i-\overline{\Phi}_i>0$. By identical logic to that above, the terms above on the LHS—especially that involving country 1's payoffs—blows up as $\delta\uparrow\theta_1^{\frac{\gamma_1}{\beta_1-1}}$. Naturally it follows that the LHS of the above inequality blows up and hence is strictly positive above a feasible cut-off discount factor $\delta(\lambda)$. Finally, identical logic as in the GGPO case shows that if the above AIC holds at time period 0 then it also holds at every other time period. That the aid induced emission reduction involves a Pareto-improvement over the GBAU follows immediately from (32). The theorem is proved. □

Remark Given that the GGPO is the maximum value of the function $a_{it}^{\beta_i}K_{it}^{\gamma_i}-\delta a_{it}\sum_j w_j$ it follows that the LHS of (32) achieves its maximum at the GGPO emission norm rather than at the emission norm $\widetilde{a}$. This might suggest that the GGPO is easiest to sustain as a norm. However, that might not be true because the payoffs on the RHS of (32) are also proportional to the size of the emission cuts and hence are highest at the GGPO emission. If that effect is stronger the cut-off discount factor $\delta(\lambda)$ might be lower for $\lambda>0$, i.e., when we sustain a norm that has a higher level of emission than the GGPO.

6 Discussion and Conclusion

To the best of our knowledge we are the first to investigate, within a fully formed model, the possibility of getting China and India to sign a climate treaty. As has been widely reported in the press, these fast growing economies are reluctant to sign onto emission caps fearing that it will compromise growth. They have also claimed that they do not have the resources to make the technological switches that are required and have pointed to the fact that the problem is not of their making. In response, Western economies have discussed various "punishment" options that range from the possibility of trade-related sanctions[22] to escalating targets on emission cuts if the first targets are not met.[23] In this paper we investigate the effectiveness of retaliatory emissions—if reductions are not made, then all countries are free to increase their emissions to BAU levels. We show that such a sanction is ineffective—the fast growing countries always have an incentive to cheat; their loss from reductions grows as quickly as their own rate of accumulation while the loss from the sanctions grow at the slower rate of others' accumulation.

We then examine a mechanism similar to the World Bank's CIF—contingent foreign aid. And we show that this is effective at getting fast (and slow) growing economies to curtail the growth of emissions. Furthermore, even inclusive of the aid given, the outcome is Pareto-superior to the BAU equilibrium.

A key simplification of our model is (1) "power functions"—the one-period payoff for each country is a Cobb-Douglas function whilst capital grows geometrically, and (2) "cost linearity"—the (incremental) damage cost is linear in the current stock of greenhouse gas. These properties of the model allows us to get closed-form solutions for the Business as Usual and Pareto optimal solutions, characterize the equilibrium subject to BAU reversion, and investigate aid-contingent equilibria. It also facilitates the possible calibration of the model, the numerical calculation of various trajectories, and sensitivity analyses. The disadvantage is that it results in a number of cases in unrealistic "unbounded" strategies, that is, strategies in which the emission rates grow infinitely large along with capital stock. In particular, a country's cost of damage due to climate change, and/or the amount of foreign aid it has promised could become unrealistically large. This aspect of the results needs to be taken "with a grain of salt." In a more realistic model, one would expect that these strategies would display a more gradual adaptation to capital growth, and capital growth might even be bounded in the long run. Our conjecture is that the analysis of the affine model yields reasonable approximations to equilibrium and optimal trajectories in the medium term. However, precise tests of this conjecture will have to await future research.

In Dutta and Radner (2006) we generalized our bench-mark model to allow for population change and in Dutta and Radner (2004) we allowed for simple technological change and presented some theoretical and numerical results on the GPO and BAU solutions. In Dutta and Radner (2007) we incorporate technical change in a more meaningful way. Eventually, we hope to develop and analyze a "complete" model that incorporates all of the above features.

---

[22]The United States House of Representatives passed a bill in June, 2009 that would place tariffs on countries that do not adjust their carbon emissions. See http://www.nytimes.com/2009/06/29/us/politics/29climate.html.

[23]Subsequent to Kyoto, at the Hague November 2000 meeting, the most popular proposal (which came from the Dutch Environment Minister Jan Pronk) was that countries would face an escalating series of target reductions in the future if they failed to comply in the current stage. A watered-down version of this proposal was adopted in Bonn in March, 2001.

The literature on (symmetric) dynamic commons games is exceedingly rich and goes back over 25 years. The earliest model was that of Levhari and Mirman (1980) who studied a particular functional representation of the neo-classical growth model with the novel twist that the capital stock could be "expropriated" by multiple players. Subsequently several authors (Sundaram 1989; Benhabib and Radner 1992; Rustichini 1992; Dutta and Sundaram 1992, 1993; Sorger 1998) studied this model in great generality and established several interesting properties relating to existence of equilibria, welfare consequences, and dynamic paths. Another variant of that model has been studied by Tornell and Velasco (1992) and, subsequently, Long and Sorger (2006).[24]

More recently in a series of papers by Dockner and his co-authors, the growth model has been directly applied to environmental problems including the problem of global warming. The paper closest to the current one is Dockner et al. (1996). It studies a model of global warming that has some broad similarities to the one we have studied here. In particular, the transition equation is identical in the two models. What is different is that they impose linearity in the emissions payoff function (whereas we have assumed it to be Cobb-Douglas and hence strictly concave) while their cost to g is strictly convex (as opposed to ours which is linear).

A large volume of literature exists that directly focuses on the economics of climate change. An excellent broad discussion is contained in the Inter-Governmental Panel on Climate Change (2007). A central question there is to determine the level of emissions that is globally optimal. An excellent example of this is Nordhaus and Boyer (2000).[25] Several of those papers, including the Nordhaus and Boyer paper, analyze only the "competitive" model, not taking strategic considerations fully into account.[26] A smaller volume of literature emphasizes the need for treaties to be self-enforcing, presenting a strategic analysis of the problem. (See Barrett 2003; Finus 2001). Where we depart from that literature is in the dynamic modelling; we allow greenhouse gases to accumulate and stay in the environment for a (possibly long) period of time. By contrast the Barrett and Finus studies restrict themselves to purely repeated games, which implies that the state variable, gas stock, remains constant over time.

What this paper does not address is a set of complementary issues regarding the economics of climate change and many of them have been addressed by other papers in this volume. These issues include whether taxes or quotas are the best instruments to achieve abatement (Karp and Zhang 2016), whether lower level

---

[24]Some of these papers allow asymmetry; however, none of them analyzes the effect of asymmetries. One significant exception is the recent paper of Long and Sorger that explicitly considers asymmetry in appropriation costs within the Tornell and Velasco model.

[25]But also see Chichilnisky (2006).

[26]To be fair, Nordhaus and Boyer (2000) and Nordhaus and Zhang 1996 do consider strategic models but restrict themselves to open-loop strategies.

"polycentric" bodies can substitute for treaty formation at national level (Ostrom 2016), whether the BAU solution can be Pareto-improved across generations by appropriate mitigation investment by existing generations (Rezai et al. 2016), and whether climate effects are mitigated if agents have preferences that value the long-run future (Asheim 2016; Figuieres and Tidball 2016; Chichilnisky 2016). Indeed this article is part of a Special Issue of Economic Theory on the topic of the Global Environment, which includes also the following articles: "Unspoken Ethical Issues in the Climate Affair Insights From a Theoretical Analysis of Negotiation Mandates" by Lecocq and Hourcade (2016), "Intergenerational equity, efficiency, and constructability", by Lauwers (2010), "Carbon Leakages: A General Equilibrium View" by Burniaux and Martins (2016), and "Detrimental Externalities, Pollution Rights, and the "Coase Theorem"" by Chipman and Tian (2016).

7 Appendix

Aggregation Lemma *The IIC above, as given by* (26) *in Sect.* 5, *hold if and only if the following Aggregate Incentive Constraints (AIC) holds*

$$\sum_{i=1}^{I}\sum_{t=\tau}^{\infty}\delta^{t-\tau}\left(\hat{\alpha}_{it}^{\beta_i}K_{it}^{\gamma_i}-\delta w_i\hat{\alpha}_{it}-\delta w_i\sum_{j\neq i}\hat{\alpha}_{jt}\right)$$
$$\geq\sum_{i=1}^{I}\sum_{t=\tau}^{\infty}\delta^{t-\tau}\left(\overline{a}_{it}^{\beta_i}K_{it}^{\gamma_i}-\delta w_i\overline{a}_{it}-\delta w_i\sum_{j\neq i}\overline{a}_{jt}\right)$$
$$+\delta\sum_{i=1}^{I}w_i\sum_{j\neq i}(\overline{a}_{j\tau}-\hat{\alpha}_{j\tau}), \textit{for all } \tau. \tag{33}$$

Proof We are required to show that (33) above implies the existence of a feasible foreign aid policy $(\Upsilon_t)_{t\geq 0}$ such that the Individual Incentive Constraints (IIC) hold for every country, i.e., that

$$\sum_{t=\tau}^{\infty}\delta^{t-\tau}\left(\hat{\alpha}_{it}^{\beta_i}K_{it}^{\gamma_i}-\delta w_i\hat{\alpha}_{it}-\delta w_i\sum_{j\neq i}\hat{\alpha}_{jt}+\Upsilon_{it}\right)$$
$$\geq\sum_{t=\tau}^{\infty}\delta^{t-\tau}\left(\overline{a}_{it}^{\beta_i}K_{it}^{\gamma_i}-\delta w_i\overline{a}_{it}-\delta w_i\sum_{j\neq i}\overline{a}_{jt}\right)$$
$$+\delta w_i\sum_{j\neq i}(\overline{a}_{j\tau}-\hat{\alpha}_{j\tau}), \textit{for all } i,\tau. \tag{34}$$

To simplify the notation let $\widehat{\alpha}_{it}^{\beta_i} K_{it}^{\gamma_i} - \delta w_i \widehat{\alpha}_{it} - \delta w_i \sum_{j \neq i} \widehat{\alpha}_{jt}$ be denoted $\widehat{u}_{it}$ and likewise let $\overline{a}_{it}^{\beta_i} K_{it}^{\gamma_i} - \delta w_i \overline{a}_{it} - \delta w_i \sum_{j \neq i} \overline{a}_{jt}$ be denoted $\overline{u}_{it}$. Fix any time-period τ and separate the group of countries into two exclusive groups where Group 1 is defined as all countries such that

$$\sum_{t=\tau}^{\infty} \delta^{t-\tau} (\widehat{u}_{it} - \overline{u}_{it}) - \delta w_i \sum_{j \neq i} (\overline{a}_{j\tau} - \widehat{\alpha}_{j\tau}) \geq 0$$

and Group 2 is made up of countries for which the inequality is reversed. For Group 2, where the IIC does not hold in the absence of foreign aid, define the life-time foreign aid receipts $\Gamma_{i\tau}$ by

$$\Gamma_{i\tau} = \left| \sum_{t=\tau}^{\infty} \delta^{t-\tau} (\widehat{u}_{it} - \overline{u}_{it}) - \delta w_i \sum_{j \neq i} (\overline{a}_{j\tau} - \widehat{\alpha}_{j\tau}) \right| .$$

Let the parameter μ_τ be defined by the following equality which ensures that the total of foreign aid grants is equal to the total of foreign aid receipts:

$$\sum_{i \in Group\ 2} \Gamma_{i\tau} = \mu_\tau \sum_{i \in Group\ 1} \left[\sum_{t=\tau}^{\infty} \delta^{t-\tau} (\widehat{u}_{it} - \overline{u}_{it}) - \delta w_i \sum_{j \neq i} (\overline{a}_{j\tau} - \widehat{\alpha}_{j\tau}) \right] . \qquad (35)$$

Note that since the AIC holds at time τ, the parameter $\mu_\tau \leq 1$. For Group 1, countries where the IIC does hold in the absence of foreign aid, define the life-time of foreign aid donations $\Gamma_{i\tau}$ by[27]

$$\Gamma_{i\tau} = -\mu_\tau \left[\sum_{t=\tau}^{\infty} \delta^{t-\tau} (\widehat{u}_{it} - \overline{u}_{it}) - \delta w_i \sum_{j \neq i} (\overline{a}_{j\tau} - \widehat{\alpha}_{j\tau}) \right] .$$

An implication of (35) is that the total life-time foreign aid $\sum_i \Gamma_{i\tau} = \sum_{i \in Group\ 1} \Gamma_{i\tau} + \sum_{i \in Group\ 2} \Gamma_{i\tau} = 0$. Finally, the lifetime aid amounts are decomposed into period by period aid amounts through the following decomposition. For every i and for every τ

$$\Upsilon_{i\tau} = \Gamma_{i\tau} - \delta \Gamma_{i\tau+1}.$$

It immediately follows that $\sum_i \Upsilon_{i\tau} = 0$ given that $\sum_i \Gamma_{i\tau} = 0$ and $\sum_i \Gamma_{i\tau+1} = 0$. So the foreign aid that is proposed aggregates to zero in every period as required. To see that (33) holds, note that for Group 1, the countries that starting at period τ are a net donor of foreign aid

[27] Recall the convention is that donations are negative while receipts are postive numbers.

$$\sum_{t=\tau}^{\infty} \delta^{t-\tau}(\widehat{u}_{it} - \overline{u}_{it}) - \delta w_i \sum_{j \neq i} (\overline{a}_{j\tau} - \widehat{\alpha}_{j\tau}) + \sum_{t=\tau}^{\infty} \delta^{t-\tau} \Upsilon_{it}$$

$$= \sum_{t=\tau}^{\infty} \delta^{t-\tau}(\widehat{u}_{it} - \overline{u}_{it}) - \delta w_i \sum_{j \neq i} (\overline{a}_{j\tau} - \widehat{\alpha}_{j\tau}) + \Gamma_{i\tau}$$

$$= (1 - \mu_\tau) \left[\sum_{t=\tau}^{\infty} \delta^{t-\tau}(\widehat{u}_{it} - \overline{u}_{it}) - \delta w_i \sum_{j \neq i} (\overline{a}_{j\tau} - \widehat{\alpha}_{j\tau}) \right] \geq 0$$

since $1 - \mu_\tau \geq 0$. For Group 2, the countries that starting at period τ are a net recipient of foreign aid,

$$\sum_{t=\tau}^{\infty} \delta^{t-\tau}(\widehat{u}_{it} - \overline{u}_{it}) - \delta w_i \sum_{j \neq i} (\overline{a}_{j\tau} - \widehat{\alpha}_{j\tau}) + \sum_{t=\tau}^{\infty} \delta^{t-\tau} \Upsilon_{it}$$

$$= \sum_{t=\tau}^{\infty} \delta^{t-\tau}(\widehat{u}_{it} - \overline{u}_{it}) - \delta w_i \sum_{j \neq i} (\overline{a}_{j\tau} - \widehat{\alpha}_{j\tau}) + \Gamma_{i\tau} = 0$$

Clearly the argument repeats at every time period τ. In other words the IIC holds (for all countries and all time-periods). Put differently, the lemma is proved. □

References

Asheim, G., Mitra, T., & Tungodden, B. (2016). Sustainable recursive social welfare functions. *Economic Theory* (reprinted in this volume).

Barrett, S. (2003). *Environment and statecraft: The strategy of environmental treaty-making*. Oxford: Oxford University Press.

Benhabib, J., & Radner, R. (1992). Joint exploitation of a productive asset. *Economic Theory*, *2*, 155–190.

Burniaux, J-M., & Martins, J. (2016). Carbon leakages: a general equilibrium view. *Economic Theory* (reprinted in this volume).

Chichilnisky, C. (2006). *Global Property Rights: The Kyoto Protocol and the Knowledge Revolution*. Discussion Paper, Paris: Institut du Development Durable et Relations Internationales (IDDRI), Ecole Polytechnique

Chichilnisky, C. (2016). Sustainable markets with short sales. *Economic Theory* (reprinted in this volume).

Chipman, J., & Tian, G. (2016). Detrimental Externalities, Pollution Rights, and the "Coase Theorem". *Economic Theory* (reprinted in this volume).

Cline, W. (1992). *The economics of global warming*. Washington, DC: Institute for International Economics.

Dockner, E., Long, N., & Sorger, G. (1996). Analysis of Nash equilibria in a class of capital accumulation games. *Journal of Economic Dynamics and Control*, *20*, 1209–1235.

Dockner, E., & Nishimura, K. (1999). Transboundary pollution in a dynamic game model. *Japanese Economic Review*, *50*, 443–456.

Dutta, P., & Radner, R. (2004). Self-Enforcing climate change treaties. *Proceedings of the National Academy of Sciences USA*, *101–14*, 5174–5179.

Dutta, P., & Radner, R. (2006). Population growth and technological change in a global warming model. *Economic Theory*, *29*, 251–270.
Dutta, P., & Radner, R. (2007). Choosing cleaner technologies: Global warming and technological change (in preparation).
Dutta, P., & Radner, R. (2009). A strategic model of global warming: Theory and some numbers. *Journal of Economic Behavior and Organization*.
Dutta, P., & Sundaram, R. (1992). Markovian equilibrium in a class of stochastic games: Existence theorems for discounted and undiscounted models. *Economic Theory*, *2*, 197–214.
Dutta, P., & Sundaram, R. (1993). How different can strategic models be? *Journal of Economic Theory*, *60*, 42–61.
Figuieres, C., & Tidball, M. (2016). Sustainable exploitation of a natural resource: A satisfying use of Chichilnisky's criterion. *Economic Theory* (reprinted in this volume).
Finus, M. (2001). *Game theory and international environmental cooperation*. Cheltenham: Edward Elgar.
Inter-Governmental Panel on Climate Change. (2007). Climate change, the synthesis report. Geneva, Switzerland: IPCC. Retrieved December 5, 2007 from. http://www.ipcc.ch/ipccreports/assessments-reports.htm.
Karp, L., & Zhang, Q. (2016). Taxes versus quantities for a stock pollutant with endogenous abatement costs and asymmetric information. *Economic Theory* (reprinted in this volume).
Lauwers, L. (2010). Intergenerational equity, efficiency, and constructability. *Economic Theory* (reprinted in this volume).
Lecocq, F., & Hourcade, J.-C. (2016). Unspoken ethical issues in the climate affair insights from a theoretical analysis of negotiation mandates. *Economic Theory* (reprinted in this volume).
Levhari, D., & Mirman, L. (1980). The great fish war: An example using a dynamic Cournot-Nash solution. *Bell Journal of Economics*, *11*, 322–334.
Long, N., & Sorger, G. (2006). Insecure property rights and growth: The role of appropriation costs, wealth effects and heterogenity. *Economic Theory*, *28*, 513–529.
Nordhaus, W., & Yang, Z. (1996). A regional dynamic general equilibrium model of alternative climate-change strategies. *American Economic Review*, *86*, 741–765.
Nordhaus, W., & Boyer, J. (2000). *Warming the world: Economic models of global warming*. Cambridge, MA: MIT Press.
Ostrom, E. (2016). Nested externalities and polycentric institutions: Must we wait for global solutions to climate change before taking actions at other scales? *Economic Theory* (reprinted in this volume).
Rezai, S., Foley, D., & Taylor, L. (2016). Global warming and economic externalities. *Economic Theory* (reprinted in this volume).
Rustichini, A. (1992). Second-best equilibria for games of joint exploitation of a productive asset. *Economic Theory*, *2*, 191–196.
Sorger, G. (1998). Markov-Perfect Nash equilibria in a class of resource games. *Economic Theory*, *11*, 79–100.
Stern, N. (2006). Review on the economics of climate change. London, UK; HM Treasury. Retrieved December 5, 2007 from. http://www.sternreview.org.uk.
Sundaram, R. (1989). Perfect equilibrium in a class of symmetric dynamic games. *Journal of Economic Theory*, *47*, 153–177.
Thomson, David J. (1997). Dependence of global temperatures on atmospheric CO_2 and solar irradiance. *Proceedings of the National Academy of Sciences USA*, *94*, 8370–8377.
Tornell, A., & Velasco, A. (1992). The tragedy of the commons and economic growth: Why does capital flow from poor to rich countries? *Journal of Political Economy*, *100*, 1208–1231.

Unspoken Ethical Issues in the Climate Affair: Insights from a Theoretical Analysis of Negotiation Mandates

Franck Lecocq and Jean-Charles Hourcade

1 Introduction

Besides controversies about the seriousness of the climatic threat, conflicting perceptions of equity between developed and developing countries are a key reason why, after Copenhagen, the international climate regime remains as "unfinished business" as ever (Jacoby et al. 1998). An archetypal exchange between the G77 and the US in Kyoto sums up the controversy over who should pay for climate mitigation. "There will be no agreement [...] until the question of emissions rights is addressed equitably", stated the G77 and China during the conference, mere months after the US Senate had unanimously voted that it would not ratify any

The authors are grateful to François Bourguignon, Graciela Chichilnisky, Sylviane Gastaldo, Roger Guesnerie, Cédric Philibert, Gilles Rotillon, Zmarak Shalizi, Tarik Tazdait, David Wheeler and anonymous referees for very constructive comments on earlier versions of this manuscript. The authors would also like to thank the participants of workshops and seminars in Kyoto, Bilbao, La Réunion, Washington D.C., and Paris for helpful discussion.

Reprinted with kind permission from the Authors: Originally published in Economic Theory, Volume 49, Number 2, February 2012.

F. Lecocq (✉)
AgroParisTech, Engref UMR 356 Economie Forestière, 54000 Nancy, France
e-mail: franck.lecocq@nancy-engref.inra.fr

F. Lecocq
INRA, UMR 356 Economie Forestière, 54000 Nancy, France

J.-C. Hourcade
Centre International de Recherche sur l'Environnement et le Développement, Nogent/Marne, France

G. Chichilnisky and A. Rezai (eds.), *The Economics of the Global Environment*, Studies in Economic Theory 29, DOI 10.1007/978-3-319-31943-8_15

Protocol without "meaningful participation" of developing countries (Bird-Hagel resolution, July 1997).

This debate presupposes that climate mitigation constitutes a *burden* to be shared among countries (an assumption we come back to later). Yet remarkably, it has mostly been framed around *ethical intuitions* about what is fair—intuitions only loosely connected to economic principles, if at all. Two such major intuitions are per capita allocation of emission rights (Agarwal and Narain 1991) which many developing countries support, and its polar opposite grandfathering which underlies the position of many developed countries.[1] The principle of "common but differentiated responsibilities" of the UN Framework Convention on Climate Change (UNFCCC, art. 4) is a rhetorical compromise between these two views, but it has no operational content per se (Stone 2004).[2]

Searching for such guidance, one strand of the literature adopts the point of view of the theory of justice (e.g., Posner and Sustein 2009; Helm and Simonis 2001; Godard et al. 2000) but it elevates the debate to a level—the choice between competing views of justice—that is hardly operational. In particular, it provides no clear judgment on pragmatic proposals for negotiating emissions quota.[3] Another strand uses models of the world economy to assess the implications of such rules (see Gupta et al. 2007 for a review). It has helped clarify the stakes for each country (e.g., Lecocq and Crassous 2003) and the rationale of alternative climate regimes (e.g., Aldy and Stavins 2007) but it has so far refrained from delivering statements about the legitimacy of different burden-sharing principles.

The caution of economists in this matter may be a symptom of their professional reflex that "it is useful to separate efficiency from equity" (Goldemberg et al. 1996) since compensating transfers can restore any income distribution judged equitable on pure political grounds. Despite the contribution of Chichilnisky (1994), who showed that the second theorem of welfare does not hold in the presence of an indivisible public good,[4] a continued deficit of economic conversation about equity principles substantially lowers the chances of finding a successor to the Kyoto Protocol. The new cycle of negotiations is indeed more complex than the pre-Kyoto one because, with

[1]The ethical justification of grandfathering is the absence of retroactive responsibility for decisions made before the reformulation of the international social contract by the climate regime.

[2]The UNFCCC is the UN body which oversees international negotiations on climate change. In fact, UNFCCC art. 4 provided a practical translation of the "common but differentiated principle" in the 1992 context by setting up emissions targets for year 2000 for developed countries only (art. 4.2), on the ground that "social development and poverty eradication are the first and overriding priorities of the developing countr[ies]" (art. 4.7). But deeper emissions cuts beyond 2012 require participation of at least major emitters among developing countries. Hence the renewed debate about who should reduce GHG emissions, and by how much.

[3]See inter alia multicriteria (Ringius et al. 1998), contraction and convergence (Meyer 2002) or historical responsibilities (Den Elzen et al. 1999).

[4]This insight was further developed in Chichilnisky and Heal (1994), and in Chichilnisky et al. (2000). In a related discussion in this volume, Chipman and Tian (2016) characterize the conditions under which, in a two-individual economy with a polluter and pollutee, the optimal level of pollution is independent of the initial assignment of property rights.

the emergence of candidate super powers in the Third World, the North/South division is no longer a pure Rich/Poor division boiling down to a distributional issue. In this context, continuing to oppose intuitions about what is 'equitable', 'fair', 'just' or 'balanced' increases the risk of endless, and ultimately self-defeating, verbal jousting with ethics as a rhetoric weapon.

Obviously, the very idea that mitigating climate change constitutes a net social cost is questionable. First, climate policies should provide a net intertemporal social *benefit* by reducing the total climate change bill (i.e., mitigation costs *plus* damages) relative to a world without mitigation but with full damages (Stern 2007; Shalizi and Lecocq 2009). It is thus possible to find Pareto-improving climate policies that make North and South better off (Chichilnisky et al. 2000). As to the fact that climate policies may still constitute a net burden for *early* generations, it can be argued that mitigation could be financed by redirecting investments instead of limiting consumption (Foley 2009; Rezai et al. 2010). And in a second-best world, removing existing barriers to development—e.g., improving energy security, accelerating technical change, or upgrading local environmental quality—may yield both lower emissions and net social benefits. In the remainder of this text however, we stick to the mainstream view that mitigation constitutes a net social cost at first period (Fisher et al. 2007). Because this may be true for very tight climate targets and once transaction costs of Pareto-improving policies are incorporated, this view still prevails among negotiators and it leads to ethical misunderstandings that need to be dispelled.

To do so, following Amartya Sen's judgment that "there is something in the methods standardly used in economics, related inter alia with its engineering aspect, that can be of use to modern ethics" (1987), we start, rather conventionally, from the social-welfare maximizing benevolent planner metaphor to represent the negotiation about the global public good problem at hand.

We show in Sect. 1 why, given the public good nature of the climate issue (Chichilnisky 1994) this model is a good caricature of the behavior of the Chair of the annual Conference of the Parties to the UNFCCC (COPs) when, in the penultimate day of the conference, she presents a take-or-leave proposal that strive to balance countries' competing demands. In Sect. 2, we show with a two-period model that a conservative but politically realistic no income redistribution constraint and the choice of the rate of pure time preference are not sufficient to fully determine the problem. Two ethical priors, so far overlooked, of the mandates that country delegates impose on the Chair have to be added. The first has to do with countries attitudes towards the evolving balance of power across nations *over time*. The second has to do with the scope of countries intergenerational solidarity (future fellow citizens vs. all future human beings).

The remainder of the paper explores the implications of four polar yet plausiblemandates deriving from these two dimensions. Section 3 shows that the burdensharing rule at first period is the same under all four mandates, with, for a largeclass of utility functions, mitigation costs in proportion of per capita income, whereas Sect. 4 reveals important disparities in the burden sharing at second period. Some mandates appear self-defeating, others are more robust to uncertainty, and all lead to different provisions of greenhouse gas (GHG) abatement at second and at first periods. Illustrative orders of magnitudes are provided with the analytic demonstrations.

2 The Benevolent Planner Model as a Metaphor of the Negotiation Process

The signature of the UNFCCC at Rio de Janeiro in 1992 initiated a very specific negotiation process in which COPs are held annually to design and adopt decisions aimed at fulfilling the overall objective of stabilizing GHG concentrations in the atmosphere.[5]

Consultations amongst Parties are conducted all year round in formal and informal settings, but negotiations culminate at the end of each year in the 2 weeks of the COP, where official decisions are adopted. At the start of the COP, draft decisions still contain many bracketed sentences that signal reservations by some Parties. The Chair of the COP acts as a facilitator to delete as many bracketed sentences as possible from the compromise text so that it can be adopted by the general assembly. This role turns out to be crucial when, just before the end of the meeting, the Chair submits a bracket-free package. Country delegates de facto treat it as a take-or-leave proposal because the few amendments they can make in the ultimate hours of the negotiation cannot substantially alter the balance of the package (Grubb et al. 1999).

Of course, an agreement at a COP is not the end of the story. Delegations may accept deals that are subsequently not ratified by their countries legislative bodies[6] and the obligation that agreements be ratified by a minimum number of countries prior to entry into force can involve additional and lengthy negotiations (eight years for the Kyoto Protocol). However, once an agreement is reached at a COP, individual countries can refuse the deal—and try to prevent its entry into force, but they cannot significantly modify it. The proposal put forward by the Chair thus remains a pivotal event.[7]

[5]The Kyoto Protocol was adopted at COP3 in 1997. The 'application decrees' of the Kyoto Protocol were adopted at COP7 (2001). Subsequent COPs like the recent COP15 in Copenhagen have focused on negotiating the post-Kyoto framework.

[6]E.g., Australia and the U.S. after Kyoto. However, Australia has since ratified the Kyoto Protocol (2007). In the U.S., the House of Representatives passed the American Clean Energy and Security Act of 2009 that would establish an economy-wide, greenhouse gas cap-and-trade system. At time of writing, the legislation is currently under debate in the U.S. Senate.

[7]The present paper focuses on the conditions for an international agreement on climate change. Yet, reaching a global agreement is a necessary but not a sufficient condition to effective mitigation as actions at different scales interact Ostrom (2016).

Historical evidence suggests that well-intentioned Chairs tend to put balanced proposals on the table to maximize the chances of acceptance by all countries, even if the proposal is not fully satisfying for each. This exercise is not very far from the program of a benevolent planner who maximizes a social welfare function written so as to translate the implicit mandate passed by the COP on to him or her. This mandate is bounded by political constraints resulting not only from Parties' differing visions of the climate change issue, but also from the existing balance of power between countries.

One such major political constraint is that the benevolent planner can compensate for undue adverse effects of climate policies but is not allowed to redistribute income across countries. This 'no-redistribution' constraint amounts to considering the current distribution of income as a given, something both ethically questionable (Azar 1999) and pragmatically defensible for reasons of political realism: Violating the 'no-redistribution' constraint would be a political non-starter for rich countries and that "it is inappropriate to redress all equity issues through climate change initiatives" (Goldemberg et al. 1996).

In analytical terms, two conditions must be met to satisfy the no-redistribution constraint. First, national contributions to the global mitigation efforts must all be non negative.[8] Second, the weights attached to individual utilities in the social welfare function must be such that the initial distribution of income is welfare maximizing. These weights are unique—up to a scale factor—and equal to the inverse of the marginal utility of income (Negishi 1960). If individual utility functions are logarithmic and if consumptions at different points in time are separable, these 'Negishi weights' are proportional to per capita income.[9]

Controversies about the measurement of GDP (PPP vs. exchange rates) notwithstanding, the Negishi weights can almost unambiguously be determined at first period. But this is no longer true at subsequent periods because future GDPs are uncertain. It is conventional modeling practice to assume time-varying Negishi weights, so that the *projected* distribution of income is also welfare maximizing at any point in time (e.g., Nordhaus and Yang 1996). But this is a pure modeling artifact which presupposes that the planner is allowed by Parties to anticipate changes in world income distribution (typically, a catching-up by poor countries), which is arguably a bold assumption about the political economy of the negotiation.

In fact, "states are cold monsters" (Machiavelli). Thus, the natural reflex of rich countries is to refuse to acknowledge ex ante a decline of their economic and

[8]In the Kyoto Protocol this condition is not met for most economies in transitions which were given more allowances than projected baseline emissions. But this situation results from tactical concessions made during the negotiations, and it is unlikely to be replicated in post-Kyoto agreements.

[9]Obviously, this results into a very unequal weighting of individual utilities. But were all the weights set to unity instead, the planner would recommend a large-scale, politically unrealistic, redistribution of wealth to achieve the equal per capita distribution of income that would maximize social welfare.

political power[10] and, on the contrary, to use this power to create institutional irreversibilities. Examples abound of long-lasting arrangements (e.g., the composition of the U.N. Security Council, or the distribution of voting shares at the IMF Executive Board before its recent reform) that mirror the balance of bargaining powers at the time they were formed—however dramatically this balance has shifted since then. In climate negotiations, country delegations range from a handful to more than one hundred in size, with a clear correlation between size and wealth, and the Chair necessarily devotes more attention to the most influential delegations at the moment of the final agreement (Grubb et al. 1999).

The key issue for our discussion is that adopting a catch up or a status quo perspective (or any combination thereof) when setting the weights attached to individual utilities at future periods affects the aggregate discounted utility of consumption and thus, for a given rate of pure time preference, influences the trade-off between the short term and the long term. Here, the choice of the weighting matters to determine the Pareto optimum because a public good is considered. Weights would not matter if a private good were considered.

Together with the no-redistribution imperative and attitudes regarding future balances of power, a third issue raised by writing a social welfare function for climate policies lies in the fact that knowing the beliefs of Parties about climate change damages does not suffice in characterizing their attitudes vis-à-vis climate change.[11] It is indeed highly likely that damages will be very unevenly distributed across countries, but climate scenarios at the regional level are at least one order of magnitude more uncertain than climate scenarios at the global level. It is thus already hard to predict the extent to which a country will suffer *directly* from climate change. And it is even harder to anticipate how individual countries will be affected *indirectly* from the propagation of impacts in other countries (via, e.g., markets, migration, transboundary impacts, common resource management, etc.). Parties attitudes vis-à-vis climate change will be very different depending on whether they disregard this uncertainty and take into account only the impacts expected to fall on their future fellow citizens—thus following a sort of *dynastic solidarity* ethics, or take into account the impacts expected to fall on their own future fellow citizens and on *all other* future individuals—thus following a *universal solidarity* ethics.[12]

[10]Developed countries may argue that convergence in per capita incomes may not occur, or at least may not *necessarily* occur because of institutional failures in developing countries or of mechanisms leading to poverty traps.

[11]For an overview of the attitudes vis-à-vis climate change damages, see Ambrosi et al. (2003).

[12]We discuss later the ethical rationale and political likelihood of this attitude, which appears in the discourses of many NGOs. For the time being, let us treat it as a pure logical possibility. Let us simply underline that the term solidarity is not synonymous of equity. Its Latin root is very suggestive: *solidus* means compact, solid, firm, while *solidarus* means with whom I consider myself to be attached.

These ethical attitudes can be translated in the language of optimization, combining them into mandates that shape the social welfare function retained by the benevolent planner. In the real world these mandates will necessarily be a composite function of many types of constraints put forward by the different delegations. However, because our purpose is to clarify the implications of various ethical postures, we will assume that the planner responds to a clear-cut mandate that combines one of two polar attitudes towards distribution of income in the future ('status quo' or 'catch-up'), and one of two polar attitudes towards damages falling on other countries ('dynastic' or 'universal'). This mandate may either reflect consensus amongst Parties or the dominant influence of a leading coalition. And the rest of the paper will demonstrate how the optimal provision of public goods differs according to the mandate, and that only the catch-up universal mandate defines a space for viable compromise at second period.

3 An Analytical Framework to Capture Three Ethical Issues

We build on Sandler and Smith's (1976) intertemporal version of the Bowen-Lindhal-Samuelson (BLS) model of the optimal provision of public goods. The world is divided in N countries, and there are two periods, present and future, the latter indexed by superscript f. At first period, we assume that climate change has no impact yet, and that the representative individual of the l_i inhabitants of country i allocates her income y_i between c_i the consumption of a composite private good chosen as numeraire, and a_i the expenses for GHG emissions abatement.

$$y_i = c_i + a_i \tag{1}$$

Let x (resp. x^f) be the amount of GHG emissions abated worldwide relative to business-as-usual. Using $x + x^f$ as an inversed index of GHG atmospheric concentration,[13] we denote $d_i\ (x + x^f)$ the damages of climate change incurred *per capita* in country i at second period.[14] Second-period budget equations are thus:

$$y_i^f - d_i\left(x + x^f\right) = c_i^f + a_i^f \tag{2}$$

---

[13]This index is a simplification of the dynamics of GHG accumulation in the atmosphere since it ignores the carbon cycle, but it is sufficient to capture the stock externality character of climate change.

[14]Damage functions d_i are twice differentiable. Damages are assumed positive ($d_i > 0$), decreasing in the amount of abatement ($d_i^{'} < 0$), but at a diminishing rate ($d_i^{''} < 0$). Finally, we assume that damages per capita in the absence of abatement remain lower than per capita income ($d_i(0) < y_i^f$).

Assuming that abatement expenditures are used efficiently, we denote $C(x)$ and $C^f(x^f)$ the worldwide abatement cost functions.[15] The total level of abatement at each period is given by:

$$\sum_i l_i a_i = C(x) \tag{3}$$

$$\sum_i l_i^f a_i^f = C^f(x^f) \tag{4}$$

To determine an abatement level for each country at both periods[16] the planner maximizes an intertemporal social welfare function W which encapsulates the no-redistribution constraint, attitudes towards distribution of income in the future, and attitudes towards damages falling on other countries. We discuss in turn each of these ethically related parameters.

First, the *no-redistribution constraint* imposes that individual contributions to abatement a_i and a_i^f be non negative, i.e.,

$$a_i \geq 0 \tag{5}$$

$$a_i^f \geq 0 \tag{6}$$

It also imposes that Negishi weights be attached to individual utilities, so as to force the planner to consider the initial distribution of income as optimal—and thus to avoid that climate policies be the occasion for income redistribution across countries. Let U_i and U_i^f denote the utility of consumption[17] of the present and future

[15]Functions C and C^f are twice differentiable, and such that $C > 0$, $C' > 0$, $C'' > 0$, and $C(0) = C'(0) = 0$ (same assumptions for C^f). One can derive aggregate abatement cost functions as follows. Let x_i be the national abatement levels relative to business-as-usual, and let $C_i(x_i)$ be the national abatement cost functions. The aggregate abatement cost function $C(x)$ is defined as: $C(x) = \mathrm{Min}_{x_i}\left\{\sum_i C_i(x_i) \middle| \sum_i x_i = x\right\}$. This is as if individual payments for mitigation were aggregated into a fund that would reduce emissions where it is the cheapest to do so. Baseline emissions and abatement costs in region i are independent from abatement in other regions. Thus, there are no leakage across regions in our model (see Burniaux and Oliveira Martins 2016, for a discussion of this issue).

[16]To ensure the existence of a solution, we also need that damages be reducible to zero if mitigation expenditures are high enough. Technically, let $\bar{x}$ be the level of abatement achieved if all available resources (short of maximum damages) were allocated to mitigation, i.e., $\bar{x} = C^{-1}\left(\sum_i l_i y_i + C^{f-1}\left(\sum_i l_i^f (a_i^f - d_i(0))\right)\right)$. We assume that for all $x \geq \bar{x}$ and all regions, $d_i(x) = 0$.

[17]Guesnerie (2004), Heal (2007) and Sterner and Persson (2008) show the importance of including a preference for the environment as an argument of the utility function to carry out a cost-benefit analysis of climate policies. In our model this inclusion is made indirectly by subtracting damages from total households' consumption.

representative individual of country i respectively.[18] The first-period component of W can be written $\sum_i \alpha_i l_i U_i$ with α_i the Negishi weights attached to individual utilities. These weights are defined by

$$\alpha_i = \frac{\alpha}{U_i'(y_i)} \tag{7}$$

With

$$\alpha = \left(\sum_i \frac{l_i}{U_i'(y_i)}\right)^{-1} \quad \left(\text{so that } \sum_i \alpha_i = 1\right) \tag{8}$$

Second, *attitudes towards the future distribution of income* can be captured via weights attached to individual utility functions at second period. Measuring relative income across countries is not easy, but the Negishi weights α_i can be reasonably derived from observable data at first period. This is not the case at second period. In addition to the intrinsic uncertainty surrounding countries growth rates, there is political uncertainty about which balancing will prevail amongst two polar alternatives:

- Weights based on first-period incomes: This option translates a configuration in which developed countries succeed in imposing on the benevolent planner that the balance of power stemming from the current distribution of income be protected over time. In this *status quo* mandate (hereafter S), the second-period component of W is $\sum_i \alpha_i l_i^f U_i^f$, where α_i are the *first-period* Negishi weights.
- Weights aligned on expected second-period incomes: In this option, developed countries are not influential enough to impose a status quo mandate on the planner or accept that developing countries will eventually catch up. But they still veto deals that would result in any wealth transfer *relative to* the future distribution of income resulting from the catch-up process (Eq. 6). In this *catch-up* mandate (hereafter C), the second-period component of W is $\sum_i \beta_i l_i^f U_i^f$, where β_i are the Negishi weights associated with the expected distribution of income at second period y_i^f[19, 20]:

$$\beta_i = \frac{\beta}{U_i^{f'}\left(y_i^f\right)} \tag{9}$$

[18] U_i and U_i^f are twice differentiable in all variables, with $U_i^{'} > 0$, $U_i^{''} < 0$, and $\frac{\partial U_i^f}{\partial d_j} < 0$.

[19] We assume here that there is common agreement about the future distribution of income.

[20] The Negishi weights β_i could be calibrated on the net per capita income at second period $y_i^f - d_i$, i.e., after accounting for the impacts of climate change. We do not retain this option to clearly distinguish between uncertainty about economic growth in the absence of climate change and uncertainty about climate change damages.

Table 1 Intertemporal welfare functions in the four planner's mandates

	Catch-up (C)	Status quo (S)
Dynastic (D)	$W_{CD}=\sum_i \alpha_i l_i U_i(c_i)+\varphi\sum_i \beta_i l_i^f U_i^f(c_i^f)$	$W_{SD}=\sum_i \alpha_i l_i U_i(c_i)+\varphi\sum_i \alpha_i l_i^f U_i^f(c_i^f)$
Universal (U)	$W_{CU}=\sum_i \alpha_i l_i U_i(c_i)+\varphi\sum_i \beta_i l_i^f U_i^f(c_i^f,d_{j\neq i})$	$W_{SU}=\sum_i \alpha_i l_i U_i(c_i)+\varphi\sum_i \alpha_i l_i^f U_i^f(c_i^f,d_{j\neq i})$

with

$$\beta=\left(\sum_i \frac{l_i^f}{U_i^{f'}\left(y_i^f\right)}\right)^{-1} \quad \left(\text{so that } \sum_i \beta_i=1\right) \tag{10}$$

Third, *attitudes towards damages falling on other countries* can be translated is the very arguments of the utility functions. Assuming that Parties share common beliefs about regional damages (Eq. 2), a 'dynastic solidarity' ethics (hereafter D) will lead Parties to include only their own descendant's consumption into the utility functions U_i^f and make it dependent upon domestic damages d_i only (via the budget Eq. 2). In a 'universal solidarity' ethic (hereafter U), on the other hand, Parties will include part of the damages falling on others into U_i^f[21]: U_i^f will be a function of both domestic *and* foreign damages d_i and d_j.[22]

This leads to four logical combinations: catch-up dynastic (CD), status quo dynastic (SD), catch-up universal (CU) and status quo universal (SU). With φ a pure time preference common to all Parties the four mandates finally write as indicated in Table 1.

In this model, climate negotiations boil down to a one-shot decision in which abatement expenditures at first and second period are decided simultaneously. This is arguably at odds with the sequential nature of real climate negotiations. However, we retain this "once for all" decision model for simplicity's sake and because it captures three key features of the real-world negotiation process.[23] First, Parties are now discussing medium-term commitments over the next two or three decades, not just commitments over the next 5 years. Second, governments will not sign agreements that have blatantly detrimental implications for their country in the medium and long

[21]Future utility functions are unobservable. We take the position that functions U_i^f represent Parties' views about their descendants' utilities.

[22]An alternative framework is possible under the ethical condition that individual agents have sustainable preferences taking into account long-term future as well. Chichilnisky (2016). demonstrates that under limited arbitrage, such preferences then lead to a sustainable market equilibrium.

[23]Because the main focus of our paper is on the relationships between *inter* and *intra* generational distributions, we adopt a two-period model (as opposed to a model with an infinite number of periods), which leaves aside critical discussions about how to include long-term sustainability concerns within intergenerational social welfare functions. The latter debate is addressed in this volume by Asheim et al. (2016), Figuières and Tidball (2016), and Lauwers (2016).

term. Third, the agreement has to incorporate some rigidity to minimize strategic games between negotiation periods and to preserve the dynamic efficiency of the regime. For example, changing rules too drastically or too often in a carbon trading system makes it difficult for agents to form expectations and might lead them to refrain from trading (OECD 1993).

4 Burden Sharing at First Period: A Simple Rule of Thumb?

At first period, all programs yield the same solution (Appendix 1): abatement expenditures should be allocated so as to equate weighted marginal utilities of consumption (WMUs). This comes back to the seminal result of Chichilnisky (1994), developed by Chichilnisky and Heal (1994), Chichilnisky et al. (2000), and more recently in Sheeran (2006). This is, *mutatis mutandis*, the BLS condition for the provision of a public good.

$$a_1 U_1^{'}(y_1 - a_1) = \cdots = a_N U_N^{'}(y_N - a_N) \tag{11}$$

Since *before-abatement* WMUs are equal by virtue of the Negishi weights, the optimal distribution of abatement costs is such that WMUs are decreased by the same amount:

$$a_1 U_1^{'}(y_1) - a_1 U_1^{'}(y_1 - a_1) = \cdots = a_N U_N^{'}(y_N) - a_N U_N^{'}(y_N - a_N) \tag{12}$$

Assuming contributions a_i remain small relative to initial revenues,[24] Eq. (12) can be approximated by:

$$-\frac{U_1^{''}}{U_1^{'}}(y_1)a_1 = \cdots = -\frac{U_N^{''}}{U_N^{'}}(y_N)a_N \tag{13}$$

For utility functions with decreasing absolute risk aversion, the optimal a_i are thus increasing with income. They are strictly proportional to it for logarithmic functions.

$$\frac{a_1}{y_1} = \cdots = \frac{a_N}{y_N} \tag{14}$$

Thus, in all mandates, the optimal distribution of abatement expenditures at first period can be encapsulated in a simple rule of thumb, i.e., mitigation costs in proportion of per capita income. This rule can be viewed as a politically astute

[24]Contributions a_i can be considered 'small' in mathematical terms if they are less than 5 % of y_i—which corresponds to an extraordinary large mitigation effort.

translation of the common but differentiated responsibilities principle of the UNFCCC.[25, 26]

It provides a basis for differentiating coordinated domestic carbon taxes (Cooper 2000) or for compensating for the uneven income effects of a unique carbon price set by a cap-and-trade regime.[27] Moreover, difficulties in comparing real incomes notwithstanding, this rule can be computed through observable parameters and independently of the optimal level of public goods $x + x^f$. But this rule does not lead to disconnecting negotiations on burden sharing from the setting of climate objectives. As we will see later indeed, the optimal level of mitigation at first period depends on the distribution of income at second period, which in turns determines the aggregate value of the future distribution of mitigation and damages.

5 Second Period: Troubling Consequences of Ethical Attitudes

At second period, the equalization of WMUs after abatement now applies to the *total* climate change bill, i.e., abatement expenditures *plus* residual damages, and it yields unexpected results. We first analyze the dynastic mandates, starting with the CD variant. Then we examine the extent to which universal solidarity mandates (CU and SU) overcome the problems identified with the dynastic mandates.

5.1 Catch-Up—Dynastic Mandate: The Winners-Losers Dilemma

In a CD mandate, the key result is that the equalization of after-abatement WMUs may not be possible under the constraint of non-negative abatement expenditures (Eq. 6).

---

[25]With a logarithmic utility function, each European should contribute 15–76 times as much as the average Indian depending on the whether purchasing power parities or current exchange rates parities are used to compare real income levels (2000 Gross National Income, World Bank 2004).

[26]For diplomatic reasons (see the last-minute statement by the G77 and China mentioned in the introduction), negotiators going to Kyoto had accepted the idea that only developed countries would take commitments. As a result, the Kyoto Protocol follows the UNFCCC (see footnote 2) and only embodies a very crude differentiation, i.e., efforts in the North, none in the South, leading to an implicit price of carbon of zero for developing countries, and the critical issue of differentiation remains unaddressed (Chichilnisky 2016).

[27]In a cap-and-trade regime, the planner can use the initial allocation of emission quotas to offset the welfare losses that might result from a uniform price of emissions allowances. In a tax system, a distinction has to be made between the industrial sector, where taxes must be equated to avoid distorting international competition, and the domestic sector, where taxes can be differentiated (Hourcade and Gilotte 2000). On the price versus quantity debate, see also in this volume Karp and Zhang (2016).

Let ψ_i be the set of shadow prices associated with inequalities (6) in each region. The general second-period equilibrium can be written as in (15) and (16) below (see Appendix 1). If all ψ_i are zero, then all second-period abatement expenditures are non-negative (all countries abate). But if one of them is positive, then that particular country does not abate. The constraint is binding and, in this corner solution, WMUs are not equated.

$$\begin{aligned} &\beta_1 U_1^{f'}(y_1^f - d_1(x + x^f) - a_1^f) \\ &\quad - \psi_1 = \cdots = \beta_N U_N^{f'}(y_N^f - d_N(x + x^f) - a_N^f) - \psi_N \end{aligned} \tag{15}$$

with

$$\begin{cases} \psi_i = 0 & \text{if} \quad a_i^f > 0 \\ \psi_i > 0 & \text{if} \quad a_i^f = 0 \end{cases} \tag{16}$$

A positive shadow price ψ_i characterize countries that suffer from damages so high that, even without contributing to mitigation at all, they face higher welfare losses than the others.

To show the plausibility of this configuration let us consider two regions, North (N) and South (S), with identical (logarithmic) utility functions, but with per capita income in N twenty-three times higher than in S at first period.[28] Let us also assume that a higher per capita income growth in S (3 % per year vs. 2.5 %) reduces this gap to eighteen times in 2050—the beginning of the second period. We posit one aggregate damage function for the World (see calibration in Appendix 2) and compute the optimal abatement expenditures, the optimal residual damages, and the optimal total climate change bills according to (15) and (16) for various *distributions* of those aggregate damages between S and N (Table 2).

The optimal distribution of abatement expenditures remains proportional to per capita income if and only if damages per capita represent the same proportion of income in N and S (scenario a in Table 2). If, as suggested by IPCC (2007), damages rip off a higher share of income in S than in N, the representative individual of N will devote a higher share of its income to abatement so that the total climate bills still represent the same proportion of the revenues in both regions. In scenario b (Table 2), for climate change bills representing the same share of incomes of N and S (1.69 %), N spends 0.75 % of its second-period revenue for mitigation, against 0.24 % for S.

But when damages become very large in S relative to N (in scenario c, for example) S faces a higher climate bill than N (1.81 % vs. 1.64 %) even with zero abatement expenditures. This corner solution is far from implausible for Africa, small Island-States, Central America or Bangladesh, and it puts the no-redistribution constraint to a serious test since, to equate the WMUs, one would

[28]Twenty-three is the 2002 gap in per capita gross national income between high-income countries, and low- and middle-income countries (World Bank 2004).

Table 2 Optimal mitigation policy at second-period (abatement expenditures, residual damages and total climate bill): catch-up—dynastic mandate

Distribution of damages			Optimal mitigation policy					
	Maximum damage N (w/o mitigation) (%)	Maximum damage S (w/o mitigation) (%)	Abatement expenditures N (a_N^f) (%)	Residual damages N (d_N) (%)	Total climate bill N $(a_N^f + d_N)$ (%)	Abatement expenditures S (a_s^f) (%)	Residual damages S (d_s) (%) es	Total climate bill S $a_s^f + d_s$ (%)
a	3	3	0.60	1.09	1.69	0.60	1.09	1.69
b	2.59	4	0.75	0.94	1.69	0.24	1.45	1.69
c	2.17	5	0.84	0.79	1.64	0.00	1.81	1.81
d	1.76	6	0.85	0.64	1.49	0.00	2.17	2.17
e	1.34	7	0.86	0.48	1.34	0.00	2.53	2.53
f	0.93	8	0.86	0.33	1.19	0.00	2.88	2.88
g	0.52	9	0.87	0.19	1.06	0.00	3.22	3.22
h	0.10	10	0.88	0.04	0.92	0.00	3.57	3.57

All figures are a percentage of second-period income per capita

Source Authors' calculation. See Appendix 2 for calibration details

need to compensate these countries for the 'excess' damages they withstand through negative abatement expenditures (i.e., direct or indirect transfers). In fact, this calls for a more lenient interpretation of the no-redistribution constraint, which applies to the *sum* of damages *and* abatement expenditures:

$$a_t^f + d_i \geq 0 \tag{17}$$

This alternative formulation would lead to higher aggregate welfare, but it contradicts the dynastic solidarity rationale. It would imply that countries likely to benefit from or not to be affected much by global warming would commit themselves to compensating countries that are severely hurt. Moreover, the amount of these compensations is almost unpredictable given the level of uncertainty surrounding climate projections at regional level (IPCC 2007).[29] Historically, uncertainties about mitigation costs have led many countries to hesitate before accepting the Kyoto Protocol (Hourcade and Ghersi 2002), and have even led some to walk away entirely. And yet, countries have some control over abatement costs, while they have basically no control over impacts. It is thus unlikely that countries will accept the risk of paying significant compensations for 'excess' damages without changing their overall attitude towards climate risks—which will be envisaged in the universal solidarity mandates.

5.2 *Status Quo—Dynastic: A Paradoxical but Meaningful Outcome*

This SD mandate faces the same problem of uneven geographical distribution of damages than the CD mandate, but, in addition, because the weights attached to individual utility functions are the Negishi weights of the *first-period* income levels, it leads to a paradoxical outcome.

The first-order condition is given in Eq. (18) below.

$$\begin{aligned} &\alpha_1 U_1^{f'}(y_1^f - d_1(x + x^f) - a_1^f) \\ &\quad - \psi_1 = \cdots = \alpha_N U_N^{f'}(y_N^f - d_N(x + x^f) - a_N^f) - \psi_N \end{aligned} \tag{18}$$

[29]Projections of future world average temperatures by global circulation models have a far higher degree of confidence than projections at local scale. And uncertainty grows by another order of magnitude when local physical impacts are translated into economic damages: Western Europe may experience either warming by 2 °C or more or cooling by several degrees depending on the evolution of the North-Atlantic thermohaline circulation. Russia is often regarded as a potential winner of global warming. But the melting of the permafrost, or dryness in the South of the country might put it among the losers.

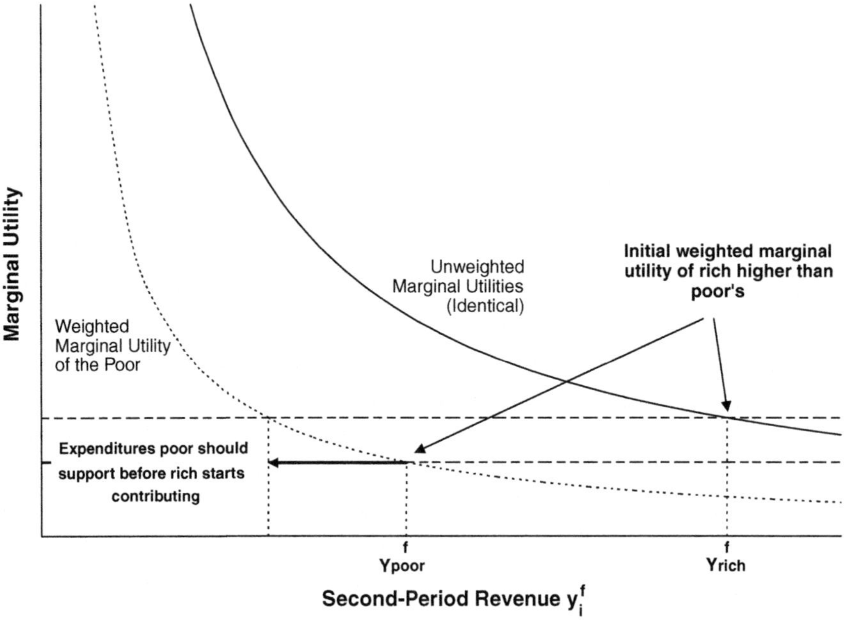

Fig. 1 Optimal abatement levels at second-period for two regions differing only by income in SD mandates

Since the weights α_i are constructed so that the distribution of *first-period* income is welfare maximizing, the vector of *second-period* income y_i^f has no reason to be welfare maximizing. In most instances, in fact, it is not, and before abatement WMUs are different. It is then optimal to charge all abatement expenditures to the country with the lowest WMU, so as to decrease its income and increase its WMU. The planner shall do so until both WMUs are equal (Fig. 1).

This paradox occurs with fairly conservative assumptions about growth rates. Let us assume that the developing world grows by 3 % annually over the next decade from now on, while rich countries grow by only 2.5 %. Then the developing world is about 5 % richer in 2019 than it would have been with a 2.5 % growth rate. This 'extra' growth amounts to about 1 % of the world GDP in 2019 and a planner following the SD mandate would allocate all the mitigation costs on the developing world, as long they do not exceed one percent of the World GDP in 2019.

This paradox is unlikely to disappear when accounting for the damages of climate change. For example, with a per capita growth in the developing world again half a point higher than in developed countries over the next 50 years, per capita GDP in 2059 is 27 % higher in the developing world than it would have been had both rates been equal. Regional damages apt to outweigh this 'extra-growth' go beyond the most pessimistic climate change scenarios.

This paradox is obviously a theoretical artifact due to the one-shot character of the model. But it helps demonstrating the consequences of ethical attitude behind the status-quo mandate, i.e., the extension of the grandfathering principle to future generations. In Eq. (18) the future inhabitants of the rich countries are de facto endowed with emissions rights based on those of their predecessors—an allocation of rights consistent with the claim by some countries that "lifestyles are not negotiable."[30] This deal is obviously not acceptable for developing countries given its so blatantly detrimental to them, and given the uncertainty and political costs of reversing any diplomatic *fait accompli*.

The above analysis thus shows that the deals resulting from mandates supported by a dynastic solidarity ethic run a risk of rejection or, if accepted, of dynamic instability as some Parties will have strong incentives to defect or to try and prevent entry into force. The CD mandates confront the reluctance of countries to commit themselves ex ante to compensating for unpredictable damages falling on other countries ex post. In addition, the SD variant is unacceptable by poor countries because it leads to a second-period equilibrium where they are asked to support most of the mitigation effort.[31]

5.3 Universal Solidarity: A Prerequisite but No Way Out?

The fact that dynastic solidarity translates the behavior of "cold monsters States" does not imply that, symmetrically, universal solidarity can derive only from a utopian *universal bonhomie*, or from Schelling's thesis (1995) that, beyond some horizon, all future individuals are indistinguishable. In fact, a universal solidarity ethic can be based on pure self-interest as well. Faced with uncertainties regarding the location of damages, Parties might refrain from indulging themselves among the 'winners' of global warming.[32] Similarly, the risks of global spillovers from local shocks (including accelerated migrations) may lead Parties to consider that any damage of climate change anywhere in the World will ultimately affect everyone's welfare and security. This is basically the key message of the Stern review (2007). In the strict etymological meaning of the term, one can show solidarity with somebody either for reasons of benevolence or because our interests stick together. In this light, even a SU mandate is plausible if it is interpreted as translating a clear selfish attitude: Conserving the current balance of powers, but taking possible negative spillovers into account.

[30]This warning, often attributed to George H. Bush Sr., would probably be endorsed, albeit in more diplomatic terms, in many quarters of the developed world.

[31]It is precisely the lack of clarification that future quotas would not be allocated through grandfathering that led the G77 to veto the rules governing carbon trading in the penultimate day of the Kyoto Conference.

[32]A situation analogous to the "veil of ignorance" of Rawls (1971).

5.3.1 A Burden-Sharing More Robust to Uncertainty?

In U mandates, both domestic and foreign damages enter into the utility functions, and this automatically makes the allocation of abatement expenditures less sensitive to the geographical distribution of damages. The question is whether that effect is sufficient to palliate the limitations of the dynastic mandates.

In the SU mandate the optimal distribution of abatement expenditures is governed by Eq. (19[33]) below, where the shadow prices associated with constraint (6) are again governed by Eq. (16).

$$\begin{aligned}\alpha_1 U_1^{f'}\left(y_1^f - d_1\left(x+x^f\right) - a_1^f, d_2, \ldots, d_N\right) - \psi_1 = \cdots \\ = \alpha_N U_N^{f'}\left(y_N^f - d_N\left(x+x^f\right) - a_N^f, d_1, \ldots, d_{N-1}\right) - \psi_N\end{aligned} \tag{19}$$

That same set of equations is valid for the CU mandate, with the only difference that coefficients β_i replace coefficients α_i:

$$\begin{aligned}\beta_1 U_1^{f'}\left(y_1^f - d_1\left(x+x^f\right) - a_1^f, d_2, \ldots, d_N\right) - \psi_1 \\ = \cdots = \beta_N U_N^{f'}\left(y_N^f - d_N\left(x+x^f\right) - a_N^f, d_1, \ldots, d_{N-1}\right) - \psi_N\end{aligned} \tag{20}$$

The comparison between U mandates and their D cousins depends on whether a universal solidarity ethics changes the perceived aggregate damages. Recognizing the existence of propagation effects across countries is consistent with the idea of a cost-multiplying effect associated with this propagation (Hallegatte et al. 2007). However, we will reason at constant aggregate damages in order to avoid controversies about the magnitude of this multiplying effect and will focus on the sole impact of changing the weights in the social welfare function. Under this assumption, what we know of the magnitude of damages is again too low to correct the likely differences in WMUs before abatement in the status-quo version of the U mandate. The SU mandate is thus likely to face the same paradox as its SD cousin. Conversely because before-abatement WMUs are equal at second period in catch-up mandates, their universal solidarity version may overcome the deadlocks of their dynastic version, provided the weight of other countries in one's utility function is large enough. This can be illustrated through a simple numerical example with the same regions and assumptions as in Sect. 5.2 (see Appendix 2). Second-period utility functions now depend both on domestic consumption and on

---

[33]The self-interest justifications for the universal mandate imposes that $\frac{\partial^2 U_i^f}{\partial c \partial d_j} \neq 0$, so that damages abroad impact on marginal utilities of consumption (directly or through compensation or security expenditures) and not only on utility levels.

aggregate damages (21), with d_i^{max} the damages that hurt region i if there were no abatement.

$$U = \ln(c)\left(1 - 0.01\frac{d_S + d_N}{d_S^{max} + d_N^{max}}\right) \tag{21}$$

Keeping aggregate damages constant, we compute the optimal mix of abatement expenditures and residual damages under several assumptions about the distribution of these damages between S and N. In Table 3, we see that a corner solution is obtained only for very uneven distribution of damages (maximum damages of 10 % in S against 0.1 % in N) in the CU mandate, whereas this corner solution is reached more easily in the CD mandate (maximum damages of 5 % in S against 2.2 % in N).

That a universal solidarity ethic enhances the chances of following a BLS-like rule at both periods in case of catch-up mandate reinforces its acceptability at first period. Its very logic indeed minimizes the consequences of mispredictions in terms of dynamic consistency of the initial agreement since part of the perceived damages is independent from where damages fall.

5.3.2 A Higher Provision of Global Public Good

In addition to easing the tensions over burden sharing the CU mandate results into a higher optimal level of abatement. The planner will indeed have to consider a higher total value of damages, even without considering that the propagation effects will increase the total level of impacts, for the reason that the rich low-impacted countries will take into account part of the damages that fall on poor countries.

To see the difference between the U and D mandates in terms of provision of global public goods, let us come back to the CD mandate and assume an interior solution. The optimal abatement levels are given by Eqs. (23) and (22) below, where ρ is the average consumption discount factor deriving from the utility discount factor φ, the economic growth rate and the shape of the utility functions (see Appendix 1 for a detailed expression of ρ).

$$C^{f'}(x^f) = -\sum_i l_i^f d_i'(x + x^f) \tag{22}$$

$$C'(x) = \rho C^{f'}(x^f) \tag{23}$$

This is the standard result that the climate should be protected up to the point where the marginal costs of mitigation equal the discounted sum of the marginal benefits in terms of avoided damages. The aggregate marginal damage function $-\sum_i l_i^f d_i'$ is the only determinant of the level of abatement, and the weights α_i and β_i do not play

Table 3 Optimal mitigation policy at second-period (abatement expenditures, residual damages and total climate bill): catch-up—universal mandate

Scenario			Optimal mitigation policy					
	Damage maximum North (%)	Damage maximum South (%)	Abatement expenditures N (a_N) (%)	Residual damages N (d_N) (%)	Total climate bill N ($a_N + d_N$) (%)	Abatement expenditures S (a_s) (%)	Residual damages S (d_S) (%)	Total climate bill S ($a_S + d_S$) (%)
a	3	3	1.28	0.59	1.87	1.28	0.59	1.87
b	2.59	4	1.36	0.51	1.87	1.09	0.78	1.87
c	2.17	5	1.44	0.43	1.87	0.89	0.98	1.87
d	1.76	6	1.52	0.35	1.87	0.69	1.18	1.87
e	1.34	7	1.61	0.26	1.87	0.49	1.38	1.87
f	0.93	8	1.69	0.18	1.87	0.30	1.57	1.87
g	0.52	9	1.77	0.10	1.87	0.10	1.77	1.87
h	0.10	10	1.81	0.02	1.83	0.00	1.96	1.96

All figures are a percentage of second-period income per capita
Source Authors' calculation. See Appendix 2 for calibration details

any role.[34] This allows for two types of separability: separability between the level of action and the distribution of abatement expenditures across countries, and separability between the level of action and the distribution of climate change impacts across countries.[35]

But this separability is lost in a CU mandate in which, again assuming an interior solution, the optimal level of abatement is given by Eqs. 24 and 25 below (see Appendix 1 for general forms). The additional term in the right-hand side of Eq. 24 translates the fact that country *i* takes into account damages falling on other countries as well as damages falling on its own citizens. Since marginal damages and partial derivatives of utility relative to damages are both negative, their product is positive. Marginal abatement costs are thus higher than they would be in the CD mandate, leading to a higher level of abatement at optimum. The extent to which abatement is higher is no longer independent from the distribution of income at second period.

$$C^{f'}(x^f) = \sum_i l_i^f \left[-d_i'(x + x^f) + \rho \sum_{j \neq i} \frac{\frac{\partial U_i^f}{\partial d_j}}{U_i^{f'}} d_j'(x + x^f) \right] \tag{24}$$

$$C'(x) = \rho C^{f'}(x^f) \tag{25}$$

For example, assuming equal damages per capita in S and N (scenario a in Table 2), the optimal level of abatement at second period is 29 % below baseline in the CD case. With utility functions of the form 21), the optimal level of abatement is 41 % in the CU case (scenario a in Table 3)—a considerable 44 % increase. A non negligible side-effect of higher levels of abatement overall is that damages in S never exceed 2 % of income, while they can reach 3.6 % in the CD case.

6 Conclusion

Although greatly influenced by political vagaries, the language of climate negotiations remains framed by an economic wisdom in which (a) the optimal level of mitigation results from priors about climate change damages and from the value of

[34]The reason is that the marginal utility of the consumption of the public goods is the marginal utility of consumption times the avoided damage. Since the Negishi weights are proportional to the inverse of the marginal utility of consumption, they cancel out, and only the sum of avoided damages remains.

[35]If the equilibrium is a corner solution, the optimal level of abatement *increases* relative to the interior solutions because some countries have a higher WMU than the others. In Table 1, for example, the optimal abatement level in scenario h—in which all damages fall on S—is 1.4 % higher than in scenario a where N and S are equally impacted.

the pure time preference, (b) intragenerational equity can be secured, for whatever mitigation objective, via appropriate transfers, and (c) it is neither legitimate nor realistic to transform climate policies into a tool for large-scale international redistribution of wealth.

This paper suggests that this framing misses two dimensions that prove critical when the climate affair is examined from a dynamic perspective. The first dimension is whether negotiations are conducted in function of the *current or future distribution of economic wealth*. The second is *whether individuals show solidarity to their own descendants only or to all the future human beings*.

The first implication of capturing these two dimensions is theoretical. The weights attached to individual utility functions define the baseline from which the no-redistribution constraint applies and affect the aggregate discounted value of future consumption and the optimal level of mitigation. Thus, (a) the pure time preference (Stern 2007; Nordhaus 2007; Heal 2007; Sterner and Persson 2008; Hourcade et al. 2009) is no longer the sole determinant of the trade-off between present and future aggregate consumption. And (b) intra- and intergenerational equity cannot be treated separately, since assumptions about future income distribution and about the scope of intergenerational solidarity impact on the value of future damages and on the optimal provisions of abatement.

The second implication concerns the conditions for a viable international climate architecture. Of the four mandates combining *status quo* or *catch-up* attitudes towards future distribution and income, and *dynastic* or *universal* solidarity towards future individuals, we show that three are dynamically inconsistent and thus self-defeating. They either result in putting the entire burden on developing countries (status quo/dynastic and status quo/universal) or prove difficult to reach because of the uncertainty about the location of damages (catch-up dynastic).

The *catch-up universal* mandate yields a stable outcome because it recognizes potential changes in world income distribution and reduces the role of uncertainty in the geographical location of impacts and in their transboundary propagation. It has two main characteristics: first, a higher optimal level of public goods relative to the others[36]; second, an optimal allocation of mitigation expenditures at first period in general proportional to per capita income[37]—which might translate of the "common but differentiated responsibilities" principle of the UNFCCC.

Although we have stuck here to the dominant view that mitigation constitutes a net burden at first period, our intuition is that these insights hold even in the

[36]Even though in a static one-period framework, a stronger preference for equity may lead to less action on climate change because low-income populations have a lower marginal utility of the environment (Tol), here, whatever the selfish or altruistic character of its motivation, a universal solidarity attitude that includes concerns about the situation of poor populations enhances the need for action.

[37]For logarithmic or power utility functions. With more general utility functions, optimal efforts are typically increasing with income.

perspective of Pareto-improving policies.[38] The political will of paying the transaction costs of deploying these policies is not unrelated to attitudes regarding solidarity in the face of climate damages or regarding the evolving discrepancies in development levels. Ultimately, the point is that negotiating the allocation of a global cap in GHGs emissions as a pure 'carbon price plus transfers' exercise, in isolation of broader issues such as the rebalancing of economic power across countries over the coming decades or the enhancement of global security, may lead to a systematic impasse as just demonstrated in Copenhagen. It is critical indeed to reflect upon the conditions under which a *status-quo universal* mandate can be adopted. We hope that further developments of the theoretical line just opened in this paper will be fruitful to scrutinize in depth these conditions and how they might open a negotiation space more conducive to a positive outcome.

Appendix 1: Model Resolution

We solve a general version of the planner's problem (26)—in which coefficients χ_i summarize both S and C mandates (27)—under constraints (3), (4), (5), and (6).

$$\mathrm{Max}_{\{a_i, a_i^f\}} \sum_i \alpha_i l_i U_i(y_i - a_i) + \varphi \sum_i \chi_i l_i^f U_i^f\left(y_i^f - a_i^f - d_i\left(x + x^f\right), d_{j \neq i}\right) \tag{26}$$

With

$$\chi_i = \begin{cases} \alpha_i = \frac{\alpha}{\mathrm{U}_{\mathrm{i}}^{'}(y_i)} \text{ in S mandates, with } \alpha = \left(\sum_i \frac{l_i}{\mathrm{U}_1^{'}(y_i)}\right)^{-1} \\ \beta_i = \frac{\beta}{\mathrm{U}_{\mathrm{i}}^{'}(y_i)} \text{ in C mandates, with } \beta = \left(\sum_i \frac{l_i^f}{\mathrm{U}_{\mathrm{i}}^{\mathrm{f}'}\left(y_i^f\right)}\right)^{-1} \end{cases} \tag{27}$$

Let λ, $\varphi\,\mu$, $l_i\,\xi_i$ and $\varphi\,l_i^f\psi_i$ be the Lagrange multipliers attached to constraints (3), (4), (5), and (6) respectively. Since (5) and (6) are inequality conditions, ξ_i and ψ_i are such that:

$$\begin{cases} \xi_i = 0 \quad \text{if} \quad a_i > 0 \\ \xi_i > 0 \quad \text{if} \quad a_i = 0 \end{cases} \tag{28}$$

[38]To identify margins of freedom for strategies with no consumption loss for the current generations of rich countries and no slowdown in the take off of developing countries, one needs a multi-goods model to study how investments can be massively redirected (200–400 G$ in 2030 according to World Bank 2009) and how those that lose from this redirection may be compensated. Using a game-theory approach, Dutta and Radner (2010) provide general conditions on how such transfers could be effective.

$$\begin{cases} \psi_i = 0 & \text{if} \quad a_i^f > 0 \\ \psi_i > 0 & \text{if} \quad a_i^f = 0 \end{cases} \tag{29}$$

With these notations, the Lagrangean of the problem becomes:

$$\begin{aligned} L = & \sum_i \alpha_i l_i U_i(y_i - a_i) + \varphi \sum_i \chi_i l_i^f U_i^f \left(y_i^f - a_i^f - d_i(x + x^f), d_{j \neq i} \right) \\ & + \lambda \left[\sum_i l_i a_i - C(x) \right] + \varphi\mu \left[\sum_i l_i^f a_i^f - C^f(x^f) \right] + \sum_i l_i \xi_i a_i + \varphi \sum_i l_i^f \psi_i a_i^f \end{aligned} \tag{30}$$

The first-order condition with regard to a_i is:

$$\frac{\partial L}{\partial a_i} = 0 \Rightarrow \alpha_i U_i'(y_i - a_i) - \xi_i = \lambda \tag{31}$$

We first demonstrate that (31) has a unique solution with strictly positive abatement levels a_i. Let us assume without loss of generality that a_1 were zero. Since $\xi_1 > 0$, $\alpha_1 U_1'(y_1 - a_1) - \xi_1 = \alpha_1 U_1'(y_1) - \xi_1 < \alpha_1 U_1'(y_1)$. If another abatement level, say a_2, were strictly positive, then $\alpha_2 U_2'(y_2 - a_2) - \xi_2 = \alpha_2 U_2'(y_2 - a_2) > \alpha_2 U_2'(y_2)$ since marginal utilities are strictly decreasing functions. Yet by definition of the Negishi weights, $\alpha_1 U_1'(y_1) = \alpha_2 U_2'(y_2)$. Thus, we would have $\alpha_1 U_1'(y_1 - a_1) - \xi_1 < \alpha_2 U_2'(y_2 - a_2) - \xi_2$, contradicting Eq. (31). Thus, if one of the abatement expenditures is zero, all abatement expenditures are zero.

But if all a_i were zero, Eq. (3) and the fact that $C(0) = 0$ would imply that $x = 0$, and thus that marginal costs of abatement C' are zero. Equalization between marginal costs of abatement and marginal damages at optimum (Eq. 36) would then imply that all marginal damages $d_i'(x + x^f)$ be zero, and thus (via Eq. 33) that the marginal costs of mitigation at second period be zero, and thus that the level of abatement at second period $x^f = 0$. But this would contradicts the assumption that no abatement leads to strictly positive marginal damages of climate change. First-period abatement levels a_i are thus all strictly positive, and since marginal utility functions are strictly decreasing, they are uniquely defined.

Similarly, derivation of L with regard to a_i^f yields:

$$\frac{\partial L}{\partial a_i^f} = 0 \Rightarrow \chi_i U_i^{f'} \left(y_i^f - a_i^f - d_i(x + x^f), d_{j \neq i} \right) - \psi_i = \mu \tag{32}$$

The second part of the argument above can be replicated to demonstrate that at least one of the second-period abatement expenditures a_i^f is strictly positive. But the first part of the argument above cannot be applied as is, and thus there is no guarantee that *all* second-period abatement expenditures be strictly positive.

In C mandates, this is because marginal utilities before abatement *and after damages* $\chi_i U_i^{f'} \left(y_i^f - d_i(x + x^f) \right)$ have no reason a priori to be equal. We only know that marginal utilities of consumption before abatement *and before damages*

$\chi_i U_i^{f'}\left(y_i^f\right)$ are equal (by construction), and thus that the *aggregate climate bills* $d_i^f + d_i$ (and not just the abatement expenditures component) are all strictly positive.

In S mandates, even that weaker property does not hold because even marginal utilities of consumption before abatement and before damages $\chi_i U_i^{f'}\left(y_i^f\right)$ have no reason a priori to be equal across regions.

Derivation of L with regard to x^f yields:

$$C^{f'}(x^f) = \sum_i l_i^f \pi_i \left[-d_i^{'}(x+x^f) + \rho \sum_{j\neq i} \frac{\frac{\partial U_i^f}{\partial d_j}}{U_i^{f'}} d_j^{'}(x+x^f) \right] \tag{33}$$

With

$$\pi_i = \frac{\chi_i}{\mu} U_i^{f'}\left(y_i^f - d_i(x+x^f) - a_i^f, d_{j\neq i}\right) = \frac{\chi_i U_i^{f'}}{\chi_i U_i^{f'} - \psi_i} \tag{34}$$

Given the assumptions made about marginal damages and marginal abatement costs, (33) has a unique solution. Coefficients π_i are the ratios of $\chi_i\, U_i^{f'}$ the weighted marginal utility of consumption of country i, and of μ the weighted marginal utility of consumption *of the countries that abate at second period* (as per Eq. 32) (μ can also be interpreted as the shadow price of abatement at second period, expressed in marginal utility terms). When country i contributes to abatement at second period, these two terms are equal and $\pi_i = 1$. When country i does not contribute to abatement—either because it does not grow rapidly enough, or because domestic damages are too high—$\pi_i > 1$.

In other words, when a country has a weighted marginal utility of consumption that is too high relative to the others, not only will it not contribute to abatement, but damages falling on this country will be weighted higher than damages falling on others because they cause higher utility losses at the margin.

When all countries contribute—which, as discussed above, occurs mostly in the C mandates—(34) simplifies in (35), which is the standard BLS condition.

$$C^{f'}(x^f) = -\sum_i l_i^f d_i^{'}(x+x^f) \tag{35}$$

Finally, derivation of L with regard to x yields:

$$C^{'}(x) = \varphi \sum_i l_i^f \omega_i \left[-d_i^{'}(x+x^f) + \rho \sum_{j\neq i} \frac{\frac{\partial U_i^f}{\partial d_j}}{U_i^{f'}} d_j^{'}(x+x^f) \right] \tag{36}$$

with

$$\omega_i = \frac{\chi_i}{\lambda} U_i^{f'} = \frac{\chi_i U_i^{f'}\left(y_i^f - d_i\left(x + x^f\right) - a_i^f, d_{j \neq i}\right)}{\alpha_i U_i^{'}(y_i - a_i)} \tag{37}$$

Coefficients ω_i capture the change in weighted marginal utility of consumptions between the first and the second period. Comparing Eqs. (33) and (36), the term φ ω_i can be interpreted as the region-specific discount rate that applies to climate change damages. In C mandates, this term becomes

$$\varphi\omega_i = \varphi \frac{\beta}{\alpha} \frac{U_i^{'}(y_i)}{U_i^{'}(y_i - a_i)} \frac{U_i^{f'}\left(y_i^f - d_i\left(x + x^f\right) - a_i^f, d_{j \neq i}\right)}{U_i^{f'}\left(y_i^f\right)} \tag{38}$$

When abatement expenditures and damages remain small with regard to income, the last two terms are close to one, and regional discount factors are all equal to $\varphi \frac{\beta}{\alpha}$. Hence coefficient ρ in Eq. (23). For example, with logarithmic utility functions, $\rho = \varphi \frac{\beta}{\alpha} = \varphi \frac{1}{(1+r)^N}$ where r is the aggregate growth rate of the economy over the first period. In D mandates, on the other hand, regional discount rates become

$$\varphi\omega_i = \varphi \frac{U_i^{f'}\left(y_i^f - d_i\left(x + x^f\right) - a_i^f, d_{j \neq i}\right)}{U_i^{'}(y_i - a_i)} \tag{39}$$

which vary depending on the regional growth rate between the two periods. For example, with logarithmic utility functions, $\rho_i = \varphi \frac{1}{(1+r_i)^N}$ where r_i are the regional growth rates over the first period.

Appendix 2: Data and Modeling Framework of Numerical Experiments

The world is divided in two regions: "North" (N) comprises high-income countries, as per World Bank (2004) definition, and "South" (S) low and middle income countries. The first period is 2000–2049, and the second 2050–2099. Initial income and population data are from World Bank (2004). Average annual economic growth in N is assumed to be 2.5 % between 2000 and 2050, against 3 % in S. World population is assumed to grow by 2 billions in that period of time, all of them in S (Table 4). For simplicity's sake, we use 2000 (resp. 2050) figures as averages for period 1 (resp. 2).

Table 4 Key assumptions in numerical example

	First period (2000–2049)		Second period (2050–2099)	
	l_i (billions)	y_i (1995 USD per capita, exchange rates)	l_i^f (billions)	y_i^f (1995 USD per capita, exchange rates)
North	0.95	26,750	0.95	91,943
South	5.11	1,160	7.11	5,085

Source World Bank (2004), authors' assumptions for 2050 figures

Without abatement, World CO_2 emissions are assumed to be 513 $GtCO_2$ during the first period, and 688 $GtCO_2$ during the second, as in the IPCC IS92a scenario (IPCC 1994).

Abatement costs at first and second period are assumed quadratic with respect to total abatement levels. We also assume that mitigating all the emissions in the world economy would cost at the margin \$1,500/tC during the first period, and \$1,000/tC during the second. After easy manipulations, Eqs. (3) and (4) become:

$$x = 1482 \times \sqrt{\frac{l_N a_N + l_S a_S}{l_N y_N + l_S y_S}} \tag{40}$$

$$x^f = 4066 \times \sqrt{\frac{l_N^f a_N^f + l_S^f a_S^f}{l_N^f y_N^f + l_S^f y_S^f}} \tag{41}$$

Damages are assumed cubic with total emissions.

$$d_i(x + x^f) = y_i^f \theta_i \left(1 - \frac{x + x^f}{1201}\right)^3 \tag{42}$$

Coefficients θ_i represent the maximum damage—expressed as a share of per capita GDP—that each region may sustain because of climate change. If there were no mitigation at all, damages would be $d_i = y_i^f \theta_i$ We use several values for coefficients θ_i to represent various assumptions about the distribution of impacts of climate change across regions—keeping the aggregate maximal damage $l_S^f y_S^f \theta_S + l_N^f y_N^f \theta_N$ constant. Finally, all utility functions are assumed logarithmic, and the rate of pure time preference φ is set to 1 %.

References

Agarwal, A., & Narain, S. (1991). *Global warming in an unequal world, a case of environmental colonialism*. Delhi: Center for Science and Environment.

Aldy, J. E., & Stavins, R. (Eds.). (2007). *Architectures for agreement: Addressing global climate change in the post-Kyoto world*. Cambridge, UK: Cambridge University Press.

Ambrosi, P., Hourcade, J.-C., Hallegatte, S., Lecocq, F., Dumas, P., & Ha-Duong, M. (2003). Optimal control models and elicitation of attitudes towards climate damages. *Environmental Modeling and Assessment, 8*, 133–147.

Asheim, G. B., Mitra, T. L., & Tungodden, B. (2016). Sustainable recursive social welfare functions. *Economic Theory* (reprinted in this volume). doi:10.1007/s00199-010-0573-7.

Azar, C. (1999). Weight factors in cost-benefit analysis of climate change. *Environmental & Resource Economics, 13*, 249–268.

Burniaux, J.-M., & Oliveira Martins, J. (2016). Carbon leakages: A general equilibrium view. *Economic Theory* (reprinted in this volume). doi:10.1007/s00199-010-0579-1.

Chichilnisky, G. (1994). The abatement of carbon emissions in industrial and developing countries. In T. Jones (Ed.), *The economics of climate change* (pp. 159–170). Paris: OECD.

Chichilnisky, G. (2016). Sustainable markets with short sales. *Economic Theory* (reprinted in this volume).

Chichilnisky, G., & Sheeran, K. A. (2009). *Saving Kyoto*. London: New Holland Publishers Ltd.

Chichilnisky, G., & Heal, G. (1994). Who should abate carbon emissions: An international viewpoint. *Economics Letters*, 443–449.

Chichilnisky, G., Heal, G., & Starrett, D. (2000). Equity and efficiency in environmental markets: Global trade in carbon emissions. In G. Chichilnisky & G. Heal (Eds.), *Environmental markets* (pp. 46–67). New York: Colombia University Press.

Chipman, J. S., & Guoqiang, T. (2016). Detrimental externalities, pollution rights, and the "Coase Theorem". *Economic Theory* (reprinted in this volume).

Cooper, R. J. (2000). International approaches to global climate change. *The World Bank Research Observer, 15*, 145–172.

Den Elzen, M., Berk, M., Schaeffer, M., Olivier, J., Hendriks, C., & Metz, B. (1999). The Brazilian proposal and other options for international burden sharing: An evaluation of methodological and policy aspects using the FAIR Model. RIVM Report No. 728001011, Bilthoven.

Dutta, P. K., & Radner, R. (2010). Capital growth in a global warming model: Will China and India sign a climate treaty? *Economic Theory* (2010, this issue).

Figuières, C., & Tidball, M. (2016). Sustainable exploitation of a natural resource: A satisfying use of Chichilnisky criterion. *Economic Theory* (reprinted in this volume). doi:10.1007/s00199-010-0579-1.

Fisher, B. S., Nakicenovic, N., Alfsen, K., Corfee Morlot, J., de la Chesnaye, F., Hourcade, J.-C., et al. (2007). Issues related to mitigation in the long term context. In: B. Metz, O. R. Davidson, P. R. Bosch, R. Dave & L. A. Meyer (Eds.), *Climate Change 2007: Mitigation. Contribution of Working Group III to the Fourth Assessment Report of the Intergovernmental Panel on Climate Change* (pp. 169–250). Cambridge: Cambridge University Press.

Foley, D. K. (2009). The economic fundamentals of global warming. In J. M. Harris & N. R. Goodwin (Eds.), *Twenty-first century macroeconomics—Responding to the climate challenge*. Northampton: Edward Elgar Publishing.

Godard, O. (2000). Sur l'éthique, l'environnement et l'économie. La justification en question. Cahier du Laboratoire d'Econométrie de l'Ecole Polytechnique. Paris: Ecole Polytechnique.

Goldemberg, J., Squitieri, R., Stiglitz, J., Amano, A., Shaoxiong, X., & Saha, R. (1996). Introduction: Scope of the assessment. In J. P. Bruce, H. Lee, & E. F. Haites (Eds.), *Climate Change 1995: Economic and Social Dimensions of Climate Change, Contribution of Working Group III to the Second Assessment Report of the Intergovernmental Panel on Climate Change* (pp. 125–144). Cambridge: Cambridge University Press.

Grubb, M., Vrolijk, C., & Brack, D. (1999). *The Kyoto protocol, a guide and an assessment*. London: Royal Institute of International Affairs.

Gupta, S., Tirpak, D. A., Burger, N., Gupta, J., Höhne, N., Boncheva, A. I., et al. (2007). Policies, instruments and co-operative arrangements. In: B. Metz, O. R. Davidson, P. R. Bosch, R. Dave & L. A. Meyer (Eds.), *Climate Change 2007: Mitigation. Contribution of Working Group III to the Fourth Assessment Report of the Intergovernmental Panel on Climate Change* (pp. 745–808). Cambridge: Cambridge University Press.

Guesnerie, R. (2004). Calcul économique et développement durable. *Rev Econ Paris, 55*(3), 363–382.
Hallegatte, S., Hourcade, J.-C., & Dumas, P. (2007). Why economic dynamics matter in assessing climate change damages: Illustration on extreme events. *Ecological Economics, 62*(2), 330–340.
Heal, G. (2007). Discounting: A review of the basic economics. *The University of Chicago Law Review, 74*, 59–77.
Helm, C., & Simonis, U. E. (2001). Distributive justice in international environmental policy: Axiomatic foundation and exemplary formulation. *Environmental Values, 10*(1), 5–18.
Hourcade, J.-C., Ambrosi, P., & Dumas, P. (2009). Beyond the stern review: Lessons from a risky venture at the limits of the cost-benefit analysis. *Ecological Economics, 68*(10), 2479–2484.
Hourcade, G. F. (2002). The economics of a lost deal: Kyoto—The Hague—Marrakesh. *The Energy Journal, 23*(3), 1–26.
Hourcade, G. L. (2000). Differentiated or uniform international carbon taxes: Theoretical evidences and procedural. In: G. Chichilnisky & G. Heal (Eds.), *Environmental markets* (pp. 135–168). New York: Colombia University Press.
Intergovernmental Panel on Climate Change (IPCC). (1994). *Radiative forcing of climate change and an evaluation of the IPCC IS92 emission scenarios*. Cambridge: Cambridge University Press.
Intergovernmental Panel on Climate Change (IPCC). (2007). Climate Change 2007: Impacts, adaptation and vulnerability. In *Contribution of Working Group II to the Fourth Assessment Report of the Intergovernmental Panel on Climate Change*. Cambridge: Cambridge University Press.
Jacoby, H. D., Prinn, R. G., & Schmalensee, R. (1998). Kyoto's unfinished business. *Foreign Affairs, 77*, 54–66.
Karp, L., & Zhang, J. (2016). Taxes versus quantities for a stock pollutant with endogenous abatement costs and asymmetric information. *Economic Theory* (reprinted in this volume). doi:10.1007/s00199-010-0561-y.
Lauwers, L. (2016). Intergenerational equity, efficiency, and constructibility. *Economic Theory* (reprinted in this volume).
Lecocq, F., & Crassous, R. (2003). International climate regime beyond 2012. Are quota allocation rules robust to uncertainty? World Bank Policy Research Working Paper 3000.
Meyer, A. (2002). *Contraction and convergence, the solution to climate change*. Schumacher Briefing No. 5, Green Books.
Negishi, T. (1960). Welfare economics and existence of an equilibrium for a competitive economy. *Metroeconomica, 12*, 92–97.
Nordhaus, W. D., & Yang, Z. (1996). A regional dynamic general-equilibrium model of alternative climate-change strategies. *American Economic Review, 86*, 741–765.
Nordhaus. (2007). A review of the stern review on the economics of climate change. *Journal of Economic Literature, 45*, 686–702.
Organization for Economic Co-operation and Development (OECD). (1993). *International economic instruments and climate change*. Paris: OECD.
Ostrom, E. (2016). Nested externalities and polycentric institutions: Must we wait for global solutions to climate change before taking actions at other scales? *Economic Theory* (reprinted in this volume). doi:10.1007/s00199-010-0558-6.
Posner, E. A., & Sunstein, C. R. (2009). Should greenhouse gas permits be allocated on a per capita basis? *California Law Review, 97*(1), 51–93.
Rawls, J. (1971). *A theory of Justice*. Cambridge: The Belknap Press of Harvard University Press.
Rezai, A., Foley, D. K., & Taylor, L. (2016). Global warming and economic externalities. *Economic Theory* (reprinted in this volume).
Ringius, L., Torvanger, A., & Holtsmark, B. (1998). Can multi-criteria rules fairly distribute climate burdens—OECD results from three burden sharing rules. *Energy Policy, 26*(10), 777–793.

Sandler, T., & Smith, K. V. (1976). Intertemporal and intergenerational pareto efficiency. *Journal of Environmental Economics and Management, 2*, 151–159.
Schelling, T. C. (1995). Intergenerational discounting. *Energy Policy, 23*, 395–401.
Sen, A. K. (1987). *On ethics and economics*. Oxford: Basil Blackwell Publishers.
Shalizi, Z., & Lecocq, F. (2009). To mitigate or to adapt: Is that the question? Observations on an appropriate response to the climate change challenge to development strategies. *The World Bank Research Observer*. doi:10.1093/wbro/lkp012.
Sheeran, K. (2006). Who should abate carbon emissions? A note. *Environmental & Resource Economics, 35*(2), 89–98.
Stern, N. (2007). *The economics of climate change: The Stern Review*. Cambridge, UK and New York: Cambridge University Press.
Sterner, T., & Persson, U. M. (2008). An even Sterner Review: Introducing relative prices into the discounting debate. *Review of Environmental Economics and Policy, 2*(1), 61–76.
Stone, C. D. (2004). Common but differentiated responsibilities in International Law. *American Journal of International Law, 98*(2), 276–301.
World Bank. (2004). *World Development Indicators*. Washington D.C: World Bank.
World Bank. (2009). *World Development Report 2009*. Washington D.C: World Bank.

Carbon Leakages: A General Equilibrium View

Jean-Marc Burniaux and Joaquim Oliveira Martins

1 Introduction

By signing the Kyoto Protocol in December 1997, a number of industrialised countries (so-called, Annex 1 group) committed themselves to reduce unilaterally their emissions of greenhouse gases (hereafter, GHG). Such unilateral action has raised concerns about its environmental effectiveness. The argument relies on the possibility that carbon abatement efforts would increase production costs and undermine the international competitiveness of Annex 1 firms. In turn, this could induce shifts in international production and additional emissions from countries that are not subject to an emission constraint. This effect has been coined to as "carbon leakages" in the literature on climate change policy (Rutherford 1992; Felder and Rutherford 1993).

The world economic context has changed a lot since the signature of the Kyoto Protocol. The scientific consensus about the influence of manmade greenhouse gases (GHGs) emissions on Earth climate has been developed since the early 2000s

The authors want to thank Graciela Chichilnisky, Jorgen Elmeskov, Peter Sturm and late Alan Manne for discussions and comments on early versions of this work. Anonymous referees also provided very useful suggestions. The views expressed are those of the authors and do not necessarily reflect those of the OECD or its Member countries.

Reprinted with kind permission from the Authors: Originally published in Economic Theory, Volume 49, Number 2, February 2012.

J.-M. Burniaux · J. Oliveira Martins (✉)
OECD, 2 Rue André Pascal, 75775 Paris Cedex 16, France
e-mail: Joaquim.Oliveira@oecd.org

J.-M. Burniaux
e-mail: Jean-Marc.Burniaux@oecd.org

G. Chichilnisky and A. Rezai (eds.), *The Economics of the Global Environment*,
Studies in Economic Theory 29, DOI 10.1007/978-3-319-31943-8_16

(IPCC 2001, 2007), and there is some evidence that earth climate is currently changing at a faster pace than was expected. On the other hand, the growth of world GHG emissions has accelerated, reflecting the economic dynamism of emerging economies and the shift of the energy mix towards coal as pressure increases on remaining oil reserves. For the first time, emissions in China are exceeding the US ones, increasing the pressure on governments of emerging economies to undertake action in the next round of negotiations. However, the existence and possible magnitude of carbon leakages is still repeatedly evoked in the current debate as an argument against an agreement among a limited group of countries to undertake carbon abatement.

Early estimates of carbon leakages with large-scale simulation models had diverged somewhat. Defining the 'leakage rate' as the ratio of the additional emissions in the non-Annex 1 countries relative to the emission reduction achieved in Annex 1 countries, the estimates ranged from around 20 % (Manne and Richels 1998; Light et al. 1999; Bollen et al. 1999) to the lower bound estimates of 2–6 % (Oliveira Martins et al. 1992; OECD 1999; McKibbin et al. 1999; Babiker and Jacoby 1999). Later work (Burniaux 2001; McKibbin and Wilcoxen 2008; OECD 2009) confirmed that leakage rates tend to be moderate, while other papers (e.g. Babiker 2005) have argued that leakage rates could even be higher than 100 %!

As a general point, the larger is the size of a country coalition to reduce carbon emissions, the smaller the potential magnitude for carbon leakages. OECD (2009) shows[1] that if the European Union was to cut its emissions unilaterally by 50 % in 2050 relative to 2005 (in line with the ambitious targets that are now considered to keep global temperature increase below 2 °C), then around 12 % of this reduction would be offset by emissions increase in the non-EU countries. However, if the same reduction is spread across all Annex 1 countries, carbon leakage becomes negligible, falling to less than 2 %. Noteworthy, the relationship between country coverage and carbon leakage is nonlinear with small increases in the size of the coalition leading to substantial reductions in leakage.

Whether the potential for carbon leakages may have been exaggerated or not, it is an important political-economy issue to the extent that it can be used to block the building-up of a coalition of acting countries at an early stage, i.e. when leakages are presumed to be higher.[2] In this context, it is unfortunate that existing econometric models disagree on the magnitude of these leakages, hence being unable to provide a clear guidance on this crucial debate. These models have different structures and are using a large number of parameters whose values often are based on little or mixed econometric evidence.

Assessing the drivers for carbon leakages is not an easy task. Indeed, they result from complex interactions between energy and non-energy markets for which there

[1]Using the OECD ENV-Linkages model (see Burniaux and Chateau 2008).

[2]The literature based on game theory shows that a high leakage rate reduces the size of a stable, self reinforcing coalition to reduce emissions (see, for instance, Carraro 1998; Botteon and Carraro 2001).

is no direct empirical evidence. The only option is to rely on model simulations. Oliveira Martins (1995)[3] showed that the values of the energy supply elasticities appear more influential in determining leakage rates than the trade elasticity of substitution for energy-intensive goods. The analysis also pointed to the possibility of *negative* leakage effects, i.e. a possible reduction of emissions in non-participating countries. This effect is due to the fall of the relative price of oil versus coal following a carbon abatement action. This fall induces a shift towards less carbon-intensive energy in non-participating countries reducing their emissions. These combined effects explain part of the low net leakage rate obtained with some large-scale models. Bollen et al. (1999) showed that another important factor determining the magnitude of carbon leakages is the substitution possibilities in the production function. Light et al. (1999) argued that the structure of the international coal market is critical to understand the leakage mechanisms; if the degree of integration of the coal international market is understated this necessarily leads to underestimation of the carbon leakages. Other factors that could potentially increase the magnitude of the leakages include the degree of international capital mobility (McKibbin and Wilcoxen 2008) or the existence of imperfect competition and firm relocation (Babiker 2005).

The aim of this article is to highlight the various factors that may lead to different leakage estimates and the parameter values that are critical in estimating the magnitude of the leakages. In other words, it is not so much to evaluate the magnitude of the leakages per se but rather to identify the modelling assumptions that are critical in making these estimates. Drawing robust conclusions in this area is made difficult by the need to take into account the interactions between different parameters over a wide range of values. In other words, sensitivity analysis cannot be carried out parameter by parameter only. Overlooking these *multidimensional* interactions can lead to misleading conclusions.

Accordingly, the paper develops a static two-country, multigood, simplified general equilibrium (GE) Model, which captures in a stylised way the main interactions between energy and non-energy markets at the world level. Compared with other work in this area, our model was designed to make tractable an extensive sensitivity analysis, in order to assess how the results are dependant on different assumptions and parameters' values. It enables to draw in a three-dimensional space the manifold representing the (equilibrium) leakage rate as a function of pairs of different parameters. Since a post-Kyoto setting has not yet emerged, our analysis uses the provisions of the Kyoto Protocol as an illustrative benchmark for the carbon abatement reduction targets used to calculate the leakage rates.

Anticipating the results of our sensitivity analysis, we did not find evidence of large leakages for a large plausible range of parameters' values. Only rather implausible values of certain parameters may generate high leakage rates. These results have important policy implications, as the argument that unilateral carbon abatement action taken by a large group of countries (such as the Annex 1 group) is

[3]Using the OECD GREEN model (see Burniaux et al. 1992).

undermined by large carbon leakages is not supported by the evidence provided in this paper. On the contrary, the likelihood of small leakages favours in fact the formation of a worldwide coalition to stabilise climate change.

Given the focus of the paper, we follow the usual approach by considering that carbon emissions affect welfare through the imposition of an aggregate emission constraint, rather than considering carbon abatement as a public good appearing directly in the utility function a question that is tackled in several other papers in the issue of Economic Theory reprinted in this volume.[4] The fact that this public good is produced in a decentralised way by the consumption and production activities of all economic agents has profound implications for the definition of an optimal carbon abatement.[5] Nonetheless, the current debate on carbon leakages is still much focused on the consequences of imposing an unilateral carbon emission constraint defined exogenously (such as the Kyoto-type targets).

The structure of the paper is as follows. We first discuss the key GE mechanisms underlying carbon leakages and present the specification of our model. A multidimensional, extensive, sensitivity analysis is then carried out in order to determine what are the critical parameters influencing the size of leakages. We found that the supply behaviour of high-carbon energy producers worldwide seems one of the most critical elements. Finally, we conclude with a summary and some policy considerations.

2 Main GE Mechanisms Underlying Carbon Leakages

What are the main general equilibrium channels generating carbon leakages? In *energy markets*, they operate in the following way. When unilateral carbon abatement is implemented in a group of countries, the reduction in world demand creates a downward pressure in the international price of the most carbon-intensive fossil fuels. This increases energy demand (and carbon emissions) in the non-participating countries. But the structure of international energy markets matters for the size and scope of this effect. While oil can be considered as an

---

[4]This article is part of a Special Issue of Economic Theory on the topic of the Global Environment, which includes also the following articles: "Unspoken Ethical Issues in the Climate Affair Insights From a Theoretical Analysis of Negotiation Mandates" by Lecocq and Hourcade, "Intergenerational equity, efficiency, and constructability" by Lauwers, "Detrimental Externalities, Pollution Rights, and the "Coase Theorem" by Chipman and Tian, "Nested externalities and polycentric institutions: must we wait for global solutions to climate change before taking actions at other scales?" by Ostrom, "Capital Growth in a Global Warming Model: Will China and India Sign a Climate Treaty?" by Dutta and Radner, "Taxes Versus Quantities for a Stock Pollutant with Endogenous Abatement Costs and Asymmetric Information" by Karp and Zhang, Sustainable recursive social welfare functions" by Asheim, Mitra, and Tungodden, "Sustainable Markets with Short Sales" by Chichilnisky, and "Sustainable Exploitation of a Natural Resource: A Satisfying Use Of Chichilnisky Criterion" by Figuières and Tidball.

[5]This insight was put forward by Chichilnisky and Heal (1994).

international commodity (or an homogenous good), the coal market is much less integrated at the world level.[6] For example, there are many varieties of coal and hence domestic demand may not shift easily across different producers. The supply response of fossil-fuel producers is also critical. The potential for reducing world carbon emissions ultimately relies upon the decision of producers to keep extracting carbon-based energy or to leave it in the ground. Given that coal is the most carbon-intensive fuel,[7] its supply elasticity can be expected to be the most influential for the size of carbon leakages.

In *non-energy markets* the channels are indirect. Unilateral carbon abatement raises production costs and reduces international competitiveness of energy-intensive industries. These industries can lose market shares in favour of producers located in countries that do not constrain their carbon emissions. This causes a corresponding shift in the production of energy-intensive goods at the world level. The intensity of this mechanism typically depends on the trade substitution elasticities (so-called Armington elasticities). The larger these elasticities, the larger the effect of price changes on market shares and production shifts at the world level. Also contributing to potential leakages is the reallocation of direct investments, which depends on the degree of international mobility of capital. In this regard, Burniaux (2001) and McKibbin and Wilcoxen (2008) found that leakage rates remain modest even in the presence of some capital mobility.

Other factors may also be important for the size of leakages. A fall in the oil price after the implementation of a unilateral carbon abatement policy could shift energy demand away from coal in the non-participating countries. This could actually induce a decrease of the carbon intensity of the large coal consumers, such as China, and generate "negative leakages". These negative leakages are more likely to materialise if the supply elasticity of oil is small and coal supply is elastic.[8] Finally, the leakages also depend on possible income losses of energy-exporting economies. By reducing their aggregate domestic demand and hence their carbon emissions, these income losses would tend to create negative leakages. Nonetheless, this is likely to be a second-order effect, at least over a medium time horizon.

Babiker (2005) finds very large leakage rates when assuming increasing returns to scale and homogenous goods in energy-intensive industries. However, in addition to the fact that these assumptions may not characterise properly the structure of

[6]High transportation costs, lack of infrastructure and other technical aspects have so far contributed to restrict coal trading to a fraction of the world coal production. Nonetheless, Light et al. (1999) argue that the international coal market is actually more integrated than it appears. More research would be needed to assess empirically this question. However, the analysis developed below shows that this result actually only holds for a narrow range of values of the coal supply elasticity. In other words, even if coal is treated as an homogenous commodity, a relatively elastic supply of coal still leads to low leakage rates.

[7]Shale oil extraction for unconventional oil production would be even more carbon-intensive, although its scope has remained limited.

[8]Note that the analysis of Light et al. (1999) was based on the assumption of a very inelastic coal supply (elasticity equal to 0.5). This assumption alone rules out the possibility of negative leakages.

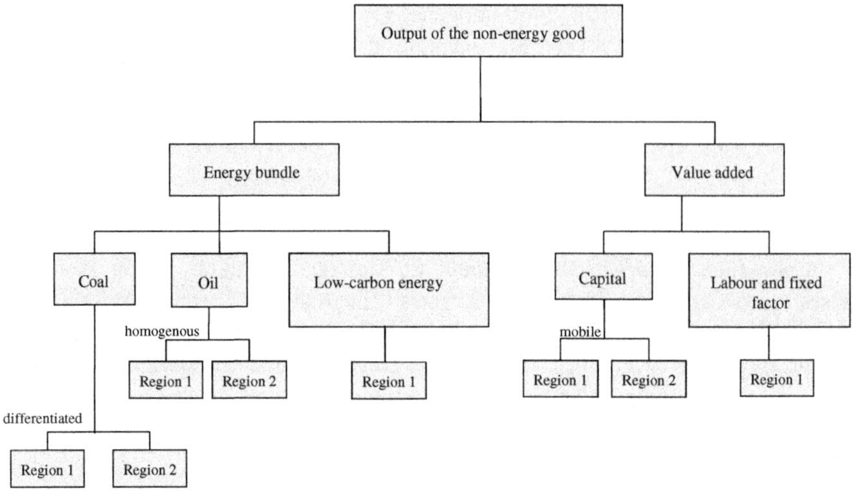

Fig. 1 Production structure of region 1 (annex 1)

all energy-intensive industries, they generate by their nature very large and unrealistic shifts in international trade shares. In contrast, Di Maria and Van der Werf (2008) argued that neglecting the effects of technological innovation leads to an overestimation of the carbon leakages. These market structure issues are more relevant at the sectoral level and therefore are not taken into account in our analysis.

3 The Simplified GE Model

Our model embodies two regions: Annex 1 and the Rest-of-the-World (or non-Annex 1). Each region uses five production inputs: a region-specific Labour and Fixed factor, Capital and three Energy inputs (Coal, Oil and a residual low-Carbon Energy source, which groups natural gas and carbon-free energy). Both regions produce all three energy sources. Oil and coal are traded commodities, with coal being differentiated by origin (Armington specification) and oil being treated as a homogenous commodity. The carbon-free energy source is region-specific (i.e. non-tradable). The final consumption good is also differentiated by region of origin. We incorporated in the model the possibility for international mobility of capital as this may be a possible channel for carbon leakages.[9] The complete production nesting corresponding to these assumptions is depicted in Fig. 1.

The demand side of the economy is represented by a single consumer in each region that maximises its utility as a function of the consumption of a non-energy

[9]This aspect has been somewhat overlooked, a significant exception being the G-Cubed model (McKibbin and Wilcoxen 1995).

(final) good, under an aggregate budget constraint. This economy could then be generically represented by an optimisation programme for the Annex 1 region as follows:

$$\begin{aligned}&\text{Max } U(C)\\&\text{subject to } g(C,K,L,E)=0\\&h(E)\leq \text{CEM}_{\text{Annex1}}\end{aligned} \tag{1}$$

where C is the consumption of the non-energy good; $g(.)$ the production frontier; K, L, E are the labour, capital and composite energy inputs, respectively; $h(.)$ is the emission generation function associated with carbon-based energy consumption; and CEM is the aggregate carbon emission constraint imposed on the Annex 1 region. It can be shown that the carbon tax correspond to the shadow price of this constraint and is equal to t_F/h, where t_F is the excise tax defined on energy consumption.

Noteworthy, if carbon abatement would have appeared directly as a public good in the utility function $U(.)$, the optimisation problem would also have been rather different,[10] with the level of carbon abatement being endogenously determined at the optimum. Nonetheless, we do not pursue that route here as the leakage problem has not been discussed in that context, but rather as an unintended consequence of an unilateral carbon abatement defined in an exogenous way (such as the Kyoto targets). For a treatment of carbon abatement or the global climate as a public good, the reader can refer to Lecocq and Hourcade (2016) or Rezai et al. (2010) in this special issue.

The numerical implementation of our specification is based on a linear approximation[11] of nested-CES functions and log-linear supply functions of fossil fuels. The capital mobility was modelled through a Constant Elasticity of Transformation (CET) function, with its transformation elasticity characterising the degree of international mobility (zero for immobile and one for perfectly mobile capital). The list of variables, parameters and details about the equations in the model are provided in the Annex. The model was calibrated on data taken from the OECD GREEN model database[12] and solved iteratively for different pairs of parameters' values.[13]

[10]For a discussion related to carbon abatement modelling, see Oliveira Martins and Sturm (2000).

[11]Based on a usual linearisation procedure of CES functions, see for example Dixon et al. (1992).

[12]The OECD GREEN Model was based on the GTAP-E data base for the year 1995 (see Hertel 1997). We could have calibrated the model with more recent data (using for example the OECD ENV-Linkages Model database), but preferred to stick with the 1995 database given that our simplified GE model uses GREEN as a benchmark. Given that this paper mainly focus on the general equilibrium mechanisms rather than an estimate of the leakage rate by itself, the flavour of the results would not have changed qualitatively.

[13]In principle, given that our model is specified in a linearised growth rate form (see Annex) it should be possible to compute algebraically the functional form of the leakage rate as a function of the key parameters. However, even this simple model turned out to be too complicated to be

Table 1 Key calibration parameters for the central case

Trade substitution elasticity for the non-energy good	4
Elasticity of transformation for capital mobility	0
(Armington) trade substitution elasticity for coal	5
Supply elasticity for coal	20
Supply elasticity for oil	1
Supply elasticity for the low-carbon energy	1
Inter-fuel (coal, oil and low-carbon energy) substitution elasticity	2
Inter-factor (K, L) substitution elasticity	0.4

Source Authors based on the calibration of the OECD GREEN Model

The values of the key parameters are given in Table 1. Most of the parameters' values are in the range of those reported in the literature (see for example Burniaux et al. 1992). The international coal market is characterised by a moderate degree of differentiation (with trade substitution elasticity equal to 5). In contrast, the coal supply elasticity is very high (equal to 20) compared with a unitary supply elasticity for the other fossil fuels. This reflects the fact that coal reserves are very large and market supply has increased steadily over the past decades without noticeable increases in international prices. A *central case* (or baseline) was calibrated with this set of parameters. We then simulated the implementation of the carbon abatement target equivalent to the Kyoto Protocol in the Annex 1 region. Under these assumptions, our GE model replicated a leakage rate comparable to those obtained with large GE models, such as the OECD GREEN or the OECD ENV-LINKAGES (i.e. below 10 %).

4 Sensitivity Analysis

The focus of this paper is the sensitivity of the results to the values of the different parameters. In line with the discussion above, the analysis considers the main parameters affecting non-energy, energy markets and changes in the mix of production inputs. We considered the joint sensitivity of different pairs of parameters in order to visualise and capture the main mechanisms underlying carbon leakages.

(Footnote 13 continued)

solved algebraically. The calculations were carried out with *Mathematica* (Wolfram 2003) and further details can be supplied upon request.

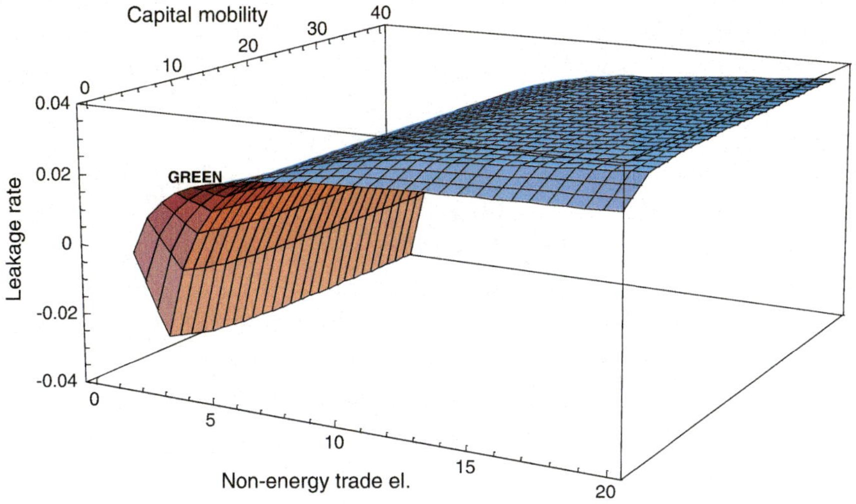

Fig. 2 Leakage rates as a function of non-energy trade substitution elasticities and international capital mobility. NB: the label GREEN in the figure correspond to the values of the parameters in the central case (see Table 1)

4.1 Parameters Characterising Non-energy Markets

Figure 2 shows the joint sensitivity of the leakage rate to the elasticity of substitution between domestic and imported non-energy goods (Armington elasticity) and the migration elasticity of capital. The first result is that whatever is the value of the trade elasticity, the leakage rate never exceeds 4 % (compared with around 2 % in the central case). The leakage rate even becomes negative when the trade elasticity is small and the capital fully mobile.[14] This suggests that the choice between the so-called Armington and Heckscher-Ohlin type assumptions[15] is not relevant for explaining large differences between leakage rates across existing models.

Contrary to possible a priori views, capital mobility has only a small impact on the leakage rate. This is because the unilateral carbon abatement in Annex 1 increases the price of energy and induces a substitution effect towards capital, increasing demand and the rate of return of capital in Annex 1 countries compared with the Rest of the World. For moderate values of the Armington elasticity (around 4–5), this implies a net inflow of capital from the rest of the World to the Annex 1

[14]Noteworthy, while the average leakage rate remains modest, the marginal leakage rates (or the incremental changes in the leakage rates) can be rather large. Important also to note, an optimal carbon tax depends on the marginal leakage rate and not on its absolute level (see Oliveira Martins 1995, footnote 11).

[15]The standard Heckscher-Ohlin model of international trade assumes that goods from different origins are homogeneous (infinite elasticity of substitution).

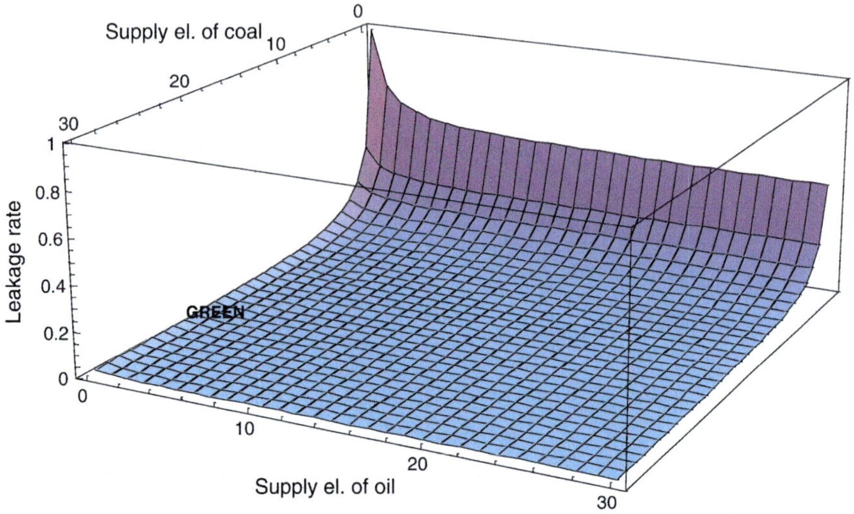

Fig. 3 Leakage rates as a function of coal and oil the supply elasticities. NB: the label GREEN in the figure correspond to the values of the parameters in the central case (see Table 1)

countries. In other words, the flow of capital goes exactly in the opposite of the expected direction, so capital mobility does not create carbon leakages.

For very low values of the trade elasticity and a high degree of capital mobility, emissions actually fall in the non-Annex 1 group creating a *negative* leakage. Only for high values of the Armington elasticities, the adverse competitiveness effects outweigh the substitution effects leading to the expected net outflow of capital from the Annex 1 to the non-Annex 1 countries. But even in this case, the leakage rate only slightly increases (Fig. 2). Given that most of the discussion on carbon leakages has focussed on heavy energy-intensive industries (steel, cement, non-ferrous), these results may highlight some compensating mechanisms for the leakages occurring in these sectors.

4.2 Parameters Characterising Energy Markets

4.2.1 Supply Elasticities of Fossil-Fuels

In Fig. 3, the leakage rate is plotted as a function of the supply elasticities of coal and oil.[16] When the supply of high-carbon fuels is totally inelastic (i.e. a zero supply elasticity), any reduction of carbon consumption in Annex 1 countries will

[16]To simplify, the values of the supply elasticities are set equal in Annex-1 and non-Annex 1 regions.

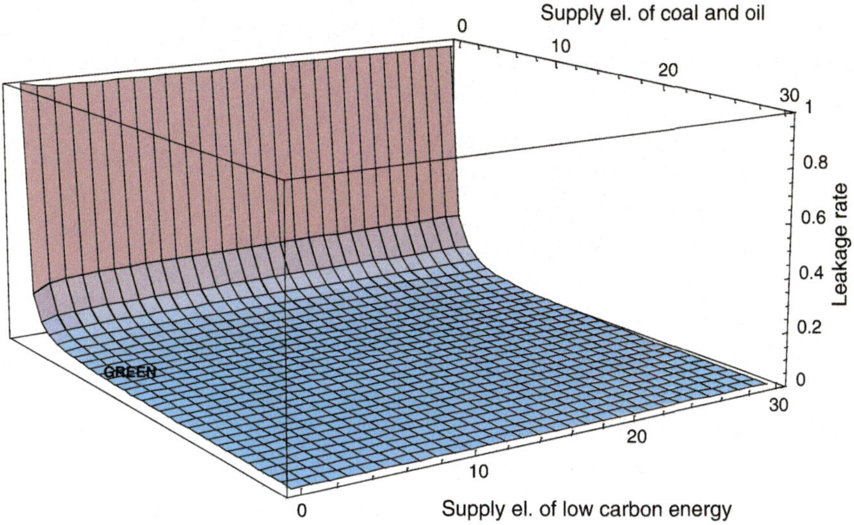

Fig. 4 Leakage rates as a function of the aggregate coal and oil supply elasticity and the low-carbon energy supply elasticity. NB: the label GREEN in the figure correspond to the values of the parameters in the central case (see Table 1)

lead to a corresponding fall in fuel prices at world level, in order to maintain the output volume unchanged. Therefore, it would impossible to reduce world emissions. Any unilateral abatement will be automatically offset by an equivalent increase of emissions in the other regions. The leakage rate would reach 100 %. For values of the coal supply elasticity lower than 2, the leakage rate is still above 20 %.

By comparison, the results are much less sensitive to the supply elasticity of oil. With a totally inelastic supply of oil but an elastic supply of coal, the leakages are small. With a fully inelastic coal supply and the supply elasticity of oil increasing to infinity, the leakage rate would stabilise at around 50 %. As it could be expected, the supply elasticity of the low-carbon energy is even less influential for the size of leakages (Fig. 4).

The bottom-line is that the size of carbon leakages depends critically on the supply response of coal producers at the world level. If one assumes an elastic supply of coal, the leakage rates would tend to be small for a large configuration of other parameters' values.

4.2.2 Degree of Integration in the International Coal Market

As suggested by Light et al. (1999), another potential channel for leakages is the degree of integration of the coal market characterised in our model by the (Armington) trade substitution elasticity: the higher this elasticity, the more integrated the market is. The joint sensitivity of the results with respect to this elasticity and to the supply elasticity of coal, which appeared above as a critical parameter, is provided in Fig. 5.

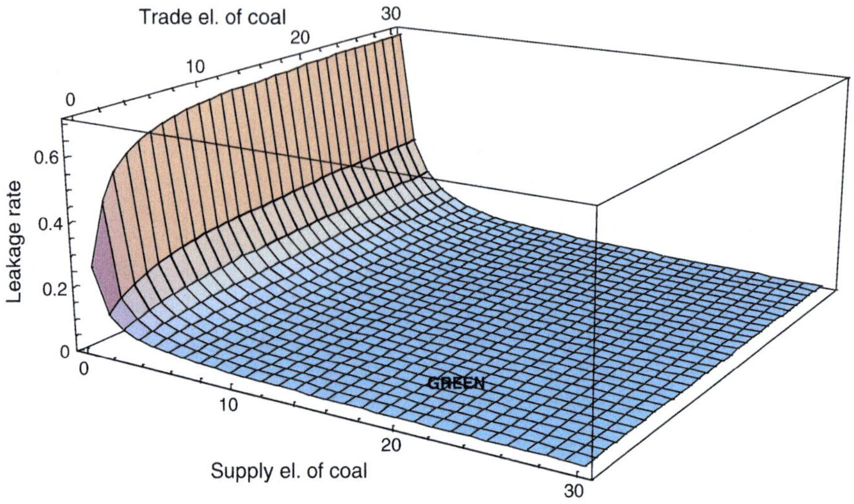

Fig. 5 Leakage rates as a function of the coal supply elasticity and the trade substitution elasticity in the coal market. NB: the label GREEN in the Figure correspond to the values of the parameters in the central case (see Table 1)

It turns out that the influence of the coal trade elasticity is strongly *conditioned* by the coal supply elasticity. With an elastic coal supply, coal prices remain relatively stable and the degree of integration in the coal market does not play a crucial role. Leakage rates remain uniformly low.[17] However, with an inelastic supply of coal, the influence of the trade elasticity can be considerable. Their joint effect pushes the leakage rate above 60 %.

How realistic is this result? For such a high leakage to occur a massive shift of coal exports would have to materialise. This outcome is nevertheless rather unlikely with high transportation costs, the difficulty of building-up infrastructure over a relatively short period, as well as the propensity of the major non-Annex 1 countries (China and India) to protect their domestic coal producers.

Unfortunately, there is little evidence in the econometric literature on the coal supply elasticity. Beck et al. (1991) derived supply elasticities based on Australian mine level cost data, which range from 2.7 to 3.5. They also estimated an aggregate coal supply elasticity ranging from 0.4 in the short term to 1.9 in the long run. This result could support the view that the value of the coal supply elasticity is low and this could induce high carbon leakages. However, Beck et al. (1991) also acknowledge the fact that their estimate only captures part of the response to the coal price. Other factors, notably price expectations can also play an important

[17]Note that the marginal leakage rate can be quite large. For instance, with a supply elasticity of coal equal to 10, the leakage rate would almost double (from 3 to 5 %) when the trade elasticity increases from 0 to 100.

role.[18] In addition, their analysis does not take into account simultaneity of supply and demand. Mellish (1998) overcame this shortcoming by estimating jointly supply and demand curves using a two-stage least squares method; his estimates imply a relatively high price elasticity of coal supply in the US market, at around 7.[19] He also argues that productivity gains were important in the coal industry and could explain why real prices have decreased while demand was steadily increasing.

4.3 The Shape of the Production Function Matters

The shape of the production function is characterised, on the one hand, by the elasticities of substitution between energy and value added (inter-factor elasticity) and, on the other hand, by the elasticity of substitution among coal, oil and low-carbon energy (inter-fuel elasticity).

Figure 6a shows the leakage rate as a function of the inter-fuel substitution elasticity, conditional on the supply elasticities of oil and low-carbon energy.[20] With energy supply elasticities equal to one (the central case), the leakage rate displays a U-shaped form. Up to a value of 5, the leakage rate is a decreasing function of the inter-fuel elasticity of substitution. Above that value, the leakages increase monotonically to reach more than 20 %. Most large GE models (like GREEN) embodies inter-fuel substitution elasticity around 2, therefore this partly explains the low leakage rates obtained with these models.

The U-shaped pattern reflects opposite mechanisms. Under relatively low substitution elasticities (downward slopping part of the U-shaped curve), the demand effect is dominant. When the coal supply is more elastic than the one of oil (central case), carbon abatement induces a relatively larger fall in the price of oil than the price of coal. This induces a shift in the non-Annex 1 countries towards less carbon-intensive fuel mix, which reduces their carbon emissions (generating marginal negative leakage rates). For high values of the inter-fuel substitution elasticity, the substitution effect becomes dominant. The demand for coal is drastically reduced in Annex 1, but the demand for oil and low-carbon energy increases substantially. As the latter are in limited supply, their world prices increases, which induces a shift in the non-Annex 1 countries towards coal consumption, thereby increasing their carbon emissions. Assuming a more elastic

---

[18]See Beck et al. 1991, p. 39.

[19]This is an approximation, as the Mellish (1998) model has not been originally set up to estimate the supply response as a function of the price, but the reverse. However, given the large R^2 of the equation both the direct and inverse specification of the supply elasticity should produce comparable results.

[20]The joint sensitivity analysis with the coal supply elasticity was less interesting as the results would be anyhow dominated by the former parameter.

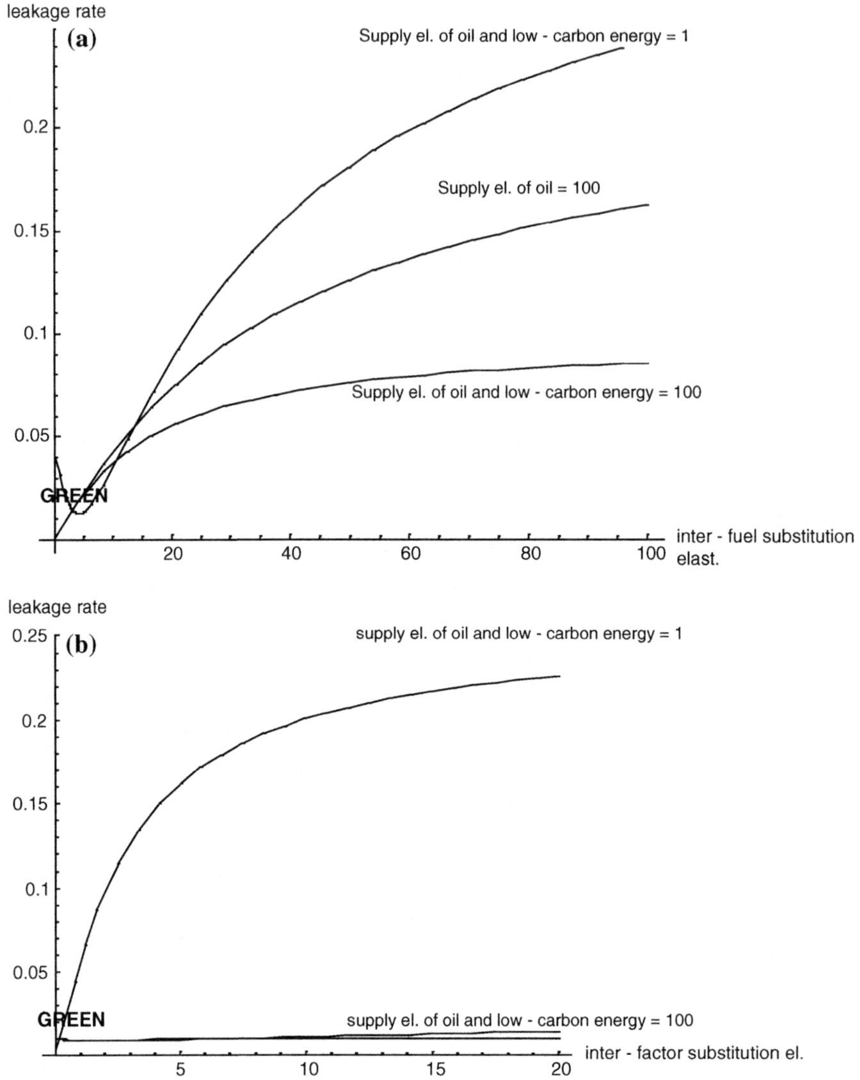

Fig. 6 **a** Leakage rates as a function of the inter-fuel substitution elasticity. **b** Leakage rates as a function of the inter-factor substitution elasticity. NB: the label GREEN in the Figure correspond to the values of the parameters in the central case (see Table 1)

supply for oil eliminates the U-shaped profile and flattens the slope of the increasing segment of the curve (Fig. 6a). Further assuming that the supply of the low-carbon fuel is elastic flattens even more the curve, leading to very low leakage rates (below 6–7 %).

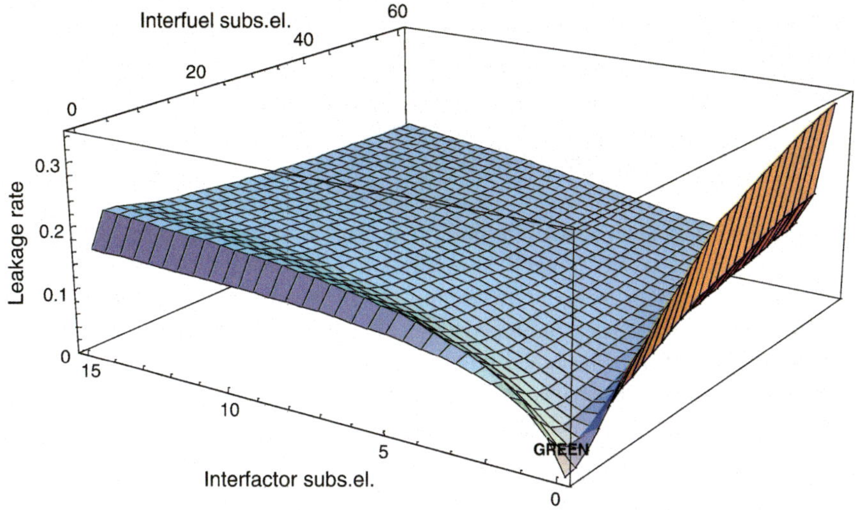

Fig. 7 Leakage rates as a function of the inter-fuel and inter-factor substitution elasticities. NB: the label GREEN in the Figure correspond to the values of the parameters in the central case (see Table 1)

Figure 6b shows the leakage rate as a function of the substitution elasticity between energy and value-added (inter-factor elasticity). The results are similar to those of the previous Figure. Under the central case, increasing inter-factor substitution elasticities induces higher leakages. Here again, the high leakage rate reflects the rigidity of the oil supply and the low-carbon fuel to adjust to demand. The explanation is the following. Following the reduction of energy demand, higher substitution possibilities imply that the demand for labour and capital increases strongly in the Annex 1 countries. As these factors are in fixed supply their prices increase. Following the increase in Annex 1export prices the demand in non-Annex 1 shifts towards domestic goods. This additional demand is met by using the factor with the most elastic supply—i.e. Coal—thus implying higher leakage. When the constraints on the supply of oil and the low-carbon fuel are relaxed this considerably reduces the scope for carbon leakages.

These results introduce some nuances with respect to the main conclusion of previous section. Even under the assumption of an elastic supply of coal, when the oil and low-carbon fuels are in restricted supply and inter-fuel and/or inter-factor substitution elasticities are higher than usually reported in the literature, there is a potential for high leakage rates. But for the substitution elasticities reported in the literature[21] the leakage rate should remain moderate. The joint sensitivity to the inter-fuel and inter-factor elasticities of substitution (Fig. 7) provides an additional

[21]See Burniaux et al. (1992) for a literature review of estimated elasticity values.

confirmation of this result. With increasing substitution elasticities (especially inter-factor substitution[22]), the carbon leakages can become very large.

5 Summary and Concluding Remarks

The complexity of general equilibrium interactions is sometimes difficult to introduce in the policy debate. Notably so when policy-makers are faced with sensitive choices and pressures from lobby groups. Such an analysis is much needed to inform the range of choices and a pedagogical effort should be carried out in order to avoid inefficient or ill-designed policies.

Along these lines, this paper provided a comprehensive analysis of the mechanisms underlying the potential carbon leakages induced by unilateral carbon emission abatement. We used a simplified static GE model, which made it possible to run extensive multidimensional sensitivity analysis over a wide range of parameters' values. The main findings are as follows:

- The leakage rate is not very sensitive to changes in the (Armington) trade elasticities, implying that competitiveness effects (in non-energy markets) have much less effect on carbon leakages than it could be expected on a priori basis. Similarly, international capital mobility does not affect leakages in a significant way. Simulations with other models are in line with this result (McKibbin et al. 1999; Babiker 2001).
- The most critical parameter for the size of leakages seems to be the supply elasticity of coal. Elasticity values above 45 yield relatively small and stable leakage rates. The result is robust to changes in the degree of integration in the coal market. However with supply elasticity below one, the leakage rates could reach 40 %. At the extreme case, with a totally inelastic supply of carbon-intensive fuels it would be impossible to reduce carbon emissions at the world level. Any change in prices in the Annex 1 would be compensated by a fall in prices in the Rest of World in order to keep the volume constant. The leakage rate could therefore reach 100 %.
- A fact that has attracted little attention in the literature, the shape of the production function matters. High inter-factor and inter-fuel substitution elasticities can generate large carbon leakages even when the supply of coal is elastic. This result is confirmed with sensitivity analysis performed with the WorldScan model (Bollen et al. 1999). However, it should be noted that such high substitution elasticities are outside the usual bounds found in the econometric literature.

[22]Note that, according to Fig. 6b, a value of 2 of the inter-factor substitution elasticity would generate a leakage rate equal to almost 10 % (against 2 % with an elasticity value of 0.4).

These results have important policy implications. The argument that unilateral carbon abatement action taken by a large group of countries (such as the Annex 1 group) is flawed because its environmental effectiveness is undermined by large carbon leakages is not supported by our sensitivity analysis over a plausible range of parameters' values. According to our analysis, only rather implausible values of certain parameters, notably low supply elasticities of high-carbon fuels, may generate high leakage rates. This also invalidates the argument that accompanying protectionist measures or tax exemptions to energy-intensive industries would need to be implemented. The likelihood of small leakages favours in fact the formation of a worldwide coalition to stabilise climate change. Nevertheless, more empirical work on the supply response of coal and oil producers to carbon abatement measures would be needed to strengthen these policy conclusions.

Annex: Specification of the GE Model

This annex provides the list of variables (Table 2), parameters (Table 3) and equations (below) of the simplified GE model used in the paper. All equations are expressed in a linearised growth rate form and variables in per cent changes, except the carbon tax, the price levels and carbon emissions in the base period (the level variables are in **bold**). The acronym 'nC' stands for low-Carbon Energy.

Energy supply

$$S_{j.r} = \varepsilon_{j.r} \cdot (P_{j.r} - PVA_r), \quad \text{for } j = \text{coal, oil, nC} \quad \text{and } r = \text{Annex1, non} - \text{Annex1}. \tag{1}$$

Inter-regional capital allocation

$$SK_r = \text{mig} \cdot (r_r - r) \quad \text{with} \quad r = \sum_r shk_r \cdot r_r \quad \text{for } r = \text{Annex1, non} - \text{Annex1}. \tag{2}$$

Consumer prices of energy (including the carbon tax).Domestic price

$$Pd_{j.r} = \left[\frac{\mathbf{Pd}^0_{j.r} \cdot (1 + P_{j.r}) + \gamma_{j.r} \cdot \mathbf{CT}_r}{\mathbf{Pd}^0_{j.r}} \right] - 1 \tag{3}$$
$$\text{for } j = \text{coal, oil, nC} \quad \text{and} \quad r = \text{Annex1, non} - \text{Annex1}.$$

Import price

Table 2 List of variables of the GE model

Equation number	Variables	Definition
1	$S_{\text{coal}.r}, S_{\text{oil}.r}, S_{nC.r}$	Supply of coal, oil and the low-carbon energy (grouping gas, nuclear and other carbon-free energy sources)
2	SK_r	Supply of capital in region r
2	r	World composite rental price of capital
3	$\mathbf{CT}_r$	Carbon tax in region r (in 1995$ per ton of carbon)
3	$Pd_{\text{coal}.r}, Pd_{\text{oil}.r}, Pd_{nC.r}$	Domestic consumer prices for coal, oil and the low-carbon energy (including the carbon tax)
4	$Pm_{\text{coal}.r}, Pm_{\text{oil}.r}$	Import consumer prices for coal and oil (including the carbon tax)
5	$PC_{\text{coal}.r}, PC_{\text{oil}.r}, PC_{nC.r}$	Composite consumer prices for coal, oil and the low-carbon energy (including the carbon tax)
6	PE_r	Composite energy price in region r
7	PVA_r	Composite factor price in region r (corresponding to the composite price of labour and capital)
8	P_r	Producer price of the non-energy good in region r
9	PC_r	Composite consumer price of the non-energy good in region r (corresponding to the composite price of domestic and imported demands for the non-energy good in region r)
9	$C_{r.r'}$	Consumption in country r of the non-energy good from region origin r'
10	X_r	Production of the non-energy good in region r
11	E_r	Total energy demand in region r
12	VA_r	Composite factor (K, L) demand in region r
13a	L_r	Labour demand in region r
13b	K_r	Capital demand in region r
14	$E_{\text{coal}.r}, E_{\text{oil}.r}, E_{nC.r}$	Demands for coal, oil and the low-carbon energy in region r
15	$Ed_{\text{coal}.r}, Em_{\text{coal}.r}$	Domestic and imported demand for coal in region r
16	$X_{\text{coal}.r}$	Coal production in region r
17	X_{oil}	World production of oil
18	CEM_r	Carbon emissions (in tons of carbon) in region r
19	$RCTAX_r$	Revenues from carbon taxes in region r (expressed in percentage of the base year GDP)
20	Y_r	Income level of region r
21–22–23	$P_{\text{coal}.r}, P_{\text{oil}.r}, P_{nC.r}$	Producer prices of coal, oil and the low-carbon energy (before tax)
24	r_r	Rental price of capital in region r
25	w_r	Wages in region r

Table 3 List of parameters of the GE model

Equation number	Parameters	Definition
1	$\varepsilon_{\text{coal}.r}, \varepsilon_{\text{oil},r}, \varepsilon_{nC.r}$	Supply elasticity for coal, oil and the low-carbon energy
2	mig	Elasticity of transformation for capital across regions
2	shk_r	Share of capital of each region in total world capital
3	$\gamma_{\text{coal}.r}, \gamma_{\text{oil}.r}, \gamma_{nC.r}$	Emission coefficient for coal, oil and the low-carbon energy (in tons of carbon per TeraJoule)
3	$\mathbf{P}^0_{\text{coal}.r}, \mathbf{Pd}^0_{\text{oil}.r}, \mathbf{Pd}^0_{nC.r}$	Levels of domestic energy prices in 1995 (in 1995$ per TeraJoule)
5	$ad_{\text{coal}.r}, ad_{\text{oil}.r}, ad_{nC.r}$	Share of domestic production in total consumption of coal, oil and the low-carbon energy. Note that $ad_{nC.r} = 1$
5	$am_{\text{coal}.r}, am_{\text{oil}.r}, am_{nC.r}$	Share of imports in total consumption of coal and oil. Note that $am_{nC.r} = 0$
6	$\alpha_{\text{coal}.r}, \alpha_{\text{oil}.r}, \alpha_{nC.r}$	Shares of coal, oil and the low-carbon energy in the total energy consumption of country r
7	$\alpha K_r, \alpha L_r$	Shares of capital and labour in total value added of region r
8	$\alpha E_r, \alpha VA_r$	Shares of energy and value-added (K, L) in total production of the non-energy good in region r
9	σ_r	Trade substitution elasticity (Armington) for the non-energy commodity in region r
9	$\beta_{r.r'}$	Shares of domestic ($r = r$) and foreign ($r \neq r$) demands in total consumption of the non-energy commodity in region r
10	$\delta_{r.r'}$	Shares of domestic demand ($\delta_{r.r}$) and exports ($\delta_{r.r'}$) in total output of the non-energy commodity produced in region r
11	κ_r	Substitution elasticity between energy and value-added (K, L) in region r
14	ϕ_r	Inter-fuel substitution elasticity in region r
15	$\sigma_{\text{coal}.r}$	Trade substitution elasticity (Armington) between domestic and imported coal in region r
16	$shd_{\text{coal}.r}, she_{\text{coal}.r}$	Shares of domestic demand ($shd_{\text{coal}.r}$) and exports ($she_{\text{coal}.r}$) of coal in total coal output in region r
17	$shd_{\text{oil}.r}$	Shares of oil demand in region r in total world oil consumption
18	$\chi_{\text{coal}.r}, \chi_{\text{oil}.r}, \chi_{nC.r}$	Energy content (in TeraJoule per 1995$) for coal, oil and the low-carbon energy in region r
19	$\mathbf{CEM}^0_r$	Level of carbon emissions in the base year 1995 in region r
19	$\mathbf{Y}^0_r$	GDP level in the base year 1995 in region r
20	shL_r, shK_r	Shares of labour and capital in total GDP of region r
20	$sh_{\text{coal}.r}, sh_{\text{oil}.r}, sh_{nC.r}$	Shares of coal, oil and the low-carbon energy in total GDP of region r
22	$shs_{\text{oil}.r}$	Shares of oil production of region r in total world production of oil

$$Pm_{j.r} = \left[\frac{\mathbf{Pd}^0_{j.r} \cdot (1 + P_{j.r'}) + \gamma_{j.r} \cdot \mathbf{CT}_r}{\mathbf{Pd}^0_{j.r}}\right] - 1$$

for $j = \text{coal, oil};\ r,\ r' = \text{Annex1, non} - \text{Annex1}$ and $r \neq r'$. (4)

Note that for oil: $P_{\text{oil.Annex1}} = P_{\text{oil.non} - Annex1}$ and $Pd_{\text{oil}.r} = Pm_{\text{oil}.r}$.

Composite energy prices

$$PC_{j.r} = \alpha d_{j.r} \,.\, Pd_{j.r} + \alpha m_{j.r} \,.\, Pm_{j.r}$$

for $j = \text{coal, oil, nC}$ and $r = \text{Annex1, non} - \text{Annex1}$. (5)

Note that for nC: $PC_{nC.r} = Pd_{nC.r}$

$$PE_r = \sum_j \alpha_{j.r} \cdot PC_{j.r} \quad \text{for } j = \text{coal, oil, nC} \quad \text{and } r = \text{Annex1, non} - \text{Annex1} \quad (6)$$

Composite factor prices

$$PVA_r = \alpha L_r.w_r + \alpha K_r.r_r \quad \text{for} \quad r = \text{Annex1, non} - \text{Annex1}. \quad (7)$$

Producer price of the non-Energy good

$$P_r = \alpha E_r.PE_r + \alpha VA_r.PVA_r \quad \text{for} \quad r = \text{Annex1, non} - \text{Annex1}. \quad (8)$$

Consumption of the non-Energy good

$$C_{r,r'} = -\sigma_r.P_{r'} + (\sigma_r - 1) \cdot PC_r + Y_r$$

for r and $r' = \text{Annex1, non} - \text{Annex1}$. (9)

and with $PC_r = \beta_{r.r}.P_r + \beta_{r.r'} \,.\, P_{r'}$

Output of the non-Energy good

$$X_r = \sum_{r'} \delta_{r.r'} \cdot C_{r.r'} \text{ for } r \text{ and } r' = \text{Annex1, non} - \text{Annex1}. \quad (10)$$

Total Energy demand

$$E_r = -\kappa_r \,.\, PE_r + \kappa_r \,.\, P_r + X_r \quad \text{for } r = \text{Annex1, non} - \text{Annex1} \quad (11)$$

Total demand of the composite Factor (K, L)

$$VA_r = -\kappa_r.PVA_r + \kappa_r\,.\,P_r + X_r \quad \text{for } r = \text{Annex1, non} - \text{Annex1} \tag{12}$$

Factor demands

$$L_r = -w_r + PVA_r + VA_r \quad \text{for } r = \text{Annex1, non} - \text{Annex1} \tag{13a}$$

$$K_r = -r_r + PVA_r + VA_r \quad \text{for } r = \text{Annex1, non} - \text{Annex1} \tag{13b}$$

Fuel specific demands

$$\begin{aligned} E_{j.r} &= -\varphi_r\,.\,PC_{j.r} + \varphi_r\,.\,PE_r + E_r \\ &\quad \text{for } j = \text{coal, oil, nC} \quad \text{and} \quad r = \text{Annex1, non} - \text{Annex1.} \end{aligned} \tag{14}$$

Demands for domestic and imported coal

$$\begin{aligned} Ed_{\text{coal}.r} &= -\sigma_{\text{coal}.r}\,.\,Pd_{\text{coal}.r} + \sigma_{\text{coal}.r} PC_{\text{coal}.r} + E_{\text{coal}.r} \\ &\quad \text{for } r = \text{Annex1, non} - \text{Annex1.} \end{aligned} \tag{15a}$$

$$\begin{aligned} Em_{\text{coal}.r} &= -\sigma_{\text{coal}.r}\,.\,Pm_{\text{coal}.r} + \sigma_{\text{coal}.r} PC_{\text{coal}.r} + E_{\text{coal}.r} \\ &\quad \text{for } r = \text{Annex1, non} - \text{Annex1.} \end{aligned} \tag{15b}$$

Coal supply

$$X_{\text{coal}.r} = shd_{\text{coal}.r}\,.\,Ed_{\text{coal}.r} + she_{\text{coal}.r}\,.\,Em_{\text{coal}.r'} \quad \text{for } r \neq r' = \text{Annex1, non} - \text{Annex1.} \tag{16}$$

World production of oil

$$X_{\text{oil}} = \sum_r shd_{\text{oil}.r}\,.\,E_{oil.r} \quad \text{for} \quad r = \text{Annex1, non} - \text{Annex1.} \tag{17}$$

Carbon emissions

$$CEM_r = \sum_j \gamma_{j.r} \cdot \chi_{j.r}\,.\,E_{j.r} \quad \text{for } j = \text{coal, oil, nC} \quad \text{and} \quad r = \text{Annex1, non} - \text{Annex1.} \tag{18}$$

Carbon tax revenues (in percentage of base-year GDP)

$$RCTAX_r = \frac{(1 + CEM_r)\,.\,\mathbf{CEM}_r^0\,.\,\mathbf{CT}_r}{\mathbf{Y}_r^0} \quad \text{for} \quad r = \text{Annex1, non} - \text{Annex1.} \tag{19}$$

Regional incomes

$$Y_r = (shL_r \, . \, w_r) + shK_r \, . \, (r_r + SK_r) + \sum_j sh_{j.r} \, . \, (P_{j.r} + S_{j.r}) + RCTAX_r$$
$$\text{for } j = \text{coal, oil, nC} \quad \text{and} \quad r = \text{Annex1, non} - \text{Annex1} \tag{20}$$

Market-clearing price of coal

$$P_{\text{coal}.r} \quad \text{such as} \quad S_{coal.r} = E_{coal.r} \quad \text{for} \quad r = \text{Annex1, non} - \text{Annex1}. \tag{21}$$

Market-clearing international price of oil

$$P_{\text{oil}} \quad \text{such as} \quad X_{\text{oil}} = \sum_r shs_{\text{oil}.r} \, . \, S_{oil.r} \quad \text{and} \quad P_{oil.r} = P_{oil}$$
$$\text{for} \quad r = \text{Annex1, non} - \text{Annex1}. \tag{22}$$

Market-clearing price of the low-carbon energy

$$P_{nC.r} \quad \text{such as} \quad S_{nC.r} = E_{nC.r} \quad \text{for} \quad r = \text{Annex1, non} - \text{Annex1}. \tag{23}$$

Market-clearing price of labour

$$w_r \quad \text{such as} \quad L_r = 0 \quad \text{for} \quad r = \text{Annex1, non} - \text{Annex1}. \tag{24}$$

Market-clearing price of Capital

$$r_r \quad \text{such as} \quad K_r = SK_r \quad \text{for} \quad r = \text{Annex1, non} - \text{Annex1}. \tag{25}$$

References

Babiker, M. (2001). Subglobal climate-change actions and carbon leakage: the implication of international capital flows. *Energy Economics, 23*, 121–139.

Babiker, M. (2005). Climate change policy, market structure and carbon leakage. *Journal of International Economics, 65*(2).

Babiker, M., & Jacoby, H. D. (1999). Developing country effects of Kyoto-type emissions restrictions, mimeo, *Joint program on the science and policy of global change*. Massachusetts Institute of Technology.

Beck, T., Jolly L., & Loncar, T. (1991). Supply response in the Australian black coal industry, Technical Paper No. 91.1, *Australian Bureau of Agricultural and Resources Economics* (ABARE).

Bollen, J., Manders, T., & Timmer, H. (1999, May 27–28). *Kyoto and carbon leakage: simulations with WorldScan*, mimeo. Paper Presented at the IPCC Working Group III Expert Meeting. The Hague.

Botteon, M., & Carraro, C. (2001). Environmental coalitions with heterogeneous countries: burden sharing and carbon leakage. In A. Ulph (Ed.), *Environmental policy, international agreements and international trade*. Oxford: Oxford University Press.

Burniaux, J. -M., Nicoletti, G., & Oliveira Martins, J. (1992). GREEN: A global model for quantifying the costs of policies to curb CO2 emissions. *OECD Economic Studies, 19* (Winter).

Burniaux, J. -M., (2001). *International trade and investment leakage associated with climate change mitigation*. Paper presented at the Fourth Annual Conference on Global Economic Analysis, available on http://www.gtap.agecon.purdue.edu.

Burniaux, J. -M. & Chateau, J. (2008). An Overview of the OECD ENV-linkages model. *OECD Economic Department Working Papers* no. 658.

Carraro, C. (1998). Beyond Kyoto: A game theoretic perspective. In *The Proceedings of the OECD Workshop on Climate Change and Economic Modelling: Background Analysis for the Kyoto Protocol*. OECD, Paris.

Chichilnisky, G., & Heal, G. (1994). Who should abate carbon emission? an international perspective. *Economics Letters, 44*, 443–449.

Di Maria, C., & Van der Werf, E. (2008). Carbon leakage revisited: unilateral climate policy with directed technical change. *Environmental and Resource Economics, 39*(2).

Dixon, P., Parmenter, B., Powell, A., & Wilcoxen, P. (1992). *Notes and problems in applied general equilibrium economics*. North-Holland.

Felder, S., & Rutherford, T. (1993). Unilateral CO2 reductions and carbon leakage: The consequences of international trade in oil and basic materials. *Journal of Environmental Economics and Management, 25*(2).

Hertel, T. (1997). *Global trade analysis: Modelling and applications*. Cambridge University Press.

IPCC (2007). *Climate change 2007: The physical science basis*. Cambridge University Press.

IPCC (2001). *Climate change 2001: The scientific basis*. Cambridge University Press.

Lecocq, F., & Hourcade, J. -C. (2016). Unspoken ethical issues in the climate affair insights from a theoretical analysis of negotiation mandates. *Economic Theory* (reprinted in this volume).

Light, M., Koldstad, C., & Rutherford, T. (1999, November). *Coal markets and the Kyoto Protocol, mimeo*.

Manne, A., & Richels, R. (1998). The Kyoto Protocol: A cost-effective strategy for meeting environmental objectives? In *Economic Modelling of Climate Change*, OECD Workshop Report, 17–18 September 1998, published in *The Energy Journal* (20), May 1999.

McKibbin, W., & Wilcoxen, P. (1995). The theoretical and empirical structure of the G-cubed model. *Brookings discussion paper in international economics* no. 118, The Brookings Institution. Washington DC.

McKibbin, W., & Wilcoxen, P. J. (2008, June 9). The *economic and environmental effects of border tax adjustments for climate policy*. Paper Prepared for the Brookings Conference on Climate Change, Trade and Competitiveness: Is a Collision Inevitable?. Washington.

McKibbin, W., Ross, M., Shackleton, R., & Wilcoxen, P. J. (1999, May 27–28). *Emissions trading, capital flows and the Kyoto Protocol*. Paper Presented at the IPCC Working Group III Expert Meeting. The Hague.

Mellish, M. (1998). *Coal pricing methodology for the Annual Energy Outlook 1998*, U.S. Energy Information Administration, Issues in Midterm Analysis and Forecasting.

OECD (1999). *Action against climate change: The Kyoto Protocol and beyond*. Paris.

OECD (2009). *The economics of climate change mitigation: Policies and options for global action beyond 2012*. Paris.

Oliveira Martins, J. (1995). Unilateral emission control, energy-intensive industries and carbon leakages. In *Global warming: Economic dimensions and policy responses*. OECD, Paris.

Oliveira Martins, J., & Sturm, P. (2000). Efficiency and distribution in computable models of carbon emission abatement. In G. Chichilnisky & G. Heal (Eds.), *Environmental markets: Equity and efficiency*. New York: Columbia University Press.

Oliveira Martins, J., Burniaux, J. -M., & Martin, J. P. (1992). Trade and the effectiveness of unilateral CO2-abatement policies: Evidence from GREEN. *OECD Economic Studies*, 19 (Winter).

Rezai, A., Foley, D. K., & Taylor, L. (2016). Global warming and economic externalities. *Economic Theory* (reprinted in this volume).

Rutherford, T. (1992). The Welfare Effects of Fossil Fuel Carbon Restrictions: results from a recursively dynamic trade model. *OECD Economics Department Working Paper*, no. 112.

Wolfram, S. (2003). The *mathematica© book.* (5th ed.). Wolfram Media.

Part IV
The Case of Climate Change

Chaos Control: Climate Stabilization by Closing the Global Carbon Cycle

Peter M. Eisenberger

1 Introduction

It has been increasingly recognized (Vincent 2001; Boccaletti 2000) that a distinctive property of a complex system is the counter intuitive notion that in fact they are easier to control than the linear complicated systems one is more familiar with. The idea captured in the growing field of Chaos Control (Vincent 2001; Boccaletti 2000) is that the feedbacks that make the future evolution of Complex Systems inherently unpredictable are also what make them easier to control. In linear complicated systems the control signal is usually the same order of magnitude as the property one wants to control, while in chaotic systems it is possible to use the feedbacks of the system itself to enable a very small feedback signal to have a very large impact on the future evolution of the system. Properly chosen, a small feedback signal can provide an efficient means to stabilize one of the many chaotic states effectively suppressing the other possible states the system might evolve into.

It has become increasingly clear that it is appropriate to think about the Earth's climate as a Complex System with chaotic dynamics, rather than a complicated system with many interacting components with predictable responses (Donner 2009). This paper and previously published articles (Eisenberger et al. 2009; Eisenberger 2013) argues that carbon cycle, specifically the CO_2 level in the atmosphere, is at the core of the feedback mechanisms that ultimately define the global climate which in turn is best characterized by a spatially averaged global temperature (Rohde et al. 2013). It is important not to interpret this in some pseudo reductionist framework that would transform the above statement into that CO_2 causes the change in climate and defines the average temperature. The main

P.M. Eisenberger (✉)
Department of Earth and Environmental Sciences, Columbia University,
110 Geoscience, 61 Route 9W, PO Box 1000, Palisades, NY 10964, USA
e-mail: petere@ldeo.columbia.edu

G. Chichilnisky and A. Rezai (eds.), *The Economics of the Global Environment*,
Studies in Economic Theory 29, DOI 10.1007/978-3-319-31943-8_17

distinctive property of chaotic systems is that states emerge out of a web of feedbacks and that it is those feedbacks rather than the direct impact that define the state that emerges (Kauffman and Stuart 1995).

Chaos control until now has been applied to very different types of complex systems. It has been successfully demonstrated in the human heart, lasers, turbulent fluids, chemical reactions and other chaotic dynamical systems (Vincent 2001; Boccaletti 2000). Our understanding of how those systems work can be translated to the climate system with very large changes in both the time and spatial dimension. This translation cannot be presented as some mathematically rigorous formulation both because the climate system equations cannot be fully specified and because experimentation is not possible. Thus the approach taken is to claim that the analysis provided below is very plausible and that using a Bayesian approach is more plausible than any other alternative given our current state of knowledge. More generally, the Bayesian approach is best matched to evaluate both the risk and likelihood of success since experimentation and thus statistical data is not available. Furthermore, one needs to use the new statistical approach proposed by Chichilnisky in support of decision analysis since it is appropriate for complex systems that can make rare but catastrophic changes (Chichilnisky et al. 2009). Most importantly, thinking about the climate as a complex system provides generally useful insights as to the nature of the climate threat and how to address it. This approach can be extended to other types of catastrophic risks in large extended systems and therefore can serve as useful example for risks that involve both human and non-human components.

The first translation of chaos control to the climate is defined by the fact that a very large system such as our planet's climate has enormous inertia. Thus, the changes will occur over long time periods compared to the systems chaos control that has been used until now. This means that the feedback control signal needs to have a long lifetime following its application so that it matches the long timescale needed to provide control. The ideal of course is to apply it once and have it last long enough without the need to continually reapply it.

The second translation is that the control signal needs to have a spatial impact that matches the size of the system. Any localized feedback signal will be inefficient if one needs to expend energy to ensure its distribution over the full spatial extent of the system in order to achieve global control. One wants the system feedbacks to provide the global impact thus enabling the feedback signal to be applied locally but have a global impact.

Of course, most importantly, the application of the feedback control must be technically feasible and economically affordable. For a very large system this could be extremely challenging, but it will be argued here that because the climate is a complex system the control can be implemented with relatively little energy and cost by having the feedbacks from the relatively small changing control signal provide a very large impact.

CO_2 in the atmosphere meets all these requirements. Its lifetime matches the timescale over which climate changes and it is uniformly distributed over the planet no matter where it is emitted. It will be shown below that the energy expended to change the CO_2 concentration in the atmosphere produces about a 50,000 times larger change in the amount of energy heating our planet. It will also be shown that the total annual energy needed for changing the CO_2 concentration in the air to stabilize the CO_2 concentration of the planet at today's, or even the preindustrial, level is less than the energy used each year to control the temperatures of all the buildings on the planet. The heat from our buildings eventually leaks out to warm the planet, yet that heat will have a negligible impact on the planet's average temperature since it is five orders of magnitude smaller than the energy input provided by the sun to our planet. The ability to use that same small amount of energy to remove CO_2 or add CO_2 to the atmosphere to stabilize the climate of the planet provides a very clear indication of the huge enhancement provided by using CO_2 to control the climate.

With respect to the spatial extent property, CO_2 is also ideal because the Earth System ensures the uniformity of the CO_2 concentration around the planet. The most important reason for this, which has other very important implications, is that CO_2 will not condense in the temperature ranges present on our planet. Thus, unlike water, a very potent Greenhouse Gas present in much greater quantities, its concentration in the atmosphere is uniform and also therefore has a very long lifetime in the atmosphere which matches the timescales over which the climate changes.

It is also expected as noted earlier that the feedback enhancement that can destabilize the system can be used to stabilize it. Thus, rather than thinking about a linear relationship between CO_2 and temperature it is in fact the feedbacks between the carbon in the atmosphere and the other major earth system components—the land, the plants, the ocean, and, of course, life—through which climate change occurs and thus through which it can be stabilized. Here it is important to make a distinction between the ease with which the feedback signal itself can be applied and the internal feedbacks within the system that responds to the feedback signal. Fixing the CO_2 concentration in the atmosphere will be shown to be relatively easy to achieve, but it is the internal feedbacks in response to a stabilized ambient CO_2 concentration that stabilizes the climate.

It is proposed based upon the analysis provided below that it is highly plausible that the CO_2 concentration in the atmosphere will make an ideal variable for global stabilization of the chaotic climate. This literally transforms the current very real risks we are facing into a historic moment and opportunity in the evolution of our planet in which the many destabilizing climate variations that have plagued life on our planet throughout its past need no longer do so. This transformation defines the positive potential of the new proposed name, Anthropocene, for our current human dominated geological era. It will be argued below that this era of humans becoming a global carbon cycle force also creates the opportunity for human stabilization of the global climate rather than leaving it to the changing CO_2 fluxes between the atmosphere land and oceans which has caused destabilizing changes in our planets past with changes of sea level in excess of 50 m (Hag et al. 1987).

It has been suggested that our use of land for agriculture and wood for our structures and fuel influenced the carbon cycle thousands of years earlier (Ruddiman 2003). Our energy use since the industrial revolution has created a much larger disturbance of the carbon cycle so for control purposes one can consider our non-energy related impacts through use of the land on the CO_2 concentration to be part of what we balance out when we Close the Global Carbon Cycle (CGCC). In general terms one would prefer our energy production not to compete with our food production but rather support it. More generally, producing carbon based energy sources via CO_2 from the air, hydrogen from water, and energy from the sun enables us to meet our energy needs and close the carbon cycle at the same time.

In applying chaos control to the climate there are two kinds of changes that can destabilize our planet that need to be stabilized. They are our own impact on the atmosphere's CO_2 concentration via our use of fossil fuel energy sources and those changes in ambient CO_2 concentrations that have been correlated with climate changes before our species existed. The latter contain both long term and more abrupt changes. The slower changes are driven by the Milankovitch Cycles (Hag et al. 1987; Ruddiman 2003). These are periodic variations in the way the sun's energy impinges on our planet. Via complicated feedbacks those changes have altered the CO_2 fluxes between the atmosphere and the planet over tens and hundreds of thousands of years which in turn have caused dramatic changes in the climate of the planet and in the sea level (Hag et al. 1987; Casebow 2008). Besides the longer term climate changes, there have been and will be more rapid changes caused by catastrophic carbon cycle events involving fluxes between the ocean, land, and the atmosphere. These are the so called methane or carbon dioxide burps (Skinner 2012) that have frequently, by geological timescales, caused catastrophic changes in the climate over the four billion years our planet has existed. The source of the carbon from the land for these burps is the permafrost and from the deep ocean is the storehouse of methane clathrates or hydrates. They store an enormous amount of carbon, more than is in the atmosphere. These burps are triggered by warming and are one way our current human heating of the planet could turn catastrophic via feedbacks. This threat can be plausibly prevented by controlling the CO_2 concentration.

These burps are another example of how the complex earth system can translate a relatively small and slow CO_2 changing concentration via feedbacks into a rapid and catastrophic change in the climate. These transitions can be associated with a Complex System making a transition from one chaotic attractor to another (NSF Advisory Committee for Environmental Research and Education 2009). They are also an example of a change that can be avoided by stabilizing the CO_2 concentration outside the region that could trigger such releases as they have in the past (Skinner 2012). It should also enable us, for example, to avoid a future in which there will be a glacier covering the New York City area as has been the case several times in the past (Sanders 1994b).

This proposed chaos control system will also involve the unique feature that human decision making based upon knowledge will be an important part of the control system. This is because in the Anthropocene era the climate dynamics are being impacted by our emissions of CO_2 and thus we need to manage them. Most importantly when human emissions dominate the changing CO_2 in the atmosphere they set the timescale for the changing CO_2 fluxes (Canadell et al. 2007); without human impact the timescale for change is much slower (Hag et al. 1987; Casebow 2008). Human control is also feasible because the large thermal mass of the planet means the temperature changes slowly allowing relatively long times to decide on the size and nature of the CO_2 flux feedback and time to implement it. This is very important because it transforms a theoretical concept into a practical approach. Our capacity to adjust the net CO_2 atmospheric flux to zero, CGCC, and stabilize the climate is a consequence of our utilizing carbon based fuels to provide the energy needed for six billion humans. It is, of course, clear that we are not currently using our capability to CGCC. It is essential that we do so because without our intervention all life on our planet will be impacted by the feedbacks of both our own emissions and those of the planet which over time have led to great, and in many cases, destructive variations from the conditions we know today.

It is expected that in the future the models we have and are developing for understanding our planets shorter term climate dynamics will improve. Those models, together with increasing data on the planets state and its changes provided by our current and growing network of global sensors, will be used to provide more detailed and quantitative basis for being able to anticipate changes in advance of them occurring. This enhanced capability will be useful for providing effective adaptive human control and stabilization of the climate. Some thoughts on what amount of advanced notice, planning horizon, that may be possible will be specified after presenting the control system and its operation. Using more technical terms, the rate at which complex systems various states diverge is characterized by its Lyapunov exponent. The application of this concept to spatially extended systems is analyzed by Paladin et al. (1994).

The same models and empirical studies, together with a rapidly growing amount of historical data about the Earth's climate history, have already shown that disruptive changes in the Earth climate system have been accompanied by temperature changes (Hag et al. 1987; Casebow 2008). They have also shown the very distinctive role played by CO_2 concentration in the atmosphere on impacting the temperature of the planet (Lacis et al. 2010). They have identified the impact on water vapor in the atmosphere as one effective positive non-linear feedback mechanism that enables relatively little energy be expended to effectively control the much larger energy that would be needed to directly control the temperature itself (Lacis et al. 2010). If the partial pressure of CO_2 in the air is changed it in turn alters plant growth, ocean chemistry and water vapor in the air.

Below it will be explained that by controlling one small component of the Earth Human System, one small component of the carbon cycle, and indeed of small component of the atmosphere itself, CO_2 at 400 ppm, one can bring long term climate stability to our planet and thus to the life it supports. It is important to be clear what can be stabilized by CGCC at a given ambient concentration of CO_2 and what cannot. This approach involves no attempt and will not directly control the weather nor many other fluctuations that will still exist within a fixed atmospheric carbon concentration with a CGCC. For example, there can still be a range of carbon fluxes between the atmosphere and the oceans and land that satisfy CGCC though they are expected to be small in magnitude if constrained by CGCC.

Thus, adaptation will be easier and much less costly if CO_2 is at a fixed level of concentration in the atmosphere. If it and the average temperature change enough to alter the sea level, and shift the local temperature extremes as well as alter the severity of floods and storms than adaptation will be much costlier. As those changes emerge they will alter the carbon cycle fluxes which can further accelerate and increase the change of the climate. For example, it is understood that the oceans take centuries to change their temperature, much longer than the response to the changing partial pressure of CO_2 in the atmosphere (Archer et al. 2009). Recently Foster concluded that CO_2 has the dominant role in determining the long term, on the scale of centuries, sea level (Foster et al. 2013). That work also concludes that by studying the planet's past that, at 400 ppm, the long term sea level is 9 m above its value today. This suggests that it may be desirable to return to pre-industrial revolution CO_2 levels of 300 ppm. Returning to 300 ppm would ensure the adequacy of the existing infrastructure and avoid the costly changes needed to strengthen the infrastructure as well as adapt to the eventual sea level rise.

This view makes clear that adaptation could be extremely costly and also futile without climate stabilization. There are other positive feedbacks of CGCC on the human system itself that were previously published (Chichilnisky et al. 2009). That paper provides the basis for asserting that by removing the negative feedbacks, so called externalities, like climate change and environmental destruction from the way we generate and utilize our energy generation, that we can create new economic dynamics that will in fact stabilize the global economic system. If in addition to using renewable energy, we produced our carbon based fuels using CO_2 from the air and hydrogen from water, our use of those carbon based fuels would close the carbon cycle like the rest of life does. While not the subject of this paper it is plausible that unlimited energy that is allocated to address the conditions that create economic catastrophes could in fact be the basis for stabilizing the complex global economic system. It is certainly true that our positive progress as a species and those of all species of life has been to convert carbon based energy to useful work and thus it is plausible that CGCC will both stabilize the climate and unleash global prosperity. This will be further discussed after the more detailed analysis of CGCC on the climate.

We will first review the crucial and distinctive role that the carbon cycle is thought to play in determining the climate of the planet. The size of the direct impact of CO_2 on temperature is heavily contested due to differing views of the effectiveness of CO_2 adsorption due to saturation effects and the various feedback mechanisms that can either enhance or reduce the CO_2 concentration in the atmosphere. A key distinction of viewing these issues from the perspective of a chaotic system is that their inherent long term unpredictability means the differing theoretical views will be hard to reconcile. Also, in a system with feedbacks the language of cause and effect appropriate for linear systems is replaced as mentioned above by the concept of the emergence of a future state as a result of the feedbacks. More generally, it is important to note that the Complex System perspective changes the framework for thinking about the threat of climate change and indeed about all catastrophic transitions in complex systems. It implies that the real danger of our current actions is that we cannot inherently predict the future because the feedbacks can dramatically alter the future resulting in catastrophic impacts for our species and the others that are flourishing in the current climate. We will show that those same feedbacks can *plausibly* provide the basis for effective control, thus catastrophe can be avoided by CGCC. The choice of the word "plausible" is important because an unavoidable consequence of the Complex System perspective, together with the inability to conduct experiments—we only have one planet—means we are limited to choosing, in the Bayesian sense, the most *plausible* approach based upon the available knowledge. Just as it is mistaken to claim that CO_2 causes climate change, it is incorrect to claim one can prove, in a reductionist sense, that fixing the CO_2 concentration will, with certainty, stabilize the climate. This important issue will be discussed further after the results of the analysis are presented.

2 The Feedback Control

The state we want to stabilize is where the concentration of CO_2 in the atmosphere is fixed. To accomplish this we need to redesign our carbon based energy system and produce carbon based fuels using CO_2 from the air and hydrogen from water using energy from the sun or other renewable or non- carbon based sources like nuclear. To reduce the CO_2 concentration and effectively cool the planet, we need to capture the CO_2 emitted when we combust our human produced carbon based energy and either store it underground or preferably as bio char or other uses that remove it from the carbon cycle for long periods of time, but provide other positive feedbacks. To increase the CO_2 concentration and effectively heat the planet we will simply release some of the previously stored CO_2. To heat the planet we could of course also burn fossil fuels as we now are doing but as will be shown below in the long run that approach is neither sustainable nor cost effective.

The carbon cycle involves complicated exchanges of CO_2 between the atmosphere, the land, and ocean. These fluxes vary with temperature, atmospheric concentration of CO_2, the PH of the ocean, weathering rate, the rate of plant growth and of course the human contributions. They have feedbacks among them that are central to the chaotic dynamics and importantly determine the net amount of CO_2 in the atmosphere over time and the lifetime of CO_2 in the atmosphere. It will be important that the rates of exchanges other than our human contributions are at time scales that vary from centuries to tens of thousands of years (Foster et al. 2013) and that the more abrupt carbon cycle changes can be suppressed by stabilizing the CO_2 concentration. For our analysis we will choose 1000 years as an average lifetime for the slow changes, a timescale consistent with the time over which climate has changed in the past. There will be both shorter and longer term variations. More generally, our focus in this paper is on the system aspects not the details since knowledge of them will not enable us to predict the longer term future dynamics anyway. The system view still allows us to come to some very important conclusions and a way to provide adaptive control to stabilize the climate.

The total solar energy, TSE, impinging on our planet is 1360 W/m^2 (NASA) which averaged over the earth's spherical surface is 340 W/m^2. This amounts to 5.5×10^{24} joules/yr calculated by multiplying 340 W/m^2 by 5.1×10^{14} m^2 the surface area of the earth and the number of seconds in a year. Of that amount, 70 % is adsorbed on the planet in the surface and the atmosphere, SEA, which is thus 0.7 TSE or 240 W/m^2. The Stefan-Boltzmann relationship determines the equilibrium average global effective black body temperature, TA, at which the earth radiates the net input energy back into space. The Stefan-Boltzmann relationship is

$$E = \sigma TA^4 \tag{1}$$

where $\sigma = 5.670373(21) \times 10^{-8}$ W m^{-2} K^{-4} where E is in W m^{-2}

for 240 W/m^2 TA = 255 K.

Now the mean effective surface temperature of the earth, TS, is measured to be 288 K (Rohde et al. 2013). The difference of 33 K between TA and TS is called the Greenhouse Effect. Interestingly, the 288 K is higher than the 279 K one would get by calculating the effective temperature if the planet was a perfect black body and adsorbed all the energy from the sun. This is because the shorter wavelength of incoming radiation that is adsorbed is reduced more than the emissivity of the longer wavelengths at which the earth radiates causing a net effective trapping of heat. This is what happens in actual greenhouses enabling the temperature inside the greenhouse to be higher than outside the greenhouse. Using Eq. 1, one can calculate the Greenhouse Energy (GE)

$$GE = \sigma(TS^4 - TA^4) \tag{2}$$

to be 150 W/m^2 making the total 390 W m^{-2} on the surface of the earth, SEA + GE, greater than the 340 W/m^2, TSE, of the incoming solar flux.

We now need to relate a change in CO_2 concentration in the atmosphere to a change in the GE and temperature T. Since this is a complex system there are many feedbacks that make it not only effectively impossible to predict the future but also to accurately estimate the expected changes in the atmospheric CO_2 concentration. In fact, the various different estimates for temperature change due to future increases in CO_2 concentration can be related to the different ways the models handle the various feedbacks including impact on water vapor, ocean, and land storage of CO_2 caused by changing the partial pressure of CO_2 in the atmosphere. For example, in (Lacis et al. 2010) they show a factor of three positive feedback factor caused by a change in CO_2 concentration and a change in water vapor. Also there has been much debate about the fact that CO_2 is already at a concentration where because of the exponential impact of concentration on absorption that changes in CO_2 concentration can at best produce a GE response proportional to the natural log of the CO_2 concentration. This has made many even more skeptical because not only is CO_2 such a small component but its impact is muted because it is already so very effective in trapping the heat. This is a very contentious issue which is discussed in a balanced way in (Best 2010).

In any case, because it is a complex system, the empirical approach followed by Muller et al. (2012) makes the most sense from a systems perspective. We have done an experiment over the past 250 years as we rapidly increased our energy production and seen what state has emerged. This can be used to estimate the change in GE that is correlated with the change in concentration of CO_2 in the atmosphere. Thus, this approach effectively is saying that it is known what emerged in the past 250 years as a result of a changing CO_2 concentration even though the impacts of all the feedbacks are hard to specify quantitatively. The whole debate using the language of cause and effect becomes from the chaotic systems perspective transformed into the analysis of the new climate that is emerging out of a set of feedbacks. In this context it is important that not only the recent data but a much greater amount of data over much longer times in our planets past also shows that temperature and CO_2 are correlated (Shakun 2012). Our unintended human experiment that Muller analyzed adds persuasive evidence to the correlation and of course led Muller to conclude that the recent CO_2 increases and temperature change were probably connected and thus human caused. As he has correctly noted, this conclusion does not prove CO_2 caused the change in temperature but it and the other correlations are the most *plausible* explanation of the linkage between temperature change and CO_2 concentration in the atmosphere. The same *plausibility* position is adopted in this paper in terms of the claim that fixing the CO_2 concentration will stabilize the climate.

Muller fit the 100 ppm change in the CO_2 concentration between 1750 and the end of the century to the temperature change that has been observed. By using the logarithm of the CO_2 concentration changes in the atmosphere for his fit he was consistent with the conclusions that the CO_2 adsorption is saturated. He also included the shorter term impact of volcanic activity which can be quite dramatic in terms of the cooling they produce. It is beyond the scope of this paper to deal with shorter term changes that might also be offset by increasing the CO_2 in the atmosphere by releasing stored CO_2. His results are reproduced here in Fig. 1 (Muller et al. 2012). The observed change over the 250-year period, 1.5 C., is due to the CO_2 variation with the volcanic activity providing shorter term variations. This empirical approach effectively includes the effects of all the feedback factors considered in the various models and its results for changing CO_2 impacts are well within the range of the various models, which indeed themselves are tested in many cases by trying to recreate the past. However, the important distinction about chaotic systems is that even though the dynamics are deterministic one cannot predict the future. That is why one needs to create an adaptive control system that can change its feedback based upon new emerging trends.

We can use Fig. 1 and the observed changes to determine the increase in GE per tonne of CO_2 added to the atmosphere. Taking the derivative of Fig. 1 with respect to temperature it is easy to show that

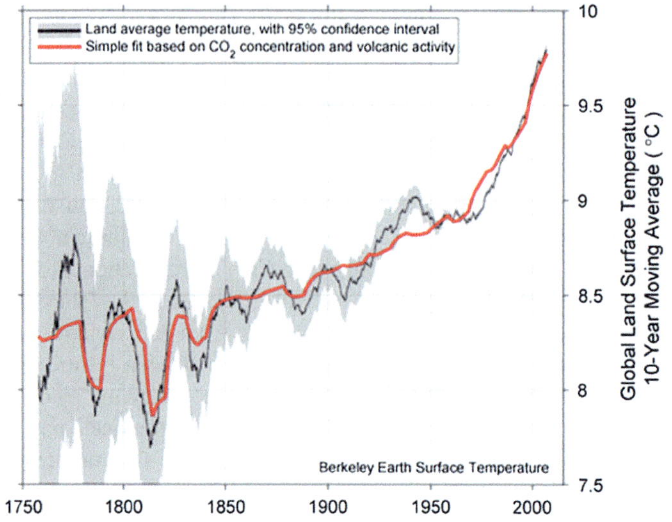

Fig. 1 Results from Muller et al. (2012)

$$\text{Delta GE} = 4\,(\text{deltaT}/288\text{K})\,(\text{SEA} + \text{GE}) = 8.1\ \text{W/m2} = 1.3 \times 10^{23}\text{joules per year} \tag{3}$$

for a deltaT of 1.5 C. This is roughly equivalent to a 2 % change in the intensity of the sun hitting our planet.

A 100 ppm change in the concentration of CO_2 in the atmosphere, 25 % of TCO_2, is equal to about 0.75 × 10^{12} tonnes of CO_2 so

$$\text{Delta GE}/0.75 \times 10^{12}\ \text{tonnes} = 1.8 \times 10^{11}\ \text{joules per tonne of } CO_2 \text{ per year} \tag{4}$$

It is worth noting that the difference between a linear dependence and a logarithmic dependence of going from 300 to 400 ppm is a factor of 500. If a planet had 100 times less CO_2 and assume for illustrative purposes that such a planet had its temperature determined by CO_2 feedbacks than from a chaotic system perspective it would be both much more sensitive by a factor of 100 to the change in CO_2 but similarly it would take 100 times less of change in CO_2 to maintain control.

To determine the effective enhancement factor of using CO_2 to control the temperature of the planet one needs to calculate the energy of combustion that puts CO_2 into the atmosphere and the energy needed to capture CO_2 from the air. The enhancement factor for warming, EFW, can be calculated as the ratio between the carbon-based energy produced for our use compared to the impact on GE. Coal will be used to make the estimate. The real situation with our diversity of energy sources is much more complicated. For coal about 1 kg of CO_2 is emitted for every kilowatt hour of useful electricity produced at about 33 % efficiency. This means we get 3.6×10^9 joules/tonne of CO_2 which means the EFW is given by

$$\text{EFW} = (1.8 \times 10^{11})/(3.6 \times 10^{9}) \times \text{YE} = 50 \times \text{YE} \tag{5}$$

where YE is the years the CO_2 emitted remains in the atmosphere.

As discussed above for this analysis we will use 1000 years which results in an enhancement of 50,000. The exact number is not important, but that it is large. This is because it both substantiates the concern about the climate impacts of our burning fossil fuels, but also substantiates the use of CO_2 to stabilize the temperature of the planet. Again here it is worth restating that there is a distinction between and effectiveness in applying the control signal and effectiveness of the internal feedbacks of the system itself to stabilize the CO_2 concentration and thus the climate.

The energy to remove CO_2 from the atmosphere and store it in biochar or in underground storage facilities can also be estimated based upon known information. The main energy to do this is to capture the CO_2 from the air. It has been shown that primary amines can effectively capture CO_2 at the low concentration in the atmosphere (Jones 2011). It has a heat of reaction that needs to be expended to remove the CO_2 from the amine that has captured it exothermically. That energy is less than 2×10^9 joules/tonne. Other energy requirements are relatively small including those for sensible heat losses and compression and to move the air over the capturing medium. Significant heat recovery options exist so we will use 3.6×10^9 joules/tonne as a conservative estimate as to what is possible and which makes our systems analysis calculations easier by using the same energy for both capture and emissions processes. The important conclusion is that both the heating and cooling of the planet with CO_2 have the large enhancement factors needed for achieving climate control with the use of a relatively small amount of energy. The value for removing CO_2 is also consistent with energy requirements for capture at the higher concentrations in the flue gas (Singh et al. 2001) since they also use amines in some of the existing capture processes. There are other proposed approaches that require much less energy that can make air capture even more energy efficient (Wang et al. 2011). They would just increase the energy effectiveness of using CO_2 to cool the planet.

But of course the removal of CO_2 from the flue gas of fossil fuel based carbon energy does not reduce the concentration in the atmosphere—it only slows down the rate of increase. This is of course desirable but it cannot prevent continued increases in the CO_2 concentration in the atmosphere at the very least because it cannot deal with the transportation sector. Thus, one really needs negative carbon (Eisenberger et al. 2009), carbon removed from the atmosphere, to have a way to stabilize the carbon cycle of the planet. It is worth noting for this analysis that the electricity production is only about 33–40 % efficient and that the CO_2 can be captured using the low temperature heat left over after electricity production (Eisenberger et al. 2009). There is three times as much heat available as needed to remove CO_2 in the case of coal and about a factor 5 in the case of natural gas. This offers energy and cost saving by using co-generation of power and heat for controlling the CO_2 concentration. The co-generation energy efficiency and cost benefits we get from providing from the same source for both electrical power and heat for our buildings can be replicated for providing our power needs and climate control for the planet. Thus, the estimates made below will significantly overestimate the actual impact on incremental human energy production needed for planetary climate control because the low temperature heat remaining after power generation or many other processes has limited utility in many circumstances and is usually removed by cooling towers or cooling water.

To calculate the enhancement factor for reducing the CO_2 concentration, EFC, one has a similar relationship if YC is the number of years between when the CO_2 that is captured is effectively re-injected into the atmosphere either intentionally to offset changes in fluxes via feedbacks between the oceans and the land and the atmosphere. Shorter times might result from the release from the biochar or leaks out of the storage site but that again is a detail which will not affect the main conclusions from this system analysis. This is clearly more under our control but to be consistent with expected variations in carbon emissions other than our own YC should have the same value in order to be able to compensate for them. Thus,

$$\mathrm{EFC} = 50 \times \mathrm{YC} = 50,000 \tag{6}$$

3 Using the Control System

For our system analysis purposes the important specifications of the carbon cycle are the total amount of CO_2 in the atmosphere, TCO_2, which today equals about 3×10^{12} tonnes, the amount we emit currently per year, HNF, which is about 3×10^{10} tonnes per yr which is offset by net storage currently of about ½ in the ocean and the land, ENF, or about 1.5×10^{10} tonnes per yr, resulting in a total net CO_2 flux, TNF, of 15 gigatonnes per yr into the atmosphere. It is important to recognize that over time the magnitude of ENF changes and can even change sign and add CO_2 to the atmosphere. In fact in the planet's past it was those changes driven by many feedbacks that impacted the average temperature of the planet and thus the climate.

To close the carbon cycle in a way that provides the desired control one needs to control the net CO_2 flux into or out of the atmosphere to provide climate stabilization by making the net flux zero. Namely,

$$\mathrm{TNF} = \mathrm{ENF} + \mathrm{HNF} = 0 \tag{7}$$

where TNF is the total net EHS CO_2 flux in or out of the atmosphere in gigatonnes per year and ENF is the net planetary flux and HNF is the human net flux also in gigatonnes per year in or out of the atmosphere. The condition for stabilizing at a given concentration becomes

$$\mathrm{HNF} = -\mathrm{ENF} \tag{8}$$

and the concentration one stabilizes at constrains climate of the planet as discussed above.

If one was only dealing with the long time changes that characterized most of the earth's temperature variations prior to any human impact we would get the full benefit of the enhancement factors in equations because the changes occur over a long period of time and our control variable CO_2 in the atmosphere has a lifetime that matches the time scale of the changes one wishes to stabilize. This means regulating in a situation where the human carbon cycle itself was closed would be relatively easy. For example assume a potential ½° centigrade change in 1,000 years for the average temperature of the planet. This is rapid compared to most changes prior to human impacts and to the recent experience in which over the 8000 years before 1750 the temperature changed less than ½°. Over the long term these changes result in periods of high glaciation or warmer periods over the tens to hundreds of thousands of years characteristic of the Malinkovitch Cycles. The carbon cycle would in that case be having changes in TNF that were either positive or negative.

A change of ½° in 1,000 years could be counteracted by changing the CO_2 concentration by about 250 gigatonnes of carbon dioxide. If one did this over 1,000 years it would require just 0.25 gigatonne per year or about 1×10^{18} joules per year additional. We currently produce 6×10^{20} joules per year so we could exercise the needed control with much less than 1 % of the energy we now produce and much less than what we currently use to provide temperature control in our homes and offices around the world which is 7×10^{19} joules per year (IEA) and much less than our growth in energy per year.

To understand how our growing use of fossil fuels and our failure to CGCC has increased the challenge to stabilize the planet it will be useful to divide our analysis of human impacts into three cases. Because of the positive correlation between our capacity to produce energy and our impact on the carbon cycle, the extra capability needed for stabilization continues to be able to be implemented with relatively little impact on our energy supply for other purposes. This follows from the basic premise of chaos control, that same feedbacks that can destabilize the system can be used to stabilize it. This is, of course, encouraging, but only if we decide to CGCC.

For the first two cases we will divide the period covered by Muller into the years 1750–1950 and between 1950 and 2010. In the years between 1750 and 1950 HNF grew to the extent that it balanced out the negative fluxes of CO_2 into the land and the oceans, ENF, reaching about 7 gigatonnes by 1950 (Canadell et al. 2007). Thus, over those 200 years our energy needs for CGCC would have been about .035 gigatonnes per year increase or about 10^{17} joules per year reaching a total of about 2×10^{19} joules per year by 1950. This is in fact roughly what happened, but with a greater increase, in the 1900s and less before. The important point is that we were roughly in balance and ENF was also low.

The interesting observation is that the developing of our energy infrastructure was the right control strategy for about 200 years and it did produce a relatively stable climate since the temperature only increased by ½° centigrade and the partial pressure of CO_2 only increased by 37 parts per million. So our first practice of control was by accident and in fact cost us nothing but it makes a big difference that we decided to grow our energy from 1950 to 2010 by burning fossil fuels rather than producing our own carbon based energy from CO_2 from the air and hydrogen from water. Thus about 75 years ago we went through the historic transition of TNF = 0 due to human impact. With hindsight that transition effectively marks the beginning of the Anthropocene Era, with HNF, human net emissions into the atmosphere, being on the scale of the net ENF changes out of the atmosphere. Until then our emissions were not on the same scale as ENF though our deforesting activities were (Canadell et al. 2007). It is important to note that our emissions are still very small, more than one order of magnitude less, compared to the flux of CO_2 between the land and the ocean with the atmosphere. They have large fluxes in both into and out of the atmosphere that produce relatively small net emissions. Our first control objective will be to again get to TNF = 0.

In the last 60 years we continued to develop our energy capability primarily using fossil fuels and now are effectively emitting 30 gigatonnes per year. We increased our energy usage and emissions in 60 years three times as much as we did in the previous 200 years. Because of feedbacks it also seems that the usually slowly changing ENF is also changing faster than usual since it has doubled since 1960 and is now removing 15 gigatonnes per year, resulting in a TNF of about 15 gigatonnes per year into the atmosphere (Canadell et al. 2007). The increasing negative feedback is presumably the result of the increased partial pressure on the pickup rates of the plants and the oceans. One less sanguine interpretation is that the partial pressure effect would have an immediate impact but that there is a lag in the temperature increase of the planet itself versus the air above it (Archer et al 2009). If this was the case, this shorter term negative feedback would be overcome by a longer term positive feedback of increased water evaporation from the oceans with the factor of three enhancement it provides triggering a more significant heating. According to Foster (Foster et al. 2013) this is in the fact the case.

The growth of TNF to 15 gigatonnes has caused an increase of 74 ppm in CO_2 and a warming of 1 degree centigrade. We can analyze the difference between whether in those 60 years we burned fossil fuels or burned carbon based fuels made from CO_2 from the air and hydrogen from water. Since burning human made carbon is carbon neutral the partial pressure would have not increased and we could have provided for our increase in energy with no need to expend any additional energy to provide any first order control. Our existing fossil fuel infrastructure would continue to offset the ENF of –7.5 gigatonnes per year. Presumably the ENF would continue to change slowly at close to the 7.5 gigatonnes per year so that any adjustment to compensate would be easily addressed as was described above. Importantly, we would be facing the expected increase in our energy use in the next

60 years of about a factor of four without the threat of climate change if we had continued to expand our energy capability using renewables and carbon neutral based fuels without any need for any capture costs. This has many positive impacts on the main challenges we are now facing in energy security and economic development as has been described elsewhere (Chichilnisky et al. 2009). More importantly, we would have already created the energy infrastructure for transitioning into long-term slow climate variations whose control should protect us for a long time to come.

Let's compare that with the situation we now are facing. We are making efforts to reduce our carbon intensity so if we only increased HNF by a factor of three instead of four we would be out of balance by 75 gigatonnes per year if ENF did not change. As discussed above we really have no idea if the past increased rate of ENF will continue so we will consider the case where it does though slightly slower rate so that TNF only increases by a factor of two. The temperature effect on the ocean component of ENF lags the partial pressure dependence and is the opposite sign. This together with the positive feedback with temperature on the amount of water vapor creates the potential of an increased rate of warming in the distant future. Limiting the net flux will suppress this positive feedback loop. For our example the TNF would be about 45 gigatonnes per year but we would have four times the energy production capability. To provide the capability to remove 45 gigatonnes in 60 years would require we increase our capability to remove CO_2 from the atmosphere on average by 0.75 gigatonnes per year reaching the 45 gigatonne stabilization condition 60 years from now. This means we need to allocate an extra 2.7×10^{18} joules per year reaching a total of 1.6×10^{20} joules per year in 60 years. To put that in perspective our use of energy would have grown to 1.8×10^{21} joules per year of which about 2×10^{20} joules per year would be allocated to providing climate control in all the buildings in the world assuming it would grow proportional to our energy use.

So we can still provide the needed control without stressing our energy system too badly. As previously mentioned, the energy need is overstated since our energy system is only about 40 % efficient, so we have much more low temperature heat available to use to remove the CO_2 than is needed. Since the power sector provides ready access to such heat and it is more than 25 % of the energy use it can provide the heat needed by co-generating power production with CO_2 removal from the air. But as was the case during the past 60 years the partial pressure of CO_2 would continue to increase during this period by almost a factor of two because of our existing installed base of fossil fuels is making HNF bigger than ENF. To prevent this we need to double our rate of CO_2 removal which is still feasible since it primarily can use low temperature heat. More detailed calculations of what is needed have been presented elsewhere (Eisenberger et al. 2009). But the fact still remains that solving the problem in the next 60 years as opposed to the previous 60 has made the energy cost significant as opposed to negligible.

4 Closing the Carbon Cycle

One can consider the above analysis of the energy needed to stabilize the carbon cycle of the planet as providing an estimate of the global operating costs for removing CO_2 from the atmosphere as the critical step in making our energy from CO_2 from the air and or capturing its emissions when combusted. The other main cost is the capital required. If we need to build 45 gigatonnes of air capture capability in 60 years, one needs on average (will clearly start slower and be at a higher rate later) 0.75 gigatonnes of new capacity each year. To avoid any contentious debate about the cost of air capture which might distract from the main point being made, one can even use a very conservative value of 600 dollars of capital per tonne of CO_2 (Eisenberger et al. 2009). Thus for each of 60 years we will construct about 0.75 gigatonne of new capacity requiring 450 billion dollars per year which is under 0.75 % of the current GDP. In 60 years we will, even at these high values for capital, have about the same installed capital cost for the heating and cooling of our homes. The specific values are not important and will become lower with time and learning. What is important is that both the energy and capital demands on the human system can be met without any catastrophic or even significant strain on our species energy and capital capabilities. In fact as shown below the real cost is much less. This correlation that exists between the rate at which we are developing as a species and, as stressed here, the ability to make the climate more stable follows naturally from the feedbacks that exist via carbon. Until the industrial revolution they were decoupled, so that at its core the Anthropocene era of the planet will be characterized by how we use our capability to manage these feedbacks.

It is important to understand the difference between Closing the Global Carbon Cycle using carbon from fossil fuel sources and carbon from the air and hydrogen from water that can be converted to a fuel using solar energy by various approaches including algae and electrolysis. We can even use CO_2 captured from the air for geothermal heat recovery in which a significant portion gets sequestered while producing electricity (Eastman and Muir 2013). If we continue to use fossil fuel sources we are taking carbon that is already effectively sequestered in the land and emitting it into the air. This also includes carbon dioxide from natural domes that is currently being used for enhanced oil recovery (DiPietro et al. 2012). While for the case of using CO_2 from the atmosphere it starts in the atmosphere and upon combustion ends back in the atmosphere. This is inherently at least carbon neutral and can is some cases be made carbon negative, reduce the CO_2 n the atmosphere. Thus if we continue to expand our use of fossil fuels to meet our energy needs we will be increasing our risk of causing destructive climate change. However, if we use energy we made ourselves using CO_2 from the air we can meet our needs without changing TNF.

We will continue for some time to use fossil fuels so we would have to first capture an equivalent of its emissions just to stay even. This will not be the case by cleaning fossil fuel plants since going beyond 90 % capture is very costly. In addition, the transportation emissions will continue to increase their amplified negative impact so we cannot provide the needed cooling using fossil fuels not to mention that the problem will continue to worsen because of the long time the CO_2 stays in the atmosphere. Using CCS the CO_2 concentration will continue to increase, though at a slower pace. Fossil fuels are only good for heating the planet exactly because they transfer CO_2 from the land to the atmosphere. In the case of using fossil fuels we will have to pay to both cleanup the fossil fuel sources and the cost of developing new carbon based energy sources as well. However, for the case of making our own carbon based energy we can provide for our energy needs either with zero impact on TNF or if desirable to add CO_2 capture and cool the planet by making TNF negative. This can also be done slowly in the case of some algae approaches that convert about 20 % of their carbon into biochar which effectively sequesters it in the land. This would enable us to drive our cars while cooling and fertilizing the planet. One could also produce electricity using CO_2 from the air as the geothermal fluid which sequesters 20–50 % of the circulating CO_2 and thus one could also turn on the lights and cool the planet (Eastman and Muir 2013). Many other products such as polymer, chemicals, and structural materials can be made from carbon from the air rather than fossil fuel sources, and provide their benefits while, in many cases, effectively sequestering the carbon. We should increase the use of carbon based structural materials as opposed to those based upon materials like steel made from material extracted from the planet.

5 Conclusions

One can draw several conclusions from the above analysis. The first is that we need to have a carbon based energy infrastructure and it should be at the scale of the changing fluxes between the land and oceans and the atmosphere. There is an enormous cost benefit of staying with carbon based fuel in addressing the challenge of climate change. First is of course that we do not have to spend the enormous amount of money to develop a new infrastructure. But even more interesting is that the same infrastructure can be used for heating the planet or cooling it: the same refinery, the same automobile, the same distribution system. All we have to do is switch from using fossil carbon to carbon from CO_2 from the air just like the plants do. We, as opposed to plants, will decide whether to emit CO_2 into the atmosphere upon combustion or to capture and store it as described above. In the future if we needed to increase the CO_2 in the atmosphere because of a changing ENF we could release the stored CO_2 at very low cost.

There are other ways one could close the carbon cycle like increasing the weathering of rocks or by fertilizing the oceans or to offset the impact on temperature of increased CO_2 in the atmosphere by the so called management of solar fluxes, but they are less desirable for many reasons. First and foremost is it transfers a problem we have created in HNF to the ENF system which at a minimum raises the risk generally of unintended consequences like other geo-engineering solutions and in particular to the living systems those carbon reservoirs support. What is unique and important from a control perspective is that using our own emissions as the control feedback to first order just balances out our perturbation on the system by making HNF close to zero. In that sense those efforts are to first order pollution control rather than geo-engineering. Thus in that sense out initial efforts are indistinguishable from pollution control except for the requirement of producing our own carbon based energy. This both prepares us for the long term control as well as provides a long term solution to our energy needs. By co-generating our energy production with our control capability we can ensure our capability to provide control feedback matches our ability to destabilize the carbon cycle like the rest of the living system that links CO_2 and energy production. It is what carbon based life does and thus would seem to remove various unintended feedbacks that might arise if one separated our energy production from our CO_2 impact.

The above analysis also makes clear that the real cost of carbon neutral or negative technology should be reduced by the extra capture cost we would need to expend to capture CO_2 from the flue gas. This in turn means that our carbon tax or market should reflect this cost. From this perspective the carbon credit should not go to the fossil fuel plant that cleans up its flue gas as it now does but rather only to those who produce energy using carbon from the atmosphere or renewable energy. In this view it makes no sense to cleanup existing carbon sources but rather that we spend our energy and capital resources for energy produced from carbon from the air. If capture from flue gas costs $50 per tonne of CO_2 and this was provided to carbon generation capability using carbon from the air it would reduce the costs significantly for generating carbon free energy. At 1 kg of CO_2 per kwhr of electricity this amounts to 5 cts per kwh. At 10 kg of CO_2 per gallon of gasoline this would be 50 cts per gallon. If with those reductions one could match the current cost of electricity and fuel production using CO_2 from the air the net incremental social cost of climate control would be zero. This does not consider the other social costs of pollution, land degradation and maintaining access to the poorly distributed fossil fuel sources which increase the social cost of our fossil fuel based energy by up to a factor of three (Greenstone and Looney 2012). This would mean we could in fact double the carbon tax or market price benefit of using carbon neutral energy and it still would cost the economy less than cleaning up fossil fuel sources. Of course one gets the increased energy security benefit at no extra cost since CO_2 from air and hydrogen from water can be produced locally essentially globally.

Finally, it makes very clear the increased cost we have paid from delaying the decision to CGCC and will pay if we delay further. For each new energy capacity that we create that is from fossil fuels, that is allowed to operate for thirty years without capture and for using fossil fuel for our cars for that time as well, will require we take out 30 times as much carbon if we decide thirty years from now to begin to approach CGCC correctly. A slightly different perspective is provided by chaotic control. Because of the difference between the human timescales of change and those of ENF, once one reaches the transition in which $TNF = 0$ we return to the need to only stabilize the slowly changing ENF for the foreseeable future with no significant additional cost in energy or capital for stabilization. So in this sense the chaos control strategy is to first get the system to TNF = 0 and then control. Once we get the control we could also slowly under our control adjust the CO_2 concentration to a different temperature if we decide we do not like the 400 ppm world of TNF = 0. This could be done by the combination of producing fuel and biochar from CO_2 from the air and hydrogen from water with little disruption and capture the additional benefit of improving the fertility of our soils globally by storing the carbon in biochar. A related conclusion is that the above analysis makes clear the increased risk we are taking during this period where our rapid changing of the CO_2 concentration is driving the climate dynamics. As noted above it appears that the ENF is also changing faster than usual in response to the changing HNF and that the shorter term negative feedback created by the increased partial pressure of CO_2 could turn and become a much larger positive impact as the oceans and land warm overtime.

Thus the future energy infrastructure that we can sustain for the life of our planet is to make our fuel from CO_2 from the air and hydrogen from water like the rest of life does. Like the rest of life we need to power our fuel making with energy from the sun or other renewable sources and like the rest of life we can store the energy produced when the sun shines in the form of fuel that can be burned when the sun does not to provide 24/7 dispatchable energy. As our energy capability continues to grow beyond the capacity needed to provide effective stabilization of the temperature of the planet, we should of course provide that energy using renewables as well. Of course we also will need to make our fuel using renewable energy so continuing to develop those sources and increasing our energy efficiency remain important objectives. In this regard nuclear becomes a very effective energy source to co-generate both liquid fuels via electrolysis and the capture of CO_2 from the air using their copious amounts of low temperature heat. Furthermore, the hydrogen they can produce can together with the CO_2 captured from air be utilized to make all the liquid fuels and other carbon based products we need.

It has, of course, not been proven here that there is not another variable we could control that might provide more effective and efficient chaos control for our climate. However, the fact that we can provide effective control as described above as part of the process of meeting our energy needs provides the unique energy efficiency and cost benefits that characterize many so called co-generation technologies. Intuitively, it makes sense and the above arguments make it *plausible* that the most effective and efficient control would be by using the carbon cycle given its critical role in supporting life. To do so wisely it will be very important to continue to understand the changing state of our planet by using sensors to make a time record of the critical parameters. It is also important to develop further our models so they can extend the time horizon over which they can reliably predict future evolution of the system. Chaos theory makes clear that the chaotic states will exponential diverge over time from a given starting state. Because of the inertia of the earth system those differing possible futures should be very similar for a reasonable time period, long enough that we will have time to react (Paladin et al. 1994). That is the case now in the sense we know the planet is warming, we know the CO_2 concentration will continue to increase because our emissions are larger than the net flux out of the atmosphere. Thus, we should act now and in doing so we should also recognize that we must produce negative carbon. Cleaning up fossil fuel plants will not solve the problem and will just drain valuable resources that could better be used for installing renewable energy and carbon neutral sources.

Our actions will shape the future evolution of our planet. In the Anthropocene era we are a global force which many view as a threat but hopefully the above will also make clear that we can protect us and other life from catastrophic climate change that has been characteristic of the past history of our planet. It is worth emphasizing again that the reason we can have such a big impact in spite of the fact that our total energy producing capability is 4 orders of magnitude less than provided by the sun is because the earth is a complex system. The enormous amplification of more than 4 orders of magnitude and the feedbacks created via the carbon cycle creates the seemingly implausible result that we can change the temperature of the planet as the evidence clearly shows we are doing. However, the basic property of the complex system that enables our actions to cause catastrophic climate change also enables us to prevent it from happening using less energy and cost than is needed to globally control the temperature of all our homes.

The very capability that has our development threatening other species with a new risk of extinction also provides us with the capability to provide increased protection to them and ourselves from such a fate in the future. It is argued here that the most *plausible* approach is to CGCC and, further, in the shorter term, we need only practice pollution control without any increased risks that geoengineering solutions cannot avoid. The above should also make clear that it really makes no difference in the end whether we are the cause of the changing the CO_2 concentration in the atmosphere or it is due to fluxes from the land or the sea as it has been in the past, they both can result in catastrophic consequences.

Carbon based life on our planet is generally characterized by strong feedbacks between energy production and species fitness. For our species this is accomplished by via meeting our needs for health, security, shelter and food, all of which can be positively impacted by using energy wisely. In becoming as successful as we have, we have become a global force able to alter the climate and ecosystems of the planet, the Anthropocene, which has connected climate stability and our welfare. But as this paper suggests we can turn this threat into an opportunity to both stabilize our planet and stabilize our own welfare. The rest of life has had no choice but to have their fate determined by the feedbacks of the chaotic earth system—we do.

Acknowledgements The author would like to acknowledge many useful discussions over the years with Graciela Chichilnisky, the author of the Carbon Market, on the critical role of CO_2. Matthew Realff helped clarify the distinction between the feedback control signal and the feedbacks that actually stabilize the planet. Roger Cohen, a climate skeptic, diligently and patiently critiqued earlier drafts of the paper and helped improve it significantly on the important distinction between correlation and causation. Klaus Lackner provided useful feedback that helped clarify some ambiguity in early formulations of this paper. Others including Wally Broecker, Tim Fox, Chris Jones, Eric Ping, and David Shearer provided useful comments on earlier drafts. The author also wishes to acknowledge that this paper was developed during preparation for a new course at

Columbia University for their new program on Carbon Management and that he is also one of the founders, together with Graciela Chichilnisky, of Global Thermostat, a company developing technology to remove CO_2 from the air. This paper expands upon an earlier published article (Eisenberger 2013).

References

Archer, et al. (2009). *Annual Review Earth Planetary Science 2009. 37*, 117–34. First published online as a Review in Advance on January 26, 2009. The Annual Review of Earth and Planetary Sciences at earth.annualreviews.org. Article doi:10.1146/annurev.earth.031208.100206.

Best, C. (2010) Doubling CO_2 and basic physics. Retrieved May 16, 2013. http://clivebest.com/blog/?p=1169.

Boccaletti, S. et al. (2000). *Physics Reports, 329*, 103–197.

Canadell, J. G., et al. (2007). *PNAS, 104*, 18866–18870.

Casebow, A. (2008). Evidence of past climate change. *La Societe Guernesiaise*. Retrieved May 16, 2013. http://www.societe.org.gg/planetguernsey/download/3_Historical.pdf.

Chichilnisky, G., et al. (2009). *International Journal Green Economics, 3*(3/4), 414–448.

DiPietro, P. et al. (2012). *SPE Economics & Management*, pp. 69–74.

Donner, R., et al (2009). *European Physics Journal Special Topics, 174*, 1–9.

Eastman, A. D., & Muir, M. P. (2013). EGS and the utilization of highly pressurized CO_2 for purposes other than power generation. GreenFire Energy. Retrieved May 16, 2013. https://pangea.stanford.edu/ERE/pdf/IGAstandard/SGW/2013/Eastman.pdf.

Eisenberger, P. (2013). Closing the Global Carbon Cycle: Climate Stabilization and Sustainable Carbon Energy, *Energy & Environmental Science*.

Eisenberger, P., et al. (2009). Environment. *Climate Change, Energy Economics and Energy Policy, Issue, 20*(6), 973–984.

Foster, G. L., et al. (2013). *PNAS, 110*(4), 1209–1214.

Greenstone, M., & Looney, A. (2012), Paying Too Much For Energy? The True Cost of Our Energy Choices. In CEEPR WP 2012-002. Retrieved May 16, 2013. http://web.mit.edu/ceepr/www/publications/workingpapers/2012-002.pdf.

Hag, B. U. et al. (1987). *Science, 235*, 1156–1167.

Jones, C. (2011). *Annual Review of Chemical and Biomolecular Engineering, 2*, 31–52.

Kauffman, S. (1995). At home in the universe: the search for laws of self-organization and complexity, *Viking*.

Lacis, A. A., et al. (2010). *Science, 330*(6002), 356–359. doi:10.1126/science.1190653.

Muller, R. A. et al. (2012). Decadal variations in the global atmospheric land temperatures. Retrieved May 16, 2013. http://berkeleyearth.org/pdf/berkeley-earth-decadal-variations.pdf.

Paladin, G., et al. (1994). Predictability in spatially extended systems. *Journal of Physics A: Mathematical and General, 27*, 4911. doi:10.1088/0305-4470/27/14/018.

Rohde, A.R. et al. (2013). Geoinfor. Geostat: An Overview 1. Retrieved May 16, 2013. http://www.scitechnol.com/GIGS/GIGS-1-101.pdf.

Ruddiman, W. F. (2003). *Climate Change, 61*, 261–293.

Sanders, J. E., & Merguerian, C. (1994b). The glacial geology of New York City and vicinity. In A.I. Benimoff (Ed.), *The Geology of Staten Island, New York, Field guide and proceedings, The Geological Association of New Jersey, XI Annual Meeting* (pp. 93–200).

Shakun, J. D. (2012). *Nature, 484*, 49–54. doi:10.1038/nature10915.

Singh, D. J. et al. (2001). CO_2 capture options for an existing coal fired power plant: O_2/CO_2 recycle combustion vs. Amine scrubbing. Retrieved May 16, 2013. http://www.netl.doe.gov/publications/proceedings/01/carbon_seq/p6.pdf.

Skinner, L. (2012). *Science, 337*, 917–919. doi:10.1126/science.1224011.
Vincent, T. L. (2001). *Nonlinear Dynamics and Systems Theory, 1*(2), 205–218.
Wang, T., et al. (2011). *Environmental Science and Technology, 45*(15), 6670–6675.
A Report by the NSF Advisory Committee for Environmental Research and Education. (2009). *Transitions and Tipping Points in Complex Environmental Systems*. http://www.nsf.gov/geo/ere/ereweb/ac-ere/nsf6895_ere_report_090809.pdf.

Climate Change, Catastrophic Risks and Social Choice Theory

Norman Schofield

1 Introduction

In this essay I shall consider what Israel (2010, 2014) calls the *Radical Enlightenment*, the program to establish rationality as the basis for society, opposed to monarchy, religion and the church. Radical enlighteners included Thomas Jefferson, Thomas Paine and James Madison. They believed that society could be based on rational constitutional principles, leading to the "probability of a fit choice." Implicit in the Radical enlightenment was the belief, originally postulated by Spinoza, that individuals could find moral bases for their choices without a need for a divine creator. An ancilliary belief was that the economy would also be rational and that the principles of the radical enlightenment would lead to material growth and the eradication of poverty and misery.[1] This enlightenment philosophy has recently had to face two troubling propositions. First are the results of Arrovian social choice theory. These very abstract results suggest that no process of social choice can be rational Second, recent events suggest that the market models that we have used to guide our economic actions are deeply flawed. Opposed to the Radical enlighteners, David Hume and Burke believed that people would need religion and nationalism to provide a moral compass to their lives. As Putnam and Campbell (2010) have noted religion is as important as it has ever been in the US. Recent models of US Elections (Schofield and Gallego 2011) show that religion is a key dimension of politics that

[1] See Pagden (2013) for an argument about the significance today of the enlightenment project, but a counter arguement by Gray (1995, 1997, 2000).

Prepared for presentation at the Climate Change Conference, Reykjavik, Iceland. June 27–28, 2013.

N. Schofield (✉)
Department of Political Science, Center in Political Economy,
Washington University, Seigle Hall Campus Box 1027, St. Louis, MO 63130, USA
e-mail: schofield.norman@gmail.com

G. Chichilnisky and A. Rezai (eds.), *The Economics of the Global Environment*,
Studies in Economic Theory 29, DOI 10.1007/978-3-319-31943-8_18

divides voters one from another. A consequence of the Industrial Revolution, that followed on from the Radical Enlightenment, has been the unintended consequence of climate change. Since this is the most important policy dimension that the world economy currently faces, this paper will address the question whether we are likely to be able to make wise social choices to avoid future catastrophe.

1.1 The Radical Enlightenment

It was no accident that the most important cosmologist after Ptolemy of Alexandria was Nicolaus Copernicus (1473–1543), born only a decade before Martin Luther. Both attacked orthodoxy in different ways.[2] Copernicus formulated a scientifically based heliocentric cosmology that displaced the Earth from the center of the universe. His book, *De revolutionibus orbium coelestium* (*On the Revolutions of the Celestial Spheres*, 1543), is often regarded as the starting point of the Scientific Revolution.

The ideas of Copernicus influenced many scholars: the natural philosopher, William Gilbert, who wrote on magnetism in De Magnete (1601); the physicist, mathematician, astronomer, and philosopher, Galileo Galilei (1564–1642); the mathematician and astronomer, Johannes Kepler (1571–1630).

Philosophiæ Naturalis Principia Mathematica (1687), by the physicist, mathematician, astronomer and natural philosopher, Isaac Newton (1642–1726) is considered to be the most influential book in the history of science.[3] Margolis (2002) argues that, after Newton, a few scholars realized that the universe exhibits laws that can be precisely written down in mathematical form. Moreover, we have, for some mysterious reason, the capacity to conceive of exactly those mathematical forms that do indeed govern reality. This mysterious connection between mind and reality was the basis for Newton's philosophy. While celestial mechanics had been understood by Ptolemy to be the domain most readily governed by these forms, Newton's work suggested that *all* reality was governed by mathematics. The influence of Newton can perhaps be detected in the work of the philosopher, mathematician, and political scientist, Marie Jean Antoine Nicolas de Caritat, Marquis de Condorcet (1743–1794), known as Nicolas de Condorcet. His work in formal social choice theory (Condorcet ([1785], 1994) was discussed in Schofield (2006) in connection with the arguments about democracy by Madison and Jefferson. The work on Moral Sentiment by the Scottish Enlightenment writers, Francis Hutcheson (1694–1746), David Hume (1711–1776), Adam Smith (1723–1790) and Adam Ferguson (1723–1816), also influenced Jefferson and Madison. Between Copernicus and Newton, the writings of Thomas Hobbes (1588–1679), René Descartes (1596–1650), John

[2]Weber (1904) speculated that there was a connection between the values of Protestantism and Capitalism. It may be that there are connections between the preference for scientific explanation and protestant belief about the relationship between God and humankind.

[3]See Feingold (2004).

Locke (1632–1704), Baruch Spinoza (1632–1677), and Gottfied Liebnitz (1646–1716) laid down foundations for the modern search for rationality in life.[4] Hobbes was more clearly influenced by the scientific method, particularly that of Galileo, while Descartes, Locke, Spinoza, and Liebniz were all concerned in one way or another with the imperishability of the soul.[5] The mathematician, Liebniz, in particular was concerned with an

> [E]xplanation of the relation between the soul and the body, a matter which has been regarded as inexplicable or else as miraculous.

Without the idea of a soul it would seem difficult to form a general scheme of ethics.[6] Indeed, the progress of science and the increasing secularization of society have caused many to doubt that our society can survive. Hawking and Mlodonow (2010) argue for a strong version of this universal mathematical principle, called *model-dependent realism*, citing its origins in Pythagoras (580–490 BCE), Euclid (383–323 BCE and Archimedes (287–212 BCE), and the recent developments in mathematical physics and cosmology.

Hawking and Mlodonow (2010) argue that it is only through a mathematical model that we can properly perceive reality. However, this mathematical principle faces two philosophical difficulties. One stems from the Godel1931-Turing 1937 undecidability theorems. The first theorem asserts that mathematics cannot be both complete and consistent, so there are mathematical principles that in principle cannot be verified. Turing's work, though it provides the basis for our computer technology also suggests that not all programs are computable.[7] The second problem is associated with the notion of *chaos* or *catastrophe*.

Since the early work of Garrett Hardin (1968) the "tragedy of the commons" has been recognised as a global prisoner' dilemma. In such a dilemma no agent has a motivation to provide for the collective good. In the context of the possibility of climate change, the outcome is the continued emission of greenhouses gases like carbon dioxide into the atmosphere and the acidification of the oceans. There has

---

[4]For Hobbes, see Rogow (1986). For Descartes, see Gaukroger (1995). For Spinoza and Liebnitz see Stewart (2006) and Goldstein (2006). See also Israel (2011) for the development of the Radical Enlightenment.

[5]It is of interest that the English word "soul" derives from Old English *sáwol* (first used in the 8th century poem, *Beowulf*.

[6]Hawking and Mlodinow (2010) assert that God did not create the Universe, perhaps implying that the soul does not exist. However they do say that they understand Isaac Newton's belief that God did "create" and "conserve" order in the universe. See other books by Dawkins (2008) and Hitchens (2007) on the same theme, as well as Wright (2009) on the evolution of the notion of God and Lilla (2007) on political theology.

[7]Tegmark (2008, 2014) suggests a version of the Hawking, Mlodonow thesis that he calls the Mathematical Universe Hypothesis, but he is aware that the Godel Turing Theorems put limits on how able we are to apprehend reality. See also Yanofsky (2013), for the limits of science and mathematics.

developed an extensive literature on the n-person prisoners' dilemma in an attempt to solve the dilemma by considering mechanisms that would induce cooperation.[8]

The problem of cooperation has also provided a rich source of models of evolution, building on the early work by Trivers (1971) and Hamilton (1964, 1970). Nowak (2011) provides an overview of the recent developments. Indeed, the last twenty years has seen a growing literature on a game theoretic, or mathematical, analysis of the evolution of social norms to maintain cooperation in prisoners' dilemma like situations. Gintis (2000, 2003), for example, provides evolutionary models of the cooperation through strong reciprocity and internalization of social norms.[9] The anthropological literature provides much evidence that, from about 500,000 years ago, the ancestors of *homo sapiens* engaged in cooperative behavior, particularly in hunting and caring for offspring and the elderly.[10] On this basis we can infer that we probably do have very deeply ingrained normative mechanisms that were crucial, far back in time, for the maintenance of cooperation, and the fitness and thus survival of early hominids.[11] These normative systems will surely have been modified over the long span of our evolution.

Current work on climate change has focussed on how we should treat the future. For example Stern (2007, 2009), Collier (2010) and Chichilnisky (2009a, b) argue essentially for equal treatment of the present and the future. Dasguta (2005) points out that how we treat the future depends on our current estimates of economic growth in the near future.

The fundamental problem of climate change is that the underlying dynamic system is extremely complex, and displays many positive feedback mechanisms.[12] The difficulty can perhaps be illustrated by Fig. 1. It is usual in economic analysis to focus on Pareto optimality. Typically in economic theory, it is assumed that preferences and production possibilities are generated by convex sets. However, climate change could create non-convexities. In such a case the Pareto set will exhibit stable and unstable components. Figure 1 distinguishes between a domain A, bounded by stable and unstable components P_1^s and P^u, and a second stable component P_2^s. If our actions lead us to an outcome within A, whether or not it is Paretian, then it is

[8]See for example Hardin (1971, 1982), Taylor (1976, 1982), Axelrod and Hamilton (1981), Axelrod (1981, 1984), Kreps et al. (1982), Margolis (1982).

[9]Strong reciprocity means the punishment of those who do not cooperate.

[10]Indeed, White et al. (2009) present evidence of a high degree of cooperation among very early hominids dating back about 4MYBP (million years before the present). The evidence includes anatomical data which allows for inferences about the behavioral characteristics of these early hominids.

[11]Gintis cites the work of Robson and Kaplan (2003) who use an economic model to estimate the correlation between brain size and life expectancy (a measure of efficiency). In this context, the increase in brain size is driven by the requirement to solve complex cooperative games against nature.

[12]See the discussion in Schofield (2011). See also Nordhaus (2013) for an economic model of climate change.

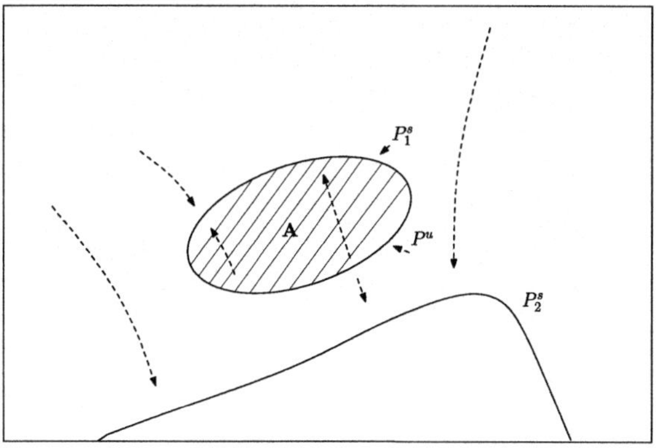

Fig. 1 Stable and unstable components of the global Pareto Set

possible that the dynamic system generated by climate could lead to a catastrophic destruction of A itself. More to the point, our society would be trapped inside A as the stable and unstable components merged together.

Our society has recently passed through a period of economic disorder, where "black swan" events, low probability occurrences with high costs, have occurred with some regularity. Recent discussion of climate change has also emphasized so called "fat-tailed climate events" again defined by high uncertainty and cost.[13] The catastrophic change implied by Fig. 1 is just such a black swan event. The point to note about Fig. 1 is everything would appear normal until the evaporation of A.

Cooperation could in principle be attained by the action of a hegemonic leader such as the United States as suggested by Kindleberger (1973) and Keohane and Nye (1977). In Sect. 2 we give a brief exposition of the prisoners' dilemma and illustrate how hegemonic behavior could facilitate international cooperation. However, the analysis suggests that in the present economic climate, such hegemonic leadership is unlikely.

Analysis of games such as the prisoner's dilemma usually focus on the existence of a Nash equilibrium, a vector of strategies with the property that no agent has an incentive to change strategy. Section 3 considers the family of equilibrium models based on the Brouwer (1912) fixed point theorem, or the more general result known as the Ky Fan theorem (Fan 1961) as well as the application by Bergstrom (1975, 1992) to prove existence of a Nash equilibrium and market equilibrium.

Section 4 considers a generalization of the Ky Fan Theorem, and argues that the general equilibrium argument can be interpreted in terms of particular properties of a preference field, H, defined on the tangent space of the joint strategy space. If this field is continuous, in a certain well-defined sense, and "*half open*" then it will exhibit a equilibrium. This half open property is the same as the non empty intersection of a

[13] Weitzman (2009) and Chichilnisky (2010, 2014). See also Chichilnisky and Eisenberger (2010) on other catastrophic events such as collision with an asteroid.

family of dual cones. We mention a Theorem by Chichilnisky (1995) that a necessary and sufficient condition for market equilibrium is that a family of dual cones also has non-empty intersection.

However, preference fields that are defined in terms of coalitions need not satisfy the half open property and thus need not exhibit equilibrium. For coalition systems, it can be shown that unless there is a collegium or oligarchy, or the dimension of the space is restricted in a particular fashion, then there need be no equilibrium. Earlier results by McKelvey (1976), Schofield (1978), McKelvey and Schofield (1987) and Saari (1997) suggested that voting can be "non-equilibriating" and indeed "chaotic."[14]

Kauffman (1993) commented on "chaos" or the failure of "structural stability" in the following way.

> One implication of the occurrence or non-occurrence of structural stability is that, in structurally stable systems, smooth walks in parameter space must [result in] smooth changes in dynamical behavior. By contrast, chaotic systems, which are not structurally stable, adapt on uncorrelated landscapes. Very small changes in the parameters pass through many interlaced bifurcation surfaces and so change the behavior of the system dramatically.

Chaos is generally understood as sensitive dependence on initial conditions whereas *structural stability* means that the qualitative nature of the dynamical system does not change as a result of a small perturbation.[15] I shall use the term *chaos* to mean that the trajectory taken by the dynamical process can wander anywhere.[16]

An earlier prophet of uncertainty was, of course, Keynes (1936) whose ideas on "speculative euphoria and crashes" would seem to be based on understanding the economy in terms of the qualitative aspects of its coalition dynamics.[17] An extensive literature has tried to draw inferences from the nature of the recent economic events. A plausible account of market disequilibrium is given by Akerlof and Shiller (2009) who argue that

> the business cycle is tied to feedback loops involving speculative price movements and other economic activity—and to the talk that these movements incite. A downward movement in stock prices, for example, generates chatter and media response, and reminds people of longstanding pessimistic stories and theories. These stories, newly prominent in their minds, incline them toward gloomy intuitive assessments. As a result, the downward spiral can continue: declining prices cause the stories to spread, causing still more price declines and further reinforcement of the stories.

---

[14]See Schofield (1977, 1980a, b). In a sense these voting theorems can be regarded as derivative of Arrow's Impossiblity Theorem (Arrow 1951). See also Arrow (1986).

[15]The theory of chaos or complexity is rooted in Smale's fundamental theorem (Smale 1966) that structural stability of dynamical systems is not "generic" or typical whenever the state space has more than two dimensions.

[16]In their early analysis of chaos, Li and Yorke (1975) showed that in the domain of a chaotic transformation f it was possible for almost any pair of positions (x, y) to transition from x to $y = f^r(x)$, where f^r means the r times reiteration of f.

[17]See Minsky (1975, 1986) and Keynes's earlier work in 1921.

It would seem reasonable that the rise and fall of the market is due precisely to the coalitional nature of decision-making, as large sets of agents follow each other in expecting first good things and then bad. A recent example can be seen in the fall in the market after the earthquake in Japan, and then recovery as an increasing set of investors gradually came to believe that the disaster was not quite as bad as initially feared.

Since investment decisions are based on these uncertain evaluations, and these are the driving force of an advanced economy, the flow of the market can exhibit singularities, of the kind that recently nearly brought on a great depression. These singularities associated with the bursting of market bubbles are time-dependent, and can be induced by endogenous belief-cascades, rather than by any change in economic or political fundamentals (Corcos 2002).

Similar uncertainty holds over political events. The fall of the Berlin Wall in 1989 was not at all foreseen. Political scientists wrote about it in terms of "belief cascades"[18] as the coalition of protesting citizens grew apace. As the very recent democratic revolutions in the Middle East and North Africa suggest, these coalitional movements are extremely uncertain.[19] In particular, whether the autocrat remains in power or is forced into exile is as uncertain as anything Keynes discussed. Even when democracy is brought about, it is still uncertain whether it will persist.[20]

Section 5 introduces the Condorcet (1994, [1795]) Jury Theorem. This theorem suggests that majority rule can provide a way for a society to attain the truth when the individuals have common goals. Schofield (2002, 2006) has argued that Madison was aware of this theorem while writing Federalist X (Madison 1999, [1787]) so it can be taken as perhaps the ultimate justification for democracy. However, models of belief aggregation that are derived from the Jury Theorem can lead to belief cascades that bifurcate the population. In addition, if the aggregation process takes place on a network, then centrally located agents, who have false beliefs, can dominate the process.[21]

In Sect. 6 we introduce the idea of a belief equilibrium, and then go on to consider the notion of "punctuated equilibrium" in general evolutionary models. Again however, the existence of an equilibrium depends on a fixed point argument, and thus on a half open property of the "cones" by which the developmental path is modeled. This half open property is equivalent to the existence of a social direction gradient defined everywhere. In Sect. 7 we introduce the notion of a "moral compass" that may provide a teleology to guide us in making wise choices for the future, by providing us with a social direction gradient. Section 8 concludes.

---

[18] Karklins and Petersen (1993), Lohmann (1994). See also Bikhchandani, Hirschleifer, and Welsh (1992).

[19] The response by the citizens of these countries to the demise of Osama bin Laden on May 2, 2011, is in large degree also unpredictable.

[20] See for example Carothers (2002) and Collier (2009).

[21] Golub and Jackson (2010).

2 The Prisoners' Dilemma, Cooperation and Morality

> For before constitution of Sovereign Power ... all men had right to all things; which necessarily causeth Warre. (Hobbes 2009 [1651]).

Kindleberger (1973) gave the first interpretation of the international economic system of states as a "Hobbesian" prisoners' dilemma, which could be solved by a leader, or "hegemon."

> A symmetric system with rules for counterbalancing, such as the gold standard is supposed to provide, may give way to a system with each participant seeking to maximize its short-term gain. ... But a world of a few actors (countries) is not like [the competitive system envisaged by Adam Smith]. ... In advancing its own economic good by a tariff, currency depreciation, or foreign exchange control, a country may worsen the welfare of its partners by more than its gain. Beggar-thy-neighbor tactics may lead to retaliation so that each country ends up in a worse position from having pursued its own gain ...
>
> This is a typical non-zero sum game, in which any player undertaking to adopt a long range solution by itself will find other countries taking advantage of it ...

In the 1970s, Robert Keohane and Joseph Nye (1977) rejected "realist" theory in international politics, and made use of the idea of a hegemonic power in a context of "complex interdependence" of the kind envisaged by Kindleberger. Although they did not refer to the formalism of the prisoners' dilemma, it would appear that this notion does capture elements of complex interdependence. To some extent, their concept of a hegemon is taken from realist theory rather than deriving from the game-theoretic formalism.

The essence of the theory of hegemony in international relations is that if there is a degree of inequality in the strengths of nation states then a hegemonic power may maintain cooperation in the context of an n-country prisoners' dilemma. Clearly, the British Empire in the 1800s is the role model for such a hegemon (Ferguson 2002).

Hegemon theory suggests that international cooperation was maintained after World War II because of a dominant cooperative coalition. At the core of this cooperative coalition was the United States; through its size it was able to generate collective goods for this community, first of all through the Marshall Plan and then in the context first of the post-world war II system of trade and economic cooperation, based on the Bretton Woods agreement and the Atlantic Alliance, or NATO. Over time, the United States has found it costly to be the dominant core of the coalition. In particular, as the relative size of the U.S. economy has declined, it would seem that international cooperation has become more difficult to maintain. Indeed, the global recession of 2008–10 suggests that problems of debt could induce "begger thy neighbor strategies", just like the 1930s.

The future utility benefits of adopting policies to ameliorate the possible climate changes in the future depend on the discount rates that we assign to the future. Dasgupta (2005) gives a clear exposition of how we might assign these discount rates. Obviously enough, different countries will in all likelihood adopt very different evaluations of the future. It is proable that developing countries like the BRICs (Brazil,

Russia, India and China) will choose growth and development now rather than choosing consumption in the future. it is true however that China seems aware of the dangers in the future, and may prove to act as a hegemon in this context.

There have been many attempts to "solve" the prisoners' dilemma in a general fashion. For example Binmore (2005) suggests that in the iterated nPD there are many equilibria with those that are *fair* standing out in some fashion. However, the criterion of "fairness" would seem to have little weight with regard to climate change. It is precisely the poor countries that will suffer from climate change, while the rapidly growing BRICs believe that they have a right to choose their own paths of development.

An extensive literature over the last few years has developed Adam Smith's ideas as expressed in the *Theory of Moral Sentiments* (1984 [1759]) to argue that human beings have an inate propensity to cooperate. This propensity may well have been the result of co-evolution of language and culture (Boyd and Richerson 2005; Gintis 2000).

Since language evolves very quickly (McWhorter 2001; Deutcher 2006), we might also expect moral values to change fairly rapidly, at least in the period during which language itself was evolving. In fact there is empirical evidence that cooperative behavior as well as notions of fairness vary significantly across different societies.[22] While there may be fundamental aspects of morality and "altruism," in particular, held in common across many societies, there is variation in how these are articulated. Gazzaniga (2008) suggests that moral values can be described in terms of various *modules*: reciprocity, suffering (or empathy), hierarchy, in-group and outgroup coalition, and purity/disgust. These modules can be combined in different ways with different emphases. An important aspect of cooperation is emphasized by Burkart, Hrdy and Van Schaik (2009) and Hrdy (2011), namely cooperation between man and woman to share the burden of child rearing.

It is generally considered that hunter-gatherer societies adopted egalitarian or "fair share" norms. The development of agriculture and then cities led to new norms of hierarchy and obedience, coupled with the predominance of military and religious elites (Schofield 2010).

North (1990), North et al. (2009) and Acemoglu and Robinson (2006) focus on the transition from such oligarchic societies to open access societies whose institutions or "rules of the game", protect private property, and maintain the rule of law and political accountability, thus faciltating both cooperation and economic development. Acemoglu et al. (2009) argue, in their historical analyses about why "good" institutions form, that the evidence is in favor of "critical junctures."[23] For example, the "Glorious Revolution" in Britain in 1688 (North and Weingast 1989), which prepared the way in a sense for the agricultural and industrial revolutions to follow (Mokyr 2005, 2010; Mokyr and Nye 2007) was the result of a sequence of historical contingencies that reduced the power of the elite to resist change. Recent work

---

[22]See Henrich et al. (2004, 2005), which reports on experiments in fifteen "small-scale societies," using the game theoretic tools of the "prisoners' dilemma," the "ultimatum game," etc.

[23]See also Acemoglu and Robinson (2008).

by Morris (2010), Fukuyama (2011), Ferguson (2011) and Acemoglu and Robinson (2011) has suggested that these fortuitous circumstances never occurred in China and the Middle East, and as a result these domains fell behind the West. Although many states have become democratic in the last few decades, oligarchic power is still entrenched in many parts of the world.[24]

At the international level, the institutions that do exist and that are designed to maintain cooperation, are relatively young. Whether they succeed in facilitating cooperation in such a difficult area as climate change is a matter of speculation. As we have suggested, international cooperation after World War II was only possible because of the overwhelming power of the United States. In a world with oligarchies in power in Russia, China, and in many countries in Africa, together with political disorder in almost all the oil producing counties in the Middle East, cooperation would appear unlikely.

To extend the discussion, we now consider more general theories of social choice.

3 Existence of a Choice

The above discussion has considered a very simple version of the prisoner's dilemma. The more general models of cooperation typically use variants of evolutionary game theory, and in essence depend on proof of existence of Nash equilibrium, using some version of the Brouwer's fixed point theorem (Brouwer 1912).

Brouwer's theorem asserts that any continuous function $f : B \to B$ from the finite dimensional ball, B (or indeed any compact convex set in $\mathbb{R}^w$) into itself, has the *fixed point property*. That is, there exists some $x \in B$ such that $f(x) = x$.

We will now consider the use of variants of the theorem, to prove existence of an equilibrium of a general choice mechanism. We shall argue that the condition for existence of an equilibrium will be violated if there are cycles in the underlying mechanism.

Let W be the set of alternatives and let X be the set of all subsets of W. A *preference correspondence*, P, on W assigns to each point $x \in W$, its *preferred set* $P(x)$. Write $P\colon W \to X$ or $P\colon W \twoheadrightarrow W$ to denote that the image of x under P is a set (possibly empty) in W. For any subset V of W, the restriction of P to V gives a correspondence $P_V\colon V \twoheadrightarrow V$. Define $P_V^{-1}\colon V \twoheadrightarrow V$ such that for each $x \in V$,

$$P_V^{-1}(x) = \{y : x \in P(y)\} \cap V.$$

$P_V^{-1}(x) = \{y : x \in P(y)\} \cap V$. The sets $P_V(x), P_V^{-1}(x)$ are sometimes called the *upper* and *lower* preference sets of P on V. When there is no ambiguity we delete the suffix V. The *choice* of P from W is the set

[24]The popular protests in N. Africa and the Middle East in 2011 were in opposition to oligarchic and autocratic power.

$$C(W,P) = \{x \in W \colon P(x) = \varnothing\}.$$

Here $\varnothing$ is the empty set. The choice of P from a subset, V, of W is the set

$$C(V,P) = \left\{x \in V \colon P_V(x) = \varnothing\right\}.$$

Call C_P a *choice function* on W if $C_P(V) = C(V,P) \neq \varnothing$ for every subset V of W. We now seek general conditions on W and P which are sufficient for C_P to be a choice function on W. Continuity properties of the preference correspondence are important and so we require the set of alternatives to be a topological space.

Definition 1 Let W, Y be two topological spaces. A correspondence $P \colon W \twoheadrightarrow Y$ is

(i) *Lower demi-continuous* (*ldc*) iff, for all $x \in Y$, the set

$$P^{-1}(x) = \{y \in W \colon x \in P(y)\}$$

is open (or empty) in W.

(ii) *Acyclic* if it is impossible to find a cycle $x_t \in P(x_{t-1}), x_{t-1} \in P(x_{t-2}), .., x_1 \in P(x_t)$.

(iii) *Lower hemi-continuous* (*lhc*) iff, for all $x \in W$, and any open set $U \subset Y$ such that $P(x) \cap U \neq \varnothing$ there exists an open neighborhood V of x in W, such that $P(x') \cap U \neq \varnothing$ for all $x' \in V$.

Note that if P is ldc then it is lhc.

We shall use lower demi-continuity of a preference correspondence to prove existence of a choice.

We shall now show that if W is compact, and P is an acyclic and ldc preference correspondence $P \colon W \twoheadrightarrow W$, then $C(W,P) \neq \varnothing$. First of all, say a preference correspondence $P \colon W \twoheadrightarrow W$ satisfies the *finite maximality property* (FMP) on W iff for every finite set V in W, there exists $x \in V$ such that $P(x) \cap V = \varnothing$.

Lemma 1 (Walker 1977).
If W is a compact, topological space and P is an ldc preference correspondence that satisfies FMP on W, then $C(W,P) \neq \varnothing$.

This follows readily, using compactness to find a finite subcover, and then using FMP.

Corollary 1 *If W is a compact topological space and P is an acyclic, ldc preference correspondence on W, then $C(W,P) \neq \varnothing$.*

As Walker (1977) noted, when W is compact and P is ldc, then P is acyclic iff P satisfies FMP on W, and so either property can be used to show existence of a choice. A second method of proof is to show that C_P is a choice function is to substitute a convexity property for P rather than acyclicity.

Definition 2

(i) If W is a subset of a vector space, then the *convex hull* of W is the set, Con$[W]$, defined by taking all convex combinations of points in W.
(ii) W is *convex* iff $W = \text{Con}[W]$. (The empty set is also convex.)
(iii) W is *admissible* iff W is a compact, convex subset of a topological vector space.
(iv) A preference correspondence $P: W \twoheadrightarrow W$ is *semi-convex* iff, for all $x \in W$, it is the case that $x \notin \text{Con}(P(x))$.

Fan (1961) has shown that if W is admissible and P is ldc and semi-convex, then $C(W, P)$ is non-empty.

Choice Theorem (Bergstrom 1975; Fan 1961).

If W is an admissible subset of a Hausdorff topological vector space, and $P: W \twoheadrightarrow W$ a preference correspondence on W which is ldc and semi-convex then $C(W, P) \neq \emptyset$.

The proof uses the KKM lemma due to Knaster et al. (1929).

The original form of the Theorem by Fan made the assumption that $P: W \twoheadrightarrow W$ was *irreflexive* (in the sense that $x \notin P(x)$ for all $x \in W$) and *convex*. Together these two assumptions imply that P is semi-convex. Bergstrom (1975) extended Fan's original result to give the version presented above.[25]

Note that the Fan Theorem is valid without restriction on the dimension of W. Indeed, Aliprantis and Brown (1983) have used this theorem in an economic context with an infinite number of commodities to show existence of a price equilibrium. Bergstrom (1992) also showed that when W is finite dimensional then the Fan Theorem is valid when the continuity property on P is weakened to lhc and used this theorem to show existence of a Nash equilibrium of a game $G = \{(P_1, W_1), .P_i, W_i), ..(P_n, W_n) : i \in N\}$. Here the ith stategy space is finite dimensional W_i and each individual has a preference P_i on the joint strategy space $P_i : W^N = W_1 \times W_2 ... \times W_n \twoheadrightarrow W_i$. The Fan Theorem can be used to show existence of an equilibrium in complex economies with externalities. Define the Nash improvement correspondence by $P_i^* : W^N \twoheadrightarrow W^N$ by $y \in P_i^*(x)$ whenever $y = (x_1, ..x_{i-1}, x_i^*, \ldots, x_n)$, $x = (x_1, .., x_{i-1}, x_i, .., x_n)$, and $x_i^* \in P_i(x)$ The joint Nash improvement correspondence is $P_N^* = \cup P_i^* : W^N \twoheadrightarrow W^N$. The Nash equilibrium of a game G is a vector $\mathbf{z} \in W^N$ such that $P_N^*(\mathbf{z}) = \emptyset$. Then the Nash equilibrium will exist when P_N^* is ldc and semi-convex and W^N is admissible.

4 Dynamical Choice Functions

We now consider a *generalized preference field* $H : W \twoheadrightarrow TW$, on a manifold W. TW is the tangent bundle above W, given by $TW = \cup\{T_x W : x \in W\}$, where $T_x W$ is the tangent space above x. If V is a neighborhood of x, then $T_V W = \cup\{T_x W : x \in V\}$ which is locally like the product space $\mathbb{R}^w \times V$. Here W is locally like $\mathbb{R}^w$.

[25] See also Shafer and Sonnenschein (1975).

At any $x \in W$, $H(x)$ is a *cone* in the tangent space T_xW above x. That is, if a vector $v \in H(x)$, then $\lambda v \in H(x)$ for any $\lambda > 0$. If there is a smooth curve, $c : [-1, 1] \to W$, such that the differential $\frac{dc(t)}{dt} \in H(x)$, whenever $c(t) = x$, then c is called an *integral curve* of H. An integral curve of H from $x = c(o)$ to $y = \lim_{t\to 1} c(t)$ is called an H *-preference curve* from x to y. In this case we write $y \in \mathbb{H}(x)$. We say y is reachable from x if there is a piecewise differentiable H−preference curve from x to y, so $y \in \mathbb{H}^r(x)$ for some reiteration r.The preference field H is called *S-continuous* iff the inverse relation $\mathbb{H}^{-1}$ is ldc. That is, if x is reachable from y, then there is a neighborhood V of y such that x is reachable from all of V. The *choice* $C(W, H)$ of H on W is defined by

$$C(W, H) = \{x \in W : H(x) = \varnothing\}.$$

Say $H(x)$ is semi-convex at $x \in W$, if either $H(x) = \varnothing$ or $0 \notin \text{Con}[H(x)]$ in the tangent space T_xW. In the later case, there will exist a vector $v' \in T_xW$ such that $(v' \cdot v) > 0$ for all $v \in H(x)$. We can say in this case that there is, at x, a *direction gradient* d in the cotangent space T_x^*W of linear maps from T_xW to $\mathbb{R}$ such that $d(v) > 0$ for all $v \in H(x)$. If H is S-continuous and half-open in a neighborhood, V, then there will exist such a continuous direction gradient $d : V \to T^*V$ on the neighborhood V[26]

We define

$$Cycle(W, H) = \{x \in W : H(x) \neq \varnothing, 0 \in \text{Con}H(x)\}.$$

An alternative way to characterize this property is as follows.

Definition 3 The *dual* of a preference field $H : W \twoheadrightarrow TW$ is defined by $H^* : W \twoheadrightarrow T^*W : x \twoheadrightarrow \{d \in T_x^*W : d(v) > 0 \text{ for all } v \in H(x) \subset T_xW\}$. For convenience if $H(x) = \varnothing$ we let $H^*(x) = T_xW$. Note that if $0 \notin \text{Con}\, H(x)$ iff $H^*(x) \neq \varnothing$. We can say in this case that the field is *half open* at x.

In applications, the field $H(x)$ at x will often consist of some family $\{H_j(x)\}$. As an example, let $u : W \twoheadrightarrow \mathbb{R}^n$ be a smooth utility profile and for any coalition $M \subset N$ let

$$H_M(u)(x) = \{v \in T_xW : (du_i(x)(v) > 0, \forall i \in M\}.$$

That is $H_M(u)(x)$ is the cone of directions that increase utilty for all members of M. If $\mathbb{D}$ is a family of *decisive* coalitions, $\mathbb{D} = \{M \subset N\}$, then we define

$$H_{\mathbb{D}}(u) = \cup H_M(u) : W \twoheadrightarrow TW$$

Then the field $H_{\mathbb{D}}(u) : W \twoheadrightarrow TW$ has a dual $[H_{\mathbb{D}}(u)]^* : W \twoheadrightarrow T^*W$ given by $[H_{\mathbb{D}}(u)]^*(x) = \cap[H_M(u)(x)]^*$ where the intersection at x is taken over all $M \in \mathbb{D}$ such

[26] *ie* $d(x)(v) > 0$ for all $x \in V$, for all $v \in H(x)$, whenever $H(x) \neq \varnothing$.

that $H_M(u)(x) \neq \varnothing$. We call $[H_M(u)(x)]^*$ the *co-cone of* M at x. It then follows that at $x \in Cycle(W, H_{\mathbb{D}}(u))$ then $0 \in \text{Con}[H_{\mathbb{D}}(u)(x)]$ and so $[H_{\mathbb{D}}(u)(x)]^* = \varnothing$. Thus

$$Cycle(W, H_{\mathbb{D}}(u)) = \{x \in W : [H_{\mathbb{D}}(u)]^*(x) = \varnothing\}.$$

The condition that $[H_{\mathbb{D}}(u)]^*(x) = \varnothing$ is equivalent to the condition that $\cap[H_M(u)(x)]^* = \varnothing$ and was called the *null dual condition* (at x). Schofield (1978) has shown that $Cycle(W, H_{\mathbb{D}}(u))$ will be an open set and contains cycles so that a point x is reachable from itself through a sequence of preference curves associated with different coalitions. This result was an application of a more general result.

Dynamical Choice Theorem (Schofield 1978).
For any S-continuous field H on compact, convex W, then

$$Cycle(W, H) \cup C(W, H) \neq \varnothing.$$

If $x \in Cycle(W, H) \neq \varnothing$ then there is a *piecewise differentiable H -preference cycle* from x to itself. If there is an open path connected neighborhood $V \subset Cycle(W, H)$ such that $H(x')$ is open for all $x' \in V$ then there is a *piecewise differentiable H -preference curve* from x to x'. □

(Here piecewise differentiable means the curve is continuous, and also differentiable except at a finite number of points). The proof follows from the previous choice theorem. The trajectory is built up from a set of vectors $\{v_1, \ldots, v_t\}$ each belonging to $H(x)$ with $0 \in \text{Con}[\{v_1, \ldots, v_t\}]$. If $H(x)$ is of full dimension, as in the case of a voting rule, then just as in the model of chaos by Li and York (1975), trajectories defined in terms of H can wander anywhere within any open path connected component of $Cycle(W, H)$.

This result has been shown more generally in Schofield (1980a) for the case that W is a compact manifold with non-zero Euler characteristic (Brown 1971). For example the theorem is valid if W is an even dimensional sphere. (The theorem is not true on odd dimensional spheres, as the clock face illustrates.)

Existence of Nash Equilibrium
Let $\{W_1, \ldots, W_n\}$ be a family of compact, contractible, smooth, strategy spaces with each $W_i \subset \mathbb{R}^w$. Consider a smooth profile $u: W^N = W_1 \times W_2 \ldots \times W_n \twoheadrightarrow \mathbb{R}^n$. Let $H_i : W_i \twoheadrightarrow TW_i$ be the induced i-preference field in the tangent space over W_i. If each H_i is S-continuous and half open in TW_i then there exists a *critical Nash equilibrium*, $\mathbf{z} \in W^N$ such that $H^N(\mathbf{z}) = (H_1 \times ..H_n)(\mathbf{z}) = \varnothing$.

This follows from the choice theorem because the product preference field, H^N, will be half-open and S-continuous. (See Schofield 2007), for an application of this technique to examine existence of *local* Nash equilibrium. With smooth utility functions, a local Nash equilibrium can be found by checking the second order conditions on the Hessians.

Example 1 To illustrate the Choice Theorem, define the preference relation $P_{\mathbb{D}}: W \twoheadrightarrow W$ generated by a family of *decisive* coalitions, $\mathbb{D} = \{M \subset N\}$, so that $y \in$

$P_{\mathbb{D}}(x)$ whenever all voters in some coalition $M \in \mathbb{D}$ prefer y to x. In particular consider the example due to Kramer (1973), with $N = \{1, 2, 3\}$ and $\mathbb{D} = \{\{1, 2\}, \{1, 3\}, \{2, 3\}\}$Suppose further that the preferences of the voters are characterized by the direction gradients

$$\{du_i(x) : i = 1, 2, 3\}$$

as in Fig. 2. In the figure, the utilities are assume to be "Euclidean," derived from distance from a preferred point, but this assumption is not important.

As the figure makes evident, it is possible to find three points $\{a, b, c\}$ in W such that

$$u_1(a) > u_1(b) = u_1(x) > u_1(c)$$
$$u_2(b) > u_2(c) = u_2(x) > u_2(a)$$
$$u_3(c) > u_3(a) = u_3(x) > u_3(b).$$

That is to say, preferences on $\{a, b, c\}$ give rise to a *Condorcet cycle*. Note also that the set of points $P_{\mathbb{D}}(x)$, preferred to x under the voting rule, are the shaded "win sets" in the figure. Clearly $x \in \text{Con}P_{\mathbb{D}}(x)$, so $P_{\mathbb{D}}(x)$ is not semi-convex. Indeed it should be clear that in *any* neighborhood V of x it is possible to find three points $\{a', b', c'\}$ such that there is *local* voting cycle, with $a' \in P_{\mathbb{D}}(b'), b' \in P_{\mathbb{D}}(c'), c' \in P_{\mathbb{D}}(a')$. We can write this as

$$a' \to c' \to b' \to a'.$$

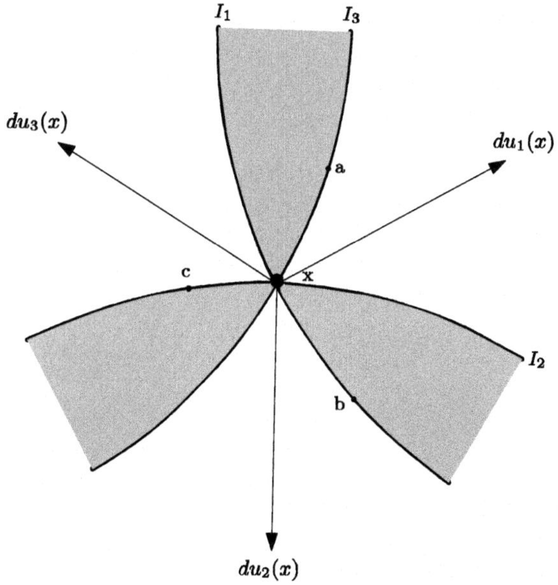

Fig. 2 Cycles in a neighborhood of x

Not only is there a voting cycle, but the Fan theorem fails, and we have no reason to believe that $C(W, P_{\mathbb{D}}) \neq \varnothing$.

We can translate this example into one on preference fields by considering the preference field

$$H_{\mathbb{D}}(u) = \cup H_M(u) : W \twoheadrightarrow TW$$

where each $M \in \mathbb{D}$.

Figure 3 shows the three difference preference fields $\{H_i : i = 1, 2, 3)$ on W, as well as the intersections H_M, for $M = \{1, 2\}$ etc.

Obviously the joint preference field $H_{\mathbb{D}}(u) = \cup H_M(u) : W \twoheadrightarrow TW$ fails the half open property at x since $0 \in \text{Con}[H_{\mathbb{D}}(u)(x)]$. Although $H_{\mathbb{D}}(u)$ is S-continuous, we cannot infer that $C(W, H_{\mathbb{D}}(u)) \neq \varnothing$.

Chichilnisky (1992, 1995, 1996a, 1997a) has obtained similar results for markets, where the condition that the dual is non-empty was termed *market arbitrage*, and defined in terms of global market co-cones associated with each player. Such a dual co-cone, $[H_i(u)]^*$ is precisely the set of prices in the cotangent space that lie in the dual of the prefered cone, $[H_i(u)]$, of the agent. By analogy with the above, she identifies this condition on non-emptiness of the intersection of the family of co-cones as one which is necessary and sufficient to guarantee an equilibrium.

Chichilnisky Theorem. (Chichilnisky 1997b)

The *limited arbitrage condition* $\cap[H_i(u)]^* \neq \varnothing$ is necessary and sufficient for existence of a competitive equilibrium. □

Chichilnisky (1993, 1997c) also defined a topological obstruction to the non-emptiness of this intersection and showed the connection with the existence of a social choice equilibrium.

Fig. 3 The failure of half-openess of a preference field

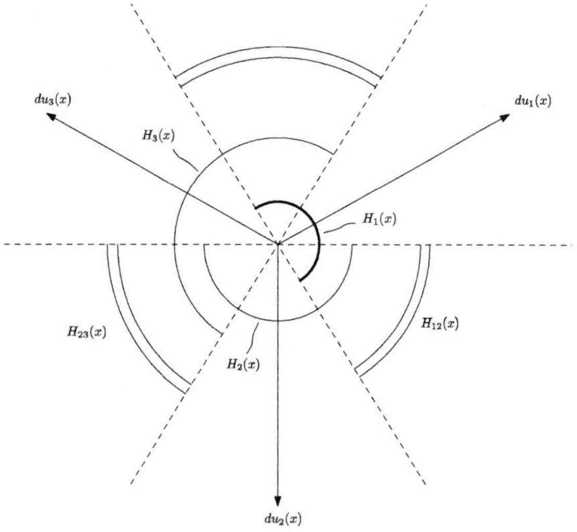

For a voting rule, $\mathbb{D}$ it is possible to guarantee that $Cycle(W, H_{\mathbb{D}}) = \varnothing$ and thus that $C(W, H_{\mathbb{D}}) \neq \varnothing$. We can do this by restricting the dimension of W.

Definition 4

(i) Let $\mathbb{D}$ be a family of decisive subsets of the finite society N of size n. If the collegium, $K(\mathbb{D}) = \cap\{M \in \mathbb{D}\}$ is non-empty then $\mathbb{D}$ is called *collegial* and the *Nakamura number* $\kappa(\mathbb{D})$ is defined to be ∞.

(ii) If the collegium $K(\mathbb{D})$ is empty then $\mathbb{D}$ is called *non-collegial*. Define the *Nakamura number* in this case to be $\kappa(\mathbb{D}) = \min\{|\mathbb{D}'| : \mathbb{D}' \subset \mathbb{D} \text{ and } K(\mathbb{D}') = \varnothing\}$.

Nakamura Theorem.

If $u \in U(W)^N$ and $\mathbb{D}$ has Nakamura number $\kappa(\mathbb{D})$ with $dim(W) \leq \kappa(\mathbb{D}) - 2$ then $Cycle(W, H_{\mathbb{D}}(u)) = \varnothing$ and $C(W, H_{\mathbb{D}}(u)) \neq \varnothing$.

Outline of proof. Consider any subfamily $\mathbb{D}'$ of $\mathbb{D}$ with cardinality $\kappa(\mathbb{D}) - 1$. Then $\cap M \neq \varnothing$, so $\cap\{[H_M(u)]^*(x) : M \in \mathbb{D}'\} \neq \varnothing$. If $[H_M(u)(x)] \neq \varnothing$, we can identify each $[H_M(u)(x)]^*$ with a non-empty convex hull generated by $(du_i(x) : i \in M\}$. These sets can be projected into $T_x W$ where they are convex and compact. Since $dim(W) \leq \kappa(\mathbb{D}) - 2$, then by Helly's Theorem, we see that $\cap\{[H_M(u)]^*(x) : M \in \mathbb{D}\} \neq \varnothing$. Thus $Cycle(W, H_{\mathbb{D}}(u)) = \varnothing$ and $C(W, H_{\mathbb{D}}(u)) \neq \varnothing$. □

See Schofield (1984b), Nakamura (1979) and Strnad (1985).

For social choice defined by voting games, the Nakamura number for majority rule is 3, except when $n = 4$, in which case $\kappa(\mathbb{D}) = 4$, so the Nakamura Theorem can generally only be used to prove a "median voter" theorem in one dimension. However, the result can be combined with the Fan Theorem to prove existence of equilibrium for a political economy with voting rule $\mathbb{D}$, when the dimension of the public good space is no more than $\kappa(\mathbb{D}) - 2$ (Konishi 1996). Recent work in political economy often only considers a public good space of one dimension (Acemoglu and Robinson 2006). Note however, that if $\mathbb{D}$ is collegial, then $Cycle(W, H_{\mathbb{D}}(u)) = \varnothing$ and $C(W, H_{\mathbb{D}}(u)) \neq \varnothing$. Such a rule can be called oligarchic, and this inference provides a theoretical basis for comparing democracy and oligarchy (Acemoglu 2008). Figure 3 showed the preference cones in a majority voting game with 3 agents and Nakamura number 3, so half openess fails in two dimensions.

Extending the equilibrium result of the Nakamura Theorem to higher dimension for a voting rule faces a difficulty caused by Bank's Theorem. We first define the *fine* C^1 topology on smooth utility functions (Hirsch 1976; Schofield 1999c, 2003).

Definition 5 Let $(U(W)^N, T_1)$ be the topological space of smooth utility profiles endowed with the the C^1–topology.

In economic theory, the existence of isolated price equilibria can be shown to be "generic" in this topological space (Debreu 1970, 1976; Smale 1974a, b). In social choice no such equilibrium theorem holds. The difference is essentially because of the coalitional nature of social choice.

Banks Theorem.

For any non-collegial $\mathbb{D}$, there exists an integer $w(\mathbb{D}) \geq \kappa(\mathbb{D})-1$ such that $dim(W) > w(\mathbb{D})$ implies that $C(W, H_{\mathbb{D}}(u)) = \varnothing$ for all u in a dense subspace of $(U(W)^N, T_1)$ so $Cycle(W, H_{\mathbb{D}}(u)) \neq \varnothing$ generically. □

This result was essentially proved by Banks (1995), building on earlier results by Plott (1967), Kramer (1973), McKelvey (1976), Schofield (1983), McKelvey and Schofield (1987). See Saari also (Saari 1985a, b, 1997, 2001a, b, 2008) for related analyses. Indeed, it can be shown that if $dim(W) > w(\mathbb{D})+ 1$ then $Cycle(W, H_{\mathbb{D}}(u))$ is generically dense (Schofield 1984c). The integer $w(\mathbb{D})$ can usually be computed explicitly from $\mathbb{D}$. For majority rule with n odd it is known that $w(\mathbb{D}) = 2$ while for n even, $w(\mathbb{D}) = 3$.

Although the Banks Theorem formally applies only to voting rules, Schofield (2010) argues that it is applicable to any non-collegial social mechanism, say $H(u)$ and can be interpreted to imply that

$$Cycle(W, H(u)) \neq \varnothing \text{ and } C(W, H(u)) = \varnothing$$

is a generic phenomenon in coalitional systems. Because preference curves can wander anywhere in any open component of $Cycle(W, H(u))$, Schofield (1979) called this *chaos*. It is not so much the sensitive dependence on initial conditions, but the aspect of indeterminacy that is emphasized.. . On the other hand, existence of a hegemon, as discussed in Sect. 2, is similar to existence of a collegium, suggesting that $Cycle(W, H(u))$ would be constrained in this case.

Richards (1993) has examined data on the distribution of power in the international system over the long run and presents evidence that it can be interpreted in terms of a chaotic trajectory. This suggests that the metaphor of the nPD in international affairs does characterise the ebb and flow of the system and the rise and decline of hegemony.

It is worth noting that the early versions of the Banks Theorem were obtained in the decade of the 1970s, a decade that saw the first oil crisis, the collapse of the Bretton Woods system of international political economy, the apparent collapse of the British economy, the beginning of social unrest in Eastern Europe, the revolution in Iran, and the second oilcrisis (Caryl 2011). Many of the transformations that have occurred since then can be seen as changes in beliefs, rather than preferences. Models of belief aggregation are less well developed than those dealing with preferences.[27] In general models of belief aggregation are related to what is now termed Condorcet's jury Theorem, which we now introduce.

[27]Results on belief aggregation include Penn (2009) and McKelvey and Page (1986).

5 Beliefs and Condorcet's Jury Theorem

The Jury theorem formally only refers to a situation where there are just two alternatives $\{1, 0\}$, and alternative 1 is the "true" option. Further, for every individual, i, it is the case that the probability that i picks the truth is ρ_{i1}, which exceeds the probability, ρ_{i0}, that i does not pick the truth. We can assume that $\rho_{i1} + \rho_{i0} = 1$, so obviously $\rho_{i1} > \frac{1}{2}$. To simplify the proof, we can assume that ρ_{i1} is the same for every individual, thus $\rho_{i1} = \alpha > \frac{1}{2}$ for all i. We use χ_i (= 0 or 1) to refer to the choice of individual i, and let $\chi = \Sigma_{i=1}^{n} \chi_i$ be the number of individuals who select the true option 1. We use Pr for the probability operator, and E for the expectation operator. In the case that the electoral size, n, is odd, then a majority, m, is defined to be $m = \frac{n+1}{2}$. In the case n is even, the majority is $m = \frac{n}{2} + 1$. The probability that a majority chooses the true option is then

$$\alpha_{maj}^{n} = \Pr[\chi \geq m].$$

The theorem assumes that voter choice is *pairwise independent*, so that $\Pr(\chi = j)$ is simply given by the binomial expression $\binom{n}{j}\alpha^{j}(1-\alpha)^{n-j}$.

A version of the theorem can be proved in the case that the probabilities $\{\rho_{i1} = \alpha_i\}$ differ but satisfy the requirement that $\frac{1}{n}\Sigma_{i=1}^{n}\alpha_i > \frac{1}{2}$. Versions of the theorem are valid when voter choices are not pairwise independent (Ladha and Miller 1996).

The Jury Theorem. If $1 > \alpha > \frac{1}{2}$, then $\alpha_{maj}^{n} \geq \alpha$, and $\alpha_{maj}^{n} \longrightarrow 1$ as $n \longrightarrow \infty$.

For both n being even or odd, as $n \longrightarrow \infty$, the fraction of voters choosing option 1 approaches $\frac{1}{n}E(\chi) = \alpha > \frac{1}{2}$. Thus, in the limit, more than half the voters choose the true option. Hence the probability $\alpha_{maj}^{n} \longrightarrow 1$ as $n \longrightarrow \infty$. □

Laplace also wrote on the topic of the probability of an error in the judgement of a tribunal. He was concerned with the degree to which jurors would make just decisions in a situation of asymmetric costs, where finding an innocent party guilty was to be more feared than letting the guilty party go free. As he commented on the appropriate rule for a jury of twelve, "I think that in order to give a sufficient guarantee to innocence, one ought to demand at least a plurality of nine votes in twelve" (Laplace Laplace 1951[1814]:139). Schofield (1972a, b) considered a model derived from the jury theorem where uncertain citizens were concerned to choose an ethical rule which would minimize their disappointment over the the likely outcomes, and showed that majority rule was indeed optimal in this sense.

Models of belief aggregation extend the Jury theorem by considering a situation where individuals receive signals, update their beliefs and make an aggregate choice on the basis of their posterior beliefs (Austen-Smith and Banks 1996). Models of this kind can be used as the basis for analysing correlated beliefs.[28] and the creation of belief cascades (Easley and Kleinberg 2010).

Schofield (2002, 2006) has argued that Condorcet's Jury theorem provided the basis for Madison's argument in Federalist X (Madison (1999) [1787]) that the judg-

[28]Schofield (1972a, b); Ladha (1992, 1993), 1995, 1996; Ladha and Miller (1996).

ments of citizens in the extended Republic would enhance the "probability of a fit choice." However, Schofield's discussion suggests that belief cascades can also fracture the society in two opposed factions, as in the lead up to the Civil War in 1860.[29]

There has been a very extensive literature recently on cascades[30] but it is unclear from this literature whether cascades will be equilibriating or very volatile. In their formal analysis of cascades on a network of social connections, Golub and Jackson (2010) use the term *wise* if the process can attain the truth. In particular they note that if one agent in the network is highly connected, then untrue beliefs of this agent can steer the crowd away from the truth. The recent economic disaster has led to research on market behavior to see if the notion of cascades can be used to explain why markets can become volatile or even irrational in some sense (Acemoglu et al. 2010; Schweitzer 2009). Indeed the literature that has developed in the last few years has dealt with the nature of herd instinct, the way markets respond to speculative behavior and the power law that characterizes market price movements.[31] The general idea is that the market can no longer be regarded as efficient. Indeed, as suggested by Ormerod (2001) the market may be fundamentally chaotic.

"Empirical" chaos was probably first discovered by Lorenz (1962, 1963) in his efforts to numerically solve a system of equations representative of the behavior of weather. A very simple version is the non-linear vector equation

$$\frac{dx}{dt} = \begin{bmatrix} \frac{dx_1}{dt} \\ \frac{dx_2}{dt} \\ \frac{dx_3}{dt} \end{bmatrix} = \begin{bmatrix} -a_1(x_1 - x_2) \\ -x_1x_3 + a_2x_1 - x_2 \\ x_1x_2 - a_3x_3 \end{bmatrix}$$

which is chaotic for certain ranges of the three constants, a_1, a_2, a_3.

The resulting "butterfly" portrait winds a number of times about the left hole (as in Fig. 3), then about the right hole,then the left, etc. Thus the "phase portrait" of this dynamical system can be described by a sequence of winding numbers ($w_l^1, w_k^1, w_l^2, w_k^2$, etc.). Changing the constants a_1, a_2, a_3 slightly changes the winding numbers. Note that the picture in Fig. 3 is in three dimensions. The butterfly wings on left and right consist of infinitely many closed loops. Figure 4 gives a version of the butterfly, namely the chaotic trajectory of the Artemis Earth Moon orbiter. The whole thing is called the Lorentz "strange attractor." A slight perturbation of this dynamic system changes the winding numbers and thus the qualitative nature of the process. Clearly this dynamic system is not structurally stable, in the sense used by Kaufmann (1993). The metaphor of the butterfly gives us pause, since all dynamic systems whether models of climate, markets, voting processes or cascades may be indeterminate or chaotic.

---

[29]Sunstein (2006, 2011) also notes that belief aggregation can lead to a situation where subgroups in the society come to hold very disparate opinions.

[30]Gleick (1987), Buchanan (2001, 2003), Gladwell (2002), Johnson (2002), Barabasi (2003, 2010), Strogatz (2004), Watts (2002, 2003), Surowiecki (2005), Ball (2004), Christakis and Fowler (2011).

[31]See, for example, Mandelbrot and Hudson (2004), Shiller (2003, 2005), Taleb (2007), Barbera (2009), Cassidy (2009), Fox (2009).

Fig. 4 The butterfly

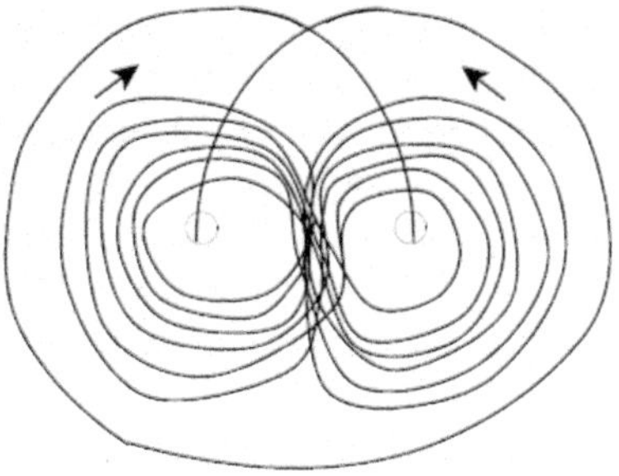

6 The Edge of Chaos

Recent work has attempted to avoid chaos by using the Brouwer fixed point theorem to seek existence of a *belief equilibrium* for a society N_τ of size $n_{\tau.}$ time τ. In this context we let

$$W_E = W_1 \times W_2 ... \times W_{n_{\tau+1.}} \times \Delta$$

be the economic product space, where W_i is the commodity space for citizen i and Δ is a price simplex. Let W_E be the economic space and $W_{\mathbb{D}}$ be a space of political goods, governed by a rule $\mathbb{D}$. At time τ, $W_\tau = W_E \times W_{\mathbb{D}}$ is the political economic space.

At τ, each individual, i, is described by a utility function $u_i : W_\tau \to \mathbb{R}$, so the population profile is given by $u : W_\tau \to \mathbb{R}^{n_\tau}$. Beliefs at τ about the future $\tau + 1$ are given by a stochastic rule, $\mathbb{Q}_\tau$, that transforms the agents' utilities from those at time τ to those at time $\tau + 1$. Thus $\mathbb{Q}_\tau$ generates a new profile for $N_{\tau+1}$ at $\tau + 1$ given by $\mathbb{Q}_\tau(u) = u' : W_{\tau+1} \to \mathbb{R}^{n_\tau+1}$. The utility and beliefs of i will depend on the various sociodemographic subgroups in the society $N_{\tau.}$ that i belongs to, as well as information about about the current price vector in Δ.

Thus we obtain a transformation on the function space $[W_\tau \to \mathbb{R}^{n_\tau}]$ given by

$$[W_\tau \to \mathbb{R}^{n_\tau}] \to {}^{\mathbb{Q}_\tau}\!\!\to [W_\tau \to \mathbb{R}^{n_\tau+1}] \to [W_\tau \to \mathbb{R}^{n_\tau}]$$

The second transformation here is projection onto the subspace $[W_\tau \to \mathbb{R}^{n_\tau}]$ obtained by restricting to changes to the original population $N_{\tau.}$ and space.

A *dynamic belief equilibrium* at τ for $N_{\tau.}$ is fixed point of this transformation. Although the space $[W_\tau \to \mathbb{R}^{n_\tau}]$ is infinite dimensional, if the domain and range of this transformation are restricted to *equicontinous* functions (Pugh 2002), then the domain and range will be compact. Penn (2009) shows that if the domain and range are convex then a generalized version of Brouwer's fixed point theorem can be applied to show existence of such a dynamic belief equilibrium. This notion of equilibrium was first suggested by Hahn (1973) who argued that equilibrium is located in the mind, not in behavior.

However, the choice theorem suggests that the validity of Penn's result will depend on how the model of social choice is constructed. For example Coros et al. (2002) consider a formal model of the market, based on the reasoning behind Keynes's "beauty contest" (Keynes 1936). There are two coalitions of "bulls" and "bears". Individuals randomly sample opinion from the coalitions and use a *critical* cutoff-rule. For example if the individual is bullish and the sampled ratio of bears exceeds some proportion then the individual flips to bearish. The model is very like that of the Jury Theorem but instead of guaranteeing a good choice the model can generate chaotic flips between bullish and bearish markets, as well as fixed points or cyclic behavior, depending on the cut-off parameters. Taleb's argument (Taleb 1997) about black swan events can be applied to the recent transformation in societies in the Middle East and North Africa that resemble such a cascade (Taleb and Blyth 2011). As in the earlier episodes in Eastern Europe, it would seem plausible that the sudden onset of a cascade is due to a switch in a critical coalition.

The notion of "criticality" has spawned in enormous literature particularly in fields involving evolution, in biology, language and culture.[32] Bak and Sneppen (1993) refer to the self organized critical state as the

> "edge of chaos" since it separates a frozen inactive state from a "hot" disordered state.
>
> The mechanism of evolution in the critical state can be thought of as an exploratory search for better local fitness, which is rarely successful, but sometimes has enormous effect on the ecosystem

Flyvbjerg et al. (1993) go on to say

> species sit at local fitness maxima..and occasionally a species jumps to another maximum [in doing so it] may change the fitness landscapes of other species which depend on it. ..Consequently they immediately jump to new maxima. This may affect yet another species in a chain reaction, a *burst* of evolutionary activity.

This work was triggered by the earlier ideas on "punctuated equilibrium" by Eldredge and Gould (1972).[33]

The point to be emphasized is that the evolution of a species involves bifurcation, a splitting of the pathway. We can refer to the bifurcation as a *catastrophe* or a *singularity*. The portal or door to the singularity may well be characterized by chaos or uncertainty, since the path can veer off in many possible directions, as suggested by the bifurcating cones in Figs. 3 and 4. At every level that we consider, the bifurcations of the evolutionary trajectory seem to be locally characterized by chaotic domains. I suggest that these domains are the result of different coalitional possibilities. The fact that the trajectories can become indeterminate suggests that this may enhance the exploration of the fitness landscape.

---

[32]See for example Cavallli-Sforza and Feldman (1981), Bowles et al. (2003).

[33]See also Eldredge (1976), Gould (1976).

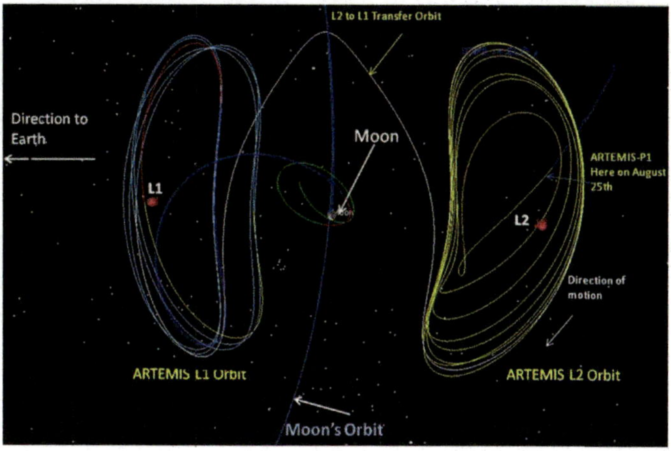

Fig. 5 A chaotic trajectory of the Artemis Earth Moon orbiter, downloaded from nasa.gov (artemis orbiter)

A more general remark concerns the role of climate change. Climate has exhibited chaotic or catastrophic behavior in the past.[34] There is good reason to believe that human evolution over the last million years can only be understood in terms of "bursts" of sudden transformations (Nowak 2011) and that language and culture co-evolve through group or coalition selection (Cavallli-Sforza and Feldman 1981). Calvin (2003) suggests that our braininess was cause and effect of the rapid exploration of the fitness landscape in response to climatic forcing. For example Fig. 5 shows the rapid changes in temperature over the last 100,000 years. It was only in the last period of stable temperature, the "holocene", the last 10,000 years that agriculture was possible.

Stringer (2012) calls the theory of rapid evolution during a period of chaotic climate change "the Social Brain hypothesis". The cave art of Chauvet, in France dating back about 36,000 years suggests that belief in the supernatural played an important part in human evolution.[35] Indeed, we might speculate that the part of our mind that enhances technological/mathematical development and that part that facilitates social/religious belief are in conflict with each other.[36] We might also speculate that market behavior is largely driven by what Keynes termed *speculation*, namely the

[34]Indeed as I understand the dynamical models, the chaotic episodes are due to the complex interactions of dynamical processes in the oceans, on the land, in weather, and in the heavens. These are very like interlinked *coalitions* of non-gradient vector fields.

[35]It is interesting to note that Alfred Wallace (1898), who developed the theory of Natural Selection at the same time as Darwin, did not believe that the theory could provide an explanation for the development of mathematical abilities and moral beliefs in humankind.

[36]This is suggested by Kahneman (2011).

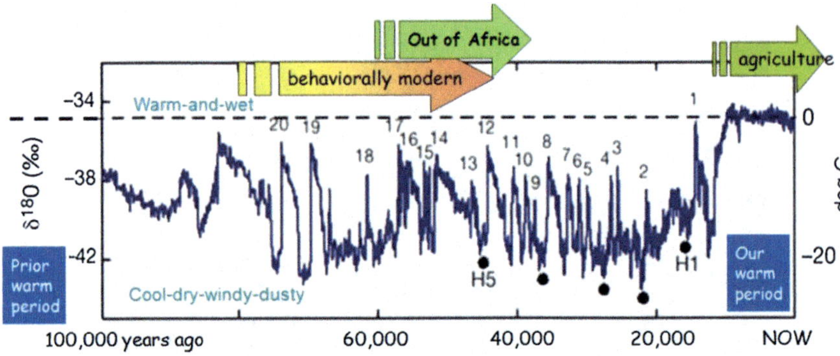

Fig. 6 Climate 100KYBP to now: chaos from 90 to 10KYBP (*Source* Global-Fever.org)

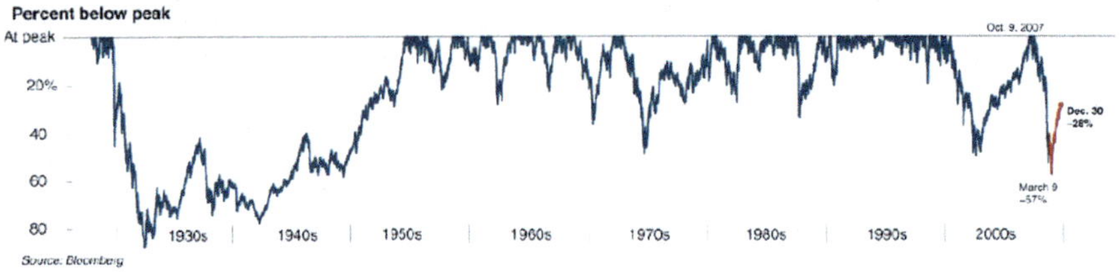

Fig. 7 Chaotic stock market prices 1930–2009 (*Source* New York Times, Dec 31, 2009)

largely irrational changes of *mood* (Casti 2010). Figure 6 gives an illustration of the swings in the US stock market over the last 80 years. While the figure may not allow us to assert that it truly chaotic, there seems no evidence that it is equilibriating (Fig. 7).

7 A Moral Compass

If we accept that moral and religious beliefs are as important as rational calculations in determining the choices of society, then depending on models of preference aggregation will not suffice in helping us to make decisions over how to deal with climate change. Instead, I suggest a moral compass, derived from current inferences made about the nature of the evolution of intelligence on our planetary home. The anthropic principle reasons that the fundamental constants of nature are very precisely tuned so that the universe contains matter and that galaxies and stars live long enough to allow for the creation of carbon, oxygen etc., all necessary for the evolution of life itself.[37] Gribbin (2011) goes further and points out that not only is the sun unusual in having the characterisics of a structurally stable system of planets,

[37]As Smolin (2007) and Rees (2001) point out, the anthropic principle has been adopted because of the experimental evidence that the expansion of the universe is accelerating. Indeed it has led to

but the earth is fortunate in being protected by Jupiter from chaotic bombardment but the Moon also stabilizes our planet's orbit.[38] In essence Gribbin gives good reasons to believe that our planet may well be the only planet in the galaxy that sustains intelligent life.[39] If this is true then we have a moral obligation to act as guardians of our planetary home. Parfit (2011) argues

> What matters most is that we rich people give up some of our luxeries, ceasing to overheat the Earth's atmosphere, and taking care of this planet in other ways, so that it continues to support intelligent life. If we are the only rational animals in the Universe, it matters even more whether we shall have descendants during the billions of years in which that would be possible. Some of our descendants might live lives and create worlds that, though failing to justify past suffering, would give us all, including those who suffered, reason to be glad that the Universe exists. (Parfit: 419)

8 Conclusion

Even if we believe that markets are well behaved, there is no reason to infer that markets are able to reflect the social costs of the externalities associated with production and consumption. Indeed Gore (2006) argues that the globalized market place, what he calls *Earth Inc* has the power and inclination to maintain business as usual. If this is so, then climate change will undoubtedly have dramatic adverse effects, not least on the less developed countries of the world.[40]

In principle we may be able to rely on a version of the jury theorem (Rae 1960; Landemore 2012; Schofield 1972a, b; Sunstein 2009), which asserts that majority rule provides an optimal procedure for making collective choices under uncertainty However, for the operation of what Madison called a "fit choice" it will be necessary to overcome the entrenched power of capital. Although we now disregard Marx's attempt at constructing a teleology of economic and political development,[41] we are in need of a more complex over-arching and evolutionary theory of political economy that will go beyond the notion of equilibrium and might help us deal with the future.

(Footnote 37 continued)
the hypothesis that there is an infinity of universes all with different laws. An alternative inference is the principle of intelligent design. My own inference is that we require a teleology as proposed in the conclusion.

[38]The work by Poincare in the late 19th century focussed on the structural stability of the solar system and was the first to conceive of the notion of chaos.

[39]See also Waltham (2014).

[40]Zhang (2007) and Hsiang et al. (2013) have provided a quantitative analysis of such adverse effects in the past. See also Parker (2013) for an historical account of the effect of climate change in early modern Europe, and Broodbank (2013) for the effects on the civilizations of the Mediterranean over a two thousand year period.

[41]See Sperber (2013) for a discussion of the development of Marx's ideas, in the context of 19th century belief in the teleology of "progress" or the advance of civilization. The last hundred years has however,made it difficult to hold such beliefs.

Gamble (1993), in his discussion of how humans colonized our planet over many thousands of years, emphasizes that they not only used reason but were driven by *purpose*, what I here refer to as teleology, not in the sense of progress, but in the sense of safeguarding our heritage.[42]

*-3pt

References

Acemoglu, D. (2008). Oligarchic versus democratic societies'. *Journal of the European Economic Association, 6*, 1–44.

Acemoglu, D., & Robinson, J. (2006). *Economic origins of dictatorship and democracy*. Cambridge: Cambridge University Press.

Acemoglu, D., & Robinson, J. (2008). Persistence of power, elites, and institutions. *American Economic Review, 98*, 267–293.

Acemoglu, D., & Robinson, J. (2011). *Why Nations Fail*. London: Profile Books.

Acemoglu, D., Johnson, S., Robinson, J., & Yared, P. (2009). Reevaluating the modernization hypothesis. *Journal of Monetary Economics, 56*, 1043–1058.

Acemoglu, D., Ozdaglar, A., & Tahbaz-Salehi, A. (2010). Cascades in networks and aggregate volatility, *NBER working paper* # 16516.

Akerlof, G. A., & Shiller, R. J. (2009). *Animal Spirits*. Princeton University Press: Princeton.

Arrow, K. J. (1951). *Social Choice and Individual Values*. New Haven: Yale University Press.

Arrow, K. (1986). Rationality of self and others in an economic system. *Journal of Business, 59*, S385–S399.

Arrow, K., & Debreu, G. (1954). Existence of an equilibrium for a competitive economy. *Econometrica, 22*, 265–290.

Austen-Smith, D., & Banks, J. (1996). Information Aggregation, Rationality, and the Condorcet Jury Theorem. *American Political Science Review, 90*, 34–45.

Axelrod, R. (1981). The emergence of cooperation among egoists. *American Political Science Review, 75*, 306–318.

Axelrod, R. (1984). *The Evolution of Cooperation*. New York: Basic.

Axelrod, R., & Hamilton, W. D. (1981). The evolution of cooperation. *Science, 211*, 1390–1396.

Bak, P. (1996). *How Nature Works: The Science of Self Organized Criticality*. Berlin: Springer.

Bak, P., & Sneppen, K. (1993). Punctuated equilibrium and criticality in a simple model of evolution. *Physics Review Letters, 71*(24), 4083–4086.

Ball, P. (2004). *Critical Mass*. New York: Ferrar, Strauss and Giroux.

Banks, J. S. (1995). Singularity theory and core existence in the spatial model. *Journal of Mathematical Economics, 24*, 523–536.

Barabasi, A.-L. (2003). *Linked*. New York: Plume.

Barabasi, A.-L. (2010). *Bursts*. New York: Dutton.

Barbera, R. (2009). *The Cost of Capitalism: Understanding Market Mayhem*. New York: McGraw Hill.

[42]The empirical work by Piketty (2014) suggests that capital is becoming predominant, and may be threatening the ability of democratic capitalism to survive. Cooper (2014) also argues against the neo-classical equilibrium model in economics, and suggests that governments engage in investment that would benefit the less well off members of society. This could be done by expanding investment in new technologies to counter climate change. Vogel (2014) sketches the increasing ability of "Big Money" to dominate politics. It is an open question whether the bad times that Randers (2014) and Turk and Rabino (2013) envision in our future will weaken the power of big money by inducing a popular backlash.

Bellah, R. N. (2011). *Religion in Human Evolution*. Cambridge MA: Belknap Press.
Berge, C. (1963). *Topological Spaces*. Edinburgh: Oliver and Boyd.
Bergstrom, T. (1975). *The existence of maximal elements and equilibria in the absence of transitivity*. Typescript, University of Michigan.
Bergstrom, T. (1992). When non-transitive relations take maxima and competitive equilibrium can't be beat. In W. Neuefeind & R. Riezman (Eds.), *Economic Theory and International Trade*. Berlin: Springer.
Bikhchandani, S., Hirschleifer, D., & Welsh, I. (1992). A theory of fads, fashion, custom, and cultural change as information cascades. *Journal of Political Economy, 100*, 992–1026.
Binmore, K. (2005). *Natural Justice*. Oxford: Oxford University Press.
Binmore, K. (2009). *Rational Decisions*. Princeton, NJ: Princeton University Press.
Black, F., & Scholes, M. (1973). The pricing of options and corporate liabilities. *Journal of Political Economy, 81*, 637–654.
Bowles, S., et al. (2003). The co-evolution of individual behaviors and socal institutions. *Journal of Theoretical Biology, 223*, 135–147.
Boyd, J., & Richerson, P. J. (2005). *The Origin and Evolution of Culture*. Oxford: Oxford University Press.
Broodbank, C. (2013). *The Making of the Middle Sea*. Oxford: Oxford University Press.
Brown, R. (1971). *The Lefschetz Fixed Point Theorem*. Glenview, IL: Scott and Foreman.
Brouwer, L. E. J. (1912). *Uber abbildung von mannigfaltikeiten. Math Analen, 71*, 97–115.
Buchanan, M. (2001). *Ubiquity*. New York: Crown.
Buchanan, M. (2003). *Nexus*. New York: Norton.
Burkhart, J. M., Hrdy, S. B., & van Schaik, C. P. (2009). Cooperative breeding and human cognitive evolution. *Evolutionary Anthropology, 18*, 175–186.
Calvin, W. H. (2003). *The Ascent of Mind*. New York: Bantam.
Carothers, T. (2002). The end of the transition paradigm. *Journal of Democracy, 13*, 5–21.
Cassidy, J. (2009). *How Markets Fail: The Logic of Economic Calamities*. New York: Farrar, Strauss and Giroux.
Cavallli-Sforza, L., & Feldman, M. (1981). *Cultural Transmission and Evolution*. Princeton, NJ: Princeton University Press.
Chichilnisky, G. (1992). Social diversity, arbitrage, and gains from trade: A unified perspective on resource allocation. *American Economic Review, 84*, 427–434.
Chichilnisky, G. (1993). Intersecting families of sets and the topology of cones in economics. *Bulletin of the American Mathematical Society, 29*, 189–207.
Chichilnisky, G. (1995). Limited arbitrage is necessary and sufficient for the existence of a competitive equilibrium with or without short sales'. *Economic Theory, 5*, 79–107.
Chichilnisky, G. (1996a). Markets and games: A simple equivalence among the core, equilibrium and limited arbitrage. *Metroeconomica, 47*, 266–280.
Chichilnisky, G. (1996b). An axiomatic approach to sustainable development. *Social Choice and Welfare, 13*, 231–257.
Chichilnisky, G. (1997a). A topological invariant for competitive markets. *Journal of Mathematical Economics, 28*, 445–469.
Chichilnisky, G. (1997b). Limited arbitrage is necessary and sufficient for the existence of a equilibrium. *Journal of Mathematical Economics, 28*, 470–479.
Chichilnisky, G. (1997c). Market arbitrage, social choice and the core. *Social Choice and Welfare, 14*, 161–198.
Chichilnisky, G. (2000). An axiomatic approach to choice under uncertainty with catastrophic risks. *Resource Energy Econ, 22*, 221–231.
Chichilnisky, G. (2009a). The topology of fear. *Journal of Mathematical Economics, 45*, 807–816.
Chichilnisky, G. (2009b). Avoiding extinction: Equal treatment of the present and the future, Working Paper: Columbia University.
Chichilnisky, G. (2010). The foundations of statistics with black swans. *Mathematical Social Science, 59*, 184–192.

Chichilnisky, G. (2014). The topology of change: foundations of statistics with black swans. Forthcoming in *Topics in Mathematical Economics: Essays in Honor of J.Marsden*, Fields Institute Communication Volume, American Math Society. Providence, Rhode Island.
Chichilnisky, G., Eisenberger, P. (2010). Asteroids: Assessing catastrophic risks, *Journal of Probability and Statistics* Article ID 954750.
Chichilnisky, G. (2011a). Catastrophic risks with finite or infinite states, *International Journal of Ecological Economics and Statistics*. 23: No. F11.
Chichilnisky, G. (2012a). Sustainable markets with short sales. *Economic Theory*, *49*, 293–307.
Chichilnisky, G. (2012b). Economic theory and the global enviroment. *Economic Theory*, *49*, 217–225.
Christakis, N., & Fowler, J. H. (2011). *Connected*. New York: Back Bay.
Collier, P. (2009). *Wars, Guns and Votes*. New York: Harper.
Collier, P. (2010). *The Plundered Planet*. Oxford: Oxford University Press.
Condorcet, N. (1994 [1785]). Essai sur l'application de l'analyse à la probabilité des décisions rendues à la pluralité des voix. Paris: Imprimerie Royale. Translated in part in: McLean, I. & Hewitt, F. (Eds.) *Condorcet:Foundations of Social Choice and Political Theory*. Aldershot UK: Edward Elgar Publishing.
Cooper, G. (2014). *Money, Blood and Revolution*. Petersfield, UK: Harriman.
Corcos, A. et al. (2002). Imitation and contrarian behavior: Hyperbolic bubbles, crashes and chaos. *Quantitative Finance*, *2*, 264–281.
Dasgupta, P. (2005). Three conceptions of intergenerational Justice. In H. Lillehammer & D. H. Mellor (Eds.), *Ramsey's Legacy*. Oxford: Clarendon Press.
Dawkins, R. (1976). *The Selfish Gene*. Oxford: Oxford University Press.
Debreu, G. (1970). Economies with a finite number of equilibria. *Econometrica*, *38*, 387–392.
Debreu, G. (1976). The application to economics of differential topology and global analysis: Regular differentiable economies. *American Economic Review*, *66*, 280–287.
Deutscher, G. (2006). *The Unfolding of Language*. New York: Holt.
Easley, D., & Kleinberg, J. (2010). *Networks, Crowds and Markets*. Cambridge: Cambridge University Press.
Edwards, P. N. (2010). *A Vast Machine: Computer models, Climate Data, and the Politics of Global Warming*. Cambridge MA: MIT Press.
Eldredge, N. (1976). Differential evolutionary rates. *Paleobiology*, *2*, 174–177.
Eldridge, N., & Gould, S. J. (1972). Punctuated equilibrium. In T. Schopf (Ed.), *Models of Paleobiology*. New York: Norton.
Fan, K. (1961). A generalization of Tychonoff's fixed point theorem. *Math Annalen*, *42*, 305–310.
Ferguson, N. (1997). Introduction. In N. Ferguson (Ed.), *Virtual history*. London: Picador.
Ferguson, N. (2002). *Empire: The Rise and Demise of the British World Order*. London: Penguin Books.
Ferguson, N. (2011). *Civilization*. London: Penguin.
Fisher, L. (2009). *The Perfect swarm*. New York: Basic.
Flyvbjerg, H., Sneppen, K., & Bak, P. (1993). A mean field theory for a simple model of evolution. *Physics Review Lettters*, *71*, 4087–4090.
Fox, J. (2009). *The Myth of the Rational Market*. New York: Harper.
Fukuyama, F. (2011). *The Origins of Political Order*. New York: Ferrar, Strauss and Giroux.
Gamble, C. (1993). *Timewalkers*. London: Penguin.
Gintis, H. (2000). Strong reciprocity and human sociality. *Journal of Theoretical Biology*, *206*, 169–179.
Gladwell, M. (2002). *The Tipping Point*. New York: Back Bay.
Gleick, J. (1987). *Chaos: Making a New Science*. New York: Viking.
Gödel, K. (1931). Uber formal unentscheidbare Satze der Principia Mathematica und verwandter Systeme. *Monatschefte fur Mathematik und Physik* 38: 173–98. Translated as On Formally Undecidable Propositions of *Principia Mathematica* and Related Systems. In van Heijenoort,

J. (Ed.) *Frege and Gödel: Two Fundamental Texts in Mathematical Logic*. Cambridge, MA, Harvard University Press.
Golub, B., & Jackson, M. (2010). Naive learning in social networks and the wisdom of crowds. *American Economic Journal: Microeconomics*, *2*, 112–149.
Gray, J. (1995). *Enlightenment's Wake*. London: Routledge.
Gray, J. (1997). *Endgames*. London: Blackwell.
Gray, J. (2000). *False Dawn*. London: New Press.
Gribbin, J. (2011). *Alone in the Universe*. NewYork: Wiley.
Hahn, F. (1973). *On the Notion of Equilibrium in Economics*. Cambridge: Cambridge University Press.
Hamilton, W. (1964). The genetical evolution of social behavior I and II. *Journal of Theoretical Biology*, *7*, 1–52.
Hamilton, W. (1970). Selfish and spiteful behavior in an evolutionary model. *Nature*, *228*, 1218–1220.
Hardin, G. (1968 [1973]). The tragedy of the commons. In H. E. Daly (Ed.), *Towards a Steady State Economy*. San Francisco: Freeman.
Hardin, R. (1971). Collective action as an agreeable prisons' dilemma. *Behavioral Science*, *16*, 472–481.
Hardin, R. (1982). *Collective Action*. Baltimore, MD: Johns Hopkins University Press.
Heinberg, R. (2011). *The End of Growth*. New York: New Society.
Henrich, J., et al. (2004). *Foundations of Human Sociality*. Oxford: Oxford University Press.
Henrich, J., et al. (2005). Economic man' in cross-cultural perspective: Behavioral experiments in 15 small-scale societies. *Behavioral Brain, Science*, *28*, 795–855.
Hinich, M. J. (2003). Risk when some states are low-probability events. *Macroeconomic Dynamics*, *7*, 636–643.
Hirsch, M. (1976). *Differential Topology*. Berlin: Springer.
Hobbes, T. (2009 [1651]). Leviathan; or the matter, forme, and power of a common-wealth, ecclesiastical and civil. In Gaskin, (Ed.). Oxford: Oxford University Press.
Hrdy, S. B. (2011). *Mothers and Others: The Evolutionary Origins of Mutual Understanding*. Cambridge MA: Harvard University Press.
Hsiang, S.M. (2013). Quantifying the influence of climate on human conflict. *Science Express* 10.1126.
Hubbard, J. H., & West, B. H. (1995). *Differential equations: A dynamical systems approach*. Berlin: Springer.
Israel, J. (2010). *Revolution of the Mind*. Princeton NJ: Princeton University Press.
Israel, J. (2002). *Radical Enlightenment*. Oxford: Oxford University Press.
Israel, J. (2006). *Enlightenment Contested*. Oxford: Oxford University Press.
Israel, J. (2014). *Revolutionary Ideas*. Princeton NJ: Princeton University Press.
Jablonka, E., & Lamb, M. J. (2006). *Evolution in four dimensions: Genetic, epigenetic, behavioral, and symbolic variation in the history of life*. Cambridge MA.: MIT Press.
Johnson, S. (2002). *Emergence*. New York: Scribner.
Kahneman, D. (2011). *Thinking Fast and Slow*. New York: Ferrar Strauss and Giroux.
Karklins, R., & Petersen, R. (1993). Decision calculus of protestors and regime change: Eastern Europe 1989. *Journal of Politics*, *55*, 588–614.
Kauffman, S. (1993). *The Origins of Order*. Oxford: Oxford University Press.
Keohane, R. (1984). *After Hegemony*. Princeton NJ: Princeton University Press.
Keohane, R., & Nye, R. (1977). *Power and Interdependence*. New York: Little Brown.
Keynes, J. M. (1921). *Treatise on Probability*. London: Macmillan.
Keynes, J. M. (1936). *The General Theory of Employment, Interest and Money*. London: Macmillan.
Kindleberger, C. (1973). *The World in Depression 1929–1939*. Berkeley, CA: University of California Press.
Kolbert, E. (2014). *The Sixth Extinction*. New York: Holt.

Knaster, B., Kuratowski, K., & Mazurkiewicz, S. (1929). Ein beweis des fixpunktsatzes fur n-dimensionale simplexe. *Fund Math*, *14*, 132–137.
Kramer, G. H. (1973). On a class of equilibrium conditions for majority rule. *Econometrica*, *41*, 285–297.
Kreps, D. M., et al. (1982). Rational cooperation in the finitely repeated prisoners' dilemma. *Journal of Economic Theory*, *27*, 245–252.
Kurz, M., & Motolese, M. (2001). Endogenous uncertainty and market volatility. *Economic Theory*, *17*, 497–544.
Ladha, K. (1992). Condorcet's jury theorem, free speech and correlated votes. *American Journal of Political Science*, *36*, 617–74.
Ladha, K. (1993). Condorcet's jury theorem in the light of de Finetti's theorem: Majority rule with correlated votes. *Social Choice and Welfare*, *10*, 69–86.
Ladha, K., & Miller, G. (1996). Political discourse, factions and the general will: correlated voting and Condorcet's jury theorem. In N. Schofield (Ed.), *Collective Decision Making*. Boston: Kluwer.
Landemore, H. (2012). *Democratic Reason*. Princeton NJ: Princeton University Press.
Laplace, P. S. (1951 [1814]). *Essai Philosophique sur les Probabilités*. Paris: Gauthiers-Villars. *A philosophical essay on probabilities*. (trans. F. Truscott and F. Emory) New York: Dover.
Li, T. Y., & Yorke, J. A. (1975). Period three implies chaos. *Mathematics Monthly*, *82*, 985–992.
Lohmann, S. (1994). The dynamics of Information cascades. *World Politics*, *47*, 42–101.
Lorenz, E. N. (1962). The statistical prediction of solutions of dynamical equations. In *Proceedings of the International Symposium om Numerical Weather Prediction, Tokyo*
Lorenz, E. N. (1963). Deterministic non periodic flow. *Journal of Atmosphere Science*, 130–141.
Lorenz, E. N. (1993). *The Essence of Chaos*. Seattle: University of Washington Press.
Madison, J. (1999 [1787]). Federalist X. In: Rakove, J. (Ed.). *Madison: Writings*. New York, Library Classics.
Mandelbrot, B., & Hudson, R. (2004). *The (Mis)behavior of Markets*. New York: Perseus.
Margolis, H. (1982). *Selfishness, Altruism and Rationality*. Cambridge: Cambridge University Press.
Margulis, L., & Sagan, D. (2002). *Acquiring Genomes*. New York: Basic.
Maynard Smith, J. (1982). *Evolution and the Theory of Games*. Cambridge: Cambridge University Press.
McKelvey, R. D. (1976). Intransitivities in multidimensional voting models and some implications for agenda control. *Journal of Economic Theory*, *12*, 472–482.
McKelvey, R. D. (1979). General conditions for global intransitivities in formal voting models. *Econometrica*, *47*, 1085–1112.
McKelvey, R. D., & Page, T. (1986). Common Knowledge, Consensus and Aggregrate Information. *Econometrica*, *54*, 109–127.
McKelvey, R. D., & Schofield, N. (1987). Generalized symmetry conditions at a core point. *Econometrica*, *55*, 923–933.
McWhorter, J. (2001). *The Power of Babel*. New York: Holt.
Meadows, D., Meadows, D. L., Randers, J., & Behrens, W. (1972). *Limits to Growth*. New York: Signet.
Meadows, D., Meadows, D. L., Randers, J. (1992). *Beyond the Limits*. White River Junction, VT.: Chelsea Green.
Merton, R. C. (1973). Theory of rational option pricing. *Bell Journal of Economics and Management Science*, *4*, 141–183.
Michael, E. (1956). Continuous selections I. *Annals of Mathematics*, *63*, 361–382.
Miller, G., & Schofield, N. (2003). Activists and partisan realignment in the U.S. *American Political Science Review*, *97*, 245–260.
Miller, G., & Schofield, N. (2008). The transformation of the Republican and Democratic party coalitions in the United States. *Perspectives on Politics*, *6*, 433–450.

Milnor, J. W. (1997). *Topology from a Differential Viewpoint*. Princeton NJ: Princeton University Press.
Minsky, H. (1975). *John Maynard Keynes*. New York: Columbia University Press.
Minsky, H. (1986). *Stabilizing an Unstable Economy*. New Haven: Yale University Press.
Mokyr, J. (2005). The intellectual origins of modern economic growth. *Journal of Economic History*, *65*, 285–351.
Mokyr, J. (2010). *The Enlightened Economy: An Economic History of Britain 1700–1850*. New Haven, CT: Yale University Press.
Mokyr, J., & Nye, V. C. (2007). Distributional coalitions, the Industrial Revolution, and the origins of economic growth in Britain. *Southern Economic Journal*, *74*, 50–70.
Morris, I. (2010). *Why the West Rules*. New York: Ferrar, Strauss and Giroux.
Nagel, T. (2012). *Mind and Cosmos*. Oxford: Oxford University Press.
Nakamura, K. (1979). The vetoers in a simple game with ordinal preference. *International Journal of Game Theory*, *8*, 55–61.
Nordhaus, W. (2013). *Climate Casino*. New Haven: Yale University Press.
North, D. C. (1990). *Institutions, Institutional Change and Economic Performance*. Cambridge: Cambridge University Press.
North, D. C., & Weingast, B. R. (1989). Constitutions and commitment: The evolution of institutions governing public choice in seventeenth century England. *Journal of Economic History*, *49*, 803–32.
North, D. C., Wallis, B., & Weingast, B. R. (2009). *Violence and Social Orders: A Conceptual Framework for Interpreting Recorded Human History*. Cambridge: Cambridge University Press.
Nowak, M. (2011). *Supercooperators*. New York: Free Press.
Ormerod, P. (2001). *Butterfly Economics*. New York: Basic.
Ostrom, E. (1990). *Governing the Commons*. Cambridge: Cambridge University Press.
Pagden, A. (2013). *The Enlightenment* New York: Random.
Pareto, V. (1935). *The Mind and Society [Trattato di Sociologia Generale]*. New York: Harcourt, Brace.
Parfit, D. (2011). *On What Matters*. Oxford: Oxford University Press.
Parker, G. (2013). *Global Crisis*. New Haven CT: Yale University Press.
Penn, E. (2009). A model of far-sighted voting. *American Journal of Political Science*, *53*, 36–54.
Piketty, T. (2014). *Capital*. Cambridge, MA: Harvard University Press.
Plott, C. R. (1967). A notion of equilibrium and its possibility under majority rule. *American Economic Review*, *57*, 787–806.
Penn, E. (2009). A model of far-sighted voting. *American Journal of Political Science*, *53*, 36–54.
Pugh, C. C. (2002). *Real Mathematical Analysis*. Berlin: Springer.
Putnam, R. D., & Campbell, D. E. (2010). *American Grace: How Religion Divides and Unites Us*. New York: Simon and Schuster.
Rader, T. (1972). *Theory of General Economic Equilibrium*. New York: Academic Press.
Randers, J. (2012). *2052: A Global forecast for the next forty years*. White River Junction, VT.: Chelsea Green.
Rees, M. (2001). *Our Cosmic Habitat*. Princeton University Press: Princeton.
Reinhart, C. M., & Rogoff, K. (2008). *This Time Is Different: Eight Centuries of Financial Folly*. Princeton: Princeton University Press.
Saari, D. (1985a). Price dynamics, social choice, voting methods, probability and chaos. In: Aliprantis, D., Burkenshaw, O., Rothman, N. J. (Eds.) *Lecture Notes in Economics and Mathematical Systems*, No. 244. Berlin, Springer.
Saari, D. (1985b). A chaotic exploration of aggregation paradoxes. *SIAM Review*, *37*, 37–52.
Saari, D. (1995). Mathematical complexity of simple economics. *Notes of the American Mathematics Society*, *42*, 222–230.
Saari, D. (1997). The Generic existence of a core for q-Rules. *Economic Theory*, *9*, 219–260.
Saari, D. (2001a). *Decisions and Elections: Explaining the Unexpected*. Cambridge: Cambridge University Press.

Saari, D. (2001b). *Chaotic Elections*. Providence, RI: American Mathematical Society.
Saari, D. (2008). *Disposing Dictators, Demystifying Voting Paradoxes*. Cambridge: Cambridge University Press.
Schofield, N. (1972a). Is majority rule special?. In: Niemi, R. G., & Weisberg, H. F. (Eds.). *Probability Models of Collective Decision-Making*. Columbus, OH: Charles E. Merrill Publishing Co..
Schofield, N. (1972b). Ethical decision rules for uncertain voters. *British Journal of Political Science*, *2*, 193–207.
Schofield, N. (1975). A game theoretic analysis of Olson's game of collective action. *Journal of Conflict Resolution*, *19*, 441–461.
Schofield, N. (1977). The Logic of catastrophe. *Human Ecology*, *5*, 261–271.
Schofield, N. (1978). Instability of simple dynamic games. *Review of Economic Studies*, *45*, 575–594.
Schofield, N. (1979). Rationality or chaos in social choice. *Greek Economic Review*, *1*, 61–76.
Schofield, N. (1980a). Generic properties of simple Bergson-Samuelson welfare functions. *Journal of Mathematical Economics*, *7*, 175–192.
Schofield, N. (1980b). Catastrophe theory and dynamic games. *Quality Quantity*, *14*, 529–545.
Schofield N. (1983a). Equilibria in simple dynamic games. In Pattanaik, P., & Salles, M. (Eds.) *Social Choice and Welfare* (pp. 269–284). Amsterdam: North Holland.
Schofield, N. (1983b). Generic instability of majority rule. *Review of Economic Studies.*, *50*, 695–705.
Schofield, N. (1984a). Existence of equilibrium on a manifold'. *Mathematics of Operations Research*, *9*, 545–557.
Schofield, N. (1984b). Social equilibrium and cycles on compact sets. *Journal of Economic Theory*, *33*, 59–71.
Schofield, N. (1984c). Classification theorem for smooth social choice on a manifold. *Social Choice and Welfare*, *1*, 187–210.
Schofield, N. (1985). Anarchy, altruism and cooperation. *Social Choice and Welfare*, *2*, 207–219.
Schofield, N., & Tovey, C. (1992). Probability and convergence for supra-majority rule with Euclidean preferences. *Mathematical Computer Modelling*, *16*, 41–58.
Schofield, N. (1999a). The heart and the uncovered set. *Journal of Economics Supplement*, *8*, 79–113.
Schofield, N. (1999b). A smooth social choice method of preference aggregation. In: Wooders, M. (Ed.). *Topics in mathematical economics and game theory: Essays in honor of R. Aumann*. Fields Institute, Providence, RI: American Mathematical Society.
Schofield, N. (1999c). The C^1–topology on the space of smooth preferences. *Social Choice and Welfare*, *16*, 445–470.
Schofield, N. (2002). Evolution of the constitution. *British Journal of Political Science*, *32*, 1–20.
Schofield, N. (2002). Madison and the founding of the two party system in the US. In Kernel S (Ed.), *James Madison: The Theory and Practise of Republican Government* (pp. 302–327). Stanford, CA: Stanford University Press.
Schofield, N. (2003). *Mathematical methods in economics and social choice*. Berlin: Springer.
Schofield, N. (2006). *Architects of Political Change*. Cambridge: Cambridge University Press.
Schofield, N. (2007). The mean voter theorem: Necessary and sufficient conditions for convergent equilibrium. *Review of Economic Studies*, *74*, 965–980.
Schofield, N. (2010). Social orders. *Social Choice and Welfare*, *34*, 503–536.
Schofield, N. (2011). Is the political economy stable or chaotic? *Czech Economic Review*, *5*, 76–93.
Schofield, N., & Gallego, M. (2011). *Leadership or Chaos*. Berlin: Springer.
Schweitzer, F., et al. (2009). Economic networks: The new challenges. *Science*, *325*, 422–425.
Shafer, W., & Sonnenschein, H. (1975). Equilibrium in abstract economies without ordered preferences. *Journal of Mathematical Economics*, *2*, 245–248.
Shiller, R. (2003). *The New Financial Order*. Princeton, NJ: Princeton University Press.
Shiller, R. (2005). *Irrational Exuberance*. Princeton, NJ: Princeton University Press.

Smale, S. (1966). Structurally stable systems are not dense. *American Journal of Mathematics*, *88*, 491–496.
Smale, S. (1974a). Global analysis and economics IIA: Extension of a theorem of Debreu. *Journal of Mathematical Economics*, *1*, 1–14.
Smale, S. (1974b). Global analysis and economics IV: Finiteness and stability of equilibria with general consumption sets and production. *Journal of Mathematical Economics*, *1*, 119–127.
Smith, A. (1984 [1759]), *The Theory of Moral Sentiments*, Indianapolis, IN: Liberty Fund
Sperber, J. (2011). *Karl Marx: A nineteenth century life*. New York: Liveright.
Smolin, L. (2007). *The Trouble with Physics*. New York: Houghton Mifflin.
Strnad, J. (1985). The structure of continuous-valued neutral monotonic social functions. *Social Choice and Welfar*, *2*, 181–195.
Strogatz, S. (2004). *Sync*. New York: Hyperion.
Stern, N. (2007). *The Economics of Climate Change*. Cambridge: Cambridge University Press.
Stern, N. (2009). *The Global Deal*. New York: Public Affairs.
Sunstein, C. R. (2006). *Infotopia*. Oxford: Oxford University Press.
Sunstein, C. R. (2009). *A Constitution of Many Minds*. Princeton, NJ: Princeton University Press.
Sunstein, C. R. (2011). *Going to Extremes*. Oxford: Oxford University Press.
Surowiecki, J. (2005). *The Wisdom of Crowds*. New York: Anchor.
Taleb, N. N. (2007). *The Black Swan*. New York: Random.
Taleb, N. N., & Blyth, M. (2011). The black swan of Cairo. *Foreign Affairs*, *90*(3), 33–39.
Taylor, C. *A Secular Age*, Cambridge MA: Belknap Press.
Taylor, M. (1976). *Anarchy and Cooperation*. London: Wiley.
Taylor, M. (1982). *Community, Anarchy and Liberty*. Cambridge: Cambridge University Press.
Tegmark, M. (2008). The mathematical universe. *Foundations of Physics*, *38*, 101–150.
Tegmark, M. (2014). *Our Mathematical Universe*. New York: Random House.
Trivers, R. (1971). The evolution of reciprocal altruism. *Quarterly Review of Biology*, *46*, 35–56.
Trivers, R. (1985). *Social Evolution*. Menlo Park, CA: Cummings.
Turing, A. (1937). On Computable Numbers with an Application to the Entscheidungs Problem. In *Proceedings of the London Mathematical Society* 42, 230–265. Reprinted in Jack Copeland, ed.I *The Essential Turing*. Oxford: The Clarendon Press.
Vogel, K. (2014). *Big Money*. New York: Public Affairs.
Walker, M. (1977). On the existence of maximal elements. *Journal of Economic Theory*, *16*, 470–474.
Wallace, A. R. (1898). *Natural Selection*. New York: Classics US.
Waltham, D. (2014). *Lucky Planet*. New York: Basic.
Watts, D. (2002). A simple model of global cascades on random networks. *Proceedings of the National Academy of Science*, *99*, 5766–5771.
Watts, D. (2003). *Six Degrees*. New York: Norton.
Weitzman, M. (2009). Additive damages, fat-tailed climate dynamics, and uncertain discounting. *Economics*, *3*, 1–22.
Yanofsky, N. S. (2013). *The Outer Limits of Reason*. Cambridge MA: MIT Press.
Zeeman, E. C. (1977). *Catastrophe Theory: Selected Papers, 1972–77*. New York: Addison Wesley.
Zeeman, E. C. (1992). Evolution and catstrophe theory. In: Bourriau (Ed.). *Understanding Catastrophe*. Cambridge: Cambridge University Press.
Zhang, D. D. (2007). Global climate change, war, and population decline in recent human history. In *Proceedings of the National Academy of Science*, *104*(49), 19214–19219.

Discounting Utility and the Evaluation of Climate Policy

Larry Karp

1 Introduction

This essay considers three aspects of the relation between utility discounting and climate policy: the sensitivity of policy recommendations to discounting assumptions, the relation between discounting and catastrophic risk, and the difference between discounting for intra- and inter-personal intertemporal transfers.

The utility discount rate converts future and current utility into the same units, thereby making inter-temporal comparisons sensible.[1] Much of our intuition about the sensitivity of climate policy to discounting comes from cost-benefit examples. In this context, we are willing to spend very little to influence a non-catastrophic event that occurs in the distant future; the amount that we are willing to spend may be sensitive to the discount rate. These conclusions might be reversed if the date of the policy outcome is random instead of deterministic (Sect. 2). Analytic examples and a review of numerical climate models show varying levels of sensitivity of optimal policy to discounting assumptions. The complexity of these models makes an explanation for these differences unattainable. But it may be interesting to note that

[1]The utility discount rate (the pure rate of time preference, or PRTP) is related to, but distinct from, the social discount rate, used to compare consumption (as distinct from utility) across two points in time. In the standard deterministic setting, the social discount rate equals the PRTP plus the growth rate multiplied by the inverse of the elasticity of intertemporal substitution. The latter provides a measure of the willingness to substitute consumption between "infinitely close" periods.

I thank an anonymous reviewer for comments on an earlier draft of this paper. The usual disclaimer applies. Sects. 3.2 and 4 of this paper are adapted from Karp (2009), and Sect. 5 is based on Karp (2013a, b).

L. Karp (✉)
Department of Agricultural and Resource Economics,
University of California, 207 Giannini Hall, Berkeley, CA 94720, USA
e-mail: karp@are.berkelely.edu

G. Chichilnisky and A. Rezai (eds.), *The Economics of the Global Environment*,
Studies in Economic Theory 29, DOI 10.1007/978-3-319-31943-8_19

when the underlying model is "more linear", the solution appears more sensitive to discounting. The definition of "more linear" is context-specific (Sect. 3).

Time-discounting and catastrophe-avoidance are logically distinct topics, but recent papers claim that the risk of catastrophe swamps any consideration of discounting. Sect. 4 explains why I am skeptical of this claim.

Most policy models use an infinitely lived agent model. That model makes no distinction between intertemporal transfers involving the same agent, e.g. a person when they are young and old, and transfers involving two different people. These two types of transfers are quite different. Even if we use a constant discount rate to evaluate each, there is no reason that we should use the same constant for the two types of transfers. A two-parameter discounting model can distinguish between these two types of transfers. A planner who gives equal weight to the welfare of all people currently living, but distinguishes between intertemporal transfers for the same person and intertemporal transfers between different people, has time inconsistent preferences. Sect. 5 illustrates the discount rate induced by such a planner in an overlapping generations setting.

2 The Cost-Benefit Setting

Although society cannot literally insure itself against climate-related events, current expenditures can reduce the probability of those events (abatement), and the cost associated with them (mitigation). I consider the extreme case where society has a binary choice, either to do nothing and face the risk (or the certainty) of the event, or to take a costly action that eliminates the risk. I refer to the action as buying insurance, and the cost of the action as the cost of the premium. The formal question is to determine how the pure rate of time preference (PRTP) affects society's maximum willingness to pay for "perfect" insurance, which eliminates the risk of climate change (Karp 2009).

Society's actual policy choice is not binary, but instead requires choosing among many different types and levels of control; and the actions we take might reduce, but cannot eliminate the risk or the consequences of climate change. However, the simplicity of the model makes it easy to see how key parameters, in particular the PRTP, affect society's willingness to incur current costs to ameliorate future damages. Abstracting from the complications of more realistic policy-driven models throws into relief certain relations between parameter assumptions and model recommendations.

I compare society's willingness to pay to eliminate the risk in two extreme cases: where the time of the event, T, is known with certainty, and where the event time, $\tilde{T}$, is a random variable. I set the expected time, $E\left(\tilde{T}\right)$, in the stochastic case, equal to the known time in the deterministic case, T, so that the two models are comparable. The two noteworthy qualitative results are the same for both zero and positive consumption growth. First, moving from a certain event time to a random event time increases the maximum premium that society is willing to pay, especially for low probability events. Second, the premium is less sensitive to the PRTP in the stochastic case, compared to the deterministic case.

The intuition for these results, based on Jensen's inequality, is quite simple. In moving from the deterministic setting, which concentrates all of the probability mass at the expected event time, to the stochastic case, we transfer some of that concentrated probability mass to earlier times, and some to later times. The higher probability of an earlier event increases the expected present value cost of the event, and the higher probability of a later event decreases that expected present value. However, because of discounting, the effect of the first change is larger than the effect of the second. Therefore, moving from a deterministic to a stochastic event time increases the premium that society is willing to pay.

The explanation for the greater sensitivity (to the PRTP) of the maximum premium in the deterministic case is only slightly more involved. The elasticity of the maximum premium, with respect to the PRTP, in the stochastic case, is a *weighted* expectation of the elasticity under certainty. This formula assigns higher weight to the elasticities corresponding to earlier event times; those elasticities are lower than their counterparts at later times.

Table 1 collects the parameter definitions used in this section.

2.1 Zero Growth

The payment of a premium and the loss associated with the event both reduce consumption, and therefore reduce utility. With zero growth, these are the only factors that affect consumption and utility. I express the costs associated with the event and with payment of the premium in units of utility.

Table 1 Parameter definitions

Parameter	Definition
ρ	Pure rate of time preference
$h = \frac{1}{T}$	Hazard (inverse of expected event time)
Δ	Fraction of income lost due to event
g	Consumption growth rate
η	Elasticity of marginal utility: $-\frac{u''(c)}{u'(c)}c$
r	Social discount rate
x	Maximum willingness to pay to avoid event (the premium)
ϕ	Elasticity of premium wrt ρ: $-\frac{dx}{d\rho}\frac{\rho}{x}$

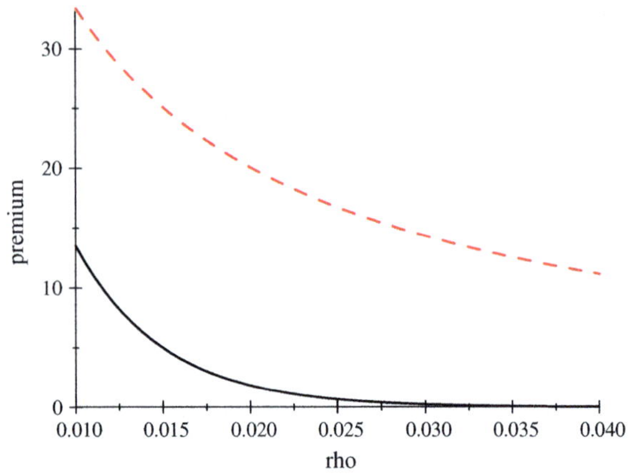

Fig. 1 Maximum premium as a percent of the flow loss when event time is $T = 200$ (*solid*) and when event time is exponentially distributed with $E(T) = 200$ (*dashed*)

2.1.1 Deterministic Event Time

For the deterministic case, suppose that by paying a premium that has a utility flow cost of x per unit of time, society insures itself against (and in that sense avoids) a utility flow loss of 100 in each period from time T to ∞. With a PRTP ρ, the present value of the utility loss that begins at T is $e^{-\rho T}\frac{100}{\rho}$, and the present value of the premium payment, beginning today, is $\frac{x}{\rho}$. Equating these expressions implies that society would be willing to pay a premium of at most $x(T,\rho) = e^{-\rho T}100$ over $(0,\infty)$. In this deterministic setting, the premium, x, is a convex function of T; this fact is key to understanding the effect of moving to an uncertain event time. If $T = 200$, the premium, x, changes by a factor of 55, ranging from 13.5 to 0.25, as ρ ranges from 0.01 to 0.03 (1 to 3 % per annum); see Fig. 1. The elasticity of x with respect to ρ is $\phi(x) = \rho T$. Thus, x is particularly sensitive to the PRTP when the event occurs in the distant future. This example illustrates the role of discounting in the simplest cost-benefit calculation.

2.1.2 Stochastic Event Time

I begin with a two point distribution to provide intuition, and then move to the exponentially distributed event time, which yields a simpler formula for the premium and its elasticity with respect to ρ. With the two-point distribution, $\tilde{T}$ takes two values, $T - \varepsilon$ and $T + \varepsilon$, each with probability 0.5, so $E(\tilde{T}) = T$. The maximum premium for the risk neutral planner, x', is the expectation over $\tilde{T}$ of $x(\tilde{T},\rho)$:

$$x' = E_{\tilde{T}}x = \left(e^{-\rho(T-\varepsilon)} + e^{-\rho(T+\varepsilon)}\right) 50 \Rightarrow$$

$$\phi\left(x'\right) = -\frac{dx'}{d\rho}\frac{\rho}{x'} = \rho\left((T-\varepsilon)\frac{e^{-\rho(T-\varepsilon)}}{e^{-\rho(T-\varepsilon)}+e^{-\rho(T+\varepsilon)}} + (T+\varepsilon)\frac{e^{-\rho(T+\varepsilon)}}{e^{-\rho(T-\varepsilon)}+e^{-\rho(T+\varepsilon)}}\right)$$

$$= \rho\left(T - \varepsilon\left(\frac{1-e^{-2\rho\varepsilon}}{1+e^{-2\rho\varepsilon}}\right)\right) < \rho T.$$

The fact, noted above, that $x(T)$ is convex, together with Jensen's inequality, implies that $E_{\tilde{T}}x\left(\tilde{T}\right) > x\left(E_{\tilde{T}}\tilde{T}\right) = x(T)$, so the first line implies that moving from a certain to a random event time increases the maximum premium. The second line shows that the elasticity of the premium, with respect to the PRTP, is a *weighted* sum of the probability-weighted elasticities; the first weight exceeds the second. The previous subsection shows that the elasticity is higher at the high event time. Thus, the elasticity in the stochastic case, $\phi\left(x'\right)$, is lower than the expectation of the deterministic elasticities, ρT.

Where $\tilde{T}$ is an exponentially distributed random variable with hazard rate h, $E\left(\tilde{T}\right) = \frac{1}{h}$. The expected present value cost of the uncertain event is $E\left(e^{-\rho T}\frac{100}{\rho}\right) = \frac{100h}{\rho(\rho+h)}$. Equating this expression to the cost of the premium, $\frac{x'}{\rho}$, gives the maximum premium that society would pay for perfect insurance, $x' = \frac{h100}{h+\rho}$. Setting $\frac{1}{h} = T$ makes the stochastic and deterministic models comparable, and yields $x' = \frac{100}{1+\rho T}$.

2.1.3 Comparison of Deterministic and Stochastic Event Time

For $T = 200$ as above, the maximum acceptable premium in the stochastic setting, x', ranges from 33.3 to 14.3 as ρ ranges from 0.01 to 0.03. A change from the deterministic to the stochastic event time increases the maximum premium by a factor that ranges from 2.5 for $\rho = 0.01$ to 57 for $\rho = 0.03$ (Fig. 1).

For general T, with $h = \frac{1}{T}$, the ratio of the maximum premium in the stochastic compared to the deterministic event time is

$$\text{premium ratio: } \frac{x'}{x} = \frac{\exp(\frac{\rho}{h})}{1 + \frac{\rho}{h}} > 1.$$

Figure 2 shows the graph of this ratio of premiums as a function of the ratio $\frac{\rho}{h}$. For example, with an annual discount rate $\rho = 0.02$, $\frac{\rho}{h} = 0.5$ corresponds to an expected event time of $T = 25$ years and a premium ratio of 1.1; $\frac{\rho}{h} = 7$ corresponds to an expected event time of 350 years and a premium ratio of 137. Thus, for low probability events, moving from a deterministic to a stochastic model increases the maximum premium by two orders of magnitude. "Low probability" means that the hazard rate is small relative to the PRTP.

Although the maximum premium is higher in the stochastic compared to the deterministic setting, the premium in the former is much less sensitive to the PRTP. The (absolute value) elasticity of x (in the deterministic case) with respect to ρ

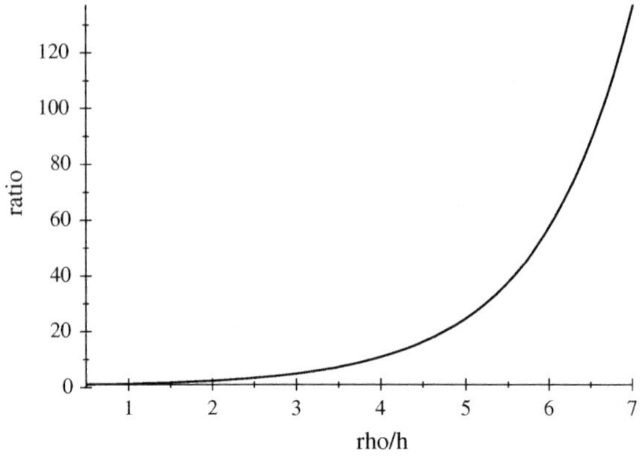

Fig. 2 The ratio of the maximum premium under stochastic to the maximum premium under deterministic event time $\left(\frac{x'}{x}\right)$ as a function of $\frac{\rho}{h}$

is $\phi(x) = \rho T = \frac{\rho}{h}$, which is linear in ρ; and the elasticity of x' (in the stochastic case) with respect to ρ is $\phi\left(x'\right) = \frac{\rho}{h+\rho}$, which decreases in ρ. The ratio of these elasticities is

$$\text{elasticity ratio: } \frac{\phi\left(x'\right)}{\phi(x)} = \frac{\frac{\rho}{h+\rho}}{\frac{\rho}{h}} = \frac{h}{h+\rho},$$

which is small for "low probability events".

This example illustrates the two features described above: moving from the deterministic to a stochastic setting increases the maximum premium, and also makes it less sensitive to the PRTP.

2.2 *Positive Consumption Growth*

If per capita income is expected to grow, future generations will be richer than current generations. With decreasing marginal utility of income, growth makes people today less willing to sacrifice to avoid future damages. In the deterministic setting, the Ramsey formula gives the social discount rate (SDR), r, as a function of the pure rate of time preference, ρ, the growth rate, g, and the elasticity of marginal utility (the inverse intertemporal elasticity of substitution), η: $r = \rho + \eta g$. With zero growth or infinite intertemporal elasticity of substitution, the social discount rate equals the pure rate of time preference. Positive growth and finite intertemporal elasticity of substitution increases the social discount rate.

I normalize consumption at $t = 0$ to 1, and assume that growth is constant, g, so potential consumption at time $t > 0$ prior to the event is e^{gt}. The event results in a permanent $\Delta \times 100\,\%$ reduction in potential consumption flow, so at a post-event time t, consumption is $c = e^{gt}(1 - \Delta)$. The insurance premium is deducted from potential consumption (not investment), so payment of the premium does not change the growth rate. The premium needed to eliminate the risk is proportional to the value-at-risk, Δe^{gt}, with proportionality factor X. If society pays the premium, consumption is $c = e^{gt}(1 - \Delta X)$. Utility (u) is isoelastic in consumption: $u(c) = \frac{c^{1-\eta}-1}{1-\eta}$.[2]

In order for the premia to be easily compared with the zero-growth analogs, I present them as a percent of Δ. In the deterministic case, the maximum premium (as a percent of Δ) that society is willing to pay for perfect insurance is

$$y = \frac{1 - \left(\left(1-(1-\Delta)^{1-\eta}\right)\left(1-e^{-\frac{(\rho+(\eta-1)g)}{h}}\right) + (1-\Delta)^{1-\eta}\right)^{\frac{1}{1-\eta}}}{\Delta} 100. \tag{1}$$

In the exponentially distributed case, the maximum premium that society is willing to pay for perfect insurance is

$$z = \frac{1 - \left(\left(1-(1-\Delta)^{1-\eta}\right)\left(1-\frac{h}{(\rho+g(\eta-1)+h)}\right) + (1-\Delta)^{1-\eta}\right)^{\frac{1}{1-\eta}}}{\Delta} 100. \tag{2}$$

As a consistency check, note that for $g = 0 = \eta$, y and z collapse to their deterministic and stochastic analogs in Sect. 2.1. Although the social discount rate, r, equals the PRTP if either $g = 0$ or $\eta = 0$, we need both of those equalities to hold in order for the premia in this section to equal their analogs in Sect. 2.1. For example, with $\eta > 0$, utility is nonlinear in consumption, so it matters whether we express costs in units of utility or of consumption, even if there is no growth.

Figure 3 graphs the premia y and z as functions of ρ, and Fig. 4 graphs the elasticity of these premia with respect to ρ.[3] With $g = 0 = \eta$, the elasticity in the deterministic case is linear, equal to $\frac{\rho}{h} = \rho T$. With $g > 0$ and $\eta > 0$ the elasticity with respect to ρ is approximately linear in ρ under the deterministic event time, ranging between 1 and 4 as ρ ranges from 0.005 to 0.02 (for $T = 200$). By this measure, it appears that optimal policy is very sensitive to discounting assumptions. However, the elasticity of the premium with respect to ρ under stochastic event time is small and insensitive to ρ, reaching only about 0.5 at $\rho = 0.02$ and scarcely increasing thereafter.

Comparing these figures to Fig. 1 shows that introducing growth ($g > 0$) and measuring costs in consumption rather than utility units ($\eta > 0$) leaves unchanged the qualitative comparisons between deterministic and stochastic event times discussed above: The maximum premium under the stochastic event time is much larger, but

---

[2]The model in this section is based on Karp and Tsur (2008). That paper considers non-constant PRTP, whereas here the PRTP is constant. In the model here, society can eliminate the risk, whereas in the Karp and Tsur model society is able only to prevent the hazard from increasing. Neither model is nested in the other. Appendix 1 derives Eqs. 1 and 2.

[3]Figures 11, and 12 in Appendix 1 shows graphs of the premia as functions of η, and g. Those parameters have the same qualitative effect as ρ on the two premia (under deterministic and stochastic event time), although of course the magnitudes of their effects are different.

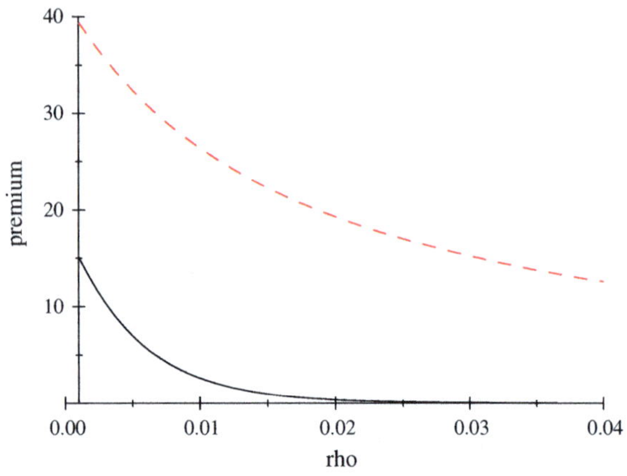

Fig. 3 Maximum premium (as a percent of cost of event) as a function of ρ, under certain event time (*solid*) and random event time (*dashed*) for $T = 200 = \frac{1}{h}$, $\Delta = 0.3$, $g = 0.01$ and $\eta = 2$

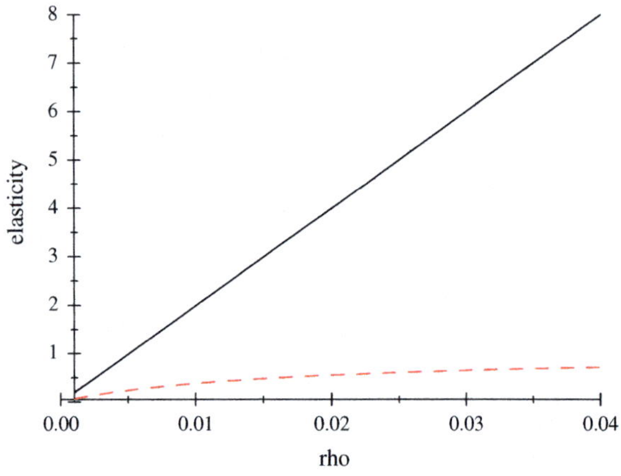

Fig. 4 The elasticity of the maximum premium under certain event time (*solid*) and under random event time (*dashed*) for $T = 200 = \frac{1}{h}$, $\Delta = 0.4$, and $\eta = 2$

also much less sensitive to the PRTP (and other discounting parameters) compared to the premium under the deterministic event time.

Arrow (2007) examines the effect of the discount rate on our willingness to avoid climate change, posing the question in terms of growth rates rather than levels of damages. His examples suggest that the benefits of significant climate policy outweighs the cost to such a large extent that the cost-benefit ratio is not sensitive to discounting assumptions.

3 Optimization Models

The comparison in Sect. 2 is based on a cost-benefit exercise. That material shows that moving from a deterministic to random event time increases the maximum acceptable insurance premium and makes that premium less sensitive to discounting assumptions. This section considers the sensitivity of policies in an optimizing framework. I first consider two analytic examples, which (together with the results above) illustrate circumstances where greater nonlinearity in the model makes the optimal policy less sensitive to the discount rate. I then assess this relation using several climate policy models.

3.1 Analytic Examples

Two familiar and tractable renewable resource models illustrate the relation between the steady state and the discount rate. In the simplest fishery model, the change in the stock, S, equals a growth function, $f(S)$, minus harvest, h; and the utility flow, $u(h)$, depends on harvest but not on the stock. The growth equation is $\frac{dS}{dt} = \dot{S} = f(S) - h$ and the payoff, evaluated at time 0, is the infinite stream of discounted utility, $\int_0^\infty e^{-\rho t} u(h_t)\, dt$, where the discount rate is ρ. The optimal (interior) steady state is the solution to $f'(S) = \rho$. The elasticity of the steady state, with respect to the discount rate, is

$$-\frac{\rho}{S_\infty}\frac{dS_\infty}{d\rho} = \frac{-1}{S_\infty}\frac{f'(S_\infty)}{f''(S_\infty)}.$$

The elasticity of the steady state has the same form as the inverse of the Arrow-Pratt risk aversion, but here applied to the growth (instead of utility) function. As the growth function becomes more steeply curved (evaluated at the steady state) the steady state becomes less sensitive to the discount rate.

For the logistic growth function, $f = \varphi S\left(1 - \frac{S}{k}\right)$; φ is the intrinsic growth rate and k is the carrying capacity. For this function, the elasticity of the steady state with respect to the discount rate is

$$\frac{-1}{S_\infty}\frac{f'(S_\infty)}{f''(S_\infty)} = \frac{\rho}{\varphi - \rho}.$$

A higher intrinsic growth rate, φ, increases the curvature of the growth function, and makes the steady state less sensitive to the discount rate. Figure 5 graphs the elasticity as a function of the PRTP for estimates of the intrinsic growth rate $\varphi = 0.71$ for Pacific Halibut and $\varphi = 0.08$ for Antarctic fin-whale (Clark 1975). Even for the slow-growing fin-whale, the elasticity is less than 1 for reasonable discount rates. For $\rho = \varphi$, where it is optimal to drive the stock to extinction, the elasticity is infinite.

The second simplest fishery model allows harvest costs (and thus the utility flow) to depend on the stock, but assumes that the flow of benefits due to harvest (like the growth function) is linear in the harvest. If the price per unit of harvest, p, is constant and the harvest costs linear in harvest, h, then the flow of benefit equals $\left(p - c(S_t)\right) h_t$. The payoff is the discounted stream of benefits, $\int_0^\infty e^{-\rho t}\left(p - c(S_t)\right) h_t dt$. In this case,

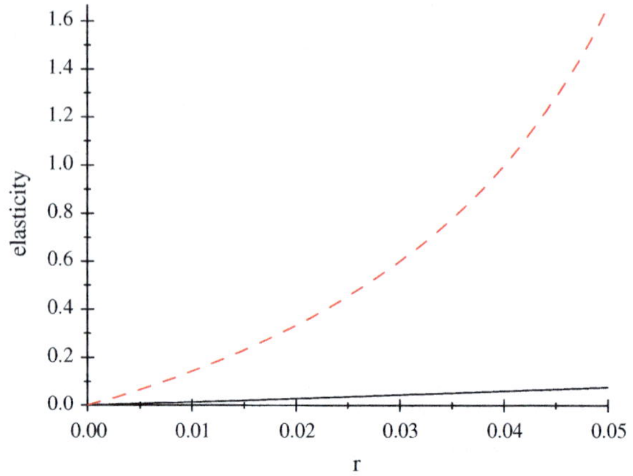

Fig. 5 The elasticity of the steady state with respect to the ρ in the zero-extraction cost model, for Pacific Halibut (*solid*, with $\varphi = 0.71$) and Antartic Fin-whale (*dashed*, with $\varphi = 0.08$)

the optimal harvest policy is bang-bang: the harvest is set to its maximum feasible level if the stock is above the steady state, and the harvest is set to 0 if stock is below the steady state; when the stock is at the steady state, the harvest maintains it at that level: $h_\infty = f\left(S_\infty\right)$. The steady state is the solution to

$$\rho = f'(S) - \frac{c'(S)f(S)}{p - c(S)}. \tag{3}$$

Using the logistic growth model and $c(S) = \frac{c}{S}$, with parameter values taken from Clark (1975), Fig. 6 (taken from Ekeland et al. 2012) shows the elasticity of the steady state stock for Pacific halibut (solid graph) and Antarctic fin-whale (dashed graph). The elasticities are non-monotonic in the discount rate, but both are well below 1. For reasonable values of ρ (i.e., values much less than 0.2, or 20 % per year), the steady state is much less sensitive to the discount rate

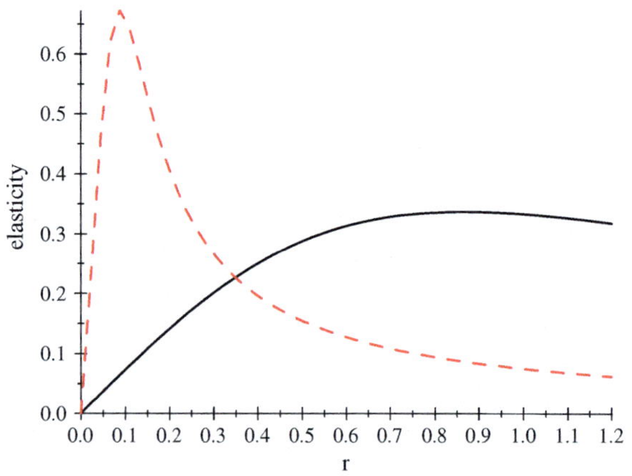

Fig. 6 Elasticity of steady state with respect to ρ for Pacific halibut (*solid graph*, using $K = 80.5 \times 10^6$ and $\frac{c}{p} = 17.7 \times 10^6$ kg) and for Antartic fin-whale (*dashed graph*, where $K = 400{,}000$ whales and $\frac{p}{c} = 40{,}000$ whales). Parameter values from Clark (1979)

for the fast-growing halibut, compared to the slow growing whales. Again, a higher intrinsic growth rate implies a more non-linear growth function, and for reasonable parameter values causes the optimal policy to be less sensitive to the discount rate.

3.2 Climate-Related Models

The Stern Review (2006) (hereafter SR) is perhaps the most widely discussed document on climate policy during the past decade. Several economists focused on the SR's discounting assumptions. SR chose a PRTP of $\rho = 0.001$, an elasticity of marginal utility of $\eta = 1$, and a growth rate of 0.013, implying a social discount rate (SDR) of $r = 0.014$ (or 1.4 %).

Nordhaus (2007) illustrates the importance of these discounting assumptions by comparing the results of three runs of the DICE model. Two of these runs use combinations of the PRTP and η consistent with a SDR of about 5.5 %, almost four times the level in the SR. With the Nordhaus values, the optimal carbon tax in the near term is approximately \$35/ton Carbon (or \$9.5/ ton CO_2), and the optimal level of abatement in the near term about 14 % of Business as Usual (BAU) emissions. A third run, using the SR's values of ρ and η (together with the DICE assumptions about growth) led to a carbon tax of \$350/ton and a 53 % level of abatement, close to the level that the SR recommends. Thus, the carbon tax increases by a factor of 10 and abatement increases by a factor of $\frac{53}{14} = 3.8$ with the decrease in the SDR. Because abatement costs are convex, the percent change in the tax (caused by changes in the discounting assumptions) is larger than the percent change in abatement. Based on these numbers, an estimate of the elasticity of the tax with respect to the discount rate is approximately $\frac{10}{4} = 2.5$ and the elasticity of abatement with respect to the tax is approximately $\frac{3.8}{4} = 0.95$. These values are in the range of elasticities of the maximum premium, in the deterministic cost-benefit setting shown in Fig. 4.

In a different context (focused on the effect of catastrophic damages) Norhaus (2009) compares optimal policy under a PRTP of 0.015 and 0.001, holding other DICE parameters (including the elasticity of marginal utility) at their baseline levels. He reports that the reduction in PRTP increases the optimal carbon tax from \$42/tC to \$102/tC. This 2.4-fold increase in the optimal tax is much less than the 10-fold increase reported in Nordhaus (2007), where both the PRTP and the elasticity are changed.[4]

Karp (2005) uses a linear-quadratic model, calibrated to reflect abatement costs and climate-related damages that are of the same order or magnitude as in DICE. In this stationary, partial equilibrium model, there is no growth, so the pure rate

[4]The increase in the tax from \$42/tC to \$102/tC leads to a fall in per capital income (during the period when it is lowest, presumably the first period) from \$6,801 to \$6,799, i.e. about 0.03 %. In view of parameter uncertainty, a 0.03 % reduction in per capita income, equivalently a \$15 billion increase in aggregate abatement costs (at Gross World Product—GWP—of \$50 trillion), is close to rounding error.

of time preference equals the social discount rate. A decrease in the discount rate from 3 to 1 % increases abatement in the first period by a factor of 2.5. Nordhaus's (2007) experiments, described above, reduce the pure rate of time preference by approximately 1.5 %, increasing abatement by a factor of 3.8. By this measure, the sensitivity of policy to the discount rate is of the same order of magnitude in the two models.

Fujii and Karp (2008) provide a more involved analysis of the role of discounting, using a one-state variable model calibrated to approximate the costs and benefits underlying the SR recommendations. In that model, Δ_t is the consumption loss due to mitigation expenditures and remaining climate-related damage, as a fraction of the no-damage no-control scenario (i.e. where there is no potential for climate damage). Reducing the discount rate increases abatement expenditures and reduces the trajectory of the damages, as expected. However, the magnitude of those induced changes was much smaller than the previous studies led us to expect. This insensitivity is probably not due to a peculiarity of our climate model, because Fujii and Karp (2006) find a similar relation using a standard renewable resource model. The analytic examples in Sect. 3.1 provide some insight into this result.

The highly nonlinear relation between expenditures and damages may explain this insensitivity. Figure 7, taken from Fujii and Karp (2008), shows the graph of the steady state climate-related costs, Δ, as a function of steady state expenditures (expressed as a fraction of income) x. This graph reaches a global minimum where climate related expenditure is 0.845 % of consumption. This is the optimal steady state level of expenditures under zero discounting. Total costs fall rapidly for smaller values of x. This graph implies that small increases in expenditure, below the global optimum, achieve significant reductions in total costs. Therefore, a very low SDR achieves nearly the global minimum, and even substantially larger SDRs take us close to the global minimum. Because initial expenditures (compared to steady state expenditures) are even less sensitive to the SDR, the entire trajectory is "relatively insensitive" to the discount rate.

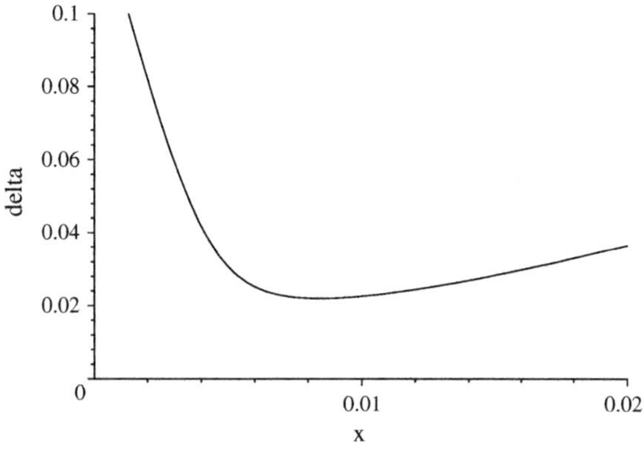

Fig. 7 The graph of steady state costs, Δ, as a function of steady state expenditures, x

Gerlagh and Liski (2012), using an extension of Golosov et al. (2013), calibrate a model to reflect climate-related damages and abatement costs similar to those in Norhaus (2009). They find that the optimal tax is very sensitive to discounting assumptions. The logarithmic utility is the inverse of the exponential damages, causing the flow payoff to be linear in the cumulative emissions (a state variable); the state dynamics are also linear in emissions. Their model is thus linear in the stock of emissions. I conjecture that this linearity contributes to the sensitivity of the tax, with respect to the discount rate(s).

This selective review shows that the discount rate matters a great deal in some, but not all, dynamic optimization models. All of this evidence comes from specific models or specific ways of presenting the tradeoff between abatement costs and avoided damages, and therefore it cannot lead to general conclusions. It provides some (but certainly not conclusive) examples where optimal policy tends to be less sensitive to the PRTP when the model is highly nonlinear.

4 Do Catastrophes Swamp Discounting?

Weitzman (2009) examines the effect of parameter uncertainty on the social discount rate. Using a two period model, representing the current period and the distant future, he calculates the marginal expected value of transferring the first unit of certain consumption from the present into an uncertain future. His chief result is open to several interpretations. In my view, a "modest" interpretation is correct and useful. A controversial interpretation is that the result undermines our ability to sensibly apply cost-benefit analysis to situations where there is uncertainty about catastrophic events. A "corollary" to this interpretation is that the recognition of catastrophic events makes discounting a second order issue. I think that both the controversial interpretation and the corollary to it are incorrect.

In order to explain these points, I consider a simplified version of his model. Let c be the known current consumption, c' the random future consumption, v the number of certain units of consumption transferred from the current period to the future, β the utility discount factor, and u the utility of consumption.[5] The social discount factor for consumption, i.e. the marginal rate of substitution between "the first" additional certain unit of consumption today and in the future, is

$$\Gamma = -\beta E_{c'} \left(\frac{\frac{du(c'+v)}{dv}}{\frac{du(c-v)}{dv}} \right)_{|v=0} . \tag{4}$$

[5]This model assumes that the sacrifice of one unit of consumption today makes one unit available in the future period. If instead, the sacrifice of v units makes $\theta(v)$ units available the future, the numerator of the right side of Eq. 4 becomes $\frac{du(c'+\theta(v))}{dv}\theta'(v)$. These and other extensions can be easily treated, but the simple model that I use is adequate for my purposes.

The chief result is that, under the assumptions of the model, $\Gamma = \infty$; Weitzman dubs this result "the dismal theorem".

The model includes a number of important features, including: (i) the uncertainty about c' is such that there is a "significant" probability that its realization is 0; (ii) the marginal utility of consumption at $c' = 0$ is infinite; and (iii) it is possible to transfer a certain unit of consumption into the future. Features (ii) and (iii) are assumptions, but (i) is an implication of the assumption that the variance of c' is an unknown parameter, and the decision-maker's subjective distribution for this parameter has "fat tails".

Any of these assumptions can be criticized, but in my view, a more fundamental issue involves the interpretation of the dismal theorem. A modest interpretation is that uncertainty about the distribution of a random variable can significantly increase "overall uncertainty" about this random variable, leading to a much higher risk premium (and therefore a much higher willingness to transfer consumption from the present into the future) relative to the situation where the distribution of the random variable is known. This modest interpretation is not controversial.

An extreme interpretation is that under conditions where the dismal theorem holds, society should be willing to make essentially any sacrifice to transfer a unit of certain consumption into the future. That interpretation is also not controversial, because it is so obviously wrong. Even with 0 discounting (in this two period stationary model with the same utility function in both periods), we would never be willing to transfer to the future more than half of what we currently have.

The controversial interpretation is that the dismal theorem substantially undermines our ability to sensibly apply cost-benefit analysis to situations with "deep uncertainty" about catastrophic risks. The basis for this claim is that in order to use the social discount rate given in Eq. (4), we need to modify the model so that Γ is finite. Weitzman suggests ways of doing this, such as truncating a distribution or changing an assumption about the utility function or its argument, in order to make Γ finite. The alleged problem is that the resulting Γ is extremely sensitive to the particular device that we use to render it finite. Because we do not have a consensus about how to achieve this finite value, we do not have a good way to select from the many extremely large and possibly very different social discount rates. In this setting, it is difficult to use cost-benefit analysis.

This controversial interpretation is not persuasive. Horowitz and Lange (2008) identify clearly the nub of the misunderstanding; I rephrase their explanation. Nordhaus (2008) also identifies this issue, and he provides numerical results using DICE to illustrate how cost-benefit analysis can be used even when damages are extremely large.[6] Millner (2013) contains a thorough discussion of the Dismal Theorem, together with extensions.

---

[6]These numerical results are interesting, but someone who accepts the controversial interpretation of the dismal theorem will not regard them as a convincing counter-argument to that interpretation. All of the numerical experiments arise in a deterministic context, and in that respect they do not really confront the dismal theorem, which makes sense only in a setting with risk and uncertainty. For this reason, I think that Horowitz and Lange's very simple treatment of the problem is particularly helpful.

The problem with the controversial interpretation is that the value of Γ in Eq. (4) is essentially irrelevant for cost-benefit analysis. This expression, which is evaluated at $v = 0$, gives the value of the "first" marginal unit transferred. The fact that the derivative may be infinite does not, of course, imply that the value of transferring one (non-infinitesimal) unit of sure consumption is infinite. If we want to approximate the value of a function, it makes no sense to use a Taylor approximation evaluated where that function's first derivative is infinite. We would make (essentially) this mistake if we were to use Eq. (4) as a basis for cost-benefit analysis with climate policy. The only information that we obtain by learning that the value of the derivative evaluated at $v = 0$ is infinite is that a non-infinitesimal policy response must be optimal. This fact is worth knowing, but it does not create problems for using cost-benefit analysis.

Although I think that the controversial interpretation is unsound, it has a "corollary" that is not so easy to dismiss. This corollary states that catastrophic risks swamps the effect of the pure rate of time preference. The idea is that because the expectation of the term in parenthesis in Eq. (4) is so large, the magnitude of β, and thus of the pure rate of time preference, is relatively unimportant. Norhaus (2009), despite his criticism of the controversial interpretation of the dismal theorem, endorses this view (emphasis added.):

> ...discounting is a second-order issue in the context of catastrophic outcomes. ... If the future outlook is indeed catastrophic, that is understood, and policies are undertaken, the discount rate has little effect on the estimate of the social cost of carbon *or to the optimal mitigation policy*.

This corollary may hold in specific settings, but it would be surprising if it is a general feature of catastrophic risk. The magnitude of the expectation of the term in parenthesis in Eq. (4), evaluated at $v = 0$, can certainly swamp the magnitude of β; but I have just noted that the former term is irrelevant for cost-benefit analysis (beyond telling us that non-infinitesimal policy is optimal).

In order to get a sense of whether the corollary is likely to hold, and also to illustrate why the controversial interpretation of the dismal theorem is not persuasive, I use an example with $u = \frac{c^{1-\eta}}{1-\eta}$. Set $\beta = \exp(-\rho T)$ and choose a unit of time equal to a century. With this choice of units, ρ is the annual pure rate of time preference expressed as a percent. Suppose that c' takes the value c with probability $1 - p$ and the value 0 with probability p. For $p > 0$ the right side of Eq. (4) is infinite, as in the dismal theorem. The optimization problem is

$$\max_{v} \left(u(c - v) + \beta \left[(1 - p)\, u(c + v) + pu(v)\right]\right).$$

Normalize by setting $c = 1$, so that v equals the fraction of current consumption that we transfer into the future. With a bit of manipulation, the first order condition for the optimal v is

$$\rho = \frac{1}{T} \ln \left((1 - p) \left(\frac{1 - v}{1 + v}\right)^{\eta} + p \left(\frac{1 - v}{v}\right)^{\eta}\right). \tag{5}$$

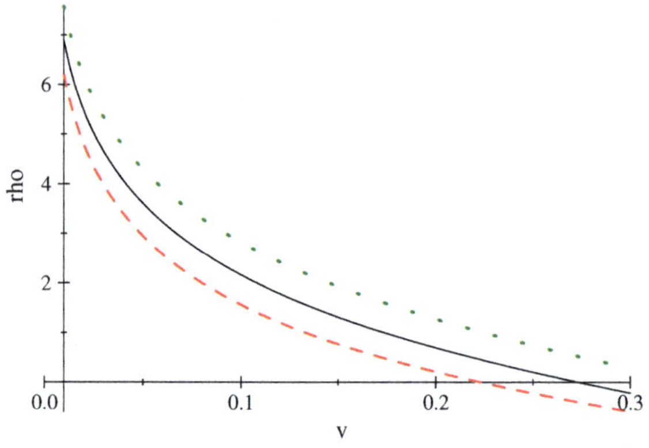

Fig. 8 The relation between the transfer, $v \in [0.01, 0.3]$, (the x axis) and the annual percentage discount rate, ρ, for $p = 0.05$ (*dashed*), $p = 0.1$ (*solid*) and $p = 0.2$ (*dotted*), with $\eta = 2$ and $T = 1$ century

Using $\eta = 2$ and $T = 1$ (so that the "future" is a century from now), Fig. 8 shows the relation between the *annual* percentage pure rate of time preference, ρ, and the optimal value of v for p equal to 0.05, 0.1 and 0.2. The figure shows that the fact that the expression in Eq. (4) is infinite does not cause any problem in determining an optimal value of the transfer. It also illustrates the less-obvious point that catastrophic risk does not swamp the effect of discounting, in determining the optimal level of the transfer.

We get a bit more insight into the relative importance of the PRTP, ρ, and the probability of catastrophe, p, by taking the ratio of the elasticities of the optimal v with respect to these variables. Define the elasticities and the ratio of elasticities as:

$$\frac{dv}{dp}\frac{p}{v} = \alpha; \; \frac{dv}{d\rho}\frac{\rho}{v} = \phi; \; \tau = -\frac{\phi}{\alpha}.$$

For $\tau > 1$, v is more sensitive to discounting than to the probability of catastrophe, and the reverse holds for $\tau < 1$. For the isoelastic example,

$$\tau = \left(1 + \frac{v^\eta}{p\left((v+1)^\eta - v^\eta\right)}\right) \ln\left(p\left(\frac{1-v}{v}\right)^\eta + (1-p)\left(\frac{1-v}{1+v}\right)^\eta\right), \tag{6}$$

where Eq. (5) implicitly defines $v = v(\rho, p, T)$. The ratio of elasticities shown in Eq. 6 depends explicitly on p and v and it depends implicitly on ρT, because that product determines the optimal value of v, given p and η.

Figure 9 shows the graphs of τ for $T = 1$ and $p = 0.05$, i.e. a 5 % chance of the catastrophe within 1 century, for $\eta = 2$ (the solid curve) and $\eta = 0.5$, the dashed curve. If $\eta = 2$, the equilibrium transfer is more sensitive to the discount rate than to the probability of catastrophe ($\tau > 1$) if $\rho > 0.74\,\%$; for $\eta = 0.5$, the transfer is more sensitive to discounting than to the probability of catastrophe if $\rho > 0.21\,\%$. Both of these critical values are much larger than the 0.1 % PRTP used in the Stern Review, but they are much smaller than PRTPs used for other studies, e.g. the various

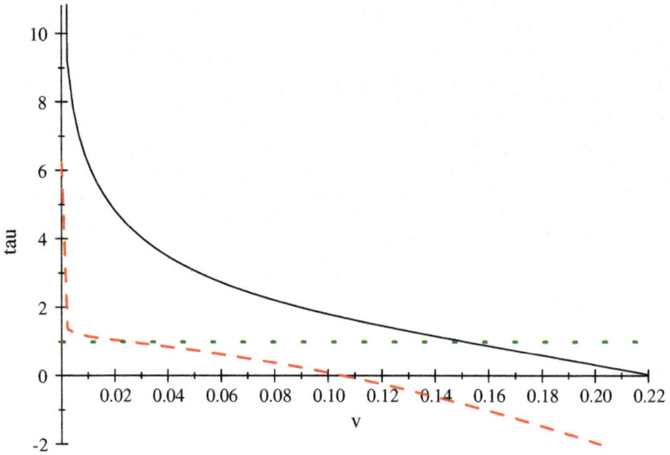

Fig. 9 The graph of τ for $\eta = 2$ (*solid*) and $\eta = 0.5$ (*dashed*), with $T = 1$ and $p = 0.05$. For $\eta = 2$, v ranges from 0.03 to 0.22 as ρ ranges from 4 to 0.1 %. For $\eta = 0.5$, v ranges from 0.06 to 9×10^{-7} asρ ranges from 4 to 0.1 %. The *dotted line* shows $\tau = 1$

incarnations of DICE. Thus, the discount rate might be either more or less important than the probability of catastrophe in determining the optimal transfer; it is likely to be less important for discount rates commonly used.

Here I explain how to interpret Fig. 9. The dotted line shows $\tau = 1$; for values of v where the graph lies above the dotted line, the equilibrium is more sensitive to the PRTP ρ, than to the risk of catastrophe, p. As noted above, for given values of η, T, p, the equilibrium value of v is a decreasing function of ρ. For $\eta = 2$, $v \in \left[3 \times 10^{-2}, 0.22\right]$ as ρ falls from a PRTP of 4 % (per year) to 0.1 %. For $\eta = 0.5$, $v \in \left[9 \times 10^{-7}, 0.06\right]$ as ρ falls from a PRTP of 4 % (per year) to 0.1 %. Given p, η, Eq. 6 determines the value of v at which $\tau = 1$. Given this value of v, and p, T, η, Eq. 5 determines the critical value of ρ at which $\tau = 1$. Those critical values are $\rho = 0.74$ % for $\eta = 2$ and $\rho = 0.21$ % for $\eta = 0.5$.

5 How Do We View the Distant Future?

The sections above examine the importance, to climate policy, of the magnitude of the PRTP. This section considers the applicability of a constant PRTP in the climate context. The PRTP measures a person's willingness to transfer utility between two points in time. Even if this person uses a constant PRTP to evaluate a utility transfer from one period to another for herself, there is no reason that society would use the same constant rate to evaluate an intertemporal transfer between two different people. If we are better able to distinguish among people who are closer to us, either in space, time, or genetically, compared to people who are further from us, then we plausibly discount hyperbolically (with respect to space, time, or genetics), not at a constant rate. More generally, if our willingness to transfer utility between two generations depends not only on the distance between those two generations, but also on their distance from us, then we discount at a non-constant rate.

Many papers discuss the plausibility of hyperbolic (more generally, nonconstant) discounting, both in the context of individual decision problems (Phelps and Pollack 1968), (Laibson 1997), (Barro 1999), (Heal 1998), (Harris and Laibson 2001) and for societal problems such as climate change (Cropper and Laibson 1999), (Karp 2005), (Fujii and Karp 2008), (Karp and Tsur 2011), (Schneider et al. 2012), (Gerlagh and Liski 2012), (Karp 2013b). The "Weber-Fechner law" states that human response to a change in stimulus, such as sound or light, is inversely proportional to the pre-existing stimulus. For example, from the standpoint of period 0, delaying utility from period 1 to period 2 represents a much larger proportional increase in delay than does a delay from period 10 to 11, even though the absolute increase in delay is the same in the two cases. Heal (2001) invokes this observation as justification for a decreasing discount rate. Applied to discounting, the "law" is consistent with a discount factor of t^{-K}, where K is a positive constant. Heal calls this "logarithmic discounting".

Ramsey (1928) remarked "My picture of the world is drawn in perspective. ... I apply my perspective not merely to space but also to time." Karp (2013c) shows that perspective applied to space corresponds to a special case of logarithmic "spatial discounting."To the extent that spatial perspective provides a useful analogy for temporal perspective, this result provides further support for the hypothesis that our view of the world corresponds to hyperbolic discounting.

A two-parameter model of preferences makes it possible to distinguish between intertemporal utility transfers for a single individual and across individuals. The PRTP, $\rho \geq 0$, measures a person's willingness to sacrifice their own future utility in order to increase their own current utility. A second discounting parameter, denoted $\lambda \geq 0$, measures a planner's willingness to transfer utility across *different people at different points in time*. I consider a pure public good (or bad), in which every person alive in a period has the same utility flow. In this setting, there is no reason to consider transfers between different people alive at a point in time. A richer model would be able to evaluate transfers across different people at different points in time, and across different people at the same point in time, but that is too much to ask of a two-parameter model. I assume that the planner gives equal weight to all people currently alive, and in that respect is utilitarian.

Consider the simplest constant-population OLG model in which agents lives for two periods. A "public project", e.g. emissions of a certain amount of greenhouse gasses, increases current aggregate utility, and reduces next-period aggregate utility by one unit. [7] What is the minimal increase in current aggregate utility needed to justify these emissions? The evaluation is complicated by the fact that it involves a utility exchange both between the current young people and their future old selves,

[7] In order to keep this example simple, suppose that current emissions affect only next period utility. Obviously, a realistic climate model has to take into account that today's emissions persist in the atmosphere, affecting utility over a long span.

and also between the next period young people and the current old people. That is the nature of climate policy.

An increase in current utility of $e^{-\rho}$ leaves the current young people indifferent. If $\lambda = \rho$, the planner is willing to transfer utility between two people one period apart at the same rate as a person would transform utility between her future and current selves. If $\lambda = 0$, the planner gives equal weight to the future young and the current old people. If $\lambda = \infty$, the planner gives zero weight to the future young. It is worth emphasizing that the value $\lambda = 0$ ($not \lambda = \rho$) implies that the planner treats currently living and not-yet-born agents symmetrically. Readers will have different views about the "reasonable" relation between ρ and λ, but it is clear that smaller values of λ imply greater weight to the not-yet-born, and in that respect correspond to greater altruism.

In this model, half of the people are young, and half are old in any period. The planner would accept the project if and only if the current increase in aggregate utility is no less than $D(1) = 0.5\left(e^{-\rho} + e^{-\lambda}\right)$, the planner's one period discount factor. This planner would accept a project that lowers aggregate utility by one unit t periods from now if it increases current aggregate utility by no less than $D(t) = 0.5\left(e^{\lambda(t-1)}e^{-\rho} + e^{-\lambda t}\right)$, the planner's t period discount factor. The term $0.5e^{\lambda(t-1)}e^{-\rho}$ equals the present value at $t-1$ to the agents born at $t-1$ of the one unit loss in utility ($e^{-\rho}$) times the planner's weight on those people's welfare, $0.5e^{\lambda(t-1)}$. The second term, $0.5e^{-\lambda t}$, equals the weight the planner puts on the utility of people born t periods in the future. Defining $\beta = \frac{e^{-\rho}+e^{-\lambda}}{2}e^{\lambda}$ and $\delta = e^{-\lambda}$, gives $D(1) = \beta\delta$ and $D(t) = \beta\delta^t$ for $t > 1$. This particular form of discounting is known as β, δ or quasi-hyperbolic discounting (Phelps and Pollack 1968). The case usually emphasized is $\lambda < \rho$, where $\beta < 1$. This model is typically used for a single-agent decision problem, to capture present bias, whereas I use it to distinguish between transfers across time for a single agent from transfers across time between different agents.

Karp (2013b) provides the formula for the discount factor in a generalization of this model, where each agent lives for T years, and time is continuous. Ekeland and Lazrak (2012) provide the discount function for a model in which agents' lifetime is exponentially distributed with mortality rate (hazard rate) θ. Setting $\theta = \frac{1}{T}$ makes the two models comparable. Figure 10, taken from Karp (2013b), shows the discount rates under exponentially distributed and finite lifetimes, for parameter values $\rho = 0.02 = \theta = \frac{1}{T}$, and for $\lambda \in \{0.01, 0.06\}$. The planner's discount rate falls if $\lambda < \rho$ (as with hyperbolic discounting), and rises for $\lambda > \rho$. In both of these cases, the planner's preferences are time inconsistent. Preferences are time consistent only for $\lambda = \rho$, where the planner makes no distinction between transferring utility across time for the same individual and between two different individuals.

For $\lambda \neq \rho$ in this two-parameter model, time consistency requires that the planner gives less weight to the old than to the young agent in a period (Karp 2013a). Obstfeld (1988) use this type of time consistent model to study fiscal policy, and (Schneider et al. 2012) use it to study climate policy. Those authors discount the utility of currently living agents back to the time of their birth. This procedure means that, for $\lambda < r$, currently living people have less weight in the planner's objective function,

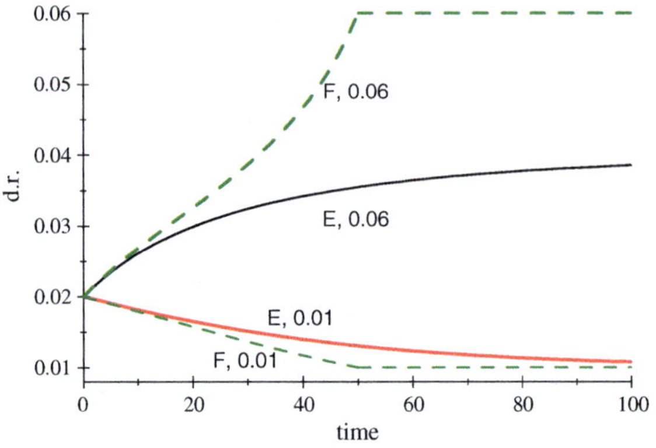

Fig. 10 Discount rates (d.r.) for $\theta = 0.02 = r = \frac{1}{T}$. *Solid curves* (labelled E) correspond to exponentially distributed lifetime and *dashed curves* (labelled F) correspond to fixed lifetime. Numerical values in label show value of λ

the older they are; with this procedure, the planner's preferences are time consistent. However, if the planner gives all currently living people equal weight, then for $\lambda \neq r$, the planner's preferences are time inconsistent.

6 Conclusion

Agents who discount the utility of future generations may be unwilling to make much of a sacrifice to avoid or ameliorate large damages that occur in the probably distant future. The degree of sacrifice we are willing to undertake may be very sensitive to discounting assumptions. Many prominent integrated assessment models illustrate these conclusions. I do not dispute their importance, but it is worth remembering that they are model-dependent. Moving from a deterministic to a random event time, in a cost-benefit setting, can weaken both conclusions, merely as a consequence of Jensen's inequality. Using analytic models and a review of numerical results, I provide examples where policy prescriptions are more sensitive to discounting assumptions the more linear is the model.

Although most models emphasize the sensitivity of optimal policy to assumptions about discounting, there is a strand in the literature that claims that discounting is relatively unimportant when discussing potentially catastrophic events. In my view, the model that has been adduced to support this conclusion in fact has little if anything to say about the conclusion. In some circumstances, catastrophic risk does swamp discounting in determining optimal policy, and in other circumstances it does not. I do not think that we currently have a basis for thinking that either tendency is more plausible.

Most climate policy models use an infinitely lived agent. These models provide a sensible starting point, because they admit normative conclusions. However, intertemporal transfers across a single individual or between different individuals are conceptually distinct. Even if, in the interest of tractability, we want to use a constant discount rate to evaluate both types of transfers, there is no logical reason for using the same constant to evaluate the two types of transfers. If we accept that different discount rates should be used to evaluate the two types of transfers, and if we also want to attach the same weight to the welfare of *currently* living agents, then the implicit social planner has time inconsistent preferences. In that case, the planner's problem becomes a sequential game instead of an optimization problem, and we lose the straightforward normative implications of the model.

Appendix 1

If society pays the premium, thus eliminating risk, the present discounted stream of utility is

$$R(X;\Delta,g,\eta)\equiv\int_0^\infty e^{-\rho t}\frac{(e^{gt}(1-\Delta X))^{1-\eta}-1}{1-\eta}dt =$$
$$\int_0^\infty \frac{e^{-(\rho+(\eta-1)g)t}((1-\Delta X))^{1-\eta}-e^{-\rho t}}{1-\eta}dt=\frac{1}{1-\eta}\left(\frac{((1-\Delta X))^{1-\eta}}{(\rho+(\eta-1)g)}-\frac{1}{\rho}\right)$$

If society does not pay the insurance premium, and the event occurs at time T, the payoff is

$$P(T;\Delta,g,\eta)\equiv\int_0^T e^{-\rho t}\frac{(e^{gt})^{1-\eta}-1}{1-\eta}d\tau+\int_T^\infty e^{-\rho t}\frac{(e^{gt}(1-\Delta))^{1-\eta}-1}{1-\eta}d\tau$$
$$=\int_0^T e^{-\rho t}\frac{((e^{gt})^{1-\eta}-1-((e^{gt}(1-\Delta))^{1-\eta}-1))}{1-\eta}d\tau+\int_0^\infty e^{-\rho t}\frac{(e^{gt}(1-\Delta))^{1-\eta}-1}{1-\eta}d\tau$$
$$=\int_0^T e^{-\rho t}\frac{e^{-g(\eta-1)t}\left(1-(1-\Delta)^{1-\eta}\right)}{1-\eta}d\tau+\int_0^\infty e^{-\rho t}\frac{(e^{gt}(1-\Delta))^{1-\eta}-1}{1-\eta}d\tau$$
$$=\frac{\left(1-(1-\Delta)^{1-\eta}\right)}{1-\eta}\frac{1-e^{-(\rho+(\eta-1)g)T}}{(\rho+(\eta-1)g)}+\frac{1}{1-\eta}\left(\frac{(1-\Delta)^{1-\eta}}{(\rho+(\eta-1)g)}-\frac{1}{\rho}\right)$$

If the event time T is exponentially distributed with hazard h, then

$$Ee^{-(\rho+(\eta-1)g)T}=\int_0^\infty e^{-(\rho+(\eta-1)g)t}he^{-ht}dt=\frac{h}{(\rho+g\eta-g+h)}. \tag{7}$$

Using this formula, we have

$$EP(T;\Delta,g,\eta)=\frac{\left(1-(1-\Delta)^{1-\eta}\right)}{1-\eta}\frac{1-\frac{h}{(\rho+g\eta-g+h)}}{(\rho+(\eta-1)g)}+\frac{1}{1-\eta}\left(\frac{(1-\Delta)^{1-\eta}}{(\rho+(\eta-1)g)}-\frac{1}{\rho}\right)$$

For the certain event time, with $T=\frac{1}{h}$, the maximum premium society would pay, X, is the solution to

$$R(X;\Delta,g,\eta)=P(T;\Delta,g,\eta)$$

or

$$\frac{1}{1-\eta}\left(\frac{((1-\Delta X))^{1-\eta}}{(\rho+(\eta-1)g)}-\frac{1}{\rho}\right)=$$
$$\frac{\left(1-(1-\Delta)^{1-\eta}\right)}{1-\eta}\frac{1-e^{-(\rho+(\eta-1)g)T}}{(\rho+(\eta-1)g)}+\frac{1}{1-\eta}\left(\frac{(1-\Delta)^{1-\eta}}{(\rho+(\eta-1)g)}-\frac{1}{\rho}\right).$$

Solving for X

$$\left(\frac{((1-\Delta X))^{1-\eta}}{(\rho+(\eta-1)g)}\right)=\left(1-(1-\Delta)^{1-\eta}\right)\frac{1-e^{-(\rho+(\eta-1)g)T}}{(\rho+(\eta-1)g)}+\left(\frac{(1-\Delta)^{1-\eta}}{(\rho+(\eta-1)g)}\right)\Rightarrow$$
$$(1-\Delta X)=\left(\left(1-(1-\Delta)^{1-\eta}\right)\left(1-e^{-\frac{(\rho+(\eta-1)g)}{h}}\right)+(1-\Delta)^{1-\eta}\right)^{\frac{1}{1-\eta}}\Rightarrow$$

Solving for X and setting $y = 100X$ gives Eq. 1. I obtain Eq. 2 by taking expectations with respect to T, using Eq. 7, and repeating the steps used to obtain Eq. 1 (Figs. 11 and 12).

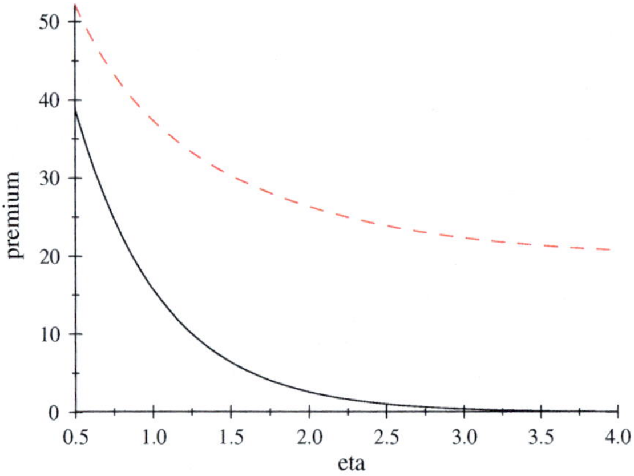

Fig. 11 Maximum premium (as a percent of cost of event) as a function of η, under certain event time (*solid*) and random event time (*dashed*) for $T = 200 = \frac{1}{h}$, $\Delta = 0.4$, $g = 0.01$ and $\rho = 0.01$

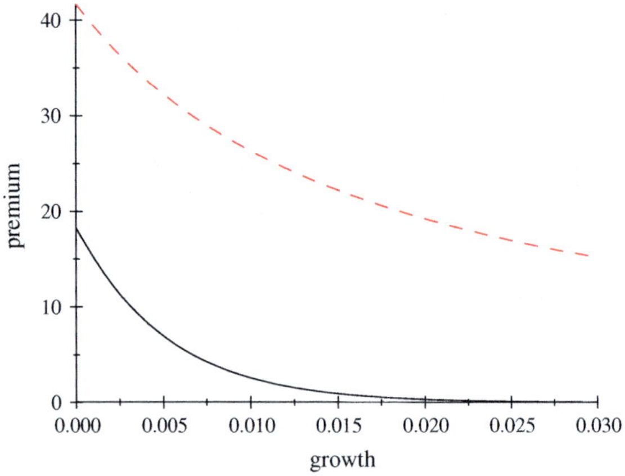

Fig. 12 Maximum premium (as a percent of cost of event) as a function of growth, under certain event time (*solid*) and random event time (*dashed*) for $T = 200 = \frac{1}{h}$, $\Delta = 0.4$, $\eta = 2$ and $\rho = 0.01$

References

Arrow, K. (2007). Globak climate change: A challange to policy. *The Economists Voice*, *4*, 1–5.

Barro, R. (1999). Ramsey meets Laibson in the neoclassical growth model. *Quarterly Journal of Economics*, *114*, 1125–52.

Calvo, G., & Obstfeld, M. (1988). Optimal time-consistent fiscal policy with finite lifetimes. *Econometrica*, *56*, 32–411.

Clark, C. W. (1975). *Mathematical bioeconomics: The optimal management of renewable resources*. Wiley.

Cropper, M., & Laibson, D. (1999). The implications of hyperbolic discounting for project evaluation. In P. R. Portney & J. Weyant (Eds.), *Discounting and Intergenerational Equity*. Resources for the Future.

Ekeland, I., Karp, L., & Sumaila, R. (2012). *Equilibrium management of fisheries with altruistic overlapping generations*. University of British Columbia Working Paper.

Ekeland, I., & Lazrak, A. (2010). The golden rule when preferences are time inconsistent. *Mathematical and Financial Economics*, *4*(1), 29–55.

Fujii, T., & Karp, L. (2008). Numerical analysis of non-constant pure rate of time preference: a model of climate policy. *Journal of Environmental Economics and Management, Forthcoming*, *56*, 83–101.

Fujii, T. (2008). Numerical analysis of non-constant pure rate of time preference: a model of climate policy. *Journal of Environmental Economics and Management*, *56*, 83–101.

Gerlagh, R., & Liski, M. (2012). *Carbon prices for the next thousand years*. Unpublished Working Paper.

Harris, C., & Laibson, D. (2001). Dynamic choices of hyperbolic consumers. *Econometrica*, *69*(5), 935–957.

Heal, G. (1998). *Valuing the future: economic theory and sustainability*. New York: Columbia University Press.

Heal, G. (2001). Intertemporal welfare economics and the environment. In K.-G. Maler & J. Vincent (Eds.), *The handbook of environmental economics*. Amsterdam: North Holland.

Horowitz, J., & Lange, A. (2008). What's wrong with infinity—a note on Weitzman's dismal theorem. University of Maryland Working Paper.

Karp, L. (2005). Global Warming and hyperbolic discounting. *Journal of Public Economics*, *89*, 261–282.

Karp, L. (2009). Sacrifice, discounting and climate policy: five questions. Giannini Foundation working paper.

Karp, L. (2013a). Overlapping generations and environmental policy: An Introduction. Giannini Foundation Working Paper.

Karp, L. (2013b). *Provision of a public good with altruistic overlapping generations and many tribes*. Giannini Foundation Working Paper.

Karp, L. (2013c). *Railroad discounting*. Giannini Foundation Working Paper.

Laibson, D. (1997). Golden eggs and hyperbolic discounting. *Quarterly Journal of Economics*, *62*, 443–78.

Nordhaus, W. D. (2007). A Review of the stern review on the economics of climate change. *Journal of Economic Literature*, *3*, 686–702.

Norhaus, W. (2009). An analysis of the dismal theorem. Cowles Foundatiuon Discussion Paper No 1686.

Phelps, E., & Pollack, R. (1968). On second-best national savings and game: equilibrium growth. *Review of Economic Studies*, *35*, 185–199.

Ramsey, F. (1928). A mathematical theory of savings. *Economic Journal*, *38*, 59–543.
Schneider, M., Traeger, C., & Winkler, R. (2012). Trading off generations: Equity, efficiency and climate change. *European Economic Review*, *56*, 1621–1644.
Stern, N. (2007). *The economics of climate change*. Cambridge University Press.
Weitzman, M. (2009). On modeling and interpreting the economics of catastrophic climate change. *Review of Economics and Statistics*, *91*(1), 1–19.

Global Warming and Economic Externalities

Armon Rezai, Duncan K. Foley and Lance Taylor

1 Introduction

Much discussion of the economics of global warming emphasizes the issue of tradeoffs in well-being between present and future generations (Nordhaus 2008; Nordhaus and Boyer 2000; Stern 2007). Specifically, is it socially beneficial for present and near future generations to sacrifice their own consumption to mitigate global warming for the benefit of generations yet to come?

In this paper we argue that the intergenerational distribution aspects of climate policy are relevant only when the externality has been corrected, and concern the distribution of welfare gains, not costs. If global warming is a negative externality, standard welfare analysis shows that all generations can benefit from its mitigation. Current generations can direct less of their foregone consumption to physical capital formation and more toward mitigation, thereby maintaining their own levels of welfare while bequeathing a better mix of conventional capital and stock of greenhouse gases (GHG) in the atmosphere to the future.

We illustrate this point by solving a *business-as-usual* economic growth model calibrated to current data for the intertemporal allocation of capital by a representative agent with an uncorrected externality and comparing the results to a solution

We received helpful comments from Larry Karp and an anonymous referee. Support from the Schwartz Center for Economic Policy Analysis is gratefully acknowledged.

Originally published in Economic Theory, Volume 49, Number 2, February 2012 DOI 10.1007/s00199-010-0573-7.

A. Rezai (✉)
Department of Socioeconomics, Vienna University of Economics and Business, Nordbergstr. 15/4, 1090 Vienna, Austria
e-mail: arezai@wu.ac.at

D.K. Foley · L. Taylor
Department of Economics, New School for Social Research, 6 East 16th Street, New York, NY 10003, USA

G. Chichilnisky and A. Rezai (eds.), *The Economics of the Global Environment*, Studies in Economic Theory 29, DOI 10.1007/978-3-319-31943-8_20

in which the externality is corrected. The results show that the correction can represent a Pareto improvement from an inefficient to an efficient growth path with higher consumption levels and lower environmental damage. There are world political efforts to implement institutions which enforce the social cost of carbon emissions on individual agents, most notably the Kyoto Protocol which created a carbon market and is international law since 2005. Our contribution provides theoretical justification for such and further measures; the existing framework only succeeds in partially internalizing the externality with the emissions restrictions set in generous manners and the majority of emissions not being subject to any caps.

In much of the literature this simple observation is obscured because the optimal path is compared to a reference path along which the externality is partially corrected. This reference path maximizes the present discounted value of the felicity of per-capita consumption subject to the constraint that mitigation expenditure is equal to zero. This *constrained optimum* implicitly includes the marginal social cost of emissions in the representative agent's production and investment decisions, thus partially internalizing the externality. Comparing this solution to the true optimum incorrectly directs attention toward "intergeneration tradeoffs" because typically the constrained optimum shows higher per capita consumption for several decades. While policy economists have persisted in presenting the global warming problem primarily as an issue of intergenerational equity, many scholars emphasize the public-good nature of the problem, which leads to the perspective developed in the present paper (Chichilnisky 1994; Chichilnisky and Heal 1994; Chichilnisky and Sheeran 2009).

When, on the other hand, the optimal and business as usual paths are compared directly, the first-order effect of optimal mitigation is a potential increase in per capita consumption in every time period. Intergenerational equity enters into the problem only as a second-order effect as the optimal program distributes the potential gains from correcting the externality across generations in accordance with the representative agent's preference for consumption smoothing. The comparison of the optimal and constrained optimal paths thus leads to an upward biased estimate of the economic costs of mitigating global warming.

This symposium includes contributions on other important aspects of the global warming problem. The most related to our are those by Karp and Zhang (2016), who discuss the effectiveness of different policy instruments, Ostrom (2016), who discusses means of introducing cost transparency at different governmental levels, and Chipman and Tian (2016), who investigate the assumptions under which markets for pollution rights lead to Pareto improvements. Burniaux and Oliveira Martins (2011) aim at identifying the sensitivity of "carbon leakage" to key parameters in a general equilibrium model of climate policy. Dutta and Radner (2016) and Lecocq and Hourcade (2016) discuss further roadblocks in the way of efficient international climate policy in a multi-region framework. The remaining contributions (Asheim et al. 2016; Chichilnisky 2016; Figuieres and Tiball 2016; Lauwers 2016) group around the ethical foundations of criteria for sustainability and their application to economics.

2 The Global Warming Problem

Human (industrial) production entails emissions of GHG. Given scientific evidence like the results presented in the 4th report of the Intergovernmental Panel on Climate Change (IPCC), such emissions impact the world climate negatively. An increase in the concentration of GHG is projected to increase the mean atmospheric temperature implying a higher frequency of disasters and natural catastrophes (such as droughts, floods, and heat waves), higher mortality rates, and a significant loss of biodiversity. These consequences have economic costs, the most apparent being a loss in the productive capacity of the world economy.[1]

The world climate is affected by the use of capital which produces a negative externality in the form of emissions. Uninternalized externalities lead to inefficiencies since economic agents do not perceive the true cost of their actions and do not equalize (social) marginal costs and benefits. In our case the representative agent is over-emitting GHG, since she perceives the marginal cost of doing so (to her as an individual) is zero. Under the perfect foresight assumption, she is able to correctly predict the path of GHG (mainly CO_2) concentrations given her (and everybody else's) consumption, production, and investment choices. Although she is aware of the collective consequences of her actions, she thinks her individual contribution to the overall result is negligible. Consequently, she will not reduce her production-related emissions, either through producing less or investing in mitigation, because she knows that nobody else will do so (as they believe their actions to be insignificant, too). All agents end up choosing the same inefficient allocation. This point was made in Foley (2008). Such socially sub-optimal outcomes are well known from simple strategic games, the most prominent being the " Prisoners' Dilemma".

Given the inefficiency of over-accumulation of GHG stock in the atmosphere as a result of capital stock accumulation, the world economy is not operating at the intertemporal production possibility frontier (PPF). Future generations would appreciate lower stocks of CO_2 which implies that current generations should accumulate less conventional capital and consume more (of it) today. There is no intergenerational trade-off despite the fact that such a trade-off is posited in most of the global warming related economic publications. The mutual gains can be illustrated by moving the economy from a point inside the PPF to its boundary. This movement to an efficient equilibrium can be achieved by cost transparency (which amounts to increasing the cost of emitting to its true value).

Creating the correct price signal for GHG emissions (by whatever means, including cap-and-trade permits, Pigouvian taxes, or direct regulation) is sufficient to internalize the negative externality of global warming. As a result our agent will start to invest into mitigation. These mitigation costs, however, are small compared with

---

[1] The worldwide economic implications of climate change are hard to quantify. This task becomes more difficult, the higher the assumed stock of carbon in the atmosphere. The report of the IPCC (2007) presents convincing evidence for the negative relationship between GDP and global warming. Tol (2009) gives an optimistic review.

the gain of obviating GHG emissions. As is shown below, averting climate change can represent a non-trivial Pareto improvement.

3 The Model

The model used here for analyzing the economic aspects of global warming is a standard Ramsey–Cass–Koopmans model of the economy extended to included GHG. In order to maximize the comparability of our results with models in the literature based on neoclassical growth theory, the economy only produces one good using a Cobb-Douglas production function, $F[K, L]$, in conventional capital and effective labor. Effective labor consists of the exogenously given growth paths of population, N, and Harrod-neutral technical change, B, (which can be translated into Hicks-neutral technical progress given the Cobb-Douglas technology), according to $L = BN$. The state equation for conventional capital, K, is in its standard form with capital increasing due to investment, I, and decreasing due to (exponential) depreciation at rate δ.

Following Nordhaus and Boyer (2000), the stock of GHG in the atmosphere, CD (for carbon dioxide, CO_2, measured in parts per million volume, ppmv), enters the model as an additional state variable. Its dynamics depend on usable output, Y, and are governed by production-related emissions, $G[Y]$, mitigation efforts, $M[m]Y$, and (exponential) depreciation at rate ϵ. m is the share of usable output invested in mitigation. Mitigation efforts, $M[m]Y$, are linear in usable output similar to the abatement cost function, $\Lambda[.]$, used in Nordhaus and Boyer (2000) and Nordhaus (2008).

In the interests of keeping the model as parsimonious as possible, temperature dynamics are omitted and CD in excess of pre-industrial levels lowers productive capacity directly via what we term a damage function, $Z[\text{CD}]$. There are no sinks and no time lags. Emissions fully affect output immediately and directly, $Y = Z[\text{CD}]F[K, L]$. Mitigation can take the form of removing existing CD from the atmosphere or by preventing current emissions. Mitigation does not alter carbon emissions intensity permanently.

Formally, the representative agent allocates shares of output to consumption, c, and investment, s, of which a certain output share, m, is invested into mitigation, in order to maximize utility, measured as the discounted present value of the felicity of per capita consumption over time. Let consumption $C[t] = (1 - s[t])Y[t - 1]$ and world output $Y[t] = Z[\text{CD}\ [t]]F[K[t], L[t]]$, then total utility is

$$U[C[t], t] = \sum_{t=1}^{T} \frac{1}{(1+\rho)^{(t-1)}} U\left[\frac{C[t]}{N[t]}\right]$$

These choices are subject to initial values $K[0]$ and $\mathrm{CD}[0]$ and the following state equations:

$$K[t+1] = (1-\delta)K[t] + (s[t+1] - m[t+1])Y[t]$$
$$\mathrm{CD}[t+1] = (1-\epsilon)\mathrm{CD}[t] + G[Y[t]] - M[m[t+1]]Y[t]$$

With $\lambda[t]$ as the shadow price of capital and $\mu[t]$ the shadow price of CO_2 in the atmosphere, both expressed in terms of undiscounted felicity in the period t, and $c[t] = \frac{C[t]}{N[t]}$ as per capita consumption in period t, the adjoined Lagrangian for the above problem is

$$\mathcal{L}[K, \mathrm{CD}, \lambda, \mu, t] = \sum_{t=1}^{T} \frac{1}{(1+\rho)^{(t-1)}}$$
$$\times (U[c[t]] + \lambda[t](K[t] - (1-\delta)K[t-1] - (s[t] - m[t])Y[t-1])$$
$$+ \mu[t](\mathrm{CD}[t] - (1-\epsilon)\mathrm{CD}[t-1] - G[Y[t-1]] + M[m[t-1]]Y[t])) \quad (1)$$

Note that $\mu[t] < 0$, as CD affects production negatively. The (negative of the) current dollar price of carbon emissions is given by $\chi[t] = \frac{\mu[t]}{\lambda[t]}$.

3.1 *The Optimal Case (OPT)*

For optimality the following first-order conditions have to hold, which simultaneously represent a social competitive equilibrium under the assumption that institutions (such as an optimal carbon tax, a universal and optimal cap-and-trade system, or optimal direct regulation) exist to impose the social costs of emission on producers and consumers:

$$\partial_\lambda \mathcal{L} = 0 \Leftrightarrow K[t] = (1-\delta)K[t-1] + (s[t] - m[t])Y[t-1] \quad (2)$$
$$\partial_\mu \mathcal{L} = 0 \Leftrightarrow \mathrm{CD}[t] = (1-\epsilon)\mathrm{CD}[t-1] + G[Y[t-1]] - M[m[t]]Y[t-1] \quad (3)$$
$$\partial_K \mathcal{L} = 0 \Leftrightarrow \lambda[t] = \frac{\lambda[t+1]}{1+\rho}\left(1 - \delta + r_K(1 - m[t+1]\right.$$
$$\left.+(G'[Y[t]] - M[m[t+1]])\chi[t+1])\right) \quad (4)$$
$$\partial_{\mathrm{CD}} \mathcal{L} = 0 \Leftrightarrow \mu[t] = \lambda[t]\chi[t] = \frac{\lambda[t+1]}{1+\rho}\left(r_{\mathrm{CD}}(1 - m[t+1])\right.$$
$$\left.+\left(1 - \epsilon + r_{\mathrm{CD}}(G'[Y[t]] - M[m[1+t]])\right)\chi[1+t]\right) \quad (5)$$
$$\partial_s \mathcal{L} = 0 \Leftrightarrow \lambda[t] = \frac{c[t]U'[c[t]]}{C[t]} \quad (6)$$
$$\partial_m \mathcal{L} = 0 \Leftrightarrow \chi[t] = -\frac{1}{M'[m[t]]} \quad (7)$$

The first two equations are simply the state equations for $K[t]$ and $\mathrm{CD}[t]$. With $r_K = (1 - Z[\mathrm{CD}])F_K[K, L]$ the marginal product of capital, the next equation tells us that in an optimal program the current value of capital must be equal to its marginal benefit, which is the discounted value of its net marginal product factoring in mitigation costs and net emissions resulting from a larger capital stock. Since the time path of $\lambda[t]$ reports the change in the shadow price of capital, its change times the discount factor yields the marginal rates of intertemporal substitution which equal the real interest rate $j[t] = \frac{\lambda[t]}{\lambda[t+1]}(1 + \rho)$.

With $r_{\mathrm{CD}} = \frac{-Z'[\mathrm{CD}]Y}{Z[\mathrm{CD}]}$ the marginal product of CD, the same has to hold in the fourth equation for CD: The price of CO_2 must be equal to the discounted value of its net marginal product, again, factoring in mitigation positively and net emissions resulting from higher output negatively. This equation thus reflects the assumption that an effective system of optimal pricing of emissions is in effect when we calculate the OPT path. The last two equations are the Euler equations and establish optimality with regard to the choice variables, $s[t]$ and $m[t]$. They tell us, first, that marginal utility of consumption per capita in period t has to be equal to cost of capital (which is measured in per capita utils per unit of capital). Through Eq. (4) the marginal utility of consumption per capita is equal to the per capita marginal benefits of accumulating more over the remaining time horizon; second, that marginal cost of mitigating has to be equal to the marginal future benefit of doing so. The capital (usually dollar) price of carbon emissions is fixed by the cost of the marginal emission reduction. This thought will be taken up later. Given the Euler equations and the co-state equations, one can derive Ramsey-Keynes rule equivalents of the system.

3.2 The Business-as-Usual Case (BAU)

We model the business-as-usual case as an equilibrium of the economy in which global warming is the outcome of a negative externality. A state variable is an externality when it has a real impact on the objective function or constraints, but no institutions exist to enforce the social price on individual agent decisions involving it. Each agent assumes that her decisions will not affect the path of the externality, but when all agents make the same decisions the path of the externality changes. On a perfect-foresight equilibrium path with an uncorrected externality, each agent is assumed to correctly forecast the path of the externality, but ignores the effect of her decisions on the path of the externality. Thus on the equilibrium path with CD as an uncorrected externality the typical agent solves the above maximization problem expecting a certain time path of for the external $\mathrm{CD}[t]$. The correct forecasting assumption amounts to the side condition that expected $\mathrm{CD}^e[t] = \mathrm{CD}[t]$, where $\mathrm{CD}[t]$ is the path of the externality corresponding to the representative agent's chosen decisions. The difference between the equilibrium path with an uncorrected externality and the optimal path is the fact that the typical agent does not adjust her controls to take account of their effect on the externality. She is not aware of the true social cost of emitting. No social institutions exist to provide the correct price signal to steer the economy. As a result, the socially competitive equilibrium and the optimum diverge. Also, the social bid price and social ask prices for the externality are not equal.

It is possible to express the second-best equilibrium path with an uncorrected externality through the Lagrangian first-order conditions. The first-order conditions with respect to the shadow prices return the real laws of motion of the system, which must be obeyed. The representative agent, however, ignores the effect of her decisions on the external state variables, which corresponds to setting the shadow price on these variables equal to zero in the first-order conditions with respect to the non-external state and choice variables. The first-order condition with respect to the external state variable then plays no active role in the solution, but does keep track of the real social value of the externality in terms of its shadow-price μ. The BAU path solves these modified first-order conditions:

$$\partial_\lambda \mathcal{L}_{\text{BAU}} = 0 \Leftrightarrow K[t] = (1-\delta)K[t-1] + s[t]Y[t-1] \tag{8}$$

$$\partial_\mu \mathcal{L}_{\text{BAU}} = 0 \Leftrightarrow \text{CD}[t] = (1-\epsilon)\text{CD}[t-1] + G[Y[t-1]] \tag{9}$$

$$\partial_K \mathcal{L}_{\text{BAU}}|_{\mu[t]=0} = 0 \Leftrightarrow \lambda[t] = \frac{\lambda[t+1]}{1+\rho}\left(1-\delta+r_K\right) \tag{10}$$

$$\partial_{\text{CD}} \mathcal{L}_{\text{BAU}}|_{\mu[t]=0} = 0 \Leftrightarrow \mu[t] = \lambda[t]\chi[t] = 0 \tag{11}$$

$$\partial_m \mathcal{L}_{\text{BAU}}|_{\mu[t]=0} = 0 \Leftrightarrow m[t] = 0 \tag{12}$$

$$\partial_s \mathcal{L}_{\text{BAU}}|_{\mu[t]=0} = 0 \Leftrightarrow \lambda[t] = \frac{c[t]U'[c[t]]}{C[t]} \tag{13}$$

It is important to notice the subtle differences between these equations and the fully optimal path equations. The state equations (8) and (9) are the same on the optimal and BAU paths. But on the BAU path the controls are optimized without taking account of the externality, since $\mu[t] = 0$ in the first-order condition. As a result $m[t] = 0$. $s[t]$ remains unaltered due to the specific form of the model ($s[t]$ is the total share of income saved). Likewise, Eq. (10) determines the shadow price of the non-external state variables with the shadow price of the externality, $\mu[t] = 0$. In the absence of the externality, the Euler equation for capital and the optimal savings decision reduce to the usual Ramsey-Keynes rule.

There are two types of misallocation on the BAU path. First, because there is no market price for carbon emissions, the typical agent allocates too little (zero) resources to mitigation. Second, the typical agent invests too much in conventional capital because she ignores the impact of increasing output on increasing climate damage.

In our calculations we keep track of how $m[t]$ and $\mu[t]$ would evolve according to their first-best first-order conditions given the other, second-best variables.

$$\partial_m \mathcal{L} = 0 \Leftrightarrow \chi[t] = M'[m[t]]$$

$$\partial_{\text{CD}} \mathcal{L} = 0 \Leftrightarrow \mu[t] = \lambda[t]\chi[t] = \frac{\lambda[t+1]}{1+\rho}\left(r_{\text{CD}}(1-m[t+1])\right.$$
$$\left. + \left(1-\epsilon+r_{\text{CD}}(G'[Y[t]] - M[m[1+t]])\right)\chi[1+t]\right)$$

3.3 The Constrained Optimal Case (COPT)

Researchers of the economic consequences of global warming (most prominently and recently Nordhaus and Boyer 2000; Nordhaus 2008) analyze an optimal path under the constraint that no mitigation is undertaken. Because this type of path implicitly partially internalizes the externality, it seems to us that it does not represent "business-as-usual", and in the remainder of this paper we call this type of path *constrained optimal*.

How exactly the constrained optimal path can be derived as an equilibrium within the representative-agent perfect foresight methodology is somewhat mysterious. Fully rational agents with perfect foresight acting with complete markets adopt the first best solution presented above as the optimal path. When markets are incomplete and there is no price signal for the marginal social value of the externality, the equilibrium is the BAU path described in the last section. The constrained optimal (COPT) path, on the other hand, represents an inconsistent mixture of assumptions about the representative agent's information. On the one hand, the representative agent on this type of path correctly estimates the marginal social cost of emissions in making her consumption, investment, and production decisions. On the other hand, she seems to ignore the availability of mitigation technologies, despite this understanding of the marginal social cost of emissions. This divergence results in a difference between the marginal social value and marginal social cost of mitigation. The agents in this mixed scenario perceive the marginal social cost of emitting as zero, the only price that justifies no mitigation. At the same time, however, the agent is confronted with the true carbon price in her decision on how much output to consume and how much to re-invest for capital formation. While this inconsistency within the perfect foresight framework is corrected for in the business-as-usual case above, we also solve for the constrained optimal case given its importance in the economic literature on global warming.

The first-order conditions of the constrained optimal case are the special case of the fully optimal with $m[t] = 0$.

$$\partial_\lambda \mathcal{L}_{\mathrm{COPT}} = 0 \Leftrightarrow K[t] = (1-\delta)K[t-1] + s[t]Y[t-1] \tag{14}$$

$$\partial_\mu \mathcal{L}_{\mathrm{COPT}} = 0 \Leftrightarrow \mathrm{CD}[t] = (1-\epsilon)\mathrm{CD}[t-1] + G[Y[t-1]] \tag{15}$$

$$\partial_K \mathcal{L}_{\mathrm{COPT}} = 0 \Leftrightarrow \lambda[t] = \frac{\lambda[t+1]}{1+\rho}\left(1-\delta + r_K(1+G'[Y[t]]\chi[t+1])\right) \tag{16}$$

$$\begin{aligned}\partial_{\mathrm{CD}} \mathcal{L}_{\mathrm{COPT}} = 0 \Leftrightarrow \mu[t] = \lambda[t]\chi[t] = {} & \frac{\lambda[t+1]}{1+\rho} \\ & \times \left(r_{\mathrm{CD}} + \left(1-\epsilon + r_{\mathrm{CD}}G'[Y[t]]\right)\chi[1+t]\right)\end{aligned} \tag{17}$$

$$\partial_s \mathcal{L}_{\mathrm{COPT}} = 0 \Leftrightarrow \lambda[t] = \frac{c[t]U'[c[t]]}{C[t]} \tag{18}$$

Notable changes include the altered state equation for CD[t] in which carbon dioxide concentration can only be changed through emissions and dissipation and the two modified co-state equations. The change in the marginal benefit of capital from the optimal to the constrained optimal path depends on the functional form of $M[m[t]]$. For any meaningful mitigation function, the marginal benefit of capital will be lower in the constrained optimal case. As a result, less capital will be accumulated. This is intuitive as a reduction in K is the only means by which CD[t] increases can be counteracted with mitigation constrained to zero. Note that while agents are deprived of the choice to mitigate, they still fully respond to changes in the price of carbon in their accumulation decisions. The same logic applies to the price of CD[t]. Lower marginal benefit will lead to higher CO_2 concentration, as CD is a "bad".

In our calculations below we retain the (for this program) superfluous optimality condition for $m[t]$ in order to see what mitigation effort would be under an optimal scenario (at that point in time of the program).

$$\partial_m \mathcal{L} = 0 \Leftrightarrow \chi[t] = -\frac{1}{M'[m[t]]}.$$

3.4 Basic Logic

The basic logic and qualitative features of the three cases can be seen even without specifying functional forms, calibrating the model to reflect current economic values, and solving it over long time horizons.

On the OPT equilibrium path agents will invest enough resources in mitigation to compensate emissions to the point where marginal cost of doing so is equal to the benefit in output due to less environmental damages. This level is defined by the damage function which is key to the quantitative outcomes of the model. The carbon price will be defined by mitigation efforts. It will be equal to the cost of marginal mitigation efforts.

On the BAU equilibrium path agents are not only deprived of the mitigation instrument, but also see themselves incapable of affecting the stock of CO_2. Their decisions are taken solely with respect to maximizing their intertemporal utility and consumption. The carbon dynamics drive the system. Emissions from rapid capital accumulation will drive up carbon dioxide levels to the point where environmental damage chokes off further accumulation due to the falling profit rate.

On the COPT equilibrium path agents are deprived of the mitigation instrument. In order to move the economy to a steady state equilibrium, the savings decisions and the capital stock will have to do all of the adjustment. It is important to note that in the COPT case (as well as in the BAU case), the climate dynamics dominate the outcome. A steady state can only be achieved when emissions equal to the dissipation of existing carbon stock in the atmosphere. This sets the level of admissible capital stock and the appropriate savings rate.

Given the logic of the three model cases, it is trivial, but nonetheless important, to note that the overall utility will be the greatest on the OPT path, followed by the COPT and BAU paths. Growth in capital stock will be highest on the OPT, followed by the BAU. Investment will be lower in the COPT than in the BAU scenario as agents are aware of the deleterious effects of their accumulation decisions and, hence, more cautious. Climate catastrophes—meaning high equilibrium levels of CO_2—are certain in the BAU and very likely in the COPT case. Over the range of all realistic mitigation functions, CO_2 levels will stabilize at low or moderate levels in the OPT case.

These characteristics are given by the model structure and are, thus, independent of the functional forms and parameter values. We hope that in this light the secondary relevance of much of the current debate on discounting factors will be apparent.

Fast convergence to the steady state implies that the steady state results will drive much of the model's behavior and its transition dynamics. As can be seen below, the optimal path reaches its steady state within 10 decades. Nordhaus (2008) and Stern (2007) do not devote much attention to the steady states of their models and the implications of steady state values on the transition dynamics.

4 Functional Forms

There are several choices that need to be made about the forms the production, damage, mitigation, and emissions functions take. The functional forms we use in the simulations reported here are as follows: Utility has its traditional iso-elastic manifestation $U[c[t]] = \frac{c[t]^{1-\eta}}{1-\eta}$, or $U[c[t]] = \mathrm{Log}[c[t]]$ when $\eta = 1$. Potential output is a Cobb-Douglas production function, $F[K[t], L[t]] = AK[t]^{\alpha}L[t]^{1-\alpha}$. Carbon-related damages are measured on a scale between 0 and 1 with zero damage at the pre-industrial level of 280 ppmv and complete output loss at a CDMax = 780 ppmv, with the damage function $Z[\mathrm{CD}[t]] = \left(1 - \left(\frac{\mathrm{CD}[t]-280}{\mathrm{CDMax}-280}\right)^{\frac{1}{\gamma}}\right)^{\gamma}$. This functional form implies that even at current CO_2 levels of 380 ppmv a certain fraction of potential output is lost due to environmental degradation. Emissions, $G[Y[t]] = \beta Y[t]$, are linear in output at a constant carbon intensity of production. The mitigation function, $M[m[t]] = \zeta\frac{1-e^{-\nu m[t]}}{\nu}$, where ζ is a scaling parameter and $0 < \nu$ is a semi-elasticity reflecting diminishing returns to m, which converts the unitless proportion of output devoted to mitigation, m, into CO_2 reduction per \$ spent on mitigation, that is ppmv/\$. In our specification of the mitigation function, we diverge from the other studies in assuming positive mitigation costs even at very low mitigation efforts $(\partial_m M[m[t]]|_{m[t]=0} = \zeta \neq 0)$. From the first-order condition for $m[t]$, $M'[m[t]]$ equals the carbon price $\chi[t]$ in the first best solution.

Population growth follows Nordhaus (2008) and UN projections in assuming that world population will rise from currently 6500 to 8600 million over the next 10–20 decades, and then stabilize at this level. Labor productivity is assumed to start at a yearly growth rate of 2 % and to flatten out at 3 times its current value after 30 decades.

Given these functional forms, the adjoined Lagrangian (with $L[t] = B[t]N[t]$) is:

$$\mathcal{L}[K[t], \mathrm{CD}[t], \lambda[t], \mu[t], t]$$

$$= \sum_{t=1}^{T} \frac{1}{(1+\rho)^{(t-1)}} \left(\frac{\left(\frac{(1-s[t])Z[\mathrm{CD}[t-1]]F[K[t-1],L[t-1]]}{N[t]}\right)^{(1-\eta)}}{1-\eta} \right.$$

$$+ \lambda[t](K[t] - (1-\delta)K[t-1]$$

$$- (s[t] - m[t])Z[\mathrm{CD}[t-1]]F[K[t-1], L[t-1]])$$

$$+ \mu[t]\left(\mathrm{CD}[t] - (1-\epsilon)\mathrm{CD}[t-1] - \left(\beta - \zeta\frac{1-e^{-vm[t]}}{v}\right)\right.$$

$$\left.\left. \times Z[\mathrm{CD}[t-1]]F[K[t-1], L[t-1]]\right)\right)$$

5 Calibration

Given the assumptions of functional forms, the actual functions need to be calibrated to match economic and physical realities. All parameters are geared towards a decadal time interval. In the benchmark case Lagrangian, the discounting factor, $\rho = 0.1$. With output measured in units current (2000–2010) $ trillion, initial capital stock is assumed to be $K_0 = 200$. CO_2 is measured as parts per million volume (ppmv). Initial CO_2 concentration $\mathrm{CD}_0 = 380$. Capital decays radioactively with $\delta = 0.7$.

Currently, 7 Gt carbon are burnt per year. This corresponds to an increase in CD of 3.37 ppmv. With a yearly world output of $60 trillion this implies a carbon dioxide emission intensity $\beta = \frac{3.37}{60} = 0.0561 \frac{\text{ppmv}}{\$\text{trillion}}$. As the actual increase in atmospheric carbon dioxide concentration is only about 2 ppmv, dissipation is 1.37 ppmv. This yields a depreciation factor $\epsilon = \frac{1.37}{380} = 0.0036$.

The marginal product of capital in the Cobb–Douglas production function α is set at 0.35 in line with standard economic research. Total factor productivity is calibrated to match current world output of $60 trillion. The elasticity parameter in the utility function η is set at 2, in our baseline simulations.

The damage function $Z[\mathrm{CD}[t]]$ also takes an iso-elastic form. This allows us to combine the apparent global warming optimism of economists towards the low damages of global warming at low carbon dioxide concentration with the serious warnings of climate scientists about severe output loss at high carbon dioxide concentration (which is set to 780 ppmv in our model). We set $\gamma = 0.5$, which is at the higher end of potential damages (Barker 2008). The calibration points usually cited are the IPCC (2007) predictions of an increase to a doubling of pre-industrial concentrations (about 280 ppmv) leading to a temperature rise of 3 °C and an increase of 4 °C leading to a potential output loss of 1–5 % of current output. Nordhaus (2008) assumes that current damages to the world economy are 0.15 % of output. This corresponds to $\gamma = 0.3$. The results for $\gamma = 0.3$ are reported in the sensitivity analysis section below. We deviate from Nordhaus as his damage function is lacking any reasonable upper limit on temperature and ultimately CO_2 concentration. Rezai (2010) shows that our parametric form of a damage function lies consistently above Nordhaus' for $\gamma = 0.5$ and below it with $\gamma = 0.3$ up to 500 ppmv (which is the relevant range for the optimal program). It is worth mentioning that the asymptotic behavior of the damage function for high concentrations has little impact on the OPT path as long as its shape for low concentrations is close to the mentioned calibration points. This holds for our range of damages.

Figure 1 plots the damage function for $\gamma = 0.5$ and a band of $0.7 \geq \gamma \geq 0.3$, around which a sensitivity analysis is carried out below. It also includes the assumptions on environmental damages from Nordhaus (2008) as a dark gray area up to CD = 580. Note that while our assumptions on γ regarding damages can be regarded as high, so are the assumptions on mitigation costs with the carbon price at \$160 (per t of C).

The parameter ζ represents the marginal reduction in CO_2 concentration (ppmv) per \$T from spending a small amount on mitigation when mitigation is zero. If it costs \$160 to remove one tonne of C at present (carbon markets suggest level between \$75 and \$125), when effectively $m = 0$, then to reduce CO_2 concentration by 1 ppmv through removing 2.07 Gt C from emissions would cost (2.07)(0.160\$ trillion) = 0.331\$ T, so we set $\zeta = 3\frac{\text{ppmv}}{\$T}$. Using this specification the carbon price is directly linked to and anchored by marginal mitigation efforts. Note that the assumption of a lower current carbon price increases ζ and makes the mitigation function more effective. Figure 2 plots the mitigation function for these parameter values.

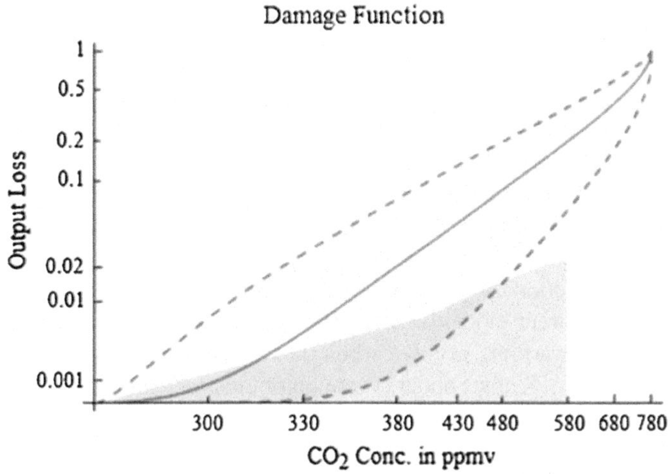

Fig. 1 Damage function with $\gamma = 0.5$

Fig. 2 Mitigation function for $\zeta = 3$ and $\nu = 6$

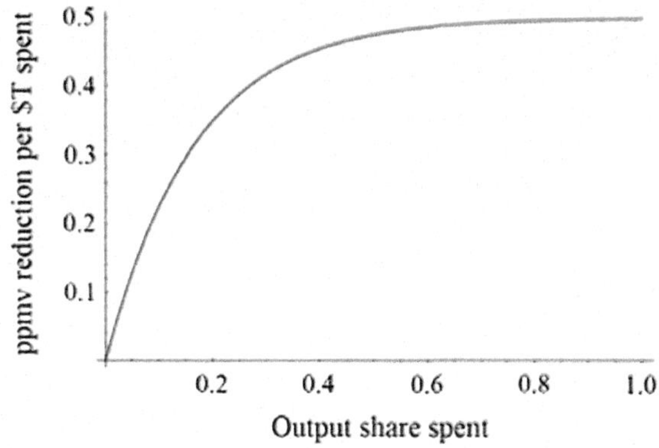

Table 1 Overview of the parameter values used in the numerical simulation

Function	Parameter	Value
$\mathcal{L}[.]$	ρ	0.1
	K_0	200
	CD_0	380
	δ	0.7
	ϵ	0.036
U[.]	η	2
F[.]	α	0.35
	A	28.4
Z[.]	γ	0.5
	CDMax	780
G[.]	β	0.056
M[.]	ζ	3
	ν	6

Table 1 provides an overview over the parameter assumption.

6 Computational Implementation

The above systems of equations can be reduced to 4 laws of motion, 2 for the state variables and 2 for the co-state variables by substituting the optimal expressions for the controls into the other equations and forming thus the maximized Lagrangian. These 4 difference equations, of which some form a subsystem in the COPT and BAU case, have to hold for $t = 1, ..., T$. In addition, initial conditions on the two state variables and terminal transversality conditions on the co-state variables have to hold. This yields $4(T+1)$ conditions to determine $4(T+1)$ variables.

In order to solve this set of equations, we make use of the program *Mathematica* and its root finding command. In this process the specification of initial search parameters is crucial. Generally, *Mathematica* proves to be quite agile in finding the equilibrium path even if state and co-state variables are persistently shifting on a steady growth path. Specifying variables in logarithmic forms some times facilitates the search routine in this case.

7 Growth Paths

The above first-order conditions are sufficient as well as necessary for a global maximum as the maximized Hamiltonian is concave in K and CD for any given λ and μ (that is, the objective function is quasi-concave and the constraints convex in K and CD). The programs below are set up as finite-horizon problems with the additional requirement of the transversality conditions $\lambda[T]K[T] = 0$ and $\mu[T]\mathrm{CD}[T] = 0$ with a fixed T at 60 decades. These terminal conditions guarantee that the calculated OPT path is a valid optimum of the primal problem over the finite time horizon. We choose the time horizon sufficiently long such that the paths to approximate the "steady-state" in the middle of the time horizon. This characteristic is known as the turnpike property which assures that finite horizon programs mimic their infinite horizon twins for sufficiently long horizons (Samuelson 1965). The paths below effectively reach their steady state within 30 decades. Solving over 60 periods becomes sufficiently close to the infinite horizon problem. Note that it would be possible to change the transversality end-point conditions to require, for example, some minimal capital and maximal CO_2 stocks at the end of the program, but it is not very easy to see how to choose those levels, or, in fact, to use this method except in the finite time horizon case. (It is not really correct to force the path to converge to the steady state at any finite time, for example.) We also emphasize the illustrative character of the simulations below, since long-run projections of any economic growth model are subject to high uncertainties in key parameters and the possible appearance of intervening factors not included in the model.

Figure 3 is a comparison of the optimal, constrained optimal and business-as-usual paths for these parameters, in terms of world per capita consumption, the damage from global warming, the implied price of carbon, and the CO_2 concentration. The finite-horizon program approaches its steady state very quickly, as to be expected from the turnpike theorem.

On the unconstrained optimal path capital accumulation combined with "small" mitigation efforts of around 1–2 % of GDP enables sustainably rising output and consumption levels. Mitigation efforts are front-loaded, meaning that most of the mitigation is done in the first few decades, such that only current CO_2 emissions have to be mitigated in later periods. The carbon price stabilizes around \$180/t which is close to zero mitigation carbon cost of \$160. In fact, atmospheric carbon concentration decreases almost to its pre-industrial level.

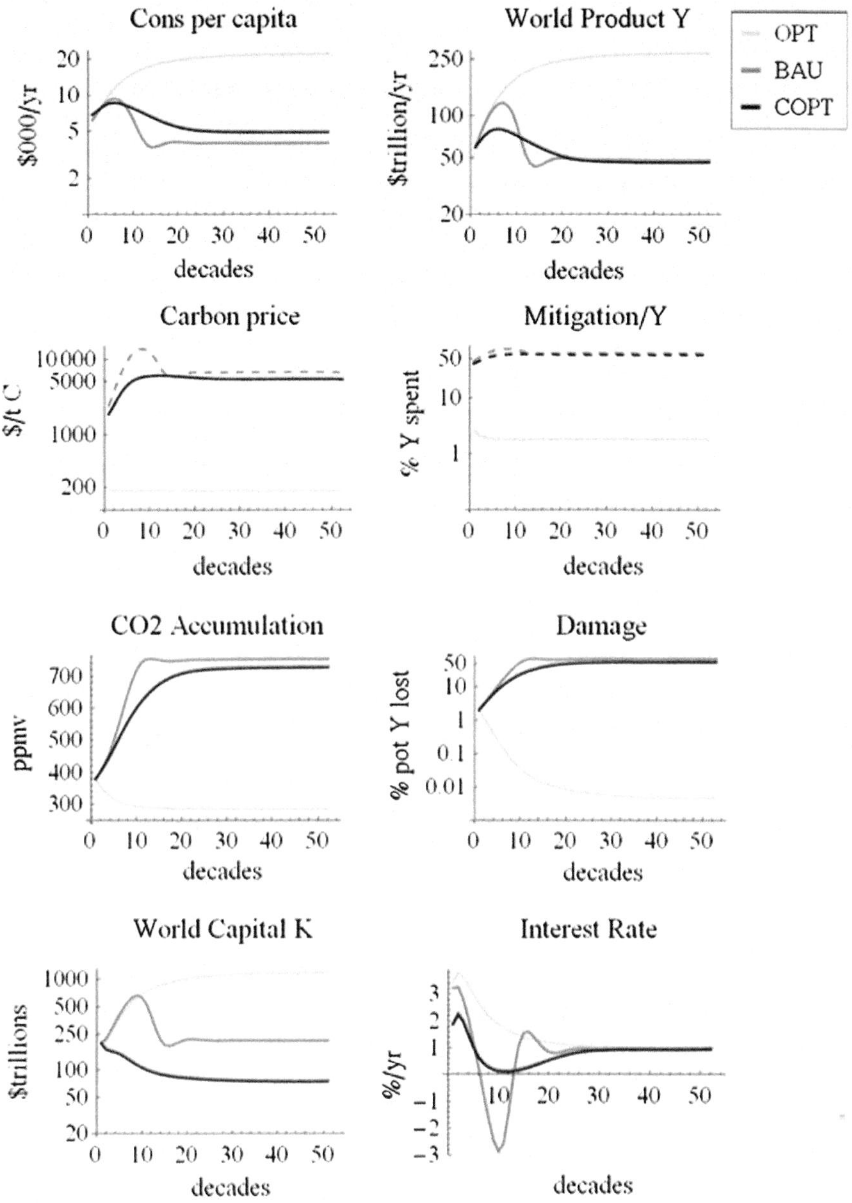

Fig. 3 The optimal equilibrium path, OPT, is plotted in *pale-gray*, the BAU equilibrium in *gray*, and the constrained equilibrium, COPT, in *dark-gray*. The carbon price on the optimal path is about \$200/t, and the damage on the optimal path is less than 1 % of potential output. The carbon price is the social marginal value of foregoing the emissions from 1t of carbon. For OPT and COPT the carbon price is effective in economic decisions, but not in BAU (and is plotted as a *dashed line*). For OPT the mitigation percentage is the proportion of world product devoted to mitigation. For COPT and BAU the mitigation percentage is the investment called for by the imputed carbon price (and is plotted as a *dashed line*), while the actual mitigation is zero

On the BAU path agents lack the correct price signals to correct the negative externality. This leads to inefficiencies in several respects. Capital rises at a rate similar to the optimal case during the first 100 years although no mitigation can be carried out; the implication of this over-accumulation is rapidly rising CO_2 concentration and environmental damage. Since the capital accumulation equation for $\lambda[t]_{BAU}$ is independent of any carbon related costs (or their price signals), accumulation continues until damages are so high that further accumulation cannot occur due to the declining productivity of capital and labor inputs. Output and consumption are bound to decline due to ever higher carbon concentration and damages until the marginal product of capital has fallen sufficiently to approach a stable equilibrium. Equilibrium output and consumption per capita are almost 20 % below current levels and less than 25 % of the OPT levels despite significant technical progress and population growth. The gray dashed lines in the carbon price and mitigation graphs depict the implied carbon price and the mitigation called for by this price. Implicit carbon price and mitigation efforts on the BAU are fifty times higher than their optimal counterparts. The inefficiency of the BAU can also be seen in the equilibrium saving rates. While higher capital stock implies higher saving to compensate the lower marginal product, the equilibrium BAU saving rate with its lower capital stock is higher in our simulations due to the (unnecessary) high carbon concentration.

The constrained optimal path does slightly better than the BAU path. Although the agents in the COPT also are confined to zero mitigation, they conceive the correct price signals and run up GHG in the atmosphere much more cautiously than in the BAU scenario. In fact, overall capital stock is decreasing on this path as current levels are suboptimally high, implying too many carbon emissions. As carbon concentration increases, so must the marginal product of capital which can only be achieved with a lower capital stock. Equilibrium GHG concentration entails a high carbon price which, again, calls for high mitigation efforts. Equilibrium carbon concentration and price are lower than on the BAU path. The carbon price is about twenty times the optimal level.

While the quantitative results of our simulations are dependent on the specific parameter assumptions, it is important to point out that the qualitative results are independent of these assumptions. Especially the finding that moving from the inefficient BAU path to the efficient OPT path through mitigation constitutes a Pareto improvement should be noted. This result implies that there is no cost to mitigation, but there are significant gains from doing so (in our simulations up to 400 % of GDP). Higher or lower discounting rates will not alter this. In light of the magnitude of the avail, we argue that questions of uncertainty and intergenerational equity which are discussed further below can be considered of second-order importance.

The high carbon prices and low, often negative real interest rates implied by these simulation results also underline the important methodological point that the present discounted value of future costs and benefits are conditional on some particular consumption path for the representative agent. On the BAU path simulated here, for example, the implicit real interest rate is negative for a significant part of the transition path and belowthe discount rate on average over the whole path. The real interest rate is negative because the representative agentwould pay a high price to move consumption from the early part of the BAU path, where capital and labor are highly productive, to the later part of the path, where productivity has been destroyed by the climate catastrophe. Even though the time horizon of the simulations is very long, the present discounted value of carbon emissions on the actual path is, as a result, quite substantial. Much of the existing literature on global warming ignores the dependence of present discounted values on the consumption path and calculates present discounted values of future costs and benefits using steady-state values of real interest rates. Since steadystate real interest rates are positive, discounting at these interest rates results in very low present discounted values even of very significant future damage, if the damage is far enough in the future (see Chichilnisky and Eisenberger 2010). The consistent present discounted values using real interest rates implied by equilibrium paths are much higher.

8 Intergenerational Equity

An important aspect of the current climate change debate in the economic literature centers on the problem of intergenerational equity. This focus on generational equity arises primarily from a failure to appreciate that the business-as-usual path with an uncorrected externality is inefficient. In particular, it is a consequence of the mistaken use of the COPT path rather than the real BAU path as the benchmark with which the OPT path is compared. The most striking difference between the COPT and OPT paths is the generational distribution of consumption. As we have explained above, the COPT path is not a theoretically relevant benchmark, because it represents an inconsistent mixture of partial internalization of the global warming externality and a failure to divert resources from conventional investment to mitigation.[2] The use of the COPT path as the benchmark comparison leads to the misleading impression that the problem of correcting the global warming externality is primarily an issue of intergenerational equity, how much to sacrifice the consumption of current generations to protect the environment for future generations. In an optimal growth framework the resolution of this trade-off depends on the discount factor, ρ, and the degree of social preference for consumption smoothing expressed in the elasticity of felicity with respect to consumption, η. But with the correct BAU benchmark the correction of the global warming externality can provide an intergenerational Pareto improvement, raising the per capita consumption and felicities of every generation. The parameters ρ and η influence the distribution of this intergenerational gain, which is a

[2]Shiell and Lyssenko (2008) offers a clear explanation of the logic supporting this point, in the context of a two-state variable model very similar to the present one. This paper also presents an ingenious alternative method to ours of computing approximate BAU paths. Shiell and Lyssenko also find that the asymptotic behavior of COPT and BAU paths is very similar. Because they do not focus on the initial transient levels of consumption on the OPT and BAU paths, however, their discussion does not bring out the critical role of the COPT path in suggesting that an intergenerational distributional tradeoff is at the center of global-warming policy evaluations.

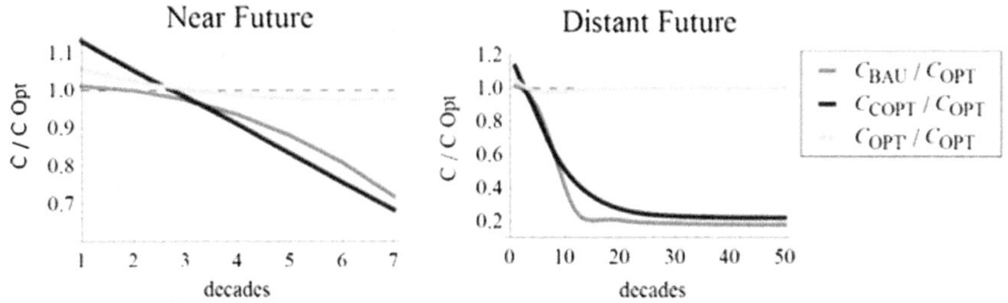

Fig. 4 Comparisons of the consumption paths for the business-as-usual, constrained optimal, and the altered optimal *palegray* OPT' (with $\eta = 3$) cases. All normalized to the optimal consumption stream

second-order consideration.[3] In fact, with our benchmark values of $\rho = 0.1/\text{decade}$ and $\eta = 2$, the OPT path exhibits slightly lower per capita consumption than BAU in the first decade. One way to understand this fact is that the representative agent in the BAU equilibrium values current consumption too highly relative to investment in mitigation; when she grasps the full social marginal value of mitigation she prefers to reduce her consumption slightly in the first decade because of the high rate of return of this investment to the consumption of future generations. If the representative agent put more weight on intergenerational smoothing of consumption, for example, if $\eta = 3$, the corresponding OPT path would dominate the BAU path for $\eta = 2$, thus yielding higher utility for any value of η and ρ and demonstrating the inefficiency of the BAU equilibrium.

In Fig. 4, below we plot the initial decades of four paths to underline these points. The COPT, and BAU paths relative to the OPT path all of which are the same as those plotted in Fig. 3 and calculated with the benchmark parameters, in particular with $\eta = 2$. The OPT' path has the same parameters for the technical side of the model, but sets $\eta = 3$, to emphasize the inefficiency of the BAU path. The figure shows that the use of the COPT equilibrium as the benchmark distorts the perception of the economic issues involved in global warming policy by incorrectly suggesting that correction of the global warming externality will depress per capita world consumption significantly for several decades.

Given the inefficiency of over-accumulation of GHG stock in the atmosphere as a result of capital stock accumulation, the world economy is not operating at the intertemporal production possibility frontier (PPF). Future generations would appreciate lower stocks of CO_2 which implies that current generations should accumulate less conventional capital and consume more (of it) today. There is no intergenerational trade-off as is posited in most of the global warming–related economic literature. The mutual gains are illustrated by moving the economy from an inefficient point inside the PPF (the BAU path) to its boundary (the OPT' path with $\eta = 3$).

---

[3]On the relation between ρ and η and intergenerational equity, see Arrow (2007), who emphasizes the public good nature of global warming, without, however, noting the inefficiency of the BAU path.

Consumption in the COPT case lies above consumption in the optimal case in the first three decades as agents who are aware of the deleterious effects of global warming wisely choose to accumulate less and consume more. This positive difference in the first few decades forms the basis for the intergenerational equity discussion since the choice of the appropriate program now depends on its parameterization (most importantly the discount factor and the elasticity of felicity with respect to per capita consumption). Nordhaus has placed much emphasis on the dominance of baseline consumption over optimal consumption in the first few decades. This implies that current generations would attain lower utility levels if they started investing into mitigation. The consumption paths of the optimal and business-as-usual scenarios virtually move together in the first few decades with the BAU dominating the OPT path for the first two decades. The OPT' path, however, shows that this effect is due to the representative agent preferring to transfer some of the gain from global-warming mitigation to future generations given the very high rate of return to mitigation investment given the initial conditions.

Figure 5 illustrates these equilibrium allocations in terms of PPFs and indifference curves. The curve closest to the origin represents all attainable allocations with $\mu = 0$ (the BAU PPF). The dark line next to it is the COPT PPF and represents all supportable allocations with $m = 0$. The internalization of the externality in the absence of mitigation allows a welfare maximizing COPT allocation with higher present and future consumption levels compared to BAU. The outmost curve represents what is commonly understood as a PPF: all technologically feasible, unconstrained first-best allocations. It is apparent that there are many allocation along the OPT PPF

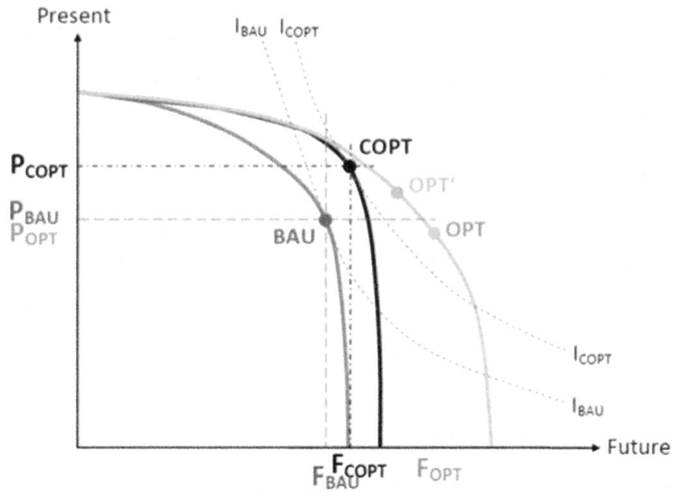

Fig. 5 Comparisons of the equilibrium allocations for the business-as-usual, constrained optimal, the optimal and the altered optimal OPT' scenarios. Given the parameter values for η and ρ), OPT has the lowest present consumption level. OPT' distributes the gains suboptimally, but such that consumption levels rise in both periods

which yield higher utility than COPT and BAU. The preference parameters ρ and η pin down the welfare-maximizing one. In our results they are such that the optimal intertemporal allocation (OPT) features the lowest present consumption level. As Fig. 5 demonstrates, the gains from moving from the BAU to the OPT allocation are much greater than from the COPT to the OPT. In particular, the set of allocations with non-decreasing consumption in either period is greater in the former than in the latter. This is due to the fact that the high saving rate in the BAU allocation provides more room to reshuffle the investment portfolio toward green investment in the form of mitigation. While the OPT' allocation does not lie on the highest indifference curve and is suboptimal, it distributes the gains from moving from the BAU to the OPT PPF such that consumption in neither period falls.

9 Parameter Sensitivity

In order to gauge the effect of parametric changes on the quantitative results of our simulations, this section explores sensitivity analysis for some model parameters. Generally, consumption smoothing dominates most of the adjustment in parametric changes. Due to spatial limitations, only the selected parameters are reported here. A complete set can be obtained from the authors. Table 2 presents an overview over the effects of parametric changes (rows) on selected variables (columns) relative to their values with standard parameters (which are reported in the bottom row). The sensitivity table reports lower and upper bounds for each parameter.

For example, the first Table in Table 2 captures the effects on the OPT path. The first two rows report changes in the discount rate of which the standard value is 0.1. The first row has $\rho = 0.05$ and the second $\rho = 0.15$. The first column reports the changes on consumption per capita. A lower time preference increases consumption by 0.5 % at the median and the 20 % quantile compared to the consumption stream under standard parameter assumptions. Consumption rises by 0.6 % above the corresponding standard consumption level at the 80 % quantile. An increase in the discount rate to $\rho = 0.15$ decreases consumption by 0.6 % at the 20 % quantile and by 0.7 % at the median and 80 % quantile. The next column reports effects of the changes on the carbon price. The change in pure time preference has virtually no effect on the carbon price. The third column reports the effects of parametric changes on mitigation efforts, the fourth on atmospheric carbon concentration, and so on. The third and fourth row the changes in the severity of damage, γ, the fifth and the sixth changes in the upper bound of atmospheric changes in CD, and so on.

The table does not report parameter changes in ν and ζ for the BAU and COPT paths, since these parameters concern the mitigation function and there is no mitigation available in these scenarios. Varying the discount factor has very different effects in the three scenarios. In the optimal case higher time preference yields higher consumption today and higher CO2 concentration and damage in the future. This is the standard result which most of the intergenerational literature rests on. In the BAU the higher consumption induced by higher time preference leads to less capital ac-

Table 2 Effects of parametric changes on time paths of selected variables relative to their standard value

		Cons per cap ($/year)			Carbon price ($/t)			Mitigation/Y (%)			CO_2 (ppmv)			K ($trillions)			Interest rate (%/year)		
		Q20	Median	Q80	Q20	Median	Q80	Q20	Median	Q80	Q20	Median	Q80	Q20	Median	Q80	Q20	Median	Q80
OPT																			
ρ	0.05	1.005	1.005	1.006	1.0	1.0	1.0	1.003	1.003	1.003	0.989	0.993	0.993	1.099	1.104	1.104	0.539	0.604	0.762
	0.15	0.992	0.993	0.993	1.0	1.0	1.0	0.997	0.997	0.997	1.007	1.008	1.011	0.911	0.911	0.914	1.209	1.38	1.439
γ	0.3	0.97	0.971	0.971	0.998	0.998	0.999	0.981	0.982	0.992	1.229	1.244	1.307	0.969	0.97	0.97	0.999	1.0	1.002
	0.7	1.086	1.086	1.086	1.0	1.001	1.001	0.997	1.007	1.007	0.949	0.979	0.982	1.086	1.086	1.086	0.999	1.0	1.0
CDMax	580	1.061	1.061	1.061	1.0	1.001	1.001	0.998	1.005	1.005	0.966	0.986	0.988	1.061	1.061	1.061	1.0	1.0	1.0
	980	0.959	0.984	0.984	1.0	1.0	1.0	0.998	0.998	1.004	1.017	1.019	1.027	0.969	0.984	0.984	1.0	1.0	1.011
ν	3	1.001	1.001	1.001	0.945	0.945	0.946	0.972	0.973	0.973	0.997	0.999	0.999	1.001	1.001	1.001	1.0	1.0	1.0
	12	0.998	0.998	0.998	1.13	1.132	1.132	1.061	1.062	1.062	1.002	1.003	1.006	0.998	0.998	0.998	1.0	1.0	1.001
ζ	1.5	0.964	0.967	0.967	2.261	2.262	2.263	2.119	2.119	2.139	1.024	1.029	1.064	0.961	0.965	0.965	1.0	1.0	1.005
	6	1.015	1.015	1.016	0.472	0.473	0.473	0.482	0.487	0.487	0.972	0.989	0.991	1.016	1.016	1.017	0.998	1.0	1.0
BAU																			
ρ	0.05	0.965	0.972	0.977	1.093	1.096	1.129	1.024	1.025	1.033	1.002	1.002	1.004	1.062	1.068	1.088	0.504	0.51	0.879
	0.15	1.015	1.025	1.028	0.848	0.913	0.915	0.963	0.976	0.976	0.994	0.998	0.998	0.923	0.94	0.945	1.099	1.467	1.477
γ	0.3	1.019	1.027	1.044	0.978	1.197	1.201	0.995	1.048	1.049	1.03	1.032	1.035	0.989	1.026	1.043	0.822	1.005	1.036
	0.7	0.939	0.955	0.971	0.855	0.858	0.91	0.958	0.959	0.974	0.944	0.95	0.951	0.949	0.956	1.0	0.568	0.987	0.997
CDMax	580	0.718	0.755	0.76	1.107	1.121	1.226	1.027	1.03	1.049	0.756	0.757	0.833	0.736	0.755	0.898	0.293	1.0	1.007
	980	1.172	1.235	1.301	0.7	0.894	0.901	0.92	0.97	0.972	1.02	1.231	1.234	0.988	1.235	1.255	0.244	0.998	1.107
COPT																			
ρ	0.05	0.98	1.019	1.042	1.122	1.137	1.15	1.032	1.036	1.039	0.969	0.985	0.991	0.725	0.803	0.837	−0.209	0.345	0.482
	0.15	0.969	0.991	1.015	0.894	0.905	0.91	0.969	0.972	0.974	1.005	1.009	1.023	1.086	1.092	1.142	1.476	1.617	2.173
γ	0.3	1.063	1.083	1.117	1.063	1.166	1.181	1.017	1.043	1.047	1.041	1.049	1.059	0.711	0.716	0.922	0.481	1.012	1.069
	0.7	0.896	0.925	0.93	0.901	0.91	0.975	0.97	0.974	0.993	0.941	0.944	0.961	0.947	1.096	1.11	0.848	0.993	1.235
CDMax	580	0.683	0.765	0.787	1.11	1.158	1.172	1.029	1.042	1.044	0.764	0.776	0.872	0.467	0.54	0.55	0.212	1.024	1.51
	980	1.199	1.228	1.298	0.814	0.892	0.915	0.942	0.968	0.976	1.072	1.201	1.226	1.474	1.496	1.597	0.8	0.976	1.541

$\alpha \to 0.35,\ \beta \to 0.056,\ \gamma \to 0.5,\ \delta \to 0.7,\ \epsilon \to 0.036,\ \zeta \to 3,\ \eta \to 2,\ \nu \to 6,\ \rho \to 0.1,\ \mathrm{A} \to 28.4,\ \mathrm{CDMax} \to 780$

cumulation and less carbon emissions which implies less carbon damage later on. This lowers the extent of the climate crisis. A lower valuation of the future, thus, leads to (marginally) higher consumption today and tomorrow as the externality is mitigated. In the constrained optimal case the increase in ρ leads to the standard intertemporal reallocation. A higher consumption level is sustained by higher accumulation with higher carbon concentrations early on. The shift also implies that the marginal rates shift. In the future a higher damage level has to be compensated with higher maintenance investments. Consumption is, consequently, lowered.

A reduction in γ to 0.3 leads to lower environmental damages. In the optimal case capital, output, and consumption can rise, while mitigation efforts fall. For marginal rates to equalize, CO_2 concentration and damages rise. In the constrained optimal case lower damages, again, lead to higher output and consumption and higher CO_2 concentrations. Damages and capital stock, however, fall. A milder damage function allows higher GHG concentrations in the atmosphere although these are not run up to the same damage level as before. Less damage requires less capital stock for the same level of consumption and less maintenance investment. The resources freed increase consumption even further. In the BAU case a milder damage function implies a milder climate crisis. More capital and CD can be accumulated. In the BAU capital accumulation is only slowed down by falling profitability due to environmental damages. A softer damage function allows a higher level of consumption to be obtained before and after a climate catastrophe. The climate crisis itself is more pronounced as can be seen in the steep fall of output and the reversal of the interest rate to almost $-6\,\%$. World output and consumption increase while CO_2 is run up, but fall more drastically thereafter.

In the optimal case agents are able to fully internalize the externality. As a result the effects of increases in the maximal permissible CO_2 level are negligible. In the BAU the increase of CDMax increases output and consumption considerably. Generally the transition to the steady state occurs much smoother which is due to the relative decrease of carbon emissions to carbon dissipation at higher CO_2 levels. This allows a higher capital stock and higher economic activity. The same rationale holds true for the constrained optimal case. Capital stock, output and consumption increase together with CD, while damages decrease.

The mitigation function is only relevant in the optimal case and consists of two parameters. ζ measures the cost of mitigation as it ties marginal mitigation to the carbon price. ν measures the effectiveness of mitigation for different levels of mitigation investment, m. An increase in the cost structure of mitigation leads to a higher carbon price and higher mitigation costs, with slight increases in CD and environmental damages to offset some of the cost increase. A decrease in the effectiveness of mitigation leads, again, to increased mitigation costs and a higher carbon price. This movement is counteracted by a slight increase in CO_2 and environmental damage. Note that front-loading of mitigation efforts is not a robust result and can change depending on parametric assumptions regarding the mitigation and damage functions. Convergence to full mitigation levels remains robust at 6 decades.

10 Conclusions

Human emissions of carbon dioxide into the atmosphere are a negative externality; despite world political efforts such as the Kyoto Protocol and its carbon market, the majority of individual producers do not take into account the impact of their emissions on atmospheric CO_2, but atmospheric concentration of CO_2 affects their (or their descendants') collective well-being through the deleterious effects of global warming. Such a negative externality leads to a market failure and an inefficient allocation of resources. In the case of global warming the inefficiency takes the form of over-accumulation of capital and under-investment in mitigation. The correction of this market failure through the implementation of institutions which enforce cost transparency represents a Pareto improvement: the current generation invests less, spending the retrenchment on consumption and mitigation such that future generations enjoy higher output, higher consumption combined with lower GHG concentration.

Our simulations show that the gains from such a movement to the intertemporal production possibility frontier are large. While this reallocation will imply significant changes of our life-style and the technology used, the mitigation costs to return to pre-industrial carbon dioxide levels are, however, relatively small with their peak at around 2 % of world product.

The question of how such cost transparency can be achieved and which mechanisms to use are important, but cannot be answered in our framework.[4] This is also true for the Pareto improvement question of who needs to compensate whom. Given our representative agent assumption, such a realistic question is clearly unanswerable. These problems will need to be solved by the world political system.[5]

If global warming is a real economic problem, there is no economic cost to correct it. In principle, the costs of reducing emissions in the current generation can be shifted to the future generations who will benefit from a cooler planet by reducing conventional investment.

References

Arrow, K. (2007, June). Global climate change: a challenge to policy. *Economists' Voice* 1–5.

Asheim, G., Mitra, T., & Tungodden, B. (2016). Sustainable recursive social welfare function. Econ Theory. reprinted in this volume.

Barker, T. (2008). The economics of dangerous climate change. *Climatic Change*, 173–194.

Burniaux, J.-M., & Oliveira Martins, J. (2011). Carbon Leakages: a general equilibrium view. *Economic Theory*. this isssue.

Chichilnisky, G. (1994). North-south trade and the global environment. *American Economic Review*, *84*, 851–874.

[4]See Ostrom (2016) for a discussion of the former and Karp and Zhang (2016) and Chipman and Tian (2016) of the latter.

[5]See Lecocq and Hourcade (2016) and Dutta and Radner (2016) for a discussion of these issues.

Chichilnisky, G. (2016). Sustainable markets with short sales. *Economic Theory*. reprinted in this volume.

Chichilnisky, G., & Eisenberger, P. (2010). Asteroids: assesing atastrophic risks. *Journal of Probability and Statistics* 954750, 15.

Chichilnisky, G., & Heal, G. (1994). Who Should abate carbon emission? An international perspective. *Economic Letters*, *44*, 443–449.

Chichilnisky, G., & Sheeran, K. (2009). *Saving kyoto*. UK: New Holland.

Chipman, J. S., & Tian, G. (2016). Detrimental externalities, pollution rights, and the "Coarse Theorem". *Economic Theory*. reprinted in this volume.

Dutta, P. K., & Radner, R.: Capital growth in a global warming model will china and india sign a climate treaty? *Economic Theory*. reprinted in this volume.

Figuieres, C., & Tidball, M. (2016). Sustainable exploitation of a natural resource: A satisfying use of chichilnisky's criterion. *Economic Theory*. reprinted in this volume.

Foley, D. K. (2008). The economic fundamentals of global warming. In: J. M. Harris, N. R. Goodwin (Eds.) Twenty-first century macroeconomics: responding to the climate challenge, ch. 5. Cheltenham and Northampton: Edward Elgar Publishing.

IPCC. (2007). *Climate Change 2007, the Fourth IPCC Assessment Report*. Retrieved August 10, 2008 from. http://www.ipcc.ch/ipccreports/ar4-syr.htm.

Karp, L., Zhang, J. (2016). Taxes versus quantities for a stock pollutant with endogenous abatement costs and asymmetric information. *Economic Theory*. reprinted in this volume.

Lauwers, L. (2016). Intergenerational equity, efficiency and constructibility. *Economic Theory*. reprinted in this volume.

Lecocq, F., Hourcade, J. -C. (2016). Unspoken ethical issues in the climate affair insights from a theoretical analysis of negotiation mandates. *Economic Theory*. reprinted in this volume.

Nordhaus, W. D. (2008). *A question of balance: weighing the options on global warming policies*. New Haven: Yale University Press.

Nordhaus, W. D., & Boyer, J. (2000). *Warming the world*. Cambridge: MIT Press.

Ostrom, E. (2016). Nested externalities and polycentric institutions: must we wait for global solutions to climate change before taking actions at other scales? *Economic Theory*. reprinted in this volume.

Rezai, A. (2010). Recast DICE and its policy implications, Macroeconomic Dynamics.

Samuelson, P. A. (1965). A catenary turnpike theorem involving consumption and the golden rule. *American Economic Review*, *55*, 486–496.

Shiell, M. L., Lyssenko, N. (2008). Computing business-as-usual with a representative agent and a pollution externality. *Journal of Economic Dynamics and Control*, *32*, 1543–1568.

Stern, N. (2007). *The Economics of climate change: The stern review*. Cambridge: Cambridge University Press.

Tol, R. (2009). The economic effects of climate change. *Journal of Economic Perspectives*, *23*, 29–51.

Part V
Economic Policy and Regulation

Detrimental Externalities, Pollution Rights, and the "Coase Theorem"

John S. Chipman and Guoqiang Tian

1 Introduction

This paper reconsiders the question of the conditions for the validity of the "Coase theorem" (Coase 1960, IV, p. 6; Stigler 1966, p. 113) according to which the equilibrium amount of pollution is independent of the assignment of legal rights as between a polluter and the pollutee. Hurwicz (1995) analyzed this problem as a two-person, two-commodity exchange equilibrium between a polluter and a pollutee exchanging "money" and pollution, and characterized the Coasian solution as one in which the set of Pareto optima in the Edgeworth box exhibits a constant level of pollution for positive money holdings of both parties, as depicted in Fig. 2 below. He recognized that a sufficient condition for this outcome is that preferences be "parallel" with respect to the x-commodity (money in this case)—a result that in fact goes back to Edgeworth (1891)—and endeavored to show that this condition is also necessary for the result. In this paper we show that this result is incorrect, but that a weaker (yet still restrictive) necessary and sufficient condition leads to the Coase result.

This paper is dedicated to the memory of our esteemed respective former colleague and former thesis advisor Leonid Hurwicz. We greatly regret not having been able to discuss the final section with him before his death. Thanks are due to Augustine Mok for his help with the diagrams.

Originally published in Economic Theory, Volume 49, Number 2, February 2012 DOI 10.1007/s00199-010-0573-7.

J.S. Chipman (✉)
Department of Economics, University of Minnesota, 4-101 Hanson Hall, 1925 4th Street South, Minneapolis, MN 55455, USA
e-mail: jchipman@umn.edu

G. Tian
Department of Economics, Texas A & M University, College Station, TX 77843, USA
e-mail: gtian@tamu.edu

G. Chichilnisky and A. Rezai (eds.), *The Economics of the Global Environment*, Studies in Economic Theory 29, DOI 10.1007/978-3-319-31943-8_21

In recent years, a very interesting literature has developed in which these concepts have been applied to *countries* as opposed to individuals, with "pollution" taking the form of emission of carbon dioxide and other greenhouse gases; cf. Chipman (1974), Chichilnisky and Heal (2000a), Sheeran (2006), and Chichilnisky et al. (2000), also (in this volume) Chichilnisky (2016), Asheim et al. (2016). An important question is then how to implement policies in the context of such detrimental externalities, as discussed in Burniaux and Martins (2016), Dutta and Radner (2016), Figuiéres and Tidball (2016), Karp and Zhang (2016), Lauwers (2016), Lecocq and Hourcade (2016), Ostrom (2016), and Rezai et al. (2016). One solution (adopted in this paper) is to follow the Coasian approach of clearly defining property rights. However, the problem of climate change differs somewhat from the Coasian one in that CO_2 is caused by breathing on the part of humans, cattle, and other animals, and only beyond a certain level (which may certainly be claimed to have been reached) by fuel combustion; but also in that it requires assumptions on preferences needed to aggregate individuals to countries (for such aggregation conditions see, e.g., Chipman 2006). It turns out that some of these assumptions are consistent with but others are incompatible with those needed to justify the Coase theorem; hence in order to avoid confusion it is safer to conduct the exposition in terms of the two-individual model. This path will be followed here.

We shall suppose that there are two individuals: individual 1 who likes to engage in an activity (e.g., smoking, blowing leaves) which is annoying to individual 2 because it produces a "detrimental externality" (smoke, noise) which may be characterized as pollution. Individual 1 will be called the polluter and individual 2 the pollutee. The cost of the activity to the polluter (e.g., purchase of cigarettes, fuel for the leaf-blower, time taken to engage in the polluting activity) will be disregarded. This externality may be "internalized" by the introduction of pollution rights or permits which can be traded. Suppose that there is a maximum amount of pollution that could be produced by individual 1 per period of time, indicated by η, and let s denote the actual amount of pollution (smoke, or noise) produced during this period of time. Then of course

$$0 \leqq s \leqq \eta. \tag{1.1}$$

Other things being equal, individual 1 will wish to increase s and individual 2 will wish to reduce it. Pollution is a public commodity in the sense of Samuelson (1954, 1955, 1969)—a public good for individual 1 and a public bad for individual 2.

Suppose a system is developed whereby a quantity η of pollution rights (permits) is made available by the government, initially allocated between the two individuals, according to

$$\eta_1 + \eta_2 = \eta, \tag{1.2}$$

where η_i is the initial allocation of pollution rights to individual i. Then individual 1 has the legal right (which we assume will be exercised) to emit η_1 units of pollution, while individual 2 has the legal right to emit $\eta_2 = \eta - \eta_1$ units of pollution, which (since we assume it will *not* be exercised) is equivalent to a right to η_2 units of *pollution avoidance*. Suppose further that the two individuals start out with amounts ξ_i

of another good which may be called "money",

$$\xi_1 + \xi_2 = \xi. \tag{1.3}$$

Letting p denote the price of a pollution right in terms of money, letting y_i denote the amount of pollution rights held by individual i, and letting x_i denote the amount of money individual i has left over after purchasing or selling pollution rights, individual i is constrained by the budget inequality

$$x_i + py_i \leqq \xi_i + p\eta_i \quad (i = 1, 2). \tag{1.4}$$

Now since it may be assumed that each individual will desire a larger final holding x_i of money than less, and for the reasons given above will also want to end up with a larger holding y_i of pollution *rights* than less, (1.4) will be an equality

$$x_i + py_i = \xi_i + p\eta_i \quad (i = 1, 2). \tag{1.5}$$

Since this equality is valid for all prices, p, summing (1.5) over the two individuals we see from (1.2) and (1.3) that[1]

$$x_1 + x_2 = \xi; \quad y_1 + y_2 = \eta. \tag{1.6}$$

We finally must relate the pollution rights (which are just pieces of paper) to the pollution itself. It would not be beneficial for individual 1 (the polluter) to hold onto y_1 pollution rights unless he or she intended to exercise them, i.e., to produce an equal amount, s, of pollution. Consequently we may assume that $y_1 = s$. Likewise, in order to limit him or herself to an amount s of pollution, individual 2 (the pollutee) will need to limit individual 1's pollution rights to $y_1 = s$ units, and will therefore need to obtain possession of the remaining $y_2 = \eta - y_1 = \eta - s$ rights.[2] Thus we have

$$y_1 = s \quad \text{and} \quad y_2 = \eta - s. \tag{1.7}$$

Hence the same piece of paper which gives individual 1 the right to emit 1 unit of pollution, if transferred to individual 2, gives individual 2 the right to 1 unit of *pollution avoidance*.

---

[1]This formulation differs from that of Hurwicz (1995) (which it otherwise follows closely), who considers pollution rights $z_1 \equiv y_1$ and $z_2 \equiv \eta - y_2$, which must satisfy $z_1 = z_2$, analogously to the theory of public goods. This corresponds to the second equation of (1.6), since $z_1 = y_1 = \eta - y_2 = z_2$. Thus the difference is largely one of notation. The present formulation, which provides a notation for individual i's final holdings of money and of pollution rights (x_i, y_i), makes it somewhat easier to interpret the diagrams in Figs. 1 and 2 below.

[2]This shows that the validity of the analysis in this paper is limited to the case of a single pollutee. The introduction of a second pollutee at once introduces a "free-rider" problem.

Now the preferences of the polluter and the pollutee may be represented by (differentiable) utility functions $U_1(x_1, s)$ and $U_2(x_2, s)$ respectively, where $\partial U_i/\partial x_i > 0$ and $\partial U_1/\partial s > 0$ but $\partial U_2/\partial s < 0$ for $(x_i, s) \in (0, \xi) \times (0, \eta)$, $i = 1, 2$.

Then the necessary first-order interior (tangency) condition for Pareto-optimality takes the form (as in the Lindahl-Samuelson public-goods condition)[3]

$$\frac{\partial U_1}{\partial s} \Big/ \frac{\partial U_1}{\partial x_1} + \frac{\partial U_2}{\partial s} \Big/ \frac{\partial U_2}{\partial x_2} = 0, \tag{1.8}$$

where $\frac{\partial U_i}{\partial s} / \frac{\partial U_i}{\partial x_i}$ is agent i's marginal rate of substitution of s for x.

2 Homothetic Preferences

Let the utility functions of the two individuals be given by

$$\begin{aligned} U_1(x_1, s) &= x_1 s, \\ U_2(x_2, s) &= x_2(\eta - s). \end{aligned} \tag{2.1}$$

For a given amount x_1 of money, individual 1 will have maximum utility when $s = \eta$ and minimum (zero) utility when $s = 0$. Likewise, for a given amount x_2 of money, individual 2 will have maximum utility when $s = 0$ and minimum (zero) utility when $s = \eta$. Because of the relations (1.7), the utility functions (2.1) may be expressed in terms of the pollution rights instead of the pollution itself:

$$\begin{aligned} U_1(x_1, y_1) &= x_1 y_1, \\ U_2(x_2, y_2) &= x_2 y_2. \end{aligned} \tag{2.2}$$

Both functions are strictly quasi-concave and increasing in both arguments. We derive the two individuals' demand functions for x_i and y_i as functions of the price, p, of the permits.

Individual i's objective is to maximize $U_i(x_i, y_i) = x_i y_i$ subject to (1.4). Equating the marginal rate of substitution to the price of a right, we have

$$\frac{\partial U_i/\partial y_i}{\partial U_i/\partial x_i} = \frac{x_i}{y_i} = p, \quad \text{hence } x_i = p y_i \quad (i = 1, 2). \tag{2.3}$$

[3]Note that the first-order necessary (Lindahl-Samuelson) condition for the Pareto optimality for public-goods economies is given by $\frac{\partial U_1}{\partial s} / \frac{\partial U_1}{\partial x_1} + \frac{\partial U_2}{\partial s} / \frac{\partial U_2}{\partial x_2} = g'(s)$, where $g(s)$ is the x-input requirement for producing s units of the public good s. But in the pollution model there is no x-input requirement to produce s, so $g(s) = 0$ identically. Hence the two marginal rates of substitution add up to zero.

Since (from (1.5)) equality must hold in (1.4), substituting (2.3) into this equality, we obtain

$$\left.\begin{array}{l} 2x_i = \xi_i + p\eta_i \text{ hence } x_i = \dfrac{\xi_i}{2} + \dfrac{\eta_i}{2}p; \\ 2py_i = \xi_i + p\eta_i \text{ hence } y_i = \dfrac{\xi_i}{2p} + \dfrac{\eta_i}{2} \end{array}\right\} \quad (i = 1, 2). \tag{2.4}$$

Summing the two individuals' demands for money and for pollution rights from (2.4) and using the facts (from (1.6), (1.3), and (1.2)) that $x_1 + x_2 = \xi_1 + \xi_2 = \xi$ and $y_1 + y_2 = \eta_1 + \eta_2 = \eta$, we have

$$\xi = x_1 + x_2 = \frac{\xi}{2} + \frac{\eta}{2}p, \quad \eta = y_1 + y_2 = \frac{\xi}{2p} + \frac{\eta}{2},$$

from either of which it follows that

$$p = \frac{\xi}{\eta}. \tag{2.5}$$

Evaluating the demand functions (2.4) at the equilibrium price (2.5) we obtain

$$x_i = \frac{\xi_i}{2} + \frac{\eta_i}{2}\frac{\xi}{\eta} \quad \text{and} \quad y_i = \frac{\xi_i}{2}\frac{\eta}{\xi} + \frac{\eta_i}{2} \quad (i = 1, 2). \tag{2.6}$$

This is the desired competitive equilibrium.

Now let us consider two cases, in both of which $\xi = \eta = 1$ and $\xi_1 = \xi_2 = \frac{1}{2}$. Then from (2.5), $p = 1$. In Case (i) the polluter (individual 1) starts out with initial holdings $(\xi_1, \eta_1) = (\frac{1}{2}, 0)$, i.e., with half the money and no pollution rights, and ends up with final holdings $(x_1, y_1) = (\frac{1}{4}, \frac{1}{4})$, i.e., with one quarter of the money (half of his initial amount, the other half of which is used to purchase pollution rights from the pollutee), and the right to pollute only one quarter of the time. On the other hand, in Case (ii) the polluter starts out with $(\xi_1, \eta_1) = (\frac{1}{2}, 1)$, i.e., with half the money and all the pollution rights, and ends up with final holdings $(x_1, y_1) = (\frac{3}{4}, \frac{3}{4})$, i.e., three quarters of the money (one quarter of which is obtained from selling pollution rights to the pollutee) and the right to pollute three quarters of the time. This violates the "Coase theorem" according to which the assignment of rights does not affect the amount of pollution.

The situation is depicted in Fig. 1, a slightly modified Edgeworth box in which individual 1's initial (ξ_1, η_1) and final (x_1, y_1) holdings of money and pollution rights are measured rightward and upward from the southwest origin $O_1 = (0, 0)$, while individual 2's initial (ξ_2, η_2) and final (x_2, y_2) holdings of money and pollution rights are measured leftward and downward from the northeast origin $O_2 = (1, 1)$. For both individuals, the amount of pollution itself (desired by individual 1 and undesired by individual 2), is measured upward from the bottom of the box. All the numbers shown in the figure are measured from O_1. With the assumed initial values $\xi = \eta = 1$,

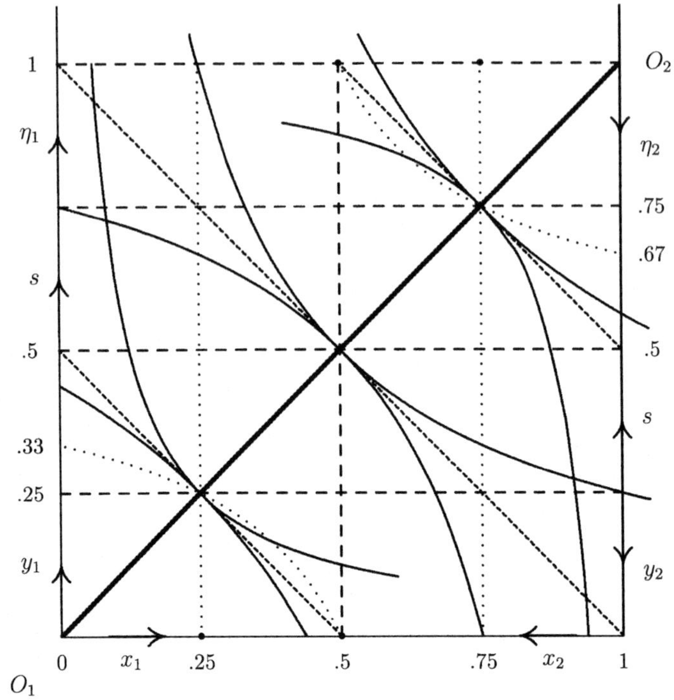

Fig. 1 Edgeworth box in the case of homothetic preferences

$\xi_1 = \xi_2 = \frac{1}{2}$, and $\eta_1 + \eta_2 = \eta$, we consider Case (i) in which the initial holdings of money and pollution rights are $(\xi_1, \eta_1) = (\frac{1}{2}, 0)$ and $(\xi_2, \eta_2) = (\frac{1}{2}, 1)$ [shown by the point $(0.5, 0)$ in the box, measured from O_1], and the final equilibrium holdings of money and pollution rights are $(x_1, y_1) = (\frac{1}{4}, \frac{1}{4})$ and $(x_2, y_2) = (\frac{3}{4}, \frac{3}{4})$ [shown by the point $(0.25, 0.25)$ in the box, measured from O_1]; and Case (ii) in which the initial holdings are $(\xi_1, \eta_1) = (\frac{1}{2}, 1)$ and $(\xi_2, \eta_2) = (\frac{1}{2}, 0)$ [shown by the point $(0.5, 1)$ in the box, measured from O_1], and the final (equilibrium) holdings are $(x_1, y_1) = (\frac{3}{4}, \frac{3}{4})$ and $(x_2, y_2) = (\frac{1}{4}, \frac{1}{4})$ [shown by the point $(0.75, 0.75)$ in the box, measured from O_1]. The tangential indifference curves of the two individuals are displayed at these two points, with the budget lines (shown as short-dashed lines) going through the points $(x_1, y_1) = (\frac{1}{2}, 0)$ and $(\frac{1}{4}, \frac{1}{4})$ in Case (i), and through the points $(x_1, y_1) = (\frac{1}{2}, 1)$ and $(\frac{3}{4}, \frac{3}{4})$ (measured from O_1) in Case (ii). (In both cases, $x_1 + x_2 = y_1 + y_2 = 1$.)

Figure 1 also displays (as dotted curves) the two individuals' offer curves. These are obtained from the two demand Eq. (2.4) for individual 1 with $\xi_1 = \frac{1}{2}$ and (in Case (i)) with $\eta_1 = 0$, to obtain

$$x_1 = \frac{1}{4}, \quad y_1 = \frac{1}{4p} \quad \text{(independently of } x_1\text{)}. \tag{2.7}$$

This is shown in the southwest part of Fig. 1 as the vertical straight line at $x_1 = \frac{1}{4}$ for individual 1. For individual 2, we eliminate the price variable from (2.4) to obtain (again in Case (i) for $\eta_1 = 0$ and $\eta_2 = 1$)

$$y_2 = \frac{1}{2} + \frac{1}{8x_2 - 2}. \tag{2.8}$$

This is shown in the southwest part of Fig. 1 as the dotted curve starting at the initial-endowment point $(\xi_2, \eta_2) = (0.5, 1)$ (measured from O_2 and corresponding to $(\xi - \xi_2, \eta - \eta_2) = (0.5, 0)$ measured from O_1), going through the equilibrium point $(x_2, y_2) = (0.75, 0.75)$ (measured from O_2 and corresponding to the equilibrium point $(1 - x_2, 1 - y_2) = (0.25, 0.25)$ measured from O_1) and ending at the point $(x_2, y_2) = (1, \frac{2}{3})$ (measured from O_2 and corresponding to the point $(1 - x_2, 1 - y_1) = (0, \frac{1}{3})$ measured from O_1).

In Case (ii), with $\eta_1 = 1$ $(\eta_2 = 0)$, we eliminate p from the demand Eq. (2.4) to obtain, for individual 1,

$$y_1 = \frac{1}{2} + \frac{1}{8x_1 - 1}, \tag{2.9}$$

a curve which is shown in the northeast part of Fig. 1 starting at $(\xi_1, \eta_1) = (\frac{1}{2}, 1)$ and going through the equilibrium point $(x_1, y_1) = (\frac{3}{4}, \frac{3}{4})$ and ending at the point $(x_1, y_1) = (1, \frac{2}{3})$. For individual 2, from the Eq. (2.4) for the case $\xi_2 = \frac{1}{2}$ and $\eta_2 = 0$ we obtain

$$x_2 = \frac{1}{4}, \quad y_2 = \frac{1}{4p} \quad \text{(independently of } x_2\text{)}, \tag{2.10}$$

showing that individual 2's offer curve is the vertical straight line $x_2 = \frac{1}{4}$ corresponding to $1 - x_2 = \frac{3}{4}$ in the diagram (measured from O_1).

This being a case of pure exchange with identical homothetic preferences, the set of Pareto optima (the "contract curve") is the dark diagonal of the box. The equilibrium amount of pollution is $y = \frac{1}{4}$ when the polluter starts out with no pollution rights, compared with $y = \frac{3}{4}$ when the polluter starts out with all the pollution rights, in contradiction to the "Coase theorem". Since the assumption of identical homothetic preferences is the leading condition making possible the aggregation of individuals into groups (cf. Chipman 1974), this raises problems with the application of Coasian economics to groups of individuals.

A third point $(x_1, y_1) = (0.5, 0.5)$ is shown in the diagram; this corresponds to the case in which not only the initial holdings of money are the same $(\xi_1 = \xi_2 = \frac{1}{2})$, but also the initial holding of pollution rights are the same $(\eta_1 = \eta_2 = \frac{1}{2})$. This point is also Pareto optimal, and no trading of pollution rights is needed to attain it.

3 Parallel Preferences

Now let the utility functions of the two individuals, expressed in terms of money and pollution, be given by

$$\begin{aligned} U_1(x_1, s) &= x_1 + \sqrt{s}, \\ U_2(x_2, s) &= x_2 + \sqrt{\eta - s}. \end{aligned} \tag{3.1}$$

As before, for a given amount x_1 of money, individual 1's utility is maximized when $s = \eta$ and minimized when $s = 0$, whereas for a given amount x_2 of money individual 2's utility is maximized when $s = 0$ and minimized when $s = \eta$. Substituting (1.7), the utility functions (3.1) may be expressed in terms of money and pollution rights:

$$\begin{aligned} U_1(x_1, y_1) &= x_1 + \sqrt{y_1}, \\ U_2(x_2, y_2) &= x_2 + \sqrt{y_2}. \end{aligned} \tag{3.2}$$

This is a case of identical "parallel" preferences.[4]

We will examine the above two cases (i) and (ii) with these utility functions in place of the utility functions (2.1).

With the utility functions (3.2), (2.3) is replaced by

$$\frac{\partial U_i/\partial y_i}{\partial U_i/\partial x_i} = \frac{1}{2\sqrt{y_i}} = p \quad \text{hence } y_i = \frac{1}{4p^2} \quad (i = 1, 2). \tag{3.3}$$

Substituting (3.3) in the budget constraint (1.4) (with equality) we obtain the two demand functions for individual i:

$$x_i = \xi_i + \eta_i p - \frac{1}{4p} \quad \text{and} \quad y_i = \frac{1}{4p^2}. \tag{3.4}$$

Thus, each individual's demand y_i for pollution rights is independent of his or her initial money holdings ξ_i or holdings of pollution rights η_i and of the budget constraint (so long as it is consistent with the budget constraint).

Now setting $x_1 + x_2 = \xi$ and $y_1 + y_2 = \eta$ we have from (3.4)

$$\xi = x_1 + x_2 = \xi + \eta p - \frac{1}{2p}, \quad \eta = y_1 + y_2 = \frac{1}{2p^2}$$

from both of which we conclude that

[4] A parallel preference ordering is one that is representable by a *quasi-linear* utility function $U(x, y) = vx + \phi(y)$ for $v > 0$ and $\phi'(y) > 0$ (cf. Hurwicz 1995, p. 55n). The term "parallel" was introduced by Boulding (1945) and followed by Samuelson (1964), though the concept goes back to Auspitz and Lieben (1889, Appendix II, Sect. 2, pp. 470–483), Edgeworth (1891, p. 237n; 1925, p. 317n), and Berry (1891, p. 550). The concept was also analyzed by Pareto (1892) and Samuelson (1942), and by Katzner (1970, pp. 23–26) who describes such preferences as "quasi-linear". See also Chipman and Moore (1976, pp. 86–91, 108–110; 1980, pp. 940–946); Chipman (2006, p. 109).

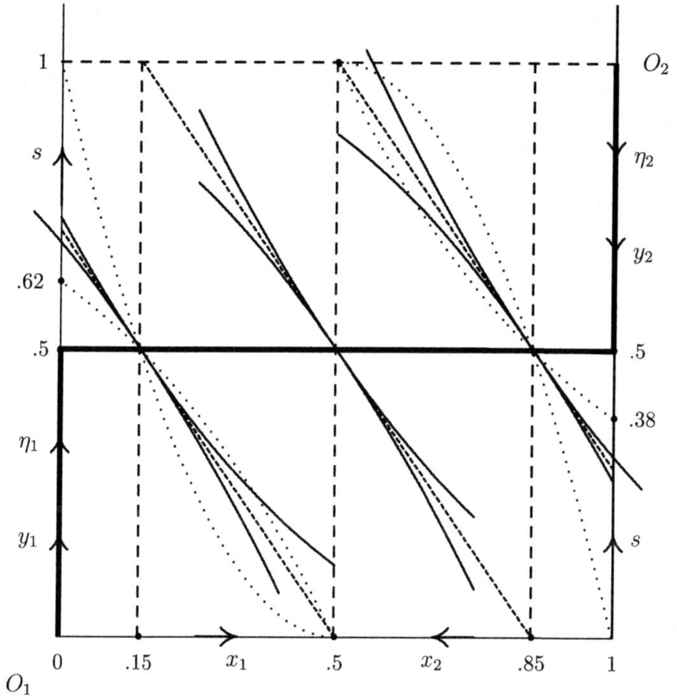

Fig. 2 Edgeworth box in the case of parallel preferences

$$p = \frac{1}{\sqrt{2\eta}}. \tag{3.5}$$

Substituting this price in the demand functions (3.4) we obtain

$$x_i = \xi_i + \frac{\eta_i}{\sqrt{2\eta}} - \frac{\sqrt{2\eta}}{4} \quad \text{and} \quad y_i = \frac{\eta}{2} \quad (i = 1, 2). \tag{3.6}$$

This is the desired competitive equilibrium.

Now let us look as before at the special case $\xi = \eta = 1$ and $\xi_1 = \xi_2 = \frac{1}{2}$, and consider two cases (see Fig. 2). When the utility functions are as in (3.1), in Case (i), while the polluter starts out without any pollution rights ($\eta_1 = 0$) and ends up with $\frac{1}{2}(1 - \frac{1}{\sqrt{2}}) = 0.14645$ of the money (less than 30 % of his initial amount $\xi_1 = \frac{1}{2}$, the other 70 % of which is used to purchase pollution rights from the pollutee), he ends up with the right to emit pollution half the time. In Case (ii) the polluter starts out with all the pollution rights ($\eta_1 = 1$) and ends up with $x_1 = \frac{1}{2}(1 + \frac{1}{\sqrt{2}}) = 0.85355$ of the money (more than 70 % of his initial amount, the extra 0.35355 coming from the sale of pollution rights to the pollutee), but the right to emit pollution only half the time. This is in accord with the "Coase theorem" which states that the initial allocation of property rights does not affect the amount of pollution. This follows from a basic property of "parallel" preferences according to which the set of Pareto optima (the "contract curve") is (for $0 < x_i < \xi$) a horizontal straight line, shown as the dark line $y_1 = 0.5$ in Fig. 2. Allowing for zero amounts of money ($x_i = 0$), the entire set of Pareto optima is the half-swastika-shaped dark line shown.[5]

In the case of the utility functions (3.1), in Case (i) when $\xi_i = \frac{1}{2}$ and $\eta_1 = 0$ ($\eta_2 = 1$), elimination of p from the demand Eq. (3.4) yields individual 1's offer function

$$y_1 = 4(\tfrac{1}{2} - x_1)^2, \tag{3.7}$$

shown as the dotted convex curve in the left part of Fig. 2 starting at $(x_1, y_1) = (\frac{1}{2}, 0)$, going through the equilibrium point $(x_1, y_1) = ((1 - 1/\sqrt{2})/2, 1/2) = (0.1464466, 0.5)$ and ending at the point $(0, 1)$. Likewise, elimination of p from the demand Eq. (3.4) yields individual 2's offer function (for $\eta_2 = 1$ in Case (i))

$$x_2 = \frac{1}{2} + \frac{1}{2\sqrt{y_2}} - \frac{\sqrt{y_2}}{2}, \tag{3.8}$$

which may be written as a quadratic equation in $\sqrt{y_2}$:

$$y_2 + (2x_1 - 1)\sqrt{y_2} - 1 = 0.$$

Taking the positive root this gives

$$y_2 = \frac{1}{4}\left(1 - 2x_2 + \sqrt{(2x_2 - 1)^2 + 4}\right)^2. \tag{3.9}$$

This is shown in the left part of Fig. 2 as the concave dotted offer curve starting at the initial-endowment point $(\xi_2, \eta_2) = (\frac{1}{2}, 1)$ (measured from O_2 and corresponding to $(\frac{1}{2}, 0)$ measured from O_1), going through the equilibrium point $(x_2, y_2) = ((1 + 1/\sqrt{2})/2, \frac{1}{2}) = (0.85355339, 0.5)$ (measured from O_2 and corresponding to $(1 - x_2, 1 - y_2) = ((1 - 1/\sqrt{2})/2, \frac{1}{2}) = (0.14644661, 0.5)$ measured from O_1), and ending at $(x_2, y_2) = (1, 0.381966)$ (corresponding to $(0, 0.618034)$ as measured from O_1).

It remains to consider Case (ii) in which $\eta_1 = 1$ ($\eta_2 = 0$). Eliminating p from Eq. (3.4) we obtain for individual 1 the offer function

---

[5]The swastika is an ancient Buddhist and Hindu symbol found on temples in central Asia. Hitler adopted it (after rotating it clockwise 45°) as the symbol of his Nazi party. As L. Hurwicz reminded the first author in a seminar presentation, the German word for swastika is *Hakenkreuz* (hook-cross); so the set of Pareto optima in the Edgeworth box with parallel preferences is one of these hooks.

$$x_1 = \frac{1}{2} + \frac{1}{2\sqrt{y_1}} - \frac{\sqrt{y_1}}{2}, \tag{3.10}$$

which may also be expressed as a quadratic equation in $\sqrt{y_1}$:

$$y_1 + (2x_1 - 1)\sqrt{y_1} - 1 = 0.$$

Taking the positive root, this yields

$$y_1 = \frac{1}{4}\left(1 - 2x_1 + \sqrt{(2x_1 - 1)^2 + 4}\right)^2. \tag{3.11}$$

This is shown in the right part of Fig. 2 by the convex dotted offer curve starting at the initial-endowment point $(\xi_1, \eta_1) = (0.5, 1)$, going through the equilibrium point $(x_1, y_1) = ((1 + 1/\sqrt{2})/2, 1/2) = (0.85355339, 0.5)$ and ending at the point $(1, (\sqrt{5} - 1)^2)/4) = (1, 0.381966)$. In the case of individual 2, with $\eta_2 = 0$, elimination of p from Eq. (3.4) yields the offer function for individual 2:

$$y_2 = 4\left(\frac{1}{2} - x_2\right)^2. \tag{3.12}$$

This is shown in the right part of Fig. 2 by the dotted concave curve starting at the initial-endowment point $(\xi_2, \eta_2) = (0.5, 0)$ (measured from O_2 and corresponding to the point (0.5, 1) as measured from O_1), going through the equilibrium point

$$(x_2, y_2) = \left(\frac{1 - 1/\sqrt{2}}{2}, \frac{1}{2}\right) = (0.14644661, 0.5)$$

(corresponding to

$$(1 - x_2, 1 - y_2) = \left(\frac{1 + 1/\sqrt{2}}{2}, \frac{1}{2}\right) = (0.85355339, 0.5),$$

as measured from O_1), and ending at the point $(x_2, y_2) = (0, 1)$ (measured from O_2 and corresponding to $(1 - x_1, 1 - y_2) = (1, 0)$ as measured from O_1).

Because of the "parallel" nature of the preferences, the set of Pareto optima is (for $0 < x_i < \xi$) the horizontal line at $y = 0.5$. In this case, the Coase theorem holds: the equilibrium amount of pollution is independent of the initial allocation of pollution rights. Note, however, that it does not hold if either the polluter has no money ($x_1 = 0, x_2 = \xi$) or the pollutee has no money ($x_1 = \xi, x_2 = 0$).

As in the previous case, a third Pareto-optimal point is shown at $(x_1, y_1) = (0.5, 0.5)$, and this is the case in which no trading is required to reach the optimum.

4 The Question of the Necessity of Parallel Preferences

The set of Pareto optima (equivalent here to the set of competitive equilibria—Edgeworth's "contract curve" (1881)) may be obtained as in Lange (1942) by maximizing the pollutee's utility $U_2(x_2, y_2)$ subject to that of the polluter $U_1(x_1, y_1)$ being constant at u_1, i.e.,

$$\text{Maximize } U_2(\xi - x_1, \eta - y_1) \text{ subject to } U_1(x_1, y_1) = u_1. \tag{4.1}$$

Setting up the Lagrangean expression

$$\mathcal{L}(x_1, y_1; \lambda) = U_2(\xi - x_1, \eta - y_1) - \lambda[U_1(x_1, y_1) - u_1] \tag{4.2}$$

and differentiating it with respect to x_1 and y_1, one obtains after eliminating λ the well-known mutual tangency condition

$$\frac{\partial U_1}{\partial y_1} \Big/ \frac{\partial U_1}{\partial x_1} = \frac{\partial U_2}{\partial y_2} \Big/ \frac{\partial U_2}{\partial x_2}, \quad \text{or} \quad \frac{\partial U_2}{\partial x_2} \cdot \frac{\partial U_1}{\partial y_1} = \frac{\partial U_2}{\partial y_2} \cdot \frac{\partial U_1}{\partial x_1} \tag{4.3}$$

as obtained by Edgeworth (1891, p. 236; 1925, p. 316). Now Edgeworth (1891, p. 237n; 1925, p. 317n) introduced the *sufficient* conditions that the marginal utilities of money of the respective individuals be positive constants $\partial U_i/\partial x_i = v_i > 0$ for $x_i > 0$, so that[6]

$$U_i(x_i, y_i) = v_i x_i + \phi_i(y_i) \quad (i = 1, 2) \tag{4.4}$$

(for $x_i > 0$), from which (4.3) reduces (together with (1.6)) to

$$\phi_1'(y_1)/v_1 = \phi_2'(\eta - y_1)/v_2. \tag{4.5}$$

Edgeworth concluded that one can solve this equation for y_1 independently of x_1,—i.e., that y_1 = constant—a conclusion which follows if it is assumed that $\phi_i' > 0$ and $\phi_i'' < 0$, as well as $\phi_2'(\eta)/v_2 < \phi_1'(0)/v_1$ and $\phi_2'(0)/v_2 > \phi_1'(\eta)/v_1$. Thus the contract curve for $0 < x_i < \xi$ is a horizontal line in the (x, y) space.[7]

It was further shown by Berry (1891, p. 550n)—see also Marshall (1961, II, pp. 793–5) Marshall—that the equilibrium price ratio is constant along this horizontal contract curve. His reasoning was that at any point (x_1, y_1) of an indifference curve

[6] The symbols x and y need to be interchanged to reconcile the present notation with Edgeworth's.

[7] Edgeworth's interest in this problem stemmed from his inquiry into the conditions under which the bargaining process introduced by Marshall in his Note on Berry (1891, pp. 395–397, 755–756; 1961, I, pp. 844–845; II, pp. 791–798), involving a succession of partial contracts at independently reached prices, with recontracting ruled out, would lead to a competitive equilibrium with in fact a uniform price, and thus a "determinate" solution.

$v_1 x_1 + \phi_1(y_1) = \bar{u}_1$, the slope is $v_1 dx_1/dy_1 + \phi_1'(y_1) = 0$; hence, along the horizontal line y_1 = constant, the slope $dx_1/dy_1 = \phi'(y_1)/v_1$ is constant.[8]

It was Hurwicz's aim to show the *necessity* of parallel preferences in order to justify Coase's result. His argument will be followed here, except that the present exposition is cast in terms of pollution *rights* rather than pollution itself, hence it applies generally to a two-agent, two-commodity model in which the marginal utilities of both commodities are nonnegative. We may characterize as *Coase's condition* the condition that the set of Pareto optima (the contract curve) in the (x, y) space for $x_i > 0$ is a horizontal line y = constant. It was shown above that a sufficient condition for this (given by Edgeworth) is the cardinal condition $\partial U_i/\partial x_i$ = constant for $i = 1, 2$. In order to investigate the necessity, we must obtain a corresponding ordinal condition.

It was shown by Hurwicz (1995, p. 67) that preferences that are parallel with respect to the x-commodity, i.e., representable by a differentiable utility function $U(x, y) = f(vx + \phi(y))$, where $v > 0$, $\phi' > 0$, and $f' > 0$, are characterized by the condition

$$\frac{\partial^2 U}{\partial x^2} \cdot \frac{\partial U}{\partial y} = \frac{\partial^2 U}{\partial y \partial x} \cdot \frac{\partial U}{\partial x}, \quad \text{or} \quad \frac{\partial^2 U}{\partial x^2} - \frac{\partial U/\partial x}{\partial U/\partial y} \cdot \frac{\partial^2 U}{\partial y \partial x} = 0. \tag{4.6}$$

This is verified immediately by performing the computations. The second condition of (4.6) is the ordinal counterpart of Edgeworth's cardinal condition $\partial^2 U/\partial x^2 = 0$.

Now we assume that the contract curve for $x_i > 0$ is a horizontal line $y_i = \bar{y}$. Accordingly, following Hurwicz (1995, p. 67), we differentiate the competitive equilibrium condition (4.3) with respect to x_1 subject to $x_2 = \xi - x_1$ and $y_i = \bar{y}$ for $i = 1, 2$. This gives

$$\frac{\partial^2 U_1}{\partial y_1 \partial x_1} \cdot \frac{\partial U_2}{\partial x_2} + \frac{\partial U_1}{\partial y_1} \cdot \frac{\partial^2 U_2}{\partial x_2^2} = \frac{\partial^2 U_1}{\partial x_1^2} \cdot \frac{\partial U_2}{\partial y_2} + \frac{\partial U_1}{\partial x_1} \cdot \frac{\partial^2 U_2}{\partial y_2 \partial x_2}. \tag{4.7}$$

Now dividing (4.7) through by $\partial U_1/\partial y_1 \cdot \partial U_2/\partial y_2$ and employing the tangency condition (4.3), we obtain Hurwicz's important formula (Hurwicz 1995, p. 67, A.3):

$$\frac{1}{\partial U_1/\partial y_1}\left[\frac{\partial^2 U_1}{\partial x_1^2} - \frac{\partial U_1/\partial x_1}{\partial U_1/\partial y_1}\frac{\partial^2 U_1}{\partial y_1 \partial x_1}\right] = \frac{1}{\partial U_2/\partial y_2}\left[\frac{\partial^2 U_2}{\partial x_2^2} - \frac{\partial U_2/\partial x_2}{\partial U_2/\partial y_2}\frac{\partial^2 U_2}{\partial y_2 \partial x_2}\right]. \tag{4.8}$$

In view of the second formula of (4.6), Hurwicz's formula (4.8) shows that, assuming (as we do) that $\partial U_i/\partial y_i > 0$ (pollution rights are positively desired by both agents), if individual 2's preferences are parallel (the bracketed term on the right is zero), so must be individual 1's, and *vice versa*. But of course this does not imply the desired converse of Edgeworth's proposition, namely that the horizontality of the contract curve implies that both individuals' preferences must be parallel.

---

[8] See Fig. 2 above, also Fig. 1 in Hurwicz (1995, p. 57). But this condition is violated in Fig. 2 of Samuelson (1969, p. 113).

In his *Sketch of proof* of the desired proposition, Hurwicz (1995, p. 71) assumed that individual 2 has a "linear preference", i.e., one representable by a utility function $U_2(x_2, y_2) = ax_2 + by_2$. But this is a special case of parallel preferences. This oversight appears to have escaped his attention. The problem of obtaining necessary conditions for the "Coase conjecture" was therefore left open. In effect, the proof consisted in showing how, starting from individual 2's linear preference (which is varied throughout the proof) one can infer that individual 1's must be parallel in order for the contract curve to be a horizontal line. But this already followed from (4.8).[9]

In fact, there are two problems in Hurwicz (1995). First, the equation in (4.8) alone cannot be used to fully characterize competitive equilibrium with $0 < x_i < \xi$ and $y_1 = y_2 = \bar{y}$. The following example shows that, although the equation in (4.8) is satisfied, it cannot guarantee that the contract curve is horizontal so that the set of Pareto optima for the utility functions need not be $y =$ constant.

Example 4.1 Suppose the initial endowments for money $\xi = 1$ and pollution rights $\eta = 1$. Let $U_i = x_i - x_i^2/2 + y_i$ which is not quasilinear in x_i. But, U_i is monotonically increasing for $(x_i, y_i) \in [0, 1] \times [0, 1]$ and concave. Moreover, it can be easily checked that (4.8) is satisfied for all $(x_i, y_i) \in [0, 1] \times [0, 1]$.

In fact, if we let

$$U_i = x_i - x_i^2/2 + \phi(y_i), \tag{4.9}$$

where ϕ is any concave and monotonically increasing function, then (4.8) is also satisfied for all $x_i, \in [0, 1]$ and $y_1 = y_2$. Thus, for the class of utility functions given by (4.9), although (4.8) is satisfied, the set of Pareto optima is not a horizontal line $y =$ constant.

To make the equation in (4.8) fully characterize competitive equilibrium with $0 < x_i < \xi$ and $y_1 = y_2 = \bar{y}$, we need to assume that the mutual tangency (first-order) condition (4.3) is also satisfied for all $x_i \in [0, \xi]$ and $y_1 = y_2 = \bar{y}$. Note that, for the class of utility functions given by (4.9), (4.3) is satisfied only for $x_i = 1/2$. This is why, even if (4.3) is satisfied for all $x_i \in [0, \xi]$ and $y_1 = y_2 = \bar{y}$, the set of Pareto optima is not a horizontal line $y =$ constant.

Secondly, Hurwicz's argument on the necessity of parallel preferences for "Coase's conjecture" is also problematic. To see this, without loss of generality,[10]

---

[9] Hurwicz mentioned (p. 68) that he had found an example (unfortunately not displayed) of cubic utility functions generating horizontal contract curves, but he dismissed this on the ground that "it must be possible to choose the two utility functions independently, while in the cubic case 'the choice of u^2 is limited by the choice of u^1.'" But formula (4.8) above, as well as the mutual tangency (4.3), shows that assuming the horizontality of the contract curve to be true, the two terms in brackets cannot be entirely unrelated. This does not imply any *psychological* dependence between the utility functions, but simply that there must be some kind of relationship, e.g. as in Edgeworth's formula (4.5), in order for the horizontality of the contract curve to be true.

[10] Since Eq. (1.7) defines a continuous one-to-one mapping between the individuals' pollution rights y_i with $y_1 + y_2 = \eta$ and allocation of pollution s, the two constrained optimization problems are equivalent. Note that through such a monotonic transformation, one may transform an original

we study the allocation of pollution s rather than the individuals' pollution rights y_i and consider the following class of utility functions $U_i(x_i, s)$ that have the functional form:

$$U_i(x_i, s) = x_i e^{-s} + \phi_i(s), \qquad i = 1, 2 \tag{4.10}$$

where

$$\phi_i(s) = \int e^{-s} b_i(s) ds. \tag{4.11}$$

$U_i(x_i, s)$ is then clearly not quasi-linear in x_i. It is further assumed that for all $s \in (0, \eta]$, $b_1(s) > \xi$, $b_2(s) < 0$, $b_i'(s) < 0$ $(i = 1, 2)$, $b_1(0) + b_2(0) \geq \xi$, and $b_1(\eta) + b_2(\eta) \leq \xi$.

We then have

$$\begin{aligned} \partial U_i/\partial x_i &= e^{-s} > 0, \quad i = 1, 2, \\ \partial U_1/\partial s &= -x_1 e^{-s} + b_1(s) e^{-s} > e^{-s}[\xi - x_1] \geq 0, \\ \partial U_2/\partial s &= -x_2 e^{-s} + b_2(s) e^{-s} < 0 \end{aligned}$$

for $(x_i, s) \in (0, \xi) \times (0, \eta)$, $i = 1, 2$. Thus, by (1.8), we have

$$0 = \frac{\partial U_1}{\partial s} \Big/ \frac{\partial U_1}{\partial x_1} + \frac{\partial U_2}{\partial s} \Big/ \frac{\partial U_2}{\partial x_2} = -x_1 - x_2 + b_1(s) + b_2(s) = b_1(s) + b_2(s) - \xi, \tag{4.12}$$

which is independent of x_i. Hence, if (x_1, x_2, s) is Pareto optimal, so is (x_1', x_2', s) provided $x_1 + x_2 = x_1' + x_2' = \xi$. Also, note that $b_i'(s) < 0$ $(i = 1, 2)$, $b_1(0) + b_2(0) \geq \xi$, and $b_1(\eta) + b_2(\eta) \leq \xi$. Then $b_1(s) + b_2(s)$ is strictly monotone and thus there is a unique $s \in [0, \eta]$, satisfying (4.12). Thus, the contract curve is horizontal even though individuals' preferences need not be parallel.

Example 4.2 Suppose $b_1(s) = (1 + s)^{\alpha} \eta^{\eta} + \xi$ with $\alpha < 0$, and $b_2(s) = -s^{\eta}$. Then, for all $s \in (0, \eta]$, $b_1(s) > \xi$, $b_2(s) < 0$, $b_i'(s) < 0$ $(i = 1, 2)$, $b_1(0) + b_2(0) > \xi$, and $b_1(\eta) + b_2(\eta) < \xi$. Thus, $\phi_i(s) = \int e^{-s} b_i(s) ds$ is concave, and $U_i(x_i, s) = x_i e^{-s} + \int e^{-s} b_i(s) ds$ is quasi-concave, $\partial U_i/\partial x_i > 0$ and $\partial U_1/\partial s > 0$, and $\partial U_2/\partial s < 0$ for $(x_i, s) \in (0, \xi) \times (0, \eta)$, $i = 1, 2$, but it is not quasi-linear in x_i.

Now we investigate the necessity for the "Coase conjecture" that the level of pollution is independent of the assignments of property rights. This reduces to developing the necessary and sufficient conditions that guarantee that the contract curve is horizontal so that the set of Pareto optima for the utility functions is s-constant. This

---

(Footnote 10 continued)
problem into a concave optimization problem, in which the object function is (quasi)concave and the constraint sets are convex. Since this technique has been widely used in the literature such as in the moral hazard model in the Principal-Agent Theory (cf. Laffont and Martimort (2002, pp. 158–159).

in turn reduces to finding the class of utility functions such that the mutual tangency (first-order) condition (4.3) does not contain x_i and consequently it is a function, denoted by $g(s)$, of s only:

$$\frac{\partial U_1}{\partial s} \Big/ \frac{\partial U_1}{\partial x_1} + \frac{\partial U_2}{\partial s} \Big/ \frac{\partial U_2}{\partial x_2} = g(s) = 0. \tag{4.13}$$

Let $F_i(x_i, s) = \frac{\partial U_i}{\partial s} / \frac{\partial U_i}{\partial x_i}$ $(i = 1, 2)$, which can be generally expressed as

$$F_i(x_i, s) = x_i h_i(s) + f_i(x_i, s) + b_i(s),$$

where the $f_i(x_i, s)$ are nonseparable and nonlinear in x_i. $h_i(s)$, $b_i(s)$, and $f_i(x_i, s)$ will be further specified below.

Let $F(x, s) = F_1(x, s) + F_2(\xi - x, s)$. Then (1.8) can be rewritten as

$$F(x, s) = 0. \tag{4.14}$$

Thus, the contract curve, i.e., the locus of Pareto-optimal allocations, can be expressed by a function $s = f(x)$ that is implicitly defined by (4.14).

Then, the Coase Neutrality Theorem, which is characterized by the condition that the set of Pareto optima (the contract curve) in the (x, s) space for $x_i > 0$ is a horizontal line $s =$ constant, implies that

$$s = f(x) = \bar{s}$$

with $\bar{s}$ constant, and thus we have

$$\frac{ds}{dx} = -\frac{F_x}{F_s} = 0$$

for all $x \in [0, \xi]$ and $F_s \neq 0$, which means that the function $F(x, s)$ is independent of x. Then, for all $x \in [0, \xi]$,

$$F(x, s) = xh_1(s) + (\xi - x)h_2(s) + f_1(x, s) + f_2(\xi - x, s) + b_1(s) + b_2(s) \equiv g(s). \tag{4.15}$$

Since the utility functions U_1 and U_2 are functionally independent, and x disappears in (4.15), we must have $h_1(s) = h_2(s) \equiv h(s)$ and $f_1(x, s) = -f_2(\xi - x, s) = 0$ for all $x \in [0, \xi]$. Therefore,

$$F(x, s) = \xi h(s) + b_1(s) + b_2(s) \equiv g(s), \tag{4.16}$$

and

$$\frac{\partial U_i}{\partial s} \Big/ \frac{\partial U_i}{\partial x_i} = F_i(x_i, s) = x_i h(s) + b_i(s) \tag{4.17}$$

which is a first-order linear partial differential equation. Then, from Polyanin et al. (2002),[11] we know that the principal integral $U_i(x_i, s)$ of (4.17) is given by

$$U_i(x_i, s) = x_i e^{\int h(s)} + \phi_i(s), \qquad i = 1, 2 \tag{4.18}$$

with

$$\phi_i(s) = \int e^{\int h(s)} b_i(s) ds. \tag{4.19}$$

The general solution of (4.17) is then given by $\bar{U}_i(x, y) = \psi(U_i)$, where ψ is an arbitrary function. Since a monotonic transformation preserves orderings of preferences, we can regard the principal solution $U_i(x_i, s)$ as a general functional form of utility functions that is fully characterized by (4.17).

Note that (4.18) is a general utility function that contains quasi-linear utility in x_i and the utility function given in (4.10) as special cases. Indeed, it represents parallel preferences when $h(s) \equiv 0$ and also reduces to the utility function given by (4.10) when $h(s) = -1$.

To make the the mutual tangency (first-order) condition (4.13) be also sufficient for the contract curve to be horizontal in a pollution economy, we assume that for all $s \in (0, \eta]$, $x_1 h(s) + b_1(s) > 0$, $x_2 h(s) + b_2(s) < 0$, $h'(s) \leq 0$, $b_i'(s) < 0$ $(i = 1, 2)$, $\xi h(0) + b_1(0) + b_2(0) \geq 0$, and $\xi h(\eta) + b_1(\eta) + b_2(\eta) \leq 0$.

We then have for $(x_i, s) \in (0, \xi) \times (0, \eta)$, $i = 1, 2$,

$$\partial U_i / \partial x_i = e^{\int h(s)} > 0, \quad i = 1, 2,$$
$$\partial U_1 / \partial s = e^{\int h(s)} [x_1 h(s) + b_1(s)] > 0,$$
$$\partial U_2 / \partial s = e^{\int h(s)} [x_2 h(s) + b_2(s)] < 0,$$

and thus

$$0 = \frac{\partial U_1}{\partial s} \Big/ \frac{\partial U_1}{\partial x_1} + \frac{\partial U_2}{\partial s} \Big/ \frac{\partial U_2}{\partial x_2} = (x_1 + x_2) h(s) + b_1(s) + b_2(s) = \xi h(s) + b_1(s) + b_2(s), \tag{4.20}$$

which does not contain x_i. Hence, if (x_1, x_2, s) is Pareto optimal, so is (x_1', x_2', s) provided $x_1 + x_2 = x_1' + x_2' = \xi$. Also, note that $h'(s) \leq 0$, $b_i'(s) < 0$ $(i = 1, 2)$, $\xi h(0) + b_1(0) + b_2(0) \geq 0$, and $\xi h(\eta) + b_1(\eta) + b_2(\eta) \leq 0$. Then $\xi h(s) + b_1(s) + b_2(s)$ is strictly monotone and thus there is a unique $s \in [0, \eta]$ that satisfies (4.20). Thus, the contract curve is horizontal even though individuals' preferences need not be parallel.

In summary, we have the following proposition.

Proposition 4.1 (COASE NEUTRALITY THEOREM) *In a pollution economy considered in this paper, suppose that the transaction cost equals zero, and that the util-*

[11] It can be also seen from http://eqworld.ipmnet.ru/en/solutions/fpde/fpde1104.pdf.

ity functions $U_i(x_i, s)$ are differentiable and such that $\partial U_i/\partial x_i > 0$, and $\partial U_1/\partial s > 0$ but $\partial U_2/\partial s < 0$ for $(x_i, s) \in (0, \xi) \times (0, \eta)$, $i = 1, 2$. Then, the level of pollution is independent of the assignment of property rights if and only if the utility functions $U_i(x, y)$, up to a monotonic transformation, have a functional form given by

$$U_i(x_i, s) = x_i e^{\int h(s)} + \int e^{\int h(s)} b_i(s) ds, \tag{4.21}$$

where h and b_i are arbitrary functions such that the $U_i(x_i, s)$ are differentiable, $\partial U_i/\partial x_i > 0$, and $\partial U_1/\partial s > 0$ but $\partial U_2/\partial s < 0$ for $(x_i, s) \in (0, \xi) \times (0, \eta)$, $i = 1, 2$.

Although the above Coase neutrality theorem covers a much wider class of preferences, it still puts a significant restriction on the domain of its validity due to the special functional forms of the utility functions.

In this paper, we only consider the economy in which one individual is the polluter and the other is the pollutee. By using the first-order conditions for Pareto optimality in an economy with negative externalities, which is developed in Tian and Yang (2009), we can also study "Coase's conjecture" in an economy with more than two individuals in which the individuals are polluters and pollute each other.

References

Asheim, G. B., Mitra, T., & Tungodden, B. (2016). Sustainable recursive social welfare functions. *Economic Theory*. reprinted in this volume

Auspitz, R., & Lieben, R. (1889). *Untersuchungen über die Theorie des Preises*. Leipzig: Verlag von Duncker & Humblot.

Berry, A. (1891). Alcune breve parole sulla teoria del baratto di A. Marshall. *Giorn Economics, Series 2*, *2*, 549–553.

Boulding, K. E. (1945). The concept of economic surplus. *American Economic Review*, *35*, 851–869.

Burniaux, J. -M., & Martins, J. O. (2016). Carbon leakage: a general equilibrium view. *Economic Theory*. reprinted in this volume.

Chichilnisky, G. (2016). Sustainable markets with short sales. *Economic Theory*. reprinted in this volume.

Chichilnisky, G., & Heal, G. (Eds.). (2000a). *Environmental markets. Equity and efficiency*. New York: Columbia University Press.

Chichilnisky, G., & Heal, G. (2000b). Markets for tradable carbon dioxide emission quotas: Principles and practice, 13–45.

Chichilnisky, G., Heal, G., & Starrett, D. (2000). Equity and efficiency in environmental markets: Global trade in carbon dioxide emissions, 46–67.

Chipman, J. S. (1974). Homothetic preferences and aggregation. *Journal of Economic Theory*, *8*, 26–38.

Chipman, J. S. (1998). A close look at the Coase theorem. In J. M. Buchanan & B. Monissen (Eds.), *The economists' vision. Essays in modern economic perspectives* (pp. 131–162). Frankfurt am Main: Campus Verlag.

Chipman, J. S. (2006). Aggregation and estimation in the theory of demand. In P. Mirowski & D. W. Hands (Eds.), *Agreement on demand: Consumer theory in the twentieth century* (pp. 106–129). Durham and London: Duke University Press.

Chipman, J. S., & Moore, J. C. (1976). The scope of consumer's surplus arguments. In A. M. Tang, F. M. Westfield & J. S. Worley (Eds.), *Evolution, welfare, and time in economics, essays in honor of Nicholas Georgescu-Roegen* (pp. 61–123). Lexington, Mass.: D. C. Heath and Company.

Chipman, J. S., & Moore, J. C. (1980). Compensating variation, consumer's surplus, and welfare. *American Economic Review*, *70*, 933–949.

Coase, R. H. (1960). The problem of social cost. *Journal of Law and Economics*, *3*, 1–44. Reprinted in Coase (1988), 95–156.

Dutta, P. K., & Radner, R. (2016). Capital Growth in a global warming model: Will China and India sign a climate treaty? *Economic Theory*. reprinted in this volume.

Edgeworth, F. Y. (1881). *Mathematical psychics*. London: C. Kegan Paul & Co.

Edgeworth, F. Y. (1891). Osservazioni sulla teoria matematica dell' -economia politica con riguardo speciale ai principi di economia di Alfredo Marshall; and Ancora a proposito della teoria del baratto. *Giorn Economics*, Series 2, *2*, 233–245, 316–318. Abridged, revised, and corrected English translation in Edgeworth (1925).

Edgeworth, F. Y. (1925). On the determinateness of economic equilibrium. In *Papers relating to political economy* 3 volumes (Vol. II, pp. 313–319). London: Macmillan and Co., Limited.

Figuiéres, C., & Tidball, M. (2016). Sustainable exploitation of a natural resource: A satisfying use of Chichilnisky criterion. *Economic Theory*. reprinted in this volume.

Hurwicz, L. (1995). What is the Coase Theorem? *Japan World Economy*, *7*, 49–74.

Karp, L., & Zhang, J. (2016). Taxes versus quantities for a stock pollutant with endogenous abatement costs and asymmetric information. *Economic Theory*. reprinted in this volume.

Katzner, D. W. (1970). *Static demand theory* (pp. 47–74). New York: The Macmillan Company.

Laffont, J. -J., & Martimort, D. (2002). *The theory of incentives: The principal-agent model*. Princeton and Oxford: Princeton University Press.

Lange, O. (1942). The foundations of welfare economics. *Econometrica*, *10*, 215–228.

Lauwers, L. (2016). Intergenerational equity, efficiency, and constructability. *Economic Theory*. reprinted in this volume.

Lecocq, F., & Hourcade, J. -C. (2016). Unspoken ethical issues in the climate affair. *Economic Theory: Insights From a Theoretical Analysis of Negotiation Mandates*. reprinted in this volume.

Marshall, A. (1891). *Principles of economics* (2nd ed.). London: Macmillan and Co. (9th (variorum) ed., Vols. 2). London: Macmillan and Co. Ltd. (1961).

Ostrom, E. (2016). Nested externalities and polycentric institutions: Must we wait for global solutions to climate change before taking action at other scales? *Economic Theory*. reprinted in this volume.

Pareto, V. (1892). Considerazioni sui principii fondamentali dell'economia politica pura. *Giorn Economics*, Series 2, *4*, 485–512, *5*, 119–157.

Polyanin, A. D., Zaitsev, V. F., & Moussiaux, A. (2002). *Handbook of first order partial differential equations*. London: Taylor & Francis.

Rezai, A., Foley, D. K., & Taylor, L. (2016). Global warming and economic externalities. *Economic Theory*. reprinted in this volume.

Samuelson, P. A. (1942). Constancy of the marginal utility of income. In O. Lange, F. McIntyre, & T. O. Yntema (Eds.), *Studies in mathematical economics and econometrics, in memory of Henry Schultz* (pp. 75–91). Chicago: The University of Chicago Press.

Samuelson, P. A. (1954). The pure theory of public expenditure. *Review of Economics and Statistics*, *36*, 387–389.

Samuelson, P. A. (1955). Diagrammatic exposition of a theory of public expenditure. *Review of Economics and Statistics*, *37*, 350–356.

Samuelson, P. A. (1964). Principles of efficiency—discussion. *American Economic Review, Papers and Proceedings*, *54*, 93–96.

Samuelson, P. A. (1969). The pure theory of public expenditure and taxation. In J. Margolis & H. Guitton (Eds.), Public economics (pp. 98–123). London: The Macmillan Press Ltd. and New York: St. Martin's Press.

Sheeran, K. A. (2006). Who should abate carbon emissions? *A Note, Environmental and Resource Economics*, *35*, 89–98.
Stigler, G. J. (1966). *The theory of price* (3rd ed.). London: Collier-Macmillan Limited.
Tian, G., & Yang, L. (2009). Theory of negative consumption externalities with applications to economics of happiness. *Economic Theory*, *39*, 399–424.

Taxes Versus Quantities for a Stock Pollutant with Endogenous Abatement Costs and Asymmetric Information

Larry Karp and Jiangfeng Zhang

1 Introduction

The danger that greenhouse gas (GHG) stocks cause environmental damage has led to a renewed interest in the problem of controlling emissions when there is asymmetric information about abatement costs. Although hybrid policies, e.g., cap and trade with a price ceiling, are more efficient than either the tax or quantity restriction (Pizer 1999), the comparison of taxes and quotas remains an important policy question. Since GHGs are a stock pollutant, the *regulator's* problem is dynamic. Most of the current literature on this dynamic problem assumes that non-strategic *firms* solve a succession of static problems. If, however, a firm's abatement costs depend on its stock of abatement capital, the firm makes a dynamic investment decision as well as the static emissions decision. We study the regulatory problem with asymmetric information when firms invest in abatement capital. We find that for general functional forms, quotas have an advantage over taxes that had not previously been recognized: quotas, but not taxes, are "information-constrained" optimal with respect to investment. We then consider a special functional form (linear-quadratic) where

Jiangfeng Zhang: The opinions expressed in this paper do not necessarily reflect the views of the Asian Development Bank.

We benefitted from comments by two anonymous referees. The usual disclaimer applies.

Reprinted with kind permission from the Authors: Originally published in Economic Theory, Volume 49, Number 2, February 2012 DOI 10.1007/s00199-010-0561-y.

L. Karp (✉)
Department of Agricultural and Resource Economics, University of California, Berkeley, USA
e-mail: karp@berkeley.edu

J. Zhang
Asian Development Bank, Manila, Philippines
e-mail: jzhang@adb.org

G. Chichilnisky and A. Rezai (eds.), *The Economics of the Global Environment*, Studies in Economic Theory 29, DOI 10.1007/978-3-319-31943-8_22

that advantage disappears, and here we find that taxes can increase abatement and welfare.

Earlier literature, e.g., Chichilnisky and Heal (1995), provided policymakers with the economic framework needed to compare different types of policies, thereby contributing to the creation of the carbon market currently used by some Kyoto signatories in the European Trading Scheme. Ellerman (2010) discusses the current state of this market. Our paper contributes to policy discussions by helping to illuminate a difference between taxes and quotas when forward-looking firms make investments that reduce abatement costs. In our setting, investment decreases costs by increasing the stock of abatement capital. We can also think of the firms' activity as investment in (excludable) R&D.

For a variety of pollution problems, capital costs comprise a large part of total abatement costs (Vogan 1991) and investment in abatement capital depends on the regulatory environment. In these cases, the endogeneity of investment is an important aspect of the regulatory problem. Several recent papers (Buonanno et al. 2001; Goulder and Schneider 1999; Goulder and Mathai 2000; Norhaus 1999) assume that the regulator can choose emissions and also induce firms to provide the first-best level of investment, e.g., by means of an investment tax/subsidy.

We consider the situation where the regulator has a single policy instrument, either a sequence of emissions taxes or a sequence of quotas. This assumption is consistent with many regulations and proposals that involve an emissions policy but ignore endogenous investment (e.g., the Kyoto Protocol). In virtually any real-world problem, the regulator is likely to have fewer instruments than targets. Our model is an example of this general disparity between the number of instruments and targets, and therefore is empirically relevant. We identify a previously unrecognized difference between taxes and quantity restrictions, and we provide a simple means of solving the regulatory problem when a certain condition holds.[1]

We now describe the problem in more detail. In each period the representative firm observes an abatement cost shock that is private information. If this cost shock is serially correlated, the regulator learns something about its current value by observing past behavior. The firm knows the current value of the cost shock and therefore is better informed than the regulator. Both the regulator and firms obtain information over time. We examine a subgame perfect equilibrium in which the regulator and firms condition their decisions on "directly payoff-relevant" information, as distinct, for example from the entire history of actions. That is, we consider a Markov Perfect Equilibrium (MPE).

[1]Jaffe et al. (2003) and Requate (2005) survey the literature on pollution control and endogenous investment. Many papers in this literature, including Biglaiser et al. (1995); Gersbach and Glazer (1999); Kennedy and Laplante (1999); Montero (2002); Fischer et al. (2003); Moledina et al. (2003); Tarui and Polasky (2005, 2006) assume that firms behave strategically with respect to the regulator: firms believe that their investment decisions will affect future regulation. Several paper, including Malueg (1989); Milliman and Prince (1989); Requate (1998); Requate and Unold (2003); Karp (2008) treat firms as non-strategic. Papers that discuss time-inconsistency arising because of the disparity between the number of targets and the number of instruments include Abrego and Perroni (2002); Marsiliani and Renstrom (2002).

For the regulator, the payoff-relevant information consists of the aggregate stock of abatement capital (which affects the industry-wide marginal abatement costs), the stock of pollution (which determines marginal damages) and the regulator's beliefs about the current cost shock (which also affects the industry's marginal abatement costs). For the firms, the payoff-relevant information consists of the current policy level (the tax or quota), the current cost shock, the individual firm's level of abatement capital, the aggregate industry capital and the pollution stock. The last two state variables are payoff-relevant for the firm because the firm understands that these variables affect the evolution of aggregate capital stock and pollution stock, and the firm understands that the regulator conditions future policy levels on future values of those two stocks.

The assumption of Markov Perfection means that the regulator cannot make binding commitments regarding future policies. Firms have rational expectations; they take the current emissions policy as given and they understand how the regulator chooses future policies. The non-atomic representative firm is not able to affect the economy-wide variables that determine future policies. The representative firm therefore behaves non-strategically, but not myopically, and also uses Markov policies.

The regulator understands that future emissions policies affect the current shadow value of abatement capital and thus affect current investment. For example, firms' anticipation that future emissions policies will be strict would increase the shadow value of abatement capital, thereby increasing the current level of investment. The regulator at the current time, t, might want to commit to a future policy, implemented at $t' > t$, as a means of affecting current investment in abatement capital. This incentive is the source of the familiar time-consistency problem. After time t, the time-t investment is predetermined, so the motivation for the time t' policy choice that was optimal at time t has changed. Our setting has the usual ingredients that lead to this problem: the regulator with a second-best instrument (the emissions tax or quota) wants to influence forward-looking agents.

Subgame perfection is stronger than time-consistency; the former requires that no agent wants to deviate from the equilibrium strategy at any possible subgame, and the latter requires only that no agent wants to deviate from the equilibrium strategy at subgames that actually occur in equilibrium. The assumption of Markov perfection (a refinement of subgame perfection) therefore implies time-consistency.

In order to understand an important difference between emissions taxes and quotas, it is useful to determine whether the restriction to time-consistency is a binding constraint. To this end, we ask whether the solution to the regulator's problem, in the absence of a time-consistency constraint, would in any case be time-consistent. If the private level of investment under the equilibrium emissions policy (the tax or quota) is socially optimal, then the regulator wants to use the emissions policy exclusively to influence emissions, without having to consider how the emissions policy affects investment. In this case, the regulator has no incentive to alter a previously announced policy rule. In this circumstance there is no time-consistency problem. Here we can obtain the MPE by solving an optimization problem that contains elements of the regulator's and the firms' problems. If, in contrast, firms' investment decisions are not socially optimal given beliefs about the emissions trajectory, then the regulator would like to choose the emissions policy partly to influence investment. In this case, the Markov Perfect restriction, which implies time-consistency, actually constrains the regulator. Here, we need to solve an equilibrium problem (a dynamic game between the regulator and non-strategic firms) rather than a relatively simple dynamic optimization problem. In other words, the type of problem that we need to solve an equilibrium problem or an optimization problem depends on whether the equilibrium level of investment is socially optimal, conditional on the emissions trajectory.

There is another way of thinking about the time-consistency problem. The only market failure is that firms do not take into account the social damages arising from emissions. If there were no cost shock, or if the regulator and firms had symmetric information, either the emissions tax or the quota would be sufficient to induce firms to emit at the optimal level. In that case, firms' investment decisions would be first best. Therefore, if in addition to the emissions policy the regulator were able to use an investment tax, the optimal level of that tax would be identically zero. However, when there is asymmetric information about abatement costs, there is no assurance that either the emissions tax or the quota leads to the first best level of emissions. Therefore, with asymmetric information, the equilibrium level of investment under the emissions policy might not be (information-constrained) socially optimal. In that case, the optimal level of an investment tax would be non-zero.

We can ask our basic question in two equivalent ways. (1) Is the optimal emissions tax or quota policy time-consistent? (2) Would a regulator who uses either the emissions tax or the quota increase welfare by additionally using an investment tax/subsidy? (In other words: Is the optimal investment tax/subsidy identically 0?).

We provide a simple answer to these questions. The optimal quota policy is time-consistent; equivalently, when the regulator can use both an emissions quota and an investment tax, the latter is identically 0. However, the optimal emissions tax policy is time-inconsistent, unless a particular "separability condition" holds; if this condition does not hold, the optimal investment tax, when used with the emissions tax, is not identically 0.

This result is useful for two reasons. First, under plausible circumstances the separability condition does not hold. In these circumstances, an emissions tax creates a secondary investment distortion, whereas the emissions quota does not. Thus, we have identified a difference between taxes and quotas that has previously been unnoticed.

Second, when the separability condition *does* hold, we can solve the dynamic game by solving a much simpler dynamic optimization problem that combines elements of the regulator's and the firms' optimization problems. The separability condition holds for an important special case, the linear-quadratic model, which has been previously used to study the problem of regulating both a flow and a stock pollutant under asymmetric information. Our separability result means that we can generalize the linear-quadratic model by including endogenous abatement capital. This generalization enables us to learn how the inclusion of endogenous abatement costs affects the ranking of the two policies. Our principal numerical finding is that, in the linear-quadratic framework, including abatement costs increases the advantage of taxes over quotas.

In summary, explicit treatment of the investment decision renders the firms' decision problems dynamic. In general, this feature favors the use of quotas, because these, unlike taxes, introduce no distortion into firms' investment decisions. However, in special cases that include the linear-quadratic model, this distortion disappears with taxes. For our calibration, the explicit treatment of the investment decision favors taxes in the linear-quadratic setting.

The next section discusses a static problem that provides the intuition for the separability condition. Subsequent sections describe the dynamic model and show the role of the separability condition. We then discuss the linear-quadratic specialization, and explain how endogenous investment affects the comparison of taxes and quotas in that setting. In closing the paper, we discuss other aspects of the tax versus quantity debate, as it applies to climate change policy.

Other papers in this Symposium discuss related aspects of the climate change problem. Asheim et al. (2016); Lauwers (2016); Figuieres and Tidball (2016); Chichilnisky (2016) consider the ethical foundations of criteria for sustainability, and the possibility of implementing such programs. Lecocq and Hourcade (2016) examine the relation between income levels and abatement expenditure in an efficient solution. Dutta and Radner (2016) show that capital accumulation can exacerbate the tragedy of the commons, and that income transfers can alleviate this problem. Rezai et al. (2016) explain why meaningful climate policy may require smaller (or even zero) sacrifice by the current generation, contrary to the conclusions of mainstream integrated assessment models. Burniaux and Martins (2016) use a computable general equilibrium model to evaluate the key parameters in determining the magnitude of "carbon leakage". Ostrom (2016) emphasizes the importance of climate change policy at sub-global levels. Chipman and Tian (2016) elaborate on the role of the Coase Theorem in climate change policy.

2 The One-Period Example

This section uses a one-period model that demonstrates, in a simple setting, the difference between taxes and quotas when abatement costs are endogenous. We show that the emissions quota is always time-consistent; equivalently, if the regulator uses an emissions quota and also is able to use an investment tax/subsidy, the optimal investment tax is 0. In contrast, the emissions tax is not time-consistent; equivalently, if the regulator uses an emissions tax and also is able to use an investment tax/subsidy, the investment tax is not 0 in general. However, the optimal emissions tax *is* time-consistent if and only if the primitive functions satisfy a particular "separability condition".

There is a simple explanation for this difference between taxes and quotas. An emissions policy chosen before investment has a "direct welfare effect" via its direct effect on emissions, and an "indirect welfare effect", via its effect on investment.

When the policy level is chosen after investment, the regulator takes into account only the direct welfare effect, since investment is fixed. The optimality condition for the policy chosen before investment requires that the sum of the direct and the indirect welfare effects is set equal to 0. The optimality condition for the policy chosen after investment requires that the direct welfare effect is set equal to 0. These two optimality conditions are equivalent if and only if the indirect effect is 0 whenever the direct effect is 0. That equivalence does not hold in general under taxes, but it does hold when the separability condition is satisfied. In contrast, the indirect welfare effect under quotas is *always* 0, so the optimal quota is the same regardless of whether it is chosen before or after taxes. All of this becomes obvious once we develop some notation and write down the optimality conditions.

We normalize the initial level of abatement capital to 0. The non-strategic but forward looking representative firm can buy k units of abatement capital at cost $c(k)$; the firm obtains benefits $B(x, k, \theta)$ by emitting x units of emissions when its stock of abatement capital is k and the cost shock is θ. We can think of the function $B(\cdot)$ as a restricted profit function in which input and output prices are suppressed. Alternatively, we can interpret $B(\cdot)$ as the amount of avoided abatement costs. For the latter interpretation, define x^b as the Business-as-Usual (BAU) level of emissions, i.e., the level of emissions under the status quo. Define $a = x^b - x$ as the level of abatement, i.e., the reduction in emissions due to a new regulatory policy. The abatement costs associated with the new regulations are $A = A(k, \theta, a)$. If x^b is a function of (k, θ), we can rewrite the abatement cost function as $A(k, \theta, a) = B(x, k, \theta)$, with $A_a(\cdot) = B_x(\cdot)$: marginal abatement costs equal the marginal benefit of emissions.

The benefit function is increasing and concave in x and k and increasing in θ ($B_k > 0$, $B_\theta > 0$, $B_x > 0$, $B_{kk} < 0$, $B_{xx} < 0$). More abatement capital decreases the marginal cost of abatement and therefore lowers the marginal benefit of pollution, so $B_{xk} < 0$. A higher cost shock increases the marginal benefits of abatement capital and emissions: $B_{k\theta} \geq 0$, $B_{x\theta} \geq 0$.

The damage from emissions (external to the firm) is $D(x)$. The regulator chooses either an emissions tax p or an emissions quota $\bar{x}$. Throughout this paper, we assume that the emissions quota is binding for all realizations of θ. Both the regulator and the firm have the same information about the distribution of θ before the firm observes its value.

Each firm has measure 0, and by choice of units the mass of firms has measure 1. With this normalization, in a symmetric equilibrium k and x represent the industry-wide capital stock and aggregate emissions, as well as the firm level values. The non-strategic firm chooses its (possibly constrained) level of k and x but takes the industry-wide levels as exogenous. In this section it is clear from the context whether we mean firm or aggregate level variables, but in a later section we modify the notation to avoid the possibility of misunderstanding.

We consider the following two time-lines:

Time Line A	Time Line B
1. The regulator chooses the policy level (p or $\bar{x}$)	1. The firm chooses investment (k)
2. The firm chooses investment (k)	2. The regulator chooses the policy level (p or $\bar{x}$)
3. Nature reveals the cost shock (θ)	3. Nature reveals the cost shock (θ)
4. The firm makes its emissions decision (x)	4. The firm makes its emissions decision (x)

With Time Line A, the emissions policy can influence both the levels of investment and emissions. With Time Line B, the emissions policy depends on the level of investment, and influences only the emissions level. In the one-period game, neither of the two time lines has a greater claim to plausibility, but the comparison of the two helps to understand the time-consistency problem in the dynamic setting.

If the optimal policy for the regulator is the same under both time lines, then it is obvious that the regulator uses that policy only to affect the emissions decision, not to influence the investment decision. In this case, the emissions policy does not create a secondary distortion in the investment decision; if we were to add a period after stage 2 and before stage 3 ("stage 2.5") to Time Line A, at which the regulator were permitted to revise the policy announced in stage 1, the regulator would not want to make a revision when policies are time-consistent.

We show that the emissions quota is always time-consistent, but the emissions tax is time-consistent if and only if a particular separability condition holds. Equivalently, if we were to add a "stage 0" to either time line, at which the regulator announces an investment tax, the optimal investment tax is always 0 when the regulator uses an emissions quota, but it is 0 when the regulator uses an emissions tax if and only if the separability condition holds. To establish this claim, we examine each policy under both time lines.

Emissions taxes Under Time Line A, p is predetermined when the individual firm chooses k. Under Time Line B, p depends on *aggregate* capital, about which firms have rational point expectations. Since firms take aggregate capital as given when making their investment decision, and since they know the relation between p and aggregate capital, it is "as if" they take p as given under Time Line B as well. In short

Remark 1 Under both time lines, the individual firm takes p as given when choosing its level of capital.

This remark simplifies comparison of the two time lines because we need only consider how the choice of time lines affects the regulator's strategic incentives; firms do not behave strategically, so the choice of time lines obviously does not affect their strategic incentives.

Consider Time Line A when the regulator uses a tax. The representative firm's payoff in stage 2 is

$$E\left[B(x,k,\theta) - px - c(k)\right],$$

where the expectation is with respect to θ. The firm chooses x in the last stage, conditional on k, p, and θ. It chooses k before it learns θ. The first-order conditions for x and k and the corresponding decision rules (denoted using $*$) in stages 4 and 2 are

$$B_x(x,k,\theta) - p = 0 \Rightarrow x = x^*(k,p,\theta). \tag{1}$$

$$E\left[B_k(x^*,k,\theta) - c'(k)\right] = 0 \Rightarrow k = k^*(p). \tag{2}$$

Note that $k^*(p)$ is independent of θ. Differentiating the first-order condition (1) gives the comparative statics result

$$\frac{\partial x^*}{\partial p} = \frac{-1}{B_{xx}(x,k,\theta)} \text{ and } \frac{\partial x^*}{\partial k} = \frac{-B_{xk}(x,k,\theta)}{B_{xx}(x,k,\theta)}. \tag{3}$$

The regulator's problem under Time Line A is

$$\max_p E\left[B(x^*,k^*,\theta) - c(k^*) - D(x^*)\right],$$

leading to the first-order condition

$$E\left\{\left[B_x(*,\theta) - D'(x^*)\right]\frac{\partial x^*}{\partial p}\right\} + E\left\{\left[\left[B_x(*,\theta) - D'(x^*)\right]\frac{\partial x^*}{\partial k} + B_k(*,\theta) - c'(k^*)\right]\frac{dk^*}{dp}\right\} = 0, \tag{4}$$

(using the notation $* = (x^*, k^*)$). Because $k^*(p)$ is independent of θ, we can take $\frac{dk^*}{dp}$ outside the expectations operator and use the firm's optimality condition (2) to write the regulator's optimality condition as

$$E\left\{\left[B_x(*,\theta) - D'(x^*)\right]\frac{\partial x^*}{\partial p}\right\} + \frac{dk^*}{dp}E\left\{\left[B_x(*,\theta) - D'(x^*)\right]\frac{\partial x^*}{\partial k}\right\} = 0, \tag{5}$$

The first term on the left side is the direct welfare effect of the tax and the second term is the indirect welfare effect, operating through investment.

The first-order condition for the firms' emissions decision is the same under the two time lines, since in both cases the firm conditions its emissions decision on predetermined values of p and k and the realized value of θ. In view of Remark 1, the first order condition to the individual firm's investment problem under Time Line B is (apart from an inessential notational difference) still given by Eq. (2). This notational difference is that instead of treating p as a predetermined variable, under Time Line B the firm treats p as a function of aggregate capital, which the firm takes as given.

The fact that (apart from the notational difference) the firm's first order conditions are the same under the two time lines means that the functions $x^* (k, p, \theta)$, and $k^* (p)$ are also the same under the two time lines, although of course the values of p in the two scenarios (and therefore the equilibrium values of k and x) might differ. This possible difference is the key to the time-consistency issue. Under Time Line B the regulator takes (aggregate) k as given, so its first-order condition for the tax is

$$E\left\{ \left[B_x(*, \theta) - D'(x^*)\right] \frac{\partial x^*}{\partial p} \right\} = 0. \tag{6}$$

Note that this optimality condition equates the expected marginal benefits and costs of the tax, not the expected marginal benefits and costs of emissions. The two need not be the same, because in general $\frac{\partial x^*}{\partial p}$ depends on θ.

Denote the optimal tax under Time Line B as $\hat{p}$. We assume that the regulator's problem is concave under both time lines, so that the solution to the respective first order condition is unique. Comparison of Eqs. (5) and (6) shows that the optimal emission tax is the same under the two time lines if and only if the indirect effect of the tax, evaluated at $p = \hat{p}$, is zero, i.e., if

$$E\left\{ \left[B_x(*, \theta) - D'(x^*)\right] \frac{\partial x^*}{\partial k} \right\}_{|p=\hat{p}} = 0. \tag{7}$$

We refer to the following as the "separability condition":

Condition 1 *(Separability) B_{xx} and B_{xk}, evaluated at the optimal x^*, are independent of θ.*

Remark 2 Equation (7) holds for all functions $B(x, k, \theta)$ if and only if the separability condition holds.

Proof In order to establish the sufficiency of Condition 1, note that it implies (using Eq. 3) that both $\frac{\partial x^*}{\partial p}$ and $\frac{\partial x^*}{\partial k}$ are independent of θ. This independence, together with Eq. (6) and the fact that $\frac{\partial x^*}{\partial p} \neq 0$ imply that $\hat{p}$ (the optimal tax under Time line B) satisfies

$$E\left[B_x(x^*(k^*, p, \theta), k^*, \theta) - D'(x^*(k^*, p, \theta))\right] = 0. \tag{8}$$

That is, under Condition 1 the tax equates expected marginal benefits of emissions with marginal damages. The independence of $\frac{\partial x^*}{\partial k}$ and θ, and the fact that $\frac{\partial x^*}{\partial k} \neq 0$, mean that Eq. (8) implies Eq. (7). Therefore, the optimal tax under the two lines is the same.

In order to establish necessity, note that if either B_{xx} or B_{xk} are not independent of θ, it is straightforward to construct examples under which the regulator's first order conditions for p differ under the two time lines. □

It is also easy to show:

Remark 3 Suppose that the regulator uses an emissions tax. If we modify either time lines by adding a stage 0 at which the regulator is able to choose an investment tax/subsidy, the optimal level of this policy is identically 0 for all functions $B(x, k, \theta)$ if and only if the separability condition holds.

We omit the proof, which parallels the proof of Remark 2.

We see that time-consistency requires that Eq. (6) implies Eq. (7). Equation (6) states that the first-order change in welfare due to a change in the tax, (holding investment fixed), is zero. Equation (7) states that the first-order change in welfare due to a change in investment (holding the tax fixed) is zero. In general, of course, there is no reason that one equation implies the other, so in general the tax chosen before investment is not time-consistent. However, the separability condition implies two things about the problem: (1) $\frac{\partial x^*}{\partial p}$ is independent of θ, so that setting the expected net marginal benefit of the tax (holding investment fixed) equal to 0 is equivalent to setting the expected net marginal benefit of emissions equal to 0; and (2) $\frac{\partial x^*}{\partial k}$ is independent of θ, so that setting the expected net marginal benefit of investment (holding the tax fixed) equal to 0 is equivalent to setting the expected marginal benefit of emissions equal to 0. When both Eqs. (6) and (7) are equivalent to setting expected net marginal benefit of emissions equal to 0, the two are equivalent to each other.

Emissions quotas Based on the same reasoning that led to Remark 1, we have

Remark 4 Under both time lines, the individual firm takes the emissions quota as given when choosing its level of capital.

If the regulator uses quotas (that by assumption are binding for all θ) the firm's emissions decision equals $\bar{x}$, and $\frac{\partial x^*}{\partial \bar{x}} = 1$. In view of Remark 4, the firm's first-order condition for the choice of k (for both of the two time lines) is

$$E\left[B_k(\bar{x}, k, \theta) - c'(k)\right] = 0 \Rightarrow k = k^*(\bar{x}). \tag{9}$$

As was the case with taxes, there is an unimportant notational issue: with Time Line A, $\bar{x}$ is literally predetermined, while with Time Line B, the firm treats $\bar{x}$ as a known function of aggregate investment, and the firm takes aggregate investment as given.

Under Time Line A, the regulator's first order condition for $\bar{x}$ is

$$\begin{aligned} E\left\{ \left[B_x(\bar{x}, k^*, \theta) - D'(\bar{x})\right] + \left[B_k(\bar{x}, k^*, \theta) - c'(k^*)\right] \frac{dk^*}{d\bar{x}} \right\} \\ = E\left\{ \left[B_x(\bar{x}, k^*, \theta) - D'(\bar{x})\right] \right\} = 0, \end{aligned} \tag{10}$$

where the first equality uses Eq. (9), the fact that k^* is independent of θ, and $\frac{dk^*}{d\bar{x}} \neq 0$. The first order condition under Time Line B is identical to the second equality in Eq. (10). Thus, when non-strategic firms have rational expectations, the optimal quota is the same under the two time lines. The regulator uses the quota to target only emissions, and the firm's investment decision is information-constrained socially optimal. There is no social value in using an investment tax when the regulator uses an emissions quota.

The time-consistency of quotas is due to the fact that the indirect welfare effect of the quota is always 0. Although the quota does affect the level of investment, the fact that the quota is always binding means that this investment does not affect the level of emissions. The only remaining indirect effect comes via the change in total cost; however, the private optimality of investment insures that it chosen so that expected marginal costs savings due to an extra unit of investment equals the marginal cost of investment. Thus, the private optimality of the investment decision insures that investment is also socially optimal.

3 Basics of the Dynamic Model

The stock of pollution at the beginning of period t is S_{t-1} and the flow of emissions in period t is x_t. The fraction $0 \leq \Delta \leq 1$ of the pollution stock lasts into the next period, so the growth equation for S_t is:

$$S_t = \Delta S_{t-1} + x_t. \tag{11}$$

The period t stock-related environmental damage equals $D_t = D\left(S_{t-1}\right)$, with $D' > 0$, $D'' > 0$.

At time t the representative firm's level of abatement capital is K_{t-1} and its cost shock is θ_t; when it emits at x_t its benefit is $B_t = B\left(K_{t-1}, \theta_t, x_t\right)$. At time t only the firm knows the value of the random cost shock θ_t; there is persistent asymmetric information. All agents know the stochastic process for the cost shock, which we assume is $AR(1)$:

$$\theta_t = \rho\theta_{t-1} + \mu_t, \qquad \mu_t \sim iid\left(0, \sigma_\mu^2\right), \qquad \forall t \geq 1, \tag{12}$$

with $-1 < \rho < 1$.[2] The sequence $\left\{\mu_t\right\}$ $(t \geq 1)$ is generated by an i.i.d. random process with zero mean and common variance σ_μ^2. At time 0 the regulator knows θ_{-1}, so the subjective expectation and variance of θ_0 is $\left(\rho\theta_{-1}, \sigma_\mu^2\right)$. This assumption about the regulator's initial priors makes the problem stationary; it has no bearing on our results, but merely simplifies the notation. At time $t \geq 1$ the regulator's variance for the current shock is σ_μ^2 provided that he has learned the value of the previous shock, θ_{t-1}.

The representative firm invests in abatement capital to reduce future abatement costs, i.e., to increase future benefits from pollution. The flow of investment in period t is I_t. The fraction of abatement capital that lasts into the next period is $0 \leq \delta \leq 1$, so the growth equation for K_t is:

---

[2]Throughout the paper we refer to θ as a "cost shock", as an abbreviation for "random cost parameter". In most economically meaningful circumstances, this parameter is positively serially correlated: $\rho > 0$.

$$K_t = \delta K_{t-1} + I_t. \tag{13}$$

The cost of investment, $C_t = C\left(I_t, K_{t-1}\right)$, is increasing and convex in I_t. This convexity means that abatement capital does not adjust instantaneously.

The endogeneity of the investment decision means that the marginal abatement cost function, $B_x(\cdot)$, changes endogenously. Slower adjustment of abatement capital means that it is optimal to adjust emissions more slowly.

4 The Game

In this section it is helpful to distinguish between the representative firm's level of capital and the aggregate level of capital. We denote the former by k and the latter by k^A. Where there is no danger of confusion, we denote both using K. Since we normalize the number of representative firms to 1, $k^A = k = K$ in a symmetric equilibrium. The representative firm understands that it controls k, and that this variable affects its payoff directly, via the function $B(\cdot)$. This firm takes the aggregate level of capital k^A as exogenous; k^A has no direct effect on the firm's payoff. However, in a Markov Perfect equilibrium, where the regulator conditions policies on payoff-relevant information, k^A affects the firm's beliefs about future policies.

To avoid a proliferation of notation, we do not distinguish between the firm's level of emissions and the aggregate level of emissions. However, it is important to bear in mind that the firm treats aggregate emissions, and therefore the aggregate pollution stock, as exogenous.

The regulator always uses taxes or always uses quotas. The period t policy is the tax p_t or the quota x_t. At time t the regulator knows the aggregate capital stock k^A_{t-1}, the pollution stock S_{t-1} and (as we explain below), the lagged cost shock θ_{t-1}. These are the payoff-relevant variables for the regulator. In a Markov Perfect rational expectations equilibrium, the representative firm takes the current level of the regulatory policy (at time t) as given; it understands that the policy at time $\tau > t$ will be a function of $\left(k^A_{\tau-1}, S_{\tau-1}, \theta_{\tau-1}\right)$. Since the firm takes these conditioning variables to be exogenous, it treats future policies as exogenous. This firm chooses investment I_t under both policies, and it chooses the level of emissions if the regulator uses a tax.

In view of the timing conventions in the model, the regulator's current (tax or quota) policy influences the firm's current emission, but not the current level of investment. Investment depends on the firm's beliefs about *future* policies (as was the case with Time Line B in Sect. 2).

4.1 The Firm's Emission and Investment Responses

The firm wants to maximize the expectation of the present value of the stream of cost saving from polluting (B) minus investment cost (C) minus pollution tax payments (under taxes). The constant discount factor is β, and we use the superscripts T and Q to distinguish functions and variables under taxes and quotas.

Taxes The firm's value function under taxes, $V^T\left(k_{t-1},\theta_t,p_t;S_{t-1},k^A_{t-1}\right)$, solves the dynamic programming equation (DPE)

$$\begin{aligned}V^T\left(k_{t-1},\theta_t,p_t;S_{t-1},k^A_{t-1}\right) = \max_{x_t,I_t}\Big\{&B\left(k_{t-1},\theta_t,x_t\right)-p_tx_t-C\left(I_t,k_{t-1}\right)\\&+\beta E_t\left[V^T\left(k_t,\theta_{t+1},p_{t+1;}S_t,k^A_t\right)\right]\Big\},\end{aligned}$$

subject to the equation of motion for the cost shock (12), the capital stock (13), and the pollution stock (11). The firm's expectation at t of θ_{t+1} and p_{t+1} is conditioned on the payoff-relevant variables $\left(k^A_{t-1},\theta_t,S_{t-1}\right)$.

The optimal level of emissions solves a static problem with the following first-order condition

$$B_x\left(k_{t-1},\theta_t,x_t\right)-p_t=0. \tag{14}$$

Solving for x, we obtain the optimal emission response

$$x_t^*=\chi\left(k_{t-1},\theta_t,p_t\right)\equiv\chi_t. \tag{15}$$

The optimal level of investment equates the marginal cost of investment and the discounted shadow value of abatement capital. Setting $k^A=k=K$, the stochastic Euler equation is[3]

$$\beta E_t\left\{B_K\left(K_t,\theta_{t+1},\chi_{t+1}\right)-C_K\left(I_{t+1},K_t\right)+\delta C_I\left(I_{t+1},K_t\right)\right\}-C_I\left(I_t,K_{t-1}\right)=0. \tag{16}$$

This second-order difference equation has two boundary conditions, the current abatement capital K_{t-1}, and the transversality condition

$$\lim_{T\to\infty}E_t\left\{\beta^{T-t}C_I\left(I_T,K_{T-1}\right)K_T\right\}=0. \tag{17}$$

Quotas Firms are homogeneous and quotas are not bankable. Thus, under a quota policy, there is no incentive to trade permits. The firm solves the DPE

$$\begin{aligned}V^Q\left(k_{t-1},\theta_t,x_t;S_{t-1},k^A_{t-1}\right)=\max\nolimits_{I_t}\Big\{&B\left(k_{t-1},\theta_t,x_t\right)-C\left(I_t,k_{t-1}\right)\\&+\beta E_t V^Q\left(k_t,\theta_{t+1},x_{t+1};S_t,k^A_t\right)\Big\}.\end{aligned}$$

Again, the firm's beliefs about the quota in the next period depend on $\left(k^A_{t-1},\theta_t,S_{t-1}\right)$.

---

[3]For all of the control problems, we merely write the Euler equation since the derivations are standard. The first-order condition of the DPE with respect to I_t provides one equation. In this first order condition, the firm's expectation of p_{t+1} is independent of its investment. This independence reflects the fact that the firm is unable to affect aggregate capital or pollution stock, and therefore cannot affect values of the variables that affect future regulation. We differentiate the DPE with respect to k_{t-1}, using the envelope theorem, to obtain a second equation. Combining these two equations gives the stochastic Euler equation.

The optimal level of investment solves the stochastic Euler equation

$$\beta E_t \left\{ B_K\left(K_t, \theta_{t+1}, x_{t+1}\right) - C_K\left(I_{t+1}, K_t\right) + \delta C_I\left(I_{t+1}, K_t\right) \right\} - C_I\left(I_t, K_{t-1}\right) = 0, \tag{18}$$

and the transversality condition (17).

The investment rule Under both taxes and quotas, the current level of investment depends on the firm's beliefs about future policy levels, but it does not depend on the current policy level. The firm has rational expectations about future policies; we discuss this policy rule in the next section. Under either taxes or quotas, the representative firm's equilibrium investment rule at time t is a function of $\left(k_{t-1}, \theta_t; S_{t-1}, k^A_{t-1}\right)$. When there is no danger of confusion, we write the investment rule as $I^j\left(K_{t-1}, \theta_t, S_{t-1}\right), j = T, Q$ (for tax or quota).

4.2 The Regulator's Problem

The regulator's payoff equals the payoff to the representative firm net of taxes, minus environmental damages. The regulator maximizes the expectation of the present discounted value of the flow of the payoff, i.e., the expectation of

$$\sum_{t=0}^{\infty} \beta^t \left(B\left(K_{t-1}, \theta_t, x_t\right) - C\left(I_t, K_{t-1}\right) - D(S_{t-1})\right).$$

His policy (always a tax or always a quota) can be a function of (only) payoff-relevant variables: the current stocks of pollution and capital, and the regulator's current information about the cost shock. Under taxes the regulator knows that Eq. (15) determines emissions. Under either policy, he knows that investment is given by $I^j\left(K_{t-1}, \theta_t, S_{t-1}\right), j = T, Q$.

The regulator takes as given the investment rule and (under taxes) the emissions rule. At time t the regulator observes the aggregate stocks S_{t-1}, K_{t-1}. If $\rho = 0$, the regulator learns nothing about the current cost shock by observing firms' past behavior. The past cost shock provides information about the current shock if and only if $\rho \neq 0$. Under taxes, the regulator learns the previous cost by observing the response to the previous tax (via Eq. (15)). With tradable quotas, the regulator learns the previous cost by observing the previous quota price.[4]

The regulator's decision rule is a function $z^j\left(K_{t-1}, \theta_{t-1}, S_{t-1}\right), j = T, Q$ that determines the current tax ($j = T$) or quota ($j = Q$) as a function of his current information, given his beliefs about the firm's decision rules.

---

[4]Under either policy, the regulator can also learn the previous cost shock by observing lagged investment, provided that $B_{K\theta} \neq 0$. From Eq. (18), $B_{K\theta} \neq 0$ means that current investment depends on the firm's beliefs about future cost shocks. When $\rho \neq 0$ these beliefs—and therefore current investment—depend on the current cost shock.

4.3 *The Equilibrium*

Both the regulator and the representative firm solve stochastic control problems; the exact problem that one agent solves depends on the solution to the other agent's problem. The rational expectations equilibrium investment rule for the firm depends on the regulator's policy rule, and that policy rule depends on the equilibrium investment rule. The investment and the regulatory decision rules generate a random sequence of pollution and capital stocks. Agents have rational expectations about these random variables.

An equilibrium consists of a (possibly non-unique) pair of decision rules $I^{j*}\left(K_{t-1}, \theta_t, S_{t-1}\right)$ and $z^{j*}\left(K_{t-1}, \theta_{t-1}, S_{t-1}\right)$ for $j = T, Q$ that are mutually consistent; the superscript "$*$" indicates equilibrium functions. Hereafter, we refer to $I^{j*}(K_{t-1}, \theta_t, S_{t-1})$, and $z^{i*}\left(K_{t-1}, \theta_{t-1}, S_{t-1}\right)$ as Markov Perfect policy rules.

Modern computational methods make it possible to (approximately) solve these kinds of dynamic equilibrium problems, i.e., to find a fixed point in function space (Judd 1998; Marcet and Marimon 1998; Miranda and Fackler 2002). These fixed point problems are not trivial, especially when the state space has more than one dimension—it has three in our problem.

5 Finding the Markov Perfect Equilibrium

In many cases, the type of model described in the previous section must be solved as an equilibrium problem rather than as an optimization problem. The next subsection explains why this complication might arise. Using an auxiliary control problem in which the regulator has two policy instruments, we then identify conditions under which the model *can* be solved as a straightforward optimization problem.

5.1 *The Time-Consistency Problem*

In general, the regulator might want to announce a rule that would determine future levels of the tax or quota. The purpose of such an announcement would be to alter the firm's investment rule—as distinct from altering a stock that appears as an argument of the investment rule. The inability to make binding commitments, and the Markov assumption, exclude this possibility. In a rational expectations equilibrium, current investment depends on beliefs about future policies, and these beliefs and policies depend on the pollution stock. By choice of the current quota or tax level, the regulator affects the future pollution stock, which can affect future investment. Under our assumptions, the only means by which *this period's* policy can influence future investment is by influencing the future level of the pollution stock.

Consider a simpler problem without asymmetric information, where a representative firm with rational expectations makes investment decisions. The firm's optimal decisions depend on its beliefs about future regulations, and the regulator wants to influence the firm's decisions. If the regulator has a first best policy (defined as one that does not cause secondary distortions), he can induce the firm to select exactly the decisions that the regulator would have used, had he been in a position to choose them directly. In that case, the regulatory problem can be solved as standard optimization problem. If, however, the regulator has only a second-best policy, the familiar time-consistency problem arises. (See Xie 1997; Karp and Lee 2003 for discussions of this problem, and references.) The Markov restriction is binding in this setting, so finding the equilibrium requires solving an equilibrium problem rather than a standard optimization problem.

The presence of asymmetric information in our model leads to the possibility of time-inconsistency of the optimal emissions tax or quota. We know from the literature on principal-agent problems that with asymmetric information, non-linear policies are generally superior to either the linear tax or the quota. We noted in Sect. 4.1 that the firm's investment depends on its beliefs about future policies. Since the regulator has two targets, (emissions and investment) and only one instrument, it appears that the regulator *might* want to use future emissions taxes or quotas to influence the firm's current investment decision. In that case, the information-constrained first best tax or quota would be time-inconsistent: the ability to make commitments about future taxes or quota decision rules would enable the regulator to achieve a higher payoff than under the Markov restriction. If this were the case, we would not be able to obtain a Markov Perfect equilibrium merely by solving a dynamic optimization problem, but would instead have to solve the equilibrium problem described in the previous section.

5.2 An Auxiliary Control Problem

This subsection describes an auxiliary control problem that helps identify conditions under which the Markov Perfect equilibrium can be obtained by solving an optimization problem. In this auxiliary control problem, in each period the regulator sets an emissions tax or quota using the same information as in the game; later in the same period he observes the current cost shock and then chooses investment directly. In contrast, in the game the regulator chooses only an emissions policy. The regulator's ability (in the auxiliary problem) to control current investment directly, knowing the current cost shock, eliminates any incentive to use future emissions policies to control current investment.

In this setting, it does not matter whether the regulator chooses investment directly (e.g., by command and control), or decentralizes this decision by means of an investment tax/subsidy. In the former case, firms make no investment decision, and in the latter case, firms merely carry out the optimal investment decision induced by the investment tax/subsidy. However, the model with the investment tax/subsidy is more helpful with intuition, so we emphasize that model. The *optimal* investment tax/subsidy is identically 0 if and only if Markov Perfect rules in the game are equivalent to the optimal policy rules in the auxiliary problem. With an identically zero investment tax, agents in the auxiliary problem have exactly the same optimization problem as in the game. It is optimal to use a non-zero investment tax/subsidy if and only if the Markov Perfect policies do not solve the auxiliary problem.

In the auxiliary problem we need to consider a two-stage optimization within each period. At the beginning of the period the regulator knows $\left(K_{t-1}, \theta_{t-1}, S_{t-1}\right)$ and chooses the emissions policy (a tax or quota); the regulator then learns θ_t and chooses the level of investment or the investment tax/subsidy. It does not matter whether this time-line is plausible. We use this problem only as a means of finding conditions under which the Markov Perfect rules can be obtained by solving a control problem.

Suppose we find that the investment tax/subsidy is identically 0 in the auxiliary control problem. In this case, the regulator is willing to allow firms to choose investment, given that the regulator chooses the emissions policy. The regulator in the game is therefore also willing to allow firms to choose investment. That is, the regulator in the game also has no wish to use an investment tax/subsidy.

5.2.1 Quotas in the Auxiliary Problem

When the regulator chooses emissions quotas in the auxiliary problem he solves the following DPE:

$$\begin{aligned} \mathcal{J}^Q\left(K_{t-1}, S_{t-1}, \theta_{t-1}\right) = \max_{x_t} E_{\theta_t|\theta_{t-1}} \{ & B\left(K_{t-1}, \theta_t, x_t\right) - D\left(S_{t-1}\right) \\ & + \max_{I_t} \left[-C\left(I_t, K_{t-1}\right) + \beta \mathcal{J}^Q\left(K_t, S_t, \theta_t\right)\right]\} \end{aligned} \tag{19}$$

subject to Eqs. (11) and (13). The first-order condition for the optimal quota is

$$E_{\theta_t|\theta_{t-1}}\{B_x\left(K_{t-1}, \theta_t, x_t\right) + \beta \mathcal{J}_S^Q\left(K_t, S_t, \theta_t\right)\} = 0 \tag{20}$$

and the Euler equation for investment under quotas is

$$\begin{aligned} &\beta E_{\theta_{t+1}|\theta_t}\left\{B_K\left(K_t, \theta_{t+1}, x_{t+1}\right) - C_K\left(I_{t+1}, K_t\right) + \delta C_I\left(I_{t+1}, K_t\right)\right\} \\ &- C_I\left(I_t, K_{t-1}\right) = 0. \end{aligned} \tag{21}$$

The transversality condition is

$$\lim_{T\to\infty} E_{\theta_T|\theta_t}\left\{\beta^{T-t} C_I\left(I_T, K_{T-1}\right) K_T\right\} = 0. \tag{22}$$

5.2.2 Taxes in the Auxiliary Problem

Using the firm's emission response function (15), the regulator in the auxiliary problem with emissions taxes solves the following DPE

$$\begin{aligned}\mathcal{J}^T\left(K_{t-1},S_{t-1},\theta_{t-1}\right) = \max\nolimits_{p_t} E_{\theta_t|\theta_{t-1}}\{B\left(K_{t-1},\theta_t,x_t^*\right) - D\left(S_{t-1}\right) \\ + \max\nolimits_{I_t}\left[-C\left(I_t,K_{t-1}\right) + \beta\mathcal{J}^T\left(K_t,S_t,\theta_t\right)\right]\}\end{aligned} \tag{23}$$

subject to Eqs. (11), (13) and (15). We use the definition

$$H_t \equiv \left[B_x\left(K_{t-1},\theta_t,x_t^*\right) + \beta\mathcal{J}_S^T\left(K_t,S_t,\theta_t\right)\right],$$

and the abbreviation $\chi_t \equiv \chi\left(K_{t-1},\theta_t,p_t\right) = x_t^*$. The function H_t is the social benefit of an additional unit of emissions, and recall that χ is the firm's decision rule for emissions under emissions taxes (Eq. (15)). With this notation, we can write the first-order condition with respect to p_t as

$$E_{\theta_t|\theta_{t-1}}\left\{H_t\frac{\partial\chi_t}{\partial p_t}\right\} = 0, \tag{24}$$

and the stochastic Euler equation for investment as

$$\begin{aligned}\beta E_{\theta_{t+1}|\theta_t}\left\{B_K\left(K_t,\theta_{t+1},x_{t+1}^*\right) - C_K\left(I_{t+1},K_t\right) + \delta C_I\left(I_{t+1},K_t\right) + H_{t+1}\tfrac{\partial\chi_{t+1}}{\partial K_t}\right\} \\ - C_I\left(I_t,K_{t-1}\right) = 0.\end{aligned} \tag{25}$$

The transversality condition is Eq. (22).

5.3 *Social Optimality of the Markov Perfect Rules*

Differentiating the first-order condition (14) implies the following:

Lemma 1 *Condition 1 is equivalent to the following two conditions: (1)* $\frac{\partial\chi(K_{t-1},\theta_t,p_t)}{\partial p_t}$ *is independent of* θ_t. *(2)* $\frac{\partial\chi(K_{t-1},\theta_t,p_t)}{\partial K_{t-1}}$ *is independent of* θ_t, *where* p_t *is the time t emissions tax.*

Our main result is the following:

Proposition 1 *(i) When the regulator uses emissions quotas, the solution to the auxiliary problem (19) is a Markov Perfect equilibrium to the original game. (ii) When the regulator uses emissions taxes, the solution to the auxiliary problem (23) is a Markov Perfect equilibrium to the original game if and only if the separability condition holds.*

The proof, contained in Appendix 1, verifies that the equilibrium conditions in the games and in the auxiliary problems are identical under the conditions stated in the Proposition.

5.3.1 Significance of the Proposition

When the regulator uses quotas to control emissions, the Markov Perfect investment rule is always information-constrained socially optimal. With emissions quotas, the ability to use an additional policy instrument to influence investment does not increase social welfare.

If the regulator uses emissions taxes to control emissions, the Markov Perfect investment rule is socially optimal if and only if Condition 1 is satisfied. This condition depends only on the benefit function $B(\cdot)$, not on the damage or the investment cost function. If the separability condition holds, the investment tax that supports optimal investment in the auxiliary problem is identically 0.

Proposition 1 identifies a previously unnoticed difference between taxes and quotas. When the separability condition does not hold, the regulator who uses an emissions tax to control pollution creates a secondary distortion in investment. In these circumstances, private investment is optimal under an emissions quota but not under an emissions tax. The emissions tax, but not the quota, creates the need for an investment tax/subsidy.

The Proposition also provides a simple way of obtaining the equilibrium for the game when the separability condition holds. This method requires only solving a dynamic optimization problem rather than a dynamic equilibrium problem.

5.3.2 Interpretation of the Separability Condition

We first identify the secondary distortion under emissions taxes, and we explain why it vanishes if the separability condition holds. This discussion also explains why emissions taxes and quotas typically have different effects, as regards the secondary distortion.

In order to identify the secondary distortion, we follow the standard procedure of computing the investment tax/subsidy that supports the information-constrained first best investment policy. Suppose that firms face an investment tax s_t, so their single period payoff is $B(\cdot) - C(\cdot) - s_t I_t - p_t x_t$. We can write the Euler equation for the capital stock corresponding to this problem, and compare it to the optimal investment policy under an emissions tax, Eq. (25). We omit the details, but the comparison implies that the investment tax supports the socially optimal level of investment if and only if[5]

$$-s_t + \beta\delta E_{\theta_{t+1}|\theta_t} s_{t+1} = \beta E_{\theta_{t+1}|\theta_t}\left\{ H_{t+1}\frac{\partial \chi_{t+1}}{\partial K_t}\right\}. \tag{26}$$

[5]The right side of Eq. (26) equals the function τ, used in the proof of Proposition 1.

The left side of Eq. (26) equals the effect of the tax sequence on the marginal incentive to invest in the current period. Under the investment tax, an additional unit of investment costs the firm s_t in the current period, but reduces the expected cost of tax payments by $\delta E_t s_{t+1}$ in the next period. The right side of Eq. (26) is the present value of the expectation of the secondary distortion. H_{t+1} is the marginal value to society of an additional unit of emissions in the next period, and $\frac{\partial \chi_{t+1}}{\partial K_t}$ equals the change in emissions in the next period caused by an additional unit of investment in the current period. Thus, the term in brackets in Eq. (26) is the value to society of the lower future emissions caused by the additional investment. This benefit is external to the firm. The optimal investment tax sequence induces the firm to internalize the present value of the expectation of this additional social benefit of investment, i.e., to internalize the externality.

The optimal emission quota does not create a secondary distortion. Under the quota, the expected social benefit of an additional unit of emissions is zero in each period (Eq. (20)). The socially optimal rule for determining investment, Eq. (21), involves only the current and future expected marginal investment and abatement costs. The socially optimal balance of these costs is identical to the balance that firms choose.

The optimal emissions tax, in contrast, requires that a marginal change in the tax has zero expected social value (Eq. (24)). This condition is not, in general, equivalent to the requirement that the expected social marginal benefit of emissions (H_t) is zero. The expected social marginal benefit of an additional unit of emissions *is* zero if and only if B_{xx} is independent of θ (equivalently, if and only if $\frac{\partial \chi}{\partial p}$ is independent of θ). This independence implies that $E_t H_t = 0$.

Even if this independence holds, H_t is a random variable, a function of θ. If $\frac{\partial \chi}{\partial K}$ is also a function of θ (i.e., if B_{xK} is not independent of θ), then the social marginal benefit of emissions is correlated with $\frac{\partial \chi}{\partial K}$. In that case, the expected marginal value to society of the lower future emissions caused by the additional investment (i.e., the secondary distortion, measured by the right side of Eq. (26)) is non-zero. Here, the investment externality is non-zero. Consequently, both B_{xx} and B_{xK} must be independent of θ in order for the investment externality to vanish under emissions taxes.

6 The Linear-Quadratic Model

The static model with a flow pollutant shows that a simple comparison of taxes and quotas requires strong functional assumptions: quadratic abatement costs and quadratic damages and additive uncertainty (the "linear-quadratic model") (Weitzman 1974). Without these assumptions, the ranking of taxes and quotas depends on parameters such as the variance of the cost uncertainty, for which we have very poor (if any) estimates. The major insight from the static linear-quadratic model is that taxes dominate quotas when the marginal abatement cost function is steeper than the marginal environmental damage function.

Analytical comparisons of the two policies in the climate change literature use the linear-quadratic model in which damages arise from the pollution stock, rather than the flow of emissions. Some commentators have claimed that for the regulation of GHGs, taxes obviously dominate quotas, because the marginal damage function for GHGs is so flat relative to the marginal abatement cost function. This reasoning is faulty, because in the dynamic setting the marginal abatement cost depends on the flow of emissions, while the marginal damage depends on the stock of pollution. The two slopes have different units in the dynamic problem, whereas they have the same units in the static problem.[6] In the dynamic setting it is not sensible to simply compare magnitudes of the two slopes. The dynamic optimization problem has to be solved in order to know how to compare these slopes, i.e., to know what constitutes a "large slope" and a "small slope" with GHGs.

This analysis has been undertaken with the competing assumptions that the regulator announces the entire sequence of future policies today (the open loop assumption) (Newell and Pizer 2003) or that the regulator conditions future policies on future information (the feedback assumption) (Hoel and Karp 2002; Karp and Zhang 2005), and under the assumption that the regulator expects to learn about a damage parameter (Karp and Zhang 2006). This analysis, together with available estimates of parameter values, supports the view that taxes dominate quotas for climate policy. Numerical results that do not use the linear-quadratic model also support this conclusion (Pizer 1999; Hoel and Karp 2001; Pizer 2002).

To determine the effect of endogenous abatement capital, it makes sense to use the linear-quadratic model. This functional form allows us to compare our results with those of earlier papers in this literature. With different functional assumptions, we would not be able to isolate the effect of endogenous capital. The linear-quadratic model satisfies the separability assumption—a fortunate circumstance, because it means that the effect of endogenous capital is not confounded by differences (under taxes and quotas) in strategic interactions. Both the information constrained optimal tax and quota are subgame perfect. Moreover, Proposition 1 enables us to obtain this equilibrium by solving the auxiliary control problem introduced in Sect. 5.2, instead of solving a dynamic game.

The representative firm's benefit function is

$$B\left(K_{t-1}, \theta_t, x_t\right) = f_0 + \left(f_1 + \psi\theta_t\right) K_{t-1} - \frac{f_2}{2} K_{t-1}^2 + \left(a - \phi K_{t-1} + \theta_t\right) x_t - \frac{b}{2} x_t^2$$

---

[6]Suppose we measure stock S in tonnes and emissions x in tonnes/year. Suppose that single period environmental damage is $a + bS^2$ and abatement cost is $c + dx^2$ and both are measured in dollars per year. Then the units of b, the slope of marginal damages, are $\frac{\$}{\text{year}\cdot(\text{tonne})^2}$ and the units of d, the slope of marginal abatement costs, are $\frac{\$\cdot\text{year}}{(\text{tonne})^2}$.

with $f_1 > 0, f_2 > 0, b > 0, \psi \geq 0, \phi \geq 0$. The function $B(\cdot)$ (which includes the rental cost of capital) satisfies the separability condition. The cost of changing the level of capital is[7]

$$C(I_t) = \frac{d}{2}(I_t)^2, \quad d > 0.$$

Environmental damages are also quadratic:

$$D(S_{t-1}) = \frac{g}{2}\left(S_{t-1} - \bar{S}\right)^2$$

where $\bar{S}$ is the stock level that minimizes damages.

The following Remark collects a number of useful facts about the comparison of policies. These results will be obvious to readers familiar with the linear-quadratic control problem, so we state them without proof:

Remark 5 In this linear-quadratic model with additive errors, the Principle of Certainty Equivalence holds. The expected trajectories of all stock and flow variables are the same under taxes and quotas. The higher moments of these trajectories differ under the two policies. Neither the policy ranking nor the magnitude of the payoff difference depends on the information state $\left(K_{t-1}, S_{t-1}, \theta_{t-1}\right)$. The magnitude (but not the sign) of the difference in payoffs depends on the variance of cost, σ_μ^2.

6.1 Regulated Emissions and Investment

For the linear-quadratic model we obtain an explicit equation for the emissions rule (Eq. 15) under taxes:

$$x_t^* = e_t - \frac{\phi}{b}K_{t-1} + \frac{\theta_t}{b}; \quad e_t \equiv \frac{a - p_t}{b}.$$

A higher cost realization increases current emissions, and a higher tax or a higher stock of abatement capital decreases emissions.

Using standard methods (e.g., Chap. 14 of Sargent 1987) we can solve the firm's Euler equation (16) under taxes and (18) under quotas) to write current investment as a linear function of current capital (K_{t-1}) and the firm's expectations of the future cost variables and policies (taxes or quotas). The optimal investment under emissions taxes is

[7] We can replace the investment cost function with a quadratic function of net rather than gross investment, so that adjustment costs are zero in the steady state. This slightly more plausible model does not lead to any interesting changes in analysis below. However, it complicates the problem of calibrating the model. Therefore we discuss only the model in which adjustment cost depends on gross investment.

$$I_t^* = \frac{\lambda\beta f_1}{d\delta(1-\lambda\beta)} + (\lambda - \delta) K_{t-1} + \frac{\lambda\beta}{d\delta} E_t \left[\left(\psi - \frac{\phi}{b} \right) \sum_{j=0}^{\infty} (\lambda\beta)^j \theta_{t+1+j} - \phi \sum_{j=0}^{\infty} (\lambda\beta)^j e_{t+1+j} \right]. \tag{27}$$

where $0 < \lambda < 1$ is the smaller root of the quadratic equation $\lambda^2 + \frac{h}{\beta}\lambda + \frac{1}{\beta} = 0$ and $h \equiv -\left[\frac{1}{\delta} + \frac{\beta}{d\delta}\left(f_2 - \frac{\phi^2}{b}\right) + \beta\delta\right]$. A lower expected future tax (i.e., a higher value of e_{t+j}) decreases current investment. A higher expected future cost shock increases (decreases) current investment if $\psi - \frac{\phi}{b}$ is positive (negative). Since $B_{K\theta} = \psi > 0$, a higher expected cost shock increases the expected marginal benefit of capital—and thus increases the marginal shadow value of capital. This effect encourages investment. However, a higher expected cost shock increases expected emissions, reducing the expected marginal benefit of capital ($B_{xK} = -\phi < 0$) and discouraging investment. These offsetting effects are exactly balanced if $\psi = \frac{\phi}{b}$, in which case the cost shock has no effect on investment, under emissions taxes.

The optimal investment under emissions quotas is[8]

$$I_t^* = \frac{\mu\beta f_1}{d\delta(1-\mu\beta)} + (\mu - \delta) K_{t-1} + \frac{\mu\beta}{d\delta} E_t \left[\psi \sum_{j=0}^{\infty} (\mu\beta)^j \theta_{t+1+j} - \phi \sum_{j=0}^{\infty} (\mu\beta)^j x_{t+1+j} \right]. \tag{28}$$

where $0 < \mu < 1$ is the smaller root of the quadratic equation $\mu^2 + \frac{w}{\beta}\mu + \frac{1}{\beta} = 0$ and $w \equiv -\left(\frac{1}{\delta} + \frac{\beta f_2}{d\delta} + \beta\delta\right)$. Higher expected quotas decrease investment, and higher expected cost shocks increase investment. With quotas, cost shocks have an unambiguous effect, because the firm treats future emissions quotas as exogenous.

6.2 A Limiting Case: Flow Externality

If $\Delta = 0$ all of the pollution stock decays in a single period, and the model collapses to the case of a flow externality. In this case, emissions in the current period cause damages only in the next period: $D(S_{t-1}) = D(x_{t-1})$.[9] By defining $\tilde{D}(x_t) = \beta D(x_t)$ we can write the difference between the benefits and costs of current emissions as $B(K_{t-1}, \theta_t, x_t) - \tilde{D}(x_t)$. This simplification eliminates a state variable (S), making it possible to obtain some analytic results. We can solve the dynamic programming equations under taxes and quotas and compare the payoffs.

[8]An algebraic proof confirms that expectated investment is the same under the rules given by Eqs. (27) and (28)—as Remark 5 states they must be.

[9]The specialization in this section simplifies the the stock pollution problem, and it is also of independent interest, because it shows how to compare taxes and quotas for a flow pollutant when abatement costs are endogenous.

We show that in two cases, the policy ranking does not depend on the parameters associated with abatement capital: (1) $\rho = 0$; or (2) $\rho \neq 0$ and $\psi \neq 0$. If neither of these two conditions hold, so that $\rho \neq 0$ and $\psi = 0$, the policy ranking does depend on the parameters associated with abatement costs.

If $\rho = 0$, or if $\rho \neq 0$ and $\psi \neq 0$, the payoff difference under taxes and quotas, is

$$\mathcal{J}^T - \mathcal{J}^Q = \frac{\sigma_\mu^2}{2b(1-\beta)} \left(1 - \frac{\beta g}{b}\right).$$

This expression reproduces a result in Weitzman (1974)'s static model and in two dynamic models (Hoel and Karp 2002; Karp and Zhang 2005).

If $\rho \neq 0$ and $\psi = 0$, the payoff difference equals

$$\mathcal{J}^T - \mathcal{J}^Q = \frac{\sigma_u^2}{2b(1-\beta)} \left[\Gamma + \left(1 - \frac{\beta g}{b}\right)\right]. \tag{29}$$

The function $\Gamma > 0$ depends on f_2, d and δ (among other parameters).

We summarize the implications of these expressions in the following:

Remark 6 For a flow pollutant ($\Delta = 0$): (i) If (a) $\rho = 0$, or if (b) $\rho \neq 0$ and $\psi \neq 0$), the policy ranking depends only on the relative slopes (appropriately discounted) of the marginal benefit and damage functions. (ii) When neither conditions (a) or (b) in part (i) hold the policy ranking also depends on the parameters associated with abatement capital.

The next section considers the problem of a stock-related pollutant with $\rho \neq 0$ and $\psi \neq 0$; there the policy ranking does depend on the parameters associated with abatement capital—in contrast to Remark 6.i. Here we explain why stock and flow pollutants have this qualitative difference.

As Remark 5 notes, the expected levels of emissions and of investment are the same under taxes and quotas. The first-order condition for investment (using Eqs. 19 or 23) is

$$-C_I\left(I_t, K_{t-1}\right) + \beta \mathcal{J}_K^i\left(K_t, S_t, \theta_t\right) = 0, \quad i = T, Q.$$

The linear-quadratic structure with additive uncertainty implies that $\mathcal{J}_K^T\left(K_t, S_t, \theta_t\right) \equiv \mathcal{J}_K^Q\left(K_t, S_t, \theta_t\right)$: the shadow value of capital and therefore the investment rules under taxes and quotas, conditional on $\left(K_{t-1}, S_t, \theta_t\right)$, are identical.

For a stock pollutant, $\mathcal{J}_{K,S}^i \neq 0$, so investment at time t depends on the pollution stock at the beginning of the next period, S_t. That pollution stock depends on current emissions; therefore, emissions in period t affect investment in period t. Conditional on the regulator's information at the beginning of a period, the current level of emissions is random under taxes and is a choice variable under quotas. Therefore, conditional on the information at the beginning of a period, the distribution function for the current level of investment differs under the two policies. The expected payoff difference therefore depends on the parameters associated with abatement capital.

In contrast, with a flow pollutant, the current level of emissions has no effect on future payoffs. The shadow value of capital $\mathcal{J}_K^i$ depends only on (K_t, θ_t). With a flow pollutant, the *current* investment and *current* emissions decisions are decoupled. Therefore, the value to the regulator of the difference in emissions under taxes and quotas does not depend on investment costs.

7 An Application to Climate Change

With a stock externality problem such as greenhouse gasses, we have three state variables (greenhouse gasses, the capital stock, and the expected cost shock) and therefore cannot obtain an analytic solution. However, using Proposition 1, it is straightforward to solve the tax and quota problems numerically. The resulting control problem is almost standard, except that new information arrives within a period, so there are two stages of optimization within a period. This fact accounts for the nested maximization in Eqs. (19) and (23). For the linear-quadratic model, we can solve each of these dynamic programming problems by solving a matrix Riccati equation.

7.1 *Model Calibration*

Table 1 describes the model. In order to calibrate the general linear-quadratic model described in the previous section, we assume that benefits are equal to the value of abatement cost that the firm avoids by increasing emissions. Abatement costs are a quadratic function of abatement, $x_t^b - x_t$ (row 6), where the BAU emissions x_t^b is a

Table 1 The model of global warming

1. Pollutant stock growth	$S_t - \bar{S} = \Delta\left(S_{t-1} - \bar{S}\right) + x_t$
2. Environmental damage	$D\left(S_{t-1}\right) = \frac{g}{2}\left(S_{t-1} - \bar{S}\right)^2$
3. Abatement capital growth	$K_t = \delta K_{t-1} + I_t$
4. Investment cost	$C\left(I_t\right) = \frac{d}{2} I_t^2$
5. "Business as usual" emissions	$x_t^b = m_0 - m_1 K_{t-1} + \tilde{\theta}_t$
6. Abatement cost	$A\left(x_t\right) = \frac{b}{2}\left(x_t^b - x_t\right)^2$
7. "General" benefit function	
$B\left(K_{t-1}, \theta_t, x_t\right) = f_0 + \left(f_1 + \psi\theta_t\right) K_{t-1} - \frac{f_2}{2} K_{t-1}^2 + \left(a - \phi K_{t-1} + \theta_t\right) x_t - \frac{b}{2} x_t^2$	
Parameter restriction	
$0 \leq \Delta \leq 1,\ g > 0,\ 0 \leq \delta \leq 1,\ d > 0,\ m_0 > 0,\ m_1 \geq 0,\ b > 0$	
Relation of parameters	
$\theta_t = b\tilde{\theta}_t, f_0 = -\frac{b}{2} m_0^2, f_1 = b m_0 m_1, f_2 = b m_1^2,\ a = b m_0,\ \phi = b m_1$, and $\psi = \frac{\phi}{b} = m_1$	

Table 2 Parameter values for the baseline model

Parameter	Note	Value
β	A continuous discount rate of 5 %	0.9512
Δ	Pollutant stock persistence	0.9917
δ	Capital stock persistence	0.85
π	The percentage loss in GWP from doubling $\bar{S}$	1.33
g	Slope of the marginal damage billion \$/(billion tons of carbon)^2	0.0022
b	Slope of the marginal abatement cost, billion \$/(billion tons of carbon)^2	26.992
d	Slope of the marginal investment cost, billion \$	703.31
m_0	Intercept of the BAU emissions, billion tons of carbon	12.466
m_1	Slope of the BAU emissions, (billion tons of carbon)/(billion \$)	0.7266
ρ	Cost correlation coefficient	0.90
σ_μ	Standard deviation of cost shock, \$/(ton of carbon)	1.7275
x_0^b	Current CO_2 emissions into the atmosphere billion tons of carbon	5.20
$\bar{S}$	Preindustrial stock, billion tons of carbon	590
S_{-1}	Current pollutant stock, billion tons of carbon	781
K_{-1}	Initial capital stock, billion \$	10

decreasing linear function of abatement capital (row 5). A higher level of abatement capital makes it cheaper to reduce emissions, and also decreases the marginal abatement costs. The cost variable $\tilde{\theta}$ (which is proportional to the random variable θ used above) changes the level of BAU emissions and therefore changes marginal abatement costs. Row 7 of Table 1 repeats the general linear-quadratic model; the final row gives the parameter restrictions under which this general model reproduces the special model described in the rows 2–6 of the table.[10] If $m_1 = 0$, capital does not affect abatement costs. This limiting case reproduces previous linear-quadratic models of a stock pollutant (Karp and Zhang 2005).

Table 2 lists baseline parameter values. In presenting the simulation results, we use the parameter π, defined as the percentage loss in Gross World Product due to a doubling of greenhouse gasses. This parameter is linearly related to g, the slope of marginal damages. Our baseline parameters assume that $\pi = 1.33$, an estimate that has been widely used. For comparison, we also discuss results when $\pi = 3.6$ (the average of expert opinions, reported in Nordhaus 1994) and $\pi = 21$ (the maximum of these expert opinions).

[10]We ignore the effect of $\tilde{\theta}$ on the constant term since the constant has no effect on the regulator's control.

Appendix 3 explains our calibration of the abatement costs (rows 3–6 of Table 1). Our companion paper (Karp and Zhang 2006)[11] describes the calibration of the growth and damage functions (rows 1 and 2 of Table 1) and of the equation for the random shock (Eq. 12).

7.2 *Numerical Results*

We begin by summarizing results from earlier static and dynamic models that exclude abatement capital. We then discuss new results—those directly related to abatement capital.

7.2.1 Previous Results

Previous papers study the relation between the policy ranking and parameters in the linear-quadratic model with additive errors (Hoel and Karp 2002; Newell and Pizer 2003; Karp and Zhang 2005). Those papers show that the difference in payoffs under optimal taxes and quotas, $\mathcal{J}^T - \mathcal{J}^Q$, is decreasing in $\frac{g}{b}$. The intuition is the same as in Weitzman (1974)'s static model. A larger value of g means that damages are more convex in S. In view of Jensen's inequality, as damages become more convex it becomes more important to control emissions exactly (as under a quota) rather than to choose only the expected value of emissions (as under a tax). A higher value of b makes it more important for the firm to be able to respond to changes in the cost variable by changing emissions. It is able to respond under a tax but not under a quota.

There is a critical value of $\frac{g}{b}$ above which quotas are preferred. This critical value is decreasing in both β and Δ. When more weight is put on future costs and benefits (higher β), or when the stock is more persistent (higher Δ), it is more important to control the exact level of emissions (as under quotas) rather than the first moment of emissions (as under taxes).

The previous papers calibrate models using parameter values that are consistent with published estimates of the abatement costs and environmental damages associated with greenhouse gasses. These studies find that taxes dominate quotas for the control of greenhouse gasses.

These qualitative results also hold for our parameterization of the model with endogenous abatement capital. This robustness is worth noting, but our analysis adds nothing to the intuition for these results, and therefore we do not discuss them

[11]That paper studies the problem in which the regulator learns about the relation between pollution stocks and environmental damages; there we ignore abatement capital. Since performing this calibration, more recent estimates of climate-related damage have been published (including Stern 2007; Intergovernmental Panel on Climate Change 2007) but these are within the range of estimates in our calibration. For this reason, and in order for the results here to be comparable to those in our earlier paper, we use the same calibration.

further. Instead, we emphasize the comparative statics and dynamics associated with endogenous abatement costs.

7.2.2 The Role of Abatement Capital

There are three important parameters related to abatement capital: δ (capital stock persistence), d (slope of marginal investment cost), and m_1 (marginal effect of capital on BAU emissions) We consider the first two briefly, and then concentrate on the third. In all cases, we perform the obvious experiment of varying one of these parameters, holding all others constant. This experiment has a shortcoming that we discuss later, where we consider a second type of experiment.

We explained why a more durable pollution stock (higher Δ) decreases the preference for taxes. However, a more durable capital stock (higher δ) increases the preference for taxes. Under taxes, the firm responds to a cost shock by changing the level of emissions. For $\rho \neq 0$,[12] the firm responds to a cost shock by changing the level of investment, thereby changing the future level of capital under both taxes and quotas. The adjustment mechanism via capital provides a partial substitute for the inability to change emissions under quotas. A large value of δ means that current investment has long-lasting effects, tending to make capital less flexible. The decreased flexibility associated with larger values of δ increases the value of being able to respond to the cost shock by changing emissions. A larger value of δ therefore increases the advantage of taxes.

A lower value of m_1 (a decrease in the marginal effect of capital on BAU emissions) or a larger value of d (an increase in the adjustment cost for abatement capital), favors quotas. Figure 1 shows the relation between the difference in payoffs (the value of using taxes minus the value of using quotas) and the parameters d and m_1 for three values of π, holding all other parameters constant. (Recall that π is the percentage loss in global world product due to a doubling of greenhouse gasses.) When environmental damages are moderate ($\pi = 1.33$ or $\pi = 3.6$) the difference in payoffs is insensitive to changes in d and m_1; for large environmental damages ($\pi = 21$) the change in either parameter has a noticeable affect on the payoff difference. Previous linear-quadratic models that do not include investment capital are a special case of the model here, obtained by letting $d \to \infty$ or $m_1 \to 0$. Those models tend to understate (slightly) the advantage of using taxes.

As d increases, capital increasingly resembles a fixed input; as m_1 decreases, abatement capital has less effect on the marginal benefit of pollution. A larger value of d or a smaller value of m_1 both imply less flexibility of marginal abatement costs. This diminished flexibility favors quotas, just as does the diminished flexibility in marginal abatement costs associated with a smaller value of b (the slope of B_x).

---

[12] If $\rho = 0$, the current cost shock provides no information about the future cost shocks. Since current investment reduces abatement costs only in future periods, the firm's investment does not depend on the current cost shock if $\rho = 0$.

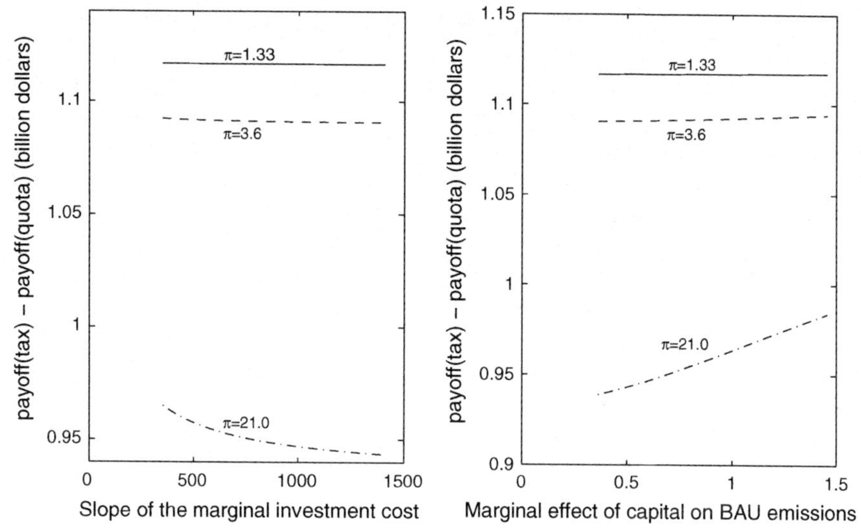

Fig. 1 Dependence of expected payoff difference on cost-related parameters

In all cases, the present discounted value of the payoff difference under taxes and quotas is approximately 1 billion dollars, implying an annualized cost of about 50 million dollars. Our parameterization of abatement costs assumes that the annualized cost of stabilizing emissions is about 1 % of income, or 290 billion dollars. Thus, the payoff difference of the two policies is less than 0.02 of the estimated costs of stabilizing emissions.

The small difference in the expected payoffs may be due largely to the Principle of Certainty Equivalence, mentioned in Sect. 5: the expected stock trajectories are identical under taxes and quotas—only higher moments differ. Uncertainty in our calibrated model (but not in the general formulation) arises only because BAU emissions are uncertain. Given the (small) magnitude of this particular type of uncertainty, the higher moments of stocks simply are not very important. Models that do not satisfy the Principle of Certainty Equivalence find a larger payoff difference under taxes and quotas (Pizer 1999; Hoel and Karp 2001).

The relations between the equilibrium decision rules and levels of the state variables are as expected. The optimal quota (which equals the expected level of emissions under the optimal tax) decreases with the level of pollution and with the capital stock and increases with the lagged cost shock (for $\rho > 0$, as in our calibration). Equilibrium investment is an increasing function of the stock of pollution and a decreasing function of capital stock. Firms understand that a higher pollution stock will lead to lower future equilibrium emissions, increasing the marginal value of investment. A higher aggregate capital stock encourages the regulator to reduce future emissions, increasing the value of investment. However, the representative firm's level of capital equals the aggregate level. For a given quota or tax, a higher capital stock reduces the marginal value of investment. The net effect of higher capital stocks is to reduce investment.

As we mentioned above, the comparative dynamics associated with a change in a single parameter value might be misleading. For example, when we decrease m_1 holding other parameters constant, we change the BAU level of emissions and the abatement costs associated with a particular emissions trajectory, in addition to changing the marginal effect of capital on abatement costs. Here we consider a slightly different experiment: When we vary m_1 we make offsetting changes in m_0 in order to maintain current BAU emissions at 5.2, and we require that the year 2100 BAU emissions are consistent with a particular IPCC scenario.

Our baseline calibration ($m_1 = 0.7266$) makes our model consistent with the *IPCC IS92a* scenario that projects BAU CO_2 stocks of 1500 GtC in the year 2100—an approximate doubling of stocks relative to pre-industrial levels. For comparison we also choose parameters that are consistent with the *IS92c* scenario of a 35 % increase in CO_2 concentration ($m_1 = 0.0416$) and with the *IS92e* scenario of a 170 % increase in CO_2 concentration ($m_1 = 1.6622$).

Figure 2 graphs optimal abatement levels, i.e., the difference in the BAU and the optimal levels of emissions (the left panel) and the difference between BAU and the regulated pollution stock (the right panel), as a function of time. The three graphs in each panel correspond to the three values of m_1. In all cases, abatement increases over time. Both the level and the change over time of abatement is greatest when abatement capital has a large effect on marginal abatement costs (m_1 is large). This result is further evidence that the consideration of endogenous investment in abatement capital increases the optimal level of abatement.

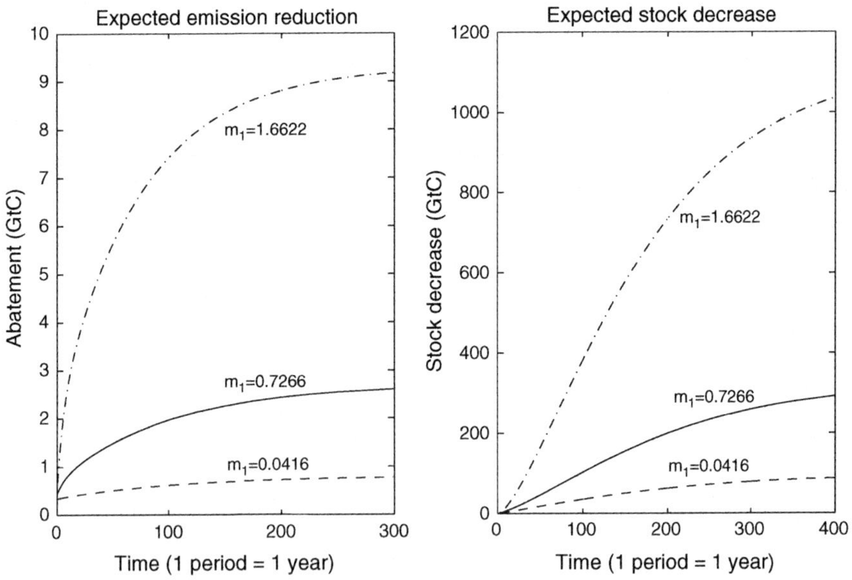

Fig. 2 Changes in expected pollution flows and stocks relative to BAU levels

8 Discussion and Conclusion

The previous literature that compares taxes and quotas assumes that firms solve a sequence of static problems. Our paper recognizes that firms also make investment decisions which affect their future abatement costs. The value of this investment depends on the severity of future environmental restrictions, so the policymaker might have an incentive to announce future environmental policies in order to influence current investment. When this incentive arises, the firms' investment decisions are not constrained optimal, so the regulator would increase welfare if he were able to use an investment tax/subsidy together with the emissions policy. We showed that for general functional forms, when the regulator uses a quota (cap and trade), the competitive firms' investment policy is information-constrained efficient. In contrast, for general functional forms, when the regulator uses an emissions tax, the firms' investment policy is not information-constrained efficient. In this sense, there is an advantage to quotas, relative to emissions taxes, that had not previously been recognized.

This particular advantage disappears under a "separability condition" on the primitive functions. The linear-quadratic model, generalized to include endogenous investment, satisfies this condition. Using a calibrated model and a numerical solution, we found that making capital more durable or more effective in reducing the cost of abatement, or reducing the marginal adjustment cost of capital, all favor the use of taxes rather than quotas. These numerical results and the previously described analytic result lead to a mixed message for the comparison of policies. Within the functional assumptions that most previous studies have used, we find that the inclusion of endogenous investment increases the advantage of taxes. However, for more general functional forms, quotas have an entirely different type of advantage. We do not know anything about the magnitude of the latter advantage; its measurement would require a more complicated (i.e., non-linear quadratic) model, which presents problems of calibration, and it would also require the solution to a dynamic game rather than an optimization problem.

We close by discussing several other views of the relative efficiency of taxes and quotas. One view is that the risk of extreme environmental damages, associated with high GHG stocks, means that over some range damages are likely to be very convex in stocks, i.e., the slope of marginal damages is actually very large. In addition, over a long enough time span, given the opportunities for the development and adoption of new technologies, the marginal abatement cost curve is actually rather flat. Based on these (in our view, plausible) observations, and reasoning from the standard static model, Dietz and Stern (2007) conclude that quantity restrictions are more efficient than taxes for climate policy. We have three reasons for doubting this conclusion. First, the use of the static framework (or the open loop assumption in a dynamic setting) is not appropriate for studying climate policy, because the current policymaker cannot choose policy levels decades into the future. More rapid adjustment of policy, i.e., a decrease in the length of period between policy adjustments, favors the use of taxes. Second, even if the possibility of extreme events makes the marginal damage function much steeper than current estimates suggest, the magnitude of the slope of damages would have to be implausibly large to favor quotas. (Hoel and Karp 2002 demonstrate both of these claims.) Third, the current paper shows that endogenous investment in abatement capital is likely to increase the advantage of taxes, given the linear-quadratic framework.

A second view, which we have heard propounded orally but not in writing, is that the existing models inaccurately describe the abatement problem and are therefore simply inappropriate for comparing policies. The objection is that firms will first exhaust the cheapest abatement opportunities; once these are used, they are unavailable in the future. There are (at least) two ways to respond to this objection. First, a stationary upward sloping marginal abatement cost curve (used in most previous analyses) is obviously consistent with the claim that firms first use the cheapest way of reducing emissions, and then use more expensive means when regulation becomes stricter. However, because abatement is a flow decision, the fact that the cheap abatement opportunities were used early in the program does not mean that they are unavailable later in the program. The firms move up their marginal abatement curves as the policy becomes stricter. A second response interprets the objection as a call to use a model in which abatement is a stock rather than a flow decision—specifically, a model with endogenous investment in abatement capital, in which there is a sequence of increasingly expensive technologies that reduce emissions. It would be fairly straightforward to produce that kind of model, using a slight modification of the model in this paper. We assumed that the cost of investment is a function of gross investment. To address the objection, we could modify the cost function so that the cost of an additional unit of capital increases with the current level of capital. With this formulation, the firms's level of capital is a proxy for it's stage of technology. Because it first adopts the cheapest (most efficient) technologies, it becomes increasingly expensive to make further reductions in abatement costs. It is not clear how this change affects the policy ranking.

There are several other model variations that would address other interesting questions. For example, network externalities may cause the productivity of a firm's capital to increase with the level of aggregate capital. Also, if we think of investment as being R&D rather than the installation of new capital, there are likely to be important spillovers. At least in the linear-quadratic framework, it would be straightforward to include such spillovers. There may be intra-firm increasing returns to scale. There might also be learning by doing, so that an increase in cumulative abatement decreases abatement costs. The inclusion of intertemporal trade (banking and borrowing) under quantity restrictions would be even more interesting. Because GHGs are a stock pollutant, the stream of damages can be sensitive to the cumulative emissions over a long period of time without being sensitive to the precise timing of emissions. Intertemporal trading allows firms to optimally allocate over time a given cumulative level of emissions. The introduction of banking and borrowing (under the quantity restriction) would likely significantly erode the advantage of taxes. The effect of banking and borrowing on the incentive to invest is not clear. These questions, and the model variations that they entail, are the subject of current research.

Appendix

The appendix consists of thee parts. Part 1 contains the proof of Proposition 1. Part 2 provides the formulae for Γ used in Eq. (29). Part 3 contains information calibration information. An additional appendix, available on request, contains additional information on calibration.

Proof of Proposition 1

We use $\mathbb{J}^j(\cdot)$ $(j = T, Q)$ to denote the regulator's value function in the dynamic game (where the regulator chooses only an emissions policy), and $\mathcal{J}^j(\cdot)$ $(j = T, Q)$ to denote the regulator's value function in the corresponding auxiliary problem (where the regulator chooses an emissions policy and then chooses investment after observing the current cost variable). We want to find conditions under which the equilibrium capital and pollution stocks are identical in the Markov Perfect equilibrium to the game and in the auxiliary problem. Equivalently, we want to find conditions under which the optimal investment tax/subsidy is identically 0 in the auxiliary problem.

(i) *Quotas.* When the regulator uses an emissions quota, the Euler equations for investment in the Markov perfect equilibrium (Eq. 18) and investment in the auxiliary problem (Eq. 21) are identical, as are the corresponding transversality conditions. We need to confirm that the Euler equations for the pollution stock are also identical in the two settings.
In the Markov Perfect equilibrium with quotas the regulator solves the following DPE:

$$\mathbb{J}^Q\left(K_{t-1}, S_{t-1}, \theta_{t-1}\right) = \max_{x_t} E_{\theta_t|\theta_{t-1}}\{B\left(K_{t-1}, \theta_t, x_t\right) - D\left(S_{t-1}\right) \\ -C\left(I_t^Q, K_{t-1}\right) + \beta\mathbb{J}^Q\left(\delta K_{t-1} + I_t^Q, \Delta S_{t-1} + x_t, \theta_t\right)\},$$

subject to the private investment rule $I_t^Q \equiv I^Q\left(K_{t-1}, \theta_t, S_{t-1}\right)$, which is independent of the current quota level x_t. The stochastic Euler equation for pollution stock is:

$$E_{\theta_t|\theta_{t-1}} B_x\left(K_{t-1}, \theta_t, x_t\right) - \beta D'\left(\Delta S_{t-1} + x_t\right) - \beta\Delta E_{\theta_{t+1}|\theta_{t-1}} \\ B_x\left(K_t, \theta_{t+1}, x_{t+1}\right) = 0.$$

The transversality condition is

$$\lim_{T\to\infty} E_{\theta_T|\theta_{t-1}}\left\{\beta^{T-t} B_x\left(K_{T-1}, \theta_T, x_T\right) S_T\right\} = 0.$$

A straightforward calculation confirms that the corresponding Euler equation and transversality condition in the auxiliary problem are identical to the last two equations. (To obtain the Euler equation in the auxiliary problem we differentiate the DPE (19) with respect to S_{t-1}, using the envelope theorem; we combine the resulting equation with the first-order condition Eq. (20).)

(ii) *Taxes* We first consider the equations that determine the evolution of capital stock. Inspection of the Euler equations for capital (Eq. (16) in the Markov Perfect equilibrium and Eq. (25) in the auxiliary problem) establishes that these are identical if and only if the function τ, defined as

$$\tau_t \equiv \beta E_{\theta_{t+1}|\theta_t} \left\{ H_{t+1} \frac{\partial \chi_{t+1}}{\partial K_t} \right\},$$

is identically 0. We therefore find necessary and sufficient conditions for $\tau_t \equiv 0$. Note that the assumptions that $B_{xK} < 0$ and $B_{KK} < 0$ imply that $\frac{\partial \chi_{t+1}}{\partial K_t} \neq 0$.

By Lemma 1, the separability condition is equivalent to

Condition 2 *(a)* $\frac{\partial \chi(K_{t-1},\theta_t,p_t)}{\partial p_t}$ *is independent of* θ_t. *(b)* $\frac{\partial \chi(K_{t-1},\theta_t,p_t)}{\partial K_{t-1}}$ *is independent of* θ_t.

We, therefore, need only show that Condition 2 is necessary and sufficient for $\tau_t \equiv 0$. We first consider sufficiency. If Condition (2a) holds, the first-order condition (24) implies

$$E_{\theta_t|\theta_{t-1}} \{H_t\} = 0, \qquad \forall t. \tag{30}$$

If Condition (2b) also holds, we can write τ_t as

$$\tau_t \equiv \beta \left(\frac{\partial \chi_{t+1}}{\partial K_t} \right) E_{\theta_{t+1}|\theta_t} \{H_{t+1}\}.$$

Using Eq. (30), the last equality implies that $\tau_t \equiv 0$. Clearly the transversality conditions in the two problems are the same.

The necessity of the separability condition follows from the previous argument. If either part of Condition 2 does not hold the function τ is not identically 0. (Of course the equality $\tau = 0$ might hold for some values of the information state, but we need the stronger condition that the equality hold identically, i.e., for all possible values of the information state.)

To complete the proof, we need only check that the Euler equations and transversality conditions for the pollution stock are also the same in the two problems. In the Markov Perfect equilibrium with taxes, the regulator solves the following DPE:

$$\begin{aligned} \mathbb{J}^T (K_{t-1}, S_{t-1}, \theta_{t-1}) = \max\nolimits_{p_t} E_{\theta_t|\theta_{t-1}} \{ & B (K_{t-1}, \theta_t, \chi_t) - D (S_{t-1}) - C (I_t^T, K_{t-1}) \\ & + \beta \mathbb{J}^T (\delta K_{t-1} + I_t^T, \Delta S_{t-1} + \chi_t, \theta_t) \}, \end{aligned} \tag{31}$$

subject to emissions χ_t given by Eq. (15), and the private investment rule $I_t^T \equiv I^T(K_{t-1}, \theta_t, S_{t-1})$. I_t^T is independent of the current tax level p_t as discussed in Sect. 4; $\frac{\partial \chi_t}{\partial p_t}$ is independent of θ_t because of Condition 1. Thus the first-order condition for the optimal tax is

$$E_{\theta_t|\theta_{t-1}}\left\{B_x\left[K_{t-1}, \theta_t, \chi\left(K_{t-1}, \theta_t, p_t\right)\right] + \beta \mathbb{J}_S^T\left[K_t, \Delta S_{t-1} + \chi\left(K_{t-1}, \theta_t, p_t\right), \theta_t\right]\right\} = 0. \quad (32)$$

Differentiating the DPE (31) with respect to S_{t-1}, using the envelope theorem, and combining the resulting equation with the first-order condition (32) gives the stochastic Euler equation for the pollution stock in the dynamic game:

$$\begin{aligned} &E_{\theta_t|\theta_{t-1}}\left\{B_x\left[K_{t-1}, \theta_t, \chi\left(K_{t-1}, \theta_t, p_t\right)\right] - \beta D'\left[\Delta S_{t-1} + \chi\left(K_{t-1}, \theta_t, p_t\right)\right]\right\} \\ &- \beta \Delta E_{\theta_{t+1}|\theta_{t-1}} B_x\left[K_t, \theta_{t+1}, \chi\left(K_t, \theta_{t+1}, p_{t+1}\right)\right] = 0. \end{aligned} \quad (33)$$

The transversality condition is

$$\lim_{T\to\infty} E_{\theta_T|\theta_{t-1}}\left\{\beta^{T-t} B_x\left[K_{T-1}, \theta_T, \chi\left(K_{T-1}, \theta_T, p_T\right)\right] S_T\right\} = 0.$$

Again, it is straightforward to obtain the Euler equation for pollution stocks in the auxiliary problem. We differentiate Eq. (23) with respect to S_{t-1}, using the envelope theorem. Combining the resulting equation with the first-order condition (30) leads to the stochastic Euler equation for the pollution stock in the auxiliary problem. This equation is identical to Eq. (33). The transversality conditions are also the same. □

Formulae for Γ

The function Γ used in Eq. (29) is

$$\Gamma = \frac{\beta^2\rho^2\phi^2 \frac{(d-\beta h)}{b\left(1+\frac{\beta g}{b}\right)^2 (d-\beta h - d\beta\rho)^2} + \frac{\beta\rho^2}{1+\frac{\beta g}{b}}}{1-\beta\rho^2} > 0$$

with

$$h = \frac{-\Xi - \sqrt{\Xi^2 + 4\beta d\left(f_2 - \frac{\phi^2}{b+\beta g}\right)}}{2\beta} < 0$$

$$\Xi \equiv \left(f_2 - \frac{\phi^2}{b+\beta g}\right)\beta - d\left(1 - \beta\delta^2\right)$$

Calibration of Abatement Costs and the Shock

We assume that abatement capital depreciates at an annual rate of 16.25 %, the mean of capital stock depreciation rates in 14 OECD countries (Cummins et al. 1996). This depreciation rate implies that $\delta = 0.85$.

A higher unit of abatement capital decreases the BAU emissions by m_1 units. When $m_1 = 0$, BAU emissions are constant, and abatement capital has no effect on the marginal benefit of pollution (i.e., on marginal abatement costs). In this special case, the firm's emission decision and investment decision are decoupled, and the firm's capital stock has no effect on the regulator's optimal policy. The restriction $m_1 = 0$ therefore reproduces the linear-quadratic models of global warming in Karp and Zhang (2006).

The dependence of adjustment costs on gross rather than net investment leads to a simple method of calibration. In the absence of additional regulation, i.e., under Business as Usual-firms never invest: $I_t^b = 0, \forall t \geq 0$. If the initial level of abatement capital is positive, the level monotonically decreases over time, so BAU emissions monotonically increase:

$$K_t^b = \delta^{t+1} K_{-1}, \quad x_t^b = m_0 - m_1 K_{t-1}^b + \tilde{\theta}_t = m_0 - m_1 \delta^t K_{-1} + \tilde{\theta}_t,$$

where $K_{-1} > 0$ is the abatement capital at the beginning of the initial period ($t = 0$). Our assumptions provide a simple way to include endogenous investment, and also to reproduce the stylized fact that BAU emissions will increase. The model is "incomplete", since it does not explain why $K_{-1} > 0$. The expected future BAU atmospheric CO_2 stock is:

$$S_t = \Delta^{t+1} S_{-1} - m_1 K_{-1} \frac{\delta^t \left[1 - \left(\frac{\Delta}{\delta} \right)^{t+1} \right]}{1 - \frac{\Delta}{\delta}} + \left[m_0 + (1 - \Delta) \bar{S} \right] \frac{1 - \Delta^{t+1}}{1 - \Delta}, \tag{34}$$

where S_{-1} is the pollutant stock at the beginning of the initial period.

The current anthropogenic fluxes of CO_2 into the atmosphere is 5.2 GtC[13] so we set $Ex_0^b = m_0 - m_1 K_{-1} = 5.2$ to obtain one calibration equation. The *IPCC IS92a* scenario projects BAU CO_2 stocks at 1500 GtC in 2100 (Intergovernmental Panel on Climate Change 1996), page 23. This estimate, Eq. (34), and the estimate of current atmospheric CO_2 concentration at $S_{-1} = 781$GtC (Keeling and Whorf 1999), gives a second calibration equation. The two equations imply

[13]We use "current" to mean the year 2000. The current total anthropogenic CO_2 emissions are about 8.12 GtC, which equals the sum of 6.518 GtC of global CO_2 emissions from fossil fuel combustion and cement production (Marland et al. 1999) and 1.6 GtC annual average net CO_2 emissions from changes in tropical land-use (Intergovernmental Panel on Climate Change 1996). We obtain the current anthropogenic fluxes of CO_2 into the atmosphere 5.20 GtC by multiplying the total anthropogenic emissions by 0.64, the marginal atmospheric retention ratio.

$$m_0 = 12.466, \qquad m_1 K_{-1} = 7.2661.$$

We do not have data on abatement capital, so we choose an arbitrary value for K_{-1}.[14] We set $K_{-1} = 10$.

We choose the baseline values of d (the slope of the marginal investment cost) and b (the slope of the marginal abatement cost) to satisfy a scenario in which firms are required to maintain emissions at the current level in each period. Firms begin with the initial abatement capital and solve an infinite horizon investment problem to minimize the present discounted sum of investment and abatement cost under emission stabilization. In order to determine the two unknown parameters, we assume:

- The annualized discounted present value of firms' total (abatement-related) costs is about 1 % of 1998 GWP (Manne and Richels 1992).[15]
- In the steady state the ratio of investment costs to total abatement costs is about 0.5 (Vogan 1991).

These two assumptions lead to the baseline parameter values: $d = 703.31$, and $b = 26.992$.

Calibration Material Not Intended for Publication

Row 1 in Table 1 is pollutant stock growth equation. We measure S_t, the CO_2 atmospheric concentration, in billions of tons of carbon equivalent (GtC). $\bar{S}$ equals 590 GtC, the preindustrial CO_2 concentration (Neftel et al. 1999). Let e_t be total anthropogenic CO_2 emissions in period t. The proportion of emissions contributing to the atmospheric stock is estimated at 0.64 (Goulder and Mathai 2000; Nordhaus 1994b). This fraction accounts for oceanic uptake, other terrestrial sinks, and the carbon cycle (Intergovernmental Panel on Climate Change 1996). The linear approximation of the evolution of the atmospheric pollutant stock is

$$S_t - 590 = \Delta\left(S_{t-1} - 590\right) + 0.64 e_t.$$

This equation states that 64 % of current emissions contribute to atmospheric CO_2, and that CO_2 stocks in excess of the preindustrial level decays naturally at an annual rate of $1 - \Delta$. We take $x_t \equiv 0.64 e_t$, the anthropogenic fluxes of CO_2 into the atmosphere, as the control variable. The stock persistence is $\Delta = 0.9917$ (an annual decay rate of 0.0083 and a half-life of 83 years) (Goulder and Mathai 2000; Nordhaus 1994b).

[14]Even for pollution problems that have been studied in more detail, data on abatement capital is difficult or impossible to obtain. For example, Becker and Henderson (1999) note the absence of estimates of abatement capital stocks associated with U.S. air quality regulation.

[15]Manne and Richels (1992) estimate that the total global costs of stabilizing CO_2 emissions at the 1990 level are about 4,560 billions of 1990 US dollars, or 20.25 % of the 1990 GWP. We take the same percentage loss and use the annuanlized value $(1 - \beta) \times 20.25\,\% = 1\,\%$.

We assume that the preindustrial CO_2 concentration has zero environmental damage. Damages from higher CO_2 concentration are $\frac{g}{2}\left(S-\bar{S}\right)^2$. (Row 2 in Table 1). For ease of interpreting the numerical values, we use π to denote the percentage loss in GWP (Gross World Product) from a doubling of the preindustrial CO_2 concentration. With the 1998 GWP of 29,185 billion dollars (International Monetary Fund, 1999) we have

$$\pi\% \cdot 29185 = g/2 \cdot 590^2 \implies g = 0.0017\pi.$$

For example, $\pi = 1.33$ which is widely used corresponds to $g = 0.0022$. For the sensitivity analysis we consider two other damage parameters, $\pi = 3.6$ and $\pi = 21.0$, the mean and the maximum of expert opinions.

Using maximum likelihood, we fit the following data generating process for global carbon emissions over the 50-year period 1947–1996 from Marland et al. (1999).

$$e_t = e_0 + nt + \varepsilon_t, \quad \varepsilon_t = \rho\varepsilon_{t-1} + \nu_t, \quad \nu_t \sim iid\, N\left(0, \sigma_\upsilon^2\right).$$

The estimates are $\rho = 0.9$ and $\sigma_\upsilon = 0.1$ GtC. We convert the emission uncertainty σ_υ into cost uncertainty σ_μ by multiplying it by 0.64 (because $x_t \equiv 0.64e_t$), and then by the slope of marginal abatement cost $b = 26.992$ (because $\theta_t \equiv b\tilde{\theta}_t$). The result is $\sigma_\mu = 0.1 \times 0.64 \times 26.992 = 1.7275\$/(\text{ton of carbon})$.

References

Abrego, L., & Perroni, C. (2002). Investment subsidies and time-consistent environmental policy. *Oxford Economic Papers*, *54*, 617–635.

Asheim, G., Mitra, T., & Tungodden, B. (2016). Sustainable recursive social welfare functions. *Economic Theory*. reprinted in this volume.

Becker, R., & Henderson, J. (1999). Costs of air quality regulation. Working paper 7308, National Bureau of Economic Research.

Biglaiser, G., Horowitz, J., & Quiggin, J. (1995). Dynamic pollution regulation. *Journal of Regulatory Economics*, *8*, 33–44.

Buonanno, P., Carraro, C., & Galeotti, M. (2001). Endogenous induced technical change and the costs of Kyoto. Fondazione ENI Enrico Mattei Nota di Lavoro, No 64-01.

Burniaux, J.-M., & Martins, J. (2016). Carbon leakages: A general equilibrium view. *Economic Theory*. reprinted in this volume.

Chichilnisky, G. (2016). Sustainable markets with short sales. *Economic Theory*. reprinted in this volume.

Chichilnisky, G., & Heal, G. (1995). Markets for tradeable C02 eission quotas: Principles and practice. OECD. http://mpra.ub.uni-muenchen.de/8486/1/MPRA_paper_8486.pdf.

Chipman, J., & Tian, G. (2016). Detrimental externalities, pollution rights, and the "Coase Theorem". *Economic Theory*. reprinted in this volume.

Cummins, J., Hassett, K., & Hubbard, R. (1996). Tax reforms and investment: A cross-country comparison. *Journal of Public Economics*, *62*(1–2), 237–273.

Dietz, S., & Stern, N. (2007). Why economic analysis supports strong action on climate change. *Review of Environmental Economics and Policy*, *2*, 94–113.

Dutta, P., & Radner, R. (2016). Capital growth in a global warming model: Will China and India sign a climate treaty?. *Economic Theory*. reprinted in this volume.

Ellerman, D. (2010). The EU emissions trading scheme: A prototype global system? In J. Aldy & R. Stavins (Eds.), *Post-Kyoto international climate policy* (pp. 88–118). Cambridge: Cambridge University Press.

Figuieres, C., & Tidball, M. (2016). Sustainable exploitation of a natural resource: A satisfying use of Chichilnisky's criterion. *Economic Theory*. reprinted in this volume.

Fischer, C., Parry, I., & Pizer, W. (2003). Instrument choice for environmental protection when technological innovation is endogenous. *Journal of Environmental Economics and Management, 45*, 523–545.

Gersbach, H., & Glazer, A. (1999). Markets and regulatory hold-up problems. *Journal of Environmental Economics and Management, 37*, 151–164.

Goulder, L., & Schneider, S. (1999). Induced technological change and the attractiveness of CO_2 abatement policies. *Resource and Energy Economics, 21*, 211–253.

Goulder, L. H., & Mathai, K. (2000). Optimal CO_2 abatement in the presence of induced technological. *Journal of Environmental Economics and Management, 39*(1), 1–38.

Hoel, M., & Karp, L. (2001). Taxes and quotas for a stock pollutant with multiplicative uncertainty. *Journal of Public Economics, 82*, 91–114.

Hoel, M., & Karp, L. (2002). Taxes versus quotas for a stock pollutant. *Resource and Energy Economics, 24*, 367–384.

Intergovernmental Panel on Climate Change. (1996). In J. T. Houghton, L. G. Meira Filho, B. A. Callander, N. Harris, A. Kattenberg & K. Maskell (Eds.) *Climate change 1995: The science of climate change*. Cambridge, UK: IPCC

Intergovernmental Panel on Climate Change. (2007). Climate Change 2007. http://www.ipcc.ch/SPM040507.pdf.

International Monetary Fund. (1999). *World Economic Outlook* International Monetary Fund (IMF), Washington, DC.

Jaffe, A., Newell, R., & Stavins, R. (2003). Technological change and the environment. In K.-G. Maler & J. Vincent (Eds.), *Handbook of environmental economics* (Vol. 1, pp. 461–516). Amstredam: North-Holland.

Judd, K. (1998). *Numerical methods in economics*. Cambridge, Massachusetts: MIT Press.

Karp, L. (2008). Correct (and misleading) arguments for using market-based pollution control technologies. http://are.berkeley.edu/~karp/marketbasedJuly08.pdf.

Karp, L., & Lee, I. H. (2003). Time consistent policies. *Journal of Economic Theory, 112*(2), 353–364.

Karp, L., & Zhang, J. (2005). Regulation of a stock pollution with correlated abatement cost. *Environmental and Resource Economics, 32*, 273–300.

Karp, L., & Zhang, J. (2006). Regulation with anticipated learning about environmental damage. *Journal of Environmental Economics and Management, 51*, 259–280.

Keeling, C., & Whorf, T. (1999). Atmospheric CO_2 records from sites in the SIO air sampling network. In *Trends: A compendium of data on global change*. U.S and Oak Ridge, Tennessee: Carbon Dioxide Information Analysis Center, Oak Ridge National Laboratory and Department of Energy.

Kennedy, P., & Laplante, B. (1999). Environmental policy and time consistency: emission taxes and emission trading. In E. Petrakis, E. Sartzetakis, & A. Xepapadeas (Eds.), *Environmental regulation and market power* (pp. 116–144). Cheltenham: Edward Elgar.

Lauwers, L. (2016). Intergenerational equity, efficiency and constructibility. *Economic Theory*. reprinted in this volume.

Lecocq, F., & Hourcade, J.-C. (2016). Unspoken ethical issues in the climate affair: Insights from a theoretical analysis of negotiation mandates. *Economic Theory*. reprinted in this volume.

Malueg, D. (1989). Emission credit trading and the incentive to adopt new pollution abatement technology. *Journal of Environmental Economics and Management, 16*, 52–57.

Manne, A., & Richels, R. (1992). *Buying greenhouse gas insurance: The economic costs of carbon dioxide emission limits*. Cambridge, Massachusetts: MIT Press.
Marcet, A., & Marimon, R. (1998). Recursive contracts. Working Paper ECO No. 98/37, European University Institute, Florence, Badia Fiesolana, Economics Department.
Marland, G., Boden, T., Andres, R., Brenkert, A., & Johnston, C. (1999). *Global, regional, and national* CO_2 *Emissions in trends: A compendium of data on global change. U.S and Oak Ridge, Tennessee: Carbon Dioxide Information Analysis Center, Oak Ridge National Laboratory and Department of Energy.
Marsiliani, L., & Renstrom, T. (2002). Time inconsistency in environmental policy: Tax earmarking as a commitement solution. *Economic Journal, 110*, C123–C128.
Milliman, S., & Prince, R. (1989). Firm incentives to promote technologicalchange in pollution control. *Journal of Environmental Economics and Management, 17*, 247–265.
Miranda, M. J., & Fackler, P. L. (2002). *Applied computational economics andfinance*. Cambridge, Massachusetts: MIT Press.
Moledina, A., Polasky, S., Coggins, J., & Costello, C. (2003). Dynamic environmental policy with strategic firms: Prices vs. quantities. *Journal of Environmental Economics and Management, 45*.
Montero, J. (2002). Permits, standards, and technology innovation. *Journal of Environmental Economics and Management, 44*.
Neftel, A., Friedli, H., Moor, E., Lötscher, H., Oeschger, H., Siegenthaler, U., et al. (1999). Historical CO_2 *R*ecord from the Siple Station Ice Core. In *Trends: A compendium of data on global change*. U.S and Oak Ridge, Tennessee: Carbon Dioxide Information Analysis Center, Oak Ridge National Laboratory and Department of Energy.
Newell, R., & Pizer, W. (2003). Regulating stock externalitiues under uncertainty. *Journal of Environmental Economics and Management, 45*, 416–432.
Nordhaus, W. (1994a). *Managing the global commons: The economics of the greenhouse effect*. Cambridge: MIT Press.
Nordhaus, W. D. (1994b). Expert opinion on climate change. *American Scientist*, 45–51.
Norhaus, W. (1999). Modeling induced innovation in climate-change policy. *Presented at the IIASA Workshop on Induced Technological Change and the Environment*, Luxenburg.
Ostrom, E. (2016). Nested externalities and polycnetric institutions: Must we wait for global solutions to climate change before taking actions at other scales? *Economic Theory*. reprinted in this volume.
Pizer, W. (1999). The optimal choice of climate change policy in the presence of uncertainty. *Resource and Energy Economics, 21*(3–4), 255–287.
Pizer, W. (2002). Combining price and quantity controls to mitigate global climate change. *Journal of Public Economics, 85*, 409–434.
Requate, T. (1998). Incentives to innovate under emissions taxes and tradeable permits. *European Journal of Political Economy, 14*, 139–165.
Requate, T. (2005). Dynamic incentives by environmental policy instruments. *Ecological Economics, 54*, 175–195.
Requate, T., & Unold, W. (2003). Environmental policy incentives to adopt advanced abatement technology: Will the true ranking please stand up? *European Economic Review, 47*, 125–146.
Rezai, A., Foley, D., & Taylor, L. (2016). Global warming and economic externalities. *Economic Theory*. reprinted in this volume.
Sargent, T. J. (1987). *Macroeconomic theory* (2nd edn.). San Diego, California: Academic Press, Inc.
Stern, N. (2007). *The economics of climate change*. Cambridge University Press.
Tarui, N., & Polasky, S. (2005). Environmental regulation with technology adoption, learning and strategic behavior. *Journal of Environmental Economics and Management, 50*, 447–467.
Tarui, N., & Polasky, S. (2006). Environmental regulation in a dynamic model with uncertainty and investment. *Presented at NBER Conference*.
Vogan, C. (1991). Pollution abatement and control expenditures. *Survey of Current Business*, 48–67.

Weitzman, M. (1974). Prices versus quantities. *Review of Economic Studies, 41*(4), 477–491.
Xie, D. (1997). On time consistency: A technical issue in Stackelberg differential games. *Journal of Economic Theory, 76*, 412–430.

Walrasian Prices in Markets with Tradable Rights

Carlos Hervés-Beloso, Francisco Martínez and Jorge Rivera

1 Introduction

Tradable-licence systems are the focus of current interest in market-based natural resources or environmental policies. For example, a system of licences is interesting as it could provide a mean to achieve decentralized solutions to set restrictions on fishing for certain fish species or in order to organize a market of emission licences or pollution rights. For general references see Ellerman et al. (2008), Joskow et al. (1998) and Newell et al. (2005).

A licence confers the agents holding it, the right to consume. In the examples above, the right to capture a certain amount of a protected species of fish or to emit pollutants at a certain rate. However, it is not always desirable to allow such rights to be transferred on a one-to-one basis. In a market system these licences should be

This work was supported by the Spanish Ministerio de Ciencia e Innovación under project ECO2009-14457-C04-01, Xunta de Galicia, FEDER under project 10PXIB300141PR, Instituto Milenio Sistemas Complejos de Ingeniería and Fondecyt # 1110716.

The authors are grateful to J. Correia da Silva, E. Moreno-García, P. Montero and J. P. Torres-Martínez for their useful comments and suggestions.

Reprinted with kind permission from the Authors: Originally published in Economic Theory, Volume 49, Number 2, February 2012.

C. Hervés-Beloso (✉)
RGEA, Facultad de Económicas, Universidad de Vigo, Vigo, Galicia, Spain
e-mail: cherves@uvigo.es

F. Martínez
Departamento de Ingeniería Civil, FCFM, Universidad de Chile, Santiago, Chile
e-mail: fmartine@ing.uchile.cl

J. Rivera
Departamento de Economía, FEN, Universidad de Chile, Santiago, Chile
e-mail: jrivera@econ.uchile.cl

G. Chichilnisky and A. Rezai (eds.), *The Economics of the Global Environment*,
Studies in Economic Theory 29, DOI 10.1007/978-3-319-31943-8_23

tradable and the desirable rule governing exchange of licences or rights should be based on a market-price system.

Models of "Cap System" with tradable licences have been analyzed by several authors during the last forty years, (see Baumol and Oates 1988; Ellerman and Joskow 2008 and Montero 2001 as general references). However, the literature on models focusing on pricing rights in a purely competitive basis, is scarce. In this issue (Burniaux and Martins 2016) analyze the consequences of imposing an unilateral carbon emission constraint defined exogenously in a general equilibrium model with two countries and (Chichilnisky 2016) studies the existence of equilibrium prices in a sustainable market.

A precise formulation of an emission licence model with a competitive basis appears in the seminal paper by Montgomery (1972). In a scenario where an exchange of such licences between polluters at different locations is considered, Montgomery shows that a market equilibrium in emission licences exists and that, with some restrictions on the initial allocation of licences, the market equilibrium is efficient. Later, Boyd and Conley (1997) were the first to directly treat the efficiency problem in presence of externalities opposed to an indirect way through *Arrovian* commodities, arguing that essential non-convexities highlighted by Starrett (1972) are due to unboundedness of the negative effects of an externality, rather than the externality itself.

Conley and Smith (2005) extended the Boyd and Conley model to allow firms to benefit from public goods and be damaged by externalities, proving the existence of a competitive equilibrium and stating a first welfare theorem. Their main result could be viewed as a type of general equilibrium Coase theorem. More recently, the paper by Mandel (2009), focuses on the influence on the general equilibrium of an economy of the opening of a licences market. Assuming there existed an equilibrium before the opening of allowances market, the paper describes the changes in the firms' behavior which guarantee that an equilibrium can be reached in the enlarged economy.

The models considered by Boyd and Conley, Conley and Smith, or Mandel imply to re-consider the pollutants as crucial consumption goods as well as key input factors for production, which drive them to the necessity of re-defining the individual preferences and production sets in order to take into account these new factors in their formulations. The problem we see in this approach is that the equilibrium solution critically depends on the assumptions on the set of properties that define the preferences (and production sets) of new goods and thus, the result is specific for those assumptions. How to model changes in preferences (and production sets) in the presence of new goods in the market is certainly an open question, for which we do not have a satisfactory answer. Complementarily, in Chipman and Guoqiang (2016) is considered the presence of tradable pollution right in the economy. However, as a crucial difference with our work, authors assume that agents' preferences depend explicitly on the pollution right.

In this paper, we consider a scenario in which limits to the consumption of certain commodities have been established exogenously and that the consumption of these commodities requires the availability of certain amount of rights or licences for its consumption. The scenario may reflect a situation where, due to binding international agreements, limits to excessive consumption of certain raw materials, or limits to the capture of protected species have been established in order to restrict the potential negative effects produced by their consumption. These negative effects may be, for instance, greenhouse effects, different types of environmental pollution, or the risk of extinction of a fish species.

Our aim is to set a simple model of an economy in order to show the existence of an equilibrium price system linking tradable licence prices with commodity prices and to highlight the immediate consequences on equilibrium prices when limits to consumption are set.

For it, we consider an exchange economy with externalities (the individual's preference depends on private consumption goods chosen by this individual and on the entire consumption plan chosen by other agents in the economy). The enforcement of licences for the consumption is exogenous; the amount of such licences is defined by an exogenous mapping that associates pollutants with consumption plans. In our model licences do not participate directly in preferences. However, the requirement of licences for the consumption of specific commodities leads to the existence of a licences market and consequently, licences become tradable modifying the budgetary constraints of agents.

The restrictions of the model primarily affect the agents' consumption sets. Agents may not consume certain quantities of specific commodities even when these form part of their endowments. Secondly, it may affect the agents' budget sets, since in order to consume they will need to have the required rights. If an agent does not have those licences, she may buy them investing part of her income coming from her endowments, or on the contrary, if she has any licences left over, she could sell them to get an additional income.

It is also assumed that the estimated negative effects, and consequently the licences required for the consumption of specific commodities, could depend not only on the quantity of those commodities but also on the entire consumption plan selected by the consumer. Our objective here is to reflect the situation in which a consumption plan entailing high technology, may involve less adverse effects, and consequently require fewer consumption licences than another less technological consumption plan.

This model assumes that each agent is endowed with a certain amount of each type of the required licences for consumption and also assumes that licences are perfectly divisible and tradable. The agent's choice of a specific consumption plan requires that she has the inherent licence for that consumption.

Our approach differs from other previous works in several aspects. Firstly, we do not explicitly consider production. In our model, agents evaluate their utility considering all the consequences involved in their consumption plan. Thus, our model is a pure exchange market in which the consumption rights or licences are traded at the same time as the commodities, that is, licences must be required at the same time that contracts for raw materials are signed, no matter the raw materials purpose. Therefore, and more importantly, we do not require to measure the actual negative effects of consumption. Instead, we suppose the existence of an external mapping which evaluates the potential negative effects derived from each contract, by mapping every consumption plan (or contract) into a theoretical amount of licences of each type. Secondly, we do not need to introduce any other *type of good* in agents preferences and neither in the production sector, which avoid us from justifying how preferences and/or production sets could be distorted by the introduction of these new goods in the market.

Due to the presence of externalities in consumption (as we setup the model in Sect. 2), we introduce the concept of *Nash–Walras equilibrium* as a competitive outcome in our framework. This concept coincides with the standard Walras notion if we were not to consider externalities.

In Sect. 3 we prove a Walras' Law for our equilibrium concept. The main result of this paper is Theorem 1 in Sect. 4, which establishes the existence of a Nash-Walras equilibrium under general conditions on the fundamentals of the economy. Finally, Sect. 5 is devoted to the conclusion remarks and further developments.

2 The Model

Following the standard Arrow-Debreu model, let us consider an economy with $m \in \mathbb{N}$ consumers and $\ell \in \mathbb{N}$ different consumption goods; the consumption set of consumer $i \in I = \{1, 2, \ldots, m\}$ is denoted by $X_i \subseteq \mathbb{R}^\ell$ and each consumer i is endowed with consumption goods denoted by $\omega_i \in X_i$. We set $\omega = \sum_{i \in I} \omega_i$, $X = \prod_{i \in I} X_i$ and given $i \in I$, we define

$$X_{-i} = \prod_{j \in I \setminus \{i\}} X_j.$$

In order to incorporate externalities in consumption, preferences of an individual $i \in I$ will be represented by a utility function

$$u_i : X_{-i} \times X_i \to \mathbb{R}.$$

We assume that limits to the consumption of certain commodities have been established exogenously due to binding international agreements established, where consumption of these commodities requires the availability of certain licences. After an exogenous *Cap-setting Process*, limits to consumption are given by the mapping

$$f : \mathbb{R}^\ell_+ \to \mathbb{R}^k_+,$$

which defines the amount of the each type of $k \in \mathbb{N}$ negative effects that could produce the consumption of the allocation $x \in \mathbb{R}^\ell_+$.

For $j \in K = \{1, \ldots, k\}$, the *Cap-setting Process* sets a limit $R_j \in \mathbb{R}_{++}$ on the total allocation of the economy; we set

$$R = (R_j) \in \mathbb{R}^k_{++}.$$

In our model, the Cap-setting process mentioned above implies that for each $j \in K$, any consumption plan $x_i \in X_i$, $i \in I$, should comply with

$$\sum_{i \in I} f_j(x_i) \leq R_j,$$

where f_j denotes the $j \in K$ component of f that defines the caps to consumption that have been exogenously established. Observe that $f_j(x_i)$ could be the amount of commodity j representing a certain raw material for which a cap has been established in order to restrict the potential negative effects that this consumption will produce. However, here we are considering a more general setting; in this model, each one of the potential negative effects and, consequently each cap, is measured globally in the sense that it depends not only on the amount of a given commodity but on the global consumption plan of the individuals. Proceeding in this way, we have in mind, for example, that a more technological consumption plan may produce less negative effects than a technologically poorer alternative.

On the other hand, we assume that for each $j \in K$ there is a type of *licence* and that each individual $i \in I$ is endowed with an amount of each of them. Formally, each agent $i \in I$ is endowed with a vector

$$r_i = (r_i^j) \in \mathbb{R}_+^k$$

in such a way that

$$\sum_{i \in I} r_i^j = R_j,\ j \in K.$$

If agent $i \in I$ decides to consume $x \in X_i$ then she must have an *amount* $f(x) \in \mathbb{R}_+^k$ of each consumption right (licence). One key assumption in our model is that consumption rights can be traded in the market and that they do not participate in the individual's preferences. The fact that licences can be traded in the market implies that any individual may exchange them with consequences on the size of her budgetary set; similar to prices of consumption goods, prices for licences will be determined endogenously as part of the equilibrium.

Thus, the difference

$$r_i - f(x) \in \mathbb{R}^k$$

defines the amount of licences that individual $i \in I$ may sell in the market (those for which the corresponding component is positive) and those she needs to buy since his initial endowment of the corresponding licence is not enough to support the consumption of x (negative components).

If the price for licences is $s \in \mathbb{R}_+^k$, then the consumption of x, as already mentioned, implies that the total wealth she can obtain (or pay if negative) from trading them in the market is:

$$s \cdot \left[r_i - f(x)\right] \in \mathbb{R}.$$

In the following, Δ denotes the Simplex in $\mathbb{R}^{\ell+k}$ and for $n \in \mathbb{N}_+$ and $x, \ y \in \mathbb{R}^n$, we say that $x \leq_n y$ iff $x_i \leq y_i$, for each $i = 1, 2, \ldots, n$, $x <_n y$ iff $x \leq_n y$ and $x \neq y$ and, $x <<_n y$ iff $x_i < y_i$, for each $i = 1, 2, \ldots, n$. Finally, 0_n is zero in $\mathbb{R}^n$.

Definition 1 For $(p, s) \in \Delta$, the budgetary set for individual $i \in I$ at prices (p, s) is defined by

$$B_i(p, s) = \left\{\xi_i \in X_i \mid p \cdot \xi_i \leq p \cdot \omega_i + s \cdot \left[r_i - f(\xi_i)\right]\right\}.$$

Definition 2 An economy with consumption rights and externalities is defined as

$$\mathcal{E}_R = (X_i, (u_i), (\omega_i), (r_i), f)_{i \in I}.$$

The corresponding economy without consumption rights ("exchange economy with externalities") is denoted by

$$\mathcal{E} = (X_i, (u_i), (\omega_i))_{i \in I}.$$

In order to define the equilibrium notion for economy $\mathcal{E}_R$, we consider feasibility in both consumption goods and consumption of licences.

Definition 3 We say that $x = (x_i) \in X$ is a feasible allocation for the economy $\mathcal{E}_R$ if

$$\sum_{i \in I} x_i \leq_\ell \omega \in \mathbb{R}^\ell_+$$

and

$$\sum_{i \in I} f(x_i) \leq_k R \in \mathbb{R}^k_+.$$

The set of feasible allocation for the economy $\mathcal{E}_R$ is denoted by $\mathcal{F}_R$.

Remark 1 Observe that the endowments $(\omega_i) \in X$ need not to be a feasible allocation for the economy $\mathcal{E}_R$. This occurs if for some j

$$\sum_{i \in I} f_j(\omega_i) > R_j.$$

More generally, for $j \in K$ suppose that f_j is a convex function and $f_j(0_\ell) = 0$; if $x = (x_1, x_2, ..., x_m) \in X$ allocates the total endowment, that is, $\sum_{i \in I} x_i = \omega$, then we have that

$$f_j(\omega/m) \leq \frac{1}{m} \sum_{i \in I} f_j(x_i) \leq \frac{1}{m} R_j.$$

Consequently, if $R_j < m f_j(\omega/m)$ the cap is effective. That is, it is not possible to allocate the total endowment of the economy.

Finally, the definition below is a natural extension of the competitive equilibrium notion we have for an exchange economy.[1]

Definition 4 We say that $((p^*, s^*), (x_i^*)) \in \Delta \times \mathbb{R}_+^{m\ell}$ is a Nash-Walras equilibrium for the economy $\mathcal{E}_R$ if

(a) $x^* = (x_i^*) \in \mathcal{F}_R$
(b) for each $i \in I$, $x_i^* \in B_i(p^*, s^*)$, and x_i^* maximizes $u_i(x_{-i}^*, \cdot)$ on $B_i(p^*, s^*)$.

3 Walras' Law and Some Direct Consequences

We begin this Section with the following straightforward lemmata, which will be useful to show the Walras' Law in our context (Proposition 1).

Lemma 1 *Suppose that $f : \mathbb{R}_+^{\ell} \to \mathbb{R}_+^k$ is continuous and that for $i \in I$ and for any $x_{-i} \in X_{-i}$, $u_i(x_{-i}, \cdot) : X_i \to \mathbb{R}$ is locally non-satiated.[2] Given $((p^*, s^*), (x_i^*))$ a Nash-Walras equilibrium of $\mathcal{E}_R$, if for $x_i \in X_i$ holds that $u_i(x^*) \le u_i(x_{-i}^*, x_i)$, then*

$$p^* \cdot x_i \ge p^* \cdot \omega_i + s^* \cdot \left[r_i - f(x_i)\right].$$

Proof Suppose that $p^* \cdot x_i < p^* \cdot \omega_i + s^* \cdot \left[r_i - f(x_i)\right]$. Since f is continuous, there exist $\epsilon > 0$ such that

$$p^* \cdot x_i' < p^* \cdot \omega_i + s^* \cdot \left[r_i - f(x_i')\right]$$

for all $x_i' \in B(x_i, \epsilon)$. Therefore, by local non-satiation, there is a point $z \in B(x_i, \epsilon)$ such that $u_i(x_{-i}^*, x_i) < u_i(x_{-i}^*, z)$ and then $u_i(x^*) < u_i(x_{-i}^*, z)$, which contradicts that (x_i^*) is the equilibrium allocation at prices (p^*, s^*). □

A direct consequence of Lemma 1 is the following proposition.

Proposition 1 **Walras' Law** *Under the conditions of Lemma 1, if $((p^*, s^*), (x_i^*))$ is a Nash-Walras equilibrium of $\mathcal{E}_R$ then*

$$p^* \cdot \left[\sum_{i \in I} x_i^* - \omega\right] = 0, \quad s^* \cdot \left[\sum_{i \in I} f(x_i^*) - R\right] = 0.$$

[1] In the following, for $x = (x_i) \in X$, we adopt the notation $u_i(x) = u_i(x_{-i}, x_i)$.
[2] That is, for any $x_{-i} \in X_{-i}$, $\epsilon > 0$ and $x_i \in X_i$, there exists $x_i' \in B(x_i, \epsilon) \cap X_i$ such that $u_i(x_{-i}, x_i) < u_i(x_{-i}, x_i')$, where $B(x_i, \epsilon)$ is the open ball with center x_i and radius ϵ.

Proof From Lemma 1, for each $i \in I, p^* \cdot x_i^* = p^* \cdot \omega_i + s^* \cdot \left[r_i - f(x_i^*)\right]$, which leads us to conclude

$$p^* \cdot \left[\sum_{i \in I} x_i^* - \omega\right] + s^* \cdot \left[\sum_{i \in I} f(x_i^*) - R\right] = 0. \tag{1}$$

Since $\sum_{i \in I} x_i^* \leq_\ell \omega$, $\sum_{i \in I} f(x_i^*) \leq_k R$ and $(p^*, s^*) \in \mathbb{R}_+^{\ell+k}$, follows that $p^* \cdot \left[\sum_{i \in I} x_i^* - \omega\right] \leq 0$ and $s^* \cdot \left[\sum_{i \in I} f(x_i^*) - R\right] \leq 0$, which along with (1) implies the desired result. □

Remark 2 For a Nash–Walras equilibrium $((p^*, s^*), (x_i^*))$, the fact that the requirement of licences may effectively restrict the consumption of a good $k \in \{1, 2, \ldots, \ell\}$ corresponds to $\sum_{i \in I} x_{ik}^* - \omega_{ik} < 0$; under this situation, the Walras' Law implies that $p_k^* = 0$. Note that this fact does not depend on the distribution of licences among individuals but only depends on the aggregate amount of licences. In this situation, as we will see in the next example, the amount of licences assigned to each individual could have consequences on their welfare in the equilibrium, allowing further analysis regarding public policy through the assignment of licences among agents.

Suppose that the amount of licences effectively restrict the consumption of a good $k \in \{1, 2, \ldots, \ell\}$ and that for some consumer i the good k is desirable, that is, for any positive λ, $u(x_i^* + \lambda e_k) > u(x_i^*)$, where e_k is the kth vector of the canonic basis of $\mathbb{R}^\ell$. From these assumptions, it immediately follows that $p_k^* = 0$ and, from the budgetary constrain, we have for some $j \in K$, $s_j^* > 0$ and $f_j(x_i^* + \lambda e_k) > f_j(x_i^*)$. Consequently, an effective cap on a commodity implies that the equilibrium price of that commodity is zero and that the price of the corresponding licence becomes the relevant price.

On the contrary, note that when the level of licences is high enough, the price of the licence becomes zero at the equilibrium and the economy becomes equivalent to a classical exchange market with externalities $\mathcal{E}$.

Example 1 In order to define economy $\mathcal{E}_R$, suppose $m = 2$, $\ell = 2$ and that individual's preferences are given by $u_1(x_1, x_2) = u_2(x_1, x_2) = x_1^\alpha x_2^{1-\alpha}$, with $0 < \alpha < 1$. Endowments of goods are $(\omega_{i2}, \omega_{i2}) \in \mathbb{R}_+^2$, $i = 1, 2$; set $\omega_j = \omega_{1j} + \omega_{2j} > 0$, $j = 1, 2$. Additionally, suppose that $K = 1$, $f(x_1, x_2) = bx_2$ (with $b > 0$) and the endowment for licences is $r_i \geq 0$, $i = 1, 2$. Set $R = r_1 + r_2 > 0$. The economy $\mathcal{E}$ is defined by u_i and $(\omega_{i1}, \omega_{i2})$, $i = 1, 2$, as before. As follows, assume that good one is the *numerary* and prices for good two and licences are denoted by p and s respectively. From the monotonicity of the involving functions, the consumer's problem for agent $i = 1, 2$ is

$$\max_{x_{i1}, x_{i2}} x_{i1}^\alpha x_{i2}^{1-\alpha} \; s.t. \; x_{i1} + px_{i2} = \omega_{i1} + p\omega_{i2} + s\left[r_i - bx_{i2}\right], \; x_{i1}, \; x_{i2} \geq 0,$$

whose unique solution is

$$x_{i1}(p, s) = \alpha\left[\omega_{i1} + p\omega_{i2} + sr_i\right], \quad x_{i2}(p, s) = (1 - \alpha)\left[\frac{\omega_{i1} + p\omega_{i2} + sr_i}{p + bs}\right], \; i = 1, 2.$$

The equilibrium conditions for goods one and two are, respectively,

$$x_{11}(p,s) + x_{21}(p,s) = \omega_1 \quad \Leftrightarrow \quad \alpha\left[\omega_1 + p\omega_2 + sR\right] = \omega_1 \tag{2}$$

$$x_{12}(p,s) + x_{22}(p,s) \leq \omega_2 \quad \Leftrightarrow \quad (1-\alpha)\left[\frac{\omega_1 + p\omega_2 + sR}{p+bs}\right] \leq \omega_2. \tag{3}$$

Combining (2), (3) and the budget constraint, for any $s \geq 0$

$$s\left[R - b\omega_2\right] \leq 0. \tag{4}$$

For the case $R > b\omega_2$, the unique equilibrium price is

$$s^c = 0, \quad p^c = \left(\frac{1-\alpha}{\alpha}\right)\frac{\omega_1}{\omega_2},$$

which coincides with the equilibrium price for the economy $\mathcal{E}$. For the case $R = b\omega_2$, there are infinite equilibrium prices $(p,s) \in \mathbb{R}^2_+$, parameterized by the relation $p + sb = p^c$.

For the case $R < b\omega_2$, from (4) we have that $s \geq 0$. However, note that $s = 0$ is not an admissible solution, since in such a case the aggregated equilibrium demand for consumption good two would be equal to those obtained for economy $\mathcal{E}$ (i.e., ω_2), which is not a feasible allocation from the side of the licences. Consequently, we may assume $s > 0$ and then, in order to preserve feasibility from the side of the licences, (3) holds that

$$b(1-\alpha)\left[\frac{\omega_1 + p\omega_2 + sR}{p+bs}\right] \leq R.$$

If we denote by R' the consumption effectively employed by agents, we have

$$b(1-\alpha)\omega_1 + p[b(1-\alpha)\omega_2 - R'] + s[b(1-\alpha)R - bR'] = 0, \tag{5}$$

from which, along with (2) we conclude that

$$p\left[\omega_2 - \frac{R'}{b}\right] + s[R - R'] = 0.$$

Since $R' \leq R < b\omega_2$, in order to obtain positive equilibrium prices we must impose $R' = R$, which lead us to conclude that the equilibrium price for good two is $p^* = 0$, and from (2), the equilibrium price for licences should be

$$s^* = \frac{(1-\alpha)\omega_1}{\alpha R}.$$

Regarding good one, the equilibrium allocation is

$$x_{i1}^r = \alpha \left[\omega_{i1} + \frac{(1-\alpha)\omega_1}{\alpha R} r_i \right] = \alpha\omega_{i1} + (1-\alpha)\omega_1 \frac{r_i}{R}, \; i = 1, 2, \tag{6}$$

which, for individual $i = 1, 2$, would be greater than those obtained in the exchange economy without consumption rights, provided that

$$\frac{r_i}{R} > \frac{\omega_{i2}}{\omega_2}.$$

Regarding good two, given $\delta = R - b\omega_2 > 0$, the aggregated demand at the equilibrium is given by

$$(1-\alpha)\left[\frac{\omega_1 + sR}{bs}\right] = \omega_2 - \frac{\delta}{b} < \omega_2. \tag{7}$$

Note that $R < b\omega_2$ implies that for some $i = 1, 2$, $r_i < b\omega_{i2}$. Thus, the initial endowment of goods and licences do not necessarily belong to the budgetary set for this individual, at any price. This fact is relevant in our model, since it implies that we cannot use standard arguments to prove the existence of equilibrium in our setting by considering an extended economy where consumption rights (licences) appear as new commodities in the market, even though they do not directly participate in agent's preferences.

Finally, from (6), the presence of consumption rights in the market imply a redistribution of good one between agents that otherwise may not be reached as a competitive outcome in the economy $\mathcal{E}$, unless a redistribution of endowments is carried out. However, from (7) we also have that the presence of consumption rights (licences) may effectively restrict the consumption of goods, implying an *excess of supply* that may not be assigned to any individual. Thus, consumption rights may not necessarily be interpreted as a tax mechanism whose role is to reach a certain point in the contract curve of the economy $\mathcal{E}$. Indeed, Karp and Zhang (2016) show the advantage of quotas over emissions taxes in a model with asymmetric information. On the other hand, from Eq. (6) we have a redistribution effect as a result the introduction of consumption rights.

4 Existence of Equilibrium

For the existence of equilibrium in our model we will consider standard hypotheses on the fundamentals of the economy. The strongest condition we are assuming for the existence of equilibrium result is **SS** (*a survival* condition).

Assumption C. For each $i \in I$, $X_i \subseteq \mathbb{R}^\ell_+$ is convex, closed and $0_\ell, \omega_i \in X_i$.

Assumption SS. For each $i \in I$, $\omega_i \in \mathbb{R}^\ell_{++}$ and $r_i \in \mathbb{R}^k_{++}$.

Assumption R. For each $j \in K$, $f_j : \mathbb{R}^\ell_+ \to \mathbb{R}_+$ is convex, continuous and $f_j(0_\ell) = 0$ (i.e., $f(0_\ell) = 0_k$).

Assumption U. For each $i \in I$, $u_i : X \to \mathbb{R}$ is continuous and for each $x_{-i} \in X_{-i}$, $u_i(x_{-i}, \cdot) : X_i \to \mathbb{R}$ is locally non-satiated and quasi-concave.

In order to facilitate the demonstration of our main result, we introduce the auxiliary economy $\mathcal{E}_R^M$, which differs from $\mathcal{E}_R$ only in the consumption sets that now, for individual $i \in I$, is defined by[3]

$$X_i^M = X_i \cap clB\left(0_\ell, M\|\omega\|\right),$$

with $M > 1$ a given constant[4] We set $X^M = \prod_{i\in I} X_i^M$ and for $i \in I$, define

$$X_{-i}^M = \prod_{j\in I\setminus\{i\}} X_j^M.$$

Lemma 2 *Under Assumptions* **C**, **SS** *and* **R**, *for* $i \in I$ *the correspondence*

$$B_i^M : \Delta \to X_i^M \mid B_i^M(p,s) = \left\{\xi_i \in X_i^M \mid p\cdot\xi_i \leq p\cdot\omega_i + s\cdot\left[r_i - f(\xi_i)\right]\right\}$$

is continuous.

Proof From Assumption **C**, it follows directly that for each $i \in I$, B_i^M is a closed correspondence. Since X_i^M is compact, it is upper semi-continuous.

Now, in order to show the lower semi-continuity of B_i^M at any point $(p_0, s_0) \in \Delta$, let G be any open set such that $B_i^M(p_0, s_0) \bigcap G \neq \emptyset$ and let ξ belonging to this set. Observe that by Assumption **SS** we have that

$$0 < p_0\cdot\omega_i + s_0\cdot\left[r_i - f(0_\ell)\right],$$

and therefore, from the convexity of f we conclude that for all $\lambda \in [0, 1)$

$$p_0\cdot\lambda\xi < p_0\cdot\omega_i + s_0\cdot\left[r_i - f(\lambda\xi)\right].$$

Let be $\lambda_0 < 1$ such that $\lambda_0\xi \in G$. Since f is continuous, there exists $\epsilon > 0$ such that $\| (p,s) - (p_0, s_0) \| < \epsilon$ implies that

$$p\cdot\lambda_0\xi < p\cdot\omega_i + s\cdot\left[r_i - f(\lambda_0\xi)\right],$$

---

[3]The closure of $A \subseteq \mathbb{R}^n$ is denoted by clA and the Euclidean norm of $x \in \mathbb{R}^n$ by $\|x\|$.

[4]Note that from feasibility condition for consumption bundles, any relevant consumption plan x_i for an individual $i \in I$ should comply with $0_\ell \leq x_i \leq_\ell \omega$ and therefore $\|x_i\| \leq \|\omega\|$.

from which we deduce that $B_i^M(p,s) \bigcap G \neq \emptyset$ for all $(p,s) \in \Delta$ such that $\| (p,s) - (p_0, s_0) \| < \epsilon$. This last assertion finally leads us to conclude that B_i^M is a continuous correspondence as required. □

Theorem 1 Existence of Equilibrium *Under Assumptions* **C**, **SS**, **R** *and* **U** *there exist a Nash-Walras equilibrium for economy* $\mathcal{E}_R$.

Proof For $i \in I$ define the function

$$u_i^* : \Delta \times X^M \times X_i^M \to \mathbb{R} \mid u_i^*((p,s),x,z) = u_i(x_{-i},z),$$

and the correspondence

$$\mathbf{B}_i^M : \Delta \times X^M \to X_i^M \mid \mathbf{B}_i^M((p,s),x) = B_i^M(p,s).$$

Note that under Assumption **U**, the demand correspondence of the auxiliary economy $\mathcal{E}_R^M$, D_i^M defined by

$$D_i^M : \Delta \times X^M \to X_i^M \mid D_i^M((p,s),x) = \{\xi_i \in \mathbf{B}_i^M((p,s),x) \mid u_i(x_{-i},\xi_i) \geq u_i(x_{-i},z),\ \forall z \in \mathbf{B}_i^M((p,s),x)\},$$

is compact and convex valued and from Lemma 2 and the Maximum Theorem (Berge 1997), it is upper semi-continuous.

Following the standard approach, for the *additional agent (the market)*, we define the function

$$u_0^* : \Delta \times X^M \times \Delta \to \mathbb{R} \ |u_0^*((p,s),x,(p',s')) = p' \cdot \left(\sum_{i \in I} x_i - \omega \right) + s' \cdot \left(\sum_{i \in I} f(x_i) - R \right),$$

and the constant correspondence

$$B_0^M : \Delta \times X^M \times \Delta \to \Delta \mid B_0^M((p,s),x,(p',s')) = \Delta.$$

The demand of the *market* is defined by the correspondence,

$$D_0^M : \Delta \times X^M \to \Delta \mid D_0^M((p,s),x) = \left\{ (p',s') \in \Delta \mid (p'-\tilde{p}) \cdot \left(\sum_{i \in I} x_i - \omega \right) + (s'-\tilde{s}) \cdot \left(\sum_{i \in I} f(x_i) - R \right) \geq 0, \forall (\tilde{p},\tilde{s}) \in \Delta \right\},$$

which is convex and compact valued and, again by the Maximum Theorem (Berge 1997), it is upper semi-continuous.

Thus, if we define

$$D^M : \Delta \times X^M \to \Delta \times X^M \mid D^M = \prod_{i=0}^{m} D_i^M,$$

follows immediately that D^M is compact and convex valued and upper semi continuous and since $\Delta \times X^M$ is convex and compact, from Kakutani's Fixed Point Theorem we conclude that there exist $((p^*, s^*), (x_i^*)) \in \Delta \times X^M$ such that

$$((p^*, s^*), (x_i^*)) \in D^M(((p^*, s^*), (x_i^*))),$$

that is,

(i) for each $i \in I$, $x_i^* \in D_i^M((p^*, s^*), (x_i^*))$,
(ii) $(p^*, s^*) \in D_0^M((p^*, s^*), (x_i^*))$.

From condition (*i*), x_i^* maximizes $u_i(x_{-i}^*, \cdot)$ on the budget set $\mathbf{B}_i^M((p^*, s^*), x^*)$. On the other hand, since $x_i^* \in \mathbf{B}_i^M((p^*, s^*), x^*)$, $i \in I$, we have that

$$p^* \cdot x_i^* \leq p^* \cdot \omega_i + s^* \cdot \left[r_i - f(x_i^*)\right],$$

which, by summing in all the agents, it leads us to conclude

$$p^* \cdot \left(\sum_{i \in I} x_i^* - \omega\right) + s^* \cdot \left(\sum_{i \in I} f(x_i^*) - R\right) \leq 0. \tag{8}$$

From condition (*ii*) and inequality (8), holds that for each $(p, s) \in \Delta$

$$p \cdot \left(\sum_{i \in I} x_i^* - \omega\right) + s \cdot \left(\sum_{i \in I} f(x_i^*) - R\right) \leq 0.$$

Taking $s = 0_k$ and letting p be each vector the canonic basis of $\mathbb{R}^\ell$, the last inequality implies that

$$\sum_{i \in I} x_i^* - \omega \leq_\ell 0_\ell.$$

In the same way, taking $p = 0_\ell$ and letting s be each vector of the canonic basis of $\mathbb{R}^k$, we conclude that

$$\sum_{i \in I} f(x_i^*) - R \leq_k 0_k.$$

Thus, all the foregoing implies that $((p^*, s^*), (x_i^*)) \in \Delta \times X^M$ is an equilibrium for economy $\mathcal{E}_r^M$. In order to show that $((p^*, s^*), (x_i^*))$ is also an equilibrium for economy $\mathcal{E}_R$, let us suppose that for some $i \in I$ there exists $\tilde{x}_i \in X_i \setminus X_i^M$ such that

(a) $u_i(x^*_{-i}, \tilde{x}_i) > u_i(x^*_{-i}, x^*_i)$,
(b) $p^* \cdot \tilde{x}_i \leq p^* \cdot \omega_i + s^* \cdot \left[r_i - f(\tilde{x}_i)\right]$.

Taking $\tilde{\lambda} \in]0, 1[$ close enough to one, Assumption **C** implies that $\tilde{\lambda}\tilde{x}_i \in X_i$ and from Assumption **U**, $u_i(x^*_{-i}, \tilde{\lambda}\tilde{x}_i) > u_i(x^*_{-i}, x^*_i)$. Moreover, condition (b) above directly implies

$$p^* \cdot (\tilde{\lambda}\tilde{x}_i) < p^* \cdot \omega_i + s^* \cdot \left[r_i - f(\tilde{x}_i)\right]. \tag{9}$$

Additionally, from Assumption **R** it is easy to check that $-f(\tilde{x}_i) \leq_k -f(\tilde{\lambda}\tilde{x}_i)$, and then, considering that $s^* \in \mathbb{R}^k_+$, inequality (9) finally implies

$$p^* \cdot (\tilde{\lambda}\tilde{x}_i) < p^* \cdot \omega_i + s^* \cdot \left[r_i - f(\tilde{\lambda}\tilde{x}_i)\right]. \tag{10}$$

For $\mu \in]0, 1[$ define

$$x_i^\mu = \mu x_i^* + (1 - \mu)\tilde{\lambda}\tilde{x}_i.$$

From (10) and Assumption **R**, holds that $p^* \cdot x_i^\mu < p^* \cdot \omega_i + s^* \cdot \left[r_i - f(x_i^\mu)\right]$, and from the quasi-concavity of $u_i(x_i^*, \cdot)$, $u_i(x^*_{-i}, x_i^\mu) \geq u_i(x^*_{-i}, x^*_i)$.

Note now that for $\mu \in]0, 1[$ close enough to one, x_i^μ belongs to X_i^M and therefore, from Assumption **U** we have that for some $\epsilon > 0$ there exists $\bar{x}_i \in X_i^M \cap B(x_i^\mu, \epsilon)$ such that

$$u_i(x^*_{-i}, \bar{x}_i) > u_i(x^*_{-i}, x_i^\mu) \geq u_i(x^*_{-i}, x^*_i).$$

Finally, choosing μ sufficiently close to 1, the continuity of f implies that

$$p^* \cdot \bar{x}_i \leq p^* \cdot \omega_i + s^* \cdot \left[r_i - f(\bar{x}_i)\right],$$

which contradicts the fact that x_i^* maximizes $u_i(x^*_{-i}, \cdot)$ on $\mathbf{B}_i^M((p^*, s^*), x^*)$. □

5 Conclusions

This paper deals with the problem of setting a price system for licences in an economy in which consumption caps exist, and in order to consume, agents are required to have the corresponding licence for consumption. This leads to the establishment of a market for rights or licences.

Examples of this situation are the European Union Emissions Trading System established in 2005 to reduce greenhouse effects under the Kyoto Protocol (see Ellerman and Joskow 2008). Also, there are other cap-and-trade systems for emissions that have been implemented in the U.S. In these kind of systems the price of licences are set, depending on the cost of controlling the negative effects.

Our model can be used not only on emission control systems, but also to deal with any other licence-based models where permits are required in advance. Such rights are for instance, aircraft landing licences, or fishing licences in a region where the amount of captures is regulated. Is it also possible to consider such model to control road congestion by distributing total transit rights for specific links such that flow capacity ratios are limited on these links.

In our approach, agents evaluate their utility considering all the consequences involved in their consumption plan and the consumption plans of the other consumers. Licences must be acquired at the same time as contracts for raw materials are signed. Thus, prices of licences are linked to prices of commodities. Our model is based on the existence of an exogenous function which evaluates the potential negative effects derived from each contract. This mapping associates to every consumption plan (contract): a theoretical amount of licences of each type and consequently, to measure the actual negative effects of consumption is not required in our model. Our aim is to analyze the immediate consequences of setting a cap with a trade system of licences in a simple model of general equilibrium.

We have shown that under standard conditions on the fundamentals of the economy, equilibrium exists. Our analysis points out that if the cap is effective for a raw material, the price of this commodity becomes irrelevant at the equilibrium and is the price of the corresponding licence that matters. Given that we deduce that the effectiveness of the cap only depends on the total amount of licences, the political welfare aspects derived from the distribution of these allowances among the agents become the relevant problem for the planner of the cap-and-trade system.

Finally, we would like to remark that in this paper we are not considering the political welfare aspects derived from the distribution of the licences among the agents. We shall focus on this problem in a future study, but it is worth remarking the existence of redistributive effects in the economy with caps for externalities and consumption rights.

References

Baumol, W., & Oates, W. (1988). *The Theory of Environmental Policy*. Cambridge: Cambridge University Press.

Berge, C. (1997). *Topological Spaces*. New York: Dover.

Boyd, J., & Conley, J. (1997). Fundamental nonconvexities in arrovian markets and a Coasian solution to the problem of externalities. *Journal of Economic Theory*, *72*, 388–407.

Burniaux, J-M., Martins, J.O. (2016). Carbon leakage: a *general equilibrium view*. *Economic Theory* (reprinted in this volume)

Chichilnisky, G. (2016). Sustainable markets with short sales. *Economic Theory*

Chipman, J., Guoqiang, T. (2016). Detrimental externalities, pollution rights, and the Coase theorem. *Economic Theory* (reprinted in this volume)

Conley, J., & Smith, S. (2005). Coasian equilibrium. *Journal of Mathematical Economics*, *41*, 687–704.

Ellerman, A., & Joskow, P. (2008). *The European Union's Emissions Trading System in Perspective*. http://www.pewclimate.org/docUploads/EU-ETS-In-Perspective-Report.pdf.

Ellerman, A., Webster, M., Parsons, J., Jacoby, H., & McGuinness, M.: *Cap-and-trade: Contributions to the design of a U.S. Greenhouse gas program*. MIT CEEPR (2008)

Joskow, P., Schmalensee, R., & Bailey, E. (1998). The market for sulfur dioxide emissions. *American Economic Review*, *88*, 669–685.
Karp, L., & Zhang, J.: Taxes versus quantities for a stock pollutant with endogenous abatement costs and asymmetric information. *Economic Theory* (2016) (reprinted in this volume)
Mandel, A. (2009). Changes in the firms behavior after the opening of markets of allowances. *Economic Theory*, *40*, 1–25.
Montero, J. (2001). Multipollutant markets. *RAND Journal of Economics*, *32*, 762–774.
Montgomery, D. (1972). Markets in licences and efficient pollution control programs. *Journal of Economic Theory*, *5*, 395–418.
Newell, R., Sanchirico, J., & Kerr, S. (2005). Fishing quota markets. *Journal of Environmental Economics and Management*, *49*, 437–462.
Starrett, D. (1972). Fundamental non-convexities in the theory of externalities. *Journal of Economic Theory*, *4*, 180–199.

Part VI
Catastrophic Risk in Economic Practice

Exploring the Role of Emotions in Decisions Involving Catastrophic Risks: Lessons from a Double Investigation

Olivier Chanel, Graciela Chichilnisky, Sébastien Massoni and Jean-Christophe Vergnaud

1 Introduction

Natural disasters due to climate change (like floods, hurricanes, heat waves or droughts) combine a risk of large losses and a low probability of occurrence. Faced with such risks, qualified as catastrophic, individuals often adopt irrational behavior, like over-coverage for the material consequences with low probability of occurrence or under-coverage for natural disaster-related risks. Conversely, when the risk affects the individual's physical well-being more directly, behavior can appear over-cautious. While these behaviors may partially be explained by psychological determinants, scant light has been shed on the subject by economic models.

Yet the stakes are high. The total cost of natural disasters worldwide was €122 billion in 2012, €50 billion of which went to losses covered by insurance (Munich RE 2013), which have been growing almost continuously since 1970. Between 1970 and 2014, of the world's ten most costly disasters, five involved flooding, two earthquakes, two hurricanes, the last being the terrorist attacks on 11 September 2001, the only non-natural disaster (Sigma 2015). Moreover,

O. Chanel (✉)
Aix-Marseille University (Aix-Marseille School of Economics), CNRS and EHESS, Marseille, France
e-mail: olivier.chanel@univ-amu.fr

G. Chichilnisky
Columbia University, New York, USA

S. Massoni
QuBE-School of Economics and Finance, Queensland University of Technology, Brisbane, Australia

J.-C. Vergnaud
CNRS - Université de Paris I, Paris, France

G. Chichilnisky and A. Rezai (eds.), *The Economics of the Global Environment*,
Studies in Economic Theory 29, DOI 10.1007/978-3-319-31943-8_24

flooding and hurricanes are major natural hazards whose intensity and frequency are likely to increase with climate change in the 21st century (IPCC 2013).

When seeking to determine the best decision to mitigate or avoid the harmful consequences of such catastrophic risks, decision theorists use the choice under uncertainty framework, especially the widely-applied Expected Utility (EU), whose axiomatic framework was formalized by von Neumann and Morgenstern (1944). However, as early as the early 1950s, several weaknesses of this model were identified, even though the original model was amended, in particular to account for choices involving catastrophic risks. Sunstein (2002) thinks, for instance, that strong emotions like fear or regret, when associated with an event such as a natural disaster or extreme weather risk, can lead to decisions that are not rational if the probabilities of occurrence of this event are negligible: the event carrying the strongest emotions prevails in the decision ("probability neglect").

The first work that took into account the fact that individuals' emotions could affect their decisions considered regret (Bell 1982; or Loomes and Sugden 1982) and disappointment (Loomes and Sugden 1986; Gul 1991). Loewenstein et al. (2001) then introduced the possibility that emotions experienced during the decision-making process could influence decisions, through their assumption of "Risks as feelings". Chichilnisky (2000, 2009) also offers an axiomatic allowing for sensitivity of preference ranking for catastrophic events, which introduces the feeling of fear as an emotion involved in the decision-making process.

But emotions also affect the decision by altering the outcome assessment, or by inducing individuals to recall previous similar situations. Ultimately, understanding the role of emotions in decision-making when catastrophic risks are involved is clearly important, especially if the goal is to understand and possibly control for the underlying psychological determinants of actual behavior, and to guide toward more rational decisions.

This paper investigates the role of emotions in individuals' choices among alternatives pertaining to catastrophic events, either artificial (laboratory experiment) or real-life (field experiment on flooding).

The experimental protocols are used to assess the role of emotions in decision-making and the formation of beliefs. These protocols allow us to use a wider than usual range of behavioral data in a controlled environment, handling via contextual devices the emotional burden experienced. The use of experimental paradigms from psychophysics allows control over the sources of uncertainty felt by the subjects. We examine in particular how insurance choices and the associated formation of beliefs are driven by anxiety.

By also conducting field surveys on populations exposed to different levels of flood risk, we seek to test assumptions related to real-life behavior. We examine whether individual psychological factors measured experimentally have predictive power regarding actual behavior. In particular, we gather and analyze data on respondents' emotions, their expectations of these emotions, their personality and psychological determinants, their symptoms before and after catastrophic events that generated emotions.

Section 2 describes how emotions are currently taken into account in models of choice under uncertainty. Section 3 proposes four additional routes as well as the emotional/psychological data required to test them. Section 4 presents the field survey on flooding and Sect. 5 presents laboratory experiment results. Section 6 concludes.

2 Emotions in Decision-Making Under Uncertainty Models

2.1 Models Without Emotions: EU

The standard approach in decision theory under risk and uncertainty considers only behavioral data concerning the choices observed between lotteries (i.e. probability distributions on consequences). Before the decision, the individual must identify possible actions (anticipated outcomes) and uncertainty (subjective probabilities).

Yet the analysis is consequential: as shown in Fig. 1, it is assumed that individuals' goals are embedded in the outcomes. The cognitive assessment phase generates immediate emotions that are only viewed as a side effect of the cognitive task. Thus, the decision is made according to a rationality criterion, almost exclusively the EU: the decision-maker has preferences over lotteries that satisfy a number of axioms (completeness, transitivity, independence and continuity) and must choose the one that maximizes his utility.

Economists however, suspecting "emotions" of being potential irrationality factors, consequently propose formalizations that introduce emotions and whose predictions are more consistent with observed behaviors.

2.2 Introducing Emotions

Let us start with some utility models alternative to the EU model, in which the decision maker is considered a "unique self", before discussing approaches in

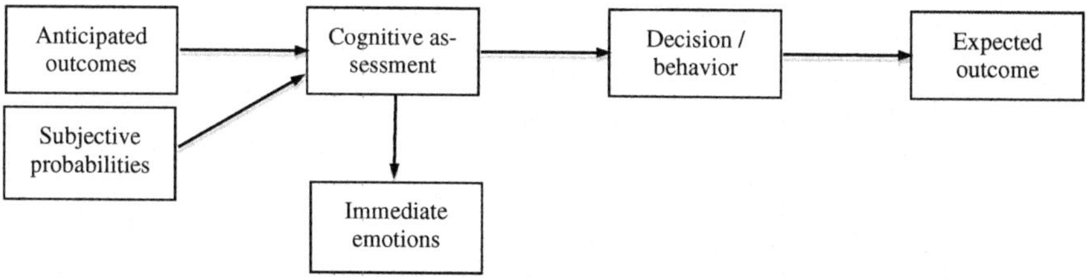

Fig. 1 Standard consequential and behavioral representation of decision under uncertainty

which the decision maker is seen as a dual system ("multiple self"). Note, however, that all these formalizations are still part of the behavioral and consequential approach.

2.2.1 Unique Self

Various models pertaining to emotions have been proposed, some based on an axiomatic change, others only offering a functional form of utility.

For the first group of models, the axioms of expected utility are modified: the independence axiom for Bell (1982), Loomes and Sugden (1982), Gul (1991) or Kahneman and Tversky (1979), the continuity axiom for Chanel and Chichilnisky (2009).

In their "prospect theory" for instance, Kahneman and Tversky (1979) partially explain the differences in the shape of the utility function and the distortion function of probabilities, as well as the discontinuity of the derivative of the utility function, by emotional considerations (loss aversion). One way to interpret this model is to suggest that there are in fact two different functional forms for utility depending on emotions (whether or not a risk of loss is felt). Emotions are therefore an intermediate product in the decision-making process, but one that is not formally represented in the decision model.

The regret theory of Loomes and Sugden (1982) is more explicit regarding emotions, since the utility function is explicitly called "regret-based": agents make a choice that is aimed at minimizing their expected regret.

Effect A in Fig. 2 reflects the fact that by taking a decision now, while the consequences of this decision will only be known in the future, the emotional state of the individual is affected by the fact that s/he imagines the emotions s/he will likely feel when the outcome is finally realized: these are *Expected (or Anticipated) emotions.*

When the choice is binary, the regret is measured for every possible contingency as a function of the outcome that would be obtained with the choice in question and the outcome that would be obtained with the second possible choice (see also Bell 1982). This theory is, however, difficult to generalize to non-binary choice settings or when there are no explicit contingencies that allow for the comparison of

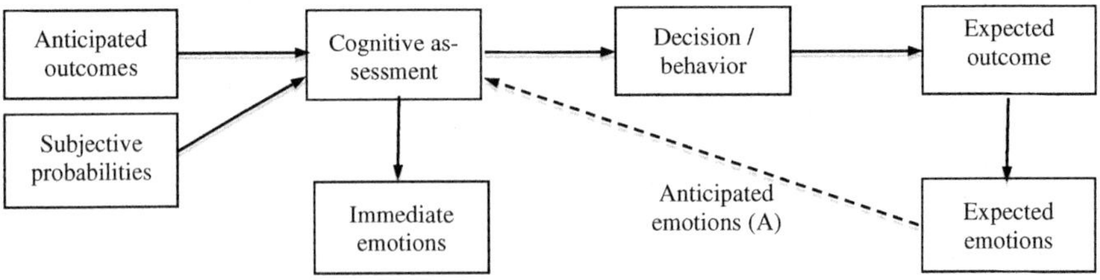

Fig. 2 Consequential and behavioral representation of decision under uncertainty with anticipated emotion

outcomes. In addition, the empirical test of the relevance of this theory still requires data collection on choices alone.

Chichilnisky (2000, 2009) proposed a model taking into account the presence of catastrophic risk, which is based on a weakening of the EU continuity axiom. The functional form proposed is expressed as a weighted sum of an EU function and a function that introduces the possibility of modulating the decision-making depending on emotional impact. Chanel and Chichinilsky (2009, 2013) presented empirical evidence supporting this possible effect of emotions, and one of the objectives of our catastrophic risk experiment is to explicitly test this hypothesis.

The models of the second group are more descriptive and only seek to explain the form of the utility function. This category includes Caplin and Leahy (2001) and Brunnermeier and Parker (2005). The basic idea is that there is an interval between the decision and the realization of the outcome, which induces emotions (stress, anxiety, fear or savoring) that are then part of the consequences of the decision.

2.2.2 Multiple Self

The idea of contrasting cognition with emotion within a multiple self has a long history in philosophy (Plato, Descartes, or Kant), is a common view in psychology (see Evans 1989), but is more recent in economics (Kahneman and Frederick 2002).

Slovic et al. (2004) refer to this dual system of risk perception by the terms "risk as feelings"—a fast, intuitive "emotional evaluation" that relies on emotions from the amygdala in the brain—and "risk as analysis"—a slower "cognitive evaluation" using logic and computation abilities from the cortex. However, they stress that a rational decision is a mixture of the two processes (cognitive and emotional). For economists, however, this is difficult to test empirically and in practice, the description of this dual system is not based on clearly-identified determinants (see Rustichini 2008; or Keren and Schul 2009).

Loewenstein (1996, 2000) and Loewenstein et al. (2001) incorporate this idea, explaining that visceral factors are the basis of a discrepancy between behavior and individual interests and that emotions alter the standard consequential evaluation.

Rick and Loewenstein (2008) and Caplin and Leahy (2001) propose two effects in addition to *Anticipated emotions* (see Fig. 3).

Effect B reflects the fact that emotions felt at the time of decision-making, when we imagine future outcomes, change the cognitive assessment of the risk and its consequences through cognitive and emotional risk assessment, thus influencing the expected utility and thereby the decision. They are referred to as *Immediate integral (or Anticipatory) emotions*.

Effect C reflects the fact that emotions immediately associated with current external factors not related to the decision (emotional state, mood, personality of the individual at the time of decision-making, their general perception of probabilities) are likely to affect the decision. The literature refers to them as *Immediate incidental emotions*.

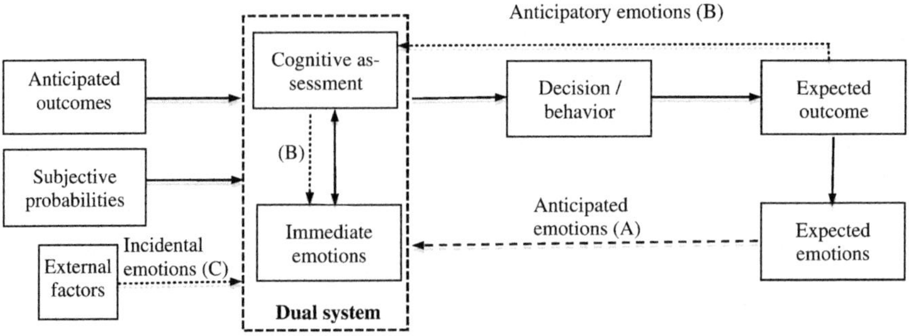

Fig. 3 Consequential representation of decision under uncertainty with the dual system and emotions (Rick and Loewenstein 2008; Caplin and Leahy 2001)

Overall, the way emotions are accounted for in economic models under uncertainty currently seems consistent with the view that the brain functions as a modular system with multiple areas dedicated to given functions (memory, vision, language, attention) that are interconnected. Emotions are also assumed to have functional roles, which probably differ depending on type of emotion.

3 Revisiting the Framework

3.1 A Wider Look at the Influence of Emotions

Given the complexity of emotional functions, the existing formalizations offered by economists are not fully convincing: no variables representing the emotions are taken into account as explanatory variables of choice, or as dependent variables. Yet emotions clearly play both roles, that of explanatory factors for behaviors and that of outcomes of the decision-making process. This raises a methodological issue when including measures related to emotions in an approach that tallies with the revealed preference approach. Reid and Gonzalez-Vallejo (2009), for example, proposed a model in which the emotions felt during the cognitive process explicitly enter the weighting function that determines behaviors, and help predict choices.

We chose to take a wider look at the various paths through which emotions, taken in a deliberately broad sense, are involved in decision-making in the face of catastrophic risks. In addition to the three above effects (A–C), we propose four other effects (D–G) presented in Fig. 4.

Effect D reflects the fact that the emotions an individual felt in the past, when s/he experienced an event similar to that s/he faces currently, affect her/his current assessment of probabilities and outcomes. The recollection of the event and the

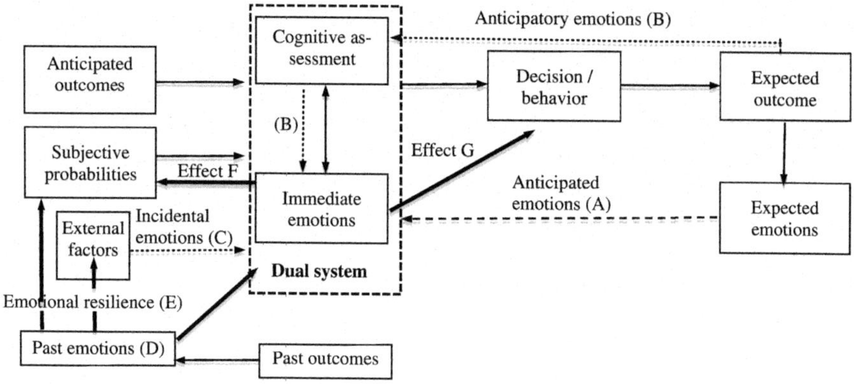

Fig. 4 Comprehensive representation of decision under uncertainty with emotions (derived from Chanel et al. 2013)

emotions felt when it is recalled are a reminder of the emotions felt at the time of the event. We refer to these emotions as *Past emotions*.

Effect E reflects the fact that emotions felt in the past have an impact on current external factors independent of the decision: on personality traits, on preferences or on the subjective perception of the probability of occurrence of the event. We call this *Emotional resilience.* Although "resilience" covers various notions, especially in psychology, we use it in its original meaning in the field of physics, i.e. the ability of a material (in this case, the respondent) subject to an impact (here a potentially traumatic event) to recover its initial state (here, the behavior of respondents who did not undergo this traumatic event).

The next two effects are explored in the laboratory experiments.

Effect F also concerns the emotions felt as they impact the subjective perception of the event that is feared, but the impact is now due to current (immediate) emotions. Typically, negative emotions activate a prevention focus attitude, with increased attention being devoted to preventing the occurrence of the negative event. We call this *Prevention focus reinforcement.* This effect can be seen as a sub-case of effect B, since it determines how emotions will affect the cognitive assessment of the risk through the subjective beliefs channel.

Finally, **effect G** captures a direct effect *of Immediate emotions* in the decision-making process. While effect B describes how emotions can affect the cognitive assessment of the risk and therefore may influence the decision, effect G reflects the fact that emotions can contribute directly to the decision through an emotional decision-making process that competes with the cognitive one. The idea is that emotions can hinder rational decision-making by triggering automatic responses in reaction to some perceived threat. We call this *Stress response.*

3.2 Measuring Emotions

To determine the paths through which emotions can affect decision-making, we collected various psychological/emotional measurements not common in standard studies, and used them to characterize the emotional dimension, both in the laboratory experiments and in the field survey on exposure to a real risk of flooding.

In both investigations, we collect data on personality traits which can be considered to capture major elements in the decision-making process. We use psychometric questionnaires to measure respondents' sense of control (Rotter 1966) and whether they are overly worried (Neuroticism) or conscientious (Conscientiousness) (using the "Big Five Inventory", John et al. 1991). Respondents are then asked about their aversion to risk in everyday life and their subjective assessment of their level of happiness and luckiness compared to the general population. All these elements reflect an individual heterogeneity on how to make decisions. They would not be used in the standard models of decision-making, which do not consider the process involved in decision-making but only its outcomes.

In the laboratory experiments, we also collect a score to measure the personality trait "worry" ("Penn State Worry Questionnaire" (PSWQ), Meyer et al. 1990). We measure stated worry each time the subjects face a new decision-making problem, thus obtaining multiple worry values for each subject. We also measure mood at the beginning of the experiment ("Brief Mood Introspection Scale" (BMIS), Mayer and Gaschke 1988), since mood has been shown to have an effect on behavior, particularly financial (Isen and Labroo 2003; Kliger and Levy 2003; Drichoutis and Nayga 2013).

In the flood field survey, we elicit personality traits indicating a possible Post-Traumatic Stress Disorder (PTSD) in respondents with a standardized questionnaire (Weathers and Ford 1996). If the score is above 44, the respondent is diagnosed as having a PTSD syndrome. S/he is also asked whether the origin of this syndrome is a past flood or not. In addition, the respondent is scored on emotions that s/he anticipates feeling in the event of future flooding (anticipated emotions).

A contingent valuation module determines respondents' willingness to participate in actions that will reduce risks and, if so, their Willingness To Pay (WTP) for protective devices (reducing the hazard) and individual insurance (reducing vulnerability). What is being measured here can be considered the value of the psychological gain related to prevention (insurance does not compensate for the psychological effects of a flood whereas the protective devices avoid them). In standard decision-making models, these anticipated emotions are only seen as one of several consequences, whereas they are potential factors in our approach.

A "flood-specific" module then elicits a subjective assessment of the risk of flooding at the place of residence in the coming year, in the next 10 years and in the next 100 years. It also elicits a score for severity of the flood (Chanel et al. 2013), which is zero for those who have never experienced flooding. People with at least one experience of flooding are questioned on number of flood experiences, on any protective action taken since and on how frequently they seek information on the risk of flooding during heavy rains.

4 Flood Survey

A quarter of the French population is at risk of flooding (MEDE 2012), which represents the major hazard in terms of number of claims paid and in terms of cost for the Cat Nat regime (the French insurance regime providing reimbursement for damage due to natural disasters). With €4.7 billion paid out between 1995 and 2006 under the natural disaster warrant (10 % for individuals, 90 % for firms), flooding accounts for 57 % of overall Cat Nat expenditure (CEPRI 2013), which justifies the choice of flood risk in our study.

4.1 Data

Four municipalities in the Provence Alpes Côte d'Azur (PACA) region (South Eastern France), within a 65 km radius, were chosen for their varying degrees of exposure to flood risk. Two municipalities have never been flooded: Miramas (25,300 inhabitants), at no risk of flooding, and Berre-l'Etang (13,800 inhabitants), located in an area with a potential risk of flooding due to torrential rivers (Arc river) and dam failure (Bimont). Two municipalities have unfortunately been flooded in the past 30 years due to flash floods. Vaison-la-Romaine (6,200 inhabitants) was flooded in September 1992 (20 years before the survey) by the Ouvèze river rising, with 37 deaths and four missing. Draguignan (36,600 inhabitants) was flooded in June 2010 (2 years before the survey) by the Nartuby river rising, with 23 deaths (12 in Draguignan itself) and two missing. These two floods currently represent the respective benchmarks for a once-a-century flood (i.e. which has a probability of occurrence of 0.01 per year) for the two municipalities.

The respondents interviewed had to live in one of the four municipalities and be older than 18 at the time of the survey. In addition, for the two flooded cities, respondents had to have been physically present when flooding occurred, and be over the age of 18 at this time. The main objective of this survey was to explore the influence of emotions on respondents' willingness to reduce their vulnerability and exposure to flooding. Although the questionnaire included eight modules, we only present below the issues pertinent to this paper (see Chanel et al. 2013; for additional results and the complete questionnaire).

The empirical analysis is based on a sample of 599 respondents interviewed at home face-to-face between 26 April and 30 June 2012 by a specialized survey institute. The average age of the sample is 51.3 years (standard deviation (s.d.)

17.02); 55.1 % are female; 36.2 % have at least one child at home; 41.8 % have at least a high school certificate; the monthly mean respondent income is €1,422 (s.d. €903); the monthly mean household income is €2,106 (s.d. €1,287); 47.6 % are owners. Section 4.2 examines the relationships between subjective probabilities, expected emotions, experience of flooding and WTP to avoid the risk (or the consequences) of a flood. Section 4.3 further examines whether the effects of emotions on decision-making described in Sect. 3 are supported by the data.

4.2 Sensitivity to Level of Flood Risk Exposure

Two types of test were used to explore sensitivity to the level of flood risk exposure. We computed standard tests of equality of means (or proportions) and, because such tests are sensitive to distributional assumptions, we also computed non-parametric and distribution-free tests for the continuous variables: the Kolmogorov Smirnov test (Chakravarti et al. 1967). The corresponding *p*-values give the level of significance and indicate whether equality with a given reference holds; if not, the sign of the difference determines the direction of the inequality (higher or lower). These two tests usually yielded similar conclusions, sometimes different conclusions but never opposite conclusions.

The top of Table 1 explores whether the experience of flood risk (four levels: not at risk—taken as reference, at risk but never flooded, flooded 20 years ago and flooded 2 years ago) is related to subjective beliefs (probabilities), anticipated emotions and willingness to protect against flood. First, we find that being at risk and recent experience of flooding have a positive effect on subjective probabilities, while experience in the more distant past has a negative effect. We also find that recent experience has a positive effect on the emotions expected in the event of a flood and no effect in the population never having experienced flooding; there is a negative effect when experience goes further back. We find that having had experience of flooding (recent or more distant) lowers willingness to participate in protection, while WTP to protect increases with recent experience (and living in a municipality at risk), although it decreases when experience goes further back.

It is interesting that although recent experience increases subjective probabilities, the anticipated negative emotions in the event of future flooding and the level of WTP, an experience going further back decreases them. Possible explanations: the population in Vaison-la-Romaine is older than in the three other municipalities on average; the population surveyed had chosen not to move after the flood, which might reveal a greater capacity to cope with this dramatic event.

The centre of Table 1 shows that a PSTD diagnosis (i.e. a PTSD score over 44), whatever the reason, is significantly related to an overestimation of subjective probabilities, higher anticipated negative emotions in the event of a future flood and

Table 1 Effects of flood experience and post-traumatic disorder on subjective probabilities, anticipated emotions and willingness to reduce flood risk (n = 599)

	Subjective probabilities of flood in the next …			Anticipated emotion in the event of flood	Willingness to participate in		Willingness to pay for …	
	… year	… 10 years	… 100 years		Prot. devices	Insurance	Prot. devices	Insurance
Experience with flooding (reference is not at risk, n = 152)								
Recent: 2 years (n = 151)	Higher (0.00011)	Higher (0.0007)	Equal (0.267)	Higher (0.0001)	Lower (0.0002)	Lower (0.0001)	Higher (0.0035)	Higher (0.0062)
	Higher (0.0001)	*Higher (0.0001)*	*Higher (0.0001)*	*Higher (0.0001)*			*Higher (0.022)*	*Higher (0.025)*
Old: 20 years (n = 149)	Equal (0.2129)	Lower (0.021)	Lower (0.0001)	Lower (0.0001	Lower (0.0017)	Lower (0.0046)	Lower (0.0001)	Lower (0.0008)
	Lower (0.026)	*Lower (0.001)*	*Lower (0.001)*	*Lower (0.0001)*			*Lower (0.0001)*	*Lower (0.0001)*
None but at risk (n = 147)	Higher (0.0002)	Higher (0.011)	Equal (0.2853)	Equal (0.2392)	Higher (0.0055)	Equal (0.7122)	Higher (0.0018)	Higher (0.0130)
	Higher (0.0001)	*Higher (0.0001)*	*Equal (0.225)*	*Equal (0.309)*			*Equal (0.108)*	*Equal (0.370)*
PTSD symptoms (reference is No, n = 535)								
PTSD (n = 64)	Higher (0.0005)	Higher (0.0001)	Higher (0.0001)	Higher (0.0273)	Equal (0.956)	Equal (0.9077)	Higher (0.0001)	Higher (0.0076)
	Higher (0.008)	*Higher (0.005)*	*Higher (0.011)*	*Higher (0.048)*			*Higher (0.003)*	*Higher (0.048)*
Severity score for flood consequences (reference is first half, n = 295)								
Fourth quartile (n = 154)	Equal (0.5247)	Equal (0.4705)	Equal (0.229)	Equal (0.7545)	Lower (0.0001)	Lower (0.0001)	Lower (0.0228)	Lower (0.0386)
	Equal (0.990)	*Equal (0.965)*	*Lower (0.014)*	*Equal (0.152)*			*Lower (0.0001)*	*Lower (0.0001)*

(continued)

Table 1 (continued)

	Subjective probabilities of flood in the next …			Anticipated emotion in the event of flood	Willingness to participate in		Willingness to pay for …	
	… year	… 10 years	… 100 years		Prot. devices	Insurance	Prot. devices	Insurance
Third quartile (n = 150)	Equal (0.241)	Equal (0.277)	Lower (0.0019)	Equal (0.9404)	Lower (0.0001)	Lower (0.0002)	Equal (0.1268)	Equal (0.2657)
	Equal (0.835)	*Equal (0.765)*	*Lower (0.063)*	*Equal (0.731)*			*Lower (0.009)*	*Lower (0.013)*

In each cell, the result of the test of equality with the reference population is given in regular and the result of equality of distribution (Kolmogorov-Smirnov test) with the reference is given in italics. *p*-values (bi-lateral if equality is found, unilateral otherwise) are given in brackets for each test

higher levels of WTP (for insurance as well as for protective devices against flooding). However, it does not significantly affect willingness to participate in protection.

The bottom of Table 1 shows that the severity of the flood experience is not significantly related to subjective probabilities and anticipated negative emotions in the event of a future flood, but is significantly related to a lower willingness to participate in protection and to lower levels of WTP for protection.

In Table 2, we find a positive relationship between anticipated negative emotions in the event of a future flood and level of subjective probabilities (for all three projected periods), on average a higher willingness to participate in protection but no significant effect on the level of WTP.

Finally, it is worth noting that having experienced flooding at least twice is positively correlated with a respondent being diagnosed with PTSD due to a flood event (p-value = 0.0179), with the anticipated emotions (p-value = 0.0069) and with the frequency of seeking information during heavy rains (p-value = 0.0752). Overall, we find evidence of relationships among emotions, flood experience and severity, a positive effect of recent experience on WTP but no clear evidence regarding how emotions impact willingness to protect.

Table 2 Effects of anticipated emotions on subjective probabilities, and willingness to reduce flood risk (n = 599)

	Subjective probabilities of flood in the next …			Willingness to participate in …		Willingness to pay for …	
	1 year	10 years	100 years	Prot. devices	Insurance	Prot. devices	Insurance
Anticipated emotion in the event of flood (reference is first quartile, n = 155)							
Fourth quartile (n = 154)	Higher (0.0001)	Higher (0.0001)	Higher (0.0003)	Equal (0.2612)	Higher (0.001)	Equal (0.2294)	Equal (0.6935)
	Higher (0.0001)	*Higher (0.0001)*	*Higher (0.0001)*			*Equal (0.471)*	*Equal (0.999)*
Third quartile (n = 134)	Higher (0.0167)	Higher (0.0016)	Higher (0.0016)	Higher (0.0259)	Equal (0.371)	Lower (0.0752)	Equal (0.469)
	Higher (0.001)	*Higher (0.0001)*	*Higher (0.0001)*			*Equal (0.648)*	*Equal (0.999)*
Second quartile (n = 156)	Higher (0.016)	Higher (0.0042)	Higher (0.027)	Higher (0.0492)	Higher (0.002)	Equal (0.7638)	Equal (0.8557)
	Higher (0.001)	*Higher (0.0001)*	*Higher (0.052)*			*Equal (0.999)*	*Equal (0.999)*

Cell contents as in Table 1

4.3 Exploring Emotion Effects

In order to explore further whether the effects previously defined are supported by the data, we compute correlation tests and study the sign and significance (p-value) of the links between the variables involved. Our flood data allow us to test five of the seven effects listed in Fig. 4: effects A, C, D, E and F, and we only discuss below correlations significantly different from zero. Note however that a correlation does not necessarily reflect a causal relationship between variables: it may be due to a third variable itself correlated with these two variables.

Effect A: Anticipated emotions. We seek to determine whether the emotions that we expect to feel in the event of flooding (anticipated emotions) influence the emotional evaluation that has prevailed in decisions to reduce risk through protective devices or to reduce vulnerability via insurance, in the frequency of seeking information about the risk of flooding during heavy rains or in the importance of protective action for those with experience of flooding. The score related to anticipated emotions is correlated neither with WTP nor with having implemented protective action following a flood. In contrast, it is very positively and significantly correlated with the decision to participate in the contingent valuation market (i.e. not to protest against the monetary evaluation exercise) for both WTP questions (p-values < 0.0005) and with the frequency of seeking information during heavy rains (p-value $= 0.0044$).

Effect C: Incidental emotions. We investigate whether external factors concerning personality and not related to the decision (including PTSD considered as a personality factor) influence five emotional evaluation measurements: anticipated emotions, WTP for protective devices or for insurance, frequency of seeking information about the risk of flooding during heavy rains or the importance of protective action for those with experience of flooding. The score for everyday risk-taking behavior aggregated over all domains is correlated with none of the five variables tested.

Neuroticism (i.e. over-anxious behavior) is strongly and positively correlated with anticipated emotions (p-value < 0.0001) and the frequency of seeking information on flooding during heavy rains (p-value $= 0.0117$), which is consistent with intuition. The feeling of control is only correlated with anticipated emotions, negatively (p-value $= 0.0493$). Conscientiousness, which affects the way we control our impulses and reflects a tendency to self-discipline, is positively correlated with WTP for insurance (p-value $= 0.0662$) and negatively with anticipated emotions (p-value $= 0.049$).

PTSD score is positively correlated with WTP for protective devices (p-value < 0.0001) and for insurance (p-value $= 0.0310$), and with the frequency of seeking information during heavy rains (p-value $= 0.0071$). The higher the PTSD score, the more willing a respondent is to reduce his/her exposure to the risk of flooding (or limit the financial consequences). It is also positively correlated with anticipated emotions (p-value = 0.0078). Being diagnosed as having a post-traumatic stress disorder (PTSD score greater than 44) is positively correlated with the two WTPs (p-values 0.0001 and 0.0274): being in a state of post-traumatic stress, whatever the cause, leads to higher WTP in all events. Self-assessing oneself as a lucky person is positively correlated with WTP for insurance (p-value = 0.043), as well as considering oneself happy (p-value = 0.0188).

Effect D: Past emotions. We investigate whether proxies of the emotions previously felt during a flood (i.e. being diagnosed with PTSD due to flooding, having experienced flooding at least twice before, having experienced flooding at least once, and the severity of the consequences) influence—through emotional evaluation—decisions to reduce risk (WTP for protective devices) and vulnerability (WTP for insurance, taking action for those with experience of flooding, frequency of seeking information about the risk of flooding during heavy rains and the level of anticipated emotions).

Having one previous experience of flooding is negatively and significantly correlated with the two WTPs (p-value of 0.0143 for protective devices, of 0.0471 for insurance), which is rather counter-intuitive. We find similar results regarding the severity of the consequences of previous flooding, which is negatively correlated with the WTPs (p-values of 0.0116 and 0.0633). However, it is positively correlated with the decision to take protective actions against flooding (p-value of 0.004), which confirms intuition.

We obtain similar results for PTSD score related to a flood event, although the p-values less strongly reject the absence of correlation (p-values between 0.0333 and 0.0704). We find no correlation between one experience of flooding and level of anticipated emotions, but level of anticipated emotions is positively correlated with having experienced flooding at least twice (p-value = 0.0069).

Effect E: Emotional resilience. We investigate whether having felt emotions in the past during a flood (i.e. diagnosed with PTSD due to a flood, having experienced flooding at least twice before, having experienced flooding at least once, and severity of the consequences) influences the personality factors that do not depend on the decision (external factors, including the PTSD score considered here as a component of personality). Having experienced a flood once before is positively correlated with the personality trait of conscientiousness (p-value < 0.0001), with a sense of control (p-value = 0.0002) and with a PTSD score related to a flood event (p-value = 0.0019). Neither having experienced flooding at least twice before, nor the severity of the last flooding event, is significantly correlated with any of the personality factors tested.

Effect F: Prevention focus reinforcement. The anticipated emotions score is positively correlated with the three subjective assessments of the probability of being flooded in the coming year (p-value = 0.0018) in the next 10 years (p-value < 0.0001) and in the next 100 years (p-value = 0.0088).

5 Experiment on Insurance Behavior Under Anxiety

The lab experiments allowed us to complement and confirm survey results through within-subjects analysis. The laboratory setting means that subjects can be observed making repeated choices, with the emotions felt being manipulated. In our experiment, subjects have to make repeated decisions on insuring against the risk of failing in a real effort task. We manipulate feelings of worry through changes in a gain/loss framework and stakes. During the experiment, in some situations subjects are confronted with a low probability of losing a high gain: these situations mimic "catastrophic" situations.

5.1 Experimental Design

We present the design of the experiment with the sequences of choices and the different characteristics used to induce an emotional component during the decision-making process. Subjects perform a succession of 64 bingo activities. A bingo activity is divided into two parts: the information and decision process and the execution of the real effort task (see Fig. 5).

We chose a perceptual numerosity task for three reasons: each trial is very quick, we can control for difficulty (we make subjects perform at an equal success rate) and we dispose of robust psychophysics models and tools (Signal Detection Theory) to treat the data. In this task, two circles, each containing a given number of dots, are briefly presented to subjects (see Fig. 5b, c). They have to find the one containing the most dots. One of the two circles always contained 50 dots while the other contained $50 + x$ dots. The difficulty level x is adjusted for each subject in a preliminary calibration phase. The trial proceeds as follows: subjects first observe two fixation crosses and start the trial at their convenience. Then the stimulus appears for 700 milliseconds (ms). Subjects give their choice (right or left) and their level of confidence from 0 to 100. The mechanism that reveals the level of confidence (trial confidence) is the matching probability (see Fig. 5b), which is a proper elicitation rule that is not biased by risk aversion (cf. Hollard et al. 2016). During the task, each trial is repeated 5 times in order to obtain a number of successes over 5 trials that will determine the accuracy of the bingo activity.

The information and decision process comes in when subjects bet on their accuracy in the perceptual task. To succeed in a bingo bet, subjects have to obtain a minimum of n successes over 5 trials, with n = 2, 3 or 4. We also vary the level of stakes that they can win: €200 in the high stake bet and €20 in the low stake bet. Finally, we frame the decision as gain or loss: in the gain frame, subjects have €0 and can win €20 or €200 if they obtain at least n successes out of 5; in the loss frame, they have €20 or €200 and they can lose everything if they fail more than 5 − n times out of 5.

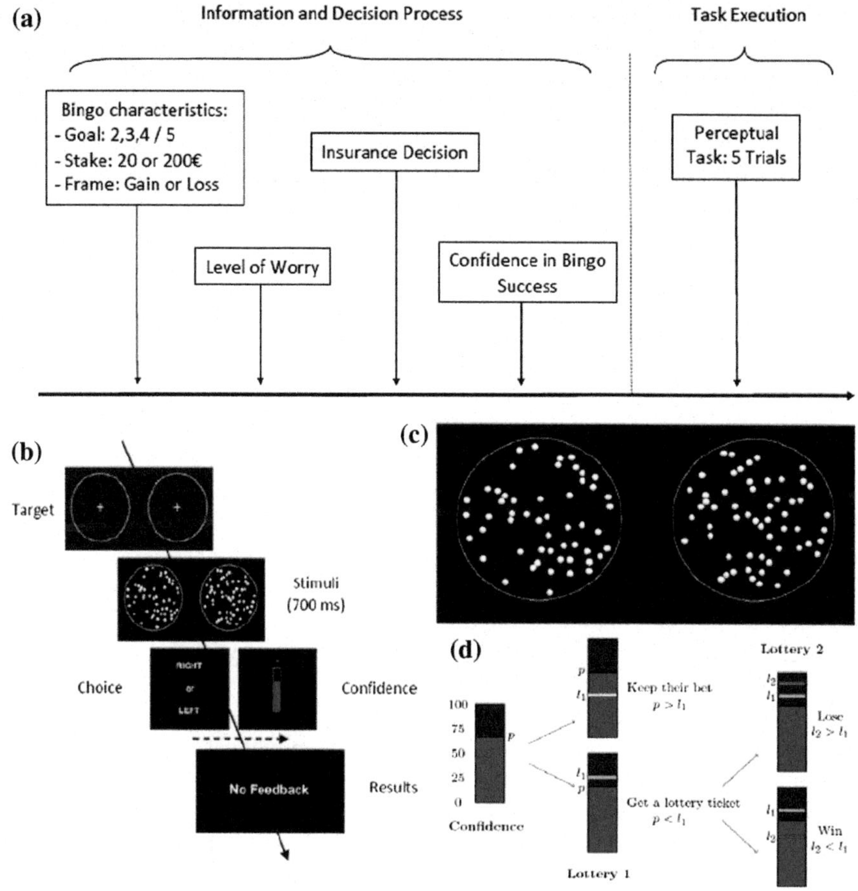

Fig. 5 Design of the experiment. **a** Represents the two main components of the experiment: first the information gathering and decision process, then execution of the perceptual task. **b** Describes a sequence of choices during one trial of the perceptual task. **c** Is an example of stimuli used during the task. **d** Details the mechanism of the matching probability used for confidence

A sequence of bingo is organized as follows (see Fig. 5a). First, we provide information to subjects about the bet: goal to reach, amount of stake and bet in gain or in loss frame. Then, subjects have to reveal their level of worry on a 10-point scale. Using a BDM mechanism (Becker et al. 1964), we then propose an insurance mechanism based on the elicitation of their certainty equivalent: on a scale from 0 to 20 or 200 (for the stake of the bet) they are asked to give the minimum amount of money they want to win for sure in the gain frame or the maximum amount they are willing to lose for sure in the loss frame. The difference between this certainty equivalent and the stake is the maximum amount they are willing to pay for full insurance (insurance premium). Finally, we ask them to reveal their level of confidence in bingo success via the matching probability mechanism (bingo confidence). After all these steps, the perceptual task begins with 5 trials. At the end of the bingo, no feedback is provided and subjects proceed with the next bingo. Subjects play a total of 64 bingos: 32 in the gain frame, 32 in the loss frame.

The experiment took place in May and July 2012 at the Laboratory of Experimental Economics in Paris (LEEP). We collected data on 98 subjects, students from all fields (25 % studying economics). The average age was 22.9 years (s.d. 0.02); 52.9 % were female.

5.2 Descriptive Results

Since the purpose of the experiment was to manipulate emotions in order to examine behavioral consequences, we first examine whether the expected effect is obtained on inducing worry.

We observe an important gender effect, as women tend to be more anxious than men on average. Regarding psychological variables, we find some expected results: a high correlation with the PSWQ, which measures the tendency to worry, and with neuroticism. Subjects that declare they have better control also declare that they are less anxious. We also find a negative correlation, but not statistically significant, with conscientiousness; there are no significant correlations with the various mood measures (BMIS).

To examine the effects of the design on worry, we first transform the reported level of worry into a normalized worry by computing individual z-scores in order to control for a potential bias in the use of the worry scale and to obtain values that are comparable across subjects. We expect to find higher stated worry in the loss frame, when the stakes are high and when the goal to reach is more difficult. We perform an ANOVA test of the different aspects of the experiment on the normalized worry and find that the expected effects are highly statistically significant. In particular, the mean normalized worry is higher in the loss frame than in the gain frame (+0.13), higher when the stake is high than when it is low (+0.14). When we perform the same ANOVA test individual-by-individual, we observe a statistically significant effect, at 10 %, of the loss/gain frame for 47 % of our sample, and of the level of stake for 38 % of our subjects.

Note that, for easy objectives ($n = 2$ out of 5), the normalized stated worry is much higher in the situations of "catastrophic" risk (loss frame and high stake) than in less extreme situations (gain frame and low stake) (+0.60).

Concerning the other behavioral data, we observe that the mean accuracy in the perceptual task is 68.3 %, which is in line with the calibration objective (71 %). The mean trial confidence is 70.1 %, which shows a moderate overconfidence. The mean success rate in the bingo is 80.0 %, with 50.1 % when the objective is difficult ($n = 4$ out of 5), 75.7 % when the objective is moderately difficult ($n = 3$ out of 5) and 95.1 % when the objective is easy ($n = 2$ out of 5). The mean bingo confidence is 75.7 %, which indicates underconfidence. The mean bingo confidence is respectively 42.1 % for $n = 4$, 78.7 % for $n = 3$ and 91.0 % for $n = 2$. The mean

Table 3 Percentages of evolution of mean values due to various aspects of the design (N = 98) obtained by an OLS regression (for accuracy we proceed by standard tests of equality of means)

	Goal: 3/5 versus 2/5	Goal: 4/5 versus 2/5	Loss versus Gain	Stake: High versus Low
Normalized worry	+0.42 (0.0001)	+0.97 (0.000)	+0.29 (0.0001)	+0.31 (0.0001)
Accuracy	+1.6 % (0.012)	+0.7 % (0.300)	−0.6 % (0.269)	+1.9 % (0.0001)
Trial confidence	+0.1 % (0.629)	−1.3 % (0.0001)	−0.5 % (0.033)	−0.0 % (0.913)
Insurance premium	+8.2 % (0.0001)	+23.7 % (0.0001)	+0.2 % (0.741)	+4.2 % (0.0001)
Bingo confidence	−12.3 % (0.0001)	−48.9 % (0.0001)	−0.9 % (0.003)	−0.9 % (0.002)

p-value are given in brackets

insurance premium corresponds to 37.0 % of the stake. In detail, the mean value ranges from 52.7 % for n = 4, to 37.2 % for n = 3 and 29.1 % for n = 2: the risk premium is very high when the probability of failure is low, while the subjects are almost risk-neutral when the probability of failure is high. We report the impacts of the design at an aggregate level in Table 3.

More specifically, for easy objectives (n = 2 out of 5), we observe that the insurance premium is 6.4 % higher in the situations of "catastrophic" risk (loss frame and high stake) than in the less extreme situations (gain frame and low stake).

5.3 *Exploring the Emotion Effects*

To examine the effects of anxiety on behavior, we perform a second transformation of the worry data, computing a level of worry controlled for the characteristics of the bingo: we retain the residual of a regression of all the characteristics of the bingo (12 dummies for 3 goals, 2 frames, 2 stakes) on the normalized worry. This residual worry captures variations in worry, which can be considered as independent of the design. The rationale for using this residual worry instead of the total worry is to reduce the risk of overestimating the effect of worry on subjective beliefs and insurance decisions. We define a bingo activity as being performed under high worry if this value is strictly positive, and under low worry if it is negative or null. Since we use the residual worry that takes the characteristics of the bingo into account, values are balanced between high and low worry.

Effect F: Prevention focus reinforcement. We examine the existence of effect F on bingo confidence, trial confidence and metacognition abilities.

Bingo confidence is significantly lower (−1.3 %) under high worry (p-value < 0.0001). Note also that subjects who are sensitive to the loss/gain framework in terms of worry (47 % of the pool) suffer from greater depressive effect: −1.7 % under high worry (vs. −0.9 % for non-sensitive subjects) and −1.4 % in the loss frame (vs. −0.5 % for non-sensitive subjects).

Trial confidence is also significantly lower: −1.6 % under high worry (p-value < 0.0001) but with no significant difference between sensitive and non-sensitive subjects. There is also no difference for the loss frame effect between sensitive and non sensitive subjects.

We expect the attentional effect of emotion to have an impact on metacognition ability in the perceptual task (see Massoni (2014) for a complete analysis of worry's effects on metacognitive abilities). We consider two metacognitive aspects: overconfidence and discrimination. Overconfidence is a measure of how close confidence judgments are on average to real success. We speak of overconfidence since usually the mean confidence is above the real success rate. Overconfidence is significantly lower under high worry (p-value $= 0.044$).

Discrimination is a measure of how well variations in confidence match variations in performance. We estimate discrimination ability by the meta-d' measure (Maniscalco and Lau 2012). Discrimination is significantly higher under high worry (p-value $= 0.032$).

Effect G: Stress response. We first consider the effects of worry on insurance decisions. The insurance premium is significantly higher (+3.2 %) under high worry (p-value < 0.0001). An OLS regression of insurance premiums on a set of variables (bingo confidence, bingo characteristics, personality measures) that also includes worry shows that worry has a positive and significant effect.

The next step is to examine whether the decision process is directly affected by the emotions. Note that, under high worry, reaction times are significantly longer for insurance (p-value $= 0.0199$), bingo confidence (p-value $= 0.0544$), trial perception (p-value $= 0.0367$) and trial confidence (p-value $= 0.014$). Conversely, the reaction times for the same behavioral data are significantly shorter (the respective p-values are 0.0173, <0.0001, <0.0001 and <0.0001) in the loss frame.

If the decision process is more driven by emotions under stress, we should observe, in contrast, that the cognitive process plays a lesser role in the insurance decision. In a rational decision-making process, subjective beliefs should be the main determinants of the insurance decision. Therefore, we investigate how insurance decisions vary with bingo confidence. We define as irrational a subject for whom the correlation between insurance premium and bingo confidence is not negative and significant at 10 %. We observe that 29 % of subjects are classified as irrational in the loss domain, against only 17 % in the gain domain. This difference of proportion is statistically significant (p-value $= 0.0309$). If we compare occurrences of irrational behavior under high and low worry, we find no difference. The results of two OLS regressions of the insurance premium, where we split the bingo confidence variable into two crossed variables with the gain/loss framework or with high/low worry, confirm these previous results: we observe that the insurance premium is more dependent on bingo confidence in the gain domain than in the loss domain, while the difference is less significant for low and high worry, although we observe a slightly higher dependence under low worry.

6 Discussion and Conclusion

Our results confirm that emotions (taken in a broad sense that includes personality traits or PTSD score) may help explain choice under uncertainty related to catastrophic risks.

First, we do find a significant and positive relationship between anticipated emotions and **formation of beliefs** in the flood survey, and this effect is also observed in the lab experiments, regardless of whether we consider the impacts of the gain/loss frame or the worry effect. The prevention focus triggered by negative emotions is a robust effect that reinforces the perception of threatening events.

Second, although emotions clearly have an effect on **decisions**, the nature of this effect appears complex and will require further elucidation. While we find that anticipated emotions impact desire to reduce the risk or vulnerability associated with a flood, they do not appear to affect the importance of this reduction. It seems that the anticipation of high negative emotions in the event of flooding induces a desire to protect without determining the intensity of this protection, probably mainly determined by other factors (income, housing characteristics). Immediate emotion (anxiety) increases the WTP for insurance but it seems that it plays a role through the cognitive evaluation channel. The loss frame seems to activate a stronger stress response with a significant change in terms of irrational behavior but no effect on the WTP for insurance. We also find evidence of a counter-intuitive relationship between past emotions and decisions: having experienced flooding before and the severity of flooding are negatively correlated with WTP. To sum up, the varying results we observe show that different yet related emotions play different roles in the decision-making process.

Third, our data also suggest that incidental emotions felt immediately (i.e. during the decision process) are associated with external factors that are not related to the decision on the flood event, and that conversely, having experienced a flood (even without severe consequences) can affect some personality traits in a durable way.

So it is too early for a verdict on whether examining the effects of emotions on decision-making can identify which emotion(s) to focus on when guiding vulnerable populations towards more rational decisions. Clearly, evidence is not fully conclusive, and this points to several valuable research avenues. The different types of negative emotions need to be better distinguished, and we need more finely-tuned analyses regarding the effect of emotions on perception of the efficiency of protection, whether via physical devices or by insurance. While our lab experiment reveals that insurance is perceived as efficient when people feel anxious, future work should assess whether this holds true in other emotional situations.

Acknowledgements We thank Marjorie Sweetko for her thorough re-reading of the English, as well as the program Riskemotion (ANR-08-RISKNAT-007-01) and both the OT-Med Labex and the A*MIDEX project (n° ANR-11-IDEX-0001-02) funded by the «Investissements d'Avenir» French Government program, managed by the French National Research Agency (ANR) for financial support.

References

Becker, G., DeGroot, M., & Marschak, J. (1964). Measuring utility by a single-response sequential method. *Behavioral Science, 9*(3), 226–232.

Bell, D. E. (1982). Regret in decision making under uncertainty. *Operations Research, 30*(5), 961–981.

Brunnermeier, M. K., & Parker, J. A. (2005). Optimal expectations. *American Economic Review, 4*, 1092–1118.

Caplin, A., & Leahy, J. (2001). Psychological expected utility theory and anticipatory feelings. *Quarterly Journal of Economics, 116*(1), 55–79.

Cepri, Centre Européen de Prévention du Risque Inondation. (2013). Indemnisation des victimes. Retrieved from November 13, 2013, http://www.cepri.net/indemnisation-des-victimes.html.

Chakravarti, I. M., Laha, R. G., & Roy, J. (Eds.). (1967). Techniques of computation, descriptive methods and statistical inference. In *Handbook of methods of applied statistics* (Vol. 1). Wiley.

Chanel, O., & Chichilnisky, G. (2013). Valuing life: Experimental evidence using sensitivity to rare events. *Ecological Economics, 85*, 198–205.

Chanel, O., Chichilnisky, G., Massoni, S., Vergnaud, J.-C., & Vincent Lyk-Jensen, S. (2013). Décision en présence d'incertitude et d'émotions face à des risques de catastrophes naturelles (RISKEMOTION). *Final Report ANR-08-RISKNAT-007-01*, Greqam, Marseille, France.

Chanel, O., & Chichilnisky, G. (2009). The influence of fear in decisions: Experimental evidence. *Journal of Risk and Uncertainty, 39*(3), 271–298.

Chichilnisky, G. (2000). An axiomatic approach to choice under uncertainty with catastrophic risks. *Resource and Energy Economics, 22*, 221–231.

Chichilnisky, G. (2009). The topology of fear. *Journal of Mathematical Economics, 45*, 807–816.

Drichoutis, A. C., & Nayga, R. M, Jr. (2013). Eliciting risk and time preferences under induced mood states. *The Journal of Socio-Economics, 45*, 18–27.

Evans, J. S. B. (Ed.). (1989). *Bias in human reasoning: Causes and consequences*. Lawrence Erlbaum Associates, Inc.

Gul, F. (1991). A theory of disappointment aversion. *Econometrica, 59*(3), 667–687.

Hollard, G., Massoni, S., & Vergnaud, J.-C. (2016). In search of good probability assessors: an experimental comparison of elicitation rules for confidence judgments. *Theory and Decision, 80*(3), 363–387.

IPCC, Intergovernmental Panel on Climate Change. (2013). Summary for policymakers. In T. F. Stocker, D. Qin, G.-K. Plattner, M. Tignor, S. K. Allen, J. Boschung, A. Nauels, Y. Xia, V. Bex & P. M. Midgley (Eds.), *Climate Change 2013: The Physical Science Basis. Contribution of Working Group I to the Fifth Assessment Report of the Intergovernmental Panel on Climate Change.* Cambridge, UK and New York, NY, USA: Cambridge University Press.

Isen, A. M., & Labroo, A. A. (2003). Some ways in which positive affect facilitates decision making and judgment. In *Emerging perspectives on judgment and decision research* (Vol. 365).

John, O. P., Donahue, E. M., & Kentle, R. L. (1991). *The big five inventory-versions 4a and 54*. Berkeley, CA, USA: University of California, Berkeley, Institute of Personality and Social Research.

Kahneman, D., & Frederick, S. (2002). Representativeness revisited: Attribute substitution in intuitive judgment. *Heuristics and Biases: The Psychology of Intuitive Judgment*, 49–81.

Kahneman, D., & Tversky, A. (1979). Prospect theory: An analysis of decision under risk. *Econometrica, 47*, 263–291.

Keren, G., & Schul, Y. (2009). Two is not always better than one: A critical evaluation of two system theories. *Perspectives on Psychological Science, 4*, 533–550.

Kliger, D., & Levy, O. (2003). Mood-induced variation in risk preferences. *Journal of Economic Behavior & Organization, 52*(4), 573–584.

Loewenstein, G. (1996). Out of control: visceral influences on behavior. *Organizational Behavior and Human Decision Processes, 65*(3), 272–292.

Loewenstein, G., Weber, E., Hsee, C., & Welch, N. (2001). Risk as feelings. *Psychological Bulletin, 127*(2), 267–286.

Loewenstein, G. (2000). Emotions in economic theory and economic behavior. *American Economic Review, 90*(2), 426–432.

Loomes, G., & Sugden, R. (1986). Disappointment and dynamic consistency in choice under uncertainty. *Review of Economic Studies, 53*, 271–282.

Loomes, G., & Sugden, R. (1982). Regret theory: An alternative theory of rational choice under uncertainty. *Economic Journal, 92*(368), 805–824.

Maniscalco, B., & Lau, H. (2012). A signal detection theoretic approach for estimating metacognitive sensitivity from confidence ratings. *Consciousness and Cognition, 21*, 422–430.

Massoni, S. (2014). Emotion as a boost to metacognition: how worry enhances the quality of confidence. *Consciousness and Cognition*, 29, 189–198.

Mayer, J. D., & Gaschke, Y. N. (1988). The experience and meta-experience of mood. *Journal of Personality and Social Psychology, 55*(1), 102–111.

Meyer, T. J., Miller, M. L., Metzger, R. L., & Borkovec, T. D. (1990). Development and validation of the Penn State Worry Questionnaire. *Behaviour Research and Therapy, 28*(6), 487–495.

MEDE, Ministère de l'Écologie, du Développement durable et de l'Énergie. (2012). Première évaluation nationale des risques d'inondation, Principaux résultats - EPRI 2011, juillet, report.

Munich RE. (2013). Natural catastrophe statistics for 2012 dominated by weather extremes in the USA. Press release, 3 Jan, Munich, Germany.

Reid, A., & Gonzalez-Vallejo, C. (2009). Emotion as a tradeable quantity. *Journal of Behavioral Decision Making, 22*, 62–90.

Rick, S., & Loewenstein, G. (2008). The role of emotion in economic behavior. In M. Lewis, J. Haviland-Jones, & L. Feldman-Barrett (Eds.), *The handbook of emotion* (3rd ed., pp. 138–156). Guilford: New York, NY, USA.

Rotter, J. (1966). Generalized expectancies for internal versus external control of reinforcements. *Psychological Monographs, 80*, Whole No. 609.

Rustichini, A. (2008). Dual or unitary system? Two alternative models of decision-making. *Cognitive, Affective, and Behavioral Neuroscience, 8*, 355–362.

Sigma. (2015). Natural catastrophes and man-made disasters in 2014, N° 2/15 (52 pp.). Swiss Reinsurance Company Ltd, order 270_0215_en.

Slovic, P., Finucane, M. L., Peters, E., & MacGregor, D. G. (2004). Risk as analysis and risk as feelings: Some thoughts about affect, reason, risk, and rationality. *Risk Analysis, 24*(2), 311–322.

Sunstein, C. (2002). Probability neglect: Emotions, worst cases, and Law. *The Yale Law Journal, 112*(1), 61–107.

Von Neumann, J., & Morgenstern, O. (1944). *Theory of games and economic behavior*. Princeton, NJ, USA: Princeton University Press.

Weathers, F., & Ford, J. (1996). Psychometric review of PTSD Checklist (PCL-C, PCL-S, PCL-M, PCL-PR). In B. H. Stamm (Ed.), *Measurement of stress, trauma, and adaptation*. Luthewille, MD, USA: Sidran Press.

How the Change of Risk Announcement on Catastrophic Disaster Affects Property Prices?

Hayato Nakanishi

1 Introduction

A number of papers have expressed concern with the expected utility theory. Some of these discussions stem from the theory's insensitivity to rare events. Chichilnisky (2009) proved that conventional expected utility is not sensitive to rare events and noted the significance of generalized expected utility, which is sensitive to rare events. Rare events with major consequences, catastrophes, have contributed particularly to researcher concerns. This is because the insensitivity of expected utility leads to the disregarding of catastrophic risk, which in turn leads to the failure of economic evaluation. Chanel and Chichilnisky (2013) observed that, with contingent valuation, experimental participant perception of a participant's own life differed from the values that were estimated by conventional expected utility theory.

However, Hausman (2012) and other authors have doubted the contingent valuation method because of hypothetical bias. A substantial number of studies have been conducted in laboratory experiments that survey the importance of real payment and contingent valuation to overcome this bias; however, researchers cannot employ these incentives for the evaluation of catastrophic risk because of ethical considerations. There are evidences from neuroscience that rare catastrophic events play a particular role in our decision processes; however, few studies offer real data. Levitt and List (2007) noted that laboratory experiments may not be generalizable to the real world, and a lack of real data since this pronouncement has amounted to a lack of convincing evidence in support of economic theories that can evaluate catastrophes. Therefore, this article presents real data as evidence that the information on catastrophic risk has an effect on economic behavior.

H. Nakanishi (✉)
Japan Society for the Promotion of Science (JSPS), Institute of Economic Research, Kyoto University, Yoshida honmachi, Sakyo-ku, Kyoto 606-8501, Japan
e-mail: nakanishi.h.aa@gmail.com

G. Chichilnisky and A. Rezai (eds.), *The Economics of the Global Environment*, Studies in Economic Theory 29, DOI 10.1007/978-3-319-31943-8_25

The Japanese government published a report on the risk of mega earthquakes and tsunamis that are believed to occur once in a millennium. An earthquake with a magnitude of nine and a subsequent tsunami in a certain part of Japan were considered. Until the report was published, these earthquakes and tsunamis had been disregarded because they are considered rare. However, as the East Japan Great Earthquake demonstrated, such a tsunami can have catastrophic consequences. The expected damage of the tsunami of this mega earthquake was far greater than had been predicted. The information concerning the risk of big earthquakes that have occurred once in 50 or 100 years was not affected by the new report. Therefore, we use this report as a treatment that affected the information on catastrophes.

For econometric analysis, we employed the hedonic method for econometric analysis because the risk of a tsunami is tied to the location of land. Rosen (1974) discussed that the marginal price of an attribute reveals the marginal willingness to pay; the housing and land price hedonic method has been widely employed for the economic evaluation of risk related to the geographical attributes of land.

Additionally, because numerous studies employ the hedonic method for evaluating the risk tied to the location of land, we can discuss the estimation results and compare them to existing hedonic research. For example, Nakagawa et al. (2007) estimated earthquake risk aversion in housing rents using a hazard map of Tokyo, Beron et al. (1997) estimated the earth quake risk before and after a massive earthquake, Gayer et al. (2002) estimated the value of reducing cancer risk. Bin et al. (2008) estimated willingness to pay for coastal amenities and risk, and Donovan et al. (2007) examined the effectiveness of a wildfire risk rating project. Although there are few papers concerning catastrophes, some discussion is required because the focus of this study is risk evaluation.

To identify the causal effect with nonexperimental data, identification strategies have been developed in the context of policy and program evaluation and treatment effect. These quasi-experimental designs include regression discontinuity (RD), instrumental variable estimation, propensity score matching, before and after (BA), and difference in differences (DD). These quasi-experimental approaches can be employed to avoid biased estimation of causal effect, which is caused by cross-sectional hedonic regression with omitted variables. With respect to the risk of natural disasters, the number of studies which employ quasi-experimental design is increasing. Beron et al. (1997) conducted a BA analysis and specified the structure that individual perception of risk affects property prices, Bin and Landry (2012) examined flood hazard effect for Pitt County, North Carolina, using multiple storm events with spatial DD regression design. With respect to the quasi-experimental approach, Lee (2005) presents a comprehensive review of literature, and an extensive review of the hedonic and quasi-experimental design is presented by Parmeter and Pope (2012).

This paper employs a DD design to identify the causal effect of a change in the predicted maximum damage of a tsunami. There are three reasons for the employment of a DD design. First, a new report changed the risk information to reflect only a certain area of Japan. Therefore, we can divide the data into a control group and a treatment group. Second, the area where the prediction was updated was decided exogenously by the government. Third, the nationwide balanced panel data is provided by the Japanese government. In DD design, balanced panel data and the differencing operations allow us to eliminate changes in land price caused by reasons other than treatment, such as depopulation if treatment and control groups are properly determined. These facts enable us to employ direct DD estimation.

The conventional expected utility derived from the objective distribution of risk predicts no effect since catastrophic events have sufficiently low probability of occurrence. That is, measure of catastrophic events is zero while expected maximum damages are updated by the announcement. Hence household's decision on landowning would not change. This predicts no effect on land prices by the announcement.

With nonparametric DD design, both positive and negative land price effects of new risk announcements are observed. Specifically, negative effects to land prices are observed in high tsunami risk areas and positive effects to land prices are observed in low tsunami risk areas. These positive or negative effects are not observed in relatively high-altitude areas, The observed result, that new risk reports on rare catastrophic assumption cause change in land prices, implies that the risk of rare catastrophic event affects land prices while conventional theory predicts no effect. However, generalized theory of Chichilnisky (2009) which includes conventional theory can explain these significant results.

The rest of the paper is structured as follows. Section 2 presents the background of our study, Sect. 3 presents the specification of the treatment effect and the estimation procedure, Sect. 4 describes the data, Sect. 5 presents the estimation results and a discussion of the treatment effect, and Sect. 6 summarizes the paper and presents the conclusions.

2 Background of the Study

Japan experiences earthquakes and tsunamis because it lies at the nexus of four tectonic plates and is surrounded by the sea. The Japanese government has researched and released reports on the expected hazard levels of earthquakes and tsunamis for those earthquake events with intervals that are understood to a certain extent, or earthquake events with catastrophic anticipated damages. They are Kuril Trench earthquakes; Japan Trench earthquakes; the Tokai, Tonankai and Nankai earthquakes; Tokyo inland earthquakes, and the Chubu and Kinki region inland earthquakes. (Our control group is not included because research concerning this region is not extensive; however, some massive tsunamis have occurred as Fig. 1 shows). These reports are openly available on the Internet and local governments compose hazard maps based on the data. As presented by Nakagawa et al. (2009, 2007) and Naoi et al. (2009), it is certain that these earthquake-related hazard maps affect property values. Therefore, these reports are considered to be influential.

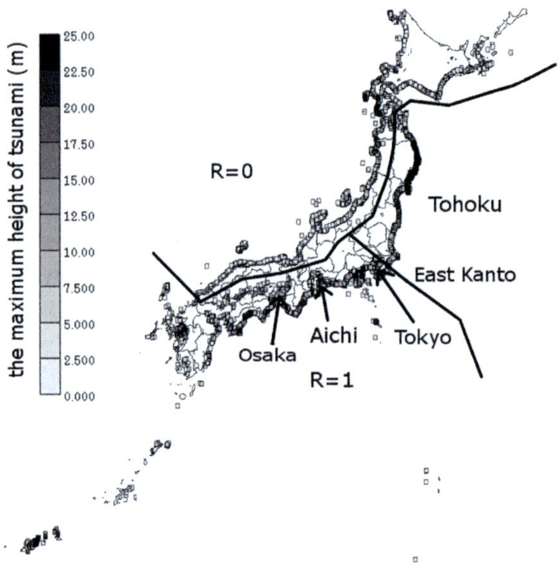

Fig. 1 Recorded Height of Japanese Tsunamis n (R = 1 indicates the region covered by the new report, R = 0 indicates the region employed as a control group in this study)

Before the East Japan Great Earthquake, expected hazard levels were calculated based on the seismic source models that can replicate "most" seismic intensities and tsunami heights previously recorded. However, if the seismic intensity or tsunami heights of an earthquake were not reproducible by the model, the earthquake was considered to have a low likelihood of occurrence, even if such an earthquake may have occurred in the past, and was disregarded from the hazard assumptions. Catastrophic earthquakes and tsunamis that occur once in a millennium such as the East Japan Great Earthquake are also included in this low likelihood category.

Therefore, in response to the severe damage caused by the East Japan Great Earthquake and the disregard of unreproduced massive earthquakes and tsunamis, the Japanese government decided to re-calculate the risk of earthquake and flooding from tsunamis under an assumption that includes previously disregarded catastrophic earthquakes and tsunamis.

Under the new assumption, the predicted maximum damage of the Nankai Trough massive earthquake was updated on March 31, 2012 and August 29, 2012 (R = 1 area in Fig. 1). The Nankai Trough massive earthquake was coincidental with the Tokai, Tonankai and Nankai earthquakes, which are replicable by conventional earthquake models. However, the Nankai Trough massive earthquake had been disregarded. In the updated report, the maximum possible damage of a tsunami is considered significantly more severe than in the conventional report.

With respect to earthquakes that are reproducible by the conventional model, the expected risks that are related to tsunamis remain unchanged following the report update because a conventional seismic model cannot reproduce only rare catastrophic earthquakes. Hence, under the assumption that property values related to a

Table 1 Time series description of official land prices and updated announcements

1 January 2012	Evaluation point for the official 2012 land prices (published on 23 March 2012)
31 March 2012	The first version of the report on the risk of the Nankai Trough massive earthquake was published
29 August 2012	The second version of the report on the risk of the Nankai Trough massive earthquake was published
1 January 2013	Evaluation point for the official 2013 land prices (published on 22 March 2013)

catastrophically severe tsunami are additively separable from property values related to the risk of a tsunami reproducible by a conventional model, the conventional risk effect can be removed using a differencing operation. The DD design is therefore suitable for identifying the effect of this risk announcement.

For the DD design risk evaluation, we employ official Japanese land price data to conduct direct nonparametric estimation. The official land prices in Japan are reported annually in March according to law. This price is calculated as the expected unit price of the land as of January 1 in a competitive market. Thus, these data are composed of balanced panel data of selected lots with no movement either in or out of the regions. This allows us to estimate the treatment effect nonparametrically. The evaluation is conducted by at least two real estate appraisers and is based on the actual trades of neighboring land. These official land price data is employed by Nakagawa et al. (2009) and Tsutsumi and Seya (2009).

Authors (e.g., Ma and Swinton 2012) question the accuracy of the assessed land value; however, the assessed price eliminates the potential problems that could bias the market as a result of special transaction. Moreover, we employ a differencing operation in this study. Therefore, these noises are differenced out in the estimation of treatment effect, whereas biases may arise in the ordinal hedonic regression (Table 1).

3 A Difference in Differences Design

A DD design has been widely employed in program evaluation literature. In this paper, because it is reasonable to assume that the “update effect” is heterogeneous (a function of altitude and the maximum height of historical tsunamis, which includes rare catastrophic events), we employ a nonparametric conditional DD design to avoid miss-specifying the heterogeneity. A general discussion of semiparametric DD is presented by Abadie (2005). However, we follow the discussion of Lee (2005) because exogenous treatment and balanced panel data provides a simpler DD estimator.

Let $P_{it}(r)$ be the potential price of land i (= 1,…N) at time t (a, b; a = 2012, b = 2013) in Region r (= 0 for control group, 1 for treatment group). $X_{it}(=X_i)$ is the maximum height of historical tsunamis, $H_{it}(=H_i)$ is the altitude of the land, $Z_{it}(=Z_i)$ is other attributes of land i, R_i is the dummy variable or indicator for treatment group (Nankai Trough area not including Tokyo, Aich, and Osaka), and τ_i is the dummy variable that indicates time is equal to b. In this study, we removed the observation that attributes of the land have varied in time period [a, b] for simplicity. Hence, index t is removed from X_{it}, Z_{it}, H_{it} This is valid if the change of land attribute is exogenous. Because the lots of officially reported land prices are determined before t = a, the changes of land attributes are considered exogenous. In this setting, observed response P_{it} can be written as

$$P_{it} = (1 - R_i\tau_t)P_{it}(0) + R_i\tau_t P_{it}(1) \tag{1}$$

Because R_i is determined exogenously by the government, $E(P_{ib}(1) - P_{ib}(0)|r = 1, x, h)$ is the treatment effect for Region 1 at time b. This is the conditional average economic value of reducing the risk level of flooding from a tsunami. However, $P_{ib}(0)$ and $P_{ib}(1)$ are not observed simultaneously in the real world. Thus, quasi-experimental design is required to identify this conditional average effect. For identification, we need to ensure a same time effect assumption.

Assumption (same time effect)

$$E(P_{ib}(0) - P_{ia}(0)|r = 1, x, h) = E(P_{ib}(0) - P_{ia}(0)|r = 0, x, h) \tag{2}$$

It is easy to verify that this same time effect condition is weaker than $E(P_{it}(1)|r = 1, x, h) = E(P_{it}(1)|x, h)$ t = a, b. (e.g., Lee 2005). This $E(P_{it}(1)|r = 1, x, h) = E(P_{it}(1)|x, h)$ assumption is imposed in the literature that employs linear DD regression.

For accurate interpretation purposes, we introduce a linear DD regression approach. One of the special cases that allows $E(P_{ib}(1) - P_{ib}(0)|r = 1, x, h)$ to be the coefficient of linear model is

$$P_{it} = c_i + \tau_t\alpha + R_i\beta + R_i\tau_t\gamma + \varepsilon_i \tag{3}$$

where α, β, and γ are the parameters that indicate time, regional, and treatment effect. c_i is the fixed effect of property i, and ε_i is the error term. In this setting, the same time effect condition is written as $E(\varepsilon_{it}|r = 1) = E(\varepsilon_{it}|r = 0)$. Under this assumption, the fixed effect linear regression can capture the causal effect of the change of the announcement. Linear specification has been widely employed as a result of this simplicity. For example, Naoi et al. (2009) estimated individual valuation of earth quake risk using the hedonic implicit price of earthquake risk and its changes following massive earthquakes. However, because γ is a one dimensional parameter, linear specification is not suitable for capturing local or heterogeneous effect.

To capture this heterogeneity and to avoid miss specification, nonparametric specification is useful. Treatment is considered to be exogenous. Moreover, because we can use balanced panel data, a direct DD estimation is possible. Under the same time effect assumption, we have

$$\begin{aligned} DD \equiv & E(P_{ib} - P_{ia}|r = 1, x, h) - E(P_{ib} - P_{ia}|r = 0, x, h) \\ = & E(P_{ib}(1) - P_{ia}(0)|r = 1, x, h) - E(P_{ib}(0) - P_{ia}(0)|r = 0, x, h) \\ = & E(P_{ib}(1) - P_{ia}(0)|r = 1, x, h) - E(P_{ib}(0) - P_{ia}(0)|r = 1, x, h) \\ & + E(P_{ib}(0) - P_{ia}(0)|r = 1, x, h) - E(P_{ib}(0) - P_{ia}(0)|r = 0, x, h) \\ = & E(P_{ib}(1) - P_{ia}(0)|r = 1, x, h) - E(P_{ib}(0) - P_{ia}(0)|r = 1, x, h) \\ = & E(P_{ib}(1) - P_{ib}(0)|r = 1, x, h). \end{aligned} \tag{4}$$

Since $E(P_{ib} - P_{ia}|r = 1, x, h)$ and $E(P_{ib} - P_{ia}|r = 0, x, h)$ are estimated nonparametrically from observable data, causal effect $E(P_{ib}(1) - P_{ia}(0)|r = 1, x, h)$ is identified by DD estimation.

For the estimation of $E(P_{ib} - P_{ia}|r = 1, x, h)$ and $E(P_{ib} - P_{ia}|r = 0, x, h)$ we employ nonparametric locally linear estimation. This is because locally linear regression reduces biases around boundary points (e.g., Li and Racine 2005; Lee 2010). This allows precise estimation of the treatment effect in a maritime area considered to be sensitive to tsunami risk. For the illustration of estimation procedure, let $v = (x, h)'$ as a two dimensional vector. Note that $E(P_{ib} - P_{ia}|r = 1, v)$ is approximated as

$$E(P_b - P_a|r, V_i) \approx \theta_{r0}(v) + \theta_{r1}(v)'(V_i - v) \tag{5}$$

when two dimensional vector $V_i = (X_i, H_i)$ is in the neighborhood of v, where $\theta_{r0}(v) = E(P_b - P_a|r, v)$ is a scalar, and $\theta_{r1}(v) = \nabla_v E(P_b - P_a|r, v)$ is a two-dimensional vector. Hence, $E(P_{ib} - P_{ia}|r = 1, x, h)$. is estimated as $\theta_{10}(v)$ of Eq. 6.

$$(\theta_{00}(v), \theta_{01}(v)) = \arg \min_{(\theta_0, \theta_1)} \sum_{i=1}^{N_1} [\bar{P}_i - \theta_0 - \theta_1(V_i - v)]^2 K\left(\frac{V_i - v}{\delta}\right) \tag{6}$$

where, empirical objective function in Eq. 6 is estimated by observations that belongs to treatment group, N_i is the number of observations that belongs to treatment group, $\bar{p}_i = p_{ib} - p_{ia}$, K(.) is a product Gaussian kernel function, δ is a vector of bandwidth. $E(P_{ib} - P_{ia}|r = 0, v)$ can be estimated from the same procedure using observations form control group. Put $(\theta_{00}(v), \theta_{01}(v))$ as the estimator of $(\theta_{00}(v), \theta_{01}(v))$. Thus DD is estimated as

$$\widehat{DD(v)} = \theta_{10}(v) - \theta_{00}(v) \tag{7}$$

4 Data Description

4.1 The Official Land Prices of Japan

Official land price data include land prices for each year, land use, address, presence or absence of electric, gas, water facilities, and the regulations of the City Planning Act such as building to land ratio. This is available at the web site of the National Land Numerical Information Download Service. As of 2013, the data of 25,983 samples are available.

To consider the heterogeneous effect on land prices of the changes in risk prediction, we utilize the data concerning the altitude of the land and the historical maximum height of tsunamis that have hit the area. This is because the risk that the lot is hit by a catastrophic tsunami that the new report is concerned with depends on the altitude and historical traces of tsunamis. Because the information on regional specific normal tsunami risk explained by a conventional seismic source model is considered to be unchanged before and after the announcement, these risk effects are removed by the differencing operation. Therefore, we do not use the variable that indicates average risk such as the average height of tsunamis. For altitude, we use the longitude and latitude data for each lot and calculate the altitude from SRTM (The Shuttle Radar Topography Mission) 3 published by NASA which is based on Farr et al. (2007). For the maximum historical tsunami height, we use the Japan Tsunami Trace Database published by Tohoku University. This database includes the data of major earthquakes and tsunami heights. The database contains the information from traces and observational data, which includes ancient documents. For the maximum tsunami height of each lot, we employ the data from the municipality in which the lot is located. The recorded height of tsunamis is presented in Fig. 1.

4.2 The Data Used in the Research

To estimate the announcement effect, we setup region 1($r = 1$) as the tsunami area that could be hit a Nankai Trough earthquake, and region 0 ($r = 0$) as other areas except Tohoku, East Kanto, Tokyo, Aichi, and Osaka. The area where $r = 1$ is determined by the report that was updated for the Nankai Trough area only by The Central Disaster Management Council. The area where $r = 0$ represents the other regions. The removal of Tohoku and East Kanto is a result of an unstable land market structure caused by the tsunami shock of the East Japan Great Earthquake and related accidents. Tokyo, Aichi, and Osaka are also removed to avoid the violation of the same time effect assumption and noted inaccuracy of the assessed prices in complex land use areas. The number of earthquakes in $r = 0$ is low compared to $r = 1$; however, Fig. 1 implies that there is no reason to assume that a catastrophic tsunami will not hit $r = 0$. Figure 1 demonstrates that the historical maximum height of a tsunami that hit the region was over 20 m. Moreover, the probability difference of earthquakes, which is explained by the conventional model, is removed by the differencing operation as fixed effect and regional effect. The DD estimation is therefore interpreted as announce effect (Table 2).

The original sample size of official land prices is 25,983. After the trimming of data for Tohoku, East Kanto Tokyo, Aichi and Osaka, and lots located in the area

Table 2 Definition of variables

Variable	Definition
BL ratio	Building to land ratio
FA ratio	Floor area ratio
CP low	Dummy variable: 1 for a lot in an area with low levels of construction (This includes two different types of low construction areas)
CP mid	Dummy variable: 1 for a lot in an area with middle levels of construction (This includes two different types of middle construction areas)
LU res	Dummy variable: 1 for a lot belonging to an area of residential land; 0 otherwise (This includes two different types of residential land use)
LU com	Dummy variable: 1 for a lot belonging to an area of commercial land; 0 otherwise (This includes two different types of commercial land use)
LU ind	Dummy variable: 1 for a lot belonging to an area of industrial land; 0 otherwise (This includes three different types of industrial land use)
Fire pro	Dummy variable: 1 for a lot belonging to a fire protection area
Dist st	Log10 of the distance from the nearest rail station
r	Dummy variable: 1 for region 1 (treatment group)
Price	Log10 of land price (Yen)
Alt	The altitude of a lot
Max	The maximum height of a tsunami that has hit the area of an existing lot

Table 3 Summary statistics

	Min	Median	Mean	Max	Std.dev
BL ratio	0	60	63.96	80	10.981
FA ratio	0	200	229.3	800	111.632
CP low	0	0	0.110	1	0.313
CP mid	0	0	0.167	1	0.372
LU res	0	0	0.259	1	0.438
LU com	0	0	0.228	1	0.420
LU ind	0	0	0.097	1	0.296
Fire pro	0	0	0.277	1	0.447
Dist st	−1	3.146	3.104	4.724	0.595
r	0	1	0.644	1	0.478
Price	3.405	4.76	4.782	6.428	0.376
Alt	−3.39	8.89	11.68	39.99	9.221
Max	0	3	5.472	57	7.972

not adjacent to the sea, the sample size of (r = 0) is 1,198 and the sample size of (r = 1) is 2,779. The final sample size used for estimation is 7,954 (= (1198 + 2779) * 2 (two time periods)). The summary statistics of the data used in the analysis are presented in Table 3.

5 Results

In the following analysis, we estimate the treatment effect of a change of announcement. For the comparison, we also estimate a familiar linear model and an extended semiparametric model as a baseline for our approach. The corresponding dependent variables of the baseline regressions are the logarithms of unit land prices.

5.1 *Linear and Semiparametric DD Regression (Baseline Result; Special Case)*

As the base line model, we report the estimation result of linear and semi-parametric specification as the baseline. First, we present the estimation result of a linear specification, Eq. 8, in Table 4:

$$P_{it} = Z_i\beta_1 + \tau_t\beta_2 + R_i\beta_3 + R_i\tau_t\beta_4 + \varepsilon_{it}. \quad (8)$$

Table 4 The estimation results of the baseline model

	Estimate	Std. error	t value	Pr(>|t|)
Intercept	4.581	0.039	117.4	0
BL ratio	0.0001	4.453	20.14	4.9E-88
FA ratio	−0.004	0.0005	−8.58	1.12E-17
CP low	0.466	0.013	33.85	9.1E-235
CP mid	0.333	0.010	31.60	1.3E-206
LU res	0.280	0.009	28.60	2.4E-171
LU com	0.337	0.014	23.27	6.00E-116
LU ind	0.218	0.011	18.24	6.4E-73
Fire pro	0.293	0.008	35.90	9.3E-262
Dist st	−0.077	0.005	−13.3	1.6E-40
r	0.233	0.008	26.78	2E-151
τ	−0.016	0.010	−1.53	0.124
$r\tau$ (treatment effect)	0.004	0.012	0.353	0.723
Sample size	7954			

In this model, treatment effect is captured by the value of. β_4.

From the OLS estimation, the majority of the estimated coefficients are reasonable. For example, the sign of building to land ratio and fire protection are positive and the log of distance from the nearest station is negative. However, the effect of treatment is not significant. This is not consistent with many other previous studies on the risks associated with flood and earthquakes (e.g., Bin et al. 2008; Nakagawa et al. 2007; Nakagawa et al. 2009). One possible reason for this insignificance is that the true treatment effect is zero because the tsunamis under consideration have a low likelihood of occurrence. This reasoning is natural from the view of conventional expected utility since measure of catastrophic events is zero. This means that household's utility maximizing problem for landowning remains same after the announcement. Another possible reason is that the specification error of functional form caused biased results. The risk of tsunami is high in low-altitude areas and low in high-altitude areas. It is also possible that the reason for this insignificance is caused by the data, which includes lots located in the areas with altitudes over 20 m, although the historic maximum height of a tsunami is at most 20 m in a low-lying area. Moreover, if land prices in a high risk area are decreasing and land prices in a low risk area are increasing after the announcement, the average treatment effect may not be significant and the effect of the report has been misread. Therefore, the treatment effect should be allowed to be heterogeneous with respect to altitude and the past maximum height of tsunamis.

5.2 *The Nonparametric Treatment Effect*

As we discussed in previous section, linear specification is friendly to the interpretation of the marginal effect of other attributes such as log of distance from the nearest station. However, linear estimation is valid only when the specifications of the remaining parts are correct. Moreover, simple linear specification cannot capture heterogeneous effects. Although nonlinear regression is a possible solution, we can control the estimation result by assuming an arbitrary functional form. Therefore, we report a direct nonparametric estimation result of the treatment effect. The result of a nonparametric estimation of the treatment effect is presented in Fig. 2.

Overall, we can observe both positive and negative treatment effect which is similar to the result of semiparametric specification. Because the treatment effect is observed to be heterogeneous with respect to the altitude and maximum past tsunami height, the standard linear specification cannot capture this effect.

In low-altitude areas, both positive and negative effects are observed. Specifically, negative effect is observed in low-altitude areas (2.5–20 m) where the historical maximum height of a tsunami is high (15–25 m). However, a positive effect is observed in mid-altitude areas (10–25 m) with a relatively low historical maximum tsunami height (0–15 m). This heterogeneity is considered to be natural because a positive treatment effect is observed in relatively low risk areas and a

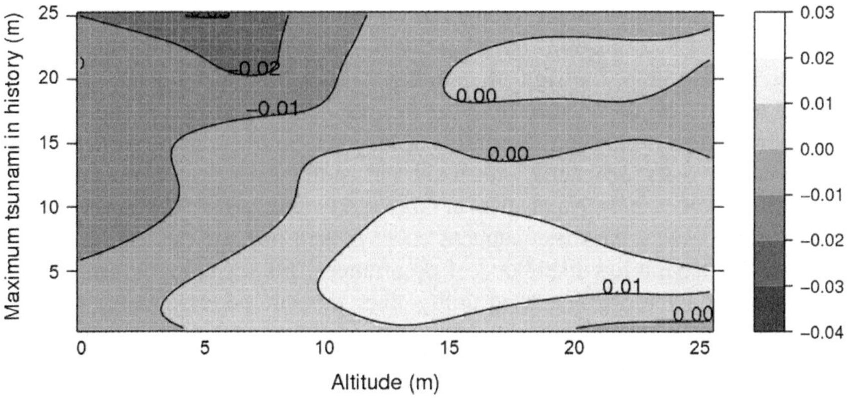

Fig. 2 Estimation result of nonparametric DD

negative treatment effect is observed in relatively high risk areas. It is possible that the announcement reduced the demand for high risk areas and that substitution of low risk areas for high risk, low-altitude areas has occurred. Additionally, the treatment effect of an announcement is almost zero or insignificant in very low-altitude (0–2.5 m) areas. This suggests that the substitution may occur only between low risk, mid-altitude areas and high risk, low-altitude areas.

For the statistical inference, we report a 90 % confidence interval. Upper and lower bounds of a 90 % confidence interval are presented in Figs. 3 and 4. Both positive and negative effect is significant in the low-altitude areas (2.5–20 m).

We impose the same time effect condition only on altitude and historical maximum height of the tsunami. Although the estimation result is valid under the same time effect assumption, the time effect may be different with respect to land use. Therefore, a more plausible result would be derived if we employ the same time effect assumption conditional on altitude, historical maximum height of tsunami, and other attributes of land.

5.3 *Theoretical Considerations*

The estimation result of the nonparametric DD demonstrated that the risk of rare catastrophic events has statistically significant effects on land pricing. This subsection discusses whether the estimated significant effect is justifiable within a conventional framework of expected utility theory.

In conventional models that consider the hedonic price of risk related to earthquakes such as Brookshire et al. (1985) and Naoi et al. (2009), it is assumed that there are two states, "w" and "o", corresponding to a "with earthquake" and "without earthquake" situation. Let π be the probability of earthquake occurrence

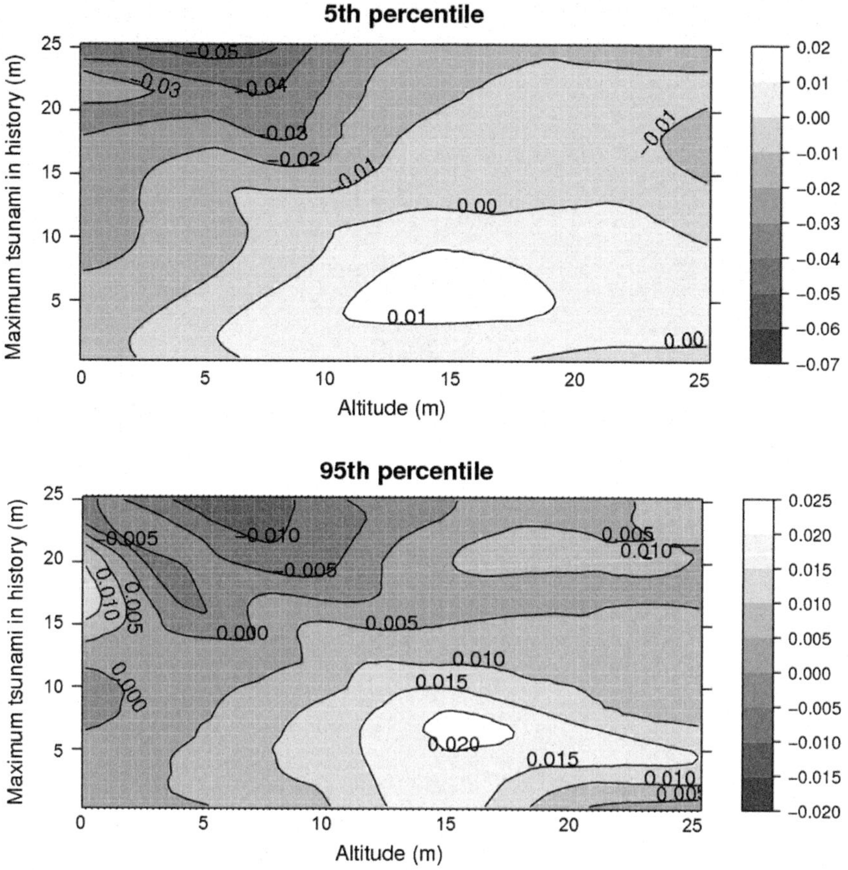

Fig. 3 Confidence interval of nonparametric DD estimation. Confidence interval is estimated by bootstrap

(i.e., the probability assigned to state w). The probability of catastrophic earthquake and tsunami occurrence is zero. The hedonic pricing method suggests that land prices can be described as a function of the locational characteristics of the land. In our context, the risk of a tsunami will be a particularly important factor influencing the market cost of land. Hence, the hedonic price function can be written as

$$p = p(z, \pi(v)) \tag{9}$$

where p is the observed land price, v is the altitude and the maximum height of a historic tsunami in the land area, and z is a vector of location specific characteristics that are not related to tsunami risk. Before the announcement which is based on catastrophic rare events, hazard maps are made based on noncatastrophic events. Therefore, it is natural to consider that the hedonic price function is not a function

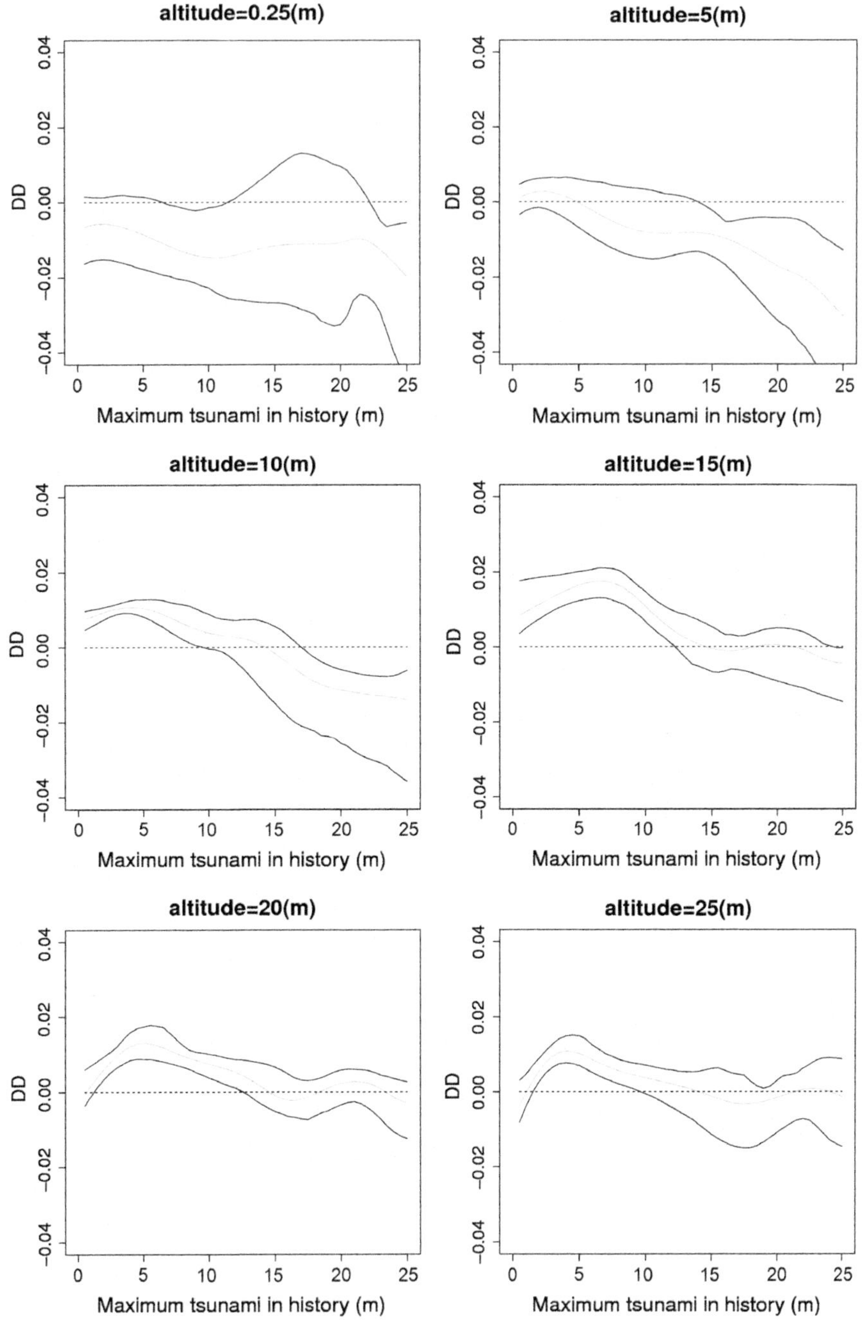

Fig. 4 The 90 % confidence intervals for fixed altitudes (0.25, 5, 10, 15, 20, 25 m)

of catastrophic risk (i.e., a function of only z and $\pi(v)$). Even after the announcement, since the probability of the events is sufficiently small to be regarded as zero, ordinal discussion suggests that land price is a function of only z and $\pi(v)$.

A "household" maximizes its expected utility by choosing a bundle of land characteristics and a level of tsunami risk. (It is possible that the purchaser is a firm; however, land is one input of the firm. We can therefore apply the same discussion when the purchaser is a firm.) Let $u(z, \mu)$ be the utility function, where μ is an amount of numeraire goods consumption, with $u_z \lesseqgtr 0$ and $u_\mu > 0$. Let y^w and y^0 be the state contingent income, where we assume $y^o > y^w$ with $L = y^o - y^w$ representing the monetary loss from the tsunami. Then, the household's budget constraint suggests that the numeraire goods' consumption under two states can be written as $\mu^o = y^o - p(z, \pi(v))$ and $\mu^w = y^w - p(z, \pi(v))$. The household's maximization of expected utility is written as:

$$\max_{z,v} EU(v, z) \tag{10}$$

Where

$$EU(v, z) = \pi(v)U(z, y^w - p(z, \pi(v))) + (1 - \pi(v))U(z, y^o - p(z, \pi(v))) \tag{11}$$

The first order condition gives us the following equilibrium conditions.

$$\frac{\partial p}{\partial z} = \frac{\pi \frac{\partial U^w}{\partial Z} + (1 - \pi) \frac{\partial U^o}{\partial Z}}{\pi \frac{\partial U^w}{\partial \mu} + (1 - \pi) \frac{\partial U^o}{\partial \mu}} \tag{12}$$

$$\frac{\partial p}{\partial \pi} = \frac{U^w - U^o}{\pi \frac{\partial U^w}{\partial \mu} + (1 - \pi) \frac{\partial U^o}{\partial \mu}} \tag{13}$$

$$\frac{\partial p}{\partial v} = \frac{\partial p}{\partial \pi} \frac{\partial \pi}{\partial v} \tag{14}$$

This model predicts the insignificant effects of the announcement on rare catastrophic events. That is, DD estimation is not significant. This is because rare events do not affect the value of $\pi(v)$, as Chichilnisky (2009) demonstrated. However, we derived significant DD estimates. This fact suggests that conventional VN-type expected utility theory may not work when rare catastrophic events are analyzed.

Hence, expected utility theory that is sensitive to rare events will be required because a significant treatment effect is observed. To explain the phenomenon and why VN-type expected utility provides a poor explanation, Chichilnisky (2009) developed expected utility theory which is sensitive to rare catastrophic events. This theory satisfies the following three axioms, whereas VN-type expected utility satisfies only Axioms 1 and 2.

- Axiom 1: Continuity
- Axiom 2: Sensitivity to frequent events
- Axiom 3: Sensitivity to rare events

Mathematical definitions of these axioms are given in Chichilnisky (2009). Under these axioms, the objective function can be written as

$$EU^* = \lambda \int U(x)\phi_1(x)dx + (1-\lambda) < U, \phi_2 > \tag{15}$$

by Theorem 2 of Chichilnisky (2009) where $\int \phi_1(x)dx = 1$, ϕ_2 is a purely finitely additive measure, and $0<\lambda<1$. We use x in Eq. 15 following the notation of Chichilnisky (2009), and this x is not related to altitude. In our discussion, since λ, ϕ_1, ϕ_2, and U can vary with geographic attributes as Eq. 11, EU^* is a function of (v, z). That is

$$\begin{aligned} EU^*(v,z) = \lambda(v)\{\pi(v)U(z, y^w - p(z, \pi(v), \lambda(v))) \\ + (1-\pi(v))U(z, y^o - p(z, \pi(v), \lambda(v)))\} + 0 \cdot (1-\lambda(v)) \end{aligned} \tag{16}$$

where, we put numeraire goods' consumption for catastrophic outcome to be zero. In this framework, land price is a function of z, $\pi(v)$, and $\lambda(v)$ since $\lambda(v)$ also affect the utility maximization problem for landowning. Hence, the household's maximization problem is written as

$$\max_{z,v} EU^*(v,z) \tag{17}$$

The first order condition gives us the equilibrium conditions presented in Eqs. 18–21.

$$\frac{\partial p}{\partial z} = \frac{\pi \frac{\partial U^w}{\partial Z} + (1-\pi)\frac{\partial U^o}{\partial Z}}{\pi \frac{\partial U^w}{\partial \mu} + (1-\pi)\frac{\partial U^o}{\partial \mu}} \tag{18}$$

$$\frac{\partial p}{\partial \pi} = \frac{U^w - U^o}{\pi \frac{\partial U^w}{\partial \mu} + (1-\pi)\frac{\partial U^o}{\partial \mu}} \tag{19}$$

$$\frac{\partial p}{\partial \lambda} = \frac{\pi U^w + (1-\pi)U^o}{\lambda\left(\pi \frac{\partial U^w}{\partial \mu} + (1-\pi)\frac{\partial U^o}{\partial \mu}\right)} \tag{20}$$

$$\frac{\partial p}{\partial v} = \frac{\partial p}{\partial \pi}\frac{\partial \pi}{\partial v} + \frac{\partial p}{\partial \lambda}\frac{\partial \lambda}{\partial v} \tag{21}$$

Because the value of $\lambda(v)$ is changed by the new risk announcement, a significant estimation result of DD design is justified in this framework. It should be noted that this expected utility is equal to VN-expected utility if there is no probability of a catastrophic event (i.e., $\lambda(v) = 1$).

As Kask and Maani (1992) and other authors suggest, our significant result can be explained if we employ subjective distribution of risk. This is because subjective distribution can change even if objective distribution does not change.

6 Discussion and Conclusion

The purpose of this paper is to examine the economic effect of a change in information concerning catastrophic risk. We used a Japanese tsunami-risk announcement that includes rare catastrophic disasters that had been disregarded until the report on the Nankai Trough massive earthquake was published. We used Japanese official land price panel data for 2012 and 2013 and the data on maximum historical heights of tsunamis. To identify the treatment effect, we used a nonparametric DD design. Estimated risk-announcement effects were heterogeneous and significant. This significant effect of the announcement suggests the need to employ expected utility, which is sensitive to rare events when catastrophic outcomes are possible.

However, other factors exist that require consideration. Mayer (1995) notes the importance of the precise specification of the time series. In this paper, this is related to the same time effect assumption imposed in Sect. 3. One possible way to overcome this problem is to condition the treatment effect with the land prices at other points in time. In this paper, we removed the data of Tohoku, East Kanto, Tokyo, Aichi, and Osaka because it was not possible to use a sufficient number of observations for nonparametric conditioning with our balanced panel data. To ensure a sufficient sample size for two-dimensional nonparametric estimation, we did not run further matching operations. However, this may not be sufficient if there are other factors that cannot be removed by the differencing operation.

It is possible to increase the length of time between time point “a” and “b” (see Sect. 3) in the observations. However, longer intervals imply a higher risk of the violation of the same time effect condition, which is central to DD identification including linear specification. Hence, we used the data for 2012 and 2013 in this study.

There is the possibility that factors other than information on catastrophic risk affects land prices. For example, Naoi et al. (2009) observed that there have been certain modifications of subjective assessments of earthquake risk after massive earthquakes. Because the report on the Nankai Trough massive earthquake was published in 2012 and the East Japan Great Earthquake occurred in 2011, the period that we studied may be included in the process in which the update of subjective risk affected land prices. Additionally, it is possible that individuals do not believe that catastrophic events such as the Nankai Trough massive earthquake have a low probability of occurring. These possibilities can be examined if we use panel data from questionnaires that Naoi et al. (2009) utilized. Hence, we leave this as a suggestion for future study.

The policy implications of our results are evident. Because it is observed that a risk-announcement based on a catastrophic disaster leads to decreasing land prices of low-altitude, high risk areas, such announcements are effective with respect to expected utility even if these event are rare. However, the price changes caused by the announcements on rare catastrophic events are predicted to be zero when researchers employ conventional VN-type expected utility. Therefore, expected utility derived from axioms that can manage these rare events should be used for the evaluation and prediction of economic phenomenon when there is a possibility of a catastrophic outcome.

Acknowledgements This work is supported by JSPS (The Japan Society for the Promotion of Science) Grant-in-Aid for JSPS Fellows Grant Number 12J09136.

References

Abadie, A. (2005). Semiparametric difference-in-differences estimator. *Review of Economic Studies, 72*, 1–19.

Beron, K. J., Murdoch, J. C., Thayer, M. A., & Vijverberg, W. P. M. (1997). An analysis of the housing market before and after the 1989 Loma Prieta Earthquake. *Land Economics, 73*(1), 101–113.

Bin, O., Crawford, T. W., Kruse, J. B., & Landry, C. E. (2008). Viewscapes and flood hazard: Coastal housing market response to amenities and risk. *Land Economics, 84*(3), 434–448.

Bin, O., & Landry, C. E. (2012). Changes in implicit flood risk premiums: Empirical evidence from the housing market. *Journal of Environmental Economics and Management, 65*(3), 361–376.

Brookshire, D. S., Thayer, M. A., Tschirhart, J., & Schulz, W. D. (1985). A test of the expected utility model: Evidence from earthquake risks. *Journal of Political Economy, 93*(2), 369–389.

Chanel, O., & Chichilnisky, G. (2013). Valuing life: Experimental evidence using sensitivity to rare events. *Ecological Economics, 85*, 198–205.

Chichilnisky, G. (2009). The topology of fear. *Journal of Mathmatical Economics, 45*, 807–816.

Donovan, G. H., Champ, P. A., & Butry, D. T. (2007). Wildfire risk and housing prices: A case study from Colorado Springs. *Land Economics, 83*(2), 217–233.

Farr, T.G., Rosen, P.A., Caro, E., Crippen, R., Duren, R., Hensley, S., Kobrick, M., Paller, M., Rodriguez, E., Roth, L., Seal, D., Shaffer, S., Shimada, J., Umland, J., Werner, M., Oskin, M., Burbank, D., & Alsdorf, D. (2007). The shuttle radar topography mission. Reviews of Geographysics *45*(2), RG2004. doi:10.1029/2005RG000183, (http://www2.jpl.nasa.gov/srtm/).

Gayer, T., Hamilton, J. T., & Viscusi, W. K. (2002). The market value of reducing cancer risk: Hedonic housing prices with changing information. *Southern Economic Journal, 69*(2), 266–289.

Hausman, J. (2012). Contingent valuation: From dubious to hopeless. *Journal of Economic Perspectives, 26*(4), 43–56.

Japan Tsunami Trace Database, School of Engeneering of Tohoku University and Japan Nuclear Energy Safety Organization (tsunami3.civil.tohoku.ac.jp/).

Kask, S. B., & Maani, S. A. (1992). Uncertainty, Information, and Hedonic Pricing. *Land Economics, 68*(2), 170–184.

Lee, M. J. (2005). *Micro econometrics for policy, program, and treatment effects*. Oxford: Oxford University Press.

Lee, M. J. (2010). Micro econometrics: Methods of moments and limited dependent variables 2nd edn.

Levitt, S. D., & List, J. A. (2007). What do laboratory experiments measuring social preferences reveal about the real world. *The Journal of Economic Perspectives, 21*(2), 153–174.

Li, Q., & Racine, J. S. (2005). *Nonparametric econometrics*. Princeton, NJ: Princeton University Press.

Ma, S., & Swinton, S. M. (2012). Hedonic valuation of farmland using sale prices versus appraised values. *Land Economics, 88*(1), 1–15.

Mayer, B. D. (1995). Natural and quasi-experiments in economics. *Journal of Bussiness and Economic Statistics, 13*(2), 151–161.

Nakagawa, N., Saito, M., & Yamaga, H. (2007). Earthquake risk and housing rents: Evidence from the Tokyo Metropolitan Area. *Regional Science and Urban Economics, 37*, 87–99.

Nakagawa, N., Saito, M., & Yamaga, H. (2009). Earthquake risk and land prices: Evidence from the Tokyo Metropolitan Area. *Japanese Economic Review, 60*(2), 208–222.

Naoi, M., Seko, M., & Sumita, K. (2009). Earthquake risk and housing prices in Japan: Evidence before and after massive earthquakes. *Regional Science and Urban Economics, 39*(6), 658–669.

Parmeter, C. F., Pope, J. C. (2012). Quasi-experiments and hedonic property value methods. Unpublished munuscript.

Rosen, S. (1974). Hedonic prices and implicit markets: Product differentiation in pure competition. *Journal of Political Economy, 82*(1), 34–55.

Tsutsumi, M., & Seya, H. (2009). Hedonic approaches based on spatial econometrics and spatial statistics: Application to evaluation of project benefits. *Geographical Systems, 11*, 357–380.

Modeling US Stock Market Volatility-Return Dependence Using Conditional Copula and Quantile Regression

Terence D. Agbeyegbe

1 Introduction

There is a growing literature in economics and finance on methods of dealing with catastrophic risks which can be seen as rare events with major consequences (see Chichilnisky (2009) and the references therein). When attention is on financial econometrics, some of these methods focus on estimating parameters of time series models using quantile regression and copula techniques (see Alexander 2008; Allen et al. Allen et al. 2009, 2012; Badshah 2012; Barnes and Hughes 2002; Bouyé and Salmon 2009; Engle and Manganelli 2004; Koenker and Xiao 2006; Kumar 2012; Patton 2004, 2006a, b, 2009; Taylor 1999; Trivedi and Zimmer 2005; Xiao 2009 among many others). In this chapter we describe the application of quantile regression and copula techniques to United States index stock market price return and volatility data. The quantile regression model we use was initially described in Koenker and Bassett (1978), and is an extension of the classical least squares estimation of the conditional mean to a collection of different conditional quantile function models. It is essentially a statistical technique intended to estimate and conduct inference about conditional quantile functions. It has the additional advantage of being robust to heteroskedasticity, skewness and leptokurtosis which are typical features of financial data.

The main purpose of this chapter is to apply quantile regression methods to investigate the relation between stock returns and implied volatilities. Though such an investigation has been done before, the analysis in this chapter differs in terms of data choice, span and the use of a GARCH filter to control for changes in the volatilities of the series. Two of the series we examine have not been investigated in the quantile regression framework: the Dow Jones Industrial Average Index and the S&P 100 Index. The other two have been examined but for a different time period. We also

---

T.D. Agbeyegbe (✉)
Department of Economics, Hunter College and the Graduate Center, City University of New York, New York City, NY 10021, USA
e-mail: tagbeyeg@hunter.cuny.edu

G. Chichilnisky and A. Rezai (eds.), *The Economics of the Global Environment*, Studies in Economic Theory 29, DOI 10.1007/978-3-319-31943-8_26

focus on tails of the distributions, which is particularly important since volatility and extreme movements are not synonymous. As noted by many others, (see Neftci 2000) the prices of two assets could exhibit the same volatility but very different patterns with regards to their extremes. For this reason, we consider methods that examine the tails of the price distributions. Quantile regression methods are of use when dealing with relationships at the tails of distributions. The relation between stock returns and implied volatilities has long been studied given its practical importance for areas such as risk management, option pricing, and event studies (see for example, the early papers by Cox and Ross 1976; Black 1976; Christie 1982). In several recent papers, the relationship was shown to be asymmetric (see for example, Badshah 2012; Dennis et al. 2006; Fleming et al. 1995; Giot 2005; Hibbert et al. 2008; Low 2004; Whaley 2000; Wu 2001; Allen et al. 2012). An asymmetric relationship means that the negative change in the stock market returns has a higher impact on the volatility index than a positive change, or vice-versa. For this reason, volatility indices are often referred to as being investors gauges of fear (see Whaley 2000). The theoretical basis for this asymmetric volatility-return relationship is the focus of two hypotheses; namely, the leverage hypothesis (see Black 1976; Christie 1982) and the volatility feedback hypothesis (see French et al. 1987; Campbell and Hentschel 1992). The leverage hypothesis states that if the stock price of a firm declines, the relative proportion of equity (debt) value to the firm value decreases (increases), which makes the firm's stock riskier and increases its volatility as a consequence. The volatility feedback hypothesis states that the negative change in expected return tends to be intensified whereas the positive change in the expected return tends to be dampened and these effects generate the asymmetric volatility phenomenon.

The plan of the chapter is as follows. Section 2 discusses quantile regression. Section 3 provides a review of some copula functions and dependence measures. Section 4 deals with non-linear quantile regressions using copula theory. Section 5 deals with the data on US equities and the results. Section 6 contains the conclusion.

2 Quantile Regression

In this section, we provide a brief discussion of quantile regression. For convenience and as a prelude to introducing the simple linear quantile regression model, we briefly discuss a simple linear regression model. A simple bivariate linear regression model may be written as:

$$y_t = \alpha + \beta x_t + \varepsilon_t \tag{1}$$

where the parameters α and β are constants and y is the independent variable, x is the dependent variable, ε is the error term and subscript t is for time period t. The standard assumptions include the provision that the errors are independent and identically distributed with mean zero and that the x is exogenous suggesting that the conditional expectation of ε_t is zero. These conditions mean we can write $E(y \mid x) = \alpha + x\beta$. Assuming further that the y and x is bivariate normal will assure that the distribution function $F(y \mid x)$ is normal and this distribution is completely specified from knowledge of the conditional mean and conditional variance equations. The ordinary least squares estimates are then the solution to the optimization problem

$$min_{\alpha\beta} \sum_t (y_t - \alpha - \beta x_t)^2 \tag{2}$$

When the joint distribution of x and y is not bivariate normal we need more than the conditional mean and conditional variance to specify the conditional distribution of the dependent variable. It is for this reason we need quantiles and by implication a quantile regression framework. The definition of Koenker and Bassett's (1978) linear quantile regression is stated in terms of an optimization problem. Let $q \in (0, 1)$ and the *qth* quantile of the error term be defined as F_ε^{-1}, where the error has a distribution function given as F_ε The simple linear quantile regression model is then given as

$$F^{-1}(q \mid x) = \alpha + x\beta + F_\varepsilon^{-1}(q). \tag{3}$$

where $F^{-1}(q \mid x)$ is the q conditional quantile of the dependent variable in the general case.

More generally, let $(y_1, y_2, \ldots, y_T)$ be a random sample on the regression process with $u_t = y_t - x_t\beta$ having distribution function F and $(x_1, x_2, \ldots, x_T)$ be a sequence of K-vectors of a known design matrix, the q-th quantile regression will be any solution to the following problem:

$$min_{\beta \varepsilon R^k} (\sum_{t \varepsilon \tau_q} q \mid y_t - x_t\beta \mid + \sum_{t \varepsilon \tau_{1-q}} (1-q) \mid y_t - x_t\beta \mid) \tag{4}$$

with $\tau_q = \{t : y_t \geq x_t\beta\}$ and τ_{1-q} is the complement.

Notice that the median (quantile) regression estimator minimizes the symmetrically weighted sum of absolute errors (where the weight is equal to 0.5). The other conditional quantile functions are estimated by minimizing an asymmetrically weighted sum of absolute errors, where the weights are now functions of the quantile of interest. The properties of the estimator is provided in Theorem 1 of Koenker and Basset (1978). As noted by Buchinsky (1998), quantile regression models have many useful features: (i) with respect to non-gaussian error terms, quantile regression estimators may be more efficient than least-square estimators, (ii) the entire conditional distribution can be characterized, (iii) different relationships between the regressor and the dependent variable may arise at different quantiles.

Whilst the modern treatment of quantile regression can be traced to Koenker and Basset (1978), the use of the classical least squares' methodology as a modern statistical framework can be traced to Galton (1886). As pointed out by Abdi (2007), Galton used it in his work on the heritability of size, which formed the foundations of correlation and (also gave the name to) regression analysis. For a fuller discussion of the history and pre-history of the classical least squares methodology, the reader is referred to Harper (1974–1976). A distinguishing feature of Galton's regression approach is the minimization of the sum of squares of residuals in order to enable one to estimate models for the conditional mean functions. The least squares methodology framework is not useful if interest is not focused on the conditional mean, to avoid this short-coming researchers developed the quantile regression method. Quantile regression methods provide a way for estimating models for the conditional median function, and the full range of other conditional quantile functions. It is capable of providing a more complete statistical analysis of the stochastic relationships among random variables by supplementing the estimation of conditional mean functions with techniques for estimating an entire family of conditional quantile functions. The estimated conditional quantile functions give a much more complete picture of the effect of covariates on the location, scale and shape of the distribution of a response variable. The method has been extended, and it has found successful application in many areas of applied econometrics. For example, in labor economics, we can find examples based on the works of: Buchinsky and Leslie (1997) who investigated wage structure; Eide and Showalter (1999) together with Buchinsky and Hunt (1999) who investigated earnings mobility; and Eide and Showalter (1998) who considered issues related to educational attainment. In financial econometrics we can find examples based on the works of: Taylor (1999) who estimated the distribution of multiperiod returns using quantile regression; Engle and Manganelli (2004) who proposed estimating value at risk (VaR) using quantile regression; Koenker and Xiao (2006) who proposed a quantile autoregression model and applied it to weekly U.S. gasoline prices; Bouyé and Salmon (2009) who developed a theory of non-linear quantile regression modeling using copula and applied the theory to examine conditional quantile dependency in the foreign exchange market; and Xiao (2009) who developed a theory for quantile cointegration and applied the proposed model to US stock index data.

It should be noted that an important generalization of the basic linear quantile regression to the non-linear case was developed by Powell (1986) using a censored regression modeling framework. The consistency of non-linear quantile regression estimation has been investigated by White (1994), Engle and Manganelli (2004) and Kim and White (2003). For an overview of quantile regression, see the guideline for empirical research by Buchinsky (1998), the surveys by Koenker and Hallock (2001) and Yu et al. (2003) together with the text by Koenker (2005).

3 Review of Copula Functions and Dependence

In this section, we state some well-known properties of copula functions and briefly discuss some measures of dependence. We start with a few definitions and introduce notation and terminology that are consistent throughout this chapter.

The interest in studying the relationship between United States index stock market price return and implied volatility data motivates the need to discuss copula functions. A full treatment of copulas and their properties can be found in Joe (1997) and Nelsen (2006). Nelsen (2006) defines copulas as "functions that join or couple multivariate distribution functions to their one-dimensional marginal distribution functions." Copula functions are particularly attractive to work with since they allow us to separately model the marginal distribution and the dependence structure. In dealing with dependence, copulas can provide us information on both the degree of dependence and the structure of dependence. In particular, copula functions contain information about the joint behavior of the random variables in the tails of the distribution and can shed light on the symmetric, or asymmetric nature of the dependence. Linear correlation is unable to shed light on tail dependence and/or the symmetry property of dependence. We now provide a definition of a two-dimensional copula and we state the most important result in copula theory, Sklar (1959)'s theorem.

Definition 1 (Nelsen (2006), p. 10) A two-dimensional copula (or 2-copula, or briefly, a copula) is a real function C with the following properties:

1. For every u, v in [0, 1],
$$C(u, 0) = 0 = C(0, v) \tag{5}$$
and
$$C(u, 1) = u, C(1, v) = v; \tag{6}$$
2. For every, u_1, u_2, v_1, v_2in [0, 1] such that $u_1 \leq u_2$ and $v_1 \leq v_2$,
$$C(u_2, v_2) - C(u_2, v_1) - C(u_1, v_2) + C(u_1, v_1)) \geq 0. \tag{7}$$

Theorem 1 (Sklar (1959)'s Theorem, Nelsen (2006), p. 18) *Let X and Y be two random variables with joint distribution F. Then, there exists a copula C such that for all x,y in* $\bar{\mathbb{R}}$ *satisfying* $F(x, y) = C(F_X(x), F_Y(y))$. *If* F_X, F_Y *are continuous, then C is unique and* F_X, F_Y *represent the marginal distributions of X and Y respectively.*

The above theorem of Sklar is very important, since it provides a way for us to analyse the dependence structure of multivariate distributions without studying marginals distributions. In the case of multivariate continuous distribution functions, the theorem allows us to view the univariate margins and the multivariate dependence structure as separate entities. The underlying dependence structure of the multivariate distribution can be represented by an adequate copula function.

Note from above, any bivariate distribution function whose margins are standard uniform distributions is a copula. Furthermore, copula functions are joint distribution functions of standard uniform random variables: $C(u, v) = Pr(U_1 \leq u, U_2 \leq v)$. They are also subjected to a version of the Fréchet-Hoeffding bounds inequality.

Theorem 2 (Fréchet-Hoeffding bounds inequality, Nelsen (2006), p. 11) *Let M(u, v) = min(u, v) and W(u, v) = max(u + v - 1.0) then for every copula C and every* $(u, v) \in [0, 1]^2$,
$$W(u, v) \leq C(u, v) \leq M(u, v). \tag{8}$$

M is referred to as the Fréchet-Hoeffding upper bound and W as the Fréchet-Hoeffding lower bound.

Definition 2 A parameter θ of a copula is called the dependence parameter if for an m-variate function F, the copula associated with F is a distribution function $C : [0,1]^m \to [0,1]$ that satisfies

$$F(y_1, y_2, \ldots, y_T) = C(F_1(y_1), \ldots, F_m(y_m); \theta).$$

The copula dependence parameter measures the dependence between the marginals and may be a vector of parameters. In bivariate applications, the dependence parameter is often represented by a scalar parameter and is the focus of estimation.

Copula theory has found successful applications in many fields. For applications and overview of copula to quantitative risk, see Embrechts et al. (2003) and Embrechts et al. (2001), among others. For applications in finance and financial time series, see Cherubini et al. (2004), and Patton (2009).

3.1 Some Dependence Concepts

In this subsection, we discuss the concept of dependence. There is a fairly large literature that deals with this concept and from what has been reported we can view dependence as falling into at least three broad classes. The first discusses dependence in terms of linear dependence relationship between variables in the center of the distribution or rank correlations if interest centers on non-linear monotonic transformation of the variables. The second considers dependence between variables in the tail of the distribution in the presence of extreme events. The third examines dependence along the whole distribution. Examples of the first approach are numerous and they are exemplified in the use of classical least-squares regression to unravel dependence between variables. Measures based on "regular" linear correlation of Pearson's ρ and the rank correlation of Kendall's τ and Spearman's ρ are often reported with this kind of analysis. Pearson's ρ deals with the linear dependence between random variables and when nonlinear transformations are applied to those random variables, linear correlation is not preserved. Instead, a rank correlation coefficient measure, such as Kendall's τ or Spearman's ρ, will be more appropriate. The rank correlations measure the degree to which large or small values of one random variable associates with large or small values of another random variable. Examples of the second approach are found in the works of Longin and Solnik (2001), Ang and Chen (2002) and (Patton 2006a, b) among many others who discuss exceedance correlation and tail dependence. One focus is to discuss dependence in terms of exceedance correlation which is defined as the correlation between two variables X and Y, conditional on both variables being above or below certain thresholds μ_1 and μ_2, respectively. The other focus is in terms of tail dependence a concept which is related to exceedance correlation but it is different. Tail dependence is a key measure for risk management, which mainly focuses on the extreme events of joint distribution. It measures the probability that both variables are simultaneously in their lower or upper tails. The lower (left) and upper (right) tail dependence coefficients, λ_l and λ_r, are defined as below.

Definition 3 $\lambda_l = lim_{u \to 0} Pr[F_Y(y) \le u \mid F_X(x) \le u] = lim_{u \to 0} \frac{C(u,u)}{u}$

Definition 4 $\lambda_r = lim_{u \to 1} Pr[F_Y(y) \ge u \mid F_X(x) \ge u] = lim_{u \to 1} \frac{1-2u+C(u,u)}{1-u}$

In both cases λ_l and $\lambda_r \in [0, 1]$. If λ_l or λ_r is positive, X or Y is said to be left (lower) or right (upper) tail dependent. Patton (2009), provide examples of analysis based on tail dependency.

Examples of the third approach can be found in many of the papers on quantile regression and some recent papers in copula quantile regression modeling. In this approach, a copula quantile regression is specified and the dependency between variables of interests are reported for different quantiles. The approach is discussed in Sect. 4.

3.2 *Some Copula Functions*

There are a large number of copulas to work with when modeling data. Each copula imposes a different dependence structure on the data. Joe (1997, Chap. 5), Nelsen (2006: 116–119) and Trivedi and Zimmer (2005) discuss a wide variety of bivariate copulas and their properties. In this sub-section, we discuss some copulas that have appeared frequently in finance applications, and we briefly describe their dependence structures.

The most common copulas can be divided into two broad types: Elliptical and Archimedean Copulas. Examples of the former being-Gaussian Copula and Student's t-Copula and of the latter being Clayton copula, Frank Copula and Gumbel copula.

3.2.1 Elliptical Copulas

(i) Gaussian Copula.
Let us define $u_i = F_i(x_i)$. The Gaussian (or normal) copula is the copula of the multivariate normal distribution. It takes the form

$$C_{Gaussian}(u_1, u_2; \rho) = \Phi_G(\Phi^{-1}(u_1), \Phi^{-1}(u_2); \rho)$$
$$= \int_{-\infty}^{\Phi^{-1}(u_1)} \int_{-\infty}^{\Phi^{-1}(u_2)} \frac{1}{2\pi(1-\rho^2)^{0.5}} e^{\left\{\frac{-(x_1^2 - 2\rho x_1 x_2 + x_2^2)}{2(1-\rho^2)}\right\}} dx_1 dx_2$$

where Φ_G is the standard bivariate normal distribution, Φ is the cumulative distribution function of the standard normal distribution, with Pearson's product moment correlation coefficient $\rho, \rho \in (-1, 1)$. The normal copula is quite flexible and allows for equal degrees of positive and negative dependence and it includes both the lower and upper Fréchet bounds in its permissible range.

(ii) Student's t-copula.
Student's t-copula is based on the multivariate t-distribution in the same way the Gaussian copula is based on the multivariate normal distribution. It adds joint fat tails to the Gaussian copula. The bivariate t-copula takes the form:

$$\begin{aligned} C_t(u_1, u_2; \nu, \rho) &= \phi_\rho(\phi_\nu^{-1}(u_1), \phi_\nu^{-1}(u_2)) \\ &= \int_{-\infty}^{t_\nu^{-1}(u_1)} \int_{-\infty}^{t_\nu^{-1}(u_2)} \frac{1}{2\pi(1-\rho^2)^{0.5}} \\ &\times \{1 + \frac{(x_1^2 - 2\rho x_1 x_2 + x_2^2)}{\nu(1-\rho^2)}\}^{-\frac{(\nu+2)}{2}} dx_1 dx_2 \end{aligned}$$

where t_ν^{-1} denotes the inverse of the cdf of the standard univariate t-distribution with ν degrees of freedom. The dependency parameters are ρ and ν with $\rho \in (-1, 1)$ and $\nu > 2$. The parameter ν controls the heaviness of the tails and when $\nu \leq 3$ the variance does not exist and when $\nu \leq 5$, the fourth moment does not exist. Large values of ν, approximate a Gaussian distribution; $C_t(u_1, u_2; \nu, \rho) \rightarrow \Phi_G(u_1, u_2; \rho)$. The t-copula is attractive because the degree of tail dependency can be set by changing the degrees of freedom. The copula is important in finance and has been recommended by a number of authors. (See, for example, Breymann et. al. 2003).

3.2.2 Archimedean Copulas

Archimedean copulas are an important class of copulas that have a wide range of applications. They are easy to construct from generators. A great variety of families of copulas belongs to this class, and they have many nice properties. (see Nelsen 2006). For a generator ϕ, the Archimedean copula can be defined as:

$$C_{Archimedean}(u_1, u_2; \alpha) = \phi^{-1}(\phi(u_1) + \phi(u_2))$$

and the density is given as:

$$c_{Archimedean}(u_1, u_2; \alpha) = \phi_{(2)}^{-1}(\phi(u_1) + \phi(u_2))\Pi_{i=1}^2 \phi'(u_i).$$

where $\phi_{(2)}^{-1}$ is the 2nd derivative of the inverse generator function, $\phi()$ is a convex decreasing function, with $\phi(1) = 0$. The function $\phi()$ depends on a single parameter α that reflects the degree of dependence. Archimedean copulas allow a wide range of dependence structure. Their mathematical and statistical properties are studied in Genest and Rivest (1993). We will discuss three members of the Archimedean families, namely Gumbel, Clayton and Frank Copula. The copula parameter α of the Archimedean copula is related to Kendall's τ coefficient of correlation which is defined as

$$\tau = \frac{2}{n(n-1)} \sum_{i}^{n} \sum_{j>1} sgn(X_i - X_j)(Y_i - Y_j) \tag{9}$$

where 'sgn' refers to the sign of the term that follows it. Genest and MacKay (1986) show that there is a relationship between τ and α. The relationship is given as $\tau = 4\int_0^1 \frac{\phi(t)}{\phi'(t)} dt + 1$

(i) Clayton copula.
The Clayton (1978) copula is also referred to as the Cook and Johnson (1981) copula and was originally studied by Kimeldorf and Sampson (1975). It takes the form

$$C_{Clayton}(u_1, u_2; \alpha) = \begin{cases} (u_1^{-\alpha} + u_2^{-\alpha} - 1)^{-\frac{1}{\alpha}}, & \alpha \in (0, \infty), \\ u_1 u_2, & \alpha = 0. \end{cases}$$

and α is the dependence parameter. As α approaches zero the marginals become independent and as it approaches infinity the copula attains the Fréchet upper bound. The Clayton copula cannot account for negative dependence, although it does exhibit strong left tail dependence and relatively weak right tail dependence. It has a tail dependence property of $\lambda_r = 0$ and $\lambda_l = 2^{-\frac{1}{\alpha}}$.

(ii) Frank copula.
The Frank copula, which appeared in Frank (1979) takes the form

$$C_{Frank}(u_1, u_2; \alpha) = -\alpha^{-1} log \left\{ 1 + \frac{(e^{-\alpha u_1} - 1)(e^{-\alpha u_2} - 1)}{(e^{-\alpha} - 1)} \right\}, \tag{10}$$

$\alpha \in (-\infty, 0) \bigcup (0, \infty)$. It has a tail dependence property of $\lambda_r = 0$ and $\lambda_l = 0$. The Frank copula is useful in financial modeling for several reasons. First, it allows for negative dependence between marginals. Second, it allows for symmetric tail dependence. Third, it is able to achieve the Fréchet-Hoeffding bounds.

(iii) Gumbel copula.
The Gumbel copula which appeared in Gumbel (1960) takes the form

$$C_{Gumbel}(u_1, u_2; \alpha) = exp(\bar{u_1}^{\alpha} + \bar{u_2}^{\alpha})^{\frac{1}{\alpha}}, \tag{11}$$

$\alpha \in [1, \infty)$ and $\bar{u_j} = -log u_j$. It has a tail dependence property of $\lambda_l = 0$ and $\lambda_r = 2^{\frac{1}{\alpha}}$. Values of 1 and ∞ correspond to independence and the Fréchet-Hoeffding upper bound. The copula does not attain the Fréchet-Hoeffding lower bound for any dependence parameter value. Also it cannot account for negative dependence. The Gumbel copula exhibits strong right tail dependence and relatively weak left tail dependence.

4 Copula Quantile Regression

Both Chen et al. (2006) and Bouyé and Salmon (2009) have built on the quantile regression work of Koenker and Basset (1978) to propose methods for estimating copula based conditional quantile models. The papers assume a correct specification of the parametric copula dependence function without specifying the underlying marginal distribution functions. Chen et al. (2006) use a rescaled empirical cumulative distribution function to obtain the marginals. After this, they employ the method of maximum likelihood to obtain the copula parameter. Their resulting conditional quantile functions are obtained by plugging in the estimated copula parameter and the empirical marginal cumulative distribution function.

The approach we follow is that of Bouyé and Salmon (2009). They estimate several distinct, non-linear quantile regression models implied by their copula specifications and gave closed forms of the quantile curve for several copulas. We begin with some definitions.

Definition 5 (Bouyé and Salmon 2009) Let $p(x, y; \theta)$ be the probability distribution of y conditional on x. Then

$$p(x, y; \theta) = Pr[Y \leq y \mid X = x] \tag{12}$$
$$= C_1[F_X(x), F_Y(y); \theta] \tag{13}$$

with $C_1(u, v; \theta) = \frac{\partial}{\partial u} C(u, v, \theta)$.

Definition 6 (Bouyé and Salmon 2009) For a parametric copula $C(.,.;\theta)$, the p-th copula quantile curve of y conditional on x is defined by the following implicit equation

$$p = C_1[F_X(x), F_Y(y); \theta] \tag{14}$$

where $\theta \in \Theta$ the set of parameters.

We give three of these copula quantile regression forms.

Normal CQR: The Normal CQR takes the form

$$y = F_Y^{-1}\left[\Phi(\rho\Phi^{-1}(F_X(x)) + \sqrt{1-\rho^2}\Phi^{-1}(q))\right] \tag{15}$$

Student-t CQR: The Student-t CQR takes the form

$$y = F_Y^{-1}\left[t_\nu(\rho t_\nu^{-1}(F_X(x)) + \sqrt{(1-\rho^2)(\nu+1)^{-1}(\nu + t_\nu^{-1}(F_X(x))^2)}) t_{\nu+1}^{-1}(q))\right] \tag{16}$$

Clayton CQR The Clayton CQR takes the form

$$y = F_Y^{-1}\left[(1 + F_X(x)^{-\alpha}(q^{-\frac{\alpha}{1+\alpha}} - 1))^{-\frac{1}{\alpha}}\right]. \tag{17}$$

In the empirical exercise, we aim to estimate a different set of copula parameters $\hat{\theta}_q$ for each quantile regression. Let $(y_1, y_2, \ldots, y_T)$ and $(x_1, x_2, \ldots, x_T)$ be a random sample, the q-th quantile regression curve will be defined as $y_t = \zeta(x_t, q; \hat{\theta}_q)$. The parameters $\hat{\theta}_q$ being any solution to the following optimization problem:

$$min_\theta(\sum_{t=1}^{T}(q - \mathbf{1}_{y_t \leq \zeta(x_t,q;\theta)})(y_t - \zeta(x_t, q; \theta)) \tag{18}$$

See Chap. 7 of Alexander (2008) and Bouyé and Salmon (2009) for details on copula quantile regression modeling.

5 Data and Empirical Estimates

In this section we present the US data and the empirical estimates.

5.1 Preliminary Analysis and Summary Statistics

We examine the return-volatility relationship for indices reported on exchanges in the United States of America. In the empirical analysis, we use daily price data for market and volatility indices of four volatility-return pairs, namely, VXD and DJIA, VIX and S&P 500 (SPX), VXO and S&P 100 (OEX), VXN and NASDAQ (NDX). The daily prices are obtained from the Chicago Board Options Exchange for a period of approximately 11 years from 2/02/2001 to 31/12/2012. For the analysis we use percentage returns computed as 100 times the logarithmic changes. The volatility indices are the VXD, VIX, VXO and the VXN and are discussed below. The CBOE DJIA Volatility Index (VXD) is based on real-time prices of options on the Dow Jones Industrial Average (DJIA), and is designed to reflect investors' consensus view of future (30-day) expected stock market volatility. The SPX VIX, is an index of implied volatility of 30-day options on the S&P 500 calculated from all available stock index option calls and puts bid and ask prices. The index, which was adopted in September 2003 provides an estimate of expected stock market volatility for the subsequent 30 days. According to Hibbert et. al. (2008), the Chicago Board Options Exchange's (CBOE) calculates the VIX from all available stock index option bid and ask prices in the tradable range of these options providing an estimate of expected stock market volatility for the subsequent 30 calendar days (about 21 trading days).

It is based on options on the S&P 500 index (SPX) and it uses options across the tradable range of all strike prices possessing both a bid and ask price; furthermore, it is independent of any option-pricing model. The new method of calculation provides a more robust measure of expected volatility along with option implied volatility skew. The OEX VXO is the original VIX version that was introduced in 1993 and is now disseminated under the ticker symbol VXO, and is based on the S&P 100 index. It considers only near-the-money options, and is calculated using the implied volatilities obtained from the Black-Scholes option-pricing model. The calculation of the CBOE NASDAQ-100 VXN Volatility Index is based on the CBOEs widely accepted VIX methodology. VXN is calculated throughout the trading day based on the near term volatility determined through pricing of NASDAQ-100 Index (NDX) option prices. Like VIX, VXN is a measure of the market's expectation of 30-day volatility, but is based on the NDX rather than the SPX. The CBOE publishes indices of these implied volatilities.

Figures 1 and 2 show the logarithmic return series of the stock return indices and the volatility indices for the period 2/2/2001–31/12/2012. The time series plot

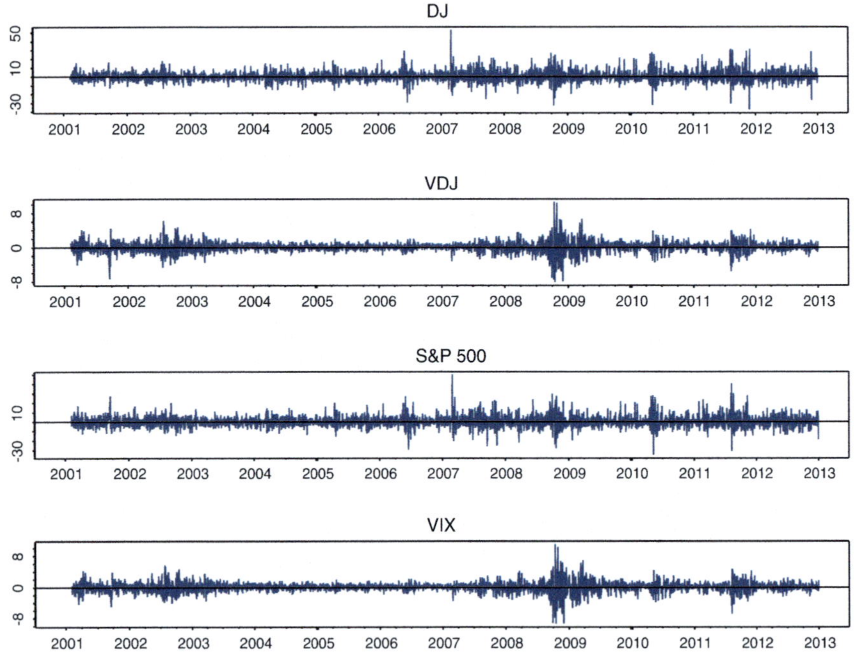

Fig. 1 Time series plot of the stock and volatility indices 2/2/2001–31/12/2012. *Notes* Daily closing percentage returns on the Dow Jones Industrial Average Index from February 2, 2001 through December 31, 2012. Daily closing percentage returns on the Dow Jones Volatility Index from February 2, 2001 through December 31, 2012. Daily closing percentage returns on the S&P 500 Index (SPX) from February 2, 2001 through December 31, 2012. Daily closing percentage returns on the VIX Index from February 2, 2001 through December 31, 2012

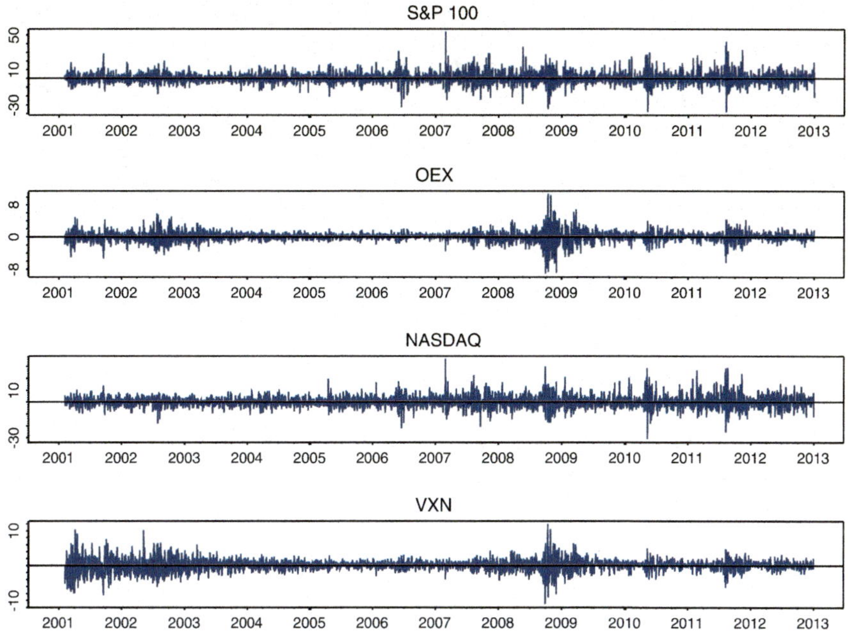

Fig. 2 Time series plot of the stock and volatility indices 2/2/2001–12/31/2012. *Notes* Daily closing percentage returns on the OEX Index from February 2, 2001 through December 31, 2012. Daily closing percentage returns on the S&P 100 Volatility Index (VXO) from February 2, 2001 through December 31, 2012. Daily closing percentage returns on the NASDAQ 100 Index from February 2, 2001 through December 31, 2012. Daily closing percentage returns on the NASDAQ 100 Volatility Index (VXN) from February 2, 2001 through December 31, 2012

seem to show that the individual volatility index changes according to the respective index return changes. Figure 3 gives the quantile-quantile plots for our series, and none of the data series shows a good fit to the normal distributions. It is well known that when the data distribution is not adequately described by a normal distribution, quantile regression (QR) can provide more efficient estimates for the return-volatility relationships (Badshah 2012). Table 1 gives the descriptive statistics for all the variables. All the variables show excess kurtosis, which indicates fat tails. Looking at the Jarque-Bera test statistics in Table 1, we see that the statistics strongly reject the presence of normal distributions in the series. Thus, we can conclude that all the return time series (both the market and the volatility series) exhibit fat tails and are not normally distributed. The reported ADF test statistics, based on an autoregression of order 8, also reject the presence of unit roots in the time series.

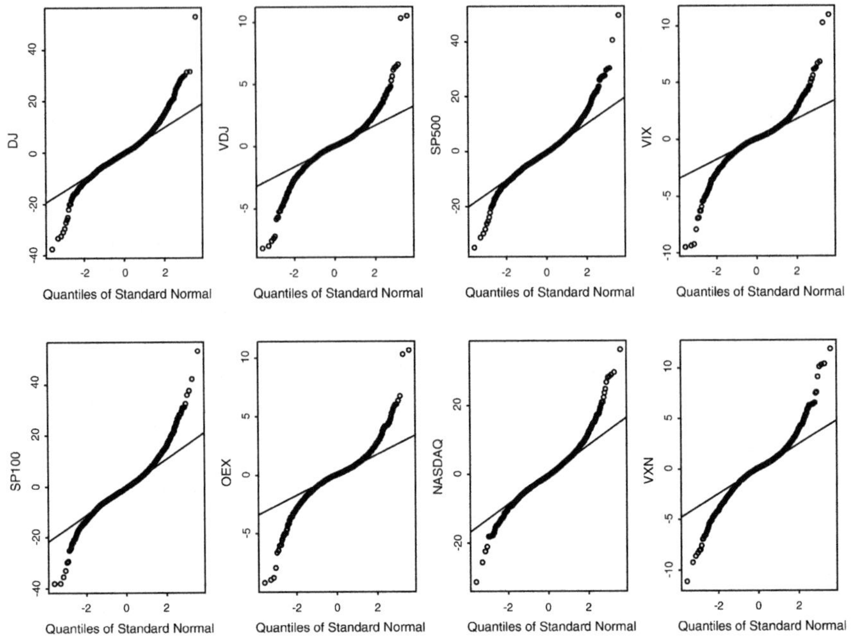

Fig. 3 Quartile-Quartile Plot of the Stock and volatility indices 2/2/2001–12/31/2012. *Notes* Normal qq-plot for Daily closing percentage returns on the Dow Jones Industrial Average Index, the Dow Jones Volatility Index, the S&P 500 Index(SPX), the VIX Index, the SP 100 Index(OEX), the SP 100 Volatility (VXO), the NASDAQ 100 Index and the NASDAQ100 Volatility Index (VXN). The data period is from February 2, 2001 through December 31, 2012

5.2 *Empirical Results Linear Quantile Regression*

Table 2 reports the point estimates of the intercept and regression coefficient for all the volatility-return pairs. The results of the regression coefficients indicate an inverse volatility return relationship. For example, if the DJ index rises by 10 %, then the VDJ will be expected to fall by 34.77 %. Similarly, if the SPX rises by 10 %, then the VIX will be expected to fall by 35.78 %.

Table 3 reports the estimates for the linear quantile regression model, with the intercept α, and the slope coefficient β. The β measures the dependence of volatility on market return. Note that as formulated, the ordinary linear regression model (OLS) is incapable of capturing both the asymmetric and tail dependence between price and implied volatility. In other words, the simple linear regression is incapable of capturing the known empirical facts that (i) volatility increases much more after a large fall in price than it decreases after a large price increase, (ii) volatility reacts more strongly to extreme price moves than normal price moves. One way of address-

Table 1 Descriptive Statistics, Stock and volatility indices 2/2/2001–12/31/2012

	VDJ	DJIA	VIX	SPX	VXO	OEX	VXN	NDX
Observations	2994	2994	2995	2995	2993	2993	2991	2991
Minimum	−37.5750	−8.2005	−35.0589	−9.4695	−38.1487	−9.1862	−31.3049	−11.1149
Quartile 1	−3.6343	−0.5405	−3.7185	−0.5792	−4.0392	−0.5872	−3.0526	−0.8198
Median	−0.3267	0.0409	−0.5013	0.0588	−0.4770	0.0494	-0.4009	0.0825
Arithmetic mean	−0.0049	0.0063	−0.0061	0.0013	−0.0086	−0.0036	−0.0361	0.0008
Quartile 3	3.0311	0.5757	3.1980	0.6130	−4.0392	0.5952	2.7605	0.8504
Maximum	52.8092	10.5084	49.6008	10.9572	53.2274	10.6551	36.2226	11.8493
SE mean	0.1160	0.0230	0.1162	0.0246	0.1287	0.0244	0.0973	0.0331
LCL mean (0.95)	−0.2324	−0.0388	−0.2339	−0.0470	−0.2609	−0.0513	−0.2269	−0.0640
UCL mean (0.95)	0.2226	0.0514	0.2216	0.0495	0.2437	0.0442	0.1547	0.6556
Variance	40.2903	1.5832	40.4082	1.8106	49.5602	1.7740	28.3248	3.2664
Standard deviation	6.3475	1.2583	6.3568	1.3456	7.0399	1.3319	5.3221	1.807
Skewness	0.5564	−0.0232	0.6526	−0.1818	0.5814	−0.1301	0.5425	0.0169
Kurtosis	4.8025	7.8877	4.2860	7.9378	4.3966	7.6032	3.7345	4.3498
ADF	−22.1359	−18.8691	−21.8574	−18.97	−22.3855	−18.9584	−21.5732	−18.3835
Jarque-Bera	3038.111	7776.073	2510.354	7893.879	2584.788	7231.034	1889.112	2363.459

Notes The table reports the various descriptive statistics. LCL is lower control limit. UCL is upper control limit. ADF is augmented Dickey Fuller.
Critical values for ADF: 10 % −3.14; 5 % −3.43; 1 % −4.00
Source for ADF is: MacKinnon (1996)

Table 2 OLS Regression: Stock and volatility indices 2/2/2001–12/31/2012

Model	α	p-value	β	p-value
VDJ-DJ	0.0169	0.841	−3.4770	<0.0001
VIX-SPX	−0.00163	0.983	−3.5778	<0.00015
VXO-OEX	−0.02295	0.787	−3.978	<0.00015
VXN-NDX	−0.034775	0.644	−1.8746	<0.00015

Notes The table reports the OLS regression results for the return volatility pairs. All the estimated β values are significant at the 1 % level

ing this limitation is to employ a linear quantile regression framework. The reported linear quantile regression results are different from those from the OLS. For example, if the DJ index rises by 10 %, then the VDJ will be expected to fall by varying amounts along the quantiles and not by 34.77 % as reported for the OLS. For example, at the 50 % quantile level, we should expect a fall of 35.32 %, and this differs from the 90 % quantile level amount of 37.3 %. Also, the results show that the estimated dependence coefficient (β) values are significant across the quantiles, and are different. Though not reported, we did perform a test to see if the slopes were the same at all the reported quantiles. For the test, we employ the anova command which produces a quantile regression analysis of variance table and is based on tests proposed by Koenker and Bassett (1982). These results indicate that the volatility-return relationship changes across the quantiles and that they are also statistically significant.

5.3 *Empirical Results Quantile Copula*

Tables 4 and 5 give estimates for the quantiles for the Normal and Student-t copulas. For the empirical analysis, we assumed the marginals for the bivariate copula quantile regression follow Normal and Student-t distributions. The univariate Student-t distributions are allowed to have different degree of freedom parameters (see Embrechts et al. 2001 or Fang and Fang 2002). Two versions of the regressions are reported. In one, we work with raw volatility and stock return series and in the second, we fit a GARCH (1, 1) with Student-t errors to the data and then work with the standardized residuals. The estimation follows the general procedure outlined by Bouyé and Salmon (2009). See also Appendix A of Koenker (2005). The rugarch package (Version 1.2-3) of Ghalanos (2013) for R is used to extract the degrees of freedom parameters and the standardized residuals of the series. The quantreg pack-

Table 3 Linear quantile regression estimates: Stock and volatility indices 2/2/2001–12/31/2012

	0.05	0.1	0.2	0.3	0.4	0.5	0.6	0.7	0.8	0.9	0.95
VDJ-DJ α	−6.642	−4.815	−3.055	−1.893	−0.981	−0.109	0.750	1.764	2.935	4.890	6.930
(p-value)	0.0	0.0	0.0	0.0	0.0	0.0	0.0	0.0	0.0	0.0	0.0
VDJ-DJ β	−3.133	−3.166	−3.374	−3.379	−3.510	−3.532	−3,552	−3.544	−3.672	−3.743	−3.795
(p-value)	0.0	0.0	0.0	0.0	0.0	0.0	0.0	0.0	0.0	0.0	0.0
VIX-SPX α	−6.033	−4.385	−3.026	−1.905	−0.999	−0.198	0.658	1.582	2.821	4.681	6.830
(p-value)	0.0	0.0	0.0	0.0	0.0	0.008	0.0	0.0	0.0	0.0	0.0
VIX-SPX β	−3.25	−3.446	−3.547	−3.583	−3.613	−3.645	−3.672	−3.643	−3.770	−3.736	−3.733
(p-value)	0.0	0.0	0.0	0.0	0.0	0.0	0.0	0.0	0.0	0.0	0.0
VXO-OEX α	−6.548	−4.738	−3.243	−2.069	−1.199	−0.283	0.703	1.719	2.955	5.225	7.569
(p-value)	0.0	0.0	0.0	0.0	0.0	0.001	0.0	0.0	0.0	0.0	0.0
VXO-OEX β	−3.706	−3.845	−3.846	−3.985	−4.020	−3.981	−3.931	−4.02	−4.006	−4.168	−4.275
(p-value)	0.0	0.0	0.0	0.0	0.0	0.0	0.0	0.0	0.0	0.0	0.0
VXN-NDX α	−6.326	−4.583	−2.912	−1.885	−1.092	−0.246	0.581	1.617	2.76	4.79	6.872
(p-value)	0.0	0.0	0.0	0.0	0.0	0.0	0.0	0.0	0.0	0.0	0.0
VXN-NDX β	−1.777	−1.72	−1.756	−1.76	−1.743	−1.733	−1.809	−1.864	−1.90	−1.90	−2.048
(p-value)	0.0	0.0	0.0	0.0	0.0	0.0	0.0	0.0	0.0	0.0	0.0

Notes The table reports the linear quantile regression results for the return volatility pairs. A p-value of ≤ 0:05 shows significance at the 5 % level
Computation is done using the R quantreg package of R. Koenker
R and the package quantreg are open-source software projects and can be freely downloaded from CRAN: http://cran.r-project.org

Table 4 Gaussian copula quantile regression coefficient estimates of volatility-return models 2/2/2001–12/31/2012

ρ	0.05	0.1	0.2	0.3	0.4	0.5	0.6	0.7	0.8	0.9	0.95
VDJ-DJ	−0.853	−0.856	−0.855	−0.823	−0.801	−0.801	−0.830	−0.854	−0.862	−0.863	−0.866
VIX-SPX	−0.879	−0.887	−0.878	−0.870	−0.868	−0.878	−0.901	−0.901	−0.888	−0.879	−0.864
VXO-OEX	−0.892	−0.890	−0.881	−0.873	−0.852	−0.873	−0.881	−0.892	−0.889	−0.8880	−0.875
VXN-NDX	−0.811	−0.811	−0.786	−0.746	−0.719	−0.712	−0.771	−0.790	−0.804	−0.806	−0.789
ρ	0.05	0.1	0.2	0.3	0.4	0.5	0.6	0.7	0.8	0.9	0.95
VDJ-DJ	−0.853	−0.851	−0.826	−0.781	−0.751	−0.746	−0.759	−0.787	−0.805	−0.817	−0.824
VIX-SPX	−0.887	−0.884	−0.870	−0.841	−0.809	−0.809	−0.830	−0.845	−0.852	−0.851	−0.839
VXO-OEX	−0.887	−0.891	−0.886	−0.857	−0.834	−0.819	−0.832	−0.854	−0.874	−0.875	−0.859
VXN-NDX	−0.795	−0.793	−0.769	−0.726	−0.705	−0.707	−0.729	−0.754	−0.759	−0.763	−0.758

Notes The table reports the Gaussian copula quantile regression coefficient estimates of volatility-return models for different quantiles. Computation is done using the R quantreg package of R. Koenker

R and the package quantreg are open-source software projects and can be freely downloaded from CRAN: http://cran.r-project.org

Table 5 Student-t copula quantile coefficient estimates of volatility-return models 2/2/2001–12/31/2012

ρ	0.05	0.1	0.2	0.3	0.4	0.5	0.6	0.7	0.8	0.9	0.95
VDJ-DJ	−0.854	−0.848	−0.838	−0.805	−0.782	−0.789	−0.815	−0.837	−0.847	−0.854	−0.859
VIX-SPX	−0.879	−0.881	−0.870	−0.86	−0.861	−0.872	−0.893	−0.895	−0.88	−0.874	−0.862
VXO-OEX	−0.89	−0.885	−0.874	−0.861	−0.85	−0.868	0.874	−0.886	−0.881	−0.878	−0.874
VXN-NDX	−0.802	−0.805	−0.779	−0.733	−0.712	−0.708	−0.76	−0.783	−0.794	−0.801	−0.785
ρ	0.05	0.1	0.2	0.3	0.4	0.5	0.6	0.7	0.8	0.9	0.95
VDJ-DJ	−0.85	−0.838	−0.802	−0.752	−0.705	−0.701	−0.712	−0.731	−0.775	−0.798	−0.822
VIX-SPX	−0.888	−0.881	−0.853	−0.826	−0.786	−0.778	−0.792	−0.816	−0.826	−0.83	−0.841
VXO-OEX	−0.89	−0.882	−0.868	−0.835	−0.799	−0.788	0.798	−0.82	−0.851	−0.857	−0.861
VXN-NDX	−0.810	−0.788	−0.751	−0.705	−0.676	−0.682	−0.69	−0.705	−0.726	−0.745	−0.764

Notes The table reports the Student-t copula quantile regression coefficient estimates of volatility-return models for different quantiles. Computation is done using the R quantreg package of R. Koenker

R and the package quantreg are open-source software projects and can be freely downloaded from CRAN: http://cran.r-project.org

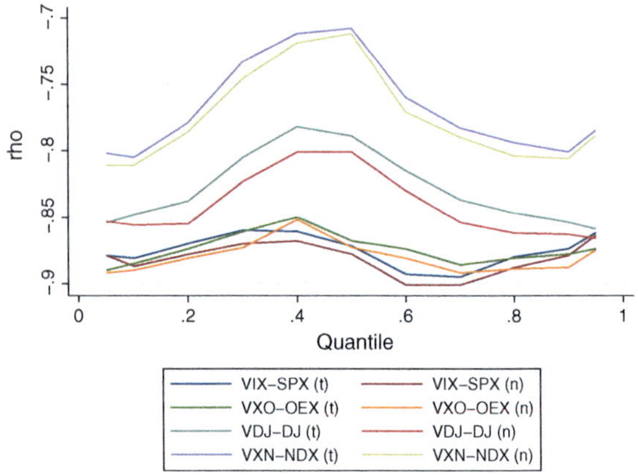

Fig. 4 Calibration of copula quantile regression of US stock volatility on return: 2/2/2001–12/31/2012. *Notes* Normal copula is (n) and t copula is (t). The data period is from February 2, 2001 through December 31, 2012 unfiltered

age (Version 5.05) of Koenker (2012) for R is used to estimate the parameters of the non-linear quantile regression. The nlrq optimization results of quantreg are dependent on the starting values of the parameters and the algorithm option chosen for optimization.The reported results here are based on using the L-BFGS-B option for the Normal copula and the Brent option for the Student-t copula. In each table, the left panel gives results for the raw data, and right panel gives results for the GARCH (1, 1) filtered data. The estimates for the Clayton CQR are not reported. The GARCH (1, 1)filter allows for control for the changes in volatility. As seen from the tables, negative dependence is greater for low and high quantiles. Furthermore, the lower tail negative dependence is higher than the upper tail negative dependence. The results reported here are similar to those of Allen et al. (2012), who used data from US and European exchanges and a different sample period and reported that for most of the pairs they investigated, the negative dependence is greater for low and high quantiles. It should be noted that they did not consider the Dow-Jones volatilitity-return pair nor the S&P 100 volatility-return pair. They also found that the lower tail negative dependence is also higher than the upper tail negative dependence. Figures 4 and 5 show the calibrated values of rho based on copula quantile regression of US stock volatility on return under both the normal and Student t copulas without and with the GARCH (1, 1) filter. The shape based on the GARCH (1,1) filtered data are much more of an inverted U-shaped as compared to the non-filtered series. Figures 6, 7, 8 and 9 show the corresponding quantile curves with the GARCH (1, 1) filter. We do not present those for the unfiltered series. It should be noted that neither Alexander (2008) nor Allen et al. (2012) used some sort of filter to control for changes in volatility. Neglecting to control for volatility changes can lead to incorrect inference in a VaR analysis. For example, suppose one is interested in a VaR analysis and estimates the 5 % quantile regression to achieve this, if one does not control for changes in the level of volatility, the 5 % quantile regression line cannot be interpreted as a true VaR measure since the probability of witnessing any particular price deviation depends crucially on the variance of the distribution.

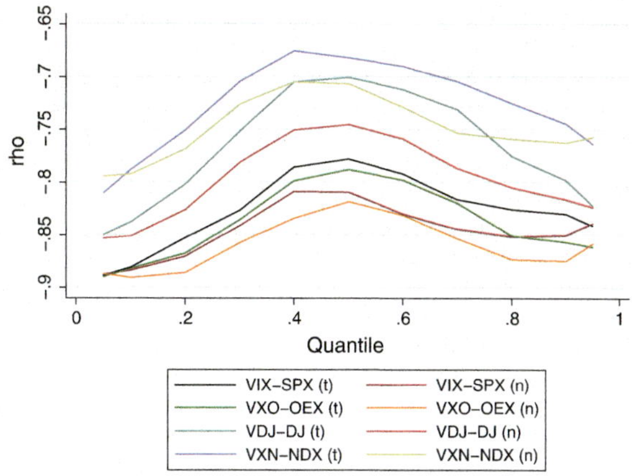

Fig. 5 Calibration of copula quantile regression of US stock volatility on return: 2/2/2001–12/31/2012. *Notes* Normal copula is (n) and t copula is (t). The data period is from February 2, 2001 through December 31, 2012 filtered with a GARCH(1, 1) specification

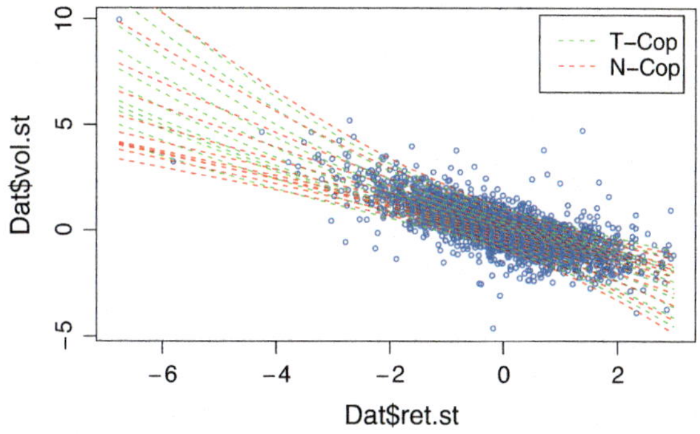

Fig. 6 DJ volatility-return quantile curves of normal and Student t copulas. *Notes* Daily closing percentage returns on the Dow Jones Industrial Average Index from February 2, 2001 through December 31, 2012. Daily closing percentage returns on the Dow Jones Volatility Index from February 2, 2001 through December 31, 2012

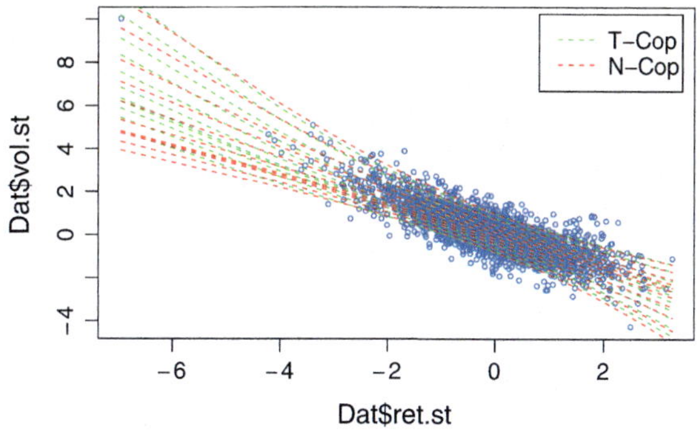

Fig. 7 S&P 500 volatility-return quantile curves of normal and Student t copulas. *Notes* Daily closing percentage returns on the S&P 500 Index (SPX) from February 2, 2001 through December 31, 2012. Daily closing percentage returns on the VIX Index from February 2, 2001 through December 31, 2012

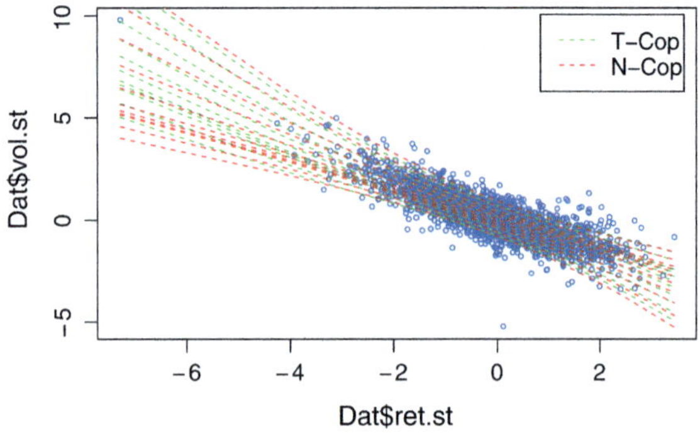

Fig. 8 S&P 100 volatility-return quantile curves of normal and Student t copulas. *Notes* Daily closing percentage returns on the OEX Index from February 2, 2001 through December 31, 2012. Daily closing percentage returns on the S&P 100 Volatility Index (VXO) from February 2, 2001 through December 31, 2012

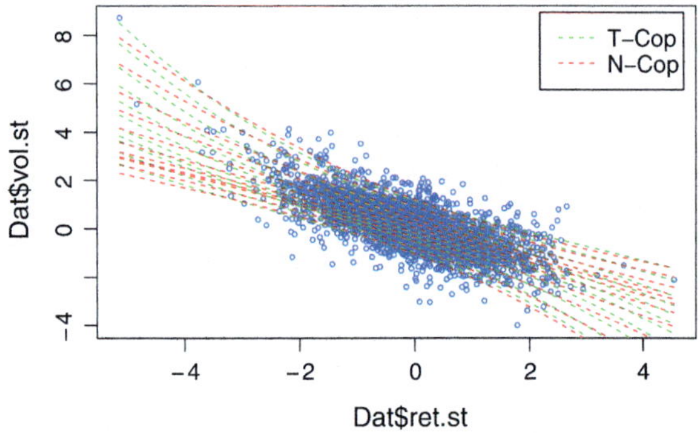

Fig. 9 NASD volatility-return quantile curves of normal and Student t copulas. *Notes* Daily closing percentage returns on the NASADAQ 100 index from February 2, 2001 through December 31, 2012. Daily closing percentage returns on the NASDAQ 100 Volatility Index (VXN) from February 2, 2001 through December 31, 2012

6 Conclusion

In this article, we have applied quantile copula regression techniques to examine the return-volatility relationship for indices reported on exchanges in the United States of America. We adopt the approach of Bouyé and Salmon (2009), which allows one to estimate copula based conditional quantile models. We utilize both linear quantile regression and copula quantile regression to evaluate the asymmetric volatility-return relationship between changes in the volatility index (VXD, VIX, VXO and VXN) and the corresponding stock index return series (DJIA, S&P 500, the S&P 100 and NASDAQ). The data period is from February 2, 2001 through December 31, 2012. We find, firstly, that the relationship between stock return and implied volatility depends on the quartile at which the relationship is being investigated. Secondly, we obtain results similar to those reported for European exchanges that show the existence of an inverted U-shaped relationship between stock return and implied volatility. This result was obtained even after controlling for changes in volatilities of return using a GARCH (1, 1) filter. This conclusion holds for all the US stock and implied volatility indices examined. Models that assumed otherwise are misspecified because ignoring the role of quartiles will result in errors in any attempt to forecast the relationship between returns and implied volatilities.

There are several issues that have not been addressed in the chapter. First, unlike Giot (2005), who examined the relationship between returns and volatility based on high volatility bull market, low volatility bull market, high volatility bear market sub-period classification, we have not concerned ourselves with such sub-period analysis in this chapter. It will be interesting to find out if the relationship is different across sub-periods. Second, the entire focus here is on the stock markets. Understanding the relationship between returns and implied volatilities for other commodities should be interesting.

Acknowledgements I like to thank Professor David Allen, Professor Graciela Chichilnisky, Professor Roger Koenker, Professor Randall Filer, a referee and the editor, for comments and suggestions. I also like to thank Lucas Dowiak for comments, suggestions and very useful computing assistance. I alone am responsible for any remaining errors.

References

Abdi, H. (2007). Method of least squares. In Neil Salkind (Ed.), *Encyclopedia of measurement and statistics*. Thousand Oaks (CA): Sage.

Alexander, C. (2008). *Market risk analysis: practical financial econometrics* (Vol. 2). Chichester: Wiley.

Allen, D. E., Gerrans, P., Singh, A. K., & Powell, R. (2009). Quantile regression and its application in investment analysis. *The Finsia Journal of Applied Finance (JASSA)*, *4*, 7–12.

Allen, D., Singh, A. K., Powell, R. J., McAleer, M., Taylor, J., & Thomas, L. (2012). *The volatility-return relationship: insights from linear and non-linear quantile regression*. Working Paper, School of Accounting Finance & Economics, Edith Cowan University, Retrieved from Complutense. http://eprints.ucm.es/16688/.

Ang, A., & Chen, J. (2002). Asymmetric correlations of equity portfolios. *Journal of Financial Economics*, *63*(3), 443–494.

Badshah, I. U. (2012). Quantile regression analysis of the asymmetric return-volatility relation. *Journal of Futures Markets*, *33*(3), 235–265.

Barnes, M. L., & Hughes, W. A. (2002). *Quantile regression analysis of the cross section of stock market returns* (Working Paper). Retrieved from Social Science Research. http://ssrn.com/abstract=458522.

Black, F. (1976). Studies of stock market volatility changes, *Proceedings of the American Statistical Association, Business and Economic Statistics Section* (pp. 177–181)

Bouyé, E., & Salmon, M. (2009). Dynamic copula quantile regressions and tail area dynamic dependence in Forex markets. *The European Journal of Finance*, *15*(7–8), 721–750.

Breymann, W., Dias, A., & Embrechts, P. (2003). Dependence structures for multivariate high-frequency data in finance. *Quantitative Finance*, *3*(1), 1–16.

Buchinsky, M. (1998). Recent advances in quantile regression models: A practical guideline for empirical research. *The Journal of Human Resources*, *33*(1), 88–126.

Buchinsky, M., & Leslie, P. (1997). *Educational attainment and the changing U.S. wage structure: Some dynamic implications* (Working Paper No. 97–13). Department of Economics, Brown University.

Buchinsky, M., & Hunt, J. (1999). Wage mobility in the united state. *The Review of Economics and Statistics*, *8*(3), 351–368.

Campbell, J. Y., & Hentschel, L. (1992). No news is good news: An asymmetric model of changing volatility in stock returns. *Journal of Financial Economics*, *31*(3), 281–318.

Chen, X., Fan, Y., & Tsyrennikov, V. (2006). Efficient estimation of semiparametric multivariate copula models. *Journal of the American Statistical Association, 101*, 1228–1240.
Cherubini, U., Luciano, E., & Vecchiato, W. (2004). *Copula methods in finance*. New York: Wiley.
Chichilnisky, G. (2009). The topology of fear. *Journal of Mathematical Economics, 45*(11–12), 807–816.
Christie, A. (1982). The stochastic behaviour common stock variances: Value, leverage and interest rate. *Journal of Financial Economics, 10*, 407–432.
Clayton, D. (1978). A model for association in bivariate life tables and its application in epidemiological studies of familial tendency in chronic disease incidence. *Biometrika, 65*, 141–151.
Cook, R. D., & Johnson, M. E. (1981). A family of distributions for modelling non-elliptically symmetric multivariate data. *Journal of Royal Statistical Society B, 43*(2), 210–218.
Cox, J., & Ross, S. (1976). The valuation of options for alternative stochastic processes. *Journal of Financial Economics, 3*, 145–166.
Dennis, P., Mayhew, S., & Stivers, C. (2006). Stock returns, implied volatility innovations, and the asymmetric volatility phenomenon. *Journal of Financial and Quantitative Analysis, 41*(2), 381–406.
Eide, E., & Showalter, M. H. (1998). The effect of school quality on student performance: A quantile regression approach. *Economics Letters, 58*(3), 345–350.
Eide, E., & Showalter, M. H. (1999). Factors affecting the transmission of earnings across generations. A quantile regression approach. *Journal of Human Resources, 34*(2), 253–267.
Embrechts, P., Höing, A., & Juri, A. (2003). Using copulae to bound the value-at- risk for functions of dependent risks. *Finance & Stochastics, 7*, 145–167.
Embrechts, P., McNeil, A., & Straumann, D. (2001). Correlation and dependence in risk management: Properties and pitfalls. In M. A. H. Dempster (Ed.), *Risk management: Value at risk and beyond* (pp. 176–223). Cambridge: Cambridge University Press.
Engle, R. F., & Manganelli, S. (2004). CAViaR: Conditional autoregressive value at risk by regression quantiles. *Journal of Business & Economics Statistics, 22*(4), 367–381.
Fang, H., Fang, K., & Kotz, S. (2002). The metaelliptical distributions with given marginals. *Journal of Multivariate Analalysis, 82*, 1–16.
Fleming, J., Ostdiek, B., & Whaley, R. E. (1995). Predicting stock market volatility: A new measure. *Journal of Futures Markets, 15*(3), 265–302.
Frank, M. J. (1979). On the simultaneous associativity of F(x, y) and x + y − F(x, y). *Aequationes Math, 19*, 194–226.
French, Kenneth R., William Schwert, G., & Stambaugh, Robert F. (1987). Expected stock returns and volatility. *Journal of Financial Economics, 19*, 3–29.
Galton, F. (1886). Regression towards mediocrity in hereditary stature. *The Journal of the Anthropological Institute of Great Britain and Ireland, 15*, 246–263.
Genest, C., & Mackay, J. (1986). The joy of copulas: Bivariate distributions with uniform marginals. *The American Statistician, 40*, 280–283.
Genest, C., & Rivest, L. P. (1993). Statistical inference procedures for bivariate Archimedean copulas. *Journal of the American Statistical Association, 88*, 1034–1043.
Ghalanos, A. (2013). rugarch: A garch r package (Version 1.2-3). Retrieved from r-project.org. http://www.r-project.org.
Giot, P. (2005). Relationships between implied volatility indices and stock index returns. *Journal of Portfolio Management, 31*, 92–100.
Gumbel, E. J. (1960) Distributions des Valeurs Extremes en Plusieurs Dimensions. Publications de lInstitute de Statistique de lUniversite de Paris 9:171–173.
Harper, H. L. (1974–1976). The method of least squares and some alternatives. Part I, II, II, IV, V, VI. International Statistical Review *42*, 147–174; *42*, 235–264; *43*, 1–44; *43*, 125–190; *43*, 269–272; *44*, 113–159.
Hibbert, A., Daigler, R., & Dupoyet, B. (2008). A behavioural explanation for the negative asymmetric return-volatility relation. *Journal of Banking and Finance, 32*, 2254–2266.
Joe, H. (1997). *Multivariate models and dependence concepts*. New York: Chapman and Hall.

Kim, T. H., & White, H. (2003). Estimation, inference, and specification testing for possibly misspecified quantile regressions. In T. Fomby & R. C. Hill (Eds.), *Maximum Likelihood Estimation of Misspecified Models: Twenty Years Later* (pp. 107–132). New York: Elsevier.
Kimeldorf, G., & Sampson, A. R. (1975). Uniform representations of bivariate distributions. *Communications in Statistics, 4*, 617–627.
Koenker, R. (2005). *Quantile regression.*, Econometric society monograph series New York: Cambridge University Press.
Koenker, R. (2012). Quantreg: A quantile regression R package (Version 5.05). Retrieved from r-project.org. http://www.r-project.org.
Koenker, R. W., & Bassett, G, Jr. (1978). Regression quantiles. *Econometrica, 46*(1), 33–50.
Koenker, R. W., & Bassett, G, Jr. (1982). Robust tests for heteroscedasticity based on regression quantiles. *Econometrica, 50*(1), 1577–1584.
Koenker, R. W., & Hallock, K. F. (2001). Quantile regression. *Journal of Economic Perspectives, 15*(4), 143–156.
Koenker, R. W., & Xiao, Z. (2006). Quantile autoregression. *Journal of the American Statistical Association, 101*(475), 980–990.
Kumar, S. (2012). A first look at the properties of India's volatility index. *International Journal of Emerging Markets, 7*(2), 160–176.
Longin, F., & Solnik, B. (2001). Extreme correlation of international equity markets. *Journal of Finance, 56*, 649–676.
Low, C. (2004). The fear and exuberance from implied volatility of S&P 100 index options. *Journal of Business, 77*, 527–546.
MacKinnon, J. G. (1996). Numerical distribution functions for unit root and cointegration tests. *Journal of Applied Econometrics, 11*, 601–618.
Neftci, S. (2000). Value at risk calculations, extreme events, and tail estimation. *Journal of Derivatives, 7*(3), 23–37.
Nelsen, R. B. (2006). *Introduction to copulas*. New York: Springer Verlag.
Patton, A. J. (2004). On the out-of-sample importance of skewness and asymmetric dependence for asset allocation. *Journal of Financial Econometrics, 2*(1), 130–168.
Patton, A. J. (2006a). Estimation of multivariate models for time series of possibly different lengths. *Journal of Applied Econometrics, 21*(2), 147–173.
Patton, A. J. (2006b). Modelling asymmetric exchange rate dependence. *International Economic Review, 47*(2), 527–556.
Patton, A. J. (2009). Copula-based models for financial time series. In T. G. Andersen, R. A. Davis, J. P. Kreiss, & T. Mikosch (Eds.), *Handbook of financial time series*. New York: Springer Verlag.
Powell, J. L. (1986). Censored regression quantiles. *Journal of Econometrics, 32*, 143–155.
Sklar, A. (1959). Fonctions de Riépartition án Dimensions et Leurs Marges. *Publications de lInstitut de Statistique de lUniversité de Paris, 8*, 229–231.
Taylor, J. (1999). A quantile regression returns. *Journal of Derivatives, 7*(1), 64–78.
Trivedi, P. K., & Zimmer, D. M. (2005). Copula modeling: An introduction for practitioners. *Foundations and Trends in Econometrics, 1*(1), 1–111.
Whaley, R. (2000). The investor fear gauge. *Journal of Portfolio Management, 26*, 12–17.
White, H. (1994). *Estimation inference and specification analysis*. New York: Cambridge University Press.
Wu, G. (2001). The determinants of asymmetric volatility. *The Review of Financial Studies, 14*(3), 837–859.
Xiao, Z. (2009). Quantile cointegrating regression. *Journal of Econometrics, 150*(2), 248–260.
Yu, K., Lu, Z., & Stander, J. (2003). Quantile regression: Applications and current research areas. Journal of the Royal Statistical Society, Series D. (The Statistician) *52*(3), 331–350.

Economic Crises: Natural or Unnatural Catastrophes?

Alan Kirman

1 Introduction

Catastrophe: "an event causing great and usually sudden damage or suffering; a disaster" *Definition Oxford English Dictionary.*

> Many disasters in anthropogenic systems should not be seen as 'bad luck', but as the results of inappropriate interactions and institutional settings. Even worse, they are often the consequences of a wrong understanding due to the counter-intuitive nature of the underlying system behaviour. Hence, conventional thinking can cause fateful decisions and the repetition of previous mistakes. This calls for a paradigm shift in thinking: systemic instabilities can be understood by a change in perspective from a component-oriented to an interaction- and network-oriented view. Helbing (2013), p. 51

The dictionary definition of catastrophe would suggest that it is an appropriate characterization of major economic crises. Yet much of the attention to catastrophic events has been devoted to "natural events" such as floods, hurricanes, earthquakes etc. These have long been recognized as being a source of economic calamities, at least locally but have, in general, been considered as exogenous. Alternatively, there has been some focus on manmade catastrophes such as those at Bhopal or Fukushima, which have important economic, societal and environmental consequences, but these have typically been regarded as technical or technological failures, and therefore, in a certain sense exogenous. Recently, however, with the recognition that we have moved into the "anthropocene" era, there has been a

I would like to thank an anonymous referee for a very helpful report and the participants in the AFOSR/CCRM Workshop on Catastrophic Risks at Stanford for their suggestions and comments and the LABEX OTMed, for funding this research.

A. Kirman (✉)
Faculte de Droit et de Sciences, GREQAM, Aix-Marseille Université, CAMS - EHESS
190-198, avenue de France, 75244 Paris Cedex 13 Aix-en-Provence, France
e-mail: alan.kirman@univ-amu.fr

G. Chichilnisky and A. Rezai (eds.), *The Economics of the Global Environment*,
Studies in Economic Theory 29, DOI 10.1007/978-3-319-31943-8_27

growing consciousness of a more general feedback from human activity, and particularly economic activity to the environment and this has called for rethinking our models of both or, at least of constructing links between, models of nature and those of economics.

In the natural sphere, particularly concerning climate, there are two approaches, one of which involves the construction of structural models which attempt to simulate the evolution of the climate and the other which essentially focuses on the analysis of the time series of such events in the past. On the one hand climatology models are insistent on the recognition that the climate is a complex system whose evolution is governed by the feed-backs between its components and, on the other, statistical analysis tries to detect some long term structure in the evolution without trying to specify an underlying mechanism. The former suggests that there is little hope for any precise forecasting but something to be gained from an understanding of the fundamental mechanisms at work.

In a sense these approaches are mirrored in economics but with a very important difference. What are the corresponding two views of the nature of in the evolution of economies and what is the place of catastrophic events? The standard approach to the structural model is to build models based on the rational actions of the individuals within them but without attributing much, if any, importance to the direct interaction between those individuals. Thus this view, while having the ambition to understand the mechanisms that govern the evolution of the economy is far from the complex system vision. Furthermore, there has been a growing conviction that it is possible to develop "scientific" models of the economy corresponding to Walras', the father of General Equilibrium Theory, vision, when he thought economics would come to have the same status as astrophysics. Worse, many economic theorists would argue that the system is inherently stable.

This is not to say that efforts have not been made by economists to examine systems in which non linear dynamics and feedbacks play a central role and in which the evolution can be "chaotic" (see e.g. Grandmont 1985 or Brock and Sayers 1988 and Chichilnisky et al. 1995). Indeed, the very term catastrophe theory based on the work of Thom (1983) gave rise to a literature on applications to economics. (see e.g. Zeeman 1974; Varian 1979 and Rosser 2007). Such efforts were rapidly pushed to one side and the view that has come to prevail, is that economies have a well defined structure and fluctuate around some "equilibrium path" and furthermore, that the fluctuations are caused by shocks which are exogenous to the economy. In this view there is some mechanism which, after such a shock, brings the economy back to its equilibrium path. The problem then becomes one of studying the distribution of the shocks and to calculate the probability of the arrival of a major one.

There is an alternative view, which has not prevailed till now, which is that sudden and major changes are intrinsic to the evolution of the economy, their repetitive appearance is a characteristic of the constant evolution and adjustment of the components of the economy. The system is never in equilibrium in the standard sense but passes through periods of stasis interspersed with endogenous upheavals. In such a view, economic catastrophes are not only endogenous but a basic characteristic of modern economies. This was the argument made by Minsky (2008) who argued that economies are intrinsically unstable. But this runs counter to a long tradition in economics of assuming that somehow the economy, particularly if left to its own devices will self organize in a stable and efficient way. This was not just the view of those who worked with mathematical models but also was considered to be a heritage from Adam Smith and Hayek referred to the notion of "spontaneous order". But since economic crises and catastrophes are difficult to reconcile with such views it is worth spending a little time on the place of stability in economic analysis. Just before doing so it is useful to look at an analogy.

2 The Antarctic

The Western Antarctic Ice Sheet is now subsiding into the sea more rapidly than previously. Furthermore, this process is now irreversible, according to two articles in Science (Joughin et al. 2014) and Geophysical Research Letters (Rignot et al. 2014). This will lead to a "short term" rise in sea level of over 1 m and a longer term rise of much greater magnitude. Anthropogenic causes are an important part of the explanation.

But why is this of interest to economists? First, because it is claimed that human behaviour has been, in part, responsible for the changes that have led to the collapse (already forecast by Mercer 1978) Secondly, because the nature of the causality is not as simple as might be thought. The obvious argument is that rising air temperatures caused by increasing CO_2 emissions have increased sea temperatures and that this has caused the melting.

In fact the mechanism is more indirect. Stronger winds have pushed warmer water which rises naturally towards the Antarctic region. These are caused, it is claimed, by global warming. This coupled with the increased Ozone hole, due in part to the emission of aerosol gases has led to the change in the ice sheet's stability.

But, and here is the important point, "There is no stabilising mechanism" as one of the authors said. Changing the things which we can control will not help now to prevent the phenomenon but could slow it. The system has self organised into an unstable state. The lessons are clear, the environment does not necessarily shift to an equilibrium as the parameters which govern it vary. Furthermore, humans are largely reduced to the role of spectators even though their activities may be, at least in part, responsible for the changes that are taking place. Within the environmental system we have come to accept that the nature of causality will be highly complex, yet for the economic system we still want to find relatively simple mechanical models which will predict the consequences of the measures that we take. My basic argument here is that we have no sound theoretical reason for arguing that the economy self organises to an efficient and stable state, even without taking into account its increasing interaction with the environment.

But it is worth reflecting on how we came in economics to the position that, if only people and markets are left to their own devices they will produce a socially satisfactory equilibrium situation. This question of the stability of the economic system and the attribution of crises to exogenous shocks is at the heart of the laissez faire tradition.

3 Stability

The idea that economies are systematically in an equilibrium state is highly counter-intuitive to non-economists. Indeed, early economists questioned the notion of an economy self-organizing into an equilibrium state. Already in 1819 Sismondi said

> Let us beware of this dangerous theory of equilibrium which is supposed to be automatically established. A certain kind of equilibrium, it is true, is reestablished in the long run, but it is after a frightful amount of suffering. de Sismondi (1819), pp. 20–21.

Again, Walras himself was convinced that economies were not perpetually in equilibrium but he did think that there was some mechanism that was constantly trying to drive it there. He said, in the Elements, that the market is:

> the market is like a lake agitated by the wind, where the water is incessantly seeking its level without ever reaching it. Walras (1954), p. 380.

He went further and then said,

> just as a lake is, at times, stirred to its very depths by a storm, so also the market is sometimes thrown into violent confusion by crises which are sudden and general disturbances of equilibrium. Walras (1954), p. 381.

Although his predecessors had doubts, Walras' quote could be seen as consistent with the assumption, which is systematically made in macroeconomics, that the economy is constantly on an equilibrium path. It is only knocked out of this state by some exogenous "storm". How did we get to this position? The historical basis for this is interesting. At one time it was thought that the problems of stability and uniqueness were closely related and that it should be possible to show that essentially proving one would guarantee, with a few restrictions, the other. For many economists it seemed to be a basic criterion for the consistency of a model that it had a unique and stable equilibrium. Scarf (1959) however, showed that one could have a unique equilibrium which was not globally stable.[1] Nevertheless, until the results of Sonnenschein (1972), Mantel (1974) and Debreu (1974), there was a persistent hope that, with the standard assumptions on individuals, one could, at least, deal with the stability problem and show that an economy starting from a disequilibrium state would tend to equilibrium, reflecting the idea expressed by Walras. Thus, those who had expressed skepticism were regarded as not having had

[1]Scarf (1969) was also a pioneer of the computational approach to finding an equilibrium rather than being satisfied with showing that one exists. But, this did not answer the question as to how one would get to such an equilibrium.

the analytical tools to show that equilibria were stable under reasonable assumptions on individuals. However, the results just mentioned were proved by three of the most sophisticated mathematical economists of their time and they showed that, even under the stringent and unrealistic assumptions made on individuals, one could not show that equilibria were stable. This led Morishima (1984) to remark,

> If economists successfully devise a correct general equilibrium model, even if it can be proved to possess an equilibrium solution, should it lack the institutional backing to realize an equilibrium solution, then the equilibrium solution will amount to no more than a utopian state of affairs which bear no relation whatsoever to the real economy. Morishima (1984), pp. 68–69.

The reaction to this could have been to study the evolution of economies in non-equilibrium states. This would have meant sacrificing the basic theorems of welfare economics and would have had profound consequences. Furthermore, the informational efficiency of the competitive allocation mechanism, long vaunted as one of its most important merits, would no longer have held. To see this, let me go back to the basic approach to theoretical economics and its notion of equilibrium. Suppose that individuals do actually satisfy the rationality axioms, and furthermore that the organization and transmission of information concerning prices is somehow achieved. Indeed, suppose, as in the standard model, that there is a single price for each good and, that it is known to everyone. Individuals simply need to know these prices and this, coupled with their income, generates the constraints that, together with their preferences, yield their demands and, of course, their excess demands for goods. The standard argument is now simple. What is needed is a vector of prices that will make these excess demands consistent, in the sense that in aggregate there is zero excess demand for all commodities. Thus all that the market mechanism has to do is to transmit the equilibrium price vector corresponding to the aggregate excess demands submitted by the individual economic agents. The information required to make this system function at equilibrium is extremely limited. In fact, a well-known result of Jordan (1982) shows that the market mechanism not only is parsimonious in terms of the information that it uses, but, moreover, it is also the only mechanism to use so little information to achieve an efficient outcome in the sense of Pareto. This extraordinary result depends, unfortunately, on one key assumption, which is that the economy is functioning *at equilibrium.*

However, as soon as one considers how the economy might function out of equilibrium the informational efficiency property is lost. What is more, if one considers how an economy might adjust to equilibrium, looking at informational efficiency provides a key to the basic problem with equilibrium theory. To see why this is so, consider one initial reaction to the stability problem which was to suggest that the problem lay with the adjustment process, the tatonnement process, that was assumed. Yet, what became immediately clear after the innovative work of Smale (1976) was that stability could only be achieved at the price of a significant increase in the amount of information needed. Smale's Global Newton Method is an extension of standard methods, which allow one to find a fixed point of a mapping, such as an aggregate excess demand function, if one starts sufficiently near the boundary of definition.[2] It has two major drawbacks. Firstly, it does not behave well in the interior of the domain, which, in the case under consideration, is the space of all strictly positive prices. Secondly, as already mentioned, it uses a great deal of information. What is needed is knowledge of all the partial derivatives of the aggregate excess demand functions and this increases the size of the message space without guaranteeing convergence from any arbitrary starting point. An additional problem is with the economic content of the process. While the original tatonnement process has a very natural interpretation this is not the case for the Newton Methods, despite the efforts of Varian (1977).

Are the informational requirements of the Newton Method a necessary evil? Saari and Simon (1978) asked the following question. Can one find what they called "Locally Effective Price Mechanisms," that is, ones that turn all economic equilibria into sinks, which use less information than the Newton Method? They proved, unhappily, that this cannot be done. One might have hoped that we had simply made the wrong choice of process, since the Generalized Newton Method has the undesirable property that it reduces excess demands monotonically and one might have hoped that, by relaxing this one could have found less informationally demanding mechanisms. Unfortunately Saari and Simon showed that any process, which would lead to equilibrium from any starting price vector, would use an infinite amount of information. Many ingenious attempts have been made to construct adjustment mechanisms, which would get around this.

However, as Jordan (1982) pointed out, all the alternative adjustment processes that had been constructed, when he wrote, had no economic interpretation. Since that time, there have been many further efforts to construct globally and universally stable price adjustment processes and, in a certain sense, Kamiya (1990), Flaschel (1991) and Herings (1997) succeeded. Yet, if one looks closely at these results there is always some feature, which is open to objection. In Kamiya's case the excess demand function is artificially defined outside the original price domain. In Flaschel's case the adjustment process depends on a parameter, which varies with the economy and indeed, he says that it is too much to hope that one would find a process that would work for all economies. Hering's mechanism has the curious feature that prices are adjusted according to the relation between current price and the starting price.

Thus it has become clear that there is no hope of finding an economically interpretable adjustment process, which will converge from any price vector independent of the economy. Had we been able to do so, this would have rehabilitated Walras' idea of the economy moving towards equilibrium even if it took an arbitrarily long time to reach it. But, to repeat, the Saari and Simon result showed that we had ended up in an impasse. Where does all this leave us? The informational requirements of adjustment processes seem to be so extreme that only economy specific processes could possibly ensure convergence. This is hardly reassuring for those who argue for the plausibility of the equilibrium notion.

[2]By this we mean starting from an initial price vector where some of the prices are close to zero.

Why dwell on the problem of stability? Precisely because the implicit assumption that the economy will always come back spontaneously to equilibrium, rules out the possibility of a sudden crisis or economic catastrophe. Yet we observe such phenomena and the whole point of this paper is to suggest that the economic system, like the environment will generate occasional large movements without any significant changes in its underlying parameters.

The alternative to taking the stability problem seriously, and to analyse what happens out of equilibrium, is the one, which has been taken in macroeconomics. This is to assume that the economy is always on or very close to an equilibrium path, thereby finessing the whole problem of stability and that any deviations from such a situation were necessarily temporary and caused by some outside perturbation.

But, the clear contradiction between the empirical evidence and the theory should have meant that the whole structure and basis of the model were thrown into question. The difficulties mentioned above reflect fundamental problems with the basic model. We somehow decided in macroeconomics to put these problems to one side and to assume that they were of no importance. This meant, to repeat, that we were in a position where any crisis or crash was simply incompatible with the model or had to be explained by extraneous causes. Endogenous collapses or booms were not part of the evolution of the economy which was assumed with the exception of a few episodes to be essentially in equilibrium. The confidence in this view was reflected in the statement by Alan Greenspan when he said:

> With notably rare exceptions (2008, for example), the global 'invisible hand' has created relatively stable exchange rates, interest rates, prices and wage rates. Alan Greenspan Former Chairman of the Federal Reserve Bank (Greenspan 2011).

As two somewhat cynical commentators observed in response:

> With notably rare exceptions, Germany remained largely at peace with its neighbours during the 20th century.

and

> With notably rare exceptions, Alan Greenspan has been right about everything. Comments on the blog Crooked Timber March 30th 2011.

Not all policymakers were so complacent however, and Adair Turner the overseer of the financial sector in the U.K said clearly:

> … there is also a strong belief, which I share, that bad or rather over-simplistic and overconfident economics helped create the crisis. There was a dominant conventional wisdom that markets were always rational and self-equilibrating, that market completion by itself could ensure economic efficiency and stability, and that financial innovation and increased trading activity were therefore axiomatically beneficial. Adair Turner, Head of the U.K. Financial Services Authority.

What is being put into question, in reality is the belief, unwarranted by theory, that markets and the economy systematically self organise into an efficient state. Hence, the only reason that it would not be in such a state must be because of some exogenous, stochastic shocks. Since the agents in the economy will be aware that there are such random events occuring, his means that the agents must take full account of the intrinsic uncertainty in the evolution in the economy. How this has been done brings me to the next topic.

4 Rational Expectations

In all that I have said up to now, the only uncertainty evoked, corresponds to shocks generated by some process external to the economy. But this is a superficial simplification. If there is uncertainty in the evolution of the economy then the individuals will make their decisions taking this uncertainty into account. This involves not only considering the expectations of individuals but also their attitudes to risk. The standard approach has come to be to assume that individuals understand the stochastic process which governs the economy and that their choices do, indeed lead the economy to behave as they assume it to do. When one considers all that is involved in such an assumption it seems, to say the least, implausible. One argument is that this is just a way of "closing the model". In other words this is a fixed point argument. If individuals were to hold such expectations then they would be self realising. But this says nothing about how they came to hold such expectations. Even if the economy was thought to be on an equilibrium path and occasionally knocked off it by exogenous shocks we have to assume that agents know the distribution of those shocks. Yet, in a world where "extreme events" are thought of as being increasingly important, it is difficult to imagine that the individuals within the model have the capacity to quantify appropriately the risks and to take account of it in making their decisions. As Bernanke (2010) one of the policymakers who became keenly aware, during the crisis, of the deficiencies of modern macroeconomic models based on rational expectations said,

> I just think it is not realistic to think that human beings can fully anticipate all possible interactions and complex developments. The best approach for dealing with this uncertainty is to make sure that the system is fundamentally resilient and that we have as many fail-safes and back-up arrangements as possible. Ben Bernanke Interview with the IHT May 17th 2010.

With all of this in mind one might rightly ask how did we arrive at the notion of rational expectations as it is currently conceived? The foundations of the modern view were laid paradoxically by John Muth and Herb Simon at the beginning of the sixties when they were colleagues at Carnegie Mellon university. The paradox lies in the fact that Herbert Simon was one of the fathers of the notion of "bounded rationality" the idea that individuals do not have the capacity to reason in the fully rational way that economists typically assume.

The problem that interested them at the time was not so much how individuals make their decisions but rather, how firms do so (see Holt et al. 1960).[3] This was, therefore not so much a question of the individual rationality that is at the basis of modern macromodels but more a question as to how well the behaviour of firms corresponds to the optimisation that economists attribute to them. Yet, given that firms change in ownership, structure and even goals over time, the task of anticipating all this is also heroic. Thus the problem to be tractable has, somehow to be simplified. Muth (1961) in an article which has become the basic reference for the rational expectations literature was explicit.

> I should like to suggest that expectations, since they are informed predictions of future events are essentially the same as the predictions of the relevant economic theory. At the risk of confusing this purely descriptive hypothesis with a pronouncement as to what firms ought to do, we call such expectations "rational". Muth Rational Expectations 1961.

Here we move to a different level of reasoning. Much of the earlier discussion in economics did not concern the specification of the uncertainty with which individuals are faced while here Muth was suggesting that, if there is a satisfactory model in economic theory which captures the evolution of the economy, people should form their expectations consistently with that model. This was, of course, extremely convenient for economists who now only had to require agents to have expectations consistent with the model that the economists proposed. In other words, Muth saw clearly that specifying the expectations as being consistent with the evolution of the economy was simply a way, as I have said earlier, of closing the model. However, Muth was well aware of what sort of assumptions were necessary for this and he questioned the empirical value of this exercise. Indeed he said,

> To make dynamic economic models complete various expectational formulae have been used. There is, however, little evidence to suggest that the presumed relations bear a resemblance to the way the economy works. Muth Rational Expectations 1961.

In other words, the rational individuals in the artificial world of the economic model would not be surprised by the fact that economic catastrophes occur, though any one of them might well constitute a surprise. This is simply because the occurrence of such events would be consistent with their view of the world, and if they waited long enough the actual distribution of these events would correspond with their prior beliefs. Simon was also working on this problem, that of incorporating people's behaviour towards the uncertainty with which they were faced into economic models, when Muth published his seminal paper in 1961. Simon actually, at one point, reached a conclusion not far from that proposed by Muth. He observed that under very strong assumptions it would be possible for individuals, without having to know the complete "true" process governing the evolution of the economy simply to substitute the expected value for the relevant stochastic

[3]As the referee pointed out to me, Muth's interest in the cyclical evolution of markets came from the idea of the cobweb model which was used to represent the markets for corn and hogs and his work followed on from that of Ezekiel (1938).

variables and then to perform their deterministic calculations to find the optimal solution. This requires the individual to have a correct view of the mean of the distribution with which he is confronted which is already an exacting requirement. However, like Muth he saw clearly that such a situation was unlikely to be an adequate characterisation of the way in which things happen in the real world. As he said,

> Of course, the solution though it provides optimal solutions for the simplified world of our assumptions, provides, at best, satisfactory solutions for the real world decision problem. In principle, unattainable optimisation is sacrificed for in practice attainable satisfaction. H. Simon "Rational decision-making in business organizations" Nobel Memorial Lecture, 8 December, 1978, p. 499.

It is interesting to observe that while Muth did not turn away from the rational expectations hypothesis, although being very reserved in his estimation of its applicability, Simon, by contrast, quickly went on to argue that it was more reasonable to assume that individuals were only "boundedly rational" and that they could well form different views of the economy's evolution. Muth, in contrast, in his seminal article was not totally convinced by the idea that differences in individual expectations might matter and thought that, in general, such differences should cancel out. He said specifically that,

> Allowing for cross-sectional differences in expectations is a simple matter, because their aggregate affect is negligible as long as the deviation from the rational forecast for an individual firm is not strongly correlated with those of the others. Modifications are necessary only if the correlation of the errors is large and depends systematically on other explanatory variables. Muth (1961, p. 321).

This is a simple appeal to the law of large numbers and argues that although there may be noise in peoples' decisions this noise would wash out. But this was, in fact, an important warning flag. As soon as peoples' expectations are influenced by each other, then the crucial independence assumption that underlies the law of large numbers is no longer valid and there is no guarantee whatsoever that, even if people converge on some common expectation, this will correspond to the "true expectation". Furthermore, this idea of a lack of correlation between people's erroneous view of the evolution of the economy in general and prices in particular had already been criticised by Poincaré (1908) who argued against the view that people observed information independently and took their actions accordingly without taking into account the actions of others. To quote him,

> Quand des hommes sont rapprochés, ils ne se décident plus au hasard et indépendamment les uns des autres; ils réagissent les uns sur les autres. Des causes multiples entrent en action, et elles troublent les hommes, les entraînent à droite et à gauche, mais il y a une chose qu'elles ne peuvent détruire, ce sont leurs habitudes de moutons de Panurge. Et c'est cela qui se conserve.
>
> (When people are in close contact with each other they no longer decide randomly and independently of each other. Many factors come into play, and they disturb people, shifting them one way and then the other, but there is one thing that they will not destroy and that is people tendency to behave like sheep. It is that which will always persist.) Poincaré (1908) Science et Methode, p. 49.

Why is this important for the present discussion? Precisely because many have used, as an explanation for the financial bubbles and crashes which have frequently led to crises in the real economy, the sort of "herding behaviour" which Poincaré thought of as being intrinsic to humans. By dismissing the possibility of such correlations Muth opened a theoretical avenue which made crises and catastrophic economic events even more difficult to incorporate in models. It is also worth recalling that the development of the "efficient markets hypothesis", which still underpins much of modern financial economics, was based on the work of Bachelier (1900) who developed the "random walk" hypothesis. Poincaré was the referee of Bachelier's thesis and his remarks above were undoubtedly influenced by what he saw as a weakness in that thesis, the idea that people act independently of each other.

For whatever reason, the early warning by Poincaré, and the caution and reservations expressed forcefully by Simon and less vigorously by Muth were essentially ignored by those who pushed for what they saw as putting macroeconomics on a sounder "scientific" footing. Worse, this move led to an even more exacting assumption on the capacity of individuals to understand the functioning of the economy. This makes this vision of how the economy functions even more implausible. This is the important point raised by Wagener (2014). As he points out, it was Lucas together with Prescott, (Lucas and Prescott 1971), who was the motor behind the adoption of rational expectations as the cornerstone of macroeconomic models. He used and extended the rational expectations hypothesis, as developed by Muth. But, and this is the crucial thing, they assumed not only as did Muth, that, on average, peoples' expectations would be correct but that for every individual this would be true. As they say,

> (...) we shall (...) go to the opposite extreme, assuming that the actual and anticipated prices have the same probability distribution, or that price expectations are rational. Lucas & Prescott, (1971, p. 260).
>
> Specifically, we assume that expectations of firms are rational, or that the anticipated price at time t is the same function of $(u_1; \ldots; u_t)$ as is the actual price. That is, we assume that firms know the true distribution of prices for all future periods. Lucas & Prescott (1971, p. 264).

Furthermore they advanced the idea that the state of the economy that would prevail would correspond to a rational expectations equilibrium, (REE), in this much stricter sense and what is more, they assumed that this equilibrium corresponded to a unique "full employment" level (potential output)—or what is referred to as a unique NAIRU or "natural" rate of unemployment. Again, few economists would seriously argue that economies during crises are at such an employment level. It would be difficult to think of an economic crisis such as the current one, as corresponding to a period of full employment, however defined. Indeed, in so doing they sidestepped an important problem. If there is more than one possible equilibrium at any time then, as Lucas (1986), himself, later admitted, the more interesting implications of the theory of rational expectations do not apply. In this case, expectations would determine the nature of the equilibrium attained, reversing the line of causation posited by rational expectations economists. Instead of being necessarily consistent with some "true process" governing the evolution of the economy, the process itself would be determined by which expectations individuals held.

Lucas suggested that this was not really a problem, since, somehow, the equilibria other than the one he was interested in would not, in fact, emerge, and he said,

> Recent theoretical work is making it increasingly clear that a multiplicity of equilibria...can arise in a wide variety of situations involving sequential trading. All but a few of these equilibria are, I believe, behaviourally uninteresting. They do not describe behaviour that collections of adaptively behaving people would ever hit on. I think that appropriate stability theory can be useful in weeding out these uninteresting equilibria... But, to be useful, stability theory must be more than simply a fancy way of saying that one does not want to think about certain equilibria. I prefer to think of it as an experimentally testable hypothesis, as a special instance of the adaptive laws that we believe govern all human behaviour. Lucas (1986), pp. 424–425.

Despite this assertion, there was no presentation of the sort of adjustment mechanism that would constitute the stability theory that Lucas envisaged. Furthermore the idea that the hypothesis is experimentally testable seems to be just wishful thinking. Two problems can therefore be clearly identified with the idea of rational expectations. Firstly, as I have said, if the process governing the evolution of the economy is not a simple stationary one then it is not clear that one can reasonably assume that an individual would have a complete understanding of the process.

Before passing to the second problem, it is worth observing that Bernanke, in his comments mentioned earlier, suggests that abandoning the hypothesis of rational expectations may open the door to considering economies with large endogenous movements. This, in turn, leads to the necessity of more regulation of the economy and he suggests that this is the way to dampen the impact of the crises which emerge in the economy. The ideological resistance to this is strong and many, despite the crisis, would like to believe that there is no alternative to free and unfettered markets. They are therefore reluctant to abandon rational expectations models which seemed to provide a theoretical justification for their beliefs.

The second problem with rational expectations is how do people come to hold them? This, as I have said, was conveniently sidestepped in Lucas' (1986) remarks. This is precisely the same problem as that of general equilibrium theory. Even if one can show that there are prices under which all markets would clear, how would these prices be established? In equilibrium theory, as I have already observed, many putative adjustment processes have been tried in order to show that prices would converge to an equilibrium, but with no success. But, is the situation any better for rational expectations equilibria?

Indeed, in a world in which the economic environment is constantly changing, it would not be, even, from an econometric point of view, rational to hold this sort of expectations. If the stochastic process governing the evolution of the economy contains what are called "structural breaks", points in time where there is a discontinuous change in the evolution of the economy, then it is not reasonable, from a purely econometric point of view, to simply condition on past information to predict the future. Yet this is what the individuals in modern macroeconomic models are assumed to do. As Hendry and Mizon (2010) two leading econometricians say,

> The mathematical derivations of dynamic stochastic general equilibrium (DSGE) models and new Keynesian Phillips curves (NKPCs), both of which incorporate 'rational expectations', fail to recognize that when there are unanticipated changes, conditional expectations are neither unbiased nor minimum mean-squared error (MMSE) predictors, and that better predictors can be provided by robust devices. Hendry and Mizon (2010).

Without going into technical details, simply conditioning on past experience is an unsatisfactory way of forming expectations. Yet the two dominant macroeconomic models, widely used as a basis for policy purposes, are based on the assumption that agents do just this. Indeed, on a practical level, the Geneva Association which involves major insurance companies in their most recent report (Geneva Association 2013) come to the conclusion that, until now, their expectations and their forecasts were too closely tied to extrapolations from the past and not sufficiently concerned with trying to anticipate the possible changes in the process governing the evolution of both the physical and economic environment, and the catastrophes that these might create.

One answer that was suggested by many authors before the rational expectations revolution and has been developed since, is to consider individuals as having "adaptive expectations". By this, it is meant that individuals revise their beliefs in function of their experience and observations, although such an approach is clearly backward looking, (see Evans and Honkapohja 2009 and Sargent 2008 for surveys of this sort of approach). To be slightly more precise, the standard approach of this general type assumes that agents' forecasts at any time t are derived from an econometric model, estimated using the data observed. up until that date. An individual forms his estimate, then as new data is revealed, revises that estimate, and then, with some rather stringent restrictions, the estimates converge.

However, as Woodford (2011) points out there is no reason to think that the agents will all believe in the same model, nor that this model should be that of the economic modeller, particularly if the latter before settling on his model had contemplated other candidate models. There is a hole in the reasoning usually employed here. One would have to show that agents who do not have the "correct" model in mind will come to believe in it. Indeed, there is, in fact, an old literature (see Bray 1982; Kirman 1975, 1983; Woodford 1990) which shows that individuals who believe in a "wrong" model, can, by behaving in a way consistent with their beliefs, produce outcomes which confirm those beliefs. In this case it is clear that there is no reason for those beliefs to be "rational".

Woodford (2013) went on to observe that neverthless, much of the literature on dynamics with learning has been concerned with the question, as to whether such learning dynamics would converge asymptotically to Rational Expectations Equilibria (REE). He says that much of the earlier literature was concerned with the foundations of the REE concept, trying to show how the postulated coincidence between subjective and objective probabilities could emerge. Obviously, as he indicates, this is only possible if the class of forecasting models that are contemplated by the participants in the economy includes a model that produces the forecasts associated with the REE. If one does not assume that economic agents are endowed with knowledge of the structural model, and hence with the information required to compute the REE, their forecasting rules might not even include, as a possibility, the precise forecasting rule implied by the REE. And Woodford concludes,

> But, if none of the models in the class contemplated, results in forecasts of this kind, convergence to REE beliefs (and hence to the REE dynamics) is obviously impossible. Woodford (2013), p. 23.

Thus, unless the individuals in the economy have taken the "real model" explicitly into account in their set of possible models of the world there is no hope that they would ever converge on a situation in which their beliefs about the world were correct. The whole issue here is that an assumption that all individuals at least contemplate the true model is a highly fragile basis on which to build our macroeconomic models. If individuals' behaviour deviates from this assumption then the dynamics of the system may exhibit all sorts of characteristics including the evolution of catastrophic crises.

5 Experiments

Indeed, one approach to examine the Rational Expectations Hypothesis has been to conduct laboratory experiments to see whether, even in a well defined and simple environment, important market fluctuations might occur. In earlier experiments it had seemed that, for example, in a double auction setting prices "converged" to equilibrium, (see e.g. Smith 1962). This appeared to take some of the sting out of the idea that agents in a market might produce fluctuating prices. Yet, even there, although the final prices were close to a market equilibrium, many of the trades were made at non-equilibrium prices. Once, uncertainty is introduced, even in markets with well defined fundamentals, prices may strongly deviate from the prices that would be implicit in those fundamentals. Hommes (2013) gives a good account of how bubbles and crashes can emerge endogenously in these experiments and one of the first major contributions in this direction was that of Smith et al. (1988).

The simplest sort of experiment is to ask individuals to bid for an asset which pays a fixed amount at each period and then can be redeemed at the last period for a fixed sum which is known from the outset. It is easy, in these circumstances to calculate what the price should be in each period. Yet, what is frequently observed is the appearance of bubbles and crashes even in such a simple market. Again, if agents are asked to forecast the price of an asset, without knowing precisely how their forecasts will determine the future price, they do coordinate in their forecasts. As Poincaré (1908) anticipated they "herd" but not necessarily on the "rational expectations" price. Whether the convergence of opinion, which almost always happens in these simple experiments, is on the equivalent of a rational expectations equilibrium or whether the price sequence exhibits waves or bubbles and crashes, is determined by whether the system has "positive or negative feedback". What is meant by this? Suppose that the system is such that when agents expect higher prices, prices do, in fact, turn out to be higher. This is what is referred to as positive feedback, or to use George Soros' term "reflexivity" This sort of self reinforcement is like the behaviour of ants who herd on one food source because each time an ant takes the trail to that source it lays more pheromone thereby attracting more ants.

Herding as a result of positive feedback was introduced in a formal way by Zeeman (1974) who modelled movements in the Hong Kong stock exchange index as a result of agents herding on different forecasting methods. He used as an example, the "chartist", and "fundamentalist" forecasting rules and thus prefigured a vast literature using that approach. Again he modelled sudden and dramatic endogenous changes in prices and it is worth repeating, that here we are only looking at what happens within the economic system. As in the environment catastrophes are an integral part of the system's evolution and do not have to be explained by exogenous shocks.

6 Tail Events

A last point is in order here. As I have explained, contrary to the models just mentioned, a convenient assumption made in macroeconomics is that the shocks to an economy can, indeed, be considered as "exogenous events". But, suppose for a moment that one accepts this idea then one has, at the very least to specify what sort of distribution these shocks have. In particular how does one deal with the extreme shocks in the tails of the distribution? There is an extensive debate as to how decisions should be made under the sort of uncertainty associated with such "tail events". How this should be taken into account, is far from clear. What a number of authors have suggested, (see e.g. Chanel and Chichilnisky 2013), is that the standard attribution of objective probabilities and the use of expected utility theory does not capture how people react to potentially catastrophic risks. As a result there is an ongoing debate in the theoretical literature as to the appropriate decision making criterion. Yet in modern macroeconomic models we assume that the behaviour of the aggregate can be assimilated to that of an average individual. Do we then assume, not only that this individual understands the process governing the evolution of the economy but that he takes his decisions according to a criterion that is not yet the subject of a consensus among theorists?

For example, when investing in assets that may be correlated in both linear and non linear ways, how does he behave? When faced with assets which are correlated in the tails of the distribution of their returns, does he have recourse to one of of the sophisticated risk measures now available, using extreme value theory (see Poon et al. 2004 or TAILCOR developed by Ricci and Veredas 2013) It seems highly unlikely that the agents the representative individual is supposed to represent, have any such idea in mind, nor that they should have arrived at such a common conclusion.

Why then, spend so much time on a notion that seems to be so full of difficulties and of little use in explaining the evolution of the economy? Precisely because this approach still underlies the modern macroeconomic models which are widely used by governments, central banks and international institutions. Yet it should be evident that the logical contradictions and the implausibility of a state in which people hold rational expectations makes models which picture the world in that way uninteresting for modelling real world phenomena. Excluding, a priori, the idea that individuals can reasonably have different views of the world and may sometimes agree on a completely erroneous picture suggests that the modellers are interested in tractability and technical sophistication but not in understanding economic phenomena. Indeed, given the nature of the models, the last thing they are capable of doing is to explain how economic catastrophes happen.

What then are the explanations advanced to explain crises? The key words used by many to explain such happenings, are contagion, panic and breakdown in trust. None of these ideas are present in the sort of "general equilibrium" macro models currently in use. General Equilibrium is in quotation marks because these models are far from those underlying the Walrasian notion of equilibrium. Could one then approach the problem differently and try to explain and understand catastrophic events in economies without having recourse to the sort of model that currently prevails?

Here I think the answer is to reduce the assumptions we make about the cognitive capacity of our individuals, to emphasise the fact that by acting according to simple rules and by directly interacting with and taking account of the actions of other individuals, they can collecively self-organise into a catastrophic situation. Crises are then emergent phenomena generated by the interaction between individuals who are not irrational in any standard sense, but are not necessarily optimisers. This is the sort of view reflected in the approach of those who regard the economy as a complex evolving system., (see e.g. Kirman 2010 and Bouchaud 2012).

To illustrate this approach, I will now present a model which portrays a market crash and in which individuals have local and limited knowledge but whose actions are governed by the choices made by those with whom they are in contact,

7 An Example

What would be an appropriate example of an endogenously generated catastrophic change in a market? The dramatic collapse of the prices of MBS, (mortgage backed securities)at the beginning of the current crisis is such an example which had profound ramifications for the economy as a whole. In Anand et al. (2013) we analyzed this event. This collapse occurred rapidly despite the fact that the

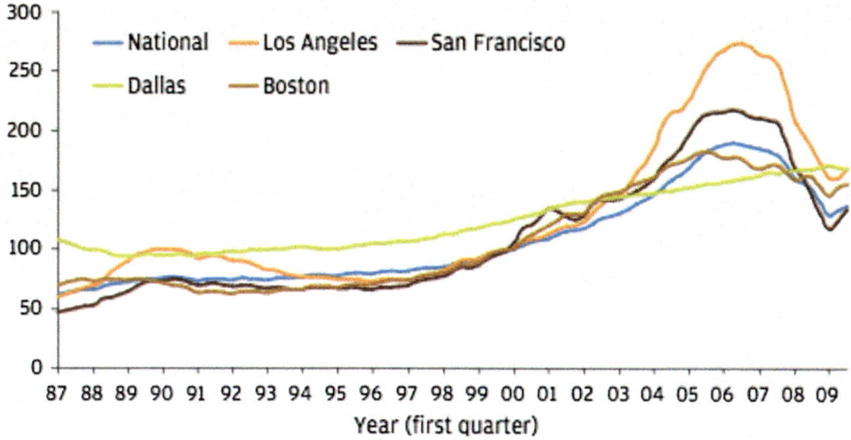

Fig. 1 The evolution of house price indices in different cities in the U.S. 1987–2009 (*Source* Case Shiller)

weakness of the assets underlying the derivatives had increased progressively over time. The instruments in question consisted of tranches of mortgages with different ratings but with a single overall rating. Figure 1 shows the evolution of house prices over 20 years and the developments from 2006 onwards should have been a first indication that defaults on the mortgages involved in the MBS were likely to rise.

The principle reason for this was that the percentage of loans that represented "positive equity", that is where the value of the house was greater than the outstanding amount of the house loan, diminished as the increase in house prices slowed down and was reversed.[4]

In Fig. 2 the default rate on mortgages issued in different years is shown for the U.S. The increase that occurred, as house prices increased more slowly, and more and more mortgages were issued on easier terms, is clear. Of mortgages issued in 2004 10 % were delinquent after 30 months whereas for those issued in 2007 10 % were delinquent after only 8 months. The evidence was public and available. Yet the evolution of the prices of MBS did not reflect this steady increase, as can be seen in Fig. 3

Prices of similarly rated assets remained stable and then suddenly collapsed. A possible explanation would be to suggest that the investors in MBS and the whole chain of actors from the mortgagor to the investor were in fact highly rational and that the constancy of the prices, before the collapse, reflected their rational expectations. Yet, Ashcraft and Schuermann (2007), right at the onset of the crisis, had already pointed out seven informational frictions in the chain of actors, from the

[4]In a number of states in the U.S. (where loans are "non-recourse") the owners of a property on which they have a loan can simply turn over the house to the bank which issued the loan without having any further financial contribution to make.

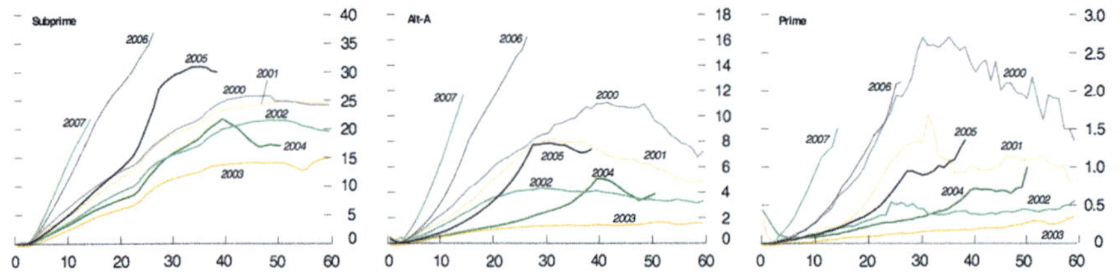

Fig. 2 Delinquency rates on mortgages, (months from origin) in the U.S. originating from 2004–2007 (*Source* Merrill Lynch and Loan Performance)

Fig. 3 The prices of MBS with different ratings

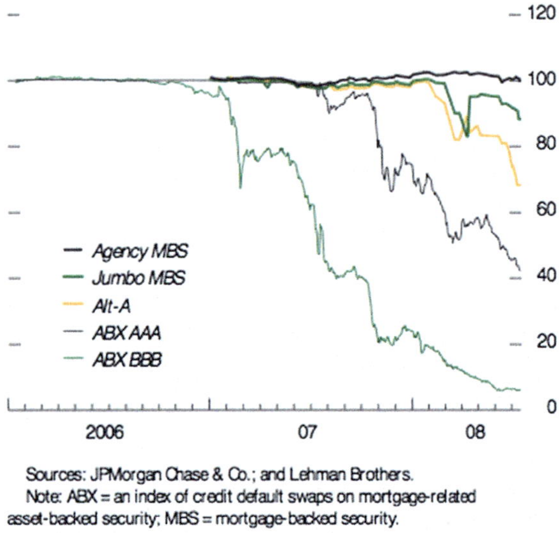

mortgagor to the investor, each of which could lead to a breakdown in the system. These frictions range from problems of moral hazard involving the payment of the credit rating agencies by asset managers, to principal agent difficulties between the investor and the asset manager, to moral hazard problems between the issuers of mortgages and the arranger who sets up a special vehicle to hold the mortgages and securitises them for eventual purchase by investors. What the authors observed is that the presence of all these frictions led to the initial breakdown of this market. Despite various pieces of legislation to control the sort of abuses that could arise from the frictions they identify, the market was very far from transparent.

Despite the fact that the delinquency of differently rated assets was evolving similarly as seen in Fig. 2 it is clear that the collapse of the better rated assets came later. This suggests that the estimates of the probability of default on these instruments were influenced by the ratings, whether or not this was justified by the underlying fundamentals. Ashcraft et al. (2010) show that prices were strongly correlated with ratings but that ratings were very poorly correlated with default rates, indicating clearly that the information provided was far from perfect. But, the essential question is, what was it in the behavior of the actors in the market, which prevented the available evidence from translating into a steady decline in the prices of MBS rather than into the sudden collapses that actually occurred?

To understand this we constructed a model of the banking system in which the participants securitise their loans and then sell them on via Special Purpose Entities, (SPE), to each other. The derivatives thus constructed were made up of bundles of parts of different underlying assets with varying degrees of riskiness. One might wonder why this happened. The reason lies in the so-called "Recourse rule" which puts a risk weight of 50 % on individual mortgages but only of 20 % on highly rated mortgage backed securities. Thus banks bundled their assets to free up capital. But, as a result, to reliably estimate the risk entailed checking on the current status of each of the assets in a bundle. In other words the banks were holding derivatives, in particular, MBS, the content of which was costly to evaluate. Worse, some of the banks knowingly misrepresented the value of the underlying assets they were selling to their investor clients as J P Morgan admitted in their recent[5] 13$ billion settlement. These investors had neither the means nor the sophistication to evaluate whether the mortgages in the MBS did, in fact, meet the underwriting standards as was claimed. Worse, the banks themselves when trading amongst themselves were not doing due diligence. Furthermore these assets were being actively traded and it is difficult to argue that the prices of the associated transactions reflected the decisions of fully informed rational agents. What I have just described is an oversimplified description of a market in which some agents had an interest in hiding or misrepresenting information and others had no incentive to go to the expense of obtaining the full information. Many of the incentives were, contrary to the conventional view of such markets, all in the wrong direction.

It was worth considering this example in some detail, since it is only by looking at the organization, incentives and the interactions between the actors that one can understand that trying to model such a framework as an anonymous market inhabited by fully informed agents with rational expectations does not capture the essence of what is happening. This market is a complex interactive network and using a standard model would not have reflected this. The model we propose focuses on the interaction between the actors in the banking network and provides an explanation as to how a sudden price collapse can occur despite the fact that the underlying fundamentals were gradually changing over a considerable period of time.

Our simple model which reflects the concerns expressed by Haldane and May (2011), describes the behavior of the participants on the market. This behavior was not irrational in the normal sense of the word but was not fully rational in the sense that that term is understood in economics. The agents in the model have short horizons and condition their evaluation of an ABS not, by always examining the

[5]22nd November 2013.

fundamentals underlying the instrument, but often on the ratings of that MBS by the Credit Rating Agencies and without analysing the general evolution of the housing market. Furthermore their willingness to buy depends on how much checking was being done by those with whom they traded.

8 The Model (Anand et al. 2013)

The system consists of $i = 1, \ldots, N$ agents, which, in the case of the sub-prime crisis, for simplicity, we can think of as the banks who were both the issuers—via SPEs—and the investors in these ABS. These banks or agents are linked in a network, corresponding to the over the counter market (OTC) and at each period an agent draws at random another agent amongst her neighbors. Each agent i is characterized by a variable $z \in \{0, 1\}$ which specifies whether she adopts ($z = 1$) or not, ($z = 0$) the following behavioral rule: purchase an ABS, relying on signals from the rating agencies, without independently evaluating the fundamental value of underlying assets

Succinctly, we write

$$z_i = \begin{bmatrix} 1 \textit{ if agent i follows the rule} \\ 0 \textit{ if agent i does risk analysis} \end{bmatrix} \tag{1}$$

The rationale for adopting the rule, as we will see, is not based on the fundamental quality of the asset but rather on the fact that others also follow the rule. If, in fact, enough other participants do so, the agent becomes convinced, not irrationally that the ABS is highly liquid and hence easy to trade.

Assume that the ABS is toxic with probability p. By toxic we mean, for example, that the underlying asset was too favourably rated by a rating agency and either that the original borrower of the loan has already defaulted, or has a higher probability of defaulting, as he is delinquent in his payments. Assume that the cost of purchasing a security is p_0 whereas the payoff from successfully re-selling the security is p_1 where $p_1 - p_0 > 0$. However, if the buyer checks and finds the ABM to be toxic the price now becomes a "fire sale" price p_2 where $p_2 - p_0 < 0$. The buyer can be sure to avoid this outcome by checking at a cost of χ drawn from a p.d.f. $\Phi(\chi)$. Now one can rescale and reduce the number of parameters by normalizing such that: $p_1 - p_2 = 1$ and $p_0 - p_2 = c$. The agent is then faced with the following problem:

	Check and toxic	Don't check
$Z(i) = 0$	$-\chi_i$	$1 - c - \chi_i$
$Z(i) = 1$	$-c$	$1 - c$

The columns represent the strategy of the buyer and the rows those of the seller. Now consider the expected pay-off to the seller of each strategy.

$$u_i(z_i = 1) = E\left[-p(1-z_j)c\right] + \left[1-p(1-z_j)\right](1-c) = 1-p(1-\overline{z_i}) - c \quad (2)$$

where $\overline{z_i} = E(z_i)$ for $j \in N_i$. That is agent i can correctly estimate the average choice of rule by his neighbors but not the choice of each individual. Thus we have,

$$z_i = \frac{1}{k_i} \sum_{j \in N_i} z_j \quad (3)$$

Now the expected pay-off from not following the rule, and choosing $z_i = 0$ that is, from checking the value of the underlying assets is:

$$u_i(z_i = 0) = (1-p)(1-c) - \chi_i \quad (4)$$

Thus if the agent checks and finds the assets to be toxic he simply incurs the cost of checking while if the asset is not toxic he obtains the difference between the selling and buying price less the checking cost. The strategy, which constitutes the best reply to the strategies of the neighbors, is then given by:

$$z_i = \Theta[u_i(1) - u_i(0)] = \Theta\left[p\left(\frac{1}{k_i}\right) \sum_{j \in N_i} z_j - c\right] \quad (5)$$

where the function Θ is defined as $\Theta(x) = 1$ if $x > 0$ and 0 otherwise. Note that the agents are assumed to know the probability of default of the underlying assets. However, in reality, the common perception of p reflected the over optimistic evaluation of the rating agencies. For low values of p there is one equilibrium in which all agents choose not to check, but, once a critical value of the commonly perceived p is passed, another equilibrium emerges in which all agents check. This is illustrated from numerical simulations, in Fig. 4.

When there are two equilibria there is no reason to believe that one or the other will be necessarily realized. However, we can introduce, as in the previous example, some noise and assume that the agents only make the best response with a certain probability. What we impose is that as the superiority of one strategy over the other increases, the probability of choosing that strategy increases. Given this, we can examine whether they coordinate on one equilibrium as p increases. We now introduce the logit rule, which has the required property. Thus, the probability of choosing $z_i = 1$ is given by:

$$P(z_i = 1) = \frac{e^{\beta u_i(1)}}{e^{\beta u_i(1)} + e^{\beta u_i(0)}} \quad (6)$$

Where β is a parameter indicating the sensitivity of the agent to the difference between the pay-offs from the two strategies. If $\beta = 0$ the agent chooses one of the

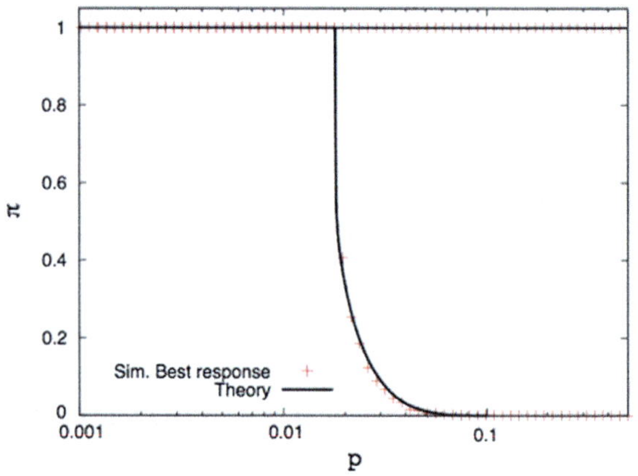

Fig. 4 The coexistence of two equilibria either all $z_i = 1$ or $z_i = 0$ (*Source* Anand et al. 2013)

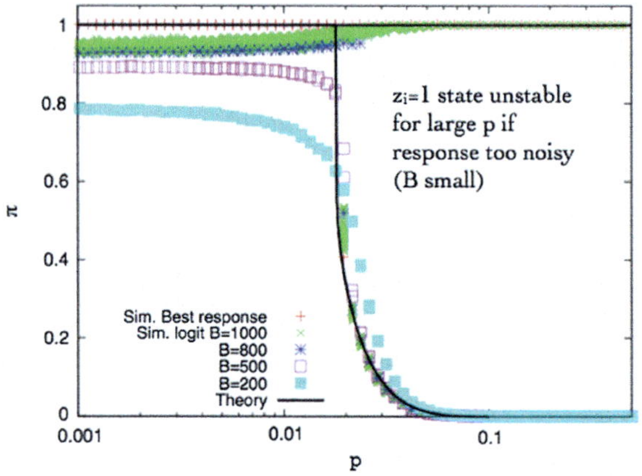

Fig. 5 The evolution of the equilibrium state as p increases when learning is noisy (*Source* Anand et al. 2013)

two strategies at random whereas if $\beta \to \infty$ then the probability of choosing the best response goes to one (Fig. 5).

With the noise in the decisions of the agents the system switches suddenly from one equilibrium to another. There are two things to observe here, firstly that a continuous evolution of p, the perceived probability of default or toxicity leads to a sudden and large change in the equilibrium state. This, in turn, provokes a sharp decline in the prices of the asset-backed security, which is just what was observed and shown in Fig. 3. The fact that the collapse occurred later for better rated MBS reflects the effect of the ratings on perceived probabilities rather than any real differences in those probabilities across assets. In fact as Ashcraft et al. (2010) say, "Our analysis also suggests MBS ratings did not fully reflect publicly available data."

The second important observation is that the existence of a certain amount of noise in the decisions of agents leads to the selection of a particular equilibrium. This view of the introduction of noise as an equilibrium selection device recalls the literature on "trembling hands" in game theory. In this context, this is a very partial equilibrium since the evolution of p has been taken as exogenous and to fully model the process this would need to be modeled also.

However, in a situation where agents are influenced by each others' decisions and where their decision making is not fully "rational" we capture some important empirical facts. The individuals involved, are far from the infinitely far sighted optimisers of standard models and are making relatively simple binary decisions, based on the actions of their partners. This can lead to major changes in the aggregate state of the market. Again this is clearly not a comprehensive model of what is, in reality, a very complicated system, but it does capture some of the characteristics which lead to major aggregate shifts without any specific major exogenous shock.

A more ambitious goal would be to build a model in which there are no equilibria and in which the market, its organization and the behavior of the agents are constantly and simulataneously evolving.

9 A More General and Important Problem

I started this paper by arguing that our models of the macroeconomy fail to capture the internal dynamics that lead to catastrophic events. I then gave a very simple example which showed how the collapse of one market could be explained as the result of the interaction between the participants. This particular collapse contributed in an important way to the current crisis. Yet this is but one dimension of a much more general problem. Much attention has been devoted to the possibility of natural catastrophes and the extent to which we can characterise the likelihood of extreme natural events, such as earthquakes, floods, and hurricanes. However, it is becoming steadily clearer that the frequency and intensity of such events is related to man's activities. This complicates the modelling problem enormously, since to begin to analyse the consequences of future climate change we have to discuss possible climatic developments and to take explicit account of the fact that we are in the anthropocene era and that man's role is important. Thus we need what one might call "integrated models" which capture the impact of man's activity on the environment, and of course, in turn, the impact of the latter on man's activity. Such models exist at the level of specific activities and a good example is given by the analysis of the exploitation of natural resources such as fisheries, (see Clark 1990 and Hommes and Rosser 2001). On the more general level there have been recurrent calls for models which seek to address directly the interdependence of the socio-economic system and the environment (see e.g. Folke et al. 2011) but what those in question have in mind, goes much further than the integrated climate change models which have been developed to date.

Indeed, recently, a number of leading economists (see Stern 2013; Pindyk 2013 and Weitzman 2013) have criticised the sort of models that have been used to analyse the consequences of climate change, as too cautious and inadequate for the purpose for which they were designed. Stern argues that the three types of models he considers, climate, impact and economic are all too narrow and too prudent. The reasons for this are, in each of the three cases, somewhat different but there is one feature which is common to all which is the desire for tractability, and this is particularly evident in the economic models. As a result Pindyk (2013) says, they create a perception of knowledge and precision, but that perception is illusory and misleading. Many of us feel that this comment could be applied much more generally to macroeconomic models. However, the weakness of the economic models is particularly disturbing. How can we model two coevolving complex systems and their impact on each other if we continue to insist that one of them is an essentially stationary equilibrium system. So far, the IAMs or Integrated Assessment Models which seek to integrate both climate and economic models and their impact on each other are extremely simplified and do not reflect the underlying mechanisms in either.

10 Conclusion

In this short paper I have first tried to show why economic crises, collapses and catastrophes are modelled as being generated by processes exogenous to the economic system. The idea that these major and often sudden events emerge intrinsically with the evolution of the economy is in direct contradiction with the underlying vision of the economy. Many economists have pointed to these difficulties in the past but, in the interest of what has been perceived to be increasing mathematical sophistication, we have continued to build models of an economy which evolves along an equilibrium path only occasionally knocked off that path by some exogenous event.

An alternative view is that the economy is a complex, evolving system which goes through "phase changes" which may have disastrous consequences. I provided a simple, even simplistic, model to explain the sort of crash that we observe in markets and to show that even by focusing on one component of the system we can generate endogenous crises. This would suggest that we should take more seriously the idea of the economy as a system in which crises emerge from the interaction of the participants in the economy.

But as I suggested in the last part of the paper, we should then have been moving towards models in which the interaction between economic activity and the environment can be thought of as the coevolution of two complex systems. Such models are unlikely to be tractable and even more unlikely to generate any sort of precise predictions. However, they would reflect our better understanding of the forces at play and the nature of the outcomes. However, sadly, rather than producing models which capture the possibly catastrophic results of this interplay between the two systems, environmental scientists have been seized by an excess of prudence in their calculations, no doubt to forestall criticism of their failure to be "fully scientific". They seem to have adopted a "Principle of Scientific Precaution" which means that one cannot make any pronouncements about something about which one is not sure. On the other hand the economic models have moved little from those of conventional economics and thus the so-called "integrated models" understate the real risks involved.

We have, at some point, to take more seriously the idea that the interaction between an economic system which is intrinsically unstable and the complex physical environment within which it operates can lead us into situations with consequences much more severe than those currently envisaged by most models. If we succeed in doing this we will no doubt have a better understanding of the evolution of the economy, but will have to make our claims as to predictions and the consequences of particular measures, much more modest.

References

Anand, K., Kirman, A., & Marsili, M. (2013). Epidemics of rules, rational negligence and market crashes. *The European Journal of Finance*, *19*(5), 438–447.

Ashcraft, A., & Schuerman, T. (2007). Understanding the securitization of subprime mortgage credit. *Working Paper Federal Reserve Bank of New York.*

Ashcraft, A. B., Goldsmith-Pinkham, P., & Vickery, J. I. (2010). *MBS ratings and the mortgage credit boom.* FRB of New York Staff Report no. 449.

Bachelier L (1900) Théorie de la Spéculation. *Annales de l'Ecole Normale Supérieure, 17*, 21–86

Bernanke B (2010) *Interview with the international Herald Tribune*, May 17th 2010.

Bouchaud, J.-P. (2012). Crises and collective socio-economic phenomena: cartoon models and challenges. arXiv:1209.0453.

Bray, M. (1982). Learning, estimation, and the stability of rational expectations. *Journal of Economic Theory, 26*, 318–339.

Brock, W. A., & Sayers, C. (1988). Is the business cycle characterized by deterministic chaos? *Journal of Monetary Economics*, 71–90 July 1988.

Chanel, O., & Chichilnisky, G. (2013). Valuing life: Experimental evidence using sensitivity to rare events. *Ecological Economics*, *85*(C), 198–205.

Chichilnisky, G. (2010). Managing catastrophic risks and climate change. *Reuters: The Great Debate UK Blog, 19 March 2010.*

Chichilnisky, G., Heal, G., & Lin, Y. (1995). Chaotic price dynamics, increasing returns and the Phillips curve. *Journal of economic behavior & organization*, *27*(2), 279–291.

Colin, C. (1990). *Mathematical bioeconomics*, 2nd edn. New York: Wiley Interscience.

Debreu, G. (1974). Excess demand functions. *Journal of Mathematical Economics, 1*, 15–23.

de Sismondi, S. (1819). *New Principles of Political Economy,* vol. 1, pp. 20–21.

Evans, G. W., & Honkapohja, S. (2009). Learning and macroeconomics. *Annual Review of Economics, 1*, 421–451.

Ezekiel, M. (1938). The Cobweb theorem. *The Quarterly Journal of Economics, 52*(2), 255–280.

Ferguson, N. (2009). *The ascent of money: a financial history of the world.* London: Penguin Press.

Flaschel, P. (1991). Stability: Independent of economic structure? *Structural change and economic dynamics, 2*(1), 9–35.

Folke, C., Jansson, Å., Rockström, J., Olsson, P., Carpenter, S., Crepín, A.-S., Daily, G., Ebbesson, J., Elmqvist, T., Galaz, V., Moberg, F. Nilsson, M., Österblom, H., Ostrom, E., Persson, Å., Polasky, S., Steffen, W., Walker, B., & Westley, F. (2011). Reconnecting to the biosphere. Working Paper No. 1. Prepared for the "3rd Nobel Laureate symposium on global

sustainability: Transforming the world in an era of global change", in Stockholm, 16–19 May 2011. Stockholm Resilience Centre, the Royal Swedish Academy of Sciences, the Stockholm Environment Institute, the Beijer Institute of Ecological Economics and the Potsdam Institute for Climate Impact.
Geneva Association. (2013). *Warming of the Oceans and Implications for the (Re)insurance Industry*. Geneva: A Geneva Association Report.
Grandmont, J. M. (1985). On endogenous competitive business cycles. Econometrica, *53*(5), 995–1045.
Greenspan, A. (2011). Dodd-Frank fails to meet test of our times. Financial Times 29th March.
Haldane, A. G., & May, R. (2011). Systemic risk in banking ecosystems. *Nature, 469*, 351–355.
Helbing, D. (2013). Globally networked risks and how to respond. *Nature, 497*, 51–59.
Hendry, D., & Mizon, G. E. (2010). On the mathematical basis of inter-temporal optimization. *Economics Series Working Papers 497*, University of Oxford, Department of Economics.
Herings, J.-J. (1997). A globally and universally stable price adjustment process. *Journal of Mathematical Economics, 27*, 163–193.
Holt, C. C., Modigliani, F., Muth, J. F., & Simon, H. A. (1960). *Planning Production, Inventories, and Work Force*. Englewood Cliffs: Prentice Hall.
Hommes, C., & Rosser, J. B., Jr. (2001). Consistent expectations equilibria and complex dynamics in renewable resource markets. *Macroeconomic Dynamics, 5*, 180–203.
Hommes, C. (2013). *Behavioural rationality and heterogeneous expectations in complex economic systems*. Cambridge: Cambridge University Press.
Jordan, J. S. (1982). The competitive allocation process is informationally efficient uniquely. *Journal of Economic Theory, 28*, 1–18.
Joughin, I., Smith, B. E., & Medley, B. (2014). Marine ice sheet collapse potentially underway for the Thwaites Glacier Basin, West Antarctica. *Science,*. doi:10.1126/science.1249055.
Kamiya, K. (1990). A globally stable price adjustment process. *Econometrica, 58*, 1481–1485.
Kirman, A. (1975). Learning by firms about demand conditions. In Day, R. (Ed.) *Adaptive economics*. Academic Press.
Kirman, A. (1983). Mistaken beliefs and resultant equilibria. In Frydman, R., Phelps, E. (Eds.) *Individual forecasting and collective outcomes*. Cambridge: Cambridge University Press.
Kirman, A. (2010). *Complex economics: individual and collective rationality*. London: Routledge.
Lucas, R. (1986). Adaptive behaviour and economic theory. *Journal of Business, 59*, 401–426.
Lucas, R., & Prescott, E. (1971). Investment under uncertainty. *Econometrica, 39*, 659–681.
Mantel, R. (1974). On the characterisation of aggregate excess demand. *Journal of Economic Theory, 7*, 348–353.
Mercer, J. H. (1978). West antarctic ice sheet and CO greenhouse effect: a threat of disaster. *Nature, 271*, 321–325.
Minsky, H. (2008). *Stabilising an unstable economy*. New York: Mcgraw Hill.
Morishima, M. (1984). The good and bad uses of mathematics. In P. J. D. Wiles & G. Routh (Eds.), *Economics in disarray*. Oxford: Basil Blackwell.
Muth, J. (1961). Rational expectations and the theory of price movements. *Econometrica, 29*(3), 315–335.
Pindyk, R. (2013). Climate change policy: What do the models tell us? *Journal of Economic Literature, 51*(3), 860–872.
Poincaré, H. (1908). *Science et méthode*. Paris: Flammarion.
Poon, S. H., Rockinger, M., & Tawn, J. (2004). Extreme value dependence in financial markets: diagnosis, models, and financial implications. *Review of Financial Studies, 17*, 581–610.
Reinhart, C. M., & Rogoff, K. S. (2010). *This time is different: a panoramic view of eight centuries of financial crises*. Princeton, NJ: Princeton University Press.
Ricci, L., & Veredas, D. (2013). TAILCOR. Working paper ECARES. Solvay Brussels School of Economics and Management, Université Libre de Bruxelles.
Rignot E, J., Mouginot, Morlighem, M., Seroussi, H., & Scheuchl, B. (2014). Widespread, rapid grounding line retreat of Pine Island, Thwaites, Smith, and Kohler glaciers, West Antarctica, from 1992 to 2011. *Geophysical research letters,* Vol. 41, Issue 10, pp. 3502–3509.

Rosser, J. B. (2007). The rise and fall of catastrophe theory, applications in economics: Was the baby thrown out with the bathwater? *Journal of Economic Dynamics and Control, 31*, 3255–3280.

Saari, D., & Simon, C. P. (1978). Effective price mechanisms. *Econometrica, 46*, 1097–1125.

Sargent, T. J. (2008). Evolution and intelligent design. *American Economic Review, 98*, 5–37.

Scarf, H .(1959). Some Examples of Global Instability of the Competitive Equilibrium. Cowles Foundation Discussion Papers 79, Cowles Foundation for Research in Economics, Yale University.

Scarf, H. (1969). An example of an algorithm for calculating general equilibrium prices. *American Economic Review, American Economic Association, 59*(4), 669–677.

Simon, H. (1979). Rational decision-making in business organizations. Nobel Memorial Lecture, 8 Dec 1978. *The American economic review 69*(4), 493–513.

Smale, S. (1976) A convergent process of price adjustment and global newton methods. *Journal of Mathematical Economics, 3*, 107–120.

Smith Vernon, L. (1962). An experimental study of competitive market behavior. *Journal of Political Economy, 70*, 111–137.

Smith, V. L., Suchanek, G. L., & Williams, A. W. (1988). Bubbles, crashes, and endogenous expectations in experimental spot asset markets. *Econometrica, 56*(5), 1119–1151.

Sonnenschein, H. (1972). Market excess demand functions. *Econometrica, 40*, 549–563.

Stern, N. (2013). The structure of economic modeling of the potential impacts of climate change: grafting gross underestimation of risk onto already narrow science models. *Journal of Economic Literature, 51*(3), 838–859.

Thom, R. (1983). *Mathematical models of morphogenesis*. Chichester: Ellis Harwood.

Turner, A. (2010). Economics, conventional wisdom and public policy. In Speech at Institute for New Economic Thinking Inaugural Conference Cambridge, April 2010.

Varian, H. R. (1977). Non-Walrasian Equilibria. *Econometrica, 45*, 573–590.

Varian, H. R. (1979). Catastrophe theory and the business cycle. *Economic Inquiry, 17*(1), 14–28.

Wagener, F. (2014). Expectations in experiments. *Annual Review of Economics, Annual Reviews, 6*(1), 421–443.

Walras, L. (1954). Elements of pure economics. (French edition 1900), Translated by W. Jaffe. London: Allen and Unwin.

Weitzman, M. (2013). Tail-hedge discounting and the social cost of carbon. *Journal of Economic Literature, 51*(3), 873–882.

Woodford, M. (1990). Learning to believe in sunspots. *Econometrica, 58*(2), 277–307.

Woodford, M. (2013). Macroeconomic analysis without the rational expectations hypothesis. *Annual Review of Economics, Annual Reviews, 5*(1), 303–346.

Zeeman, E. Christopher. (1974). on the unstable behavior of the stock exchanges. *Journal of Mathematical Economics, 1*(1), 39–44.

Printed by Printforce, the Netherlands